Lecture Notes in Mathematics 2235

More information about this series at http://www.springer.com/series/304

Pietro Aiena

Fredholm and Local Spectral Theory II

With Application to Weyl-type Theorems

 Springer

Pietro Aiena
Dipartimento di Matematica
Università degli Studi di Palermo
Palermo, Italy

ISSN 0075-8434   ISSN 1617-9692   (electronic)
Lecture Notes in Mathematics
ISBN 978-3-030-02265-5   ISBN 978-3-030-02266-2   (eBook)
https://doi.org/10.1007/978-3-030-02266-2

Library of Congress Control Number: 2018962161

Mathematics Subject Classification (2010): 47A05, 47A10, 47A11, 47A53

This Springer imprint is published by the registered company Springer Nature Switzerland AG
The registered company address is: Gewerbestrasse 11, 6330 Cham, Switzerland

A mia moglie Maria, ai miei figli Marco, Caterina, Adriana, e a tutti i miei cari familiari

Preface

The main focus of this book is on bounded linear operators on complex infinite-dimensional Banach spaces and their spectral properties. In recent years spectral theory, which has numerous and important applications in many parts of modern analysis and physics, has witnessed considerable development. In this book the reader is assumed to be familiar with the basic notions of linear algebra, functional analysis and complex analysis. Essentially, the aim of this book is to give an idea of the relationship between local spectral theory and Fredholm theory. The deep and elegant interaction between Fredholm theory and local spectral theory becomes evident when one considers the so-called single-valued extension property (SVEP). This property, which dates back to the early days of local spectral theory, was first introduced in 1952 by Dunford and has received a systematic treatment in Dunford and Schwartz's classical book [143]. It also plays an important role in the book of Colojoară and Foiaş [98], in Vasilescu [292], and in the more recent books of Laursen and Neumann [216] and Aiena [1].

H. Weyl's pioneering work [296] showed that the spectra of self-adjoint operators on Hilbert spaces have a very nice structure, which was later also observed for several classes of operators defined on Banach spaces. Nowadays, these operators are said to satisfy Weyl's theorem. A general theory of these operators has never been systematically organized in a monograph, only a few results may be found in the books by Aiena [1, Chapter 3, §8] and Lee [223]. The purpose of this book is to provide the first general treatment of the theory of operators for which Weyl-type or Browder-type theorems hold, taking into account the more recent developments. A localized version of the single-valued extension property will be a very useful tool for studying these theorems, so this monograph may also be thought of as a substantial attempt to show how the local spectral theory and Fredholm theory interact.

We describe in more detail the architecture of this monograph. This book consists of six chapters. The first chapter is mainly dedicated to some classes of operators originating from Fredholm theory. All these classes of operators are strictly related to the classical semi-Fredholm operators. In the second chapter we introduce the basic tools of local spectral theory, and relate some aspects of Fredholm theory to a localized version of the single-valued extension property. The third chapter plays a central role in this book, since it concerns the relationships between the several spectra originating from the classes of operators introduced in Chap. 1. In particular, these spectra are studied under suitable perturbations. Furthermore, in this chapter we consider some special classes of operators having a nice spectral structure: Riesz operators, meromorphic operators, algebraic operators and other classes of operators. We also see that the spectral theorem holds for many of the spectra which originate from Fredholm theory.

In Chap. 4 we introduce the classes of polaroid-type operators. The polaroid-type operators are those for which the isolated points of the spectrum, or also the isolated points of the approximate point spectrum, are poles or left poles of the resolvent, respectively. These classes of operators are very large, and contain many important classes of operators acting on Hilbert spaces which generalize, in some sense, the properties of normal operators, for instance hyponormal and paranormal operators. Other examples of polaroid operators are the isometries, the convolution operators on group algebra, the analytic Toeplitz operators defined on Hardy spaces, the semi-shift operators and weighted unilateral shifts.

The ideas developed in the first four chapters are then applied to the last two chapters. In particular, Chap. 5 is dedicated to the classes of operators which satisfy Browder-type theorems, in their classical form or in their generalized form, according to the B-Fredholm theory.

The last chapter concerns Weyl-type theorems, also in their classical or generalized form. We also see that Weyl's theorem holds for many important classes of operators, such as Toeplitz operators in Hardy spaces, symmetrizable operators, symmetries and shift operators. In the last part of this chapter we also present a theoretical and general framework from which all Weyl-type theorems may be promptly established for many classes of operators introduced in Chap. 4.

The content of this book is essentially new, and is the result of intensive research done during the last 10 years. Many results appear here for the first time in a monograph, and this in a sense is an attempt to organize the available material, most of which can only be found in research papers. Of course, it is not possible to make a presentation such as this one entirely self-contained, but in general most of the material needed is developed within the framework of the book itself, and only a few results are mentioned with suitable reference. A rudimentary knowledge of functional analysis is quite necessary for understanding the material of the book. However, some basic facts of functional analysis are reassumed in the Appendix.

Anyone who writes a book of this sort accumulates, of course, many outstanding debts. There are several friends and colleagues to whom I am indebted for their suggestions and ideas. These ideas constitute a great part of the material presented in this book. In particular, I thank Manolo Gonzalez, Vladimir Müller, Bhaggy Duggal, Salvatore Triolo, and all my PhD students. Finally, I have a special debt of thanks to the anonymous copy editor, who reads the manuscript and who provided me with corrections and useful comments.

Palermo, Italy   Pietro Aiena

Contents

Chapter 1
Fredholm Theory

The purpose of this chapter is to provide an introduction to some classes of operators which have their origin in the classical Fredholm theory of bounded linear operators on Banach spaces. The presentation is rather expository in style, and only a few results are mentioned here with suitable reference. The first three sections address some preliminary and basic notions, concerning some important invariant subspaces, such as the hyper-range, the hyper-kernel, and the analytic core of an operator. The importance of these subspaces will become evident when we study the special classes of operators treated in successive sections. The first class introduced is that of semi-regular operators, which is treated in the fourth section. We also consider, in the fourth section, an important decomposition property, the generalized Kato decomposition for operators, originating from the classical treatment of perturbation theory by Kato [195], which showed that this decomposition holds for semi-Fredholm operators. This decomposition motivates the definition of two other classes of operators which contain all semi-Fredholm operators, the class of all operators of Kato-type, and the class of essentially semi-regular operators, introduced by Müller [243].

The fifth section is devoted to the class of operators having topological uniform descent, introduced by Grabiner [162]. We also give some perturbation results concerning this class of operators that will be used in the sequel. This class properly contains the class of quasi-Fredholm operators on Banach spaces, which was first introduced by Labrousse [208] for operators acting on Hilbert spaces. The last two sections of this chapter concern a generalization of semi-Fredholm operators, the class of semi B-Fredholm operators introduced by Berkani et al. [62, 64, 71] and the class of all Drazin invertible operators. We also introduce the classes of left and right Drazin invertible operators and relate these operators to the other classes of operators previously introduced.

© Springer Nature Switzerland AG 2018

P. Aiena, *Fredholm and Local Spectral Theory II*, Lecture Notes in Mathematics 2235, https://doi.org/10.1007/978-3-030-02266-2_1

1.1 Operators with Closed Range

Since this book concerns the spectral theory of bounded linear operators, we will always assume that the Banach spaces are complex and infinite-dimensional, although many of the results in these notes are still valid for real Banach spaces. If X, Y are Banach spaces, by $L(X, Y)$ we denote the Banach space of all bounded linear operators from X into Y. Recall that if $T \in L(X, Y)$, the norm of T is defined by

$$\|T\| := \sup_{x \neq 0} \frac{\|Tx\|}{\|x\|}.$$

If $X = Y$ we write $L(X) := L(X, X)$. By $X^* := L(X, \mathbb{C})$ we denote the *dual* of X. If $T \in L(X, Y)$ by $T^* \in L(Y^*, X^*)$ we denote the *dual operator* of T, defined by

$$(T^* f)(x) := f(Tx) \quad \text{for all } x \in X, \ f \in Y^*.$$

The identity operator on X will be denoted by I_X, or simply I if no confusion can arise. Given a bounded operator $T \in L(X, Y)$, the *kernel* of T is the set

$$\ker T := \{x \in X : Tx = 0\},$$

while the *range* of T is denoted by $T(X)$. In the sequel, for every bounded operator $T \in L(X, Y)$, we shall denote by $\alpha(T)$ the *nullity* of T, is defined as $\alpha(T) := \dim \ker T$, while the *deficiency* $\beta(T)$ of T is the dimension of the *cokernel* of $T(X)$, i.e., $\beta(T) := \dim Y / T(X) = \operatorname{codim} T(X)$. The *spectrum* of $T \in L(X)$ defined as

$$\sigma(T) := \{\lambda \in \mathbb{C} : \lambda I - T \text{ is not bijective}\}.$$

It is well known that the spectrum is a compact subset of $\mathbb{C}$ and $\sigma(T) = \sigma(T^*)$ for all $T \in L(X)$. If X is a complex Banach space then every $T \in L(X)$ has non-empty spectrum. The complement $\rho(T) := \mathbb{C} \setminus \sigma(T)$ is called the *resolvent* of T.

In this book by $\mathbb{D}(\lambda_0, \varepsilon)$ we always denote the open disc in $\mathbb{C}$ centered at λ_0 with radius ε. The closed unit disc will be denoted by $\mathbf{D}(\lambda_0, \varepsilon)$.

The property of $T(X)$ being closed is an important property in operator theory. Indeed, if T is a bounded linear operator defined on a normed space X, then $T(X)$ is closed if and only if T is *normally solvable*, which means that the equation $Tx = y$, $x, y \in X$, possesses a solution exactly if $f(y) = 0$ for all $f \in \ker T^*$, see [179, §29]. If $K : C[a, b] \to C[a, b]$ is the integral compact operator defined on the Banach space $C[a, b]$ of all complex-valued continuous functions on the interval $[a, b]$ by

$$(Kx)(s) := \int_a^b k(s, t)x(t)\mathrm{d}t,$$

where $k(s, t)$ is continuous on the square $[a, b] \times [a, b]$, then $\lambda I - K$ has closed range for every $\lambda \neq 0$, and hence is normally solvable. This fact leads, jointly with the property $\alpha(\lambda I - K) = \beta(\lambda I - K) < \infty$, to the classical *Fredholm alternative criterion* of the Fredholm integral equation (of the second kind).

The property of $T(X)$ being closed may be characterized by means of a suitable number associated with T, defined as follows.

Definition 1.1 If $T \in L(X, Y)$, X, Y Banach spaces, the *reduced minimal modulus* of T is defined to be

$$\gamma(T) := \inf_{x \notin \ker T} \frac{\|Tx\|}{\text{dist}(x, \ker T)}.$$

Formally, we set $\gamma(0) := \infty$. It easily seen that if T is bijective then $\gamma(T) = \frac{1}{\|T^{-1}\|}$. In fact, if T is bijective then

$$\text{dist}(x, \ker T) = \text{dist}(x, \{0\}) = \|x\|,$$

thus if $Tx = y$

$$\gamma(T) = \inf_{x \neq 0} \frac{\|Tx\|}{\|x\|} = \left(\sup_{x \neq 0} \frac{\|x\|}{\|Tx\|} \right)^{-1}$$

$$= \left(\sup_{y \neq 0} \frac{\|T^{-1}y\|}{\|y\|} \right)^{-1} = \frac{1}{\|T^{-1}\|}.$$

Theorem 1.2 *Let $T \in L(X, Y)$, X and Y Banach spaces. Then we have*

(i) $\gamma(T) > 0$ *if and only if $T(X)$ is closed.*
(ii) $\gamma(T) = \gamma(T^*)$.

Proof

(i) The statement is clear if $T = 0$. Suppose that $T \neq 0$. Let $\overline{X} := X/\ker T$ and denote by $\overline{T} : \overline{X} \to Y$ the continuous injection corresponding to T, defined by

$$\overline{T}\,\overline{x} := Tx \quad \text{for every } x \in \overline{x}.$$

It is easily seen that $\overline{T}(\overline{X}) = T(X)$. But, by a classical result of functional analysis, it is known that $\overline{T}(\overline{X})$ is closed if and only if $\overline{T}$ admits a continuous inverse, i.e., there exists a constant $\delta > 0$ such that $\|\overline{T}\,\overline{x}\| \geq \delta\|\overline{x}\|$, for every $x \in X$. From the equality

$$\gamma(T) = \inf_{\overline{x} \neq 0} \frac{\|\overline{T}\,\overline{x}\|}{\|\overline{x}\|}$$

we then conclude that $\overline{T}(\overline{X}) = T(X)$ is closed if and only if $\gamma(T) > 0$.

(ii) The assertion is obvious if $\gamma(T) = 0$. Suppose that $\gamma(T) > 0$. Then $T(X)$ is closed. If $\overline{T}_0 : \overline{X} \to T(X)$ is defined by $\overline{T}_0\overline{x} := Tx$ for every $x \in \overline{x}$, then $\gamma(T) = \gamma(\overline{T}_0)$ and $T = J\overline{T}_0 Q$, where $J : T(X) \to Y$ denotes the natural embedding and $Q : X \to \overline{X}$ is the canonical projection defined by $Qx = \overline{x}$. Clearly, $\overline{T}_0$ is bijective, and from $T = J\overline{T}_0 Q$ it then follows that $T^* = Q^*(\overline{T}_0)^* J^*$. From this we easily obtain that

$$\gamma(T) = \frac{1}{\|(\overline{T}_0)^{-1}\|} = \frac{1}{\|(\overline{T}_0^{\,*})^{-1}\|} = \gamma(T^*).$$

∎

Corollary 1.3 *Let* $T \in L(X, Y)$, X *and* Y *Banach spaces. Then* $T(X)$ *is closed if and only if* $T^*(X^*)$ *is closed.*

In the sequel we give some sufficient conditions that ensure that the range $T(X)$ is closed.

Lemma 1.4 *Let* $T \in L(X)$ *and suppose that* $T(X)$ *is closed and that* Y *is a closed subspace of* X *such that* $\ker T \subseteq Y$. *Then* $T(Y)$ *is closed.*

Proof Consider the induced operator $\tilde{T} : X/\ker T \to T(X)$. By the open mapping theorem $\tilde{T}$ is an isomorphism, in particular an open mapping. Since $\ker T \subseteq Y$, the space $Y/\ker T$ is closed in $X/\ker T$ and this easily implies that the image $\tilde{T}(Y/\ker T) = T(Y)$ is closed in X. ∎

Lemma 1.5 *Let* $T \in L(X)$ *and suppose that* $T(X)$ *is closed. If* Y *is a (not necessarily closed) subspace of* X *for which* $Y + \ker T$ *is closed, then* $T(Y)$ *is closed.*

Proof We have $T(Y) = T(Y + \ker T)$, so the statement follows from Lemma 1.4. ∎

Theorem 1.6 *Let* $T \in L(X)$, X *a Banach space, and suppose that there exists a closed subspace* Y *of* X *such that* $T(X) + Y$ *is closed and* $T(X) \cap Y = \{0\}$. *Then the subspace* $T(X)$ *is also closed.*

Proof Consider the product space $X \times Y$ under the norm

$$\|(x, y)\| := \|x\| + \|y\| \quad x \in X, y \in Y.$$

Then $X \times Y$ is a Banach space and the continuous map $S : X \times Y \to X$ defined by $S(x, y) := Tx + y$ has range $S(X \times Y) = T(X) \oplus Y$, which is closed by assumption. Hence

$$\gamma(S) := \inf_{(x,y) \notin \ker S} \frac{\|S(x, y)\|}{\mathrm{dist}((x, y), \ker S)} > 0.$$

Moreover, ker $S = $ ker $T \times \{0\}$, so

$$\text{dist}((x, 0), \text{ker } S) = \text{dist}(x, \text{ker } T),$$

and hence

$$\|Tx\| = \|S(x, 0)\| \geq \gamma(S) \, \text{dist}((x, 0), \text{ker } S))$$
$$= \gamma(S) \, \text{dist}(x, \text{ker } T).$$

From this it follows that $\gamma(T) \geq \gamma(S) > 0$, and this implies that T has closed range. ∎

It is obvious that the sum $M + N$ of two linear subspaces M and N of a vector space X is again a linear subspace. If $M \cap N = \{0\}$ then this sum is called the *direct sum* of M and N and will be denoted by $M \oplus N$. In this case for every $z = x + y$ in $M + N$ the components x, y are uniquely determined. If $X = M \oplus N$ then N is called an *algebraic complement* of M. In this case the (Hamel) basis of X is the union of the basis of M with the basis of N. It is obvious that every subspace of a vector space admits at least one algebraic complement. The *codimension* of a subspace M of X is the dimension of every algebraic complement N of M, or equivalently the dimension of the quotient X/M. Note that codim $M = \dim M^{\perp}$, where

$$M^{\perp} := \{f \in X^* : f(x) = 0 \text{ for every } x \in M\}$$

denotes the *annihilator* of M. Indeed, by the Annihilator theorem (see Appendix A) we have:

$$\text{codim } M = \dim X/M = \dim (X/M)^* = \dim M^{\perp}.$$

Theorem 1.6 then directly yields the following important result:

Corollary 1.7 *Let $T \in L(X)$, X a Banach space, and Y a finite-dimensional subspace of X such that $T(X) + Y$ is closed. Then $T(X)$ is closed. In particular, if $T(X)$ has finite codimension then $T(X)$ is closed.*

Proof Let Y_1 be any subspace of Y for which $Y_1 \cap T(X) = \{0\}$ and $T(X) + Y_1 = T(X) + Y$. From the assumption we infer that $T(X) \oplus Y_1$ is closed, so $T(X)$ is closed by Theorem 1.6. The second statement is clear, since every finite-dimensional subspace of a Banach space X is always closed. ∎

A very important class of operators is given by the class of injective operators having closed range.

Definition 1.8 An operator $T \in L(X)$ is said to be *bounded below* if T is injective and has closed range.

Lemma 1.9 $T \in L(X, Y)$ *is bounded below if and only if there exists a $\delta > 0$ such that*

$$\|Tx\| \geq \delta \|x\| \quad \text{for all } x \in X. \tag{1.1}$$

Proof Indeed, if $\|Tx\| \geq \delta\|x\|$ for some $\delta > 0$ and all $x \in X$ then T is injective. Moreover, if (x_n) is a sequence in X for which (Tx_n) converges to $y \in X$, then (x_n) is a Cauchy sequence and hence, by completeness, it converges to some $x \in X$. Since T is continuous then $Tx = y$ and therefore $T(X)$ is closed.

Conversely, if T is injective and $T(X)$ is closed then, from the open mapping theorem, it easily follows that there exists a $\delta > 0$ for which the inequality (1.1) holds. ∎

The quantity

$$j(T) := \inf_{\|x\|=1} \|Tx\| = \inf_{x \neq 0} \frac{\|Tx\|}{\|x\|}$$

is called the *injectivity modulus* of T and, obviously, from (1.1) we have

$$T \text{ is bounded below} \Leftrightarrow j(T) > 0, \tag{1.2}$$

and in this case $j(T) = \gamma(T)$. The next result shows that the properties of being bounded below or being surjective are dual each other.

Theorem 1.10 *Let $T \in L(X)$, X a Banach space. Then:*

(i) *T is surjective (respectively, bounded below) if and only if T^* is bounded below (respectively, surjective);*
(ii) *If T is bounded below (respectively, surjective) then $\lambda I - T$ is bounded below (respectively, surjective) for all $|\lambda| < \gamma(T)$.*

Proof

(i) Suppose that T is surjective. Trivially T has closed range and therefore T^* also has closed range. From the equality $\ker T^* = T(X)^\perp = X^\perp = \{0\}$ we conclude that T^* is injective.

Conversely, suppose that T^* is bounded below. Then T^* has closed range and hence, by Theorem 1.2, the operator T also has closed range. From the equality $T(X) =^\perp \ker T^* =^\perp \{0\} = X$ we then conclude that T is surjective.

The proof of T being bounded below if and only if T^* is surjective is analogous.
(ii) Suppose that T is injective with closed range. Then $\gamma(T) > 0$ and from the definition of $\gamma(T)$ we obtain

$$\gamma(T) \cdot \text{dist}(x, \ker T) = \gamma(T)\|x\| \leq \|Tx\| \quad \text{for all } x \in X.$$

From that we obtain

$$\|(\lambda I - T)x\| \geq \|Tx\| - |\lambda|\|x\| \geq (\gamma(T) - |\lambda|)\|x\|,$$

and hence, by Lemma 1.9, the operator $\lambda I - T$ is bounded below for all $|\lambda| < \gamma(T)$.

The case of surjective T now follows easily by considering the adjoint T^*. $\blacksquare$

Two important parts of the spectrum $\sigma(T)$ are defined as follows: The *approximate point spectrum* of $T \in L(X)$ defined as

$$\sigma_{ap}(T) := \{\lambda \in \mathbb{C} : \lambda I - T \text{ is not bounded below}\}$$

and the *surjectivity spectrum* of T defined as

$$\sigma_s(T) := \{\lambda \in \mathbb{C} : \lambda I - T \text{ is onto}\}.$$

By Theorem 1.10 we have

$$\sigma_{ap}(T) = \sigma_s(T^*) \quad \text{and} \quad \sigma_{ap}(T^*) = \sigma_s(T).$$

Remark 1.11 In the case of Hilbert space operators $T \in L(H)$ instead of the dual T^* it is more appropriate to consider the Hilbert adjoint T'. By means of the classical *Fréchet–Riesz representation theorem* if U is the conjugate-linear isometry that associates to each $y \in H$ the linear form $x \to (x, y)$ we have that

$$\bar{\lambda} I - T' = (\lambda I - T)' = U^{-1}(\lambda I - T)^* U, \tag{1.3}$$

where $\bar{\lambda}$ is the conjugate of $\lambda \in \mathbb{C}$. The equality (1.3) then easily implies that

$$\sigma_{ap}(T') = \overline{\sigma_{ap}(T^*)} = \overline{\sigma_s(T)}$$

and

$$\sigma_s(T') = \overline{\sigma_s(T^*)} = \overline{\sigma_{ap}(T)}.$$

Theorem 1.12 *Let $T \in L(X)$. Then we have:*

(i) $\lambda \in \sigma_{ap}(T)$ *if and only if there exists a sequence of unit vectors (x_n) such that $(\lambda I - T)x_n \to 0$, as $n \to \infty$.*

(ii) *Both spectra $\sigma_{ap}(T)$ and $\sigma_s(T)$ are closed subsets of $\sigma(T)$ that contain the boundary $\partial\sigma(T)$. Moreover, if $X \neq \{0\}$ then $\sigma_{ap}(T)$ and $\sigma_s(T)$ are non-empty.*

Proof

(i) By Lemma 1.9 if $\lambda \in \sigma_{ap}(T)$ then there exists, for each $n \in \mathbb{N}$, a unit vector x_n such that $\|(\lambda I - T)x_n\| < 1/n$, so $(\lambda I - T)x_n \to 0$ as $n \to \infty$. Conversely, the existence of a sequence of unit vectors from which $(\lambda I - T)x_n \to 0$ implies, again by Lemma 1.9, that $\lambda I - T$ is not bounded below.

(ii) We show first that $\sigma_{ap}(T)$ is closed. Let (λ_n) be a sequence in $\sigma_{ap}(T)$ which converges to $\lambda \in \mathbb{C}$, and choose a sequence (x_n) of unit vectors such that $\|(\lambda_n I - T)x_n\| \leq 1/n$ for all $n \in \mathbb{N}$. Then

$$\|(\lambda I - T)x_n\| \leq |\lambda - \lambda_n| + \frac{1}{n}$$

for all $n \in \mathbb{N}$, from which we obtain that $\lambda \in \sigma_{ap}(T)$. Therefore $\sigma_{ap}(T)$ is closed for every operator, and by duality $\sigma_s(T) = \sigma_{ap}(T^*)$ is also closed.

To show the inclusion $\partial \sigma(T) \subseteq \sigma_{ap}(T)$, let $\lambda \in \partial \sigma(T)$ and consider a sequence (λ_n) of points of $\rho(T)$ which converges to λ. The well-known estimate

$$\|(\mu I - T)^{-1}\| \operatorname{dist}[\mu, \sigma(T)] \geq 1 \quad \text{for all } \mu \in \rho(T)$$

then implies that $\|(\lambda_n I - T)^{-1}\| \to \infty$ as $n \to \infty$. For each $n \in \mathbb{N}$, let y_n be a unit vector such that

$$\delta_n := \|(\lambda_n I - T)^{-1} y_n\| \geq \|(\lambda_n I - T)^{-1}\| - \frac{1}{n}.$$

Put

$$x_n := (1/\delta_n)(\lambda_n I - T)^{-1} y_n.$$

Then $\|x_n\| = 1$ and $\|(\lambda I - T)x_n\| \to 0$, so $\lambda \in \sigma_{ap}(T)$. The inclusion $\partial \sigma(T) \subseteq \sigma_s(T)$ easily follows by duality and the last statement is an obvious consequence of the fact that $\partial \sigma(T)$ is non-empty, since the spectrum is a non-empty closed subset of $\mathbb{C}$.   ∎

1.2 Ascent and Descent

The kernels and the ranges of the iterates T^n, $n \in \mathbb{N}$, of a linear operator T defined on a vector space X form two increasing and decreasing chains, respectively, i.e., the chain of kernels

$$\ker T^0 = \{0\} \subseteq \ker T \subseteq \ker T^2 \subseteq \cdots$$

and the chain of ranges

$$T^0(X) = X \supseteq T(X) \supseteq T^2(X) \cdots.$$

The subspace

$$N^\infty(T) := \bigcup_{n=1}^{\infty} \ker T^n$$

is called the *hyper-kernel* of T, while

$$T^\infty(X) := \bigcap_{n=1}^{\infty} T^n(X)$$

is called the *hyper-range* of T. Note that both $N^\infty(T)$ and $T^\infty(X)$ are T-invariant linear subspaces of T, i.e.

$$T(N^\infty(T)) \subseteq N^\infty(T) \quad \text{and} \quad T(T^\infty(X)) \subseteq T^\infty(X).$$

The following elementary lemma will be useful in the sequel.

Lemma 1.13 *Let X be a vector space and T a linear operator on X. If p_1 and p_2 are relatively prime polynomials then there exist polynomials q_1 and q_2 such that*

$$p_1(T)q_1(T) + p_2(T)q_2(T) = I.$$

Proof If p_1 and q_1 are relatively prime polynomials then there are polynomials such that $p_1(\mu)q_1(\mu) + p_2(\mu)q_2(\mu) = 1$ for every $\mu \in \mathbb{C}$.   ∎

The next result establishes some basic properties of the hyper-kernel and the hyper-range of an operator.

Theorem 1.14 *Let X be a vector space and T a linear operator on X. Then we have:*

(i) $(\lambda I + T)(N^\infty(T)) = N^\infty(T)$ *for every* $\lambda \neq 0$;
(ii) $N^\infty(\lambda I + T) \subseteq (\mu I + T)^\infty(X)$ *for every* $\lambda \neq \mu$.

Proof

(i) The equality will be proved if we show that $(\lambda I + T)(\ker T^n) = \ker T^n$ holds for every $n \in \mathbb{N}$ and $\lambda \neq 0$. Clearly, the inclusion $(\lambda I + T)(\ker T^n) \subseteq \ker T^n$ holds for all $n \in \mathbb{N}$. By Lemma 1.13 we know that there exist two polynomials p and q such that

$$(\lambda I + T)p(T) + q(T)T^n = I.$$

If $x \in \ker T^n$ then $(\lambda I + T)p(T)x = x$ and since $p(T)x \in \ker T^n$ we then obtain $\ker T^n \subseteq (\lambda I + T)(\ker T^n)$, so the equality (i) is proved.

(ii) Put $S := \lambda I + T$ and write

$$\mu I + T = (\mu - \lambda)I + \lambda I + T = (\mu - \lambda)I + S.$$

By assumption $\mu - \lambda \neq 0$, so, by part (i), we have

$$(\mu I + T)(\mathcal{N}^\infty(\lambda I + T)) = ((\mu - \lambda)I + S)(\mathcal{N}^\infty(S)) = \mathcal{N}^\infty(\lambda I + T).$$

From this it easily follows that

$$(\mu I + T)^n(\mathcal{N}^\infty(\lambda I + T)) = \mathcal{N}^\infty(\lambda I + T) \quad \text{for all } n \in \mathbb{N},$$

so $\mathcal{N}^\infty(\lambda I + T) \subseteq (\mu I + T)^n(X)$ for all $n \in \mathbb{N}$, from which we conclude that $\mathcal{N}^\infty(\lambda I + T) \subseteq (\mu I + T)^\infty(X)$. ∎

Lemma 1.15 *For every linear operator T on a vector space X we have*

$$T^m(\ker T^{m+n}) = T^m(X) \cap \ker T^n \quad \text{for all } m, n \in \mathbb{N}.$$

Proof If $x \in \ker T^{m+n}$ then $T^m x \in T^m(X)$ and $T^n(T^m x) = 0$, so that $T^m(\ker T^{m+n}) \subseteq T^m(X) \cap \ker T^n$.

Conversely, if $y \in T^m(X) \cap \ker T^n$ then $y = T^m(x)$ and $x \in \ker T^{m+n}$, so the opposite inclusion is verified. ∎

In the next result we give some useful connections between the kernels and the ranges of the iterates T^n of an operator T on a vector space X.

Theorem 1.16 *For a linear operator T on a vector space X the following statements are equivalent:*

(i) $\ker T \subseteq T^m(X)$ *for each $m \in \mathbb{N}$;*
(ii) $\ker T^n \subseteq T(X)$ *for each $n \in \mathbb{N}$;*
(iii) $\ker T^n \subset T^m(X)$ *for each $n \in \mathbb{N}$ and each $m \in \mathbb{N}$;*
(iv) $\ker T^n = T^m(\ker T^{m+n})$ *for each $n \in \mathbb{N}$ and each $m \in \mathbb{N}$.*

Proof The implications (iv) $\Rightarrow$ (iii) $\Rightarrow$ (ii) are trivial.

(ii) $\Rightarrow$ (i) If we apply the inclusion (ii) to the operator T^m we then obtain $\ker T^{mn} \subseteq T^m(X)$ and consequently $\ker T \subseteq T^m(X)$, since $\ker T \subseteq \ker T^{mn}$.

(i) $\Rightarrow$ (iv) If we apply the inclusion (i) to the operator T^n we obtain

$$\ker T^n \subseteq (T^n)^m(X) \subseteq T^m(X).$$

By Lemma 1.15 we then have

$$T^m(\ker T^{m+n}) = T^m(X) \cap \ker T^n = \ker T^n,$$

so the proof is complete. ∎

Corollary 1.17 *Let T be a linear operator on a vector space X. Then the statements of Theorem 1.16 are equivalent to each of the following inclusions:*

(i) $\ker T \subseteq T^\infty(X)$;
(ii) $\mathcal{N}^\infty(T) \subseteq T(X)$;
(iii) $\mathcal{N}^\infty(T) \subseteq T^\infty(X)$.

We now introduce two important notions in operator theory.

Definition 1.18 A linear operator T on a vector space X is said to have *finite ascent* if $\mathcal{N}^\infty(T) = \ker T^k$ for some positive integer k. Clearly, in such a case there is a smallest positive integer $p := p(T)$ such that $\ker T^p = \ker T^{p+1}$. The positive integer p is called the *ascent* of T. If there is no such integer we set $p(T) := \infty$. Analogously, T is said to have *finite descent* if $T^\infty(X) = T^k(X)$ for some positive integer k. In such a case there is a smallest positive integer $q = q(T)$ such that $T^{q+1}(X) = T^q(X)$. The positive integer q is called the *descent* of T. If there is no such integer we set $q(T) := \infty$.

Clearly $p(T) = 0$ if and only if T is injective and $q(T) = 0$ if and only if T is surjective. The following lemma establishes useful and simple characterizations of operators having finite ascent and finite descent.

Lemma 1.19 *Let T be a linear operator on a vector space X. For a positive natural m, the following assertions hold:*

(i) *$p(T) \leq m < \infty$ if and only if for every $n \in \mathbb{N}$ we have $T^m(X) \cap \ker T^n = \{0\}$;*
(ii) *$q(T) \leq m < \infty$ if and only if for every $n \in \mathbb{N}$ there exists a subspace $Y_n \subseteq \ker T^m$ such that $X = Y_n \oplus T^n(X)$.*

Proof

(i) Suppose $p(T) \leq m < \infty$ and n is any natural number. Consider an element $y \in T^m(X) \cap \ker T^n$. Then there exists an $x \in X$ such that $y = T^m x$ and $T^n y = 0$. From that we obtain $T^{m+n} x = T^n y = 0$ and therefore $x \in \ker T^{m+n} = \ker T^m$. Hence $y = T^m x = 0$.

Conversely, suppose $T^m(X) \cap \ker T^n = \{0\}$ for some natural m and let $x \in \ker T^{m+1}$. Then $T^m x \in \ker T$ and therefore

$$T^m x \in T^m(X) \cap \ker T \subseteq T^m(X) \cap \ker T^n = \{0\}.$$

Hence $x \in \ker T^m$. We have shown that $\ker T^{m+1} \subseteq \ker T^m$. Since the opposite inclusion is satisfied for all operators, we then conclude that $\ker T^m = \ker T^{m+1}$.

(ii) Let $q := q(T) \leq m < \infty$ and Y be a complementary subspace to $T^n(X)$ in X. Let $\{x_j : j \in J\}$ be a basis of Y. Since

$$T^q(Y) \subseteq T^q(X) = T^{q+n}(X),$$

for every element x_j of the basis there exists an element $y_j \in X$ such that $T^q x_j = T^{q+n} y_j$. Set $z_j := x_j - T^n y_j$. Then

$$T^q z_j = T^q x_j - T^{q+n} y_j = 0.$$

Therefore, the linear subspace Y_n generated by the elements z_j is contained in $\ker T^q$ and hence is contained in $\ker T^m$. From the decomposition $X = Y \oplus T^n(X)$ we obtain for every $x \in X$ a representation of the form

$$x = \sum_{j \in J} \lambda_j x_j + T^n y = \sum_{j \in J} \lambda_j (z_j + T^n y_j) + T^n y = \sum_{j \in J} \lambda_j z_j + T^n z,$$

thus $X = Y_n + T^n(X)$. We show that this sum is direct. Indeed, suppose that $x \in Y_n \cap T^n(X)$. Then

$$x = \sum_{j \in J} \mu_j z_j = T^n v,$$

for some $v \in X$, and

$$\sum_{j \in J} \mu_j x_j = \sum_{j \in J} \mu_j T^n y_j + T^n v \in T^n(X).$$

From the decomposition $X = Y \oplus T^n(X)$ we then obtain that $\mu_j = 0$ for all $j \in J$ and hence $x = 0$. Therefore Y_n is a complement of $T^n(X)$ contained in $\ker T^m$. Conversely, if for $n \in \mathbb{N}$ the subspace $T^n(X)$ has a complement $Y_n \subseteq \ker T^m$ then

$$T^m(X) = T^m(Y_n) + T^{m+n}(X) = T^{m+n}(X),$$

and hence $q(T) \leq m$. ∎

Theorem 1.20 *If both $p(T)$ and $q(T)$ are finite then $p(T) = q(T)$.*

Proof Set $p := p(T)$ and $q := q(T)$. Assume first that $p \leq q$, so that the inclusion $T^q(X) \subseteq T^p(X)$ holds. Obviously we may assume $q > 0$. From part (ii) of Lemma 1.19 we have $X = \ker T^q + T^q(X)$, so every element $y := T^p(x) \in T^p(X)$ admits the decomposition $y = z + T^q w$, with $z \in \ker T^q$. From $z = T^p x - T^q w \in T^p(X)$ we then obtain that $z \in \ker T^q \cap T^p(X)$ and the last intersection is $\{0\}$, by part (i) of Lemma 1.19. Therefore $y = T^q w \in T^q(X)$ and this shows the equality $T^p(X) = T^q(X)$, from which we obtain $p \geq q$, and hence $p = q$.

Assume now that $q \leq p$ and $p > 0$, so that $\ker T^q \subseteq \ker T^p$. From part (ii) of Lemma 1.19 we have $X = \ker T^q + T^p(X)$, so an arbitrary element x of $\ker T^p$ admits the representation $x = u + T^p v$, with $u \in \ker T^q$. From $T^p x = T^p u = 0$ it then follows that $T^{2p} v = 0$, so $v \in \ker T^{2p} = \ker T^p$. Therefore, $T^p v = 0$ and, consequently, $x = u \in \ker T^q$. This shows that $\ker T^q = \ker T^p$, hence $q \geq p$ from which we conclude that $p = q$. ∎

Lemma 1.21 *Let T be a linear operator on a linear vector space X. If $\alpha(T) < \infty$ then $\alpha(T^n) < \infty$ for all $n \in \mathbb{N}$. Analogously, if $\beta(T) < \infty$ then $\beta(T^n) < \infty$ for all $n \in \mathbb{N}$.*

Proof We use an inductive argument. Suppose that $\dim \ker T^n < \infty$. Since $T(\ker T^{n+1}) \subseteq \ker T^n$ then the restriction

$$T_0 := T | \ker T^{n+1} : \ker T^{n+1} \to \ker T^n$$

has kernel equal to $\ker T$, so the canonical mapping $\hat{T} : \ker T^{n+1} / \ker T \to \ker T^n / \ker T$ is injective. Therefore we have

$$\dim \ker T^{n+1} / \ker T \leq \dim \ker T^n / \ker T \leq \dim \ker T^n < \infty$$

and since $\dim \ker T < \infty$ we then conclude that $\dim \ker T^{n+1} < \infty$.

Suppose now that $\beta(T^n) < \infty$. Since the map

$$\tilde{T} : T^n(X)/T^{n+1}(X) \to T^{n+1}(T^n(X)/T^{n+2}(X),$$

defined by

$$\tilde{T}(z + T^{n+1}(X)) = Tz + T^{n+2}(X), \quad z \in T^n(X),$$

is onto, $\dim T^{n+1}(X)/T^{n+2}(X) \leq \dim T^n(X)/T^{n+1}(X)$. This easily implies that $\beta(T^{n+1}) < \infty$. ∎

In the next theorem we establish the basic relationships between the quantities $\alpha(T)$, $\beta(T)$, $p(T)$ and $q(T)$.

Theorem 1.22 *If T is a linear operator on a vector space X then the following properties hold:*

(i) *If $p(T) < \infty$ then $\alpha(T) \leq \beta(T)$.*
(ii) *If $q(T) < \infty$ then $\beta(T) \leq \alpha(T)$.*
(iii) *If $p(T) = q(T) < \infty$ then $\alpha(T) = \beta(T)$ (possibly infinite).*
(iv) *If $\alpha(T) = \beta(T) < \infty$ and if either $p(T)$ or $q(T)$ is finite then $p(T) = q(T)$.*

Proof

(i) Let $p := p(T) < \infty$, i.e., $\ker T^p = \ker T^{p+n}$ for all $n \in N$. Obviously if $\beta(T) = \infty$ there is nothing to prove. Assume that $\beta(T) < \infty$. By Lemma 1.19, part (i), we have $\ker T \cap T^p(X) = \{0\}$. Since $\beta(T^n) < \infty$, by Lemma 1.21, this implies that $\alpha(T) < \infty$, so T has finite deficiency. According to the index theorem (see Appendix A) we then obtain for all $n \geq p$ the following equality:

$$n \cdot \text{ind } T = \text{ind } T^n = \alpha(T^n) - (\beta T^n) = \alpha(T^p) - \beta(T^n).$$

Now suppose that $q := q(T) < \infty$. For all integers $n \geq \max\{p, q\}$ the quantity $n \cdot \text{ind } T = \alpha(T^p) - \beta(T^p)$ is then constant, so that ind $T = 0$, i.e., $\alpha(T) = \beta(T)$. Consider the other case $q = \infty$. Then $\beta(T^n) \to 0$ as $n \to \infty$, so $n \cdot \text{ind } T$ becomes eventually negative, and hence ind $T < 0$. Therefore, in this case we have $\alpha(T) < \beta(T)$.

(ii) Let $q := q(T) < \infty$. Also here we can assume that $\alpha(T) < \infty$, otherwise there is nothing to prove. Consequently, as is easy to check, also $\beta(T^n) < \infty$ and by part (ii) of Lemma 1.19 $X = Y \oplus T(X)$ with $Y \subseteq \ker T^q$. From this it follows that

$$\beta(T) = \dim Y \leq \alpha(T^q) < \infty.$$

If we use, with appropriate changes, the index argument used in the proof of part (i) then we obtain that $\beta(T) = \alpha(T)$ if $p(T) < \infty$, and $\beta(T) < \alpha(T)$ if $p(T) = \infty$.

(iii) This is clear from part (i) and part (ii).

(iv) This is an immediate consequence of the equality $\alpha(T^n) - \beta(T^n) = \text{ind } T^n = n \cdot \text{ind } T = 0$, valid for every $n \in \mathbb{N}$. ∎

Given $n \in \mathbb{N}$, we denote by $T_n : T^n(X) \to T^n(X)$ the restriction of $T \in L(X)$ on the subspace $R(T^n) := T^n(X)$. Observe that

$$R(T_n{}^m) = R(T^{m+n}) = R(T_m^n)$$

for all $m, n \in \mathbb{N}$.

Lemma 1.23 *Let T be a linear operator on a vector space X. Then the following statements are equivalent:*

(i) *$p(T) < \infty$;*

(ii) *there exists a $k \in \mathbb{N}$ such that T_k is injective;*

(iii) *there exists a $k \in \mathbb{N}$ such that $p(T_k) < \infty$.*

Proof (i) $\Leftrightarrow$ (ii) If $p := p(T) < \infty$, then by Lemma 1.19 $\ker T_p = \ker T \cap T^p(X) = \{0\}$. Conversely, suppose that $\ker T_k = \{0\}$, for some $k \in \mathbb{N}$. If $x \in \ker T^{k+1}$ then $T(T^k x) = 0$, so

$$T^k x \in \ker T \cap T^k(X) = \ker T_k = \{0\}.$$

Hence $x \in \ker T^k$. This shows that $\ker T^{k+1} \subseteq \ker T^k$. The opposite inclusion is true for every operator, thus $\ker T^{k+1} = \ker T^k$ and consequently $p(T) \leq k$.

(ii) $\Leftrightarrow$ (iii) The implication (ii) $\Rightarrow$ (iii) is obvious. To show the opposite implication, suppose that $v := p(T_k) < \infty$. By Lemma 1.19 we then have:

$$\{0\} = \ker T_k \cap R(T_k^{\,v}) = (\ker T \cap R(T^k)) \cap R(T_k^{\,v}) = (\ker T) \cap R(T_k^{\,v})$$
$$= \ker T \cap R(T^{v+k}) = \ker T_{v+k},$$

so that the equivalence (ii) $\Leftrightarrow$ (iii) is proved.   ∎

A dual result holds for the descent:

Lemma 1.24 *Let T be a linear operator on a vector space X. Then the following statements are equivalent:*

(i) *$q(T) < \infty$;*
(ii) *there exists a $k \in \mathbb{N}$ such that T_k is onto;*
(iii) *there exists a $k \in \mathbb{N}$ such that $q(T_k) < \infty$.*

Proof (i) $\Leftrightarrow$ (ii) Suppose that $q := q(T) < \infty$. Then

$$T^q(X) = T^{q+1}(X) = T(T^q(X)) = R(T_q),$$

hence T_q is onto. Conversely, if T_k is onto for some $k \in \mathbb{N}$ then

$$T^{k+1}(X) = T(T^k(X)) = R(T_k) = T^k(X),$$

thus $q(T) \le k$.

The implication (ii) $\Rightarrow$ (iii) is obvious. We show (iii) $\Rightarrow$ (i). Suppose that $v := q(T_k) < \infty$ for some $k \in \mathbb{N}$. Then $T_k^{\,v}(X) = T_k^{\,v+1}(X)$, so $T^{k+v}(X) = T^{v+k+1}(X)$, hence $q(T) \le k + v$.   ∎

Remark 1.25 As observed in the proof of Lemma 1.23, if $p := p(T) < \infty$ then $\ker T_p = \{0\}$ and hence $\ker T_j = \{0\}$ for all $j \ge p$. Conversely, if $\ker T_k = \{0\}$ for some $k \in \mathbb{N}$ then $p(T) < \infty$ and $p(T) \le k$. Hence, if $p(T) < \infty$ we have

$$p(T) = \inf\{k \in \mathbb{N} : T_k \text{ is injective}\}.$$

Analogously, if $q := q(T) < \infty$ then T_j is onto for all $j \ge q$. Conversely, if T_k is onto for some $k \in \mathbb{N}$ then $q(T) \le k$, so that

$$q(T) = \inf\{k \in \mathbb{N} : T_k \text{ is onto}\}.$$

We shall often use the following basic results:

Lemma 1.26 *Suppose that $T \in L(X)$ has closed range. Then $p(T) = q(T^*)$ and $q(T) = p(T^*)$.*

Proof If T has closed range then T^n has closed range for all $n \in \mathbb{N}$ or equivalently, by Corollary 1.3, $(T^*)^n$ has closed range. From the closed range theorem (see

Appendix A) it then follows that $\ker T^n =^{\perp} [(T^*)^n(X^*)]$ and $[(T^*)^n(X^*)] = \ker T^n$ for all $n \in \mathbb{N}$. These equalities easily imply that $p(T) = q(T^*)$ and $q(T) = p(T^*)$ ∎

Remark 1.27 It is easily seen from equality (1.3) that T^* has closed range if and only if the Hilbert adjoint T' has closed range, and $p(\lambda I - T^*) = p(\bar{\lambda} I - T')$ and $q(\lambda I - T^*) = q(\bar{\lambda} I - T')$.

Lemma 1.28 *Let T be a linear operator on a vector space X. Suppose that $X = M \oplus N$, M and N closed T-invariant subspaces. Then*

(i) $\alpha(T) = \alpha(T|M) + \alpha(T|N)$ and $\beta(T) = \beta(T|M) + \beta(T|N)$.
(ii) $p(T) = p(T|M) + p(T|N)$ and $q(T) = q(T|M) + q(T|N)$.
(iii) *If $T \in L(X)$, X a Banach space, then $T(X)$ is closed if and only if $T(M)$ is closed in M and $T(N)$ is closed in N.*
(iv) $\sigma_{\mathrm{ap}}(T) = \sigma_{\mathrm{ap}}(T|M) \cup \sigma_{\mathrm{ap}}(T|N)$ and $\sigma_{\mathrm{s}}(T) = \sigma_{\mathrm{s}}(T|M) \cup \sigma_{\mathrm{s}}(T|N)$.

Proof The equalities in (i) and (ii) follow immediately from the equality $\ker T = \ker T|M \oplus \ker T|N$ and $T(X) = T(M) \oplus T(N)$. To show (iii) denote by P the projection of X onto M along N. Clearly, $PT = TP$. If $T(X)$ is closed then $T(M) = T(P(X)) = P(T(X)) = T(X) \cap M$, so $T(M)$ is closed, and analogously $T(N)$ is closed. Conversely, suppose that $T(M)$ is closed in M and $T(N)$ is closed in N. The mapping $\Psi : M \times N \to M \oplus N$, defined by $\Psi(x, y) := x + y$ is a topological isomorphism, so the image $\Psi(T(M) \times T(N)) = T(M) \oplus T(N) = T(X)$ is closed in X.

The equalities (iv) are obvious consequences of (i) and (iii).   ∎

Recall that a bounded operator $T \in L(X)$ is said to be *relatively open* if it is open as mapping from X onto its range $T(X)$. An application of the open mapping theorem shows that T is relatively open if and only if $T(X)$ is closed, see [179, Theorem 3.21].

Lemma 1.29 *Suppose that $T, S \in L(X)$, X a Banach space, satisfy $T(X) \cap S(X) = \{0\}$ and the sum $T(X) + S(X)$ is closed. Then $T(X)$ and $S(X)$ are closed.*

Proof Define $U : X \times X \to X$ by means of $U(x, y) := Tx + Sy$ for all $(x, y) \in X \times Y$. Since U has closed range, from the open mapping theorem we have that U is relatively open. Because $T(X) \cap S(X) = \{0\}$, it easily follows that both T and S are open operators. This is equivalent to saying that T and S have closed ranges. ∎

Theorem 1.30 *If $T \in L(X)$ is an operator for which $T(X) \cap \ker T = \{0\}$ and $T(X) + \ker T$ is closed in X, then $T(X)$ is closed. In particular, if $p(T) \le 1$ and $T(X) + \ker T$ is closed in X, then $T(X)$ is closed.*

Proof The first assertion immediately follows from Lemma 1.29 applied to the operator T and the natural inclusion mapping from $\ker T$ into X. The second assertion is clear, since, by Lemma 1.19, $p(T) \le 1$ entails that $\ker T \cap T(X) = \{0\}$.
                                                                                ∎

1.3 The Algebraic and Analytic Core

In this section we shall introduce some important T-invariant subspaces, the algebraic core and its analytic counterpart, the analytic core. The first was introduced by Saphar [270].

Definition 1.31 Given a linear operator T defined on a vector space X, the *algebraic core* $C(T)$ of T is defined to be the largest linear subspace M such that $T(M) = M$.

It is easy to prove that $C(T)$ is the set of all $x \in X$ such that there exists a sequence $(x_n)_{n=0,1,\ldots}$ such that $x_0 = x$, $Tx_{n+1} = x_n$ for all $n = 0, 1, 2, \ldots$.

Trivially, if $T \in L(X)$ is surjective then $C(T) = X$. Clearly, for every linear operator T we have $C(T) = T^n(C(T)) \subseteq T^n(X)$ for all $n \in \mathbb{N}$. From that it follows that $C(T) \subseteq T^\infty(X)$. The next result shows that under certain purely algebraic conditions the algebraic core and the hyper-range of an operator coincide.

Lemma 1.32 *Let T be a linear operator on a vector space X. Suppose that there exists an $m \in \mathbb{N}$ such that*

$$\ker T \cap T^m(X) = \ker T \cap T^{m+k}(X) \quad \text{for all integers } k \geq 0.$$

Then $C(T) = T^\infty(X)$.

Proof We have only to prove that $T^\infty(X) \subseteq C(T)$. We show that $T(T^\infty(X)) = T^\infty(X)$. Evidently the inclusion $T(T^\infty(X)) \subseteq T^\infty(X)$ holds for every linear operator, so we need only to prove the opposite inclusion.

Let $Y := \ker T \cap T^m(X)$. Obviously we have

$$Y = \ker T \cap T^m(X) = \ker T \cap T^\infty(X).$$

Let us now consider an element $y \in T^\infty(X)$. Then $y \in T^n(X)$ for each $n \in \mathbb{N}$, so there exists an $x_k \in X$ such that $y = T^{m+k}x_k$ for every $k \in \mathbb{N}$. If we set

$$z_k := T^m x_1 - T^{m+k-1}x_k \quad (k \in \mathbb{N}),$$

then $z_k \in T^m(X)$ and since

$$Tz_k = T^{m+1}x_1 - T^{m+k}x_k = y - y = 0$$

we also have $z_k \in \ker T$. Thus $z_k \in Y$, and from the inclusion

$$Y = \ker T \cap T^{m+k}(X) \subseteq \ker T \cap T^{m+k-1}(X)$$

it follows that $z_k \in T^{m+k-1}(X)$. This implies that

$$T^m x_1 = z_k + T^{m+k-1}x_k \in T^{m+k-1}(X)$$

for each $k \in \mathbb{N}$, and therefore $T^m x_1 \in T^\infty(X)$. Finally, from

$$T(T^m x_1) = T^{m+1} x_1 = y$$

we may conclude that $y \in T(T^\infty(X))$. Therefore $T^\infty(X) \subseteq T(T^\infty(X))$, so the proof is complete. ∎

Theorem 1.33 *Let T be a linear operator on a vector space X. Suppose that one of the following conditions holds:*

(i) $\alpha(T) < \infty$;
(ii) $\beta(T) < \infty$;
(iii) $\ker T \subseteq T^n(X)$ *for all $n \in \mathbb{N}$.*

Then $C(T) = T^\infty(X)$.

Proof

(i) If $\ker T$ is finite-dimensional then there exists a positive integer m such that

$$\ker T \cap T^m(X) = \ker T \cap T^{m+k}(X)$$

for all integers $k \geq 0$. Hence it suffices to apply Lemma 1.32.

(ii) Suppose that $X = F \oplus T(X)$ with $\dim F < \infty$. If we let $Y_n := \ker T \cap T^n(X)$ then we have $Y_n \supseteq Y_{n+1}$ for all $n \in \mathbb{N}$. Suppose that there exist k distinct subspaces Y_n. There is no loss of generality in assuming $Y_j \neq Y_{j+1}$ for all $j = 1, 2, \ldots k$. Then for every one of these j we can find an element $w_j \in X$ such that $T^j w_j \in Y_j$ and $T^j w_j \notin Y_{j+1}$. By means of the decomposition $X = F \oplus T(X)$ we also find $u_j \in F$ and $v_j \in T(X)$ such that $w_j = u_j + v_j$.

We claim that the vectors $u_1, \cdots, u_k$ are linearly independent. To see this, let us suppose $\sum_{j=1}^k \lambda_j u_j = 0$. Then

$$\sum_{j=1}^k \lambda_j w_j = \sum_{j=1}^k \lambda_j v_j$$

and therefore from the equalities

$$T^k w_1 = \cdots = T^k w_{k-1} = 0$$

we deduce that

$$T^k \Big(\sum_{j=1}^k \lambda_j w_j \Big) = \lambda_k T^k w_k = T^k \Big(\sum_{j=1}^k \lambda_j v_j \Big) \in T^k(T(X)) = T^{k+1}(X).$$

From $T^k w_k \in \ker T$ we obtain $\lambda_k T^k w_k \in Y_{k+1}$, and since $T^k w_k \notin Y_{k+1}$ this is possible only if $\lambda_k = 0$. Analogously we have $\lambda_{k-1} = \cdots = \lambda_1 = 0$, so the vectors $u_1, \ldots, u_k$ are linearly independent. Consequently, k is smaller than or equal to the dimension of F. But then for a sufficiently large m we obtain that

$$\ker T \cap T^m(X) = \ker T \cap T^{m+j}(X)$$

for all integers $j \geq 0$. So we are again in the situation of Lemma 1.32.

(iii) Obviously, if $\ker T \subseteq T^n(X)$ for all $n \in \mathbb{N}$, then

$$\ker T \cap T^n(X) = \ker T \cap T^{n+k}(X) = \ker T$$

for all integers $k \geq 0$. Hence also in this case we can apply Lemma 1.32. ∎

The finiteness of $p(T)$ or $q(T)$ also has some remarkable consequences on $T \,|\, T^\infty(X)$, the restriction of T on $T^\infty(X)$.

Theorem 1.34 *Let T be a linear operator on the vector space X. We have:*

(i) *If either $p(T)$ or $q(T)$ is finite then $T \,|T^\infty(X)$ is surjective. Indeed, $T^\infty(X) = C(T)$.*

(ii) *If either $\alpha(T) < \infty$ or $\beta(T) < \infty$ then*

$$p(T) < \infty \Leftrightarrow T | T^\infty(X) \text{ is injective.}$$

Proof

(i) The assertion follows immediately from Lemma 1.32, because if $p = p(T) < \infty$ then, by Lemma 1.19,

$$\ker T \cap T^p(X) = \ker T \cap T^{p+k}(X) \quad \text{for all integers } k \geq 0;$$

while if $q = q(T) < \infty$ then

$$\ker T \cap T^q(X) = \ker T \cap T^{q+k}(X) \quad \text{for all integers } k \geq 0.$$

(ii) Assume that $p(T) < \infty$. We have $C(T) = T^\infty(X)$ and hence $T(T^\infty(X)) = T^\infty(X)$. Let $\widetilde{T} := T|T^\infty(X)$. Then $\widetilde{T}$ is surjective, thus $q(\widetilde{T}) = 0$. From our assumption and from the equality $\ker \widetilde{T}^n = \ker T^n \cap T^\infty(X)$ we also obtain $p(\widetilde{T}) < \infty$. From Theorem 1.20 we then conclude that $p(\widetilde{T}) = q(\widetilde{T}) = 0$, and therefore the restriction $\widetilde{T}$ is injective.

Conversely, if $\widetilde{T}$ is injective then $\ker T \cap T^\infty(X) = \{0\}$. By assumption $\alpha(T) < \infty$ or $\beta(T) < \infty$, and this implies (see the proof of Theorem 1.33) that $\ker T \cap T^n(X) = \{0\}$ for some positive integer n. By Lemma 1.19 it then follows that $p(T) < \infty$. ∎

The finiteness of the ascent and the descent of a linear operator T is related to a certain decomposition of X.

Theorem 1.35 *Suppose that T is a linear operator on a vector space X. If $p := p(T) = q(T) < \infty$ then we have the decomposition*

$$X = T^p(X) \oplus \ker T^p.$$

Conversely, if for a natural number m we have the decomposition $X = T^m(X) \oplus \ker T^m$ then $p(T) = q(T) \leq m$. In this case $T|T^p(X)$ is bijective.

Proof If $p < \infty$ and assuming, as we may, that $p > 0$, then the decomposition $X = T^p(X) \oplus \ker T^p$ immediately follows from Lemma 1.19. Conversely, if $X = T^m(X) \oplus \ker T^m$ for some $m \in \mathbb{N}$ then $p(T), q(T) \leq m$, again by Lemma 1.19, and hence $p(T) = q(T) < \infty$ by Theorem 1.20.

To verify the last assertion observe that $T^\infty(X) = T^p(X)$, so, from Theorem 1.34, $\tilde{T} := T|T^p(X)$ is onto. On the other hand,

$$\ker \tilde{T} \subseteq \ker T \subseteq \ker T^p,$$

but also $\ker \tilde{T} \subseteq T^p(X)$, so the decomposition $X = T^p(X) \oplus \ker T^p$ entails that $\ker \tilde{T} = \{0\}$.   ∎

Let λ_0 be an isolated point of the spectrum and let Γ be a closed positively oriented contour in $\rho(T) = \mathbb{C} \setminus \sigma(T)$ which separates λ_0 from the rest of the spectrum $\sigma(T)$. The spectral projection associated with the spectral set $\{\lambda_0\}$ is defined by

$$P_0 := \frac{1}{2\pi i} \int_\Gamma (\lambda I - T)^{-1} d\lambda,$$

see Appendix A. The subspaces $P_0(X)$ and $\ker P_0$ are invariant under T and $\sigma(T \, |P_0(X)) = \{\lambda_0\}$, while $\sigma(T \, |\ker P_0) = \mathbb{C} \setminus \{\lambda_0\}$.

The algebraic concepts of ascent and descent are intimately related to the analytic concept of a *pole of the resolvent*. Indeed, we have (see Heuser [179, Proposition 50.2]):

Theorem 1.36 *If $T \in L(X)$ then $\lambda_0 \in \sigma(T)$ is a pole of R_λ if and only if $0 < p(\lambda_0 I - T) = q(\lambda_0 I - T) < \infty$. Moreover, if $p := p(\lambda_0 I - T) = q(\lambda_0 I - T)$ then p is the order of the pole. In this case λ_0 is an eigenvalue of T, and if P_0 is the spectral projection associated with $\{\lambda_0\}$ then*

$$P_0(X) = \ker(\lambda_0 I - T)^p, \quad \ker P_0 = (\lambda_0 I - T)^p(X).$$

The following subspace was introduced by Vrbová [293] and studied in several papers by Mbekhta [229, 230, 232]. It is a natural analytic counterpart of the algebraic core $C(T)$ introduced before.

Definition 1.37 Let X be a Banach space and $T \in L(X)$. The *analytic core* of T is the set $K(T)$ of all $x \in X$ such that there exists a sequence $(u_n) \subset X$ and a constant $\delta > 0$ such that:

(a) $x = u_0$, and $Tu_{n+1} = u_n$ for every $n \in \mathbb{Z}_+$;
(b) $\|u_n\| \leq \delta^n \|x\|$ for every $n \in \mathbb{Z}_+$.

In the following theorem we collect some elementary properties of $K(T)$.

Theorem 1.38 *If* $T \in L(X)$ *the following statements hold:*

(i) $K(T)$ *is a linear subspace of* X;
(ii) $T(K(T)) = K(T)$;
(iii) $K(T) \subseteq C(T)$;
(iv) *If* $\lambda \neq 0$ *then* $\ker(\lambda I - T) \subseteq K(T)$.

Proof (i) It is evident that if $x \in K(T)$ then $\lambda x \in K(T)$ for every $\lambda \in \mathbb{C}$. We show that if $x, y \in K(T)$ then $x + y \in K(T)$. If $x \in K(T)$ there exists a $\delta_1 > 0$ and a sequence $(u_n) \subset X$ satisfying the condition (a) and which is such that $\|u_n\| \leq \delta_1{}^n \|x\|$ for all $n \in \mathbb{Z}_+$. Analogously, since $y \in K(T)$ there exists a $\delta_2 > 0$ and a sequence $(v_n) \subset X$ satisfying condition (a) of the definition of $K(T)$ and such that $\|v_n\| \leq \delta_2^n \|y\|$ for every $n \in \mathbb{N}$.

Let $\delta := \max\{\delta_1, \delta_2\}$. We have

$$\|u_n + v_n\| \leq \|u_n\| + \|v_n\| \leq \delta_1^n \|x\| + \delta_2^n \|y\| \leq \delta^n(\|x\| + \|y\|).$$

Trivially, if $x + y = 0$ there is nothing to prove since $0 \in K(T)$. Suppose then $x + y \neq 0$ and set

$$\mu := \frac{\|x\| + \|y\|}{\|x + y\|}.$$

Clearly $\mu \geq 1$, so $\mu \leq \mu^n$ and therefore

$$\|u_n + v_n\| \leq \delta^n \mu \|x + y\| \leq (\delta\mu)^n \|x + y\| \quad \text{for all } n \in \mathbb{Z}_+,$$

which shows that property (b) of the definition of $K(T)$ is satisfied for every sum $x + y$, with $x, y \in K(T)$. Hence $x + y \in K(T)$, and consequently $K(T)$ is a linear subspace of X.

The proof of (ii) is rather simple, while (iii) is a trivial consequence of (ii) and the definition of $C(T)$. To prove (iv), suppose that $\lambda \neq 0$ and $x \in \ker(\lambda I - T)$. If we set $u_0 = x$ and $u_n := \frac{x}{\lambda^n}$ for $n = 1, \ldots$, then the sequence (u_n) satisfies (a) and (b) of the definition of $K(T)$. Hence $x \in K(T)$. ∎

Observe that in general neither $K(T)$ nor $C(T)$ are closed. The next result shows that $K(T)$ contains every closed subspace F for which the equality $T(F) = F$ holds.

Theorem 1.39 *Suppose that $T \in L(X)$. Then we have*

(i) *If F is a closed subspace of X such that $T(F) = F$ then $F \subseteq K(T)$.*
(ii) *If $C(T)$ is closed then $C(T) = K(T)$.*

Proof

(i) Let $T_0 : F \rightarrow F$ denote the restriction of T on F. By assumption F is a Banach space and $T(F) = F$, so, by the open mapping theorem, T_0 is open. This means that there exists a constant $\delta > 0$ with the property that for every $x \in F$ there is a $u \in F$ such that $Tu = x$ and $\|u\| \leq \delta \|x\|$.

Now, if $x \in F$, define $u_0 := x$ and consider an element $u_1 \in F$ such that

$$T u_1 = u_0 \quad \text{and} \quad \|u_1\| \leq \delta \|u_0\|.$$

By repeating this procedure, for every $n \in \mathbb{N}$ we find an element $u_n \in F$ such that

$$T u_n = u_{n-1} \quad \text{and} \quad \|u_n\| \leq \delta \|u_{n-1}\|.$$

From the last inequality we obtain the estimate

$$\|u_n\| \leq \delta^n \|u_0\| = \delta^n \|x\| \quad \text{for every } n \in \mathbb{N},$$

so $x \in K(T)$. Hence $F \subseteq K(T)$.
(ii) Suppose that $C(T)$ is closed. Since $C(T) = T(C(T))$ the first part of the theorem shows that $C(T) \subseteq K(T)$, and hence, since the reverse inclusion is always true, $C(T) = K(T)$.   ∎

1.4   Semi-Regular Operators

In this section we first introduce a class of operators which are related to semi-Fredholm operators.

Definition 1.40 A bounded operator $T \in L(X)$, X a Banach space, is said to be *semi-regular* if T has closed range $T(X)$ and ker $T \subseteq T^n(X)$ for every $n \in \mathbf{N}$.

Note that the condition ker $T \subseteq T^n(X)$ is equivalent to the conditions listed in Theorem 1.16 and Corollary 1.17. Clearly, bounded below, as well as surjective operators, are semi-regular.

Lemma 1.41 *Suppose that $T \in L(X)$ is semi-regular. Then $\gamma(T^n) \geq \gamma(T)^n$.*

Proof We proceed by induction. The case $n = 1$ is trivial.

Suppose that $\gamma(T^n) \geq \gamma(T)^n$. If $x \in X$ and $u \in \ker T^{n+1}$, we have

$$\text{dist}(x, \ker T^{n+1}) = \text{dist}(x - u, \ker T^{n+1})$$
$$\leq \text{dist}(x - u, \ker T).$$

By Theorem 1.16 we also have $\ker T = T^n(\ker T^{n+1})$ and hence

$$\text{dist}(T^n x, \ker T) = \text{dist}(T^n x, T^n(\ker T^{n+1}))$$
$$= \inf_{u \in \ker T^{n+1}} \|T^n(x - u)\|$$
$$\geq \gamma(T^n) \cdot \inf_{u \in \ker T^{n+1}} \text{dist}(x - u, \ker T^n)$$
$$\geq \gamma(T^n) \, \text{dist}(x, \ker T^{n+1}).$$

From this estimate it then follows that

$$\|T^{n+1} x\| \geq \gamma(T) \, \text{dist}(T^n x, \ker T) \geq \gamma(T) \, \gamma(T^n) \cdot \text{dist}(x, \ker T^{n+1}).$$

Consequently, from our inductive assumption we obtain that

$$\gamma(T^{n+1}) \geq \gamma(T) \gamma(T)^n = \gamma(T)^{n+1},$$

so the proof is complete. ∎

Theorem 1.42 *If $T \in L(X)$ is semi-regular and $x \in X$, then $Tx \in C(T)$ if and only if $x \in C(T)$.*

Proof Clearly the equality $T(C(T)) = C(T)$ implies that $Tx \in C(T)$ for every $x \in C(T)$. Conversely, let $Tx \in C(T)$. By Theorem 1.33 we then have that $C(T) = T^\infty(X)$, and consequently for each $n \in \mathbb{N}$ there exists a $y_n \in X$ such that $T^{n+1} y_n = Tx$. Hence $z := x - T^n y_n \in \ker T \subseteq T^n(X)$. Then $x = z + T^n x \in T^n(X)$ for each $n \in \mathbb{N}$, and consequently $x \in C(T)$.

Theorem 1.43 *$T \in L(X)$ is semi-regular if and only if $T^* \in L(X^*)$ is semi-regular.*

Proof Suppose first that T is semi-regular. Then, by part (i) of Theorem 1.2, $T(X)$ is closed so $\gamma(T) > 0$. Theorem 1.41 entails that $\gamma(T^n) \geq \gamma(T)^n > 0$ and this implies, again by part (i) of Theorem 1.2, that $T^n(X)$ is closed for every $n \in \mathbb{N}$. By part (ii) of Theorem 1.2 the same argument also shows that $T^{n*}(X^*) = T^{*n}(X^*)$ is closed for every $n \in \mathbb{N}$. Therefore,

$$\ker T^{n\perp} = T^{*n}(X^*) \quad \text{and} \quad {}^\perp\ker T^{*n} = T^n(X). \tag{1.4}$$

Now, since T is semi-regular we have $\ker T \subseteq T^n(X)$ for every $n \in \mathbb{N}$, and consequently

$$T^n(X)^\perp \subseteq \ker T^\perp = T^*(X^*).$$

Finally, from the second equality of (1.4) we have $\ker T^{*n} = T^n(X)^\perp$, so $\ker T^{*n} \subseteq T^\star(X^*)$ holds for every $n \in \mathbb{N}$. Since $T^\star(X^*)$ is closed, it then follows that T^* is semi-regular.

A similar argument shows that if $T^\star$ is semi-regular then T is also semi-regular. ∎

Theorem 1.44 *Let $T \in L(X)$ be semi-regular. Then we have*

(i) T^n *is semi-regular for all $n \in \mathbb{N}$.*
(ii) $C(T)$ *is closed and $C(T) = K(T) = T^\infty(X)$.*
(iii) $\lambda I - T$ *is semi-regular for all $|\lambda| < \gamma(T)$, where $\gamma(T)$ denotes the reduced minimal modulus.*

Proof

(i) If T is semi-regular then, by Lemma 1.41, $S := T^n$ has closed range. Furthermore, $S^\infty(X) = T^\infty(X)$ and hence, by Theorem 1.16, $\ker S \subseteq T^\infty(X) = S^\infty(X)$. From Corollary 1.17 we then conclude that T^n is semi-regular.
(ii) The semi-regularity of T gives, by definition, $\ker T \subseteq T^n(X)$ for all $n \in \mathbb{N}$. Hence, by Theorem 1.33, we have $T^\infty(X) = C(T)$. But T^n is semi-regular for all $n \in \mathbb{N}$, so $T^n(X)$ is closed for all $n \in \mathbb{N}$, so $T^\infty(X) = \bigcap_{n=1}^\infty T^n(X)$ is closed. By part (ii) of Theorem 1.39 we then conclude that $K(T) = C(T)$.
(iii) First we show that $C(T) \subseteq C(\lambda I - T)$ for all $|\lambda| < \gamma(T)$. Let $T_0 : C(T) \to C(T)$ denote the restriction of T to $C(T)$. From part (ii) we know that $C(T)$ is closed and T_0 is surjective. Thus, by Lemma 1.10, the equalities

$$(\lambda I - T_0)(C(T)) = (\lambda I - T)(C(T)) = C(T)$$

hold for all $|\lambda| < \gamma(T_0)$.

On the other hand, T is semi-regular, so by Theorem 1.33,

$$\ker\ T \subseteq T^\infty(X) = C(T).$$

This easily implies that $\gamma(T_0) \geq \gamma(T)$, and hence

$$(\lambda I - T)(C(T)) = C(T) \quad \text{for all } |\lambda| < \gamma(T).$$

Note that this last equality implies that

$$C(T) \subseteq C(\lambda I - T) \quad \text{for all } |\lambda| < \gamma(T). \tag{1.5}$$

Moreover, for every $\lambda \neq 0$ we have $T(\ker(\lambda I - T)) = \ker(\lambda I - T)$, so, from part (ii) and Theorem 1.39, we have

$$\ker(\lambda I - T) \subseteq C(T) \quad \text{for all } \lambda \neq 0.$$

From the inclusion (1.5) we then conclude that the inclusions

$$\ker(\lambda I - T) \subseteq C(\lambda I - T) \subseteq (\lambda I - T)^n(X) \tag{1.6}$$

hold for all $|\lambda| < \gamma(T)$, $\lambda \neq 0$ and $n \in \mathbb{N}$. Of course, this is still true for $\lambda = 0$ since T is semi-regular, so the inclusions (1.6) are valid for all $|\lambda| < \gamma(T)$.

To prove that $\lambda I - T$ is semi-regular for all $|\lambda| < \gamma(T)$, it only remains to show that $(\lambda I - T)(X)$ is closed for all $|\lambda| < \gamma(T)$. Note that, as a consequence of Lemma 1.10, we need only consider the case $C(T) \neq \{0\}$ and $C(T) \neq X$. Indeed, if $C(T) = \{0\}$ then $\ker T \subseteq C(T) = \{0\}$, and hence T is bounded below, while in the other case $C(T) = X$ the operator T is surjective.

Let $\overline{X} := X/C(T)$, and let $\overline{T} : \overline{X} \to \overline{X}$ be the quotient map defined by $\overline{T}\,\overline{x} := \overline{Tx}$, where $x \in \overline{x}$. Clearly $\overline{T}$ is continuous. Moreover, $\overline{T}$ is injective since from $\overline{T}\,\overline{x} = \overline{Tx} = \overline{0}$ we have $Tx \in C(T)$, and this implies, by Theorem 1.42, that $x \in C(T)$, which yields $\overline{x} = \overline{0}$.

We show now that $\overline{T}$ is bounded below. We only need to prove that $\overline{T}$ has closed range. To see this we first show the inequality $\gamma(\overline{T}) \geq \gamma(T)$. In fact, for each $x \in X$ and each $u \in C(T)$ we have, since $\ker T \subseteq C(T)$,

$$\|\overline{x}\| = \text{dist}(x, C(T)) = \text{dist}(x - u, C(T))$$

$$\leq \text{dist}(x - u, \ker T) \leq \frac{1}{\gamma(T)}\|Tx - Tu\|.$$

From the equality $C(T) = T(C(T))$ we then obtain that

$$\|\overline{Tx}\| = \inf_{u \in C(T)} \|Tx - Tu\| \quad \text{for all } u \in C(T),$$

hence $\|\overline{x}\| \leq 1/\gamma(T)\|\overline{Tx}\|$, from which we obtain that $\gamma(\overline{T}) \geq \gamma(T)$.

Hence $\overline{T}$ is bounded below. By Lemma 1.10 then $\lambda \overline{I} - \overline{T}$ is bounded below for all $|\lambda| < \gamma(\overline{T})$, and hence for all $|\lambda| < \gamma(T)$.

Finally, to show that $(\lambda I - T)(X)$ is closed for all $|\lambda| < \gamma(T)$, let us consider a sequence (x_n) in $(\lambda I - T)(X)$ which converges to $x \in X$. Clearly, the sequence $(\overline{x}_n)$ converges to $\overline{x}$ and $\overline{x}_n \in (\lambda \overline{I} - \overline{T})(\overline{X})$. The last space is closed for all $|\lambda| < \gamma(T)$, and hence $\overline{x} \in (\lambda \overline{I} - \overline{T})(\overline{X})$. Let $\overline{x} = (\lambda \overline{I} - \overline{T})\overline{v}$ and $v \in \overline{v}$. Then

$$x - (\lambda I - T)v \in C(T) \subseteq (\lambda I - T)(C(T)) \quad \text{for all } |\lambda| < \gamma(T),$$

and so there exists a $u \in C(T)$ for which $x = (\lambda I - T)(v + u)$, hence $x \in (\lambda I - T)(X)$ for all $|\lambda| < \gamma(T)$. Therefore, $(\lambda I - T)(X)$ is closed for all $|\lambda| < \gamma(T)$, and, consequently, $\lambda I - T$ is semi-regular for all $|\lambda| < \gamma(T)$.   ∎

Semi-regular operators may be characterized in the following way:

Theorem 1.45 *An operator $T \in L(X)$ is semi-regular if and only if there exists a T-invariant closed subspace Y such that the restriction $T|Y$ is onto and the operator $\tilde{T} : X/Y \to X/Y$ induced by T is bounded below. For the subspace Y we can take $Y = T^{\infty}(X)$.*

Proof Let T be semi-regular and set $Y := T^{\infty}(X)$. Then Y is closed and $T(Y) = Y$, by Theorem 1.44, and, clearly, the operator $\tilde{T} : X/Y \to X/Y$ induced by T is injective. To show that $\tilde{T}$ has closed range, observe that $T(X)$ is closed by assumption, and $Y \subseteq T(X)$. We show that $\tilde{T}(X/Y) = T(Y) + Y$ is closed. Indeed, if $Tx_n + Y \to x + Y$ in X/Y then there exists $y_n \in Y$ such that $Tx_n + y_n \to x$, thus $x \in T(X)$ and $x + Y \in \tilde{T}(X/Y)$. Therefore, $\tilde{T}$ is bounded below.

Conversely, suppose that for a closed subspace Y invariant under T, $T|Y$ is onto and the operator $\tilde{T} : X/Y \to X/Y$ is bounded below. Since $T(Y) = Y$ we have $Y \subseteq K(T) \subseteq T^{\infty}(X)$, by Theorem 1.39. If $x \in \ker T$ then $\tilde{T}(x + Y) = 0$ and since $\tilde{T}$ is injective we then have $x \in Y$. Thus, $\ker T \subseteq Y \subseteq T^{\infty}(X)$.

To conclude the proof we need only to show that $T(X)$ is closed. Let $\pi : X \to X/Y$ be the canonical homomorphism. We claim that $T(X) = \pi^{-1}(\tilde{T}(X/Y))$. To see this, observe that if $y \in T(X)$ then $y = Tx$ for some $x \in X$, so

$$\pi(y) = Tx + Y = \tilde{T}(x + Y) \in (\tilde{T}(X/Y)),$$

and hence

$$T(X) \subseteq \pi^{-1}(\tilde{T}(X/Y)).$$

If $y \in X$ and $Qy \in \tilde{T}(X/Y)$, then $y + Y = Tx + Y$ for some $x \in X$, hence $y \in Tx + Y \subseteq T(X)$, since $Y \subseteq T(X)$. Thus, $T(X) = \pi^{-1}(\tilde{T}(X/Y))$, and this subspace is closed, since π is continuous and $\tilde{T}$ has closed range.   ∎

The spectrum of a bounded linear operator $T \in L(X)$ can be divided into subsets in many different ways. Another important part of the spectrum is given by the *semi-regular spectrum* defined as

$$\sigma_{\mathrm{se}}(T) := \{\lambda \in \mathbb{C} : \lambda I - T \text{ is not semi-regular}\}.$$

In the literature $\sigma_{\mathrm{se}}(T)$ is sometimes called the *Kato spectrum* or the *Apostol spectrum*. From Theorem 1.44 we see that $\rho_{\mathrm{se}}(T) := \mathbb{C} \setminus \sigma_{\mathrm{se}}(T)$ is an open subset of $\mathbb{C}$, so $\sigma_{\mathrm{se}}(T)$ is a closed subset of $\mathbb{C}$. Since a bounded below operator, as well as a surjective operator, is semi-regular, we also have $\sigma_{\mathrm{se}}(T) \subseteq \sigma_{\mathrm{ap}}(T)$ and $\sigma_{\mathrm{se}}(T) \subseteq \sigma_{\mathrm{s}}(T)$. Later, in Chap. 2, we shall prove that $\sigma_{\mathrm{se}}(T)$ is non-empty, since it contains the boundary $\partial \sigma(T)$ of the spectrum.

Theorem 1.46 *Let M and N be two closed T-invariant subspaces of X and $X = M \oplus N$. Then T is semi-regular if and only if both $T|M$ and $T|N$ are semi-regular. Consequently,*

$$\sigma_{se}(T) = \sigma_{se}(T|M) \cup \sigma_{se}(T|N).$$

Proof Observe first that $\ker T|M = M \cap \ker T$. We also have $T(M) = M \cap T(X)$. The inclusion $T(M) \subseteq M \cap T(X)$ is obvious. Conversely, if $y \in M \cap T(X)$ then $y \in M$ and $y = Tx$. Write $x = x_1 + x_2$, with $x_1 \in M$ and $x_2 \in N$. Then $y = Tx = Tx_1 + Tx_2$ and since $Tx_1 \in M$ we have $Tx_2 = y - Tx_1 \in M \cap N = \{0\}$, so $y = Tx_1 \in T(M)$.

By induction we have $(T|M)^n(M) = T^n(M) = M \cap T^n(X)$ for every $n \in \mathbb{N}$. Assume now that T is semi-regular. Then

$$\ker T|M = M \cap \ker T \subseteq M \cap T^n(X) = (T|M)^n(M),$$

for every $n \in \mathbb{N}$. Moreover, $(T|M)(M) = M \cap T(X)$ is closed and hence $T|M$ is semi-regular. In the same way we obtain that $T|N$ is semi-regular.

Conversely, if $T|M$ and $T|N$ are both semi-regular then $T(X) = T(M) \oplus T(N)$ is closed and $\ker T|M \subseteq T^n(M)$ and $\ker T|N \subseteq T^n(N)$ for all $n \in \mathbb{N}$, so

$$\ker T = \ker T|M \oplus \ker T|N \subseteq T^n(M) \oplus T^n(N) = T^n(X),$$

and hence T is semi-regular. ∎

The open set $\rho_{se}(T)$, called the *semi-regular resolvent*, can canonically be decomposed into (maximal, open, connected, pairwise disjoint) non-empty components. We want show now that the analytic cores are locally constant on each component Ω of $\rho_{se}(T)$. To do this we need first to introduce the notion of the gap between closed linear subspaces and prove some preliminary results.

Let M, N denote two closed linear subspaces of a Banach space X and define

$$\delta(M, N) := \sup\{\text{dist}(u, N) : u \in M, \|u\| = 1\} \quad \text{if } M \neq \{0\},$$

otherwise we define $\delta(\{0\}, N) = 0$ for any subspace N.

The *gap* between M and N is then defined by

$$\widehat{\delta}(M, N) := \max\{\delta(M, N), \delta(N, M)\}.$$

It is easily seen that the function $\widehat{\delta}$ is a metric on the set $\mathcal{C}(X)$ of all linear closed subspaces of X, see Kato [195, §2, Chapter IV] and the convergence $M_n \to M$ is obviously defined by $\widehat{\delta}(M_n, M) \to 0$ as $n \to \infty$. We recall that for two closed linear subspaces M and N of X we have

$$\delta(M, N) = \delta(N^\perp, M^\perp) \quad \text{and} \quad \widehat{\delta}(M, N) = \widehat{\delta}(N^\perp, M^\perp),$$

see again Kato's book [195, Theorem 2.9, Chapter IV]. From these equalities it easily follows, as $n \to \infty$, that $M_n \to M$ if and only if $M_n^{\perp} \to M^{\perp}$. Moreover, by Corollary 2.6 of [195, §2, Chapter IV], we have

$$\widehat{\delta}(M, N) < 1 \Rightarrow \dim M = \dim N. \tag{1.7}$$

In the sequel we shall need the following technical lemmas.

Lemma 1.47 *Let* $T \in L(X)$ *and consider two arbitrary points* $\lambda, \mu \in \mathbb{C}$. *Then we have:*

(i) $\gamma(\lambda I - T) \cdot \delta(\ker(\mu I - T), \ker(\lambda I - T)) \leq |\mu - \lambda|$;
(ii) $\min\{\gamma(\lambda I - T), \gamma(\mu I - T)\} \cdot \widehat{\delta}(\ker(\lambda I - T), \ker(\mu I - T)) \leq |\mu - \lambda|$.

Proof The statement is obvious for $\lambda = \mu$. Suppose that $\lambda \neq \mu$ and consider an element $0 \neq x \in \ker(\mu I - T)$. Then $x \notin \ker(\lambda I - T)$, so

$$\gamma(\lambda I - T)\mathrm{dist}\,(x, \ker(\lambda I - T)) \leq \|(\lambda I - T)x\|$$
$$= \|(\lambda I - T)x - (\mu I - T)x\|$$
$$= |\lambda - \mu|\,\|x\|.$$

From this estimate we obtain, if $Y := \{x \in \ker(\mu I - T) : \|x\| \leq 1\}$, that

$$\gamma(\lambda I - T) \cdot \sup_{x \in Y} \mathrm{dist}\,(x, \ker(\lambda I - T)) \leq |\lambda - \mu|,$$

and hence

$$\gamma(\lambda I - T) \cdot \delta(\ker(\lambda I - T), \ker(\mu I - T)) \leq |\mu - \lambda|.$$

(ii) The inequality follows from (i) by interchanging λ and μ. ∎

Lemma 1.48 *For every* $x \in X$ *and* $0 < \varepsilon < 1$ *there exists an* $x_0 \in X$ *such that* $x - x_0 \in M$ *and*

$$\mathrm{dist}(x_0, N) \geq \left((1 - \varepsilon)\frac{1 - \delta(M, N)}{1 + \delta(M, N)}\right)\|x_0\|. \tag{1.8}$$

Proof Evidently, if $x \in M$ it suffices to put $x_0 = 0$. Therefore, we can assume that $x \notin M$. Let $\widehat{X} := X/M$ denote the quotient space and put $\widehat{x} := x + M$. Clearly, $\|\widehat{x}\| = \inf_{z \in \widehat{x}} \|z\| > 0$. We show that there exists an element $x_0 \in X$ such that

$$\|\widehat{x_0}\| = \mathrm{dist}(x_0, M) \geq (1 - \varepsilon)\|x_0\|.$$

Indeed, were it not so, then

$$\|\widehat{x}\| = \|\widehat{z}\| < (1 - \varepsilon)\|z\| \quad \text{for every } z \in \widehat{x}$$

and consequently

$$\|\widehat{x}\| \leq (1 - \varepsilon) \inf_{z \in \widehat{x}} \|z\| = (1 - \varepsilon)\|\widehat{x}\|.$$

But this is impossible since $\|\widehat{x}\| > 0$. Define

$$\mu := \text{dist}(x_0, N) = \inf_{u \in N} \|x_0 - u\|.$$

Clearly, there exists a $y \in N$ such that

$$\|x_0 - y\| \leq \mu + \varepsilon\|x_0\|.$$

From that we then obtain

$$\|y\| \leq (1 + \varepsilon)\|x_0\| + \mu.$$

On the other hand, because $\text{dist}(y, M) \leq \delta(N, M) \cdot \|y\|$, we have

$$(1 - \varepsilon)\|\widehat{x_0}\| \leq \text{dist}(x_0, M) \leq \|x_0 - y\| + \text{dist}(y, M)$$
$$\leq \mu + \varepsilon\|x_0\| + \delta(N, M) \cdot \|y\|$$
$$\leq \mu + \varepsilon\|x_0\| + \delta(N, M)[(1 + \varepsilon)\|x_0\| + \mu],$$

from which we deduce that

$$\mu \geq \left[\frac{1 - \varepsilon - \delta(N, M)}{1 + \delta(N, M)} - \varepsilon \right] \|x_0\|.$$

The inequality (1.8) is then clear, since $\varepsilon > 0$ is arbitrary. ∎

Theorem 1.49 *Suppose that $T \in L(X)$ is semi-regular. Then*

$$\gamma(\lambda I - T) \geq \gamma(T) - 3|\lambda| \quad \text{for every } \lambda \in \mathbb{C}. \tag{1.9}$$

Proof Clearly, for every $T \in L(X)$ and $|\lambda| \geq \gamma(T)$ we have

$$\gamma(\lambda I - T) \geq 0 \geq \gamma(T) - 3|\lambda|,$$

so we need to prove (1.9) only in the case when $\lambda < \gamma(T)$.

Since T is semi-regular we have $C(T) = T^\infty(X)$, by Theorem 1.44. If $C(T) = \{0\}$ then $\ker T \subseteq T^\infty(X) = \{0\}$, so T is injective, and since $T(X)$ is closed it then follows that T is bounded below. From a closer look at the proof of Lemma 1.10 we can then conclude that

$$\gamma(\lambda I - T) \geq \gamma(T) - |\lambda| \geq \gamma(T) - 3|\lambda|$$

for every $|\lambda| < \gamma(T)$. The case $C(T) = X$ is trivial, since in such a case T is onto, hence T^* is bounded below, and consequently

$$\gamma(\lambda I - T) = \gamma(\lambda I^* - T^*) \geq \gamma(T^*) - 3|\lambda| = \gamma(T) - 3|\lambda|.$$

It remains to prove the inequality (1.9) in the case when $C(T) \neq \{0\}$ and $C(T) \neq X$. Assume that $|\lambda| < \gamma(T)$ and let $x \in C(T) = T(C(T))$. Then there exists a $u \in C(T)$ such that $x = Tu$ and hence

$$\text{dist}(u, \ker T) \leq (\gamma(T))^{-1} \|Tu\| = (\gamma(T))^{-1} \|x\|.$$

Let $\varepsilon > 0$ be arbitrary and choose $w \in \ker T$ such that

$$\|u - w\| \leq [(1 - \varepsilon)\gamma(T)]^{-1} \|x\|.$$

Let

$$u_1 := u - w \quad \text{and} \quad \mu := (1 - \varepsilon)\gamma(T).$$

Clearly, $u_1 \in C(T)$, $Tu_1 = x$ and $\|u_1\| \leq \mu^{-1} \|x\|$. Since $u_1 \in C(T)$, by repeating the same procedure we then obtain a sequence (u_n), where $u_0 := x$ and

$$u_n \in C(T), \quad Tu_{n+1} = u_n \quad \text{and} \quad \|u_n\| \leq \mu^{-n} \|x\|.$$

Let us consider the function $f : \mathbb{D}(0, \mu) \to X$ defined as

$$f(\lambda) := \sum_{n=0}^{\infty} \lambda^n u_n.$$

Clearly, $f(0) = x$ and $f(\lambda) \in \ker(\lambda I - T)$ for all $|\lambda| < \mu$. Moreover,

$$\|x - f(\lambda)\| = \|\sum_{n=1}^{\infty} \lambda^n u_n\| \leq \frac{|\lambda|}{\mu - |\lambda|}.$$

Consequently,

$$\text{dist}(x, \ker(\lambda I - T)) \leq \frac{|\lambda|}{\mu - |\lambda|},$$

so

$$\delta(\ker T, \ker(\lambda I - T)) \leq \frac{|\lambda|}{\mu - |\lambda|} = \frac{|\lambda|}{(1 - \varepsilon)\gamma(T) - |\lambda|}$$

for every $|\lambda| < \mu$. Since ε is arbitrary we conclude that

$$\delta(\ker T, \ker(\lambda I - T)) \leq \frac{|\lambda|}{\gamma(T) - |\lambda|} \quad \text{for every } |\lambda| < \gamma(T). \qquad (1.10)$$

Let $\delta := \delta(\ker T, \ker(\lambda I - T))$. By Lemma 1.48 we can correspond to the element u and $\varepsilon > 0$ an element $v \in X$ such that $z := u - v \in \ker(\lambda I - T)$ and

$$\text{dist}(v, \ker T) \geq \frac{1 - \delta}{1 + \delta}(1 - \varepsilon)\|v\|.$$

From this estimate it then follows that

$$\|(\lambda I - T)u\| = \|(\lambda I - T)v\| \geq \|Tv\| - |\lambda|\|v\|$$
$$\geq \gamma(T) \cdot \text{dist}(v, \ker T) - |\lambda|\|v\|$$
$$\geq \gamma(T)\frac{1 - \delta}{1 + \delta}(1 - \varepsilon)\|v\| - |\lambda|\|v\|.$$

By using inequality (1.10) we then obtain

$$\|(\lambda I - T)u\| \geq [(1 - \varepsilon)(\gamma(T) - 2|\lambda|) - |\lambda|]\|v\|$$
$$\geq [(1 - \varepsilon)(\gamma(T) - 2|\lambda|) - |\lambda|]\|u - z\|$$
$$\geq [(1 - \varepsilon)(\gamma(T) - 2|\lambda|) - |\lambda|] \cdot \text{dist}(u, \ker(\lambda I - T)).$$

From the last inequality we easily obtain that

$$\gamma(\lambda I - T) \geq (1 - \varepsilon)(\gamma(T) - 2|\lambda|) - |\lambda|,$$

and since ε is arbitrary we then conclude that inequality (1.9) holds. ∎

In the following result we show that the subspaces $K(\lambda I - T)$ are constant as λ ranges through a component Ω of the semi-regular resolvent $\rho_{se}(T)$.

Theorem 1.50 Let $T \in L(X)$ and consider a connected component Ω of $\rho_{se}(T)$. If $\lambda_0 \in \Omega$ is arbitrarily fixed then

$$K(\lambda I - T) = K(\lambda_0 I - T) \text{ for every } \lambda \in \Omega.$$

Proof By Theorem 1.44,

$$C(\lambda I - T) = K(\lambda I - T) = (\lambda I - T)^{\infty}(X) \quad \text{for all } \lambda \in \rho_{\text{se}}(T).$$

By the first part of the proof of part (iii) of Theorem 1.44 we have $K(T) \subseteq K(\mu I - T)$ for every $|\mu| < \gamma(T)$. Now, take $|\mu| < \frac{1}{4}\gamma(T)$ and define $S := \mu I - T$. From Theorem 1.49 we have

$$\gamma(S) = \gamma(\mu I - T) \geq \gamma(T) - 3|\mu| > |\mu|,$$

hence, according to the observation above,

$$K(\mu I - T) = K(S) \subseteq K(\mu I - S) = K(T).$$

From this it then follows that $K(\mu I - T) = K(T)$ when μ is sufficiently small. Take two arbitrary points λ_1, λ_2 in Ω. Write

$$\lambda_1 I - T = (\lambda_1 - \lambda_2)I - (T - \lambda_2 I).$$

If we choose λ_1, λ_2 sufficiently near to each other, the previous argument shows that

$$K(\lambda_1 I - T) = K((\lambda_1 - \lambda_2)I - (T - \lambda_2 I)) = K(\lambda_2 I - T).$$

The following standard compactness argument proves that $K(\lambda I - T) = K(\mu I - T)$ for all $\lambda, \mu \in \Omega$. In fact, join a fixed point $\lambda_0 \in \Omega$ with an arbitrary point $\lambda \in \Omega$ by a polygonal line $P \subset \Omega$. Associate with each point in P a disc in which the analytic core is constant. By the classical Heine–Borel theorem already finitely many of these discs cover P, so $K(\lambda_0 I - T) = K(\lambda I - T)$. Thus, the subspaces $K(\lambda I - T)$ are constant on Ω. ■

By Theorem 1.44 if $\lambda I - T$ is a semi-regular operator then $K(\lambda I - T) = (\lambda I - T)^{\infty}(X)$, so the statement of Theorem 1.50 is equivalent to saying that the subspaces $(\lambda I - T)^{\infty}(X)$ are constant as λ ranges through a component Ω of $\rho_{\text{se}}(T)$.

The semi-regularity of an operator may be characterized in terms of the continuity of certain mappings.

Theorem 1.51 *If $T \in L(X)$ and $\lambda_0 \in \mathbb{C}$, then the following statements are equivalent:*

(i) *$\lambda_0 I - T$ is semi-regular;*
(ii) *$\gamma(\lambda_0 I - T) > 0$ and the mapping $\lambda \to \gamma(\lambda I - T)$ is continuous at the point λ_0;*
(iii) *$\gamma(\lambda_0 I - T) > 0$ and the mapping $\lambda \to \ker(\lambda I - T)$ is continuous at λ_0 in the gap metric;*
(iv) *The range $(\lambda I - T)(X)$ is closed in a neighborhood of λ_0 and the mapping $\lambda \to (\lambda I - T)(X)$ is continuous at λ_0 in the gap metric.*

Proof There is no loss of generality if we assume that $\lambda_0 = 0$.

(i) $\Rightarrow$ (ii) Since T has closed range we have $\gamma(T) > 0$. Moreover, for every $|\lambda| < \gamma(T)$, the operator $\lambda I - T$ is semi-regular, by Theorem 1.44. Consider $|\lambda| < \gamma(T)$ and $|\mu| < \gamma(T)$. By Theorem 1.49 we have

$$|\gamma(\lambda I - T) - \gamma(\mu I - T)| \le 3|\lambda - \mu|,$$

and this obviously implies the continuity of the mapping $\lambda \to \gamma(\lambda I - T)$ at the point 0.

(ii) $\Rightarrow$ (iii) The continuity of the mapping $\lambda \to \gamma(\lambda I - T)$ at 0 implies that there exists a neighborhood $\mathcal{U}$ of 0 for which

$$\gamma(\lambda I - T) \ge \frac{\gamma(T)}{2} \quad \text{for all } \lambda \in \mathcal{U}.$$

From Lemma 1.47 we then have that

$$\widehat{\delta}(\ker(\mu I - T), \ker(\lambda I - T)) \le \frac{2}{\gamma(T)}|\lambda - \mu| \quad \text{for all } \lambda, \mu \in \mathcal{U},$$

and in particular,

$$\widehat{\delta}(\ker T, \ker(\lambda I - T)) \le \frac{2}{\gamma(T)}|\lambda| \quad \text{for all } \lambda \in \mathcal{U}.$$

From this estimate we deduce that $\ker(\lambda I - T))$ converges in the gap metric to $\ker T$, as $\lambda \to 0$, and consequently the mapping $\lambda \to \ker(\lambda I - T)$ is continuous at 0.

(iii) $\Rightarrow$ (i) It is clear that $\ker(\lambda I - T) \subseteq T^n(X)$ for every n. For every $x \in \ker T$, $n \in \mathbb{N}$ and $\lambda \ne 0$, we then have

$$\text{dist}(x, T^n(X)) \le \text{dist}(x, \ker(\lambda I - T)) \le \delta(\ker T, \ker(\lambda I - T)) \cdot \|x\|.$$

This estimate implies that

$$\text{dist}(x, T^n(X)) \le \widehat{\delta}(\ker T, \ker(\lambda I - T)) \cdot \|x\|.$$

The continuity at 0 of the mapping $\lambda \to \ker(\lambda I - T)$ entails that $x \in \overline{T^n(X)}$ for every n. Hence $\ker T \subseteq \overline{T^n(X)}$ for every $n = 1, \cdots$.

To prove the semi-regularity of T it suffices to prove that $T^n(X)$ is closed for $n \in \mathbb{N}$. We proceed by induction.

The case $n = 1$ is obvious from the assumption. Assume that $\overline{T^n(X)}$ is closed. Then $\ker T \subseteq \overline{T^n(X)} = T^n(X)$, hence $\ker T + T^n(X) = T^n(X)$ is closed. By Lemma 1.5 we then have that $T(T^n(X)) = T^{n+1}(X)$ is closed. Therefore, (i), (ii) and (iii) are equivalent.

(i) $\Rightarrow$ (iv) If T is semi-regular and $\mathbb{D}(0, \gamma(T))$ is the open disc centered at 0 with radius $\gamma(T)$ then $(\lambda I - T)$ is semi-regular for all $\lambda \in \mathbb{D}(0, \gamma(T))$, by Theorem 1.44. In particular, $(\lambda I - T)(X)$ is closed, and hence $\ker(\lambda I - T^*)^{\perp} = (\lambda I - T)(X)$ for all $\lambda \in \mathbb{D}(0, \gamma(T))$.

Now, by Theorem 1.43 T^* is semi-regular, and by the first part of the proof, this is equivalent to the continuity at 0 of the mapping

$$\lambda \to \ker(\lambda I - T^*) = (\lambda I - T)(X)^{\perp}.$$

Since

$$\widehat{\delta}(T(X)^{\perp}, (\lambda I - T)(X)^{\perp}) = \widehat{\delta}(T(X), (\lambda I - T)(X)),$$

we then conclude that the mapping $\lambda \to (\lambda I - T)(X)$ is continuous at 0.

(iv) $\Rightarrow$ (iii) Let $\mathcal{U}$ be a neighborhood of 0 such that $(\lambda I - T)(X)$ is closed for every $\lambda \in \mathcal{U}$. Then $\gamma(T) > 0$, so

$$\widehat{\delta}(\ker T^*, \ker(\lambda I - T^*)) = \widehat{\delta}(^{\perp}\ker T^*, {}^{\perp}\ker(\lambda I - T^*))$$

$$= \widehat{\delta}(T(X), (\lambda I - T)(X)).$$

Hence the mapping $\lambda \to \gamma(\lambda I - T^*) = \gamma(\lambda I - T)$ is continuous at 0.   ∎

Theorem 1.52 *Let Ω be a connected component of $\rho_{\mathrm{se}}(T)$ and fix $\lambda_0 \in \Omega$. Then*

$$K(\lambda_0 I - T) = \bigcap_{n=0}^{\infty}(\lambda_n I - T)(X) = \bigcap_{n=1}^{\infty}(\lambda_n I - T)(X), \qquad (1.11)$$

where (λ_n) is a sequence of distinct points of Ω which converges to λ_0.

Proof We first show the second equality in (1.11). Trivially, the inclusion

$$\bigcap_{n=0}^{\infty}(\lambda_n I - T)(X) \subseteq \bigcap_{n=1}^{\infty}(\lambda_n I - T)(X)$$

holds for every $T \in L(X)$.

To show the opposite inclusion, suppose that $x \in \bigcap_{n=1}^{\infty}(\lambda_n I - T)(X)$. Then

$$\mathrm{dist}(x, (\lambda_0 I - T)(X)) \leq \widehat{\delta}((\lambda_n I - T)(X), (\lambda_0 I - T)(X)) \cdot \|x\|$$

for every $n \in \mathbb{N}$. Since $\lambda_n \to \lambda_0$, from Theorem 1.51 it then follows that $x \in \overline{(\lambda_0 I - T)(X)} = (\lambda_0 I - T)(X)$. Therefore, the equality

$$\bigcap_{n=0}^{\infty}(\lambda_n I - T)(X) = \bigcap_{n=1}^{\infty}(\lambda_n I - T)(X) \qquad (1.12)$$

is proved.

It remains to prove the first equality in (1.11). By Theorem 1.50 we have

$$K(\lambda_0 I - T) = K(\lambda_n I - T) \subseteq (\lambda_n I - T)(X), \quad \text{for all } n \in \mathbb{N},$$

and hence

$$K(\lambda_0 I - T) \subseteq \bigcap_{n=1}^{\infty} (\lambda_n I - T)(X).$$

Conversely, let $x \in \bigcap_{n=1}^{\infty} (\lambda_n I - T)(X)$. From equality (1.12) we know that $x \in (\lambda_0 I - T)(X)$, so there exists an element $u \in X$ such that $x = (\lambda_0 I - T)u$. Write

$$x = (\lambda_n I - T)u + (\lambda_0 - \lambda_n)u.$$

Since $x \in (\lambda_n I - T)(X)$ for every $n \in \mathbb{N}$, we have that $(\lambda_0 - \lambda_n)u$ belongs to $\bigcap_{n=1}^{\infty} (\lambda_n I - T)(X)$. Now, by assumption, $\lambda_n \neq \lambda_0$ for every $n \in \mathbb{N}$, so $u \in \bigcap_{n=1}^{\infty} (\lambda_n I - T)(X)$. This shows that

$$x = (\lambda_0 I - T)u \in (\lambda_0 I - T)(\bigcap_{n=1}^{\infty} (\lambda_n I - T)(X)),$$

from which the inclusion

$$\bigcap_{n=1}^{\infty} (\lambda_n I - T)(X) \subseteq (\lambda_0 I - T)(\bigcap_{n=1}^{\infty} (\lambda_n I - T)(X))$$

follows. The opposite inclusion is clearly satisfied. By Theorem 1.39 we then obtain that

$$\bigcap_{n=1}^{\infty} (\lambda_n I - T)(X) \subseteq K(\lambda_0 I - T),$$

which concludes the proof.   ∎

The following example shows that the product of two semi-regular operators, also commuting semi-regular operators, need not be semi-regular.

Example 1.53 Let H denote a Hilbert space with an orthonormal basis $(e_{i,j})$ where i, j are integers for which $i \times j \leq 0$. Define $T, S \in L(H)$ and $S \in L(H)$ by the assignment:

$$T e_{i,j} := \begin{cases} 0 & \text{if } i = 0, \ j > 0 \\ e_{i+1,j} & \text{otherwise,} \end{cases}$$

and

$$Se_{i,j} := \begin{cases} 0 & \text{if } j = 0, \ i > 0 \\ e_{i,j+1} & \text{otherwise.} \end{cases}$$

Then

$$T Se_{i,j} = S T e_{i,j} = \begin{cases} 0 & \text{if } i = 0, \ j \geq 0, \text{ or } j = 0, \ i \geq 0, \\ e_{i+1,j+1} & \text{otherwise.} \end{cases}$$

Hence $TS = ST$ and, as is easy to verify,

$$\ker T = \bigvee_{j>0} \{e_{i,0}\} \subset T^{\infty}(H),$$

where $\bigvee_{j>0}\{e_{0,j}\}$ denotes the linear subspace of H generated by the set $\{e_j : j > 0\}$. Analogously we have

$$\ker S = \bigvee_{i>0} \{e_{i,0}\} \subset S^{\infty}(H).$$

Moreover, both T and S have closed range, so T and S are semi-regular. On the other hand, $e_{0,0} \in \ker TS$ and $e_{0,0} \notin (TS)(H)$, thus TS is not semi-regular.

The next example shows that the set of all semi-regular operators need not be an open subset of $L(X)$.

Example 1.54 Let H be a Hilbert space with an orthonormal basis $(e_{i,j})$ where i, j are integers and $i \geq 1$. Let T be defined by:

$$T e_{i,j} := \begin{cases} e_{i,j+1} & \text{if } j \neq 0, \\ 0 & \text{if } j = 0. \end{cases}$$

Clearly $T(H)$ is closed and

$$\ker T = \bigvee_{i>1} \{e_{0,j}\} \subset T^{\infty}(H),$$

thus T is semi-regular.

Now let $\varepsilon > 0$ be arbitrarily given and define $S \in L(H)$ by

$$Se_{i,j} := \begin{cases} \dfrac{\varepsilon}{i} e_{i,0} & \text{if } j = 0, \\ 0 & \text{if } j \neq 0. \end{cases}$$

It is easy to see that $\|S\| = \varepsilon$. Moreover, since S is a compact operator having an infinite-dimensional range, the range $S(H)$ is not closed. Let M denote the subspace generated by the set $\{e_{i,0} : i \geq 1\}$. Then the subspace $T(H)$ is orthogonal to M and hence is orthogonal to $S(H)$, since $S(H) \subseteq M$. Moreover, $(T + S)(H) = T(H) + S(H)$, from which we deduce that $(T + S)(H)$ is not closed, and hence $T + S$ is not semi-regular.

Theorem 1.55 *Let $T, S \in L(X)$ be commuting operators such that TS is semi-regular. Then both T and S are semi-regular. In particular, if T^n is semi-regular then T is semi-regular.*

Proof It suffices only to show that one of the two operators, say T, is semi-regular. From the semi-regularity of TS we obtain

$$\ker T \subseteq \ker (TS) \subseteq \bigcap_{n=1}^{\infty}(T^n S^n)(X) \subseteq \bigcap_{n=1}^{\infty} T^n(X). \tag{1.13}$$

We show now that $T(X)$ is closed. Let $(y_n) := (Tx_n)$ be a sequence of $T(X)$ which converges to some y_0. Then

$$Sy_n = STx_n = TSx_n \in (TS)(X)$$

and (Sy_n) converges to Sy_0. Since by assumption $(TS)(X)$ is closed, we have $Sy_0 \in (TS)(X) = (ST)(X)$. Consequently, there exists an element $z_0 \in X$ such that $Sy_0 = STz_0$ and hence

$$y_0 - Tz_0 \in \ker S \subseteq \ker (TS).$$

From (1.13) we deduce that

$$y_0 - Tz_0 \in \bigcap_{n=1}^{\infty} T^n(X) \subseteq T(X).$$

From this we then obtain that $y_0 \in T(X)$, so $T(X)$ is closed and T is semi-regular. ∎

The class of all *upper semi-Fredholm* operators is defined by

$$\Phi_+(X) := \{T \in L(X) : \alpha(T) < \infty \text{ and } T(X) \text{ isclosed}\}$$

and the class of all *lower semi-Fredholm* operators is defined by

$$\Phi_-(X) := \{T \in L(X) : \beta(T) < \infty\}.$$

The class of all *semi-Fredholm operators* is defined as $\Phi_\pm(X) := \Phi_+(X) \cup \Phi_-(X)$, while the class of the *Fredholm* operators is defined as $\Phi(X) := \Phi_+(X) \cap \Phi_-(X)$. The *index* of $T \in \Phi_\pm(X)$ is defined by

$$\text{ind}(T) := \alpha(T) - \beta(T).$$

An easy example of a semi-regular operator which is not Fredholm is given by an injective finite-rank operator T acting on an infinite-dimensional X. Indeed, T is obviously semi-regular, since $T(X)$ is finite-dimensional and hence closed, and obviously $\beta(T) = \text{codim}(T) = \infty$.

The classes $\Phi_+(X)$, $\Phi_-(X)$, and $\Phi(X)$ are open subsets of $L(X)$ (see Appendix A) and give rise to the following spectra. The *upper semi-Fredholm spectrum*, defined as

$$\sigma_{\text{usf}}(T) := \{\lambda \in \mathbb{C} : \lambda I - T \notin \Phi_+(X)\},$$

the *lower semi-Fredholm spectrum*, defined as

$$\sigma_{\text{lsf}}(T) := \{\lambda \in \mathbb{C} : \lambda I - T \notin \Phi_-(X)\},$$

and the *semi-Fredholm spectrum* (also known in the literature as the *Wolf spectrum*), defined as

$$\sigma_{\text{sf}}(T) := \{\lambda \in \mathbb{C} : \lambda I - T \notin \Phi_\pm(X)\}.$$

The *semi-Fredholm region* of T is defined as $\rho_{\text{sf}}(T) := \mathbb{C} \setminus \sigma_{\text{sf}}(T)$. If $\rho_{\text{usf}}(T) := \mathbb{C} \setminus \sigma_{\text{usf}}(T)$ and $\rho_{\text{lsf}}(T) := \mathbb{C} \setminus \sigma_{\text{lsf}}(T)$ then $\rho_{\text{sf}} = \rho_{\text{usf}}(T) \cup \rho_{\text{lsf}}(T)$. The *essential spectrum* (also called the *Fredholm spectrum*) is defined as

$$\sigma_e(T) := \{\lambda \in \mathbb{C} : \lambda I - T \notin \Phi(X)\}.$$

The Fredholm region of T is the set $\rho_{\text{f}}(T) := \mathbb{C} \setminus \sigma_e(T)$.

Remark 1.56 All the spectra defined above are closed subsets of $\mathbb{C}$. Moreover, these spectra are non-empty if X is an infinite-dimensional Banach space. That $\sigma_e(T)$ is non-empty is an easy consequence of the *Atkinson characterization of Fredholm operators*, which says that $T \in L(X)$ is a Fredholm operator if and only if the residual class $\hat{T} := T + K(X)$ is invertible in the Calkin algebra $\mathcal{L} := L(X)/\mathcal{K}(X)$, where $K(X)$ denotes the two-sided ideal of all compact operators, see Appendix A. Consequently, $\sigma_e(T) = \sigma(\hat{T})$ is non-empty if X is infinite-dimensional. Also $\sigma_{usf}(T)$ and $\sigma_{lsf}(T)$ are non-empty; this will be proved in Chap. 2.

Theorem 1.57 *If $T \in L(X)$ then the boundary $\partial\sigma_e(T)$ of the essential spectrum is contained in $\sigma_{usf}(T) \cap \sigma_{lsf}(T)$.*

Proof If $\lambda_0 \in \partial\sigma_e$ then there exists a sequence $\{\lambda_n\}$ which converges to λ_0 such that $\lambda_n I - T \in \Phi(X)$. By Atkinson's characterization of Fredholm operators the residual class $\lambda_n \hat{I} - \hat{T}$ is invertible for each $n \in \mathbb{N}$ in the Calkin algebra $\mathcal{L}$. But $\lambda_0 I - T \notin \Phi(X)$, so $\lambda_0 \hat{I} - \hat{T}$ is not invertible, and since $\lambda_n \hat{I} - \hat{T} \to \lambda_0 \hat{I} - \hat{T}$, as $n \to \infty$, it then follows that $\lambda_0 \hat{I} - \hat{T}$ is neither injective nor surjective. Hence $\lambda_0 I - T \notin \Phi_+(X) \cup \Phi_-(X)$, thus $\lambda_0 \in \sigma_{usf}(T) \cap \sigma_{lsf}(T)$. ∎

The punctured neighborhood theorem for semi-Fredholm operators (see Appendix A) establishes that if $T \in \Phi_+(X)$ then there exists an $\varepsilon > 0$ such that $\lambda I + T \in \Phi_+(X)$ and $\alpha(\lambda I + T)$ is constant on the punctured neighborhood $0 < |\lambda| < \varepsilon$. Moreover,

$$\alpha(\lambda I + T) \le \alpha(T) \quad \text{for all } |\lambda| < \varepsilon , \tag{1.14}$$

and

$$\text{ind}\,(\lambda I + T) = \text{ind}\,T \quad \text{for all } |\lambda| < \varepsilon.$$

Analogously, if $T \in \Phi_-(X)$ then there exists an $\varepsilon > 0$ such that $\lambda I + T \in \Phi_-(X)$ and $\beta(\lambda I + T)$ is constant on the punctured neighborhood $0 < |\lambda| < \varepsilon$. Moreover,

$$\beta(\lambda I + T) \le \beta(T) \quad \text{for all } |\lambda| < \varepsilon, \tag{1.15}$$

and

$$\text{ind}\,(\lambda I + T) = \text{ind}\,T \quad \text{for all } |\lambda| < \varepsilon.$$

Definition 1.58 Let $T \in \Phi_\pm(X)$, X a Banach space. Let $\varepsilon > 0$ be as in (1.14) or in (1.15). If $T \in \Phi_+(X)$ the *jump* of T is defined by

$$\text{jump}\,(T) := \alpha(T) - \alpha(\lambda I + T), \quad 0 < |\lambda| < \varepsilon,$$

while if $T \in \Phi_-(X)$, the jump is defined by

$$\text{jump}\,(T) := \beta(T) - \beta(\lambda I + T), \quad 0 < |\lambda| < \varepsilon.$$

The continuity of the index ensures that both definitions of the jump coincide whenever T is a Fredholm operator.

We have seen in Example 1.54 that the set of semi-regular operators $\mathcal{SR}(X)$ is, in general, not open. Since the sets $\Phi(X)$, $\Phi_+(X)$ and $\Phi_-(X)$ are open, an obvious consequence is that $\mathcal{SR}(X)$ does not coincide with one of these sets. In the sequel, we set $T_\infty := T|T^\infty(X)$.

Lemma 1.59 *If $T \in \Phi_+(X)$ then T_∞ is a Fredholm operator.*

Proof Since $\alpha(T) < \infty$, from Theorem 1.33 we have $\beta(T_\infty) = 0$, and from the inclusion ker $T_\infty \subseteq$ ker T it then follows that ker T is finite-dimensional, thus T_∞ is a Fredholm operator. ∎

Theorem 1.60 *A semi-Fredholm operator $T \in L(X)$ is semi-regular precisely when T has* jump $(T) = 0$.

Proof Since $T(X)$ is closed it suffices to show the equivalence

$$\text{jump}\,(T) = 0 \Leftrightarrow \mathcal{N}^\infty(T) \subseteq T^\infty(X)\,.$$

Assume first $T \in \Phi_+(X)$ and $\mathcal{N}^\infty(T) \subseteq T^\infty(X)$. Observe that

$$\alpha(\lambda I + T) = \alpha(\lambda I + T_\infty) \quad \text{for all } \lambda \in \mathbb{C}.$$

For $\lambda = 0$ this is clear, since ker $T \subseteq \mathcal{N}^\infty(T) \subseteq T^\infty(X)$ implies that ker $T =$ ker T_∞. For $\lambda \neq 0$ we have, by part (ii) of Theorem 1.14,

$$\text{ker } T \subseteq \mathcal{N}^\infty(\lambda I + T) \subseteq T^\infty(X),$$

so that ker $(\lambda I + T =$ ker $(\lambda I + T_\infty)$.

Now, by Theorem 1.33 we know that $\beta(T_\infty) = 0$ and hence, by Theorem 1.10, there exists an $\varepsilon > 0$ such that $\beta(\lambda I + T_\infty) = 0$ for all $|\lambda| < \varepsilon$. By Lemma 1.59 we know that T_∞ is Fredholm, so we can assume ε is such that

$$\text{ind }(\lambda I + T_\infty) = \text{ ind }(T_\infty) \quad \text{for all } |\lambda| < \varepsilon.$$

Therefore $\alpha(\lambda I + T_\infty) = \alpha(T_\infty)$ for all $|\lambda| < \varepsilon$ and hence $\alpha(\lambda I + T) = \alpha(T)$ for all $|\lambda| < \varepsilon$, thus jump $(T) = 0$.

Conversely, suppose that $T \in \Phi_+(X)$ and jump $(T) = 0$, namely there exists an $\varepsilon > 0$ such $\alpha(\lambda I + T)$ is constant for $|\lambda| < \varepsilon$. Then

$$\alpha(T_\infty) \leq \alpha(T) = \alpha(\lambda I + T) = \alpha(\lambda I + T_\infty) \quad \text{for all } 0 < |\lambda| < \varepsilon.$$

But T_∞ is Fredholm, by Lemma 1.59, and hence, from the punctured neighborhood theorem, we can choose $\varepsilon > 0$ such that $\alpha(\lambda I + T_\infty) \leq \alpha(T_\infty)$ for all $|\lambda| < \varepsilon$. This shows that $\alpha(T_\infty) = \alpha(T)$ and consequently, $\mathcal{N}^\infty(T) \subseteq T^\infty(X)$.

To conclude the proof, we need to consider the case $T \in \Phi_-(X)$ and jump $(T) = 0$. In this case, $T^* \in \Phi_+(X^*)$ and jump $(T) = $ jump $(T^*) = 0$. From the first part of the proof we deduce that $\mathcal{N}^\infty(T^*) \subseteq T^{*\infty}(X^*)$. From Corollary 1.17 it then follows that ker $T^{*n} \subseteq T^*(X^*)$ for all $n \in \mathbb{N}$, or equivalently $T^n(X)^\perp \subseteq$ ker $T^\perp$ for all $n \in \mathbb{N}$. Since all these subspaces are closed then $T^n(X) \supseteq$ ker T for all $n \in \mathbb{N}$, so, by Corollary 1.17 we conclude that $\mathcal{N}^\infty(T) \subseteq T^\infty(X)$. ∎

1.5   The Kato Decomposition

The following definition comes from Kato's classical treatment [195] of the
perturbation theory of semi-Fredholm operators.

Definition 1.61 An operator $T \in L(X)$, X a Banach space, is said to admit a
generalized Kato decomposition, abbreviated as GKD, if there exists a pair of T-
invariant closed subspaces (M, N) such that $X = M \oplus N$, the restriction $T|M$ is
semi-regular and $T|N$ is quasi-nilpotent.

Evidently, every semi-regular operator has a GKD $M = X$ and $N = \{0\}$ and a
quasi-nilpotent operator has a GKD $M = \{0\}$ and $N = X$.

A relevant case is obtained if we assume in the definition above that $T|N$ is
nilpotent, i.e. there exists a $d \in \mathbb{N}$ for which $(T|N)^d = 0$. In this case the operator
T is said to be of *Kato-type of order d*.

Evidently, if $\lambda_0 I - T$ admits a generalized Kato decomposition then $\lambda_0 I^\star -
T^\star$ also admits a generalized Kato decomposition. More precisely, if T admits a
$GKD(M, N)$ then the pair $(N^\perp, M^\perp)$ is a GKD for $\lambda_0 I^\star - T^\star$.

An operator $T \in L(X)$ is said to be *essentially semi-regular* if it admits a GKD
(M, N) such that N is finite-dimensional. Note that if T is essentially semi-regular
then $T|N$ is nilpotent, since every quasi-nilpotent operator on a finite-dimensional
space is nilpotent.

Hence we have the following implications:

$$T \text{ semi-regular} \Rightarrow T \text{ essentially semi-regular} \Rightarrow T \text{ of Kato-type}$$

$$\Rightarrow T \text{ admits a GKD}.$$

In the sequel we reassume some results concerning essentially semi-regular
operators. The reader may find a well-organized exposition of the basic results
concerning this class of operators in the book of Müller [243, §21], where the
essentially semi-regular operators are called *essentially Kato operators*.

(i) $T \in L(X)$ is essentially semi-regular if and only if $T(X)$ is closed and there
   exists a finite-dimensional subspace F of X such that $\ker T \subseteq T^\infty(X) + F$.

(ii) If $T \in L(X)$ is essentially semi-regular then T^n is essentially semi-regular for
   every $n \in \mathbb{N}$.

(iii) $T \in L(X)$ is essentially semi-regular if and only if $T^* \in L(X^*)$ is essentially
   semi-regular.

(iii) If T and S commutes and TS is essentially semi-regular then both T and S are
   essentially semi-regular.

(iv) If $T \in L(X)$ is essentially semi-regular then there exists an $\varepsilon > 0$ such that
   $T + S$ is essentially semi-regular for every $S \in L(X)$ such that $ST = TS$ and
   $\|S\| < \varepsilon$.

(v) If $T \in L(X)$ is essentially semi-regular then $T + K$ is essentially semi-regular
   for every finite-rank operator $K \in L(X)$.

We have already observed that a semi-Fredholm operator is, in general, not semi-regular. The following important result was first observed by Kato [195], for a simpler proof see also Müller [243, Theorem 16.21].

Theorem 1.62 *Every semi-Fredholm operator $T \in L(X)$ is essentially semi-regular, in particular is of Kato-type.*

In the sequel we see that some of the properties already observed for semi-regular operators may be extended to operators which admit a GKD.

Theorem 1.63 *Suppose that (M, N) is a GKD for $T \in L(X)$. Then we have:*

(i) $K(T) = K(T|M)$ *and $K(T)$ is closed;*
(ii) $\ker T|M = \ker T \cap M = K(T) \cap \ker T$.

Proof

(i) To show the equality $K(T) = K(T|M)$, we need only prove that $K(T) \subseteq M$. Let $x \in K(T)$ and, according to the definition of $K(T)$, choose a sequence (u_n) in X, and $\delta > 0$ such that

$$x = u_0, \quad Tu_{n+1} = u_n, \quad \text{and} \quad \|u_n\| \leq \delta^n \|x\| \quad \text{for all } n \in \mathbb{N}.$$

Obviously, $T^n u_n = x$ for all $n \in \mathbb{N}$. From the decomposition $X = M \oplus N$ we know that $x = y + z$, $u_n = y_n + z_n$, with $y, y_n \in M$ and $z, z_n \in N$. Then

$$x = T^n u_n = T^n y_n + T^n z_n,$$

and hence, by the uniqueness of the decomposition, $y = T^n y_n$ and $z = T^n z_n$ for all n. Let P denote the projection of X onto N along M. From the estimate

$$\|((T|N)P)^n\|^{1/n} \leq \|(T|N)^n\|^{1/n} \|P^n\|^{1/n} = \|(T|N)^n\|^{1/n} \|P\|^{1/n},$$

we deduce that $(T|N)P$ is also quasi-nilpotent, since, by assumption, $T|N$ is quasi-nilpotent. Therefore, if $\varepsilon > 0$, there is a positive integer n_0 such that

$$\|(TP)^n\|^{1/n} = \|((T|N)P)^n\|^{1/n} < \varepsilon,$$

for all $n > n_0$. We have

$$\|z\| = \|T^n z_n\| = \|T^n P u_n\| = \|(TP)^n u_n\| \leq \varepsilon^n \delta^n \|x\|, \qquad (1.16)$$

for all $n > n_0$. Since ε is arbitrary, the last term of (1.16) converges at 0, so $z = 0$ and this implies that $x = y \in M$.

The last assertion is a consequence of part (ii) of Theorem 1.44, since the restriction $T|M$ is semi-regular.

(ii) From part (i) we have $K(T) \subseteq M$ and, since $T|M$ is semi-regular, from Theorem 1.33 and part (i) we also have that

$$\ker(T|M) \subseteq (T|M)^\infty(M) = K(T|M) = K(T).$$

From this we then conclude that

$$K(T) \cap \ker T = K(T) \cap M \cap \ker T = K(T) \cap \ker(T|M) = \ker(T|M),$$

so part (ii) is also proved. ∎

For operators of Kato-type the hyper-range and the analytic core coincide:

Theorem 1.64 *Let $T \in L(X)$, X a Banach space, be of Kato-type of order d, with a GKD (M, N). Then we have:*

(i) $K(T) = T^\infty(X) = (T|M)^\infty(M)$;
(ii) $\ker(T|M) = \ker T \cap T^\infty(X) = \ker T \cap T^n(X)$ *for all natural $n \geq d$;*
(iii) *We have $T(X) + \ker T^n = T(M) \oplus N$ for every natural $n \geq d$. Moreover, $T(X) + \ker T^n$ is closed in X for every natural $n \geq d$.*

Proof

(i) We have $(T|N)^d = 0$. For $n \geq d$ we have

$$T^n(X) = T^n(M) \oplus T^n(N) = T^n(M) \tag{1.17}$$

and hence $T^\infty(X) = (T|M)^\infty(M)$. From part (ii) of Theorem 1.44, the semi-regularity of $T|M$ implies that $(T|M)^\infty(M) = K(T|M)$ and the last set, by Theorem 1.63, coincides with $K(T)$.
(ii) Let $n \geq d$. Clearly, $T^n(X) = T^n(M)$. From the equalities (1.17) and part (ii) of Theorem 1.63 we obtain

$$\ker(T|M) = \ker T \cap K(T) \subseteq \ker T \cap T^n(X) = \ker T \cap T^n(M)$$
$$\subseteq \ker T \cap M = \ker(T|M).$$

Hence for all $n \geq d$, $\ker(T|M) = \ker T \cap T^n(X)$.
(iii) It is obvious that if $n \geq d$ then $N \subseteq \ker T^n$, so

$$T(M) \oplus N \subseteq T(X) + \ker T^n.$$

Conversely, if $n \geq d$ then

$$\ker T^n = \ker(T|M)^n \oplus \ker(T|N)^n = \ker(T|M)^n \oplus N$$

and from the semi-regularity of $T|M$ it then follows that $\ker T^n \subseteq T(M) \oplus N$.

Because $T(X) = T(M) \oplus T(N) \subseteq T(M) \oplus N$, we then have

$$T(X) + \ker T^n \subseteq T(M) \oplus N.$$

Hence $T(X) + \ker T^n = T(M) \oplus N$.

To complete the proof we show that $T(M) \oplus N$ is closed. Let $M \times N$ be provided with the canonical norm

$$\|(x, y)\| := \|x\| + \|y\| \quad (x \in M, y \in N). \tag{1.18}$$

Clearly, $M \times N$ with respect to this norm is complete. Let $\Psi : M \times N \to M \oplus N = X$ denote the topological isomorphism defined by

$$\Psi(x, y) := x + y \quad \text{for every } x \in M, y \in N. \tag{1.19}$$

We have $\Psi(T(M), N) = T(M) \oplus N$ and hence, since $(T(M), N)$ is closed in $M \times N$, the subspace $T(M) \oplus N$ is closed in X. ∎

If $T \in L(X)$ is of Kato-type then $\lambda I - T$ is semi-regular on a punctured disc centered at 0:

Theorem 1.65 *If $T \in L(X)$ is of Kato-type then there exists an $\varepsilon > 0$ such that $\lambda I - T$ is semi-regular for all $0 < |\lambda| < \varepsilon$.*

Proof Let (M, N) be a GKD for T such that $T|N$ is nilpotent.

First we show that $(\lambda I - T)(X)$ is closed for all $0 < |\lambda| < \gamma(T|M)$, where $\gamma(T|M)$ denotes the minimal modulus of $T|M$. Since $T|N$ is nilpotent, the restriction $\lambda I - T|N$ is bijective for every $\lambda \neq 0$, thus $N = (\lambda I - T)(N)$ for every $\lambda \neq 0$, and therefore

$$(\lambda I - T)(X) = (\lambda I - T)(M) \oplus (\lambda I - T)(N) = (\lambda I - T)(M) \oplus N$$

for every $\lambda \neq 0$. By assumption $T|M$ is semi-regular, so by Theorem 1.44 $(\lambda I - T)|M$ is semi-regular for every $|\lambda| < \gamma(T|M)$, and hence, for these values of λ, the set $(\lambda I - T)(M)$ is a closed subspace of M.

We show now that $(\lambda I - T)(X)$ is closed for every $0 < |\lambda| < \gamma(T|M)$. Consider the Banach space $M \times N$ provided with the canonical norm defined in (1.18) and let $\Psi : M \times N \to M \oplus N = X$ denote the topological isomorphism defined as in (1.19). Then for every $0 < |\lambda| < \gamma(T|M)$ the set

$$\Psi[(\lambda I - T)(M) \times N] = (\lambda I - T)(M) \oplus N = (\lambda I - T)(X)$$

is closed since the product $(\lambda I - T)(M) \times N$ is closed in $M \times N$.

We show now that there exists an open disc $\mathbb{D}(0, \varepsilon)$ such that

$$\mathcal{N}^\infty(\lambda I - T) \subseteq (\lambda I - T)^\infty(X) \quad \text{for all } \lambda \in \mathbb{D}(0, \varepsilon) \setminus \{0\}.$$

Since T is of Kato-type, the hyper-range is closed and coincides with $K(T)$, by Theorems 1.63 and 1.64. Therefore, $T(T^\infty(X)) = T^\infty(X)$. Let $T_0 := T|T^\infty(X)$. The operator T_0 is onto and hence, by part (ii) of Lemma 1.10, $\lambda I - T_0$ is also onto for all $|\lambda| < \gamma(T_0)$, so

$$(\lambda I - T)(T^\infty(X)) = T^\infty(X) \quad \text{for all } |\lambda| < \gamma(T_0).$$

Since the hyper-range $T^\infty(X)$ is closed, by Theorem 1.39, we then have

$$T^\infty(X) \subseteq K(\lambda I - T) \subseteq (\lambda I - T)^\infty(X) \quad \text{for all } |\lambda| < \gamma(T_0).$$

From Theorem 1.14, part (ii), we then conclude that

$$\mathcal{N}^\infty(\lambda I - T) \subseteq T^\infty(X) \subseteq (\lambda I - T)^\infty(X) \quad \text{for all } 0 < |\lambda| < \gamma(T_0). \tag{1.20}$$

Since $(\lambda I - T)(X)$ is closed for all $0 < |\lambda| < \gamma(T|M)$, from the inclusions (1.20) we then deduce the semi-regularity of $\lambda I - T$ for all $0 < |\lambda| < \varepsilon$, where $\varepsilon := \min\{\gamma(T_0), \gamma(T|M)\}$. ∎

Some special classes of semi-Fredholm operators are given by the class $B_+(X)$ of all *upper semi-Browder operators,* defined as

$$B_+(X) := \{T \in \Phi_+(X) : p(T) < \infty\},$$

and by $B_-(X)$ the class of all *lower semi-Browder operators,* defined as

$$B_-(X) := \{T \in \Phi_-(X) : q(T) < \infty\}.$$

The class of all *Browder operators* is defined by $B(X) = B_+(X) \cap B_-(X)$. Clearly,

$$B_+(X) := \{T \in \Phi(X) : p(T) = q(T) < \infty\}.$$

These classes, together with some other related classes of operators, will be treated in more detail in Chap. 3. It is easy to see that if $T \in B_-(X)$ then the subspace M in the Kato decomposition (M, N) is uniquely determined and $M = T^\infty(X)$, thus $T|M$ is onto.

Lemma 1.66 *If $T \in L(X)$ is essentially semi-regular then the operator $\tilde{T} : X/T^\infty(X) \to X/T^\infty(X)$ is upper semi-Browder.*

Proof Let (M, N) be the corresponding Kato decomposition for which $X = M \oplus N$, $T|M$ is semi-regular, and $T|N$ is nilpotent with $\dim N < \infty$. Clearly,

$$T^\infty(X) = (T|M)^\infty(M) \subseteq M.$$

Moreover, $T^\infty(X)$ is closed and $T(T^\infty(X)) = T^\infty(X)$, by Theorems 1.63 and 1.64.

Let $k \geq 1$ and $x = x_1 \oplus x_2$, $x_1 \in M$ and $x_2 \in N$, satisfy $T^k x \in T^\infty(X)$. Then $(T|M)x_2 \in T^\infty(X)$, thus $x \in N + T^\infty(X)$ and dim ker $T^k \leq$ dim N. Consequently, $N^\infty(\tilde{T}) \leq$ dim $N < \infty$. Let $\pi : X \to X/T^\infty(X)$ be the canonical projection. As $T^\infty(X) \subseteq T(X)$, and since the range of $\tilde{T}$ is the set

$$\{Tx + T^\infty(X), x \in X\} = \pi(T(X)),$$

then the range of $\tilde{T}$ is closed, hence $\tilde{T}$ is upper semi-Browder.   ∎

In the sequel we give further characterizations of essentially semi-regular operators. These characterizations, in a sense, are a natural extension of the result of Theorem 1.45, established for semi-regular operators, to the case of essentially semi-regular operators.

Theorem 1.67 *For a bounded operator $T \in L(X)$, the following conditions are equivalent:*

  (i) *T is essentially semi-regular;*
 (ii) *there exists a closed T-invariant subspace Y of X such that the restriction $T|Y$ is lower semi-Fredholm and the induced operator $\tilde{T} : X/Y \to X/Y$ is upper semi-Fredholm;*
(iii) *there exists a closed T-invariant subspace Y of X such that the restriction $T|Y$ is lower semi-Browder and the induced operator $\tilde{T} : X/Y \to X/Y$ is upper semi-Browder;*
(iv) *there exists a closed T-invariant subspace Y of X such that the restriction $T|Y$ is onto and the induced operator $\tilde{T} : X/Y \to X/Y$ is upper semi-Browder;*
 (v) *there exists a closed T-invariant subspace Y of X such that the restriction $T|Y$ is lower semi-Browder and the induced operator $\tilde{T} : X/Y \to X/Y$ is bounded below.*

Proof By Lemma 1.66 we have (i) $\Rightarrow$ (iv), and the implications (iv) $\Rightarrow$ (iii) $\Rightarrow$ (ii) and (v) $\Rightarrow$ (ii) are obvious. We prove (ii) $\Rightarrow$ (i).

Suppose that statement (ii) holds. We show first that $T(X)$ is closed. Let $\pi : X \to X/Y$ be the canonical projection. If $z \in T(X)$ then $z = Tx$ for some $x \in X$. Then $\pi(z) = Tx + Y = \tilde{T}(x + Y)$ belongs to the range $R(\tilde{T})$ of $\tilde{T}$, so that $T(X) \subseteq \pi^{-1}(R(\tilde{T}))$. Let $z \in X$ such that $\pi(y) \in R(\tilde{T})$, i.e., $y + Y = Tx + Y$. Then, for some finite-dimensional subspace F of Y we have

$$y \in T(X) + Y \subseteq T(X) + (F + T(Y)) \subseteq T(X) + F,$$

thus

$$\pi^{-1}(R(\tilde{T})) \subseteq T(X) + F \subseteq \pi^{-1}(R(\tilde{T})) + F.$$

Evidently $\pi^{-1}(R(\tilde{T})) + F$ is closed, since π is continuous, $R(\tilde{T})$ is closed and F is finite-dimensional, so $T(X) + F$ is closed and hence $T(X)$ is also closed.

To show that T is essentially semi-regular, we only need to prove that $\ker T \subseteq G + T^\infty(X)$ for some finite-dimensional subspace G. As $\pi(\ker T) \subseteq \ker \tilde{T}$ and $\ker \tilde{T}$ is finite-dimensional, there exists a finite-dimensional subspace $G_1 \subseteq \ker T$ such that $\ker T \subseteq G_1 + \ker T|Y$. The operator $T|Y$ is lower semi-Fredholm and hence essentially semi-regular, so there exists a finite-dimensional subspace G_2 of Y such that

$$\ker T|Y \subseteq G_2 + (T|Y)^\infty(Y).$$

Therefore, if $G := G_1 + G_2$, we have

$$\ker T \subseteq G_1 + \ker T|Y \subseteq G_1 + G_2 + (T|Y)^\infty(Y) \subseteq G + T^\infty(X),$$

so T is essentially semi-regular.

(i) $\Rightarrow$ (v) Let (M, N) be the corresponding Kato decomposition for T, i.e., $X = M \oplus N$, $T|M$ semi-regular, $T|N$ nilpotent and $\dim N < \infty$. Set

$$Y := N \oplus (T|M)^\infty(M) = N \oplus T^\infty(X),$$

see Theorem 1.64. Evidently, Y is closed and because $T^\infty(X) = K(T)$, by Theorem 1.64, we have $T(T^\infty(X)) = T^\infty(X)$. This implies that the restriction $T|Y$ is a lower semi-Browder operator. Denote by $\tilde{T} : X/Y \to X/Y$ the operator induced by T. If $x = x_1 \oplus x_2$, $x_1 \in M$, $x_2 \in N$, satisfies $Tx \in Y$ then $Tx_1 \in (T|M)^\infty(M)$, so that $x_1 \in (T|M)^\infty(M)$ and $x \in Y$. Hence $\ker \tilde{T} = \{0\}$. We show now that the range $R(\tilde{T})$ of $\tilde{T}$ is closed. Suppose that in the topology of $X|Y$ we have $Tz_k + M \to z + M$ as $k \to \infty$, with $z, z_k \in X$. Then $z \in \overline{T(X) + Y} = T(X) + Y$, since $Y \subseteq T(X) + N$. Consequently, $x + Y \in R(\tilde{T})$, hence $R(\tilde{T})$ is closed and $\tilde{T}$ is bounded below.   ∎

1.6   Operators with a Topological Uniform Descent

We start this section with some purely algebraic lemmas, which will be used in the sequel. For abbreviation we use the symbol $X_1 \cong X_2$ to denote that the two linear spaces X_1 and X_2 are isomorphic.

Lemma 1.68 *Let U, V and W be linear subspaces of a vector space X and E a linear subspace of a vector space Y. For every linear operator $T : X \to Y$ we have:*

(a) *If $U \subseteq W$ then $[U + V] \cap W = U + (V \cap W)$.*
(b) *$U/(U \cap V) \cong (U + V)/V$.*
(c) *$T^{-1}(T(U)) = \ker T + U$.*
(d) *$T(U \cap T^{-1}(E)) = T(U) \cap E$.*

Proof The assertion (a) is the so-called modular law, see [209]. To show (b), let $[x]$ denote a coset in the quotient $(U + V)/V$. Define for every $u \in U$ $Ju := [u]$. Then J is a linear mapping from U into $(U + V)/V$. If $[x]$ is an element of $(U + V)/V$, then $x = u + v$ with $u \in U$ and $v \in V$, hence $[x] = [u]$. Therefore, J is onto. Since the kernel of J is $U \cap V$, (b) is proved. The equalities (c) and (d) follow from easy calculations.   ∎

If two quotient spaces are linearly isomorphic under an isomorphism induced by the identity, as in (b), we say that these quotient spaces are *naturally isomorphic*. In the following lemma we establish some other isomorphisms, which will be useful in the sequel.

Lemma 1.69 *Let T be a linear operator in the linear space X. Then we have:*

(i) *For every $k = 0, 1, \ldots$ and $n = 0, 1, \ldots$ we have*

$$\frac{T^n(X)}{T^{n+k}(X)} \cong \frac{X}{T^k(X) + \ker T^n}.$$

(ii) *For $n = 0, 1, \ldots$ we have*

$$\frac{\ker T^{n+1}}{[\ker T^n + T(X)] \cap \ker T^{n+1}} \cong \frac{\ker T \cap T^n(X)}{\ker T \cap T^{n+1}(X)}.$$

Proof

(i) Let $[y]$ denote any coset in the quotient $\frac{T^n(X)}{T^{n+k}}$, and for every $x \in X$ define $J : X \to \frac{T^n(X)}{T^{n+k}}$ by $Jx := [T^n x]$. If $Jx = 0$, then $T^n x = T^{n+k} z$ for some $z \in X$, and hence $x = T^k z \in \ker T^n$. This shows the inclusion

$$\ker J \subseteq T^k(X) + \ker T^n.$$

Conversely, if $x \in T^k(X) + \ker T^n$, then $T^n x \in T^{n+k}(X)$, and hence $Jx = 0$. This shows the equality $\ker J = T^k(X) + \ker T^n$, from which we obtain

$$\frac{T^n(X)}{T^{n+k}(X)} \cong \frac{X}{\ker J}.$$

(ii) Let $[y]$ denote any coset in the quotient $\frac{\ker T \cap T^n(X)}{\ker T \cap T^{n+1}}$ and define the map

$$J : \ker T^{n+1} \to \frac{\ker T \cap T^n(X)}{\ker T \cap T^{n+1}(X)}$$

by

$$Jx := [T^n x] \quad \text{for each } x \in \ker T^{n+1}.$$

In order to prove our isomorphism it suffices to show that J is onto, and that

$$\ker J = [\ker T^n + T(X)] \cap \ker T^{n+1}.$$

Evidently, if

$$[y] \in \frac{\ker T \cap T^n(X)}{\ker T \cap T^{n+1}(X)},$$

then $y = T^n x \in \ker T$ for some $x \in X$. Clearly, $x \in \ker T^{n+1}$ and $Jx = [T^n x] = [y]$, so J is onto. Take $x \in \ker J$. Clearly, $T^n x \in \ker T \cap T^{n+1}(X)$, and $T^n x = T^{n+1} z$ for some $z \in X$. But then, $x - Tz \in \ker T^n$, so $x \in \ker T^n + T(X)$. This shows the inclusion

$$\ker J \subseteq [\ker T^n + T(X)] \cap \ker T^{n+1}.$$

We show the opposite inclusion. Let $x \in [\ker T^n + T(X)] \cap \ker T^{n+1}$. Then $x = u + Tz$ for some $u \in \ker T^n$ and $z \in X$, and hence

$$T^n x = T^n u + T^{n+1} z = T^{n+1} z,$$

from which we conclude that $Jx = 0$. Therefore

$$\ker J = [\ker T^n + T(X)] \cap \ker T^{n+1},$$

and this completes the proof.   ∎

Since for every n we have $\ker T^n \subseteq \ker T^{n+1}$ we can consider, for every n, the mapping

$$\Phi_n : \ker T^{n+2}/\ker T^{n+1} \to \ker T^{n+1}/\ker T^n$$

induced by T and defined as

$$\Phi_n(z + \ker T^{n+1}) := Tz + \ker T^n \quad z \in \ker T^{n+2}.$$

Analogously, since $T^{n+1}(X) \subseteq T^n(X)$, we can consider, for every n, the sequence of mappings

$$\Psi_n : T^n(X)/T^{n+1}(X) \to T^{n+1}(X)/T^{n+2}(X)$$

defined as

$$\Psi_n(z + T^{n+1}(X)) := Tz + T^{n+2}(X), \quad z \in T^n(X).$$

Evidently, every map Φ_n is onto, while Ψ_n is injective.

For every n let us denote by $k_n(T)$ the dimension of the kernel of the map Φ_n. In the following theorem we give useful information on the kernel of Φ_n, and the cokernel of Ψ_n.

Theorem 1.70 *Let T be a linear operator on a vector space X, and n a nonnegative integer.*

(i) *the kernel of Ψ_n is naturally isomorphic to the quotient*

$$\frac{\ker T \cap T^n(X)}{\ker T \cap T^{n+1}(X)}.$$

(ii) *The cokernel of Φ_n is naturally isomorphic to the quotient*

$$\frac{\ker T^{n+1} + T(X)}{\ker T^n + T(X)}.$$

(iii) $k_n(T) = \dim \frac{\ker T \cap T^n(X)}{\ker T \cap T^{n+1}(X)}.$

(iii) $k_n(T)$ *is equal to the codimension of the image of the linear mapping Ψ_n. More precisely,*

$$k_n(T) = \dim \frac{\ker T^{n+1} + T(X)}{\ker T^n + T(X)}.$$

Proof Using Lemma 1.68 we have

$$\ker \Phi_n = \frac{[T^{n+1}(X) + \ker T] \cap T^n(X)}{T^{n+1}(X)}$$

$$= \frac{T^{n+1}(X) + [\ker T \cap T^n(X)]}{T^{n+1}(X)},$$

which is naturally isomorphic to $\frac{\ker T \cap T^n(X)}{\ker T \cap T^{n+1}(X)}$, so part (i) is proved. The proof of part (ii) is similar, so it is omitted.

(iii) $k_n(T) = \dim \ker \Phi_n$, by definition, so the equality follows from part (i).

(iv) The cokernel U of Φ_n is the quotient

$$\frac{\ker T^{n+1}}{[\ker T^{n+1} \cap T(X)] + \ker T^n}.$$

Another application of Lemma 1.68 shows that T^n induces an isomorphism from the cokernel U of Ψ_n onto the quotient

$$\frac{\ker T \cap T^n(X)}{\ker T \cap T^{n+1}(X)},$$

and the latter space is linearly isomorphic to ker Ψ_n. Therefore, by part (ii), we have

$$k_n(T) = \dim \ker \Phi_n = \dim U = \dim \frac{\ker T^{n+1} + T(X)}{\ker T^n + T(X)}.$$

∎

For every linear operator T on a vector space X define

$$c_n(T) := \dim \frac{T^n(X)}{T^{n+1}(X)}$$

and

$$c'_n(T) := \dim \frac{\ker T^{n+1}}{\ker T^n}.$$

By Theorem 1.70, we have

$$c_0(T) \geq c_1(T) \geq \cdots$$

and, analogously,

$$c'_0(T) \geq c'_1(T) \geq \cdots.$$

Moreover, by Theorem 1.70, we also have

$$c_n(T) < \infty \Rightarrow k_n(T) = c_n(T) - c_{n+1}(T)$$

and

$$c'_n(T) < \infty \Rightarrow k_n(T) = c'_n(T) - c'_{n+1}(T).$$

Note that it is possible that $k_n(T) < \infty$, while both $c_n(T)$ and $c'_n(T)$ are infinite.

Lemma 1.71 *Let T be a linear operator on a vector space X. Then $c_n(T) = $ codim $[T(X) + \ker T^n]$.*

Proof $\frac{T^n(X)}{T^{n+1}(X)}$ is isomorphic to $\frac{X}{T(X)+\ker T^n}$, by part (i) of Lemma 1.69. ∎

Definition 1.72 Let T be a linear operator on a vector space X and let d be a nonnegative integer. T is said to have *uniform descent for $n \geq d$* if $k_n(T) = 0$ for all $n \geq d$. We say that T has *almost uniform descent* if $k(T) := \sum_{n=0}^{\infty} k_n(T) < \infty$.

Note that the condition $k_n(T) = 0$ means that

$$\ker T = \ker T \cap (X) = \ker T \cap T^2(X) = \cdots = \ker T \cap T^\infty(X),$$

so that $k_n(T) = 0$ for all $n \in \mathbb{N}$ if and only if $\ker T \subseteq T^\infty(X)$.

If M and N are (not necessarily closed) subspaces of a Banach space X, then we write $M \subseteq^e N$ (M is said to be *essentially contained* in N) if there exists a finite-dimensional subspace F of X such that $M \subseteq N + F$. Clearly, $M \subseteq^e N$ if and only if the quotient $M/M \cap N$ is finite-dimensional. If $M \subseteq^e N$ and $N \subseteq^e M$ we write $M =^e N$. Evidently, $k(T) < \infty$ means that there exists a $d \in \mathbb{N}$ such that

$$\ker T =^e \ker T \cap T(X) =^e \cdots =^e \ker T \cap T^d(X) = \ker T \cap T^\infty(X).$$

Thus, $k(T) < \infty$ if and only if $\ker T \subseteq^e T^\infty(X)$, while the condition $k_n(T) < \infty$ for all $n \in \mathbb{N}$ is equivalent to saying that $\ker T^m \subseteq^e T^n(X)$ for all $m, n \in \mathbb{N}$.

Corollary 1.73 *Suppose that T is a linear operator on a vector space X.*

(i) *If T has finite nullity $\alpha(T)$, or finite defect $\beta(T)$, then T has uniform descent for $n \geq 1$.*

(ii) *If T has finite descent p then T has uniform descent for $n \geq p$.*

(iii) *If T has finite descent q then T has uniform descent for $n \geq q$.*

Proof

(i) Suppose that $\alpha(T) < \infty$. Since $\alpha(T^n) < \infty$ for every $n \geq 0$, by Lemma 1.21, the quotient spaces $\ker T^{n+2}/\ker T^{n+1}$ are all finite-dimensional, so the maps Φ_n are isomorphisms for all $n \geq 0$. Analogously, if $\beta(T) < \infty$ then $\beta(T^n) < \infty$ and hence the maps Ψ_n are isomorphisms for all $n \geq 0$.

(ii) If $n \geq p$ we have $\ker T^n = \ker T^{n+1}$, so $\ker T^{n+1}/\ker T^n = \{0\}$.

(iii) If $n \geq q$ we have $T^n(X) = T^{n+1}(X)$, so $T^n(X)/T^{n+1}(X) = \{0\}$.   ∎

The operators which have topological uniform descent may be characterized in several ways:

Theorem 1.74 *If T is a linear operator on a vector space X and d is a fixed nonnegative integer, then the following statements are equivalent:*

(i) *T has uniform descent for each $n \geq d$;*

(ii) *The sequence of subspaces $\{\ker T \cap T^n(X)\}$ is constant for $n \geq d$;*

(iii) *$\ker T \cap T^d(X) = \ker T \cap T^\infty(X)$;*

(iv) *The maps induced by T from $\ker T^{n+2}/\ker T^{n+1}$ to $\ker T^{n+1}/\ker T^n$ are isomorphisms for $n \geq d$;*

(v) *The sequence of subspaces $\{\ker T^n + T(X)\}$ is constant for $n \geq d$;*

(vi) *$\ker T^d + T(X) = \mathcal{N}^\infty(T) + T(X)$.*

Proof The equivalence of (i), (ii), (iv) and (v) follows from Theorem 1.70 and, clearly, (iii) implies (ii) and (vi) implies (v). If (ii) holds then $T^d(X) \cap \ker T \subseteq T^n(X)$ for all $n \geq d$, so that $T^d(X) \cap \ker T \subseteq T^\infty(X)$ from which the equality (iii) easily follows. The proof of the implication (iv) $\Rightarrow$ (v) is similar.   ∎

Let $T \in L(X)$, X a Banach space. The *operator range topology* on $T(X)$ is the topology induced by the norm $\| \cdot \|_T$ defined by:

$$\|y\|_T := \inf_{x \in X}\{\|x\| : y = Tx\}.$$

For a detailed discussion of operator ranges and their topology we refer the reader to [147] and [159].

To study the topological properties of the maps induced by T^n, we will always assume that in $T^n(X)$ is given the unique operator range topology under which $T^n(X)$ becomes a Banach space continuously imbedded in X. Then all the restrictions of bounded linear operators to maps between operator ranges are continuous, by the closed graph theorem, see Grabiner [159, pp. 1433–1444] or Filmore and Williams [147, pp. 255–257].

We shall need in the sequel the following two lemmas.

Lemma 1.75 *Suppose that $T \in L(X, Y)$ has closed range and $E \subseteq X$, $F \subseteq Y$ are linear subspaces such that* ker $T \subseteq E$ *and* $F \subseteq T(X)$. *Then*

(i) $T(\overline{E}) = \overline{T(E)}$.
(ii) $T^{-1}(\overline{F}) = \overline{T^{-1}(F)}$.

Proof The statements (i) and (ii) immediately follow once we observe that T induces a linear homomorphism from $X/\ker T$ onto $T(X)$, and that a subspace Z containing ker T is closed if and only if $Z/\ker T$ is a closed subspace of $X/\ker T$. $\blacksquare$

In the following lemma we give some equivalences in the operator range topology.

Lemma 1.76 *Let $T \in L(X)$ and n be a nonnegative integer. If the map $\hat{T}$: $T^n(X)/T^{n+1} \to T^{n+1}(X)/T^{n+2}$ induced by T has a finite-dimensional kernel, then the following statements are equivalent:*

(i) $T^{n+1}(X)$ *is closed in the operator range topology on $T^n(X)$;*
(ii) $T^{n+2}(X)$ *is closed in the operator range topology on $T^{n+1}(X)$;*
(iii) $T^{n+2}(X)$ *is closed in the operator range topology on $T^n(X)$.*

Proof Since $T^{n+1}(X)$ is continuously embedded in $T^n(X)$ it suffices to prove the equivalence (i) $\Leftrightarrow$ (ii). Denote by $T_n : T^n(X) \to T^{n+1}(X)$ the restriction $T|T^n(X)$. Obviously, T_n is onto, and by Lemma 1.75, $T^{n+2}(X)$ is closed in the topology of $T^{n+1}(X)$ if and only if $T_n^{-1}(T^{n+2}(X))$ is closed in the topology of $T^n(X)$. By assumption $T^{n+1}(X)$ is an operator range which has finite codimension in $T_n^{-1}(T^{n+2}(X))$. By using the fact that operator ranges of finite codimension in a Banach space are closed, it then follows that $T^{n+1}(X)$ is closed in the topology of $T^n(X)$ if and only if $T_n^{-1}(T^{n+2}(X))$ is closed in this topology. $\blacksquare$

Definition 1.77 An operator $T \in L(X)$, X a Banach space, is said to have *topological uniform descent for $n \geq d$* if T has uniform descent for $n \geq d$ and $T^n(X)$ is closed in the operator range topology of $T^d(X)$ for each $n \geq d$.

Theorem 1.78 *If $T \in L(X)$, X a Banach space, has uniform descent for $n \geq d$, then the following assertions are equivalent:*

(i) *T has topological uniform descent for $n \geq d$;*
(ii) *There is an integer $n \geq d$ and $k \in \mathbb{N}$ such that $T^{n+k}(X)$ is closed in the operator range topology on $T^n(X)$;*
(iii) *For each $n \geq d$ and $k \in \mathbb{N}$, $T^{n+k}(X)$ is closed in the operator range topology on $T^n(X)$;*
(iv) *There is an $n \geq d$ and $k \in \mathbb{N}$ such that $\ker T^n + T^k(X)$ is closed in X;*
(v) *For all $n \geq d$ and for all $k \in \mathbb{N}$, $\ker T^n + T^k(X)$ is closed in X. This is also true for $k = \infty$.*

Proof The equivalence of (i), (ii), and (iii) is immediate from Theorem 1.74. Now, for each fixed n and k, T^n induces a bounded operator from X to $T^n(X)$, with respect to the operator range topology. Hence, from Lemma 1.75, $T^{n+k}(X)$ is closed in the range operator topology on $T^n(X)$ if and only if $T^{-n}[T^{n+k}(X)] = \ker T^n + T^k(X)$ is a closed subspace in the topology of X. This completes the proof. ∎

The following theorem is the major result on the structure of operators with topological uniform descent. In the sequel if E is a subspace of $T^d(X)$, then $\overline{E}^d$ denotes the closure of E in the operator range topology.

Theorem 1.79 *Let $T \in L(X)$ be with topological uniform descent for $n \geq d$. Then we have:*

(i) *The restriction of T to $T^\infty(X)$ is onto.*
(ii) *The map induced by T on $T^d(X)/T^\infty(X)$ is bounded below.*
(iii) *The restriction of T to $T^d(X) \cap \overline{\mathcal{N}^\infty(T)}$ is onto.*
(iv) *The map $\hat{T} : X/\overline{\mathcal{N}^\infty(T)} \to X/\overline{\mathcal{N}^\infty(T)}$ defined by $\hat{T}[x] = [Tx]$ is bounded below.*

Proof Let $Y := T^d(X)$ and denote by S the restriction $T|Y$. Then $S \in L(Y)$ has topological uniform descent for $n \geq 0$ and has closed range. From Theorem 1.74, part (ii), we have that $\ker S \subseteq S^n(Y)$ for all n, so that

$$S^{-1}(S^\infty(Y)) = \bigcap_{n=1}^\infty S^{-1}(S^{n+1}(Y)) = \bigcap_{n=1}^\infty (S^n(Y) + \ker S) = S^\infty(Y).$$

A simple application of Lemma 1.68, part (d), yields $S(S^\infty(Y)) = S^\infty(Y)$. Now, $S^\infty(Y) = T^\infty(X)$, so the restriction of T to $T^\infty(X)$ is onto. Because

$$S^{-1}(S^\infty(Y)) = T^d(X) \cap T^{-1}(T^\infty(X)),$$

the map induced by T on $T^d(X)/T^\infty(X)$ is injective, and since S has closed range, this induced map also has closed range. This proves (i) and (ii).

Clearly, for every operator T we have $T^{-1}(\mathcal{N}^\infty(T)) = \mathcal{N}^\infty(T)$. Moreover, S has uniform descent for $n \geq 0$ and since, by part (v) of Theorem 1.74, we have $\mathcal{N}^\infty(S) \subseteq S(Y)$, then $S(\mathcal{N}^\infty(S)) = \mathcal{N}^\infty(S)$. A direct application of Lemma 1.75 gives $S(\overline{\mathcal{N}^\infty(S)}^d) = \overline{\mathcal{N}^\infty(S)}^d$ and $S^{-1}(\overline{\mathcal{N}^\infty(S)}^d) = \overline{\mathcal{N}^\infty(S)}^d$.

It is easily seen that $S^\infty(Y) = T^d(X) \cap \mathcal{N}^\infty(T)$, hence $T^{-d}(\mathcal{N}^\infty(S)) = \mathcal{N}^\infty(T)$. Applying Lemma 1.75, part (b), to the map induced by T^d from X onto $T^d(X)$ we then obtain $\overline{\mathcal{N}^\infty(T)} = T^{-d}(\overline{\mathcal{N}^\infty(S)}^d)$, so we have

$$T^{-1}(\overline{\mathcal{N}^\infty(T)}) = T^{-1}T^{-d}(\overline{\mathcal{N}^\infty(S)}^d) = T^{-d}T^{-1}(\overline{\mathcal{N}^\infty(S)}^d)$$
$$= T^{-d}(\overline{\mathcal{N}^\infty(S)}^d) = \overline{\mathcal{N}^\infty(T)}.$$

Hence the map $\hat{T}$ induced by T on $X/\overline{\mathcal{N}^\infty(T)}$ is one-to-one. From part (vi) and part (v) of Theorem 1.74 it then follows that $T(X) + \mathcal{N}^\infty(T)$ is closed, and hence $T(X) + \overline{\mathcal{N}^\infty(T)} = T(X) + \mathcal{N}^\infty(T)$ is closed, so (iv) is proved.

From $\overline{\mathcal{N}^\infty(S)}^d = \overline{T^d(X) \cap \mathcal{N}^\infty(T)}^d$ we also have

$$T(\overline{T^d(X) \cap \mathcal{N}^\infty(T)}^d) = \overline{T^d(X) \cap \mathcal{N}^\infty(T)}^d,$$

so, in order to complete the proof of part (iii), it suffices to prove the equality

$$T^d(X) \cap \overline{\mathcal{N}^\infty(T)} = \overline{T^d(X) \cap \mathcal{N}^\infty(T)}^d. \tag{1.21}$$

Since $T^{-d}(\overline{\mathcal{N}^\infty(T)}) = \overline{\mathcal{N}^\infty(T)}$, the left-hand side of equality (1.21) is $T^d(\overline{\mathcal{N}^\infty(T)})$. An application of part (i) of Lemma 1.75 to the map induced by T^d from X onto $T^d(X)$ then shows that the right-hand side of equality (1.21) coincides with $T^d(\overline{\mathcal{N}^\infty(T)})$, hence equality (1.21) is proved and this completes the proof of the theorem. ∎

The following lemma concerns some identities involving ranges and kernels of operators having topological uniform descent that will be needed in the sequel.

Lemma 1.80 *Let $T \in L(X)$ be with topological uniform descent for $n \geq d$. Then we have:*

(i) $T^\infty(X) + \ker T^d = T^\infty(X) + \mathcal{N}^\infty(T) = T^\infty(X) + \overline{\mathcal{N}^\infty(T)}$.

(ii) $T^d(X) \cap \mathcal{N}^\infty(T) = T^\infty(X) \cap \mathcal{N}^\infty(T)$.

(iii) $T^d(X) \cap \overline{\mathcal{N}^\infty(T)} = T^\infty(X) \cap \overline{\mathcal{N}^\infty(T)}$.

(iv) $\overline{[T^\infty(X) \cap \mathcal{N}^\infty(T)]} = \overline{[T^\infty(X) \cap \overline{\mathcal{N}^\infty(T)}]}$.

Proof

(i) From Theorem 1.74, part (iii), we know that, for each $n \geq d$ we have $T^n(X) \cap$ ker $T \subseteq T^\infty(X)$. Applying T^{-n} to both sides of the latter inclusion, and by using part (c) of Lemma 1.68, we then obtain ker $T^{n+1} \subseteq T^\infty(X) + $ ker T^n, or equivalently,

$$T^\infty(X) + \text{ker } T^{n+1} = T^\infty(X) + \text{ker } T^n.$$

By induction we then deduce that

$$T^\infty(X) + \text{ker } T^d = T^\infty(X) + \mathcal{N}^\infty(T).$$

Since, $T^\infty(X) + $ ker T^d is a closed subspace, by part (v) of Theorem 1.78 we then conclude that

$$T^\infty(X) + \mathcal{N}^\infty(T) = T^\infty(X) + \overline{\mathcal{N}^\infty(T)}.$$

(ii) The proof follows similarly to part (i), by applying T^n to both sides of the set inclusion $\mathcal{N}^\infty(T) \subseteq T(X) + $ ker T^n for $n \geq d$.

(iii) Using part (ii), equality (1.21), and the fact that $T^\infty(X)$ is closed in the operator range topology on $T^d(X)$, we obtain

$$T^\infty(X) \cap \overline{\mathcal{N}^\infty(T)} \subseteq T^d(X) \cap \overline{\mathcal{N}^\infty(T)} = \overline{(T^\infty(X) \cap \mathcal{N}^\infty(T)}^d$$
$$\subseteq T^\infty(X) \cap \overline{\mathcal{N}^\infty(T)},$$

from which the equality of part (iii) follows directly.

(iv) This follows by taking the closure in the topology of X of the equality (iii) proved above. ∎

The quantity $k(T)$ defined in Definition 1.72 may be characterized in several ways as follows:

Theorem 1.81 *If $T \in L(X)$ then $k(T)$ coincides with each of the following quantities:*

(a) $\sup\{\dim [\text{ker } T/(\text{ker } T \cap T^n(X))]\};$
(b) $\dim [\text{ker } T/ \text{ker } T \cap T^\infty(X)];$
(c) $\sup\{\dim [T(X) + \text{ker } T^n)/T(X)]\};$
(d) $\dim [(T(X) + \mathcal{N}^\infty(T))/T(X)].$

Proof From part (i) of Theorem 1.70 it follows that the dimension of the quotient ker $T/(\text{ker } T \cap T^n(X))$ for every $n \in \mathbb{N}$ is equal to $k_0(T) + k_1(T) \cdots + k_{n-1}(T)$, so, from the definition of $k(T)$, we conclude that $k(T)$ coincides with the quantity (a). A similar argument, by using part (ii) and (iii) of Theorem 1.70, shows that $k(T)$ coincides with the quantity in (c).

Evidently, the quantities in (a) and (b) coincide when $k(T) = \infty$, since ker $T \cap T^\infty(X) \subseteq$ ker $T \cap T^n(X)$. Suppose now that $k(T) < \infty$. From part (a) it then follows that for some integer d we have

$$\text{ker } T \cap T^d(X) = \text{ker } T \cap T^n(X) \subseteq T^\infty(X) \quad for \ n \geq d.$$

From this it then follows that the quantities in (a) and (b) also coincide when $k(T) < \infty$. The equality of the quantities in (c) and (d) may be proved in a similar way. ∎

The following characterization of operators having almost uniform descent, and closed ranges $T^k(X)$ for all $k \in \mathbb{N}$, follows immediately from Lemma 1.76.

Theorem 1.82 *Suppose that for $T \in L(X)$ we have $k(T) < \infty$. Then the following statements are equivalent:*

(i) $T^k(X)$ *is closed for some $k \in \mathbb{N}$;*
(ii) $T^k(X)$ *is closed for each $k \in \mathbb{N}$;*
(iii) T *has topological uniform descent for $n \geq k(T)$.*

Essentially semi-regular operators have topological uniform descent. More precisely, we have:

Corollary 1.83 *Let $T \in L(X)$. We have:*

(i) *T is a semi-regular operator if and only if T has topological uniform descent for $n \geq 0$.*
(ii) *If T is a essentially semi-regular operator then T has topological uniform descent for $n \geq d := k(T)$. Furthermore, every Kato-type operator of order d has topological uniform descent for $n \geq d$.*

Proof

(i) By Theorem 1.44, if T is semi-regular then T^k is semi-regular for every $k \in \mathbb{N}$, thus $T^k(X)$ is closed. Since ker $T \subseteq T^k(X)$ for every $k \in \mathbb{N}$ it then follows that ker $T + T^k(X) = T^k(X)$ is closed, so $T \in L(X)$ has topological uniform descent for $n \geq 0$, by part (iv) of Theorem 1.78.
(ii) We know that T is essentially semi-regular if $T(X)$ is closed and there exists a finite-dimensional subspace F for which ker $T \subseteq T^\infty(X) + F$. By Theorem 1.81 it then follows that the essentially semi-regular are precisely those operators T for which $T(X)$ is closed and $k(T) < \infty$. Finally, if T is essentially semi-regular then T^n is essentially semi-regular for each n, so $T^n(X)$ is closed for each n, and hence, by Theorem 1.82, T has topological uniform descent for $n \geq k(T)$. The second assertion follows from part (iii) of Theorems 1.64 an 1.78. ∎

The next theorem describes the quantity $k(T)$ in terms of the nullity and the defect of maps induced by T.

Theorem 1.84 *Let $T \in L(X)$. Then we have*

(i) $k(T)$ *is the dimension of the kernel of the map induced by T on $X/T^\infty(X)$.*
(ii) $k(T)$ *is the dimension of the range of the restriction $T|\mathcal{N}^\infty(T)$.*
(iii) *If T has closed range then $k(T)$ is the codimension of the range of the restriction $T|\overline{\mathcal{N}^\infty(T)}$.*

Proof We only prove statement (iii), omitting the proofs of part (i) and part (ii), which are similar and simpler.

Suppose first that $k(T) < \infty$. From part (iv) of Theorem 1.79 we have $T^{-1}(\overline{\mathcal{N}^\infty(T)}) = \overline{\mathcal{N}^\infty(T)}$. Using Lemma 1.68 and part (i) of Lemma 1.80, we see that the cokernel of the restriction $T|\overline{\mathcal{N}^\infty(T)}$ is

$$\frac{\overline{\mathcal{N}^\infty(T)}}{T(X) \cap \overline{\mathcal{N}^\infty(T)}} \cong \frac{T(X) + \overline{\mathcal{N}^\infty(T)}}{T(X)} = \frac{T(X) + \mathcal{N}^\infty(T)}{T(X)},$$

and the dimension of these quotients is exactly $k(T)$, by Theorem 1.81.

To conclude the proof, consider the case where $k(T) = \infty$. Statement (iii) then follows similarly from part (d) of Theorem 1.81, together with the fact that $T(\overline{\mathcal{N}^\infty(T)}) \subseteq T(X) \cap \overline{\mathcal{N}^\infty(T)}$, so the proof is complete. ∎

In the remaining part of this section we give some perturbation results of operators T having topological uniform descent. We consider bounded operators S which commute with T for which $T - S$ is "sufficiently small" in the sense of the following definition.

Definition 1.85 Suppose that $T \in L(X)$, X a Banach space, has topological uniform ascent for $n \geq d$, and let $S \in L(X)$ be an operator which commutes with T. We say that $S - T$ is *sufficiently small* if the norm of the restriction $(S - T)|T^d(X)$ is less than the reduced minimum modulus $\gamma(T|T^d(X))$.

Note that if $T^d(X)$ is closed in X and is given the restriction norm, it is easily seen that $\|S - T\|$ is no greater than the norm of its restriction to $T^d(X)$, so the definition above is essentially a restriction of $\|S - T\|$.

The next theorem gives some information on sufficiently small perturbations when T has closed range and has topological uniform descent for $n \geq 0$. This in particular applies to semi-Fredholm operators.

Theorem 1.86 *Suppose that $T \in L(X)$ has closed range and has topological uniform descent for $n \geq 0$. Suppose that $ST = TS$ and $S - T$ is sufficiently small. Then*

(i) $S(X)$ *is closed and S has topological uniform descent for $n \geq 0$.*
(ii) $\dim(S^n(X)/S^{n+1}(X)) = \dim(X/T(X))$ *for all $n \geq 0$.*
(iii) $\dim(\ker S^{n+1}/\ker S^n) = \dim \ker T$ *for all $n \geq 0$.*
(iv) $S^\infty(X) = T^\infty(X)$.
(v) $\overline{\mathcal{N}^\infty(S)} = \overline{\mathcal{N}^\infty(T)}$.

Proof Let Y be any of the four Banach spaces $T^\infty(X)$, $X/T^\infty(X)$, $\overline{\mathcal{N}^\infty(T)}$ and $X/\overline{\mathcal{N}^\infty(T)}$. The proof is based on considering the maps induced by T and S on these spaces. Let $\hat{T}$ and $\hat{S}$ be the maps induced by T and S on Y. By Theorem 1.79 we know that $\hat{T}$ is either bounded below or onto, and obviously $\|\hat{S} - \hat{T}\| \le \|S - T\|$. Since both $T^\infty(X)$ and $\overline{\mathcal{N}^\infty(T)}$ contain ker T and are contained in $T(X)$, by using part (iii) and part (iv) of Theorem 1.74, we easily obtain $\gamma(\hat{T}) \ge \gamma(T)$. Hence, if $S - T$ is sufficiently small, then $\|\hat{S} - \hat{T}\| \le \gamma(\hat{T})$. Suppose now that $\hat{T}$ is bounded below. A trivial calculation yields that $\hat{S}$ is also bounded below and

$$\gamma(\hat{S}) \ge \gamma(\hat{T}) - \|\hat{S} - \hat{T}\|. \tag{1.22}$$

Analogously, by duality, if $\hat{T}$ is onto then $\hat{S}$ is onto and inequality (1.22) also holds.

To show assertion (i), let us consider the cases $Y = T^\infty(X)$ and $Y = X/\overline{\mathcal{N}^\infty(T)}$. Then $S(T^\infty(X)) = T^\infty(X)$ and $S^{-1}(\overline{\mathcal{N}^\infty(T)}) \subseteq \overline{\mathcal{N}^\infty(T)}$, thus, by using Lemma 1.80, part (i), we obtain

$$\text{ker } S \subseteq \overline{\mathcal{N}^\infty(S)} \subseteq \overline{\mathcal{N}^\infty(T)} \subseteq T^\infty(X) \subseteq S^\infty(X) \subseteq S(X), \tag{1.23}$$

which, by Theorem 1.74, implies that S has uniform descent for $n \ge 0$.

(ii) Consider the case $Y = X/T^\infty(X)$. Then $\hat{T}$ and $\hat{S}$ are both bounded below. Moreover, $Y/\hat{T}(Y)$ and $Y/\hat{S}(Y)$ both have the same dimension (a proof of this may be found in [156, Corollary V.1.3, p. 111]). From (1.23) it then follows that $S(X)$ is closed and that the quotients $X/S(X)$ and $X/T(X)$ have the same dimension. The assertion (ii) then follows from the fact that S has uniform descent for $n \ge 0$, by part (i).

(iv) To prove the equality $S^\infty(X) = T^\infty(X)$, we use the maps induced on $X/T^\infty(X)$. For each $0 \le \lambda \le 1$, define $S_\lambda := T + \lambda(S - T)$. We need only to show that the hyper-range $S_\lambda^\infty(X)$ is locally constant. Since each S_λ has closed range and uniform descent for $n \ge 0$, it will be enough to prove that if $\|S - T\| \le \frac{\gamma(T)}{2}$, then $S^\infty(X) = T^\infty(X)$. By (1.23) we need only to show that $S^\infty(X) \subseteq T^\infty(X)$. As we observed above we have $\|\hat{S} - \hat{T}\| \le \|S - T\|$ and $\gamma(\hat{T}) \ge \gamma(T)$, so from (1.22) we have $\|\hat{T} - \hat{S}\| \le \gamma(\hat{S})$. Therefore, by using the inclusions (1.23), with S replaced by $\hat{T}$ and T replaced by $\hat{S}$, we obtain that the hyper-range of $\hat{S}$ is contained in the hyper-range of $\hat{T}$ and this is equal to 0. Hence $S^\infty(X) \subseteq T^\infty(X)$, thus equality (iv) is proved.

The proofs of parts (iii) and (v) are omitted, since they follow by using an argument similar to that given above, where the maps induced on $X/T^\infty(X)$ are replaced by the maps induced on $\overline{\mathcal{N}^\infty(T)}$. ∎

Let us consider the case when T has topological uniform descent $n \ge d \ne 0$. The previous theorem describes the restriction of the perturbed operator S to $T^d(X)$. In the sequel we need the following technical result, which provides a useful tool for studying S in terms of its restrictions to $T^d(X)$. We omit the proof, which is very similar to the proof of Theorem 1.70.

Lemma 1.87 *Suppose that T and S are commuting linear operators on a vector space X. If $n, d \in \mathbb{N}$ are nonnegative integers, we have:*

(i) *The mapping Φ induced by T^d from the quotient $S^n(X)/S^{n+1}(X)$ to the quotient $(S^n T^d)(X)/(S^{n+1} T^d)(X)$ is onto. The kernel of Φ is naturally isomorphic to the space $(S^n(X) \cap \ker T^d)/(S^{n+1}(X) \cap \ker T^d)$.*

(ii) *The mapping Ψ induced by the identity on X from the quotient $(\ker S^{n+1} \cap T^d(X))/(\ker S^n \cap T^d(X))$ to $\ker S^{n+1}/\ker S^n$ is injective. Furthermore, the cokernel of the mapping Ψ is naturally isomorphic to the quotient $(\ker S^{n+1} + T^d(X))/(\ker S^n + T^d(X))$.*

(iii) *T^d induces an isomorphism from $\ker T^{d+n+1}/\ker T^{d+n}$ onto the quotient $(\ker T^{n+1} \cap T^d(X))/(\ker T^n \cap T^d(X))$.*

Now we consider arbitrary small commuting perturbations of operators having topological uniform descent.

Theorem 1.88 *Suppose that $T \in L(X)$ has topological uniform descent for $m \geq d$, and that $S \in L(X)$ commutes with T. If $S - T$ is sufficiently small, then:*

(i) *$\dim(T^m(X)/T^{m+1}(X)) \leq \dim(S^n(X)/T^{n+1}(X))$ for all $n \geq 0$ and $m \geq d$.*

(ii) *$\dim(\ker T^{m+1}/\ker T^m)) \leq \dim(\ker S^{n+1}/\ker S^n)$ for all $n \geq 0$ and $m \geq d$*

(iii) *$T^\infty(X) \subseteq S^\infty(X) \subseteq T^\infty(X) + \mathcal{N}^\infty(T)$.*

(iv) *$T^\infty(X) \cap \overline{\mathcal{N}^\infty(T)} \subseteq \overline{\mathcal{N}^\infty(S)} \subseteq \overline{\mathcal{N}^\infty(T)}$.*

Proof Since T has topological uniform descent for $m \geq d$, the dimension of $T^m(X)/T^{m+1}(X)$, as well as the dimension of $\ker T^{m+1}/\ker T^m$, is constant for $m \geq d$. Therefore, parts (i) and (ii) follow directly from Theorem 1.86 and Lemma 1.87.

For the proof of (iii), let us denote by $\widehat{S}$ the restriction of S to $Y := T^d(X)$. By Theorem 1.86, part (iv), we have $\widehat{S}^\infty(Y) = T^\infty(X)$, and since $\widehat{S}^\infty(Y) \subseteq S^\infty(X)$ we then have $T^\infty(X) \subseteq S^\infty(X)$. Now, applying T^{-d} to both sides of the identity $T^\infty(X) = \bigcap_{n=1}^\infty T^d S^n(X)$, and using part (c) of Lemma 1.68 and part (a) of Theorem 1.79, we then obtain

$$T^\infty(X) + \ker T^d = \bigcap_{n=1}^\infty [S^n(X) + \ker T^d] \supseteq S^\infty(X).$$

The assertion (iii) then follows directly from part (i) of Lemma 1.80.

To prove (iv), observe that from part (v) of Theorem 1.86 we obtain

$$\overline{T^d(X) \cap \mathcal{N}^\infty(T)}^d = \overline{T^d(X) \cap \mathcal{N}^\infty(\widehat{S})}^d \subseteq \overline{S^\infty(X)}.$$

The first inclusion of part (iv) now follows from the equality (1.21) and part (iii) of Lemma 1.80. Finally, as observed in the proof of Theorem 1.86, the map induced by S on $\overline{T^d(X)/T^d(X) \cap \mathcal{N}^\infty(T)}^d$ is one-to-one. Hence, as in the proof of part (iv) of Theorem 1.79, we can conclude that $S^{-1}(\overline{\mathcal{N}^\infty(T)}) \subseteq \overline{\mathcal{N}^\infty(T)}$. This obviously implies $\overline{\mathcal{N}^\infty(S)} \subseteq \overline{\mathcal{N}^\infty(T)}$, so the proof is complete. ∎

We now consider the important case, when $S - T$ is invertible. This case, of course, subsumes the case $S = \lambda I - T$, when $\lambda \neq 0$.

Theorem 1.89 *Suppose that $T \in L(X)$ has topological uniform descent for $n \geq d$, and that $S \in L(X)$ commutes with T. If $S - T$ is sufficiently small and is invertible, then:*

(i) *S has closed range and topological uniform descent for $n \geq 0$.*
(ii) *$\dim(S^n(X)/S^{n+1}(X)) = \dim(T^d(X)/T^{d+1}(X))$, for all $n \geq 0$.*
(iii) *$\dim(\ker T^{n+1}/\ker T^n) = \dim(\ker T^{d+1}/\ker T^d)$ for all $n \geq 0$.*
(iv) *$S^\infty(X) = T^\infty(X) + \mathcal{N}^\infty(T)$.*
(v) *$\overline{\mathcal{N}^\infty(S)} = \overline{[T^\infty(X) \cap \mathcal{N}^\infty(T)]}$.*

Proof We first show the inclusions

$$\mathcal{N}^\infty(S) \subseteq T^\infty(X) \quad \text{and} \quad \mathcal{N}^\infty(T) \subseteq S^\infty(X).$$

Let us consider the invertible operator $U := S - T$. Since S and T commute, it follows from the binomial theorem that for each fixed k, there is a bounded operator W, depending on k, for which

$$U^{-k}S^k = I - TW.$$

Hence, if $x \in \ker S^k$ then $x = T^n W^n x \in T^n(X)$ for all n. Since k is arbitrary, we then have $\mathcal{N}^\infty(S) \subseteq T^\infty(X)$, and by interchanging the roles of S and T we also deduce the second inclusion $\mathcal{N}^\infty(T) \subseteq S^\infty(X)$.

(i) By Theorem 1.86, part (i), we know that $ST^d(X)$ is closed in the topology of $T^d(X)$, so that

$$T^{-d}(ST^d(X)) = S(X) + \ker T^d = S(X)$$

is closed in X. The inclusion $\mathcal{N}^\infty(S) \subseteq T^\infty(X)$, together with part (iii) of Theorem 1.88, yields $\mathcal{N}^\infty(S) \subseteq S^\infty(X)$, which implies, by Theorem 1.74, that S has topological uniform descent for $n \geq 0$.

The equalities in (ii) and (iii) easily follow from Theorem 1.86, part (ii) and part (iii), and Theorem 1.88, part (i) and part (ii).

(iv) The equality easily follows from the inclusion $\mathcal{N}^\infty(T) \subseteq S^\infty(X)$, together with Theorem 1.88, part (iv).

(v) By Theorem 1.88, part (iv), we obtain $T^\infty(X) \cap \overline{\mathcal{N}^\infty(T)} \subseteq \overline{\mathcal{N}^\infty(S)}$ and $\mathcal{N}^\infty(S) \subseteq T^\infty(X) \cap \overline{\mathcal{N}^\infty(T)}$. By taking closures, and applying part (v) of Lemma 1.80, we then conclude the proof of (iv). ∎

Since an operator which has topological uniform descent $n \geq 0$ is semi-regular we easily have:

Corollary 1.90 *Suppose that $T \in L(X)$ has topological uniform descent for $n \geq d$. If $S \in L(X)$ is an invertible operator which commutes with T and is sufficiently small, then $T + S$ is semi-regular. Moreover,*

$$(T + S)^\infty(X) = T^\infty(X) + \mathcal{N}^\infty(T) \quad and \quad \overline{\mathcal{N}^\infty(T + S)} = \overline{T^\infty(X) \cap \mathcal{N}^\infty(T)}.$$

Theorem 1.89 has some important consequences, if we assume further properties on ker $T \cap T^d(X)$ and $T(X) + $ ker T^d.

Corollary 1.91 *Suppose that $T \in L(X)$ has topological uniform descent for $n \geq d$, and that $S \in L(X)$ commutes with T. If $S - T$ is sufficiently small and is invertible, then the following assertion holds:*

(i) *If ker $T \cap T^d(X)$ has finite dimension, then S is upper semi-Fredholm and $\alpha(S) = \dim(\ker T \cap T^d(X))$.*
(ii) *If $T(X) + $ ker T^d has finite codimension, then S is lower semi-Fredholm and $\beta(S) = \operatorname{codim}(T(X) + \ker T^d)$.*

Proof We know that $S(X)$ is closed, by Theorem 1.89. Furthermore,

$$\dim \frac{\ker T^{n+1}}{\ker T^n} = \dim(\ker T \cap T^d(X)),$$

and

$$\dim \frac{T^n(X)}{T^{n+1}(X)} = \operatorname{codim}(T(X) + \ker T^d).$$

Again, by Theorem 1.89 we have that $\alpha(S) = \dim(\ker T \cap T^d(X))$ and $\beta(S) = \operatorname{codim}(T(X) + \ker T^d)$, from which we deduce that (i) and (ii) hold.   ∎

The next corollary is an immediate consequence of part (i) and part (ii) of Theorem 1.89 applied to some special classes of operators having topological uniform descent.

Corollary 1.92 *Suppose that $T \in L(X)$ has topological uniform descent for $n \geq d$, and that $S \in L(X)$ commutes with T. If $S - T$ is sufficiently small and is invertible, then:*

(i) *S has infinite ascent or descent if and only if T does.*
(ii) *S cannot have finite ascent $p(S) > 0$, or finite descent $q(S) > 0$.*
(iii) *S is onto if and only if T has finite descent.*
(iv) *S is injective (or also bounded below) if and only if T has finite descent.*
(v) *S is invertible if and only if $p(T) = q(T) < \infty$.*

The following corollary is just a special case of Corollary 1.92, part (v).

Corollary 1.93 *Suppose that λ belongs to the boundary of the spectrum $\partial\sigma(T)$, and $\lambda I - T$ has uniform descent. Then $p(\lambda I - T) = q(\lambda I - T) < \infty$, so λ is a pole of the resolvent.*

A more precise result than that of Corollary 1.93 will be obtained in Chap. 2, by using the localized SVEP. We conclude this section by proving that the dual of an operator having topological uniform descent may have not topological uniform descent.

Example 1.94 We first give an example for which $T^2(X) = T(X)$ and $T(X)$ is not closed. Let X be a Hilbert space with an orthonormal basis $(e_{i,j})$. Let T be defined by:

$$Te_{i,j} := \begin{cases} 0 & \text{if } j = 1, \\ \frac{1}{i}e_{i,1} & \text{if } j = 2, \\ e_{i,j-1} & \text{otherwise.} \end{cases}$$

Let M_1 denote the subspace generated by the set $\{e_{i,j} : j \geq 2, i \geq 1\}$, and M_2 the subspace generated by the set $\{e_{i,2} : i \geq 1\}$. It is easily seen that $T^2(X) = T(X) = M_1 + T(M_2)$. Further, if M_3 denotes the subspace generated by the set $\{e_{i,1} : i \geq 1\}$, the intersection $T(X) \cap M_3$ is not closed, from which we deduce that $T(X)$ is not closed. Therefore, $q(T) \leq 1$ so that T has topological uniform descent. We show that T^* does not have uniform topological descent. In fact, suppose that T^* has topological uniform descent $n \geq d$. Then $T^*(X^*) + \ker T^{*d}$ is closed, by part (v) of Theorem 1.78. Since $T^2(X) = T(X)$ we then have $\ker T^* = \ker T^{*2}$, so $p(T^*) \leq 1$. By Theorem 1.19 we also have $T^*(X^*) \cap \ker T^{*d} = \{0\}$ and this, by Theorem 1.30, implies that T^* has closed range, or equivalently, $T(X)$ is closed, and this is impossible. Therefore, T^* does not have uniform topological descent.

1.7 Quasi-Fredholm Operators

The class of quasi-Fredholm operator was first introduced by Labrousse [208], who considered this class in the case of Hilbert space operators. Consider the set

$$\Delta(T) := \{n \in \mathbb{N} : m \geq n, m \in \mathbb{N} \Rightarrow T^n(X) \cap \ker T \subseteq T^m(X) \cap \ker T\}.$$

The *degree of stable iteration* is defined as $\operatorname{dis}(T) := \inf \Delta(T)$ if $\Delta(T) \neq \emptyset$, while $\operatorname{dis}(T) = \infty$ if $\Delta(T) = \emptyset$.

Definition 1.95 $T \in L(X)$ is said to be *quasi-Fredholm of degree d* if there exists a $d \in \mathbb{N}$ such that:

(a) $\text{dis}(T) = d$,
(b) $T^n(X)$ is a closed subspace of X for each $n \geq d$,
(c) $T(X) + \ker T^d$ is a closed subspace of X.

Evidently, condition (a) entails that the quantity $k_n(T)$, introduced in the previous section, is equal to 0 for every $n \geq d$, hence, from part (ii) of Theorem 1.74, we see that every quasi-Fredholm operator has uniform descent for $n \geq d$. In the sequel, by $QF(d)$ we denote the class of all quasi-Fredholm operators of degree d.

Theorem 1.96 *If $T \in L(X)$ then the following implications hold:*

$$T \in \Phi_{\pm}(X) \;\Rightarrow\; T \text{ quasi-Fredholm } \;\Rightarrow\; T \text{ has topological uniform descent.}$$

In particular, if $T \in QF(d)$ then T has topological uniform descent for $n \geq d$.

Proof Every semi-Fredholm operator T has topological uniform descent for $n \geq 0$, by Corollary 1.83, so, by Theorem 1.74, $T^n(X) + \ker T$ is constant for all $n \geq 1$. Moreover, T^n is semi-Fredholm for all $n \in \mathbb{N}$, hence all $T^n(X)$ are closed. Condition (c) is trivially satisfied, by Theorem 1.78. This shows the first implication. As observed before, T has uniform descent. Since condition (iv) of Theorem 1.78 is satisfied by part (c) of the definition of quasi-Fredholm operators, it then follows that T has topological uniform descent for $n \geq d$. ∎

Remark 1.97 The converse of the second implication in Theorem 1.96 does not hold in general. The operator defined in Example 1.94 has topological uniform descent but is not quasi-Fredholm since $T^n(X) = T(X)$ is not closed for all $n \in \mathbb{N}$.

In the sequel we will need the following lemmas.

Lemma 1.98 *Let $T \in L(X)$ and let $m \in \mathbb{N} \cup \{0\}$, and $n \geq k \geq 1$. If $T^n(X) + \ker T^m$ is closed then $T^{n-k}(X) + \ker T^{m+k}$ is closed. In particular, if $T^n(X)$ is closed then $T(X) + \ker T^{n-1}$ is closed.*

Proof We first show the equality

$$T^{n-k}(X) + \ker T^{m+k} = T^{-k}[T^n(X) + \ker T^m]. \tag{1.24}$$

To show this equality it suffices to prove the inclusion $\supseteq$, since the opposite inclusion is trivial. Let $z \in T^{-k}[T^n(X) + \ker T^m]$, i.e.,

$$T^k z \in T^n(X) + \ker T^m.$$

Then

$$T^k z = T^n x + u \quad \text{for some } x \in X, u \in \ker T^m,$$

from which we obtain

$$u = T^k z - T^k (T^{n-k} x) \in T^k(X).$$

Let $v \in X$ be such that $u = T^k v$. Clearly,

$$T^{m+k} v = T^m T^k v = T^m u = 0,$$

from which we obtain $v \in \ker T^{m+k}$. Then we have

$$T^k(z - T^{n-k} x - v) = T^k z + T^n x - T\grave{\imath}kv = T^k z - T^n x - u = 0,$$

thus, $z - T^{n-k} x - v \in \ker T^k$ and hence

$$z \in T^{n-k}(X) + \ker T^{m+k} + \ker T^k.$$

Because $\ker T^n \subseteq \ker T^{n+k}$ we then conclude that $z \in T^{n-k}(X) + \ker T^{m+k}$, so the equality (1.24) is proved. Since $T^n(X) + \ker T^m$ is closed by assumption, this equality entails that $T^{n-k}(X) + \ker T^{m+k}$ is closed.

The last assertion is immediate, by taking $m = 0$ and $n = k + 1$. ∎

Lemma 1.99 *Let $T \in L(X)$ and let $n \in \mathbb{N} \cup \{0\}$. If $T^n(X)$ is closed and $T(X) + \ker T^n$ is closed then $T^{n+1}(X)$ is closed.*

Proof Let (y_k) be a sequence of elements of $T^{n+1}(X)$ which converges to some $z \in X$. Then there exist $u_k \in X$ such that $y_k = T^{n+1} u_k$, and since $T^{n+1}(X) \subseteq T^n(X)$, we have that $y_k \in T^n(X)$, and hence $z \in T^n(X)$. Let $u \in X$ be such that $z = T^n u$. Obviously, $T^n(u - Tu_k) \to 0$ as $k \to \infty$. Consider the operator $\overline{T^n} : X/\ker T^n \to X$ induced by T^n. Evidently, $\overline{T^n}$ is bounded below, and $\overline{T^n}(u - Tu_k + \ker T^n) \to 0$ as $k \to \infty$, so the quotient classes $u - Tu_k + \ker T^n \to 0$ in $X/\ker T^n$. Therefore there exists for each k an element $v_k \in \ker T^n$ such that $Tu_k + v_k \to u \in T(X) + \ker T^n$, from which we conclude $z \in T^{n+1}(X)$. ∎

The following result plays a crucial role in the characterization of quasi-Fredholm operators.

Theorem 1.100 *Let $T \in L(X)$, $d \in \mathbb{N}$ and suppose that $k_i(T) < \infty$ for all $i \geq d$. Then the following statements are equivalent:*

(i) *there exists an $n \geq d + 1$ such that $T^n(X)$ is closed;*
(ii) *$T^j(X)$ is closed for every $j \geq d$;*
(iii) *$T^j(X) + \ker T^m$ is closed for all $m, j \in \mathbb{N}$ with $m + j \geq d$.*

Proof The implications (iii) ⇒ (ii) ⇒ (i) are clear, while the implication (ii) ⇒ (iii) follows from Lemma 1.98. It only remains to show that (i) ⇒ (ii).

Suppose $T^n(X)$ is closed and let us consider first the case $n \geq d + 1$. Then, by Lemma 1.98, $T(X) + \ker T^{n-1}$ is closed. Since $k_i(T) < \infty$ for all $j \geq d$ we have

$$T(X) + \ker T^{n-1} \subset^e T(X) + \ker T^n \subset^e T(X) + \ker T^{n+1} \subset^e \cdots$$

and this gives that $T(X) + \ker T^j$ is closed for every $j \geq n$. By Lemma 1.99, inductively we obtain that $T^j(X)$ is closed for every $j \geq n$.

Consider the other case $d \leq j \leq n$. We show that $T^{n-1}(X)$ is closed. Observe first that

$$T(X) + \ker T^{n-1} = T^{-(n-1)}(T^n(X)),$$

so $T(X) + \ker T^{n-1}$ is closed. Further, $T^n(X) \cap \ker T$ is closed and it has finite codimension in $T^{n-1}(X) \cap \ker T$, and hence is closed by Theorem 1.70. By the Neubauer lemma (see Appendix A), we then conclude that $T^{n-1}(X)$ is closed. By repeating these arguments, we conclude that $T^j(X)$ is also closed for every $d \leq j \leq n$.   ∎

Theorem 1.100 in particular applies to operators having finite ascent $p := p(T)$ (in this case, by Corollary 1.73, $k_i(T) = 0$ for all $i \geq p$) or applies to operators having finite descent $q := q(T) < \infty$ (in this case, $k_i(T) = 0$ for all $i \geq q$).

Corollary 1.101 *Let $T \in L(X)$, and suppose that T has finite ascent $p := p(T) < \infty$. Then the following statements are equivalent:*

(i)   *there exists an $n \geq p + 1$ such that $T^n(X)$ is closed;*
(ii) *$T^j(X)$ is closed for every $j \geq p$;*
(iii) *$T^j(X) + \ker T^m$ is closed for all $m, j \in \mathbb{N}$ with $m + j \geq p$.*
      Analogous statements hold if T has finite descent $q := q(T) < \infty$.

Dealing with quasi-Fredholm operators, another application of Theorem 1.100 gives a characterization of these operators:

Theorem 1.102 *$T \in L(X)$ is quasi-Fredholm if and only if there exists a $p \in \mathbb{N}$ such that $T(X) + \ker T^p = T(X) + \mathcal{N}^\infty(T)$ and $T^{p+1}(X)$ is closed.*

Proof If $T \in QF(d)$, then T has uniform topological descent for $n \geq d$, so, by Theorem 1.74, the sequence $(T(X) + \ker T^n)$ is constant for $n \geq d$. Hence, for $n \geq d$ we have

$$T(X) + \ker T^d = T(X) + \ker T^n = T(X) + \mathcal{N}^\infty(T).$$

Moreover, $T^n(X)$ is closed for all $n \geq d$, by definition of the class $QF(d)$. Conversely, suppose that there exists a $p \in \mathbb{N}$ such that $T(X) + \ker T^p = T(X) + \mathcal{N}^\infty(T)$ and $T^{p+1}(X)$ is closed. Then $T(X) + \ker T^n = T(X) + \ker T^p$ for all $n \geq p$, and hence the sequence of subspaces $(T(X) + \ker T^n)$ is constant for all $n \geq p$, or equivalently, by Theorem 1.74, $k_n(T) = 0$ for all $n \geq p$. By Theorem 1.100, then $T^n(X)$ is closed for all $n \geq p$. Moreover, $T(X) + \ker T^p$ is closed, since it coincides with the set $T^{-p}(T^{p+1}(X))$. Hence, T is quasi-Fredholm of degree p.   ∎

The operators $QF(d)$ may be characterized in terms of semi-regularity as follows:

Theorem 1.103 *Suppose that $T \in L(X)$ and $d \in \mathbb{N}$. Then the following statements are equivalent:*

(i) $T \in QF(d)$;
(ii) *there exists an integer $n \geq 0$ such that $T^n(X)$ is closed and the restriction $T_n := T|T^n(X)$ is semi-regular.*

Proof (i) $\Rightarrow$ (ii) We know that $T^d(X)$ is closed. Consider the restriction $T_d := T|T^d(X)$. Then the range of T_d is $T^{d+1}(X)$, which is closed, and the equalities

$$\ker T_d = \ker T \cap T^d(X) = \ker T \cap T^m(X)$$

hold for all $m \geq d$. The latter intersection is obviously contained in the range of T_d^m, so the operator T_d is semi-regular.

(ii) $\Rightarrow$ (i) Suppose that $T^n(X)$ is closed and that the restriction $T_n := T|T^n(X)$ is semi-regular. Observe first that the range of T_n^k is $T^{n+k}(X)$. Since T_n is semi-regular, T_n^k is semi-regular for each $k = 1, 2, \ldots$, so $T^{n+k}(X)$ is closed for each $k = 1, 2, \ldots$. The semi-regularity of T_n gives that $\ker T_n = \ker T \cap T^n(X) \subseteq T^{n+k}(X)$, from which we obtain $\ker \cap T^n(X) \subseteq \ker T \cap T^{n+k}(X)$. On the other hand, $T^{n+k}(X) \subseteq T^n(X)$, so $\ker T \cap T^n(X) = \ker T \cap T^{n+k}(X)$ for all $k = 1, 2, \ldots$. By Theorem 1.74 it then follows that

$$\ker T^n + T(X) = \ker T^{n+k} + T(X) = \mathcal{N}^\infty(T) + T(X).$$

But $T^{n+1}(X)$ is closed, thus T is quasi-Fredholm, by Theorem 1.102.   ∎

Evidently, by part (iii) of Theorem 1.103,

$$T \text{ semi-regular} \Rightarrow T \text{ quasi-Fredholm}.$$

The dual of a quasi-Fredholm operator is also quasi-Fredholm:

Theorem 1.104 $T \in QF(d)$ *if and only if* $T^* \in QF(d)$.

Proof Suppose that $T \in QF(d)$. Then $T^n(X)$, and hence $T^{*n}(X^*)$, is closed for all $n \geq d$. We show first that for $n \geq d$ we have

$$\ker T^* \cap T^{*n}(X^*) = [T(X) + \ker T^n]^\perp. \tag{1.25}$$

To see this note that since $\ker T^* \cap T^{*n}(X^*)$ is closed, we have

$$^\perp[\ker T^* \cap T^{*n}(X^*)]^\perp = \ker T^* \cap T^{*n}(X^*).$$

It is easily seen that $T(X) + \ker T^n \subseteq^{\perp} [\ker T^* \cap T^{*n}(X^*)]$, so that

$$\ker T^* \cap T^{*n}(X^*) =^{\perp} [\ker T^* \cap T^{*n}(X^*)]^{\perp} \subseteq [T(X) + \ker T^n]^{\perp}.$$

On the other hand, from $T(X) \subseteq T(X) + \ker T^n$ and $\ker T^n \subseteq T(X) + \ker T^n$ we deduce that

$$[T(X) + \ker T^n]^{\perp} \subseteq T(X)^{\perp} + (\ker T^n)^{\perp} \subseteq \ker T^* \cap T^{*n}(X^*),$$

hence the proof of equality (1.25) is complete. Now, since T has uniform descent, the sequence $\{T(X) + \ker T^n\}$ is constant for $n \geq d$, by Theorem 1.74, and hence the sequence $\{\ker T^* \cap T^{*n}(X^*)\}$ is also constant for $n \geq d$, i.e., $\operatorname{dis}(T^*) = d$. Since $T^{*d+1}(X^*)$ is closed, $T^* \in QF(d)$.

Conversely, suppose that T^* is quasi-Fredholm of degree d. By Theorem 1.100, $T^{*j}(X^*)$ is closed for all $j \geq d$, hence $T^j(X)$ is closed for all $j \geq d$, and by the Sum theorem (see Appendix A) we have

$$T^{*j}(X^*) \cap \ker T^* = (\ker T^j)^{\perp} \cap T(X)^{\perp} = (\ker T^j + T(X))^{\perp}, \qquad (1.26)$$

for all $j \geq d$. Moreover, from the equality

$$T^{-j}(T^{j+1}(X)) = \ker T^j + T(X) \quad \text{for all } j \geq d,$$

we deduce that $\ker T^j + T(X)$ is closed for all $j \geq d$. Since $k_j(T^*) = 0$ for all $j \geq d$, from Eq. (1.26) we then obtain

$$\ker T^j + T(X) = {}^{\perp}[(\ker T^j + T(X))^{\perp}] = {}^{\perp}[(\ker T^d + T(X))^{\perp}] = \ker T^d + T(X),$$

for all $j \geq d$. Therefore, $\ker T^d + T(X) = \mathcal{N}^{\infty}(T) + T(X)$, and since $T^{d+1}(X)$ is closed, by Theorem 1.102, T is quasi-Fredholm. ∎

A natural question is: when is a quasi-Fredholm operator of Kato-type? We shall prove that this is always the case for Hilbert space operators, while for Banach space operators it is true under some additional conditions. We first need a preliminary lemma.

Lemma 1.105 *Suppose that $T \in L(X)$ has uniform descent $n \geq d$ and $j \geq 1$. Then* $\ker T^j \cap T^d(X) \subseteq T^{\infty}(X)$.

Proof We prove the statement by induction on j. We have

$$\ker T \cap T^d(X) = \ker T \cap T^n(X) \subseteq T^n(X) \quad \text{for all } n \geq d,$$

so ker $T \cap T^d(X) \subseteq T^\infty(X)$. Suppose that the statement is true for some $j \geq 1$. Let $x \in \ker T^{j+1} \cap T^d(X)$ and $n \geq d$. Then

$$Tx \in \ker T^j \cap T^d(X) \subseteq T^{n+1}(X),$$

so $Tx = T^{n+1}y$ for some $y \in X$. Therefore, $x - Ty \in \ker T$ and $x = T^n y + u$ for some $u \in \ker T$. Clearly, also $u \in T^d(X)$, and hence

$$x \in T^n(X) + (\ker T \cap T^d(X)) \subseteq T^n(X)$$

holds for all $n \geq d$. Therefore, $x \in T^\infty(X)$. ∎

Theorem 1.106 *Let $T \in L(X)$, X a Banach space, be quasi-Fredholm of degree d. Suppose that $T(X) + \ker T^d$ and $\ker T \cap T^{d+1}(X)$ are complemented. Then T is of Kato-type.*

Proof If $T \in QF(d)$ then $T^d(X)$ is closed, and, by Lemma 1.105,

$$\ker T^i \cap T^d(X) \subseteq T^\infty(X) \subseteq T^j(X) \quad \text{for all } i, j \geq 0.$$

If $d = 0$ then $T_0 = T$ is semi-regular, and hence of Kato-type. Suppose $d \geq 1$ and let L be a closed subspace for which $X = (\ker T \cap T^d(X)) \oplus L$. Consider the following subspaces N_j, inductively defined for each $j = 0, 1, \ldots, d$ as follows: $N_0 = 0$, while

$$N_{j+1} := T^{-1}(N_j) \cap L \quad \text{for all } j < d.$$

Clearly, $T(N_{j+1}) \subseteq N_j \cap T(X)$. On the other hand, if $x \in N_j \cap T(X)$, then $x = Tu$ for some $u \in X$, and writing $u = v + w$, with $v \in \ker T \cap T^d(X)$ and $w \in L$, we have $u - v = w \in L$ and $T(u - v) = Tu = x$. Thus $u - v \in N_{j+1}$ and $x \in T(N_{j+1})$, so $N_j \cap T(X) \subseteq T(N_{j+1})$. Therefore

$$N_j \cap T(X) = T(N_{j+1}), \quad \text{for all } j < d.$$

We prove by induction on j that $N_j \subseteq N_{j+1}$ for all $0 \leq j < d$. This inclusion is clear for $j = 0$. Suppose that $N_j \subseteq N_{j+1}$ for $j > 0$, and let $x \in N_{j+1}$. Then $Tx \in N_j \subseteq N_{j+1}$ and hence $x \in T^{-1}(N_{j+1})$. Since $x \in N_{j+1} \subseteq L$, we have $x \in N_{j+2}$, so

$$N_j \subseteq N_{j+1} \quad \text{for all } 0 \leq j < d.$$

It is easily seen that $N_j \subseteq \ker T^j$ for all j. We now prove by induction on j the following inclusion

$$\ker T^j \subseteq N_j + \ker T^j \cap T^d(X). \tag{1.27}$$

This inclusion is obvious for $j = 0$. For $j = 1$ we have

$$\ker T = \ker T \cap L + (\ker T \cap T^d(X)) = N_1 + \ker T \cap T^d(X).$$

Let $j \geq 1$ and suppose $\ker T^j \subseteq N_j + \ker T^j \cap T^d(X)$. If $x \in \ker T^{j+1}$ then $Tx \in \ker T^j$, and so $Tx = v_1 + v_2$ for some $v_1 \in N_j$ and

$$v_2 \in \ker T^j \cap T^d(X) = \ker T^j \cap T^{d+1}(X) = T(\ker(T^{j+1}) \cap T^d(X).$$

Thus, $v_1 \in N_j \cap T(X) = T(N_{j+1})$ and

$$
\begin{aligned}
x &\in N_{j+1} + \ker T^{j+1} \cap T^d(X) + \ker T \\
&= N_{j+1} + \ker T^{j+1} \cap T^d(X) + (\ker T \cap L) + (\ker T \cap T^d(X)) \\
&= N_{j+1} + \ker T^{j+1} \cap T^d(X),
\end{aligned}
$$

from which the inclusion (1.27) follows. We show now by induction that

$$N_j \cap T^d(X) = \{0\} \quad \text{for all } j \leq d. \tag{1.28}$$

For $j = 0$ it is clear. Let $j > 0$ and $N_j \cap T^d(X) = \{0\}$. If $x \in N_{j+1} \cap T^d(X)$ then $Tx \in N_j \cap T^d(X)$ and by the induction assumption we deduce that $Tx = 0$. Thus, $x \in \ker T \cap T^d(X)$, and $x \in N_{j+1} \subseteq L$, hence $x = 0$, which proves (1.28).

We show now that T is of Kato-type. Set $N := N_d$. Then N is T-invariant and $N \subseteq \ker T^d$, $(T|N)^d = 0$, so $T|N$ is nilpotent. Furthermore, $\ker T^d \subseteq N + T^d(X)$ and $N \cap T^d(X) = \{0\}$. Note also that the space $N + T^d(X) = \ker T^d + T^d(X)$ is closed, by Theorem 1.78. Since $T^* \in QF(d)$, by Theorem 1.104, we can use the same construction as above for T^*. By assumption $T(X) + \ker T^d$ is complemented, and $\ker T^* \cap T^{*d}(X^*) = [T(X) + \ker T^d]^\perp$, so we can choose a w^*-closed subspace L' of X^* such that

$$X^* = (\ker T^* \cap T^{*d}(X^*)) \oplus L'.$$

As before, for all $0 \leq i < d$ we can construct subspaces M_i of X^* by $M_0 = \{0\}$ and $M_{i+1} = T^{*-1}M_i \cap L'$. Note that all the subspaces M_i are w^*-closed. We have

$$T^*(M_d) \subseteq M_d \subseteq \ker T^{*d}.$$

Moreover, $M_d \cap T^{*d}(X^*) = \{0\}$ and

$$\ker T^{*d} \subseteq M_d + T^{*d}(X^*),$$

where the latter space is closed.

Finally, set $M := {}^{\perp}M_d$. Clearly, $T(M) \subseteq M$ and

$$M = {}^{\perp}M_d \supseteq {}^{\perp}\ker T^{*d} = T^d(X).$$

We have

$$M + \ker T^d = {}^{\perp}M_d + {}^{\perp}T^{*d}(X^*) = {}^{\perp}[M_d \cap T^{*d}(X^*)] = X,$$

where the equality

$${}^{\perp}M_d + {}^{\perp}T^{*d}(X^*) = {}^{\perp}[M_d \cap T^{*d}(X^*)]$$

follows by the Sum theorem (see Appendix A), since the space $M_d + T^{*d}(X^*)$ is closed. Furthermore,

$$T^d(X) = {}^{\perp}\ker T^{*d} \supseteq {}^{\perp}[M_d + T^{*d}(X^*)] = {}^{\perp}M_d \cap {}^{\perp}T^{*d}(X^*) = M \cap \ker T^d.$$

Thus,

$$M + N \supseteq M + T^d(X) + N \supseteq M + \ker T^d = X.$$

But,

$$M \cap N \subseteq M \cap \ker T^d \cap N \subseteq T^d(X) \cap N = \{0\},$$

thus $X = M \oplus N$. As proved before, $T|N$ is nilpotent, so we have only to show that $T|M$ is semi-regular. Let $T_M = T|M$. If $x \in \ker T_M$ then

$$x \in \ker T \cap M \subseteq \ker T^d \cap M \subseteq M \cap \ker T^d \cap T^d(X)$$
$$\subseteq M \cap T^{\infty}(X) = T_M^{\infty}(M).$$

Further, $T_M^d(M) = T^d(X)$ and hence $T_M^d(M)$ is closed. Therefore, T_M^d is semi-regular, and hence $T|M$ is also semi-regular. ∎

Since every closed subspace of a Hilbert space is complemented, by Theorem 1.106 we have:

Corollary 1.107 *Every quasi-Fredholm operator acting on a Hilbert space is of Kato-type.*

We now want to show that a finite-dimensional perturbation of a quasi-Fredholm operator is also quasi-Fredholm. We first need to do some preliminary work.

Lemma 1.108 *If $T, K \in L(X)$ and $K \in L(X)$ is finite-dimensional, then $(T + K)^n(X) =^e T^n(X)$.*

Proof We have

$$(T + K)^n - T^n = \sum_{j=0}^{n-1}[T^j(T + K)^{n-j} - T^{j+1}(T + K)^{n-j-1}]$$

$$= \sum_{j=0}^{n-1} T^j K(T + K)^{n-j-1},$$

from which we see that $(T + K)^n - T^n$ is finite-dimensional. Consequently, $(T + K)^n(X) =^e T^n(X)$. ∎

Lemma 1.109 *Let $T \in QF(d)$ and let $K \in L(X)$ be one-dimensional. If $T_d := T|T^d(X)$ and $x_0 \in \ker T_d$ then, for every $n \in \mathbb{N}$, there exist vectors $x_1, x_2, \dots x_n$ in the hyper-range of T_d such that $Tx_j = x_{j-1}$, with $x_i \in \ker K$ for all $j = 1, 2, \dots n$.*

Proof Let $M := T^d(X)$ and set $T_d := T|T^d(X)$. We proceed by induction on n. For $n = 0$ the statement is obvious. Suppose that the statement is true for n, i.e., there exist $x_1, x_2, \dots x_n$ in the hyper-range $T_d^\infty(M)$ of T_d such that $Tx_j = x_{j-1}$, with $x_i \in \ker K$ for all $j = 1, 2, \dots n$. Since T_d is semi-regular, see the proof of Theorem 1.106, $T_d^\infty(M) = K(T_d)$, by Theorem 1.44, so there exists a vector $x_{n+1} \in T_d^\infty(M)$ such that $T_d x_{n+1} = x_n$. If $Kx_{n+1} = 0$ then the statement holds for $n + 1$. Suppose that $Kx_{n+1} \neq 0$, and let k be the smallest integer for which $\ker T_d^k$ is not contained in $\ker K$. Clearly, $k \leq n+1$, since $x_{n+1} \in \ker T_d^{n+1} \setminus \ker K$. Since K is one-dimensional, we can find $z \in \ker T_d^k \subseteq T_d^\infty(M)$ such that $K(x_{n+1} - z) = 0$. Set

$$y_{n+1} := x_{n+1} - z, \quad y_n := T_d\, y_{n+1}, \quad y_{n-1} := T_d^2\, y_{n+1},$$

and

$$y_{n-1-k} := T_d^k\, y_{n+1} = T_d^k\, x_{n+1} = x_{n+1-k},$$

$$y_{n-k} = x_{n-r}, \dots, \ y_1 = x_1.$$

Evidently, $y_j \in T_d^\infty(M)$, and $T_d\, y_j = y_{j-1}$ for $j = 1, \dots n + 1$. Moreover, $Ky_{n+1} = 0$ and $Kx_j = 0$ for $1 \leq j \leq n + 1 - k$. If $n + 2 - k \leq j \leq n$ then

$$Ky_j = K(y_j - x_1) + Kx_j = K(y_j - x_j) = 0,$$

since $y_j - x_j \in \ker T_d^{k-1}$ by definition of k. ∎

Quasi-Fredholm operators are stable under (not necessarily commuting) finite-dimensional perturbations:

Theorem 1.110 *If* $T \in L(X)$ *is quasi-Fredholm and* $K \in L(X)$ *is finite-dimensional then* $T + K$ *is quasi-Fredholm.*

Proof Suppose that $T \in QF(d)$. It is sufficient to prove the case where K is one-dimensional. Since, by Lemma 1.108, $(T + K)^n(X) =^e T^n(X)$ for every $n \in \mathbb{N}$, $(T + K)^n(X)$ is closed if and only if $T^n(X)$ is closed. To show that $T + K$ is quasi-Fredholm it is sufficient to prove, by Theorem 1.102, that

$$(T + K)(X) + \ker T^d = (T + K)(X) + \mathcal{N}^\infty(T + K).$$

Since $T \in QF(d)$ then T has topological uniform descent for $n \geq d$, so, by Lemma 1.105, $\ker T \cap T^d(X) \subseteq T^\infty(X)$ and $T^d(X), T^{d+1}(X)$ are both closed. If $M := T^d(X)$ and $T_d; = T|M$, then T^d is semi-regular, by Theorem 1.103. We claim that $\ker T_d \subseteq (T + K)^\infty(X)$. Indeed, since

$$\ker T_d = \ker T \cap T^d(X) =^e \ker(T + K) \cap (T + K)^d(X),$$

we have

$$\ker(T + K) \cap (T + K)^d(X) \subseteq^e (T + K)^\infty(X).$$

This means that

$$\ker(T + K) \cap (T + K)^n(X) = \ker(T + K) \cap (T + K)^\infty(X)$$

for some $n \geq d$. By taking x_n as in Lemma 1.109 we have

$$(T + K)^n x_n = (T + K)^{n-1} x_{n-1} = \cdots = (T + K) x_1 = x_0,$$

thus $x_0 \in (T + K)^n(X)$ for each n, and hence $\ker T_d \subseteq (T + K)^\infty(X)$, which concludes the proof. ∎

The quasi-Fredholm spectrum is defined as

$$\sigma_{qf}(T) := \{\lambda \in \mathbb{C} : \lambda I - T \text{ is quasi-Fredholm}\}.$$

The spectrum $\sigma_{qf}(T)$ is a closed subset of $\mathbb{C}$, which may be empty. This is the case, for instance, for the null operator 0. A less trivial example may be obtained as follows. The spectrum $\sigma_{qf}(T)$ is a subset of the Drazin spectrum, which will be studied later, and the Drazin spectrum can be empty, as in the case of an algebraic operator. Note that, by Theorem 1.110, we have $\sigma_{qf}(T) = \sigma_{qf}(T + K)$ for all finite-dimensional operators K.

1.8   Semi B-Fredholm Operators

This section concerns a class of operators, introduced by Berkani et al., which extends the class of semi-Fredholm operators. For every $T \in L(X)$ and nonnegative integer n, let us denote by T_n the restriction of T to $T^n(X)$ viewed as a map from the space $T^n(X)$ into itself (we set $T_0 = T$).

Definition 1.111 An operator $T \in L(X)$, X a Banach space, is said to be *B-Fredholm*, (respectively, *semi B-Fredholm, upper semi B-Fredholm, lower semi B-Fredholm*), if for some integer $n \geq 0$ the range $T^n(X)$ is closed and T_n is a Fredholm operator (respectively, semi-Fredholm, upper semi-Fredholm, lower semi-Fredholm).

It is easily seen that every nilpotent operator, as well as any idempotent bounded operator, is B-Fredholm. Therefore the class of B-Fredholm operators contains the class of Fredholm operators as a proper subclass.

Theorem 1.112 *Let $T \in L(X)$ and suppose that $T^n(X)$ is closed and T_n is a Fredholm operator (respectively, semi-Fredholm, upper semi-Fredholm, lower semi-Fredholm). For every $m \geq n$, $T^m(X)$ is closed and T_m is a Fredholm operator (respectively, semi-Fredholm, upper semi-Fredholm, lower semi-Fredholm), with ind $T_m =$ ind T_n.*

Proof Suppose that $T_n : T^n(X) \to T^n(X)$ is Fredholm and let $m \geq n$. Then T_n^{m-n} is a Fredholm operator on $T^n(X)$, so that $T_n^{m-n}(T^n(X)) = T^m(X)$ is a closed subspace of $T^n(X)$, and hence is closed in X. Clearly,

$$\ker T_m = \ker T \cap T^m(X) \subseteq \ker T \cap T^n(X) = \ker T_n,$$

so $\alpha(T_m) < \infty$. Since $\beta(T_n) < \infty$ and the range of T_n is $T^{n+1}(X)$, there exists a finite-dimensional subspace U of $T^n(X)$ such that $T^n(X) = U + T^{n+1}(X)$. Then

$$T^m(X) = T^{m-n}(X) + T^{m+1}(X),$$

and T_m has range $T^{m+1}(X)$ of finite codimension in $T^m(X)$. Consequently, $\beta(T_m) < \infty$, so the restriction T_m is Fredholm. By Lemma 1.69 we have

$$\dim \frac{\ker T \cap T^n(X)}{\ker T \cap T^{n+1}(X)} = \dim \frac{\ker T^{n+1} + T(X)}{\ker T^n + T(X)},$$

and, by Lemma 1.71, we also have

$$\dim \frac{T^n(X)}{T^{n+1}(X)} = \dim \frac{X}{T(X) + \ker T^n},$$

and analogously,

$$\dim \frac{T^{n+1}(X)}{T^{n+2}(X)} = \dim \frac{X}{T(X) + \ker T^{n+1}}.$$

Therefore, $\alpha(T_n) - \alpha(T_{n+1}) = \beta(T_n) - \beta(T_{n+1})$, so $\operatorname{ind} T_{n+1} = \operatorname{ind} T_{[n]}$. An inductive argument then proves that $\operatorname{ind} T_m = \operatorname{ind} T_n$ for all $m \geq n$.

The assertions concerning semi-Fredholm operators are proved in a similar way. ∎

The previous theorem has a crucial role, since it permits us to extend the concept of index to semi B-Fredholm operators. Indeed, the index of semi B-Fredholm operators is independent of the choice of the integer n which appears in the following definition. It is clear that in the case of semi-Fredholm operators we recover the usual notion of index.

Definition 1.113 Let $T \in L(X)$ be semi B-Fredholm and let $n \in \mathbb{N}$ be such that T_n is a Fredholm operator. Then the *index* $\operatorname{ind} T$ of T is defined as the index of T_n.

The upper semi-Fredholm operators (respectively, the lower semi-Fredholm operators) are exactly the upper semi B-Fredholm operators (respectively, the lower semi B-Fredholm operators) for which we have $\alpha(T) < \infty$ (respectively, $\beta(T) < \infty$):

Theorem 1.114 *Let $T \in L(X)$. Then we have:*

(i) *T is upper semi B-Fredholm and $\alpha(T) < \infty \Leftrightarrow T \in \Phi_+(X)$.*
(ii) *T is lower semi B-Fredholm and $\beta(T) < \infty \Leftrightarrow T \in \Phi_-(X)$.*

Proof (i) If T is upper semi B-Fredholm then there exists an $n \in \mathbb{N}$ such that $T^n(X)$ is closed and T_n is upper semi-Fredholm. Since $\alpha(T) < \infty$ then $\alpha(T^n) < \infty$, hence T^n is upper semi-Fredholm. From the classical Fredholm theory then T is also upper semi-Fredholm. The converse is obvious.

Part (ii) may be proved in a similar way. ∎

The next corollary is an obvious consequence of Theorem 1.114, since every semi-Fredholm operator has closed range.

Corollary 1.115 *If $T \in L(X)$ is injective and upper semi B-Fredholm then T is bounded below.*

Every semi B-Fredholm operator T has topological uniform descent. Indeed, we show now that every semi B-Fredholm operator is quasi-Fredholm.

Theorem 1.116 *Every semi B-Fredholm operator is quasi Fredholm. More precisely, we have:*

(i) *T is upper semi B-Fredholm if and only if there is an integer d such that $T \in QF(d)$ and $\ker T \cap T^d(X)$ has finite dimension.*

(ii) *T is lower semi B-Fredholm if and only if there is an integer d such that $T \in QF(d)$ and $T(X) + \ker T^d$ has finite codimension.*

(iii) *T is B-Fredholm if and only if there is an integer d such that $T \in QF(d)$, $\ker T \cap T^d(X)$ has finite dimension and $T(X) + \ker T^d$ has finite codimension.*

Proof

(i) Let $n \in \mathbb{N}$ such that $T^n(X)$ is closed and T_n is upper semi-Fredholm. Then $\alpha(T_n) < \infty$ and since $\ker T \cap T^m(X) \subseteq \ker T \cap T^n(X) = \ker T_n$ for each $m \geq n$, it then follows that the sequence of subspaces $\{\ker T \cap T^k(X)\}$ becomes stationary for k large enough. This shows, by Theorem 1.74, that T has uniform descent $d = \mathrm{dis}(T)$ and $\dim(\ker T \cap T^d(X)) < \infty$. Since, by Theorem 1.112, $T_{[m]}$ is upper semi-Fredholm for all $m \geq n$, $T^m(X)$ is closed for all $m \geq n$. By using Theorem 1.100, we then conclude that $T^m(X)$ is closed for all $m \geq d$. Moreover,

$$T(X) + \ker T^d = (T^d)^{-1}(T^{d+1}(X))$$

is a closed subspace of X, so $T \in QF(d)$.

   Conversely, suppose $T \in QF(d)$ and $\dim(\ker T \cap T^d(X)) < \infty$. Then $T^n(X)$ is closed for all $n \geq d$. Moreover,

$$\alpha(T_d) = \dim(\ker T \cap T^d(X)) < \infty,$$

thus T_d is upper semi-Fredholm.

(ii) The proof is similar to that of part (i). If T is lower semi B-Fredholm and T_n is lower semi-Fredholm on the Banach space $T^n(X)$, then, by Lemma 1.71,

$$\beta(T_n) = \dim \frac{T^n(X)}{T^{n+1}(X)} = \dim \frac{X}{T(X) + \ker T^d},$$

hence the sequence $(T(X) + \ker T^k)_k$ becomes stationary for k large enough. Since

$$\dim \frac{\ker T \cap T^k(X)}{\ker T \cap T^{k+1}(X)} = \dim \frac{\ker T^{k+1} + T(X)}{\ker T^k + T(X)},$$

we then conclude, by Theorem 1.74, that T has uniform descent $d = \mathrm{dis}(T)$ and $T(X) + \ker T^d$ has finite codimension. Moreover, by Theorem 1.112, T_m is upper semi-Fredholm for all $m \geq n$, so $T^m(X)$ is closed for all $m \geq n$. Using Theorem 1.100, we then conclude that $T^m(X)$ is closed for all $m \geq d$. Hence $T \in QF(d)$.

Conversely, suppose that $T \in QF(d)$ and that $T(X) + \ker T^d$ has finite codimension. Then $T^n(X)$ is closed for all $n \geq d$. Moreover, by Lemma 1.71, we have

$$\beta(T_d) = \dim \frac{T^d(X)}{T^{d+1}(X)} = \dim \frac{X}{T(X) + \ker T^d} < \infty,$$

from which we see that T_d is lower semi-Fredholm.

(iii) This easily follows from part (i) and (ii). ∎

The following punctured disc theorem is obtained by combining Corollary 1.91 and Theorem 1.116, in the special case when $S = \lambda I - T$, with λ sufficiently small.

Theorem 1.117 *Suppose that $T \in L(X)$ is upper semi B-Fredholm. Then there exists an open disc $\mathbb{D}(0, \varepsilon)$ centered at 0 such that $\lambda I - T$ is upper semi-Fredholm for all $\lambda \in \mathbb{D}(0, \varepsilon) \setminus \{0\}$ and*

$$\operatorname{ind}(\lambda I - T) = \operatorname{ind}(T) \quad \text{for all } \lambda \in \mathbb{D}(0, \varepsilon).$$

Moreover, if $\lambda \in \mathbb{D}(0, \varepsilon) \setminus \{0\}$ then

$$\alpha(\lambda I - T) = \dim(\ker T \cap T^d(X)) \quad \text{for some } d \in \mathbb{N},$$

so that $\alpha(\lambda I - T)$ is constant as λ ranges over $\mathbb{D}(0, \varepsilon) \setminus \{0\}$ and

$$\alpha(\lambda I - T) \leq \alpha(T) \quad \text{for all } \lambda \in \mathbb{D}(0, \varepsilon).$$

Analogously, if $T \in L(X)$ is lower semi B-Fredholm then there exists an open disc $\mathbb{D}(0, \varepsilon)$ centered at 0 such that $\lambda I - T$ is lower semi-Fredholm for all $\lambda \in \mathbb{D}(0, \varepsilon) \setminus \{0\}$ and

$$\operatorname{ind}(\lambda I - T) = \operatorname{ind}(T) \quad \text{for all } \lambda \in \mathbb{D}(0, \varepsilon).$$

Moreover, if $\lambda \in \mathbb{D}(0, \varepsilon) \setminus \{0\}$ then

$$\beta(\lambda I - T) = \operatorname{codim}(\ker T^d + T(X)) \quad \text{for some } d \in \mathbb{N},$$

so that $\beta(\lambda I - T)$ is constant as λ ranges over $\mathbb{D}(0, \varepsilon) \setminus \{0\}$ and

$$\beta(\lambda I - T) \leq \beta(T) \quad \text{for all } \lambda \in \mathbb{D}(0, \varepsilon).$$

Theorem 1.117 may be viewed as an extension of the classical punctured neighborhood theorem for semi-Fredholm operators (see Appendix A) to semi B-Fredholm operators.

Theorem 1.118 *If $T \in L(X)$ is quasi-Fredholm then there exists an $\varepsilon > 0$ such that $\mathcal{N}^{\infty}(\lambda I - T) \subseteq (\lambda I - T)^{\infty}(X)$ for all $0 < |\lambda| < \varepsilon$. If T is semi B-Fredholm then $\lambda I - T$ is semi-regular in a suitable punctured open disc centered at 0.*

Proof Observe first that if T is quasi-Fredholm of degree d then $T^n(X)$ is closed for all $n \geq d$, so $T^{\infty}(X)$ is closed. Furthermore, by Theorem 1.79 the restriction $T|T^{\infty}(X)$ is onto, so $T(T^{\infty}(X)) = T^{\infty}(X)$. Let $T_0 := T|T^{\infty}(X)$. Clearly, T_0 is onto and hence, by Theorem 1.10, $\lambda I - T$ is onto for all $|\lambda| < \varepsilon$, where $\varepsilon := \gamma(T_0)$ is the minimal modulus of T_0. Therefore,

$$(\lambda I - T)(T^{\infty}(X)) = T^{\infty}(X) \quad \text{for all } |\lambda| < \varepsilon.$$

Since $T^{\infty}(X)$ is closed, by Theorem 1.39 it then follows that

$$T^{\infty}(X) \subseteq K(\lambda I - T) \subseteq (\lambda I - T)^{\infty}(X) \quad \text{for all } |\lambda| < \varepsilon.$$

By part (ii) of Theorem 1.14 we have $\mathcal{N}^{\infty}(\lambda I - T) \subseteq T^{\infty}(X)$ for all $\lambda \neq 0$, so we conclude that

$$\mathcal{N}^{\infty}(\lambda I - T) \subseteq (\lambda I - T)^{\infty}(X) \quad \text{for all } 0 < |\lambda| < \varepsilon,$$

and the first assertion is proved.

To show the second assertion, suppose that T is semi B-Fredholm. By Theorem 1.117, there exists an open disc $\mathbb{D}$ centered at 0 such that $\lambda I - T$ is semi-Fredholm for all $\lambda \in \mathbb{D} \setminus \{0\}$. Since semi-Fredholm operators have closed range, the last assertion easily follows. ∎

The B-Fredholm operators on Banach spaces may be characterized through a decomposition which is similar to the Kato decomposition (but recall that a Fredholm operator may be non-semi-regular):

Theorem 1.119 *Let $T \in L(X)$, X a Banach space. Then*

(i) *T is B-Fredholm if and only if there exist two closed invariant subspaces M and N such that $X = M \oplus N$, $T|M$ is Fredholm and $T|N$ is nilpotent.*

(ii) *T is B-Fredholm of index 0 if and only if there exist two closed invariant subspaces M and N such that $X = M \oplus N$, $T|M$ is Fredholm having index 0 and $T|N$ is nilpotent.*

Proof

(i) If T is B-Fredholm then, by Theorem 1.116, there is an integer d for which $T \in QF(d)$, $\ker T \cap T^d(X)$ has finite dimension and $T(X) + \ker T^d$ has finite codimension, and hence both subspaces are complemented. By Theorem 1.106, T is of Kato-type, so there exist two closed T-invariant subspaces M and N such that $X = M \oplus N$, $T|N$ is nilpotent of order d, and $T|M$ is semi-regular.

Define $\alpha_n(T) := \alpha(T_n) = \dim(\ker T \cap T^n(X)) < \infty$ and $\beta_n(T) := \dim \frac{T^n(X)}{T^{n+1}(X)} < \infty$. Clearly,

$$\alpha_d(T|M) = \alpha_d(T|N) + \alpha_d(T|M) = \alpha_d(T) < \infty,$$

and

$$\beta_d(T|M) = \beta_d(T|N) + \beta_d(T|M) = \beta_d(T) < \infty.$$

Since $k_j(T|M) = 0$ for all j, we conclude that $\alpha_0(T|M) = \alpha_d(T|M) < \infty$ and $\beta_0(T|M) = \beta_d(T|M) < \infty$, so $T|M$ is Fredholm.

(ii) The proof is clear from part (i). ∎

According to Caradus [87] an operator $T \in L(X)$ is said to be *generalized Fredholm* if there exists an operator $S \in L(X)$ such that $TST = T$ and $I - ST - TS \in \Phi(X)$. Examples of generalized Fredholm operators are projections and finite-dimensional and Fredholm operators. This class of operators has been studied in several papers by Schmoeger [275, 277–281], and a remarkable result is that an operator T is generalized Fredholm if and only if there exist two closed invariant subspaces M and N such that $X = M \oplus N$, $T|M$ is Fredholm and $T|N$ is a finite rank nilpotent operator, see [280, Theorem 1.1]. Therefore, by Theorem 1.119, every generalized Fredholm operator is B-Fredholm, but the converse is not true. For instance, a nilpotent operator with non-closed range is a B-Fredholm operator but not a generalized Fredholm operator. The relationship between B-Fredholm operators and generalized Fredholm operators is fixed by the following theorem.

Theorem 1.120 $T \in L(X)$ is B-Fredholm if and only if there exists an $n \in \mathbb{N}$ such that T^n is a generalized Fredholm operator.

Proof As observed above, every generalized Fredholm operator is B-Fredholm. Suppose that T is B-Fredholm. Then, by Theorem 1.119, $T = T_1 \oplus T_2$, where T_1 is Fredholm and T_2 is nilpotent. Let $T_2^n = 0$. Then

$$T^n = T_1^n \oplus T_2^n = T_1^n \oplus 0,$$

and since T_1^n is Fredholm we then conclude that T^n is generalized Fredholm. ∎

The following concept of invertibility was introduced in [267] and was inspired by the work of Drazin [123].

Definition 1.121 Let $\mathcal{A}$ be an algebra with unit e. An element $a \in \mathcal{A}$ is said to be a *Drazin invertible element of degree n* if there is an element $b \in \mathcal{A}$ such that

$$a^n ba = a^n, \quad bab = b, \quad ab = ba. \tag{1.29}$$

The element b is called the *Drazin inverse* (of degree n). If $a \in \mathcal{A}$ satisfies the equalities (1.29) for $k = 1$ then a is said to be *group invertible*.

Theorem 1.122 *An element a has at most one Drazin inverse. If it exists, the Drazin inverse of a belongs to the second commutant of a, i.e., the Drazin inverse of a commutes with every $x \in \mathcal{A}$ which commutes with a.*

Proof Suppose that b_1 and b_2 are Drazin inverses of a, with corresponding integers m_1 and m_2 in the first equality of (1.29) and let $m := \max\{m_1, m_2\}$. Then $b_1 a^{m+1} = a^m = a^{m+1} b_2$ and $b_1 = b_1^2 a$ and $b_2 = a b_2^2$. By induction, we easily have $b_1 = b_1^{k+1} a^k$ and $b_2 = a^k b_2^{k+1}$ for $k = 1, 2, \ldots$. In particular $b_1 = b_1^{m+1} a^m$ and $b_2 = a^m b_2^{m+1}$. Hence

$$b_1 = b_1 = b_1^{m+1} a^m = b_1^{m+1} a^{m+1} b_2 = b_1 a b_2$$

and similarly we obtain $b_2 = b_2 a b_1$, so that $b_1 = b_2$.

To prove the second assertion, assume that $ax = xa$ and let b be the Drazin inverse of a. We have

$$ba^m x = bxa^m = bxa^{m+1} b = ba^{m+1} xb = a^m xb.$$

Hence $b^{m+1} a^m x = a^m x b^{m+1}$, and because $b = b^{m+1} a^m$ we then obtain

$$bx = b^{m+1} a^m x = a^m x b^{m+1} = x b^{m+1} a^m = xb,$$

which concludes the proof. ∎

Let $a \in \mathcal{A}$ be Drazin invertible. The *index* $i(a)$ of a is the least non-negative integer n for which the equations (1.29) have a solution.

Drazin [123] proved a number of interesting properties which we now mention, in the following remark, without proof. We shall also reassume some other basic properties of group invertible elements.

Remark 1.123 Let $\mathcal{A}$ denote a Banach algebra with unit e.

(a) An element $a \in \mathcal{A}$ is Drazin invertible of degree n if and only if a^n is group invertible in $\mathcal{A}$, see [267, Lemma 1 and Corollary 5].

(b) An element $a \in \mathcal{A}$ is group invertible if and only if a admits a commuting *generalized inverse*, i.e. there is a $b \in \mathcal{A}$ such that $ab = ba$ and $aba = a$, or equivalently there exists a $b \in \mathcal{A}$ such that $aba = a$ and $e - ab - ba$ is invertible in $\mathcal{A}$, see [275, Theorem 3.3 and Proposition 3.9].

(c) If $a, b \in \mathcal{A}$ are two commuting Drazin invertible elements then ab is Drazin invertible, see [72, Proposition 2.6].

(d) If $a \in \mathcal{A}$ has a Drazin inverse b, then a^k has Drazin inverse b^k for all $k \in \mathbb{N}$.

(e) The Drazin inverse b of an element $a \in \mathcal{A}$, if it exists, is also Drazin invertible. The index of b is equal to 1 and the Drazin inverse of b is $a^2 b$.

(f) If $a \in \mathcal{A}$ has Drazin inverse b then the Drazin inverse of b is a if and only if a has index 1.

(g) If a has Drazin inverse b and b has Drazin inverse c then the Drazin inverse of c is a.

If $a \in \mathcal{A}$ is Drazin invertible, denote by a^D its Drazin inverse.

Theorem 1.124 *Let $\mathcal{A}$ denote a Banach algebra with unit e and $a, b \in \mathcal{A}$. Then ab is Drazin invertible if and only if ba is Drazin invertible. Moreover,*

$$(ba)^D = b((ab)^D)^2a \quad and \quad (ab)^D = a((ba)^D)^2b, \tag{1.30}$$

and $|i(ba) - i(ab)| \leq 1$.

Proof Let d be the Drazin inverse of ab and let $k := i(ab)$. Then $d(ab) = (ab)d$, $d(ab)d = d$, and

$$(ab)^k d(ab) = (ab)^k.$$

Let $u := bd^2a$. We have

$$u(ba) = bd^2aba = bd^2aba = bda,$$

and

$$(ba)b = (ba)bd^2a = (ba)babd^3a = bababd^3a = bda,$$

from which we obtain that $u(ba) = (ba)u$. From $u(ba) = bda$ it then follows that

$$u(ba)u = (bda)bd^2a = (bdab)abd^2(da) = (bdab)abd^2(da)$$
$$= bd^2a = u,$$

and

$$(ba)^{k+1}u(ba) = (ba)^{k+1}bda = (ba)^{k+1}b(abd^2)a = b(ab)^k ab(abd^2)a$$
$$= b(ab)^{k+1}abd^2a = b(ab)^{k+1}da = b(ab)^k a = (ba)^{k+1}.$$

Therefore, ba is Drazin invertible and its Drazin inverse is $(ba)^D = b((ab)^D)^2a$. Evidently, $i(ba) \leq i(ab) + 1$.

A similar argument shows that if ba is Drazin invertible, then ab is Drazin invertible with Drazin inverse

$$(ab)^D = a((ba)^D)^2b$$

and $i(ab) \leq i(ba) + 1$. ∎

Let $\mathcal{F}(X)$ denote the two-sided ideal of all finite-dimensional operators in $L(X)$, and denote by $\mathcal{L}$ the normed algebra $L(X)/\mathcal{F}(X)$ provided with the canonical quotient norm. Let $\pi : L(X) \rightarrow \mathcal{L}$ be the canonical quotient mapping. The well-known Atkinson's theorem says that $T \in \Phi(X)$ if and only if $\pi(T)$ is invertible in $\mathcal{L}$. A version of Atkinson's theorem may be stated for B-Fredholm theory as follows:

Theorem 1.125 $T \in L(X)$ is B-Fredholm if and only if $\pi(T)$ is a Drazin invertible element of the algebra $\mathcal{L}$.

Proof We know that T is generalized Fredholm if and only if $\pi(T)$ is group invertible in the Calkin algebra $\mathcal{L}$. Using Theorem 1.120 we see that T is B-Fredholm precisely when $\pi(T^n) = \pi(T)^n$ is group invertible for some $n \in \mathbb{N}$. By using (b) of Remark 1.123 we then conclude that T is B-Fredholm if and only if $\pi(T)$ is Drazin invertible in $\mathcal{L}$. ∎

The class of B-Fredholm operators is stable under finite-dimensional perturbation:

Theorem 1.126 *If* $T, S \in L(X)$ *are* B-*Fredholm operators. Then we have:*

(i) *If* $TS = ST$ *then* ST *is* B-*Fredholm. Moreover,* $\mathrm{ind}\,(ST) = \mathrm{ind}\,(S) + \mathrm{ind}\,(T)$.
(ii) *If* $K \in \mathcal{F}(X)$ *then* $T + K$ *is* B-*Fredholm and* $\mathrm{ind}\,(T + K) = \mathrm{ind}\,T$.

Proof

(i) We have $\pi(ST) = \pi(S)\pi(T) = \pi(S)\pi(T)$. By using (c) of Remark 1.123 we then deduce that $\pi(ST)$ is Drazin invertible in $\mathcal{L}$, hence ST is B-Fredholm by Theorem 1.125. Furthermore, according to Theorem 1.117, choose $n_0 \in \mathbb{N}$ such that $\frac{1}{n}I - T$ and $\frac{1}{n}I - S$ are Fredholm operators for $n \geq n_0$, $\mathrm{ind}\,(\frac{1}{n}I - T) = \mathrm{ind}\,T$ and $\mathrm{ind}\,(\frac{1}{n}I - S) = \mathrm{ind}\,S$. Choosing n large enough, the difference

$$ST - \left(\frac{1}{n}I - S\right)\left(\frac{1}{n}I - T\right) = \frac{1}{n}\left(\frac{1}{n}I - (T + S)\right)$$

is of small norm. As both ST and $\left(\frac{1}{n}I - S\right)\left(\frac{1}{n}I - T\right)$ are B-Fredholm then, by Theorem 1.117,

$$\mathrm{ind}\,(ST) = \mathrm{ind}\,\left[\left(\frac{1}{n}I - S\right)\left(\frac{1}{n}I - T\right)\right].$$

But both $\frac{1}{n}I - S$ and $\frac{1}{n}I - T$ are Fredholm, so

$$\mathrm{ind}\,\left[\left(\frac{1}{n}I - S\right)\left(\frac{1}{n}I - T\right)\right] = \mathrm{ind}\,\left(\frac{1}{n}I - S\right) + \mathrm{ind}\,\left(\frac{1}{n}I - T\right),$$

and this implies $\mathrm{ind}\,(ST) = \mathrm{ind}\,(S) + \mathrm{ind}\,(T)$.
(ii) Since $\pi(T + K) = \pi(T)$ then $T + K$ is B-Fredholm, again by Theorem 1.125. By Theorem 1.117 we may choose $n_0 \in \mathbb{N}$ such that $\frac{1}{n}I - T$, $\frac{1}{n}I - (T + K)$ are Fredholm, and $\mathrm{ind}\,(\frac{1}{n}I - T) = \mathrm{ind}\,(T)$ and $\mathrm{ind}\,(\frac{1}{n}I - (T + K)) = \mathrm{ind}\,(T + K)$.

Since $\frac{1}{n}I - T$ is Fredholm then ind $(\frac{1}{n}I - (T + K)) = $ ind $(\frac{1}{n}I - T)$, thus ind $(T + K) = $ ind T. ∎

Theorem 1.127 *Let $T \in L(X)$ and $X = M \oplus N$, where M and N are two closed subspaces of X. If T is B-Fredholm then $T|M$ and $T|N$ are B-Fredholm.*

Proof We show that $U := T|N$ is B-Fredholm. Let P be the projection of X onto N along M. Clearly R is B-Fredholm and commutes with T, so by part (i) of Theorem 1.126, TP is B-Fredholm. Consequently, there exists an integer $n \in \mathbb{N}$ such that $(TP)^n(X)$ is closed and the restriction $(TP)_n = TP|(TP)^n(X)$ is Fredholm. Since $(TP)^n(X) = U(X)$ and $(TP)_n = U$, then U is B-Fredholm. ∎

Consider now the *Calkin algebra* $L(X)/\mathcal{K}(X)$ provided with the usual quotient norm, where $\mathcal{K}(X)$ denotes the two-sided ideal in $L(X)$ of all compact operators. Let $\tau : L(X) \to L(X)/\mathcal{K}(X)$ be the canonical quotient mapping. By Atkinson's characterization of Fredholm operators we know that $T \in \Phi(X)$ if and only if $\tau(T)$ is invertible in $L(X)/\mathcal{K}(X)$. The following example shows that the analogue of Theorem 1.125 is not true if we replace $L(X)/\mathcal{F}(X)$ with $L(X)/\mathcal{K}(X)$.

Example 1.128 Let (λ_n) be a sequence in $\mathbb{C}$ which converges to 0 and assume that $\lambda_n \neq 0$ for all n. Let $\ell^2(\mathbb{N})$ denote the space of all 2-summable sequences of complex numbers, and denote by T the operator on the Hilbert space $\ell^2(\mathbb{N})$ defined by:

$$T(x_1, x_2, \dots) := (\lambda_1 x_1, \lambda_2 x_2, \dots) \quad \text{for all } (x_n) \in \ell^2(\mathbb{N}).$$

Then

$$T^n(x_1, x_2, \dots) := (\lambda_1{}^n x_1, \lambda_2{}^n x_2, \dots).$$

Since $\lambda_k{}^n \neq 0$ for all k and $\lambda_k{}^n \to 0$, as $k \to \infty$, $T^n \in \mathcal{K}(X)$ and T^n is not a finite-dimensional operator for every n. Hence, $T^n(X)$ is not closed for all n, thus T is not B-Fredholm. Obviously, the image $\tau(T)$ in the Calkin algebra $L(X)/\mathcal{K}(X)$ is 0, so $\tau(T)$ is Drazin invertible. This example also shows that the class of B-Fredholm operators is not stable under compact perturbations. Indeed, 0 is B-Fredholm while $0 + T = T$ is not B-Fredholm.

Theorem 1.129 *If $T \in L(X)$ then the following assertions are equivalent:*

(i) *T is B-Fredholm;*
(ii) *T^m is B-Fredholm for all integers $m > 0$;*
(iii) *T^m is B-Fredholm for some integer $m > 0$.*

Proof The implication (i) $\Rightarrow$ (ii) follows from Theorem 1.126. The implication (ii) $\Rightarrow$ (iii) is obvious. (iii) $\Rightarrow$ (i) Suppose that T^m is B-Fredholm for some integer $m > 0$. Then there exist an integer n such that $T^{nm}(X)$ is closed, and $T_n^m : T^{nm}(X) \to T^{nm}(X)$ is Fredholm. Since

$$\ker T \cap T^{nm}(X) \subseteq \ker T^m \cap T^{nm}(X),$$

it then follows that ker $T \cap T^{nm}(X)$ has finite dimension, by Theorem 1.116. Moreover, the inclusions

$$T^{(n+1)m}(X) \subseteq T^{nm+1}(X) \subseteq T^{nm}(X)$$

entail that $T^{nm+1}(X)$ has finite codimension in $T^{nm}(X)$. Consequently, T_{nm} is a Fredholm operator and hence T is B-Fredholm.

The next result shows that a characterization similar to that of Theorem 1.119 established for Fredholm operators on Banach spaces also holds for semi B-Fredholm operators in the case of *Hilbert space operators*.

Theorem 1.130 *If $T \in L(H)$, H a Hilbert space, then T is semi B-Fredholm (respectively, upper semi B-Fredholm, lower semi B-Fredholm) if and only if there exist two closed T-invariant subspaces M and N such that $X = M \oplus N$, $T|M$ is semi-Fredholm (respectively, upper semi-Fredholm, lower semi Fredholm) and $T|N$ is nilpotent.*

Proof Suppose that T is semi B-Fredholm. Then T is quasi-Fredholm, so, by Corollary 1.107, T is of Kato-type, i.e. there exist closed invariant subspaces M and N such that $X = M \oplus N$, $T|M$ is semi-regular and $T|N$ nilpotent. Let $(T|N)^d = 0$. From parts (ii) and (iii) of Theorem 1.64 we know that ker $T|M = $ ker $T \cap T^d(X)$ and $T_M(M) = T(M) \oplus N = T(X) + $ ker T^d. Now, if T is upper semi B-Fredholm then ker $T|M = $ ker $T \cap T^d(X)$ has finite dimension and since the range of $T|M$ is closed, since $T|M$ is semi-regular, $T|M$ is upper semi-Fredholm. If T is lower semi B-Fredholm, then there exists a finite-dimensional subspace Y of H such that $H = Y \oplus T(M) \oplus N$. Let P_M denote the projection of H onto M along N. Then $M = P_M(Y) + T(M)$, from which we deduce that $T(M)$ has finite codimension on M. Therefore, $T|M$ is lower semi-Fredholm.

Conversely, suppose that there exist closed invariant subspaces M and N such that $H = M \oplus N$, $T|M$ is semi-Fredholm and $T|N$ is nilpotent of order d. Then $(T|M)^d$ is semi-Fredholm, and hence $T^d(H) = T^d(M)$ is a closed subspace of M, so is a closed subspace of H. Evidently, $T_d = (T|M)_d$, so, using Theorem 1.112, and the fact that $T|M$ is semi-Fredholm, we conclude that T_d is semi-Fredholm, hence T is semi B-Fredholm.

The proofs of the assertions concerning upper semi B-Fredholm and lower semi B-Fredholm operators are omitted.   ■

We also omit the proof of the following theorem, since it is similar to Theorem 1.129, and may be easily obtained by using Theorem 1.130 instead of Theorem 1.119.

Theorem 1.131 *If $T \in L(H)$, H a Hilbert space. Then the following assertions are equivalent:*

(i) *T is upper semi B-Fredholm, (respectively, lower semi B-Fredholm);*

(ii) T^m is upper semi B-Fredholm, (respectively, lower semi B-Fredholm), for all integer $m > 0$;
(iii) T^m is upper semi B-Fredholm, (respectively, lower semi B-Fredholm), for some integer $m > 0$.

1.9 Drazin Invertible Operators

A bounded operator T which is Drazin invertible in the Banach algebra $\mathcal{A} := L(X)$ is simply said to be Drazin invertible. Drazin invertibility for operators may be characterized in several ways:

Theorem 1.132 *If $T \in L(X)$ then the following statements are equivalent:*

(i) *T is Drazin invertible, i.e. there is an $S \in L(X)$ such that $TS = ST$, $STS = S$ and $T^n ST = T^n$ for some $n \in \mathbb{N}$;*
(ii) *there is an $S \in L(X)$ which commutes with T and $n \in \mathbb{N}$ such that $T^{n+1}S = T^n$;*
(iii) *$p(T) = q(T) \leq n$;*
(iv) *$T = T_1 \oplus T_2$, where T_1 is nilpotent and T_2 is invertible;*
(v) *Either $0 \notin \sigma(T)$ or $0 \in \mathrm{iso}\,\sigma(T)$, and the restriction of T onto the range $P(X)$ of the spectral projection associated at $\{0\}$ is nilpotent.*

Proof (i) $\Rightarrow$ (ii) Suppose that there exists an $S \in L(X)$ which commutes with T such that $STS = S$ and $T^n ST = T^n$. Then

$$T^{n+1}S = TT^n S = T^n ST = T^n.$$

(ii) $\Rightarrow$ (iii) We have

$$T^n(X) = [T^{n+1}S](X) \subseteq T^{n+1}(X),$$

hence $q(T) \leq n$. If $x \in \ker T^{n+1}$ then $T^n x = ST^{n+1}x = 0$, so $x \in \ker T^n$ and, consequently, $p(T) < \infty$.

(iii) $\Rightarrow$ (iv) If $p := p(T) = q(T)$ then, by Theorem 1.35, $X = \ker T^p \oplus T^p(X)$ and, by Theorem 1.36, 0 is a pole of the resolvent of T and $T^p(X)$ is closed, since it coincides with the kernel of the spectral projection associated with 0. Let $T_1 := T|\ker T^p$ and $T_2 := T|T^p(X)$. Then $T_1{}^p = 0$ and because $\ker T^p = \ker T^{p+1}$ and $T^p(X) = T^{p+1}(X)$ it is easily seen that T_2 is invertible.

(iv) $\Leftrightarrow$ (v) If $X = X_1 \oplus X_2$ with $T_1 = T|X_1$ nilpotent and $T_2 = T|X_2$ invertible then either $0 \notin \sigma(T)$ or $0 \in \mathrm{iso}\,\sigma(T)$. Moreover, X_1 coincides with the range of the spectral projection associated with 0, hence (iv) $\Rightarrow$ (v). The reverse implication (v)$\Rightarrow$ (iv) is clear, so (iv) and (v) are equivalent.

(iv) $\Rightarrow$ (i) Define $S := 0 \oplus T_2^{-1}$ and let $n \in \mathbb{N}$ be such that $T_1{}^n = 0$. Then S satisfies (i). ∎

From Theorem 1.119 we immediately have

Corollary 1.133 *Every Drazin invertible operator is B-Fredholm.*

In the sequel we denote by S the Drazin inverse of a Drazin invertible operator, i.e.,

$$T^n S T = T^n, \quad S T S = S, \quad T S = S T. \tag{1.31}$$

The next result shows that the index of $T \in L(X)$, where T is regarded as an element of the Banach algebra $A := L(X)$, coincides with the order of the pole $p := p(T) = q(T)$ of the resolvent of T.

Theorem 1.134 *If $T \in L(X)$ is Drazin invertible of index $i(T) = k$ then $p(T) = q(T) = k$.*

Proof Let S be the Drazin inverse of T. We have

$$T S = S T \quad S T S = S \quad T^k S T = T^k, \tag{1.32}$$

and hence, by Theorem 1.132, $p(T) = q(T) \leq k$. On the other hand, if $p := p(T) = q(T)$ then $X = \ker T^p \oplus T^p(X)$ and if we denote by P the spectral projection associated with $\{0\}$ then $T_1 := T | \ker P$ is invertible and the operator $S := T_1^{-1}(I - P)$ is well-defined, since $(I - P)(X) = \ker P$. It is easily seen that $T S = S T = I - P$. Moreover,

$$S T S = T S^2 = (I - P) S = S,$$

since $S(X) = (I - P)(X)$, and

$$T^{p+1} S = T^p T S = T^p (I - P) = (I - P) T^p = T^p,$$

since $T^p(X) = (I - P)(X))$. Therefore, $i(T) \leq p$. ∎

By Theorem 1.132 we know that $T \in L(X)$ is Drazin invertible if and only if there exist two closed invariant subspaces Y and Z such that $X = Y \oplus Z$ and, with respect to this decomposition,

$$T = T_1 \oplus T_2, \quad \text{with } T_1 := T | Y \text{ nilpotent and } T_2 := T | Z \text{ invertible.} \tag{1.33}$$

Moreover, by Theorem 1.122, the Drazin inverse S of T, if it exists, is uniquely determined, and in the proof of Theorem 1.132, it has been observed that S may be represented, with respect to the decomposition $X = Y \oplus Z$, as the directed sum

$$S := 0 \oplus S_2 \quad \text{with } S_2 := T_2^{-1}. \tag{1.34}$$

Theorem 1.135 *Suppose that* $T \in L(X)$ *is Drazin invertible with* $p := p(T) = q(T) \geq 1$. *If* P *is the spectral projection associated with* $\{0\}$ *and* S *is the Drazin inverse of* T, *then*

(i) $P = I - TS$, $\ker S = \ker (T^p) = P(X)$.
(ii) $S(X) = \ker P = T^p(X)$.
(iii) S *is Drazin invertible and* $p(S) = q(S) = 1$. *The Drazin inverse of* S *is* TST.

Proof

(i) The equality $P = I - TS$ has been observed in the proof of Theorem 1.134. If $p := p(T) = q(T)$ then $P(X) = \ker T^p$ and $\ker P = T^p(X)$, by Theorem 1.36. If $x \in X$ then $Sx = 0$ if and only if $Px = x$, from which we obtain that $\ker S = P(X)$.

(ii) From $P = I - TS = I - ST$ it is easily seen that $S(X) = \ker P$.

(iii) Evidently, S is Drazin invertible, since from the decomposition (1.34) we know that S is the direct sum of a nilpotent operator and an invertible operator. Let $U := TST$. Then

$$S^2 U = SUS = STSTS = S.$$

Furthermore,

$$SU = STST = TSTS = US,$$

and

$$USU = TSTSTST = TSTST = TST = U.$$

Therefore, $U = TST$ is the Drazin inverse of S, and obviously $i(S) = 1$. ∎

A simple consequence of the spectral mapping theorem for the spectrum shows that if $T \in L(X)$ is invertible then the points of the spectrum of the inverse T^{-1} are the reciprocals of the spectrum $\sigma(T)$, i.e.

$$\sigma(T^{-1}) = \left\{ \frac{1}{\lambda} : \lambda \in \sigma(T) \right\}.$$

The relationship of "reciprocity" observed above, between the nonzero parts of the spectrum, is also satisfied when we consider Drazin invertibility.

Theorem 1.136 *Suppose that* $T \in L(X)$ *is Drazin invertible with Drazin inverse* S. *Then*

$$\sigma(S) \setminus \{0\} = \left\{ \frac{1}{\lambda} : \lambda \in \sigma(T) \setminus \{0\} \right\}. \tag{1.35}$$

Proof If T is invertible then $S = T^{-1}$, so (1.35) is clear. Suppose that $0 \in \sigma(R)$. By Theorem 1.132 $T = T_1 \oplus T_2$, where T_1 is nilpotent, T_2 is invertible, and $S := 0 \oplus S_2$, with $S_2 := T_2^{-1}$. Clearly,

$$\sigma(T) = \sigma(T_1) \cup \sigma(T_2) = \{0\} \cup \sigma(T_2) \quad \text{and } \sigma(S) = \{0\} \cup \sigma(S_2),$$

where $0 \notin \sigma(T_2)$ and $0 \notin \sigma(S_2)$. Since S_2 is the inverse of T_2, we then conclude that the nonzero part of the spectrum of S is given by the reciprocals of the nonzero points of the spectrum of T.   ∎

The spectral mapping theorem also holds for the approximate point spectrum, so, by using similar arguments as above, we obtain

$$\sigma_a(S) \setminus \{0\} = \left\{ \frac{1}{\lambda} : \lambda \in \sigma_a(T) \setminus \{0\} \right\}. \tag{1.36}$$

An operator $T \in L(X)$ is said to be *relatively regular* if there is an $S \in L(X)$ such that $TST = T$. S is also called a *pseudo inverse* of T. An operator T is relatively regular if and only if $\ker T$ and $T(X)$ are both complemented, see Appendix A. In general, a relatively regular operator admits infinite pseudo-inverses, in fact, if S is a pseudo-inverse then all operators of the form $STS + U - STUTS$, with $U \in L(X)$ arbitrary, are pseudo-inverses of T, see [87, Theorem 2]. The next example shows that for a pseudo-inverse formula (1.35), in general, does not hold.

Example 1.137 Let T be the unilateral right shift in $\ell_2(\mathbb{N})$, defined as

$$T(x_1, x_2, \dots) := (0, x_1, x_2, \dots) \quad \text{for all } (x_n) \in \ell_2(\mathbb{N}),$$

then T is relatively regular. Indeed, the range of T is closed and hence complemented in the Hilbert space $\ell_2(\mathbb{N})$, and the same happens for the kernel $\ker T$, so T is relatively regular. We have $\sigma(T) = \mathbf{D}(0, 1)$. Consequently, the points of $\sigma(S) \setminus \{0\}$, for any pseudo-spectral inverse S of T, cannot be the reciprocals of $\sigma(T) \setminus \{0\}$, otherwise $\sigma(S)$ would be unbounded.

Remark 1.138 The Drazin inverse S of a Drazin invertible operator is relatively regular, since $STS = S$, and T is a pseudo-inverse of S. But generally a Drazin invertible operator T may have no pseudo-inverse. For instance, if $V \in L(X)$ does not have a closed range then the operator $T := \begin{pmatrix} 0 & 0 \\ V & 0 \end{pmatrix}$ is nilpotent, since $T^2 = 0$, and hence is Drazin invertible, but T has no pseudo-inverse, since its range is not closed.

An interesting generalization of Drazin invertibility has been introduced by Koliha [201], see also [202]. In the case of the Banach algebra $L(X)$, an operator $T \in L(X)$ is said to be *generalized Drazin invertible* if $0 \notin \text{acc } \sigma(T)$, i.e. either T

is invertible or 0 is an isolated point of $\sigma(T)$. This is equivalent to saying that there exists an operator $B \in L(X)$, called the generalized Drazin inverse of T, such that

$$STS = S, \quad TS = ST \quad \text{and } T(I - TS) \text{ is quasi-nilpotent,}$$

see, for instance, Djordjević and Rakočević [114].

1.10 Left and Right Drazin Invertible Operators

The concept of Drazin invertibility suggests the following definition:

Definition 1.139 $T \in L(X)$, X a Banach space, is said to be *left Drazin invertible* if $p := p(T) < \infty$ and $T^{p+1}(X)$ is closed, while $T \in L(X)$ is said to be *right Drazin invertible* if $q := q(T) < \infty$ and $T^q(X)$ is closed.

The operator considered in Example 1.94 shows that the condition $q = q(T) < \infty$ does not entail that $T^q(X)$ is closed. Clearly, $T \in L(X)$ is both right and left Drazin invertible if and only if T is Drazin invertible. In fact, by Theorem 1.36, if $0 < p := p(T) = q(T)$ then $T^p(X) = T^{p+1}(X)$ is the kernel of the spectral projection P_0 associated with the spectral set $\{0\}$. We show now that the concepts of Drazin invertibility may be characterized in a very simple way by means of the restrictions $T_n := T|T^n(X)$.

Theorem 1.140 *If $T \in L(X)$ then we have:*

(i) *T is left Drazin invertible if and only if there exists a $k \in \mathbb{N}$ such that $T^k(X)$ is closed and T_k is bounded below. In this case $T^j(X)$ is closed and T_j is bounded below for all natural numbers $j \geq k$.*

(ii) *T is right Drazin invertible if and only if there exists a $k \in \mathbb{N}$ such that $T^k(X)$ is closed and T_k is onto. In this case $T^j(X)$ is closed and T_j is onto for all natural numbers $j \geq k$.*

(iii) *T is Drazin invertible if and only if there exists a $k \in \mathbb{N}$ such that $T^k(X)$ is closed and T_k is invertible. In this case $T^j(X)$ is closed and T_j is invertible for all natural numbers $j \geq k$.*

Proof

(i) Suppose $p := p(T) < \infty$ and that $T^{p+1}(X)$ closed. Then T_p is injective and $R(T_p) = T^{p+1}(X)$ is closed. Conversely, if $T_{[k]}$ is bounded below for some $k \in \mathbb{N}$ then, by Lemma 1.23, $p := p(T) < \infty$ and by Remark 1.25 we have $p \leq k$, and hence $p + 1 \leq k + 1$. Since $R(T_k) = T^{k+1}(X)$ is closed then, by Lemma 1.100, $T^{p+1}(X)$ is closed and, consequently, T is left Drazin invertible. The last assertion is clear, by Remark 1.25, T_j is injective for all $j \geq k$ and $T^j(X)$ is closed, again by Lemma 1.100.

(ii) Suppose that $q := q(T) < \infty$ and $T^q(X)$ is closed, then $R(T_q) = T^{q+1}(X) = T^q(X)$, so T_q is onto. Conversely, suppose that $T^k(X)$ is closed and T_k is onto for some $k \in \mathbb{N}$. Then, by Lemma 1.23, $q = q(T) < \infty$ and $q + 1 \le k + 1$. By Lemma 1.100, $T^q(X)$ is closed, and hence T is right Drazin invertible. By Lemma 1.100, $T^j(X)$ is closed for all $j \ge k$, and, by Remark 1.25, T_j is onto for all $j \ge k$.

(iii) Clear.   ∎

Theorem 1.141 *Every left Drazin invertible operator T is upper semi B-Fredholm with* ind $T \le 0$; *every right Drazin invertible T operator is lower semi B-Fredholm with* ind $T \ge 0$. *Every Drazin invertible operator T is a B-Fredholm operator having index* ind $T = 0$.

Proof Suppose that T is left Drazin invertible. By Theorem 1.140, T_n is bounded below for some $n \in \mathbb{N}$, in particular T_n is upper semi-Fredholm. Hence T is upper semi B-Fredholm, and, by part (i) of Theorem 1.22, ind $T_n \le 0$. Analogously, if T is right Drazin invertible, then, by Theorem 1.140, T_n is onto for some $n \in \mathbb{N}$, in particular T_n is lower semi B-Fredholm. Again, part (ii) of Theorem 1.22 implies that ind $T_n \ge 0$. The last assertion is clear.   ∎

The next result characterizes the left Drazin invertible and the right Drazin invertible operators among the operators which have topological uniform descent.

Theorem 1.142 *Suppose that $T \in L(X)$. Then the following statements are equivalent:*

 (i) *T is left Drazin invertible;*
 (ii) *T is quasi-Fredholm and has finite ascent;*
(iii) *T has topological uniform descent and has finite ascent.*
        Dually, the following statements are equivalent:
(iv) *T is right Drazin invertible;*
 (v) *T is quasi-Fredholm and has finite descent:*
(vi) *T has topological uniform descent and has finite descent.*

Proof The implications (i) $\Rightarrow$ (ii) $\Rightarrow$ (iii) are obvious, so, in order to show the equivalences of the statements (i), (ii), and (iii), it suffices to prove the implication (iii) $\Rightarrow$ (i).

Suppose that $p := p(T) < \infty$ and T has topological uniform descent for $n \ge p$. By Theorem 1.74 we know that ker $T^j + T^k(X)$ is closed in X for all $k \in \mathbb{N}$ and $j \ge p$. In particular, ker $T^p + T^k(X)$ is closed in X for all integers $k \ge 0$. By Corollary 1.101 it then follows that $T^{p+1}(X)$ is closed, and hence T is left Drazin invertible.

The implications (iv) $\Rightarrow$ (v) $\Rightarrow$ (vi) are obvious, so, in order to show the equivalences of the statements (iv), (v), and (vi), we only need to prove the implication (vi) $\Rightarrow$ (iii).

Suppose that $q := q(T) < \infty$ and T has topological uniform descent for $n \ge q$. By Lemma 1.19 we know that the condition $q = q(T) < \infty$ entails that for every $n \in \mathbb{N}$ and $m \ge q$ the subspace $T^n(X)$ admits a complementary subspace $Y_n \subseteq$ ker T^m, hence $X = $ ker $T^m + T^n(X)$ for all $n \in \mathbb{N}$, $m \ge q$. In particular, the sums ker $T^m + T^n(X)$ for all $n \in \mathbb{N}$ and $m \geq q$ are closed. From Corollary 1.101 it then follows that $T^q(X)$ is closed. Therefore T is right Drazin invertible. ∎

By Theorem 1.22, if T is a Fredholm operator with index 0 and either $p(T)$ or $q(T)$ is finite then T is Browder. This result, in the framework of B-Fredholm theory, may be generalized as follows:

Theorem 1.143 *For an operator $T \in L(X)$ the following statements hold:*

(i) *If T is upper semi B-Fredholm with index* ind $T \leq 0$ *and $q(T) < \infty$, then T is Drazin invertible.*

(ii) *If T is lower semi B-Fredholm with index* ind $T \geq 0$ *and $p(T) < \infty$, then T is Drazin invertible.*

(iii) *If T is B-Fredholm with index* ind $T = 0$ *and has either ascent or descent finite, then T is Drazin invertible.*

Proof

(i) By Theorem 1.142 we know that T is right Drazin invertible, thus ind $T \geq 0$ by Theorem 1.141. Hence ind $T = 0$, so that there exists an $n \in \mathbb{N}$ such that $T^m(X)$ is closed and T_m is Fredholm with ind $T_m = \alpha(T_m) - \beta(T_m) = 0$ for all $m \geq n$. Now, since $q := q(T) < \infty$, it is evident that the range $R(T_m)$ of T_m is $T^{m+1}(X) = T^m(X)$ for all $m \geq q$, thus T_m is onto. If we choose $m \geq \max\{n, q\}$ then T_m is both onto and Fredholm with index 0, so $\alpha(T_m) = \beta(T_m) = 0$, i.e., T_m is invertible. Consequently, T_m^k is invertible for all $k \in \mathbb{N}$, from which we deduce that

$$\ker T^k \cap T^m(X) = \ker T_m^k = \{0\} \quad \text{for all } k \in \mathbb{N},$$

and this implies that $p(T) < \infty$, by Lemma 1.19.

(ii) By Theorem 1.142 we know that T is left Drazin invertible, thus ind $T \leq 0$ by Theorem 1.141. Hence ind $T = 0$, so for some $n \in \mathbb{N}$, $T^m(X)$ is closed and T_m is Fredholm with ind $T_m = \alpha(T_m) - \beta(T_m) = 0$ for all $m \geq n$. By Lemma 1.19, the assumption $p := p(T) < \infty$ entails that

$$\ker T_m = \ker T \cap T^m(X) = \{0\} \quad \text{for all natural } m \geq p.$$

Choosing $m \geq \max\{n, p\}$ then T_m is both injective and is Fredholm with index ind $T_m = -\beta(T_m) = 0$, hence T_m is invertible. Consequently,

$$T^{m+1}(X) = R(T_m) = T^m(X),$$

and hence $q(T) < \infty$.

(iii) This is evident from part (i) and (ii). ∎

We now show that the concepts of left and right Drazin invertibility are dual to each other:

Theorem 1.144 *For every $T \in L(X)$ the following equivalences hold:*

(i) *T is left Drazin invertible $\Leftrightarrow$ T^* is right Drazin invertible.*
(ii) *T is right Drazin invertible $\Leftrightarrow$ T^* is left Drazin invertible.*
(iii) *T is Drazin invertible if and only if T^* is Drazin invertible.*

Proof

(i) Suppose that T is left Drazin invertible. Then $p := p(T) < \infty$ and, by Corollary 1.73, T has uniform descent for $n \geq p$. Since $T^p(X)$ is closed, Theorem 1.100 entails that $T^{p+1}(X)$ is also closed, and hence $T^{*p+1}(X^*)$ is closed. From the equality ker $T^p = $ ker T^{p+1} and from the classical closed range theorem (see Appendix A) we then deduce that

$$T^{*p}(X^*) = [\text{ker } T^p]^\perp = [\text{ker } T^{p+1}]^\perp = T^{*p+1}(X^*).$$

This shows that T^* has finite descent $q := q(T^*) \leq p$ and since $T^{*q}(X^*) = T^{*p}(X^*)$ is closed it then follows that T^* is right Drazin invertible.

   Conversely, suppose that T^* is right Drazin invertible. Then $q := q(T^*) < \infty$ and $T^{*q}(X^*)$ is closed. From the equality $T^{*q}(X^*) = T^{*q+1}(X^*)$ and from the closed range theorem we then obtain:

$$\text{ker } T^q = {}^\perp[T^{*q}(X^*)] = {}^\perp[T^{*q+1}(T^*)] = \text{ker } T^{q+1},$$

and hence $p := p(T) \leq q$. Since $T^{*q+1}(X^*)$ is closed then $T^{q+1}(X)$ is also closed, again by Theorem 1.100 it then follows that $T^{p+1}(X)$ is closed, so T is left Drazin invertible.

(ii) This may be proved in a similar way to part (i).

(iii) This is obvious, since both left and right Drazin invertibility entail Drazin invertibility. ∎

   The concepts of Drazin invertibility may be relaxed as follows. For every linear operator T on a vector space X consider the quantities already introduced: $c_n(T) = \dim \frac{T^n(X)}{T^{n+1}(X)}$ and $c'_n(T) = \dim \frac{\text{ker } T^{n+1}}{\text{ker } T^n}$. Evidently, the descent $q(T)$ is the infimum of the set $\{n \in \mathbb{N} : c_n(T) = 0\}$ (also here the infimum of an empty set is defined to be ∞), while the ascent is the infimum of the set $\{n \in \mathbb{N} : c'_n(T) = 0\}$.

Definition 1.145 For a linear operator T on a vector space X the *essential descent* is defined as

$$q_e(T) := \inf\{n \in \mathbb{N} : c_n(T) < \infty\},$$

while the *essential ascent* is defined as

$$p_e(T) := \inf\{n \in \mathbb{N} : c'_n(T) < \infty\}.$$

If $T \in L(X)$ is such that $p_e(T) < \infty$ and $T^{p_e(T)+1}(X)$ is closed then T is said to be *essentially left Drazin invertible*, while $T \in L(X)$ is said to be *essentially right Drazin invertible* if $q_e(T) < \infty$ and $T^{q_e(T)}(X)$ is closed.

Note that if $q := q_e(T) < \infty$ then $T^q(X) =^e T^n(X)$ for all $n \geq q$, and T has finite essential descent if and only if there exists a $d \in \mathbb{N}$ such that $c_n(T) < \infty$ for all $n \geq d$.

Remark 1.146 In the sequel we list some important properties of the essential ascent and descent:

(i) If T has finite ascent $p(T)$ then $p_e(T) = p(T)$. Analogously, if T has finite descent $q(T)$ then $q_e(T) = q(T)$.

(ii) If both $p_e(T)$, $q_e(T)$ are finite then $p_e(T) = q_e(T)$ and $T^{p_e(T)}(X)$ is a complemented subspace of X, see [245, Chap. III, Lemma 11]. T has both $p_e(T)$, $q_e(T)$ finite if and only if T is B-Fredholm, see [63, Corollary 3.7] or [245, Chap. III, Theorem 12].

(iii) If T has essential finite ascent then $T^n(X)$ is closed for some $n > p_e(T)$ if and only if $T^n(X)$ is closed for all $n \geq p_e(T)$, see [163].

(iv) Every essentially left or right Drazin invertible operator is quasi-Fredholm. More precisely, T is essentially left Drazin invertible if and only if T is quasi-Fredholm and $T_d := T|T^d(X)$ is upper semi-Fredholm (as usual, $d = \mathrm{dis}(T)$ is the degree of stable iteration). Analogously, T is essentially right Drazin invertible if and only if T is quasi-Fredholm and $T_d := T|T^d(X)$ is lower semi-Fredholm, [63, Corollary 3.5].

(v) T is essentially left Drazin invertible if and only if T is upper semi B-Fredholm and analogously, T is essentially right Drazin invertible if and only if T is lower semi B-Fredholm, [63, Theorem 3.6].

(vi) It is easily seen that T is essentially left (respectively, right) Drazin invertible if and only if its dual T^* is essentially right (respectively, left) Drazin invertible.

(vii) The essential ascent spectrum and the essential descent spectrum are compact subsets of $\mathbb{C}$, see [150] and [85]

(viii) Evidently, every lower semi-Fredholm operator has finite essential descent. As observed in Lemma 1.19, $T \in L(X)$ has finite descent if and only if $X = T(X) + \ker T^d$ for some integer $d \geq 0$. Similarly, $T \in L(X)$ has finite essential descent if and only if $T(X) + \ker T^d$ has finite codimension for some integer $d \geq 0$, see [163].

1.11   Comments

The concept of the algebraic core of an operator was introduced by Saphar [270], while the analytic core was introduced by Vrbová [294] and further studied by Mbekhta [233]. Theorem 1.33 is due to Aiena and Monsalve [19].

The theory of semi-regular operators has its origins in Kato's classical treatment [195] of perturbation theory, but these operators did not originally go by this name. Kato's pioneering work led many other authors to study this class of operators, in particular it was extensively studied by Mbekhta [229, 230, 233, 234], Mbekhta and Ouahab [236], and Schmoeger [273]. Originally the semi-regular spectrum was investigated for operators acting on Hilbert spaces by Apostol [51], and for this reason this spectrum was sometimes called the *Apostol spectrum*. This spectrum was defined, for Hilbert space operators, as the set of all complex λ such that either $\lambda I - T$ is not closed or λ is a discontinuity point for the function $\lambda \to (\lambda I - T)^{-1}$, see Theorem 1.51. Later the results of Apostol were generalized to Banach space operators by Mbekhta [229, 230], and Mbekhta and Ouahab [236], see also Harte [172] and Müller [243]. The methods adopted in the fourth section are strongly inspired by the work of Mbekhta and Ouahab [236], and Schmoeger [272]. Example 1.53 and Theorem 1.55 are taken from Müller [243].

The generalized Kato decomposition was studied in several papers by Mbekhta [230, 231] and [232]. The particular case of essentially semi-regular operators was systematically investigated by Müller and Rakočević [263]. The material presented here is completely inspired by Müller in [243], and Theorem 1.67 is taken from Kordula, Müller and Rakočević [205]. The fundamental result that a semi-Fredholm operator is essentially semi-regular was proved by Kato [195]. The section concerning the operators having uniform descent is entirely based on the work by Grabiner [162], while the sections relative to quasi-Fredholm operators are modeled after Labrousse [208], Mbekhta [230] and Berkani [62]. Theorem 1.110 was proved by Mbekhta and Müller [235].

The work of Berkani et al. [62–64, 71, 72] also inspired the section concerning B-Fredholm theory. Theorem 1.106 is taken from an unpublished paper of Müller, which also proved, for Banach space operators, Theorem 1.130 previously proved by Berkani [62] for Hilbert space operators. The concept of Drazin invertibility in associative rings and semigroups were introduced by Drazin [123], and has been investigated by several authors, see for instance Patricio and Hartwig [255], Wang and Chen [295], Roch and Silbermann [267].

The formula (1.30) for the Drazin inverse of ba is due to Cline [96] and it is a useful tool to finding the Drazin inverse of a sum of two elements and that of a block matrix, see [255]. Numerous mathematicians have studied the analogue of Cline's formula for other kinds of "invertibility", see Barnes [56], Corach et al. [102]. The last section concerning Drazin invertible operators is patterned after Caradus [87], and Caradus et al. [88]. Theorem 1.118 is taken from [38], while Theorem 1.142 is taken from Aiena and Triolo [26].

Chapter 2
Local Spectral Theory

In this chapter we shall introduce an important property, defined for bounded linear operators on complex Banach spaces, the so-called *single-valued extension property* (SVEP). This property dates back to the early days of local spectral theory and appeared first in Dunford [140] and [141]. Subsequently, this property has received a more systematic treatment in the classical texts by Dunford and Schwartz [143], as well as those by Colojoară and Foiaş [98], by Vasilescu [292] and, more recently, by Laursen and Neumann [216], and Aiena [1].

The single-valued extension property has a basic importance in local spectral theory since it is satisfied by a wide variety of linear bounded operators in the spectral decomposition problem. An important class of operators which enjoys this property is the class of all decomposable operators on Banach spaces, which includes normal operators on Hilbert spaces, Riesz operators and more generally operators which have a discrete spectrum. Another class of operators, not necessarily decomposable, is given by the class of all multipliers defined on a semi-prime commutative Banach algebra, and in particular by the class of all convolution operators of group algebras $L_1(G)$, where G is a locally compact abelian group. Other classes of operators which satisfy the SVEP are obtained by the $H(p)$-operators, which will be studied in Chap. 4, and these classes include several kinds of operators on Hilbert spaces, obtained by weakening, in some way, the condition of normality. As we shall see in Chaps. 5 and 6, a localized version of the single-valued extension property will be a very precious tool for establishing Browder-type and Weyl-type theorems.

In the first section we introduce the basic tools of local spectral theory, as the local spectrum and the local spectral subspace $X_T(F)$ associated to a subset $F \subseteq \mathbb{C}$, and we give an important characterization of the analytic core $K(T)$ as a special case of a local spectral subspace. In the second section, we introduce the concept of a glocal local subspace, a somewhat more useful variant of the concept of an analytic local subspace, which is better suited for operators without the single-valued extension property. The third section also deals with another important subspace in local

© Springer Nature Switzerland AG 2018   95
P. Aiena, *Fredholm and Local Spectral Theory II*, Lecture Notes in Mathematics
2235, https://doi.org/10.1007/978-3-030-02266-2_2

spectral theory: the quasi-nilpotent part of an operator. This subspace, which is the glocal spectral subspace associated with the set $\{0\}$, also has a relevant role in Fredholm theory.

The fourth section is devoted to the study of a localized version of the single-valued extension property, in particular we shall employ the basic tools of local spectral theory to establish a variety of conditions that ensure the single-valued extension property at a point λ_0. We shall see that the relative positions of all the subspaces introduced in the previous section are intimately related to the SVEP at a point. Most of the conditions which entail the localized SVEP at a point λ_0 involve the kernel type and the range type of subspaces previously introduced, as well as the analytic core and the quasi-nilpotent part of $\lambda_0 I - T$. Furthermore, in this section we shall give results concerning the particular case that λ_0 is an isolated point of the spectrum, or is an isolated point of the approximate point spectrum $\sigma_{\mathrm{ap}}(T)$, or, dually, λ_0 is an isolated point of the surjectivity spectrum $\sigma_{\mathrm{s}}(T)$.

In the fifth section we shall show that the localized single-valued extension property behaves canonically under the Riesz functional calculus, while in the sixth section we shall see that for operators having topological uniform descent, in particular for a quasi-Fredholm operator T, the SVEP at 0 is equivalent to several other conditions, some of them concerning the quasi-nilpotent part and the analytic core of T.

The seventh section deals with the preservation of the localized SVEP under some commuting perturbations, in particular we show that the localized SVEP is preserved under Riesz commuting perturbations, while the global SVEP is preserved under algebraic commuting perturbations. We shall also study the preservation of the localized SVEP under quasi-nilpotent equivalence. In this chapter we will introduce another local spectral property, the Dunford property (C), which is stronger than the SVEP. The last part of the chapter is centred around the spectral properties of the product of operators RS and SR, and we shall show that these products share many spectral properties, as well as many local spectral properties. We shall also consider the case when $RSR = R^2$ and $SRS = S^2$. The last section concerns the transmission of many local spectral properties from a Drazin invertible operator R to its Drazin inverse S.

2.1   The Single-Valued Extension Property

Before introducing the typical tools of local spectral theory, and in order to give a first motivation, we begin with some considerations on spectral theory. It is well known that the resolvent function $R(\lambda, T) := (\lambda I - T)^{-1}$ of $T \in L(X)$, X a Banach space, is an analytic operator-valued function defined on the resolvent set $\rho(T)$. Setting

$$f_x(\lambda) := R(\lambda, T)x \quad \text{for any } x \in X,$$

the vector-valued analytic function $f_x : \rho(T) \to X$ trivially satisfies the equation

$$(\lambda I - T) f_x(\lambda) = x \quad \text{for all } \lambda \in \rho(T). \tag{2.1}$$

Suppose that $T \in L(X)$ is a bounded operator on a Banach space X such that the spectrum $\sigma(T)$ has an isolated point λ_0, and let P_0 denote the spectral projection of T associated with λ_0. It is known that the spectrum of the restriction $T_0 := T|P_0(X)$ is $\{\lambda_0\}$, thus $\lambda I - T_0$ is invertible for all $\lambda \neq \lambda_0$. Let $x \in P_0(X)$. Obviously, Eq. (2.1) has the analytic solution

$$g_x(\lambda) := (\lambda I - T_0)^{-1} x \quad \text{for all } \lambda \in \mathbb{C} \setminus \{\lambda_0\}.$$

This shows that it is possible to find analytic solutions of the equation $(\lambda I - T) f_x(\lambda) = x$ for some, and sometimes even for all, values of λ that are in the spectrum of T.

These considerations property lead to the following concepts:

Definition 2.1 Given an arbitrary operator $T \in L(X)$, X a Banach space, let $\rho_T(x)$ denote the set of all $\lambda \in \mathbb{C}$ for which there exists an open neighborhood $\mathcal{U}_\lambda$ of λ in $\mathbb{C}$ and an analytic function $f : \mathcal{U}_\lambda \to X$ such that the equation

$$(\mu I - T) f(\mu) = x \quad \text{holds for all } \mu \in \mathcal{U}_\lambda. \tag{2.2}$$

If the function f is defined on the set $\rho_T(x)$ then f is called a *local resolvent function* of T at x. The set $\rho_T(x)$ is called the *local resolvent* of T at x. The *local spectrum* $\sigma_T(x)$ of T at the point $x \in X$ is defined to be the set

$$\sigma_T(x) := \mathbb{C} \setminus \rho_T(x).$$

Evidently $\rho_T(x)$ is the open subset of $\mathbb{C}$ given by the union of the domains of all the local resolvent functions. Moreover,

$$\rho(T) \subseteq \rho_T(x) \quad \text{and} \quad \sigma_T(x) \subseteq \sigma(T).$$

It is immediate to check the following elementary properties of $\sigma_T(x)$:

(a) $\sigma_T(0) = \emptyset$;
(b) $\sigma_T(\alpha x + \beta y) \subseteq \sigma_T(x) \cup \sigma_T(y)$ for all $x, y \in X$;
(c) For every $F \subseteq \mathbb{C}$, $\sigma_{\lambda I+T}(x) \subseteq F$ if and only if $\sigma_T(x) \subseteq F - \{\lambda\}$. In particular, $\sigma_{(\lambda I-T)}(x) \subseteq \{0\}$ if and only if $\sigma_T(x) \subseteq \{\lambda\}$.

The next example shows that $\sigma_T(x)$ may also be empty for some $x \neq 0$.

Example 2.2 Let R and L denote the right shift and the left shift on $\ell^2(\mathbb{N})$, respectively. Obviously, $LR = I$ and L is the adjoint of R. Let $\{e_k : k \geq 0\}$ be the canonical orthonormal basis of $\ell_2(\mathbb{N})$ and set

$$x := \sum_{k=1}^{\infty} \frac{1}{k} e_k.$$

Then $Lx = \frac{x}{2}$. Now, for $|\lambda| > \frac{1}{2}$, an easy computation shows that the function

$$f(\lambda) := \sum_{k=0}^{\infty} \frac{L^k x}{\lambda^{k+1}}$$

satisfies the equality $(\lambda I - L) f(\lambda) = x$. Also, if we set

$$g(\lambda) := \sum_{k=0}^{\infty} R^{k+1} \lambda^k x,$$

then it is easily seen that

$$(\lambda I - L) g(\lambda) = \sum_{k=1}^{\infty} R^{k+1} \lambda^{k+1} x - \sum_{k=1}^{\infty} R^k \lambda^k x = x \quad \text{for all } |\lambda| \leq 1,$$

in particular for all $|\lambda| \leq \frac{1}{2}$. Hence $\sigma_L(x) = \emptyset$.

Lemma 2.3 *If $T, S \in L(X)$ commutes then $\sigma_T(Sx) \subseteq \sigma_T(x)$.*

Proof Let $f : \mathcal{U} \to X$ be an analytic function on the open set $\mathcal{U} \subseteq \mathbb{C}$ for which $(\mu I - T) f(\mu) = x$ holds for all $\mu \in \mathcal{U}$. If $TS = ST$ then the function $S \circ f : \mathcal{U} \to X$ is also analytic and satisfies the equation

$$(\mu I - T)(S \circ f)(\mu) = S((\mu I - T) f(\mu)) = Sx \quad \text{for all } \mu \in \mathcal{U}.$$

Therefore $\rho_T(x) \subseteq \rho_T(Sx)$ and hence $\sigma_T(Sx) \subseteq \sigma_T(x)$.   ∎

It is well known that given two operators $R, S \in L(X)$, the spectra $\sigma(RS)$ and $\sigma(SR)$ may differ only by the inclusion of 0, see also the next Corollary 2.150. The following theorem shows a local spectral version of this result.

Theorem 2.4 *Let X and Y be Banach spaces and consider two operators $S \in L(X, Y)$ and $R \in L(Y, X)$. Then we have:*

(i) *For every $x \in X$ the following inclusions hold:*

$$\sigma_{SR}(Sx) \subseteq \sigma_{RS}(x) \subseteq \sigma_{SR}(Sx) \cup \{0\}.$$

If S is injective then $\sigma_{RS}(x) = \sigma_{SR}(Sx)$ for all $x \in X$.

(ii) *For every $y \in Y$ the following inclusions hold:*

$$\sigma_{RS}(Ry) \subseteq \sigma_{SR}(y) \subseteq \sigma_{RS}(Ry) \cup \{0\}.$$

If R is injective then $\sigma_{RS}(Ry) = \sigma_{SR}(y)$ for all $y \in Y$.

Proof

(i) Let $\lambda \notin \sigma_{RS}(x)$ and $f : \mathcal{U} \to X$ be an analytic function defined in a neighborhood $\mathcal{U}$ of λ such that $(\mu I - RS)f(\mu) = x$ for all $\mu \in \mathcal{U}$. Then

$$S(\mu I - RS)f(\mu) = (\mu I - SR)Sf(\mu) = Sx,$$

hence $\lambda \notin \sigma_{SR}(Sx)$ since $Sf(\mu)$ is analytic on $\mathcal{U}$, so the first inclusion in (i) is proved. To show the second inclusion, let $\lambda \notin \sigma_{SR}(Sx) \cup \{0\}$ and denote by $g(\mu)$ a Y-valued analytic function defined on a neighborhood $\mathcal{U}$ of λ such that $(\mu I - SR)g(\mu) = Sx$ for all $\mu \in \mathcal{U}$. If we set

$$h(\mu) := \frac{1}{\mu}(x - Rg(\mu)),$$

it is easy to check that $(\mu I - RS)h(\mu) = x$, so $\lambda \notin \sigma_{RS}(x)$.

To show the second statement, assume that $\lambda \notin \sigma_{SR}(Sx)$. There is no harm in assuming $\lambda = 0$. Thus, assume $0 \notin \sigma_{SR}(Sx)$, and let $g(\mu)$ be a Y-valued analytic function defined in a neighborhood $\mathcal{U}$ of 0 such that $(\mu I - SR)g(\mu) = Sx$. For $\mu = 0$ we have $-SRg(0) = Sx$ and from the injectivity of S it follows that $x = Rg(0)$. Moreover,

$$\mu g(\mu) = Sx + SRg(\mu) = S(x + Rg(\mu))$$

so

$$g(\mu) = S\left[\frac{1}{\mu}(x + Rg(\mu))\right].$$

Note that

$$SRg'(0) = \lim_{\mu \to 0} \frac{SRg(\mu) - SRg(0)}{\mu} = \lim_{\mu \to 0} \frac{SRg(\mu) + Sx}{\mu}$$

$$= \lim_{\mu \to 0} g(\mu) = g(0).$$

Set

$$h(\mu) := \begin{cases} \dfrac{1}{\mu}(x - Rg(\mu)) & \text{if } \mu \neq 0, \\ Rg'(0) & \text{if } \mu = 0. \end{cases}$$

We have

$$S[(\mu I - RS)h(\mu) - x] = 0 \quad \text{for all } \mu \in U.$$

Indeed, we have seen in the first part of the proof that for $\mu \neq 0$ we have

$$(\mu I - RS)h(\mu) - x = 0,$$

while for $\mu = 0$ we have

$$S[-RSRg'(0) - x] = -SR(SRg'(0) - Sx = -SRg(0) - Sx$$

$$= Sx - Sx = 0.$$

Since S is injective we have $(\mu I - RS)h(\mu) = x$ for all $\mu \in U$, hence $0 \notin \sigma_{RS}(x)$.

(ii) The proof is analogous. ∎

For an injective operator $T \in L(X)$ the local spectra of Tx and x coincide:

Corollary 2.5 *Let* $T \in L(X)$ *and* $x \in X$. *Then we have*

(i) $\sigma_T(Tx) \subseteq \sigma_T(x) \subseteq \sigma_T(Tx) \cup \{0\}$.
(ii) *If* T *is injective then* $\sigma_T(Tx) = \sigma_T(x)$.

Proof Take $S = T$ and $R = I$ in Theorem 2.4. ∎

The following example shows that if S is not injective we may have $\sigma_{RS}(x) \neq \sigma_{SR}(Sx)$.

Example 2.6 Let S denote the shift operator defined in the usual Hardy space $H^2(\mathbb{D})$ of all analytic functions $f : \mathbb{D} \to \mathbb{C}$, on the open unit disc $\mathbb{D}$, for which

$$\sup \left\{ \int_{-\pi}^{\pi} |f(re^{i\theta})|^2 d\theta : 0 \leq r < 1 \right\} < \infty,$$

and let $R := S^*$ be the adjoint of S. Then, RS is the identity operator, while SR is the projection of X onto the range $S(H)$. In particular, $\sigma_{RS}(x) = \{1\}$ for all $0 \neq x \in H$, $\sigma_{SR}(x) = \{1\}$ if $x \in S(H)$, $\sigma_{SR}(x) = \{0\}$ if $x \in \ker R$ and $\sigma_{SR}(x) = \{0, 1\}$ otherwise. In this case, $\sigma_{RS}(Sx)$ is strictly contained in $\sigma_{SR}(x)$.

Now we consider the case when $S, R \in L(X)$ satisfy the operator equation $RSR = R^2$. Evidently, the operator equation $RSR = R^2$ implies that $(SR)^2 = SR^2$. Examples of operators which satisfy this equation are given by $R = PQ$, where P and Q are idempotents, see Vidav [293].

Lemma 2.7 *Suppose that* $R, S \in L(X)$ *satisfy* $RSR = R^2$. *Then we have*

$$\sigma_R(Rx) \subseteq \sigma_{SR}(x) \quad \text{and} \quad \sigma_{SR}(SRx) \subseteq \sigma_R(x), \tag{2.3}$$

for all $x \in X$.

Proof To show the first inclusion, suppose that $\lambda_0 \notin \sigma_{SR}(x)$, i.e. $\lambda_0 \in \rho_{SR}(x)$. Then there exists an open neighborhood $\mathcal{U}_0$ of λ_0 and an analytic function $f : \mathcal{U}_0 \to X$ such that

$$(\lambda I - SR)f(\lambda) = x \quad \text{for all } \lambda \in \mathcal{U}_0.$$

From this it then follows that

$$Rx = R(\lambda I - SR)f(\lambda) = (\lambda R - RSR)f(\lambda)$$
$$= (\lambda R - R^2)f(\lambda) = (\lambda I - R)(Rf)(\lambda),$$

for all $\lambda \in \mathcal{U}_0$. Obviously, $Rf : \mathcal{U}_0 \to X$ is analytic, so $\lambda_0 \in \rho_R(Rx)$ and hence $\lambda_0 \notin \sigma_R(Rx)$. This shows the first inclusion.

To show the second inclusion, let $\lambda_0 \notin \sigma_R(x)$. Then $\lambda_0 \in \rho_R(x)$ and hence there exists an open neighborhood $\mathcal{U}_0$ of λ_0 and an analytic function $f : \mathcal{U}_0 \to X$ such that

$$(\lambda I - R)f(\lambda) = x \quad \text{for all } \lambda \in \mathcal{U}_0.$$

Consequently,

$$SRx = SR(\lambda I - R)f(\lambda) = (\lambda SR - SR^2)f(\lambda)$$
$$= (\lambda SR - (SR)^2)f(\lambda) = (\lambda I - SR)(SRf)(\lambda),$$

for all $\lambda \in \mathcal{U}_0$, and since $(SR)f$ is analytic we then obtain $\lambda_0 \in \rho_{SR}(SRx)$, i.e. $\lambda_0 \notin \sigma_{SR}(SRx)$. Hence the second inclusion is also proved. ∎

Example 2.8 A very important example of a local spectrum is given in the case of multiplication operators on the Banach algebra $C(\Omega)$ of all continuous complex-valued functions on a compact Hausdorff space Ω, endowed with point-wise operations and supremum norm. Let $T_\varphi \in L(C(\Omega))$ denote the operator of multiplication by an arbitrary function $\varphi \in C(\Omega)$. We show that

$$\sigma_{T_\varphi}(f) = \varphi(\text{supp } f) \quad \text{for all } f \in C(\Omega),$$

where the *support* of f is defined by

$$\text{supp } f := \overline{\{\lambda \in \Omega : f(\lambda) \neq 0\}}.$$

Indeed, we easily have that $f(\mu) = 0$ for all $\mu \in \Omega$ with $\varphi(\mu) \in \rho_T(f)$, so that

$$\varphi(\{\mu \in \Omega : f(\mu) \neq 0\}) \subseteq \sigma_{T_\varphi}(f),$$

and hence, since φ is continuous, $\varphi(\text{supp } f) \subseteq \sigma_{T_\varphi}(f)$. To prove the opposite inclusion, let $\lambda \notin \varphi(\text{supp } f)$ and let $\mathbf{D}(\lambda, \delta)$ be a closed disc centered at λ with radius δ such that

$$\mathbf{D}(\lambda, \delta) \cap \varphi(\text{supp } f) = \emptyset.$$

If $\mu \in \mathbb{D}(, \lambda, \delta)$, where $\mathbb{D}(\lambda, \delta)$ denotes the open disc centered at λ with radius δ, and $\omega \in \Omega$, define

$$g_\mu(\omega) := \begin{cases} \dfrac{f(\omega)}{(\varphi(\omega) - \mu)} & \text{if } \varphi(\omega) \notin \mathbf{D}(\lambda, \delta), \\ 0 & \text{if } \varphi(\omega) \notin \varphi(\text{supp } f). \end{cases}$$

Clearly, for fixed $\mu \in \mathbb{D}(\lambda, \delta)$ we have $g_\mu \in C(\Omega)$ and, evidently, the equation $(T_\varphi - \mu)g_\mu = f$ holds. Since g_μ is analytic on $\mathbb{D}(\lambda, \delta)$ it then follows that $\lambda \in \rho_{T_\varphi}(f) := \mathbb{C} \setminus \sigma_{T_\varphi}(f)$.

The next theorem shows that the local resolvent functions preserve the local spectrum.

Theorem 2.9 *Let $T \in L(X)$, $x \in X$ and $\mathcal{U}$ be an open subset of $\mathbb{C}$. Suppose that $f : \mathcal{U} \to X$ is an analytic function for which $(\mu I - T)f(\mu) = x$ for all $\mu \in \mathcal{U}$. Then $\mathcal{U} \subseteq \rho_T(f(\lambda))$ for all $\lambda \in \mathcal{U}$. Moreover,*

$$\sigma_T(x) = \sigma_T(f(\lambda)) \quad \text{for all } \lambda \in \mathcal{U}. \tag{2.4}$$

Proof Let λ be arbitrarily chosen in $\mathcal{U}$. Define

$$h(\mu) := \begin{cases} \dfrac{f(\lambda) - f(\mu)}{\mu - \lambda} & \text{if } \mu \neq \lambda, \\ -f'(\lambda) & \text{if } \mu = \lambda, \end{cases}$$

for all $\mu \in \mathcal{U}$. Clearly, the function h is analytic, and it is easily seen that $(\mu I - T)h(\mu) = f(\lambda)$ for all $\mu \in \mathcal{U} \setminus \{\lambda\}$. By continuity the last equality is also true for $\mu = \lambda$, so

$$(\lambda I - T)h(\lambda) = f(\lambda) \quad \text{for all } \mu \in \mathcal{U}.$$

This shows that $\lambda \in \rho_T(f(\lambda))$. Since λ is arbitrary in $\mathcal{U}$, $\mathcal{U} \subseteq \rho_T(f(\lambda))$ for all $\lambda \in \mathcal{U}$.

To show the identity (2.4), we first prove the inclusion $\sigma_T(f(\lambda)) \subseteq \sigma_T(x)$, or equivalently, $\rho_T(x) \subseteq \rho_T(f(\lambda))$ for all $\lambda \in \mathcal{U}$. If $\omega \in \mathcal{U}$ then $\omega \in \rho_T(f(\lambda))$ for all $\lambda \in \mathcal{U}$, by the first part of the proof. Suppose that $\omega \in \rho_T(x) \setminus \mathcal{U}$. Since $w \in \rho_T(x)$, there exist an open neighborhood $\mathcal{W}$ of w such that $\lambda \notin \mathcal{W}$ and an analytic function

$g : \mathcal{W} \to X$ such that $(\mu I - T)g(\mu) = x$ for all $\mu \in \mathcal{W}$. Define

$$k(\mu) := \frac{f(\lambda) - g(\mu)}{\mu - \lambda} \quad \text{for all } \mu \in \mathcal{W},$$

then, as is easy to verify, $(\mu I - T)k(\mu) = f(\lambda)$ holds for all $\mu \in \mathcal{W}$. This shows that $\omega \in \rho_T(x)$, and hence $\sigma_T(x) \subseteq \sigma_T(f(\lambda))$.

It remains to prove the opposite inclusion $\sigma_T(f(\lambda)) \subseteq \sigma_T(x)$.

Let $\eta \notin \sigma_T(f(\lambda))$ and hence $\eta \in \rho_T(f(\lambda))$. Let $h : \mathcal{V} \to X$ be an analytic function defined on the open neighborhood $\mathcal{V}$ of η for which the identity $(\mu I - T)h(\mu) = f(\lambda)$ is satisfied for all $\mu \in \mathcal{V}$. Then

$$(\mu I - T)(\lambda I - T)h(\mu) = (\lambda I - T)(\mu I - T)h(\mu) = (\lambda I - T)f(\lambda) = x,$$

for all $\mu \in \mathcal{V}$, so that $\eta \in \rho_T(x)$ and hence $\eta \notin \sigma_T(x)$. ∎

We have seen that the uniqueness of the analytic solution of Eq. (2.2) is a non-trivial issue. With this in mind we make the following definition:

Definition 2.10 Let X be a complex Banach space and $T \in L(X)$. The operator T is said to have *the single-valued extension property* (SVEP) at $\lambda_0 \in \mathbb{C}$ if for every neighborhood $\mathcal{U}$ of λ_0 the only analytic function $f : \mathcal{U} \to X$ which satisfies the equation

$$(\lambda I - T)f(\lambda) = 0$$

is the constant function $f \equiv 0$.

The operator T is said to have the SVEP if T has the SVEP at every $\lambda \in \mathbb{C}$.

Remark 2.11 In the sequel we collect some basic properties of the single-valued extension property.

(a) The SVEP ensures the consistency of the local solutions of Eq. (4.13), in the sense that if $x \in X$ and T has the SVEP at $\lambda_0 \in \rho_T(x)$ then there exists a neighborhood $\mathcal{U}$ of λ_0 and a *unique* analytic function $f : \mathcal{U} \to X$ satisfying the equation $(\lambda I - T)f(\lambda) = x$ for all $\lambda \in \mathcal{U}$.

The SVEP also ensures the existence of a maximal analytic extension $\tilde{f}$ of $R(\lambda, T)x := (\lambda I - T)^{-1}x$ to the set $\rho_T(x)$ for every $x \in X$. This function identically satisfies the equation

$$(\mu I - T)\tilde{f}(\mu) = x \quad \text{for every } \mu \in \rho_T(x)$$

and, obviously,

$$\tilde{f}(\mu) = (\mu I - T)^{-1}x \quad \text{for every } \mu \in \rho(T).$$

(b) It is immediate to verify that the SVEP is inherited by the restrictions on invariant subspaces, i.e., if $T \in L(X)$ has the SVEP at λ_0 and M is a closed T-invariant subspace, then $T|M$ has the SVEP at λ_0. Moreover,

$$\sigma_T(x) \subseteq \sigma_{T|M}(x) \quad \text{for every } x \in M.$$

(c) Obviously, an operator $T \in L(X)$ has the SVEP at every point of the resolvent $\rho(T) := \mathbf{C} \setminus \sigma(T)$. From the identity theorem for analytic functions it easily follows that an operator always has the SVEP at every point of the boundary $\partial\sigma(T)$ of the spectrum $\sigma(T)$. In particular, T has the SVEP at every isolated point of the spectrum $\sigma(T)$.

(d) Let $\sigma_p(T)$ denote the *point spectrum* of $T \in L(X)$, i.e.,

$$\sigma_p(T) := \{\lambda \in \mathbb{C} : \lambda \text{ is an eigenvalue of } T\}.$$

It is easy to see that if $\sigma_p(T)$ has empty interior then T has the SVEP, in particular every operator with real spectrum has the SVEP. A rather immediate argument shows the following implication:

$$\sigma_p(T) \text{ does not cluster at } \lambda_0 \Rightarrow T \text{ has the SVEP at } \lambda_0.$$

Indeed, suppose that $\sigma_p(T)$ does not cluster at λ_0. Then there is an open neighborhood $\mathcal{U}$ of λ_0 such that $\lambda I - T$ is injective for every $\lambda \in \mathcal{U}$, $\lambda \neq \lambda_0$. Let $f : \mathcal{V} \to X$ be an analytic function defined on another neighborhood $\mathcal{V}$ of λ_0 such that the equation $(\lambda I - T)f(\lambda) = 0$ holds for every $\lambda \in \mathcal{V}$. We may assume that $\mathcal{V} \subseteq \mathcal{U}$. Then $f(\lambda) \in \ker(\lambda I - T) = \{0\}$ for every $\lambda \in \mathcal{V}$, $\lambda \neq \lambda_0$, so $f(\lambda) = 0$ for every $\lambda \in \mathcal{V}$, $\lambda \neq \lambda_0$. Since f is continuous at λ_0 we then conclude that $f(\lambda_0) = 0$. Hence $f \equiv 0$ in $\mathcal{V}$ and therefore T has the SVEP at λ_0.

It should be noted that T may have the SVEP, although $\sigma_p(T) \neq \emptyset$. For instance, if X is the Banach algebra $B(\Omega)$ of all bounded complex-valued functions on a compact Hausdorff space Ω, endowed with point-wise operations and supremum norm, the operator $T \in L(X)$, defined by the assignment

$$(Tf)(\lambda) := \lambda f(\lambda) \quad \text{for all } \lambda \in \Omega,$$

has $\sigma_p(T) \neq \emptyset$, while T has the SVEP since the ascent $p(\mu I - T) \leq 1$ for all $\mu \in \mathbb{C}$, and this, as we shall see later, entails the SVEP of T at μ.

(e) A similar argument to that of part (d) shows that

$$\sigma_{ap}(T) \text{ does not cluster at } \lambda_0 \Rightarrow T \text{ has the SVEP at } \lambda_0$$

and dually,

$$\sigma_s(T) \text{ does not cluster at } \lambda_0 \Rightarrow T^* \text{ has the SVEP at } \lambda_0.$$

(f) Evidently, T has the SVEP if $\sigma_{ap}(T)$ is contained in the boundary of the spectrum $\partial\sigma(T)$, while T^* has the SVEP if $\sigma_s(T)$ is contained in $\partial\sigma(T)$.

(g) The SVEP is transmitted under translations, i.e. $T \in L(X)$ has the SVEP if and only if $\lambda I - T$ has the SVEP.

Now we introduce an important class of subspaces which play an important role in local theory.

Definition 2.12 For every subset F of $\mathbb{C}$ the *local spectral subspace* of an operator $T \in L(X)$ associated with F is the set

$$X_T(F) := \{x \in X : \sigma_T(x) \subseteq F\}.$$

Obviously, if $F_1 \subseteq F_2 \subseteq \mathbb{C}$ then $X_T(F_1) \subseteq X_T(F_2)$ and

$$X_T(F) = X_T(F \cap \sigma(T)).$$

Indeed, $X_T(F) \cap \sigma(T)) \subseteq X_T(F)$. Conversely, if $x \in X_T(F)$ then $\sigma_T(x) \subseteq F \cap \sigma(T)$, and hence $x \in X_T(F \cap \sigma(T))$. Moreover, from the basic properties of the local spectrum it is easily seen that $X_{\lambda I+T}(F) = X_T(F \setminus \{\lambda\})$. Further basic properties of local spectral subspaces are collected in the sequel.

Theorem 2.13 *Let $T \in L(X)$ and F a subset of $\mathbb{C}$. Then the following properties hold:*

(i) *$X_T(F)$ is a linear T-hyper-invariant subspace of X, i.e., for every bounded operator S such that $TS = ST$ we have $S(X_T(F)) \subseteq X_T(F)$.*

(ii) *If $\lambda \notin F$, then $(\lambda I - T)(X_T(F)) = X_T(F)$.*

(iii) *Suppose that $\lambda \in F$ and $(\lambda I - T)x \in X_T(F)$ for some $x \in X$. Then $x \in X_T(F)$.*

(iv) *For every family $(F_j)_{j \in J}$ of subsets of $\mathbb{C}$ we have*

$$X_T\left(\bigcap_{j \in J} F_j\right) = \bigcap_{j \in J} X_T(F_j).$$

(v) *If Y is a T-invariant closed subspace of X for which $\sigma(T\,|Y) \subseteq F$, then $Y \subseteq X_T(F)$. In particular, $Y \subseteq X_T(\sigma(T\,|Y))$ holds for every closed T-invariant closed subspace of X.*

(vi) $\ker(\lambda I - T)^n \subseteq X_T(\{\lambda\}$ *for all $\lambda \in \mathbb{C}$ and $n \in \mathbb{N}$.*

Proof

(i) Evidently the set $X_T(F)$ is a linear subspace of X, since the inclusion $\sigma_T(\alpha x + \beta y) \subseteq \sigma_T(x) \cup \sigma_T(y)$ holds for all $\alpha, \beta \in \mathbb{C}$ and $x, y \in X$.

Suppose now that $x \in X_T(F)$ and $TS = ST$. Then $\sigma_T(Sx) \subseteq \sigma_T(x) \subseteq F$, hence $Sx \in X_T(F)$.

(ii) The operators $\lambda I - T$ and T commute, so from part (i) it follows that $(\lambda I - T)(X_T(F)) \subseteq X_T(F)$ for all $\lambda \in \mathbb{C}$. Let $\lambda \notin F$ and consider an element $x \in X_T(F)$. Then $\sigma_T(x) \subseteq F$ and hence $\lambda \in \rho_T(F)$. Therefore, there exist an open neighborhood $\mathcal{U}$ of λ and an analytic function $f : \mathcal{U} \to X$ for which $(\mu I - T)f(\mu) = x$ for all $\mu \in \mathcal{U}$. In particular, $(\lambda I - T)f(\lambda) = x$. By Theorem 2.9 we obtain $\sigma_T(f(\lambda)) = \sigma_T(x) \subseteq F$, and hence $f(\lambda) \in X_T(F)$, from which we conclude that

$$x = (\lambda I - T)f(\lambda) \in (\lambda I - T)(X_T(F)).$$

(iii) Suppose that $(\lambda I - T)x \in X_T(F)$, $\lambda \in F$. We need to show that $\sigma_T(x) \subseteq F$, or equivalently, $\mathbb{C} \setminus F \subseteq \rho_T(x)$. Take $\mu \notin F$. By assumption $\mathbb{C} \setminus F \subseteq \rho_T((\lambda I - T)x)$, so there is an analytic function $f : \mathcal{U}_\mu \to X$ defined on some open neighborhood $\mathcal{U}_\mu$ of μ such that $\lambda \notin \mathcal{U}$ and $(\omega I - T)f(\omega) = (\lambda I - T)x$ for all $\omega \in \mathcal{U}_\mu$. Define $g : \mathcal{U}_\mu \to X$ by

$$g(\omega) := \frac{x - f(\omega)}{\omega - \lambda} \quad \text{for all } \omega \in \mathcal{U}_\mu.$$

Clearly the analytic function g satisfies the equality $(\omega I - T)g(\omega) = x$ for all $\omega \in \mathcal{U}_\mu$, so $\mu \in \rho_T(x)$. Therefore $\mathbb{C} \setminus F \subseteq \rho_T(x)$.

(iv) This easily follows from the definition.

(v) From $\sigma(T \mid Y) \subseteq F$ we obtain $\mathbb{C} \setminus F \subseteq \rho(T \mid Y)$, hence for any $y \in Y$ we have

$$(\lambda I - T)(\lambda I - T \mid Y)^{-1}y = y \quad \text{for all } \lambda \in \mathbb{C} \setminus F.$$

Clearly, $f(\lambda) := (\lambda I - T \mid Y)^{-1}y$ is analytic for all $\lambda \in \mathbb{C} \setminus F$, hence $\mathbb{C} \setminus F \subseteq \rho_T(y)$, and consequently $\sigma_T(y) \subseteq F$.

(vi) This is obvious from part (iii).   ∎

We have already observed that 0 has an empty local spectrum. The next result shows that if T has the SVEP then 0 is the *unique* element of X having empty local spectrum. In other words, this property characterizes the SVEP.

Theorem 2.14 *If $T \in L(X)$ the following statements are equivalent:*

(i) *T has the SVEP;*
(ii) *$X_T(\emptyset) = \{0\}$, i.e. $\sigma_T(x) = \emptyset$ if and only if $x = 0$;*
(iii) *$X_T(\emptyset)$ is closed.*

Proof (i) $\Leftrightarrow$ (ii) Suppose that T has the SVEP and $\sigma_T(x) = \emptyset$. Then $\rho_T(x) = \mathbb{C}$, so there exists an analytic function $f : \mathbb{C} \to X$ such that $(\lambda I - T)f(\lambda) = x$ for every $\lambda \in \mathbb{C}$. If $\lambda \in \rho(T)$ we have $f(\lambda) = (\lambda I - T)^{-1}x$, and hence, since $\|(\lambda I - T)^{-1}\| \to 0$ as $|\lambda| \to +\infty$, $f(\lambda)$ is a bounded function on $\mathbb{C}$. By Liouville's theorem $f(\lambda)$ is then constant, and therefore, since $(\lambda I - T)^{-1}x \to 0$ as $|\lambda| \to +\infty$, f is identically 0 on $\mathbb{C}$. This proves that $x = 0$. Since $0 \in X_T(\emptyset)$ we then conclude that $X_T(\emptyset) = \{0\}$.

Conversely, let $\lambda_0 \in \mathbb{C}$ be arbitrary and suppose that for every $0 \neq x \in X$ we have $\sigma_T(x) \neq \emptyset$. Consider any analytic function $f : \mathcal{U} \to X$ defined on a neighborhood $\mathcal{U}$ of λ_0 such that the equation $(\lambda I - T)f(\lambda) = 0$ holds for every $\lambda \in \mathcal{U}$. From the equality

$$\sigma_T(f(\lambda)) = \sigma_T(0) = \emptyset,$$

see Theorem 2.9, we deduce that $f \equiv 0$ on $\mathcal{U}$ and therefore T has the SVEP at λ_0. Since λ_0 is arbitrary, T has the SVEP.

(ii) $\Rightarrow$ (iii) Trivial.

(iii) $\Rightarrow$ (ii) Suppose that $X_T(\emptyset)$ is closed. From part (iii) of Theorem 2.13 we deduce that

$$(\lambda I - T)(X_T(\emptyset)) = X_T(\emptyset) \quad \text{for every } \lambda \in \mathbb{C}.$$

Now, let S denote the restriction $T \, |X_T(\emptyset)$. The operator $\lambda I - S$ is surjective and therefore $\lambda I - S^*$ is bounded below for all $\lambda \in \mathbb{C}$, i.e. $\sigma_{\mathrm{ap}}(S^*) = \emptyset$. This implies, by Theorem 1.12, that the dual of $X_T(\emptyset)$ is trivial. A standard consequence of the Hahn–Banach theorem then implies that $X_T(\emptyset) = \{0\}$. ∎

Theorem 2.14 implies that a left shift L does not have the SVEP. Indeed, in Example 2.2 we have shown that there is $0 \neq x \in \ell_2(\mathbb{N})$ such that $\sigma_L(x) = \emptyset$.

Note that $X_T(\Omega)$ need not be closed for a closed subset $\Omega \subseteq \mathbb{C}$. Indeed, Theorem 2.14 shows that $X_T(\emptyset)$ is not closed if T does not have the SVEP.

Later, Example 2.33 will show that for a closed subset $\Omega \subseteq \mathbb{C}$ the local spectral subspaces $X_T(\Omega)$ need not be closed even in the case when T has the SVEP.

Theorem 2.15 *Suppose that $T_i \in L(X_i)$, $i = 1, 2$, where X_i are Banach spaces. Then $T_1 \oplus T_2$ has the SVEP at λ_0 if and only if both T_1, T_2 have the SVEP at λ_0. If T_1, T_2 have the SVEP then*

$$\sigma_{T_1 \oplus T_2}(x_1 \oplus x_2) = \sigma_{T_1}(x_1) \cup \sigma_{T_2}(x_2). \tag{2.5}$$

Proof First, suppose that T_1 and T_2 have the SVEP at λ_0 and let $f = f_1 \oplus f_2 : \mathcal{U} \to X_1 \oplus X_2$ be analytic on a neighborhood $\mathcal{U}$ of λ_0, where $f_i : \mathcal{U} \to X_i, i = 1, 2$, are also analytic on $\mathcal{U}$. Obviously, for every $\lambda \in \mathcal{U}$ the condition $(\lambda I - T_1 \oplus T_2)f(\lambda) = 0$ implies that $(\lambda I - T_i)f_i(\lambda) = 0, i = 1, 2$. The SVEP of T_1 and T_2 then entails that $f_1 \equiv 0$ and $f_2 \equiv 0$ on $\mathcal{U}$. Thus $f \equiv 0$ on $\mathcal{U}$.

Conversely, assume that $T_1 \oplus T_2$ satisfies the SVEP at λ_0 and let $f_i : \mathcal{U} \to X_i$ be two analytic functions, defined on a neighborhood $\mathcal{U}$ of λ_0, which satisfy for $i = 1, 2$, the equations $(\lambda I - T_i)f_i(\lambda) = 0$ for all $\lambda \in \mathcal{U}$. For all $\lambda \in \mathcal{U}$ we then have

$$0 = (\lambda I - T_1) f_1(\lambda) \oplus (\lambda I - T_2) f_2(\lambda) = (\lambda - T_1 \oplus T_2)[f_1(\lambda) \oplus (f_2(\lambda)].$$

The SVEP of $T_1 \oplus T_2$ at λ_0 then implies $f_1(\lambda) \oplus (f_2(\lambda) \equiv 0$ on $\mathcal{U}$, and hence $f_i \equiv 0$ on $\mathcal{U}$ for $i = 1, 2$.

To show equality (2.5), suppose that $T_1 \oplus T_2$ has the SVEP. Assume that $\lambda \in \rho_{T_1 \oplus T_2}(x_1 \oplus x_2)$. Then there exists an open neighborhood $\mathcal{U}$ of λ and an analytic function $f := f_1 \oplus f_2 : \mathcal{U} \to X_1 \oplus X_2$, with f_1 and f_2 analytic, such that

$$(\lambda I - T_1) f_1(\lambda) \oplus (\lambda I - T_2) f_2(\lambda) = (\lambda I - T_1 \oplus T_2) f(\lambda) = x_1 \oplus x_2.$$

Therefore, $(\lambda I - T_i) f_i(\lambda) = x_i, i = 1, 2$, so $\lambda \in \rho_{T_1}(x_1) \cap \rho_{T_2}(x_2)$. This shows the inclusion $\sigma_{T_1}(x_1) \cup \sigma_{T_2}(x_2) \subseteq \sigma_{T_1 \oplus T_2}(x_1 \oplus x_2)$. The opposite inclusion has a similar proof.   ∎

As an immediate consequence of Theorem 2.15 we have:

Corollary 2.16 *Suppose that $T \in L(X)$ has the SVEP and $X = M \oplus N$, where M and N are two closed and invariant subspaces. If $T_1 := T \mid M$ and $T_2 := T \mid N$, then we have $X_T(\Omega) = M_{T_1}(F) \oplus N_{T_2}(F)$ for all closed $F \subseteq \mathbb{C}$.*   ∎

Theorem 2.17 *Suppose that $T \in L(X)$ admits, with respect to a decomposition $X = M \oplus N$, the representation $T = \begin{pmatrix} T_1 & T_2 \\ 0 & T_3 \end{pmatrix}$, where T_3 is nilpotent. Then T has the SVEP if and only if T_1 has the SVEP.*

Proof Suppose that T_1 has the SVEP. Fix arbitrarily $\lambda_0 \in \mathbb{C}$ and let $f : U \to X$ be an analytic function defined on an open disc U centered at λ_0 such that $(\lambda I - T) f(\lambda) = 0$ for all $\lambda \in U$. Set $f(\lambda) := f_1(\lambda) \oplus f_2(\lambda)$ on $X = M \oplus N$. Then we can write

$$0 = (\lambda I - T) f(\lambda) = \begin{pmatrix} \lambda I - T_1 & -T_2 \\ 0 & -\lambda I - T_3 \end{pmatrix} \begin{pmatrix} f_1(\lambda) \\ f_2(\lambda) \end{pmatrix}$$

$$= \begin{pmatrix} (\lambda I - T_1) f_1(\lambda) - T_2 f_2(\lambda) \\ (\lambda I - T_3) f_2(\lambda) \end{pmatrix}.$$

Then $(\lambda I - T_3) f_2(\lambda) = 0$ and

$$(\lambda I - T_1) f_1(\lambda) - T_2 f_2(\lambda) = 0.$$

Since a nilpotent operator has the SVEP, $f_2(\lambda) = 0$, and consequently $(\lambda I - T_1) f_1(\lambda) = 0$. But T_1 has the SVEP at λ_0, so $f_1(\lambda) = 0$ and hence $f(\lambda) = 0$ on U. Thus, T has the SVEP at λ_0. Since λ_0 is arbitrary then T has the SVEP.

Conversely, suppose that T has the SVEP. Since T_1 is the restriction of T to M and the SVEP from T is inherited by the restriction to closed invariant subspaces, then T_1 has the SVEP. ∎

In the sequel we shall need the following lemma.

Lemma 2.18 *Let $T \in L(X)$ and $K \subset \mathbb{C}$ be a compact set and let Γ be a contour in the complement $\mathbb{C} \setminus K$ that surrounds K. Suppose that $f : \mathbb{C} \setminus K \to X$ is an analytic function which satisfies $(\lambda I - T)f(\lambda) = x$ for all $\lambda \in \mathbb{C} \setminus K$. Then*

$$x = \frac{1}{2\pi i} \int_\Gamma f(\lambda)\, d\lambda.$$

Proof Let $\mathcal{U} := \mathbb{C} \setminus K$. Observe first that we may suppose that Γ is contained in the unbounded connected component of the open set $\mathcal{U}$. Indeed, f is analytic on $\mathcal{U}$ and by Cauchy's theorem only the part of Γ which lies in the unbounded component of $\mathcal{U}$ contributes to the integral $\int_\Gamma f(\lambda)\, d\lambda$.

Let Λ be the boundary, positively oriented, of a disc centered at 0 and having radius large enough to include in its interior both the sets Γ and $\sigma(T)$. From Cauchy's theorem we have

$$\int_\Gamma f(\lambda) d\lambda = \int_\Lambda f(\lambda) d\lambda.$$

Now, if $\lambda \in \Lambda$ we have $\lambda \in \rho(T)$ and

$$f(\lambda) = (\lambda I - T)^{-1} x = \sum_{n=0}^\infty \lambda^{-n-1} T^n x.$$

A simple calculation then gives that $\int_\Lambda f(\lambda) d\lambda = 2\pi i x$. ∎

Next we want to establish a local decomposition property that will be needed later.

Theorem 2.19 *Suppose that $T \in L(X)$ has the SVEP. If F_1 and F_2 are two closed and disjoint subsets of $\mathbb{C}$ then*

$$X_T(F_1 \cup F_2) = X_T(F_1) \oplus X_T(F_2),$$

where the direct sum is in the algebraic sense.

Proof The inclusion $X_T(F_1) \oplus X_T(F_2) \subseteq X_T(F_1 \cup F_2)$ is obvious.

To show the opposite inclusion, observe first that we may assume that F_1 and F_2 are both compact, since, by part (ii) of Theorem 2.13, we have $X_T(F) = X_T(F \cap \sigma(T))$ for all subsets F of $\mathbb{C}$. Now, if $x \in X_T(F_1 \cup F_2)$, the SVEP for T entails that there is an analytic function $f : \mathbb{C} \setminus (F_1 \cup F_2) \to X$ such that

$$(\lambda I - T)f(\lambda) = x \quad \text{for all } \lambda \in \mathbb{C} \setminus (F_1 \cup F_2).$$

Let Λ_1, Λ_2 be two compact disjoint sets such that Λ_i, for $i = 1, 2$, is a neighborhood of F_i and the boundary Γ_i of Λ_i is a contour surrounding F_i. From Lemma 2.18 we have $x = x_1 + x_2$, where

$$x_i := \frac{1}{2\pi i} \int_{\Gamma_i} f(\lambda) \, d\lambda \quad \text{for } i = 1, 2.$$

We show now that $x_i \in X_T(\Lambda_i)$. Set

$$g_i(\mu) := \frac{1}{2\pi i} \int_{\Gamma_i} \frac{f(\lambda)}{\mu - \lambda} \, d\lambda \quad \text{for all } \mu \in \mathbb{C} \setminus \Lambda_i.$$

Clearly, the functions $g_i(\lambda) : \mathbb{C} \setminus \Lambda_i \to X$ are analytic. Furthermore, for every $\mu \in \mathbb{C} \setminus \Lambda_i$ we have

$$
\begin{aligned}
(\mu I - T)g_i(\mu) &= \frac{1}{2\pi i} \int_{\Gamma_i} (\mu - \lambda + \lambda - T) \frac{f(\lambda)}{\mu - \lambda} \, d\lambda \\
&= \frac{1}{2\pi i} \int_{\Gamma_i} (\lambda - T) \frac{f(\lambda)}{\mu - \lambda} \, d\lambda + \frac{1}{2\pi i} \int_{\Gamma_i} f(\lambda) \, d\lambda \\
&= \frac{1}{2\pi i} \int_{\Gamma_i} \frac{x}{\mu - \lambda} \, d\lambda + x_i.
\end{aligned}
$$

But, from Cauchy's theorem, we have that

$$\frac{1}{2\pi i} \int_{\Gamma_i} \frac{x}{\mu - \lambda} \, d\lambda = 0 \quad \text{for all } \mu \in \mathbb{C} \setminus \Lambda_i,$$

thus $(\mu I - T)g_i(\mu) = x_i$ for $i = 1, 2$ and this implies $x_i \in X_T(\Lambda_i)$.

To conclude the proof observe that, again by Cauchy's theorem, the definition of x_1 and x_2 does not depend on the particular choice of Λ_1 and Λ_2, with the properties required above. This implies that $x_i \in X_T(\Lambda_i)$ for every compact neighborhood Λ_i of F_i, hence $x_i \in X_T(F_i)$ for $i = 1, 2$. Therefore, $x = x_1 + x_2$, where $x_i \in X_T(F_i)$. To see that the sum is direct, observe that since $F_1 \cap F_2 = \emptyset$, from part (v) of Theorem 2.13, we have

$$X_T(F_1) \cap X_T(F_2) = X_T(\emptyset) = \{0\},$$

because, by assumption, T has the SVEP.   ∎

We now turn to an important local spectral characterization of the analytic core $K(T)$.

Theorem 2.20 *For every $T \in L(X)$ we have*

$$K(T) = X_T(\mathbb{C} \setminus \{0\}) = \{x \in X : 0 \notin \sigma_T(x)\}.$$

Proof Let $x \in K(T)$. We can suppose that $x \neq 0$. According to the definition of $K(T)$, let $\delta > 0$ and $(u_n) \subset X$ be a sequence for which

$$x = u_0, \ Tu_{n+1} = u_n, \quad \|u_n\| \leq \delta^n \|x\| \quad \text{for every } n = 0, 1, \ldots.$$

Then the function $f : \mathbb{D}(0, 1/\delta) \to X$, where $\mathbb{D}(0, 1/\delta)$ is the open disc centered at 0 with radius $1/\delta$, defined by

$$f(\lambda) := -\sum_{n=1}^{\infty} \lambda^{n-1} u_n \quad \text{for all } \lambda \in \mathbb{D}(0, 1/\delta),$$

is analytic and satisfies the equation $(\lambda I - T)f(\lambda) = x$ for every $\lambda \in \mathbb{D}(0, 1/\delta)$. Consequently $0 \in \rho_T(x)$.

Conversely, if $0 \in \rho_T(x)$ then there exists an open disc $\mathbb{D}(0, \varepsilon)$ and an analytic function $f : \mathbb{D}(0, \varepsilon) \to X$ such that

$$(\lambda I - T)f(\lambda) = x \quad \text{for every } \lambda \in \mathbb{D}(0, \varepsilon). \tag{2.6}$$

Since f is analytic on $\mathbb{D}(0, \varepsilon)$ there exists a sequence $(u_n) \subset X$ such that

$$f(\lambda) = -\sum_{n=1}^{\infty} \lambda^{n-1} u_n \quad \text{for every } \lambda \in \mathbb{D}(0, \varepsilon). \tag{2.7}$$

Clearly $f(0) = -u_1$, and taking $\lambda = 0$ in (2.6) we obtain

$$Tu_1 = -T(f(0)) = x.$$

On the other hand

$$x = (\lambda I - T)f(\lambda) = Tu_1 + \lambda(Tu_2 - u_1) + \lambda^2(Tu_3 - u_2) + \cdots$$

for all $\lambda \in \mathbb{D}(0, \varepsilon)$. Since $x = Tu_1$ we conclude that

$$Tu_{n+1} = u_n \quad \text{for all } n = 1, 2, \cdots.$$

Hence letting $u_0 = x$ the sequence (u_n) satisfies for all $n = 0, 1, \ldots$ the first of the conditions which define $K(T)$.

It remains to prove the condition $\|u_n\| \leq \delta^n \|x\|$ for a suitable $\delta > 0$ and for all $n = 0, 1, \ldots$. Take $\mu > 1/\varepsilon$. Since the series (2.7) converges we have $|\lambda|^{n-1} \|u_n\| \to 0$ as $n \to \infty$ for all $\|\lambda\| < \varepsilon$ and, in particular, $1/\mu^{n-1} \|u_n\| \to 0$, so there exists a $c > 0$ such that

$$\|u_n\| \leq c \, \mu^{n-1} \quad \text{for every } n \in \mathbb{N}. \tag{2.8}$$

From the estimates (2.8) we easily obtain

$$\|u_n\| \leq \left(\mu + \frac{c}{\|x\|} \right)^n \|x\|$$

and therefore $x \in K(T)$.   ∎

The following theorem shows that the surjectivity spectrum $\sigma_s(T)$ of an operator is closely related to the local spectra.

Theorem 2.21 *For every operator $T \in L(X)$ we have*

$$\sigma_s(T) = \bigcup_{x \in X} \sigma_T(x).$$

The set $\{x \in X : \sigma_T(x) = \sigma_s(T)\}$ is of the second category in X.

Proof If $\lambda \notin \bigcup_{x \in X} \sigma_T(x)$ then $\lambda \in \rho_T(x)$ for every $x \in X$ and hence, directly from the definition of $\rho_T(x)$, we conclude that the equation $(\lambda I - T)y = x$ always admits a solution for every $x \in X$, hence $\lambda I - T$ is surjective and $\lambda \notin \sigma_s(T)$.

Conversely, suppose $\lambda \notin \sigma_s(T)$. Then $\lambda I - T$ is surjective and therefore $X = K(\lambda I - T)$. From Theorem 2.20 it follows that $0 \notin \sigma_{\lambda I - T}(x)$ for every $x \in X$, and consequently $\lambda \notin \sigma_T(x)$ for every $x \in X$.

To show the second assertion, let Λ denote a countable subset of $\sigma_s(T)$. Then $(\lambda I - T)(X) \neq X$ for all $\lambda \in \Lambda$, and the range $(\lambda I - T)(X)$ is of the first category in X (see Appendix A). Consequently, the subspace

$$M := \bigcap_{\lambda \in \Lambda} (\lambda I - T)(X)$$

is of the first category in X. Since X is of the second category, the complement $X \setminus M$ is also of the second category in X. Now, if $x \in X \setminus M$ then $\Lambda \subseteq \sigma_T(x)$, and hence

$$\sigma_s(T) \subseteq \overline{\Lambda} \subseteq \sigma_T(x),$$

so the second assertion follows.   ∎

2.2 Glocal Spectral Subspaces

We now introduce a variant of the concept of an analytic subspace $X_T(\Omega)$. These subspaces are more appropriate, for certain general questions of local spectral theory, than the analytic subspace $X_T(F)$, and in particular these subspaces are more useful in the case when T does not have the SVEP.

Definition 2.22 Let $F \subseteq \mathbb{C}$ be a closed subset. If $T \in L(X)$ the *glocal spectral subspace* $\mathcal{X}_T(F)$ is defined as the set of all $x \in X$ such that there is an analytic function $f : \mathbb{C} \setminus F \to X$ such that

$$(\lambda I - T)f(\lambda) = x \quad \text{for all } \lambda \in \mathbb{C} \setminus F.$$

It is easy to verify that $\mathcal{X}_T(F)$ is a linear subspace of X. Clearly

$$\mathcal{X}_T(F) \subseteq X_T(F) \quad \text{for every closed subset } F \subseteq \mathbb{C}. \tag{2.9}$$

In the following theorem we give a few basic properties of the glocal subspaces. Some of these properties are rather similar to those of local spectral subspaces. The interested reader may find further results on glocal spectral subspaces in Laursen and Neumann [216].

Theorem 2.23 *For an operator $T \in L(X)$, the following statements hold:*

(i) $\mathcal{X}_T(\emptyset) = \{0\}$ *and* $\mathcal{X}_T(\sigma(T)) = X$;

(ii) $\mathcal{X}_T(F) = \mathcal{X}_T(F \cap \sigma(T))$ *and* $(\lambda I - T)\mathcal{X}_T(F) = \mathcal{X}_T(F)$ *for every closed set $F \subseteq \mathbb{C}$ and all $\lambda \in \mathbb{C} \setminus F$;*

(iii) *If $(\lambda I - T)x \in \mathcal{X}_T(F)$ for some $\lambda \in F$, then $x \in \mathcal{X}_T(F)$;*

(iv) $\mathcal{X}_T(F_1 \cup F_2) = \mathcal{X}_T(F_1) + \mathcal{X}_T(F_2)$ *for all disjoint closed subsets F_1 and F_2 of $\mathbb{C}$;*

(v)   *T has the SVEP if and only if $\mathcal{X}_T(F) = X_T(F)$, for every closed subset $F \subseteq \mathbb{C}$, and this happens if and only if $\mathcal{X}_T(F) \cap \mathcal{X}_T(G) = \{0\}$ for all disjoint closed subsets F and G of $\mathbb{C}$.*

Proof (i) Suppose that $x \in \mathcal{X}_T(\emptyset)$ and let $f : \mathbb{C} \to X$ be an analytic function such that $(\lambda I - T)f(\lambda) = x$ for all $\lambda \in \mathbb{C}$. Then $f(\lambda)$ coalesces with the resolvent function $R(\lambda, T) := (\lambda I - T)^{-1}$ on $\rho(T)$, so $f(\lambda) \to 0$ as $|\lambda| \to \infty$. By the vector-valued version of Liouville's theorem $f \equiv 0$, and therefore $x = 0$. The second equality of part (i) is straightforward.

The proof of (ii) easily follows from Theorem 2.9. To show the assertion (iii) let $f : \mathbb{C} \setminus F \to X$ be an analytic function which satisfies $(\mu I - T)f(\mu) = (\lambda I - T)x$ for all $\mu \in \mathbb{C} \setminus F)$. Define

$$g(\mu) := \frac{f(\mu) - x}{\mu - \lambda} \quad \text{for all } \mu \in \mathbb{C} \setminus F.$$

Clearly, g is analytic and satisfies $(\mu I - T)g(\mu) = x$ for all $\mu \in \mathbb{C} \setminus F$, so $x \in \mathcal{X}_T(F)$.

The proof of the decomposition (iv) is similar to the proof of that given for spectral local subspaces.

(v) Evidently, if T has the SVEP and $F \subseteq \mathbb{C}$ is closed then $\mathcal{X}_T(F) = X_T(F)$.

Conversely, if $X_T(F) = \mathcal{X}_T(F)$ for all closed sets $F \subseteq \mathbb{C}$ then

$$X_T(\emptyset) = \mathcal{X}_T(\emptyset) = \{0\},$$

so, by Theorem 2.14, T has the SVEP. To show the second equivalence, assume that T has the SVEP. Since $F \cap G = \emptyset$ we then have $\{0\} = X_T(\emptyset) = X_T(F) \cap X_T(G) = \mathcal{X}_T(F) \cap \mathcal{X}_T(G)$. Conversely, suppose that $\mathcal{X}_T(F) \cap \mathcal{X}_T(G) = \{0\}$ for all disjoint closed subsets F and G and assume that T does not have the SVEP. Then there exists a non-trivial analytic function $f : \mathcal{O} \to X$ on an open set $\mathcal{O}$ such that $(\lambda I - T) f(\lambda) = 0$ for all $\lambda \in \mathcal{O}$. Let $\lambda_0 \in \mathcal{O}$ be such that $f(\lambda_0) \neq 0$, and set $F := \{\lambda_0\}$ and $G := \mathbb{C} \setminus \mathcal{O}$. The subsets F and G are closed and disjoint, moreover $f(\lambda_0) \in \ker(\lambda_0 I - T)$, and hence $f(\lambda_0) \in X_T(F)$, by part (iii) of Theorem 2.23. Define $g : \mathcal{O} \to X$ as follows

$$g(\lambda) := \begin{cases} \dfrac{f(\lambda) - f(\lambda_0)}{\lambda - \lambda_0} & \text{for all } \lambda \neq \lambda_0 \\ f'(\lambda_0) & \text{for } \lambda = \lambda_0. \end{cases}$$

Then we have $(\lambda I - T) g(\lambda) = f(\lambda_0)$ for all $\lambda \in \mathcal{O}$ and hence $f(\lambda_0) \in X_T(G)$. Therefore, $0 \neq f(\lambda_0) \in \mathcal{X}_T(F) \cap \mathcal{X}_T(G)$, a contradiction. Hence T has the SVEP.   ∎

If $T \in L(X)$ is onto then, by Theorem 2.20, $X = K(T) = X_T(\mathbb{C} \setminus \{0\})$. Actually, if T is onto, X is the local spectral subspace associated with a smaller set than $\mathbb{C} \setminus \{0\}$.

Theorem 2.24 *If $T \in L(X)$ is onto then there exists a $\delta > 0$ such that $X = \mathcal{X}_T(\mathbb{C} \setminus \mathbb{D}(0, \delta)) = X_T(\mathbb{C} \setminus \mathbb{D}(0, \delta))$.*

Proof From the open mapping theorem we know that there exists a $\delta > 0$ such that, for every $u \in X$, there is a $v \in X$ for which $Tv = u$ and $\delta \|v\| \leq \|u\|$. For an arbitrary $x \in X$, set $x_0 := x$ and define a sequence (x_n) such that $T x_n := x_{n-1}$ and $\delta \|x_n\| \leq \|x_{n-1}\|$ for all $n \in \mathbb{N}$. Since $\|x_n\| \leq \frac{1}{\delta^n} \|x\|$, the series defined as

$$f(\mu) := \sum_{n=0}^{\infty} \mu^n x_{n+1}$$

converges locally uniformly on the open disc $\mathbb{D}(0, \delta)$, and hence defines an analytic function $f : \mathbb{D}(0, \delta) \to X$ for which

$$(\mu I - T) f(\mu) = \sum_{n=0}^{\infty} \mu^n x_n - \sum_{n=0}^{\infty} \mu_{n+1} x_{n+1} = x$$

for all $\mu \in \mathbb{D}(0, \delta)$. Therefore, $x \in \mathcal{X}_T(\mathbb{C} \setminus \mathbb{D}(0, \delta))$. This shows that $X = \mathcal{X}_T(\mathbb{C} \setminus \mathbb{D}(0, \delta))$. Obviously, $X = X_T(\mathbb{C} \setminus \mathbb{D}(0, \delta))$, since $\mathcal{X}_T(\mathbb{C} \setminus \mathbb{D}(0, \delta)) \subseteq X_T(\mathbb{C} \setminus \mathbb{D}(0, \delta))$.   ∎

Theorem 2.25 *If $T \in L(X)$ is of Kato-type, then there exists a $\delta > 0$ for which*

$$K(T) = \mathcal{X}_T(\mathbb{C} \setminus \mathbb{D}(0, \delta)) = X_T(\mathbb{C} \setminus \mathbb{D}(0, \delta)). \qquad (2.10)$$

Proof By Theorem 1.63 $K(T)$ is closed, and, by Theorem 1.64, we have $K(T) = T^\infty(X)$. From $T(K(T)) = K(T)$ we see that $T|T^\infty(X)$ is onto. Furthermore, by Theorem 2.24, we have $T^\infty(X) \subseteq \mathcal{X}_T(\mathbb{C} \setminus \mathbb{D}(0, \delta))$ for δ small enough. From Theorem 2.20 we also deduce that

$$\mathcal{X}_T(\mathbb{C} \setminus \mathbb{D}(0, \delta)) \subseteq X_T(\mathbb{C} \setminus \mathbb{D}(0, \delta)) \subseteq X_T(\mathbb{C} \setminus \{0\} = K(T),$$

so,

$$K(T) = \mathcal{X}_T(\mathbb{C} \setminus \mathbb{D}(0, \delta)) \subseteq X_T(\mathbb{C} \setminus \mathbb{D}(0, \delta))$$

for some sufficiently small $\delta > 0$, so equality (2.10) is proved. ∎

Part (iv) of Theorem 2.23 also shows that for an operator without the SVEP the direct sum proved in Theorem 2.19 may fail.

The next result shows that if $T \in L(X)$ has a disconnected spectrum then X may be decomposed as the topological direct sum of closed glocal spectral subspaces.

Theorem 2.26 *Suppose that for $T \in L(X)$ we have $\sigma(T) = F_1 \uplus F_2$, where F_1 and F_2 are disjoint closed subsets of $\mathbb{C}$. Then the subspaces $\mathcal{X}_T(F_i)$, $i = 1, 2$, are closed and $X = \mathcal{X}_T(F_1) \oplus \mathcal{X}_T(F_2)$.*

Proof From the Riesz functional calculus we know that there is a decomposition $X = X_1 \oplus X_2$, with X_1 and X_2 closed and T-invariant subspaces. Moreover, $\sigma(T|X_i) = F_i$ for $i = 1, 2$, and it is immediate that $X_i \subseteq \mathcal{X}_T(F_i)$, so $X = \mathcal{X}_T(F_1) + \mathcal{X}_T(F_2)$.

To show that this sum is direct, let $x \in \mathcal{X}_T(F_1) \cap \mathcal{X}_T(F_2)$. Then, for each $i = 1, 2$, there exists an analytic function $f_i : \mathbb{C} \setminus F_i \to X$ such that $(\lambda I - T)f_i(\lambda) = x$ for all $\lambda \in \mathbb{C} \setminus F_i$. For every

$$\lambda \in (\mathbb{C} \setminus F_1) \cap (\mathbb{C} \setminus F_2) = \mathbb{C} \setminus \sigma(T) = \rho(T),$$

we have

$$f_1(\lambda) = (\lambda I - T)^{-1}x = f_2(\lambda).$$

Therefore, f_1 and f_2 must coincide whenever they are both defined. Because $(\mathbb{C} \setminus F_1) \cup (\mathbb{C} \setminus F_2) = \mathbb{C}$, there exists an analytic function f defined on $\mathbb{C}$ for which $(\lambda I - T)f(\lambda) = x$ holds for all $\lambda \in \mathbb{C}$. From part (i) of Theorem 2.23 we then conclude that $x = 0$. Hence, $X = \mathcal{X}_T(F_1) \oplus \mathcal{X}_T(F_2)$ and $\mathcal{X}_T(F_i) = X_i$, in particular both $\mathcal{X}_T(F_i)$, $i = 1, 2$, are closed. ∎

Evidently, $\mathcal{X}_T(F) = X$ for a closed subset F of $\mathbb{C}$ implies that $\sigma_s(T) \subseteq F$. The next result shows that the reverse implication holds. This result is based on a deep result of Leiterer [226] and we refer to [216, Theorem 3.3.12] for a proof.

Theorem 2.27 *If $T \in L(X)$ and $F \subseteq \mathbb{C}$ is closed then the following assertions hold:*

(i) $\mathcal{X}_T(F) = X$ *if and only if $\sigma_s(T) \subseteq F$.*
(ii) $\mathcal{X}_T(F) = \{0\}$ *if and only if $\sigma_{\mathrm{ap}}(T) \cap F = \emptyset$.*

Remark 2.28 Let $\mathcal{U}$ denote an open neighborhood of $\sigma(T)$ and $f : \mathcal{U} \to \mathbb{C}$ be an analytic function. Define $g : \mathcal{U} \times \mathcal{U} \to \mathbb{C}$ as follows

$$g(\mu, \lambda) := \begin{cases} \dfrac{f(\mu) - f(\lambda)}{\mu - \lambda} & \text{for all } \lambda, \mu \in \mathcal{U}, \lambda \neq \mu, \\ f'(\lambda) & \text{for } \mu = \lambda. \end{cases}$$

Clearly, g is analytic in each of the two variables μ and λ, and satisfies $f(\mu) - f(\lambda) = (\mu - \lambda)g(\mu, \lambda)$ for all $\mu, \lambda \in \mathcal{U}$. From the Riesz functional calculus we have $f(T) - f(\lambda) = (T - \lambda I)g(T, \lambda)$ for all $\lambda \in \mathbb{U}$. Furthermore, it is easily seen, from the integral formula for the Riesz functional calculus, that the function $\lambda \to g(T, \lambda)$ is an analytic operator function from $\mathcal{U}$ into $L(X)$.

The next theorem shows that the glocal spectral subspaces behave canonically under the functional calculus.

Theorem 2.29 *If $T \in L(X)$, and $f : \mathcal{U} \to X$ is analytic on an open neighborhood of the spectrum $\sigma(T)$, then*

$$\mathcal{X}_{f(T)}(F) = \mathcal{X}_T(f^{-1}(F)) \quad \text{for all closed subsets } F \subseteq \mathbb{C}.$$

Proof Let F be a closed subset of $\mathbb{C}$, and $x \in \mathcal{X}_{f(T)}(F)$. Choose an analytic function $h : \mathbb{C} \setminus F \to X$ such that $(\mu I - f(T))h(\mu) = x$ for all $\mu \in \mathbb{C} \setminus F$. Then, for every $\lambda \in \mathcal{U} \setminus f^{-1}(F) = f^{-1}(\mathbb{C} \setminus F)$, we have $(\lambda I - f(T))h(f(\lambda)) = x$. From Remark 2.28 we have that

$$(\lambda I - T)g(T, \lambda)h(f(\lambda)) = x$$

holds for all $\lambda \in \mathcal{U} \setminus f^{-1}(F)$. From Remark 2.28 we also have that the function $\lambda \to g(T, \lambda)h(f(\lambda))$ is analytic on the set $\mathcal{U} \setminus f^{-1}(F)$, hence $x \in \mathcal{X}_T(f^{-1}(F)) \cup (\mathbb{C} \setminus \mathcal{U}))$. But $\sigma(T) \subseteq \mathcal{U}$, so, by part (ii) of Theorem 2.23, we conclude that $x \in \mathcal{X}_T(f^{-1}(F))$. Therefore, $\mathcal{X}_{f(T)}(F) \subseteq \mathcal{X}_T(f^{-1}(F))$.

To show the reverse inclusion, observe that if $K := f^{-1}(F) \cap \sigma(T)$ then, by part (ii) of Theorem 2.23, it suffices to prove that $\mathcal{X}_T(K) \subseteq \mathcal{X}_T(F)$. Since K is a compact subset of the open set $\mathcal{U}$, there exists a closed disc $\mathbf{D}(0, \varepsilon_0)$ such that

$K + \mathbf{D}(0, \varepsilon_0) \subseteq \mathcal{U}$. The compactness of K and the continuity of f entails that

$$\bigcap_{0 < \varepsilon < \varepsilon_0} f(K + \mathbf{D}(0, \varepsilon_0)) \subseteq f(K) \subseteq F,$$

and hence

$$\mathbb{C} \setminus F \subseteq \bigcup_{0 < \varepsilon < \varepsilon_0} (\mathbb{C} \setminus f(K + \mathbf{D}(0, \varepsilon_0))).$$

Now, let $x \in \mathcal{X}_T(K)$ and let $h : \mathbb{C} \setminus K \to X$ be an analytic function for which $(\lambda I - T)h(\lambda) = x$ holds for all $\lambda \in \mathbb{C} \setminus K$. For $0 < \varepsilon < \varepsilon_0$, consider the open disc $\mathbb{D}(0, \varepsilon)$ and denote by Γ_ε a contour in $K + \mathbb{D}(0, \varepsilon)$ which surrounds K. Set $V_\varepsilon := \mathbb{C} \setminus f(K + \mathbb{D}(0, \varepsilon))$. Clearly, $f(\lambda) \neq V_\varepsilon$ for all $\lambda \in \Gamma_\varepsilon$. Define

$$g_\varepsilon(\mu) := \frac{1}{2\pi i} \int_{\Gamma_\varepsilon} \frac{h(\lambda)}{\mu - f(\lambda)} d\lambda \quad \text{for all } \mu \in V_\varepsilon.$$

Clearly, $g_\varepsilon : V_\varepsilon \to X$ is analytic, and, by our Remark 2.28 and by Lemma 2.18, for every $\mu \in V_\varepsilon$, we have

$$\begin{aligned}
(\mu I - f(T))g_\varepsilon(\mu) &= \frac{1}{2\pi i} \int_{\Gamma_\varepsilon} (\mu I - f(T)) \frac{h(\lambda)}{\mu - f(\lambda)} d\lambda \\
&= \frac{1}{2\pi i} \int_{\Gamma_\varepsilon} (f(\lambda)I - f(T)) \frac{h(\lambda)}{\mu - f(\lambda)} d\lambda - \frac{1}{2\pi i} \int_{\Gamma_\varepsilon} h(\lambda) d\lambda \\
&= \frac{1}{2\pi i} \int_{\Gamma_\varepsilon} g(T, \lambda) \frac{x}{\mu - f(\lambda)} d\lambda + x.
\end{aligned}$$

By Cauchy's integral theorem the last integral is 0, since for arbitrary $\mu \in V_\varepsilon$ the function $\lambda \to (\mu - f(\lambda))^{-1} g(T, \lambda)x$ is analytic on $K + \mathbb{D}(0, \varepsilon)$. Hence,

$$(\mu I - f(T))g_\varepsilon(\mu) = x \quad \text{for all } \mu \in V_\varepsilon.$$

Another application of Cauchy's theorem gives that g_ε is the restriction of every g_δ to V_ε for all $0 < \delta < \varepsilon < \varepsilon_0$. Since $\mathbb{C} \setminus F \subseteq \bigcap_{0 < \varepsilon < \varepsilon_0} V_\varepsilon$, the family of functions $\{g_\varepsilon\}$ leads to a well defined analytic function $g : \mathbb{C} \setminus F \to X$ which satisfies $(\mu I - f(T))g(\mu) = x$ for all $\mu \in \mathbb{C} \setminus F$, so $x \in \mathcal{X}_{f(T)}(F)$, and hence $\mathcal{X}_T(f^{-1}(F)) \subseteq \mathcal{X}_{f(T)}(F)$. ∎

The glocal spectral subspace $\mathcal{X}_T(\mathbf{D}(0, \varepsilon))$ associated with the closed disc $\mathbf{D}(0, \varepsilon)$ may be characterized as follows:

Theorem 2.30 *If $T \in L(X)$ then*

$$\mathcal{X}_T(\mathbf{D}(0, \varepsilon)) = \left\{ x \in X : \limsup_{n \to \infty} \|T^n x\|^{1/n} \leq \varepsilon \right\}. \tag{2.11}$$

Proof For every $x \in X$ suppose that $\eta_T(x) := \limsup_{n\to\infty} \|T^n x\|^{1/n} \leq \varepsilon$. Clearly, the series

$$f(\lambda) := \sum_{n=1}^{\infty} \lambda^{-n} T^{n-1} x, \quad \lambda \in \mathbb{C} \setminus \mathbf{D}(0.\varepsilon)$$

converges locally uniformly, so it defines an X-valued function on the set $\mathbb{C} \setminus \mathbf{D}(0, \varepsilon)$. Clearly,

$$(\lambda I - T)f(\lambda) = x \quad \text{for all } \lambda \in \mathcal{X}_T(\mathbf{D}(0, \varepsilon)),$$

so $x \in \mathcal{X}_T(\mathbf{D}(0, \varepsilon))$.

Conversely, assume that $x \in \mathcal{X}_T(\mathbf{D}(0, \varepsilon))$ and consider an analytic function $f : \mathbb{C} \setminus \mathbf{D}(0, \varepsilon) \to X$ such that $(\lambda I - T)f(\lambda) = x$ holds for all $\lambda \in \mathbb{C} \setminus \mathbf{D}(0, \varepsilon)$. If $|\lambda| > \max\{\varepsilon, \|T\|\}$ we have

$$f(\lambda) = (\lambda I - T)^{-1} x = \sum_{n}^{\infty} \lambda^{-n} T^{n-1} x,$$

and consequently $f(\lambda) \to 0$ as $|\lambda| \to \infty$. Consider now the open disc $\mathbb{D}(0, 1/\varepsilon)$ of $\mathbb{C}$ centered at 0 with radius $1/\varepsilon$. The analytic function $g : \mathbb{D}(0, 1/\varepsilon) \to X$ defined by

$$g(\mu) := \begin{cases} f\left(\dfrac{1}{\lambda}\right) & \text{if } 0 \neq \mu \in \mathbb{D}(0, 1/\varepsilon), \\ 0 & \text{if } \mu = 0, \end{cases}$$

satisfies the equality

$$g(\mu) = \sum_{n=1}^{\infty} \mu^n T^{n-1} x \quad \text{for all } |\mu| < \frac{1}{\max\{\varepsilon, \|T\|\}}. \tag{2.12}$$

Since g is analytic on $\mathbb{D}(0, 1/\varepsilon)$, from Cauchy's integral formula (proceeding exactly as in the scalar setting) we conclude that equality (2.12) holds even for all $\mu \in \mathbf{D}(0, 1/\varepsilon)$. This shows that the radius of convergence of the power series representing $g(\mu)$ is greater then $1/\varepsilon$. The standard formula for the radius of convergence of a vector-valued power series then implies that $\eta_T(x) < \varepsilon$ and hence equality (2.11) holds. ∎

2.3 The Quasi-Nilpotent Part of an Operator

We now introduce an important subspace in Fredholm theory and local spectral theory.

Definition 2.31 The *quasi-nilpotent part* $H_0(T)$ of an operator $T \in L(X)$ is defined as $H_0(T) = \mathcal{X}_T(\{0\})$.

The quasi-nilpotent part of an operator may be characterized as follows:

Theorem 2.32 *For every* $T \in L(X)$ *we have*

$$H_0(T) = \{x \in X : \limsup_{n \to \infty} \|T^n x\|^{1/n} = 0\}. \tag{2.13}$$

Moreover, if T has the SVEP then $H_0(T) = X_T(\{0\})$.

Proof Clearly, the equality (2.13) is obtained by taking $\varepsilon = 0$ in Theorem 2.30. If T has the SVEP then $\mathcal{X}_T(\{0\}) = X_T(\{0\})$, by part (iv) of Theorem 2.23. ∎

It should be noted that in general the limit $\lim_{n \to \infty} \|T^n x\|^{1/n}$ does not exist. In fact it has been observed, by Daneš in [107], that the set of all accumulation points of the sequence $(\|T^n x\|^{1/n})$ is the whole interval (a, b) where

$$a := \liminf_{n \to \infty} \|T^n x\|^{1/n} \quad \text{and} \quad b := \limsup_{n \to \infty} \|T^n x\|^{1/n}.$$

The quantity

$$r_x(T) := \limsup_{n \to \infty} \|T^n x\|^{1/n}$$

is called the *local spectral radius* of T at x. It should be noted that in general,

$$r_x(T) \leq \max\{|\lambda| : \lambda \in \sigma_T(x)\}.$$

If T has the SVEP then

$$r_x(T) = \max\{|\lambda| : \lambda \in \sigma_T(x)\}.$$

For a proof, see Laursen and Neumann [216, Proposition 3.3.13].

The following example shows that the quasi-nilpotent part $H_0(T)$ need not be closed even if T has the SVEP.

Example 2.33 Let $X := \ell_2 \oplus \ell_2 \cdots$ be provided with the norm

$$\|x\| := \left(\sum_{n=1}^{\infty} \|x_n\|^2 \right)^{1/2} \quad \text{for all } x := (x_n) \in X,$$

and define

$$T_n e_i := \begin{cases} e_{i+1} & \text{if } i = 1, \cdots, n, \\ \dfrac{e_{i+1}}{i - n} & \text{if } i > n. \end{cases}$$

It is easily seen that

$$\|T_n^{n+k}\| = \frac{1}{k!} \quad \text{and} \quad \left(\frac{1}{k!}\right)^{\frac{1}{n+k}} \to 0 \text{ as } k \to \infty,$$

from which we obtain that $\sigma(T_n) = \{0\}$. Moreover, every T_n is injective and the point spectrum $\sigma_p(T_n) = \emptyset$, thus T_n has the SVEP.

Let us define $T := T_1 \oplus \cdots \oplus T_n \oplus \cdots$. From the estimate $\|T_n\| = 1$ for every $n \in \mathbb{N}$, we easily obtain $\|T\| = 1$. Moreover, since $\sigma_p(T_n) = \emptyset$ for every $n \in \mathbb{N}$, it also follows that $\sigma_p(T) = \emptyset$, hence T has the SVEP.

Now, let us consider the sequence $x = (x_n) \subset X$ defined by $x_n := \frac{e_1}{n}$ for every $n \in \mathbb{N}$. We have

$$\|x\| = \left(\sum_{n=1}^{\infty} \frac{1}{n^2}\right)^{\frac{1}{2}} < \infty,$$

which implies that $x \in X$. Moreover,

$$\|T^n x\|^{1/n} \ge \|T_n^n \frac{e_1}{n}\|^{1/n} = \left(\frac{1}{n}\right)^{\frac{1}{n}}$$

and the last term does not converge to 0. From this it follows that $\sigma_T(x)$ properly contains $\{0\}$ and therefore, $x \notin X_T(\{0\}) = H_0(T)$.

Finally,

$$\ell_2 \oplus \ell_2 \cdots \oplus \ell_2 \oplus \{0\} \cdots \subset H_0(T),$$

where the non-zero terms are n. This holds for every $n \in \mathbb{N}$, so $H_0(T)$ is dense in X. Since $H_0(T) \ne X$ it then follows that $H_0(T)$ is not closed.

In the following we collect some basic properties of $H_0(T)$.

Lemma 2.34 *For every $T \in L(X)$, X a Banach space, we have:*

(i) $\ker(T^m) \subseteq \mathcal{N}^\infty(T) \subseteq H_0(T)$ *for every* $m \in \mathbb{N}$;
(ii) $x \in H_0(T) \Leftrightarrow Tx \in H_0(T)$;
(iii) $\ker(\lambda I - T) \cap H_0(T) = \{0\}$ *for every* $\lambda \ne 0$;
(iv) $H_0(T) \subseteq (\lambda I - T)(X)$ *for all* $\lambda \ne 0$;
(v) *If* $TS = ST$ *then* $H_0(T) \subseteq H_0(TS)$.

Proof

(i) If $T^m x = 0$ then $T^n x = 0$ for every $n \ge m$.
(ii) If $x_0 \in H_0(T)$ then from the inequality $\|T^n Tx\| \le \|T\|\|T^n x\|$ it easily follows that $Tx \in H_0(T)$. Conversely, if $Tx \in H_0(T)$, from

$$\|T^{n-1} Tx\|^{1/n-1} = (\|T^n x\|^{1/n})^{n/n-1}$$

we conclude that $x \in H_0(T)$.

(iii) If $x \neq 0$ is an element of $\ker (\lambda I - T)$ then $T^n x = \lambda^n x$, so

$$\lim_{n \to \infty} \|T^n x\|^{1/n} = \lim_{n \to \infty} |\lambda| \|x\|^{1/n} = |\lambda|$$

and therefore $x \notin H_0(T)$.

(iv) If $\lambda \neq 0$ then $\{0\} \subseteq \mathbb{C} \setminus \{\lambda\}$, and from Theorem 2.20 we obtain

$$H_0(T) = \mathcal{X}_T(\{0\}) \subseteq X_T(\{0\}) \subseteq X_T(\mathbb{C} \setminus \{\lambda\}) = K(\lambda I - T).$$

The inclusion (iv) then follows from $K(\lambda I - T) \subseteq (\lambda I - T)(X)$.

(v) This follows easily from definition.   ∎

Theorem 2.35 $T \in L(X)$ *is quasi-nilpotent if and only if* $H_0(T) = X$.

Proof If T is quasi-nilpotent then $\lim_{n \to \infty} \|T^n\|^{1/n} = 0$, so that from $\|T^n x\| \leq \|T^n\| \|x\|$ we obtain that $\lim_{n \to \infty} \|T^n x\|^{1/n} = 0$ for every $x \in X$.

Conversely, assume that $H_0(T) = \mathcal{X}_T(\{0\} = X$. If $x \in \mathcal{X}_T(\{0\} = X$ then there is an analytic function $f : \mathbb{C} \setminus \{0\} \to X$ such that

$$(\lambda I - T) f(\lambda) = x \quad \text{for all } \lambda \neq 0,$$

thus $(\lambda I - T)$ is surjective for all $\lambda \neq 0$. On the other hand, for every $\lambda \neq 0$ we have that

$$\{0\} = \ker (\lambda I - T) \cap H_0(T) = \ker (\lambda I - T) \cap X = \ker (\lambda I - T),$$

which shows that $\lambda I - T$ is invertible and therefore $\sigma(T) = \{0\}$.   ∎

We now describe the quasi-nilpotent part of an operator T which admits a generalized Kato decomposition. We start with an elementary lemma.

Lemma 2.36 *Assume that* $T \in L(X)$ *admits a GKD* (M, N). *Then*

$$H_0(T) = H_0(T|M) \oplus H_0(T|N) = H_0(T|M) \oplus N.$$

Proof From Theorem 2.35 we know that $N = H_0(T|N)$. The inclusion $H_0(T) \supseteq H_0(T|M) + H_0(T|N)$ is clear. In order to show the opposite inclusion, let us consider an arbitrary element $x \in H_0(T)$ and set $x := u + v$, with $u \in M$ and $v \in N$. Evidently, $N = H_0(T|N) \subseteq H_0(T)$, thus $u = x - v \in H_0(T) \cap M = H_0(T|M)$ and hence $H_0(T) \subseteq H_0(T|M) + H_0(T|N)$. Clearly the sum $H_0(T|M) + N$ is direct since $M \cap N = \{0\}$.   ∎

The next result shows that for a semi-regular operator T, or if T is a semi-Fredholm operator, the quasi-nilpotent part and the hyper-kernel of T have the same closure.

Theorem 2.37 *For every bounded operator $T \in L(X)$ we have:*

(i) $H_0(T) \subseteq {}^{\perp}K(T^{\star})$ and $K(T) \subseteq {}^{\perp}H_0(T^{\star})$.

(ii) *If T is semi-regular or semi-Fredholm, then*

$$\overline{H_0(T)} = \overline{\mathcal{N}^{\infty}(T)} = {}^{\perp}K(T^{\star}) \quad \text{and} \quad K(T) = {}^{\perp}H_0(T^{\star}). \tag{2.14}$$

(iii) *If T is semi-regular then $\overline{H_0(T)} \subseteq K(T)$.*

Proof

(i) Let $u \in H_0(T)$ and $f \in K(T^{\star})$. From the definition of $K(T^{\star})$ we know that there exists a $\delta > 0$ and a sequence (g_n), $n \in \mathbb{Z}_+$, of $X^{\star}$ such that

$$g_0 = f, \quad T^{\star}g_{n+1} = g_n \quad \text{and} \quad \|g_n\| \leq \delta^n \|f\|$$

for every $n \in \mathbb{Z}_+$. These equalities imply that $f = (T^{\star})^n g_n$ for every $n \in \mathbb{Z}_+$, and hence

$$f(u) = (T^{\star})^n g_n(u) = g_n(T^n u) \quad \text{for every } n \in \mathbb{Z}_+.$$

From this it then follows that $|f(u)| \leq \|T^n u\| \|g_n\|$ for every $n \in \mathbb{Z}_+$, and consequently

$$|f(u)| \leq \delta^n \|f\| \|T^n u\| \quad \text{for every } n \in \mathbb{Z}_+. \tag{2.15}$$

Since $u \in H_0(T)$ we then obtain that $\lim_{n \to \infty} \|T^n u\|^{1/n} = 0$ and hence, by taking the n-th root in (2.15), we conclude that $f(u) = 0$. Therefore $H_0(T) \subseteq {}^{\perp}K(T^{\star})$.

The inclusion $K(T) \subseteq {}^{\perp}H_0(T^{\star})$ may be proved in a similar way.

(ii) Assume that T is semi-regular. Then, by Theorem 1.43, $T^{\star}$ is semi-regular and hence, by Theorem 1.44, $(T^{\star})^n$ is semi-regular, so $T^{\star n}(X^{\star})$ is closed for all $n \in \mathbb{N}$. From the first part we also know that

$$\overline{\mathcal{N}^{\infty}(T)} \subseteq \overline{H_0(T)} \subseteq \overline{{}^{\perp}K(T^{\star})} = {}^{\perp}K(T^{\star}),$$

since ${}^{\perp}K(T^{\star})$ is closed.

To show the first two equalities of (2.14) we need only to show the inclusion ${}^{\perp}K(T^{\star}) \subseteq \overline{\mathcal{N}^{\infty}(T)}$. For every $T \in L(X)$ and every $n \in \mathbb{N}$ we have $\ker T^n \subseteq \mathcal{N}^{\infty}(T)$, and hence

$$\mathcal{N}^{\infty}(T)^{\perp} \subseteq \ker T^{n\perp} = T^{\star n}(X^{\star})$$

because the last subspaces are closed for all $n \in \mathbb{N}$.

From this we easily obtain that

$$\overline{\mathcal{N}^\infty(T)}^\perp \subseteq T^{\star\infty}(X^\star) = K(T^\star),$$

where the last equality holds by Theorem 1.64. Consequently $^\perp K(T^\star) \subseteq \overline{\mathcal{N}^\infty(T)}$, and hence the equalities (2.14) are proved. The equality $K(T) = {}^\perp H_0(T^\star)$ is proved in a similar way.

The proof in the case where T is semi-Fredholm is analogous.

(iii) The semi-regularity of T entails that $\mathcal{N}^\infty(T) \subseteq T^\infty(X) = K(T)$, where the last equality follows from Theorem 1.64. Consequently from part (ii) it follows that

$$\overline{H_0(T)} = \overline{\mathcal{N}^\infty(T)} \subseteq \overline{K(T)} = K(T),$$

since $K(T)$ is closed, by Theorem 1.64.   ∎

Corollary 2.38 *Let $T \in L(X)$ be semi-regular. Then $T(H_0(T)) = H_0(T)$.*

Proof Clearly by (ii) of Lemma 2.34 it suffices to show the inclusion $H_0(T) \subseteq T(H_0(T))$. Let $x \in H_0(T)$. From part (iii) of Theorem 2.37 we have $x \in K(T) = T(K(T))$, so $x = Ty$ for some $y \in X$ and from part (ii) of Lemma 2.34 we conclude that $y \in H_0(T)$. Hence $H_0(T) \subseteq T(H_0(T))$.   ∎

Theorem 2.39 *Suppose $H_0(T)$ closed or $H_0(T) \cap K(T)$ is closed. Then $H_0(T) \cap K(T) = \{0\}$.*

Proof Assume first that $H_0(T)$ is closed. Let $\tilde{T}$ denote the restriction of T to the T-invariant subspace $H_0(T)$. Clearly, $H_0(T) = H_0(\tilde{T})$, thus $\tilde{T}$ is quasi-nilpotent. Therefore $K(\tilde{T}) = \{0\}$. On the other hand it is easily seen that $H_0(T) \cap K(T) = K(\tilde{T}) = \{0\}$.

Assume that $Y := H_0(T) \cap K(T)$ is closed. Clearly, Y is invariant under T, so we can consider the restriction $\widehat{T} := T|Y$. If $y \in Y$ then

$$\|\widehat{T}^n y\|^{\frac{1}{n}} = \|T^n y\|^{\frac{1}{n}} \to \quad \text{as } n \to \infty,$$

hence $y \in H_0(\widehat{T})$ and consequently, $H_0(\widehat{T}) = Y$. This shows that $\widehat{T}$ is quasi-nilpotent and hence $K(\widehat{T}) = \{0\}$. We claim that $Y = K(\widehat{T})$.

Choose $y \in Y = H_0(T) \cap K(T)$. From the definition of $K(T)$ there exists a sequence $(y_n)_{n=0,1,\dots}$ in X and a $\delta > 0$ such that

$$y_0 = y, \, T y_n = y_{n-1} \quad \text{and} \quad \|y_n\| \leq \delta^n \|y\| \text{ for all } n = 0, 1, \dots.$$

Since $y \in Y \subseteq H_0(T)$, according to Corollary 2.38, we have $y_n \in H_0(T)$ for all $n \in \mathbb{N}$. Furthermore, since $y \in K(T) = X_T(\mathbb{C} \setminus \{0\})$, from part (ii) of Theorem 2.13 we also have $y_n \in K(T)$ for all n, so $y_n \in Y$ and hence $y \in K(\widehat{T})$. Therefore,

$Y \subseteq K(\widehat{T})$. The opposite inclusion is clear, since $K(\widehat{T}) = K(T) \cap Y \subseteq Y$. Thus,

$$Y = H_0(T) \cap K(T) = K(\widehat{T}) = \{0\}.$$

Corollary 2.40 *If T is semi-regular and either $H_0(T)$ or $H_0(T) \cap K(T)$ are closed then $H_0(T) = \{0\}$.*

Proof If T is semi-regular then $T(H_0(T)) = H_0(T)$, by Corollary 2.38. Suppose first that $H_0(T)$ is closed. Then $H_0(T) \subseteq K(T)$ and hence, by Theorem 2.39, we have that $H_0(T) = H_0(T) \cap K(T) = \{0\}$.

Suppose that $H_0(T) \cap K(T)$ is closed. Since T is semi-regular, $\ker T \subseteq T^n(X)$ for every $n \in \mathbb{N}$ and this is equivalent to saying that $\mathcal{N}^\infty(T) \subseteq T^\infty(X)$, by Corollary 1.17. Moreover, by Theorem 1.44, $K(T) = T^\infty(X)$ is closed. The semi-regularity of T also implies, from part (ii) of Theorem 2.37, that

$$H_0(T) \subseteq \overline{H_0(T)} = \overline{\mathcal{N}^\infty(T)} \subseteq \overline{T^\infty(X)} = \overline{K(T)} = K(T),$$

and hence $H_0(T) \cap K(T) = H_0(T)$ is closed. From the first part of the proof it then follows that $H_0(T) \cap K(T) = \{0\}$.

We now give a characterization of the isolated points of $\sigma_s(T)$.

Theorem 2.41 *If $T \in L(X)$ then $X = H_0(\lambda I - T) + K(\lambda I - T)$ if and only if $\sigma_s(T)$ does not cluster at λ. In particular, if $\lambda \in \sigma_s(T)$ then $X = H_0(\lambda I - T) + K(\lambda I - T)$ if and only if $\lambda \in \mathrm{iso}\, \sigma_s(T)$.*

Proof We can take $\lambda = 0$. The equivalence is obvious if $0 \notin \sigma_s(T)$, since $K(\lambda I - T) = X$ in this case. Suppose that $0 \in \sigma_s(T)$. By Theorems 2.27 and 2.23 we have

$$X = \mathcal{X}_T(\sigma_s(T)) = \mathcal{X}_T(\{0\}) + \mathcal{X}_T(\sigma_s(T) \setminus \{0\}).$$

But by Theorem 2.20 we have

$$\mathcal{X}_T(\sigma_s(T) \setminus \{0\}) \subseteq \mathcal{X}_T(\mathbb{C} \setminus \{0\} = K(T)),$$

from which we obtain $H_0(T) + K(T) = X$.

Conversely, suppose that $0 \in \sigma_s(T)$ and $H_0(T) + K(T) = X$. Then every $x \in X$ may be written as $x = x_1 + x_2$, where $x_1 \in H_0(T)$ and $x_2 \in K(T)$. Clearly, from the definition of $H_0(T)$, we have $\sigma_T(x_1) \subseteq \{0\}$, while $0 \notin \sigma_T(x_2)$, by Theorem 2.23. Therefore,

$$\sigma_T(x) \subseteq \sigma_T(x_1) \cup \sigma_T(x_2) \subseteq \{0\} \cap \sigma_T(x_2),$$

and this implies, since $\sigma_T(x_2)$ is closed, that 0 is isolated in $\sigma_T(x)$. By Theorem 2.21 there exists an $x_0 \in X$ for which $\sigma_T(x_0) = \sigma_s(T)$, so we can conclude that 0 is isolated in $\sigma_s(T)$.

The next corollary is an obvious consequence of Theorem 2.41, once we observe the equality $\sigma_{ap}(T) = \sigma_s(T^*)$.

Corollary 2.42 *If $T \in L(X)$ then $X^* = H_0(\lambda I - T^*) + K(\lambda I - T^*)$ if and only if $\sigma_{ap}(T)$ does not cluster at λ.*

Theorem 2.41 has some other interesting consequences:

Corollary 2.43 *If $T \in L(X)$ the following assertions hold:*

(i) $X = H_0(\lambda I - T) + K(\lambda I - T)$ *if and only if $\sigma_T(x)$ does not cluster at λ for every $x \in X$.*
(ii) $X = H_0(\lambda I - T) + K(\lambda I - T)$ *for all $\lambda \in \mathbb{C}$ if and only if $\sigma(T)$ is finite.*

Proof

(i) The direct implication is clear from the proof of Theorem 2.41. For the converse, note that if $\lambda \notin \sigma_s(T)$ then $K(\lambda I - T) = X$. Moreover, $\sigma_T(x) \subseteq \sigma_s(T)$ for all $x \in X$, by Theorem 2.21. The converse implication is then a direct consequence of Theorem 2.41.
(ii) Since a compact set consisting of isolated points is a finite set, Theorem 2.41 entails that $\sigma_s(T)$ is finite, and hence $\sigma(T)$ is also finite. ∎

Remark 2.44 Since the condition $X = H_0(\lambda I - T) + K(\lambda I - T)$ may be thought of as being dual to the condition $H_0(\lambda I - T) \cap K(\lambda I - T) = \{0\}$, see Theorem 2.37, one is tempted to conjecture that λ is isolated in $\sigma_{ap}(T)$ if and only if $H_0(\lambda I - T) \cap K(\lambda I - T) = \{0\}$ and $H_0(\lambda I - T) + K(\lambda I - T)$ is closed. The following example shows that this is not true. Set

$$i(T) := \lim_{n \to \infty} j(T^n)^{1/n},$$

where $j(T)$ is the injectivity modulus of $T \in L(X)$, defined in Chap. 1. Let us consider the *weighted right shift* $S \in L(X)$, where $X = \ell^2(\mathbb{N})$, defined by $Se_n := s_n e_{n+1}$, where (e_n) is the canonical basis of $\ell^2(\mathbb{N})$, and (s_n) is a given weight sequence, with $0 < s_n \leq 1$. We may choose the sequence (s_n) such that $i(S) = 0$ and $r(S) > 0$, $r(S)$ the spectral radius of S, see [216, Chap. 1, §1.6]. By [216, Prop. 1.6.15] we have

$$\sigma_{ap}(S) = \{\lambda \in \mathbb{C} : i(S) \leq |\lambda| \leq r(S)\},$$

so 0 is not isolated in $\sigma_{ap}(S)$. Moreover, by [216, Prop. 1.6.16], the subspaces $\mathcal{X}_S(F)$ are closed for all closed $F \subseteq \mathbb{C}$, in particular $H_0(S) = \mathcal{X}_S(\{0\})$ is closed, and

$$K(S) = \bigcap_{n=0}^{\infty} S^n(X) = \{0\}.$$

Hence, $H_0(S) \cap K(S) = \{0\}$ and $H_0(S) + K(S) = H_0(S)$ is closed, but 0 is not an isolated point of $\sigma_{ap}(S)$.

For an isolated point λ_0 of $\sigma(T)$ the quasi-nilpotent part $H_0(\lambda_0 I - T)$ and the analytic core $K(\lambda_0 I - T)$ may be precisely described as a range or a kernel of the spectral projection P_0 associated with the spectral subset $\{\lambda_0\}$.

Theorem 2.45 *Let* $T \in L(X)$ *and suppose that* λ_0 *is an isolated point of* $\sigma(T)$. *If* P_0 *is the spectral projection associated with* $\{\lambda_0\}$, *then:*

(i) $P_0(X) = H_0(\lambda_0 I - T)$.
(ii) $\ker P_0 = K(\lambda_0 I - T)$.

   Therefore, $X = H_0(\lambda_0 I - T) \oplus K(\lambda_0 I - T)$.

Proof

(i) Since λ_0 is an isolated point of $\sigma(T)$ there exists a positively oriented circle $\Gamma := \{\lambda \in \mathbb{C} : |\lambda - \lambda_0| = \delta\}$ which separates λ_0 from the remaining part of the spectrum. We have

$$(\lambda_0 I - T)^n P_0 x = \frac{1}{2\pi i} \int_\Gamma (\lambda_0 - \lambda)^n (\lambda I - T)^{-1} x \, d\lambda \quad \text{for all } n = 0, 1, \cdots.$$

Now, assume that $x \in P_0(X)$. We have $P_0 x = x$ and it is easy to verify the following estimate:

$$\|(\lambda_0 I - T)^n x\| \leq \frac{1}{2\pi} 2\pi \delta^{n+1} \max_{\lambda \in \Gamma} \|(\lambda I - T)^{-1}\| \|x\|.$$

Obviously this estimate also holds for some $\delta_0 < \delta$ (since Γ lies in $\rho(T)$), and consequently

$$\limsup \|(\lambda_0 I - T)^n x\|^{1/n} < \delta. \tag{2.16}$$

This proves the inclusion $P_0(X) \subseteq H_0(\lambda_0 I - T)$.

Conversely, assume that $x \in H_0(\lambda_0 I - T)$ and hence that the inequality (2.16) holds. Let $S \in L(X)$ denote the operator

$$S := \frac{\lambda_0 I - T}{\lambda_0 - \lambda}.$$

Evidently the Neumann series

$$\sum_{n=0}^\infty S^n x = \sum_{n=0}^\infty \left(\frac{\lambda_0 I - T}{\lambda_0 - \lambda} \right)^n x$$

converges for all $\lambda \in \Gamma$. If y_λ denotes its sum for every $\lambda \in \Gamma$, from a standard argument of functional analysis we obtain that $(I - S)y_\lambda = x$. A simple calculation also shows that $y_\lambda = (\lambda - \lambda_0)R_\lambda x$ and therefore

$$R_\lambda x = -\sum_{n=0}^{\infty} \frac{\lambda_0 I - T)^n x}{(\lambda_0 - \lambda)^{n+1}} \quad \text{for all } \lambda \in \Gamma.$$

A term by term integration then yields

$$P_0 x = \frac{1}{2\pi i} \int_\Gamma R_\lambda x \, d\lambda = -\frac{1}{2\pi i} \int_\Gamma \frac{1}{(\lambda_0 - \lambda)} x \, d\lambda = x,$$

so $x \in P_0(X)$ and this proves the inclusion $H_0(\lambda_0 I - T) \subseteq P_0(X)$, so the proof of (i) is complete.

(ii) There is no harm in assuming that $\lambda_0 = 0$. We have $\sigma(T|P_0(X)) = \{0\}$, and $0 \in \rho(T|\ker P_0)$. From the equality $T(\ker P_0) = \ker P_0$ we obtain $\ker P_0 \subseteq K(T)$, see Theorem 1.39.

It remains to prove the reverse inclusion $K(T) \subseteq \ker P_0$. To see this we first show that $H_0(T) \cap K(T) = \{0\}$. This is clear because $H_0(T) \cap K(T) = K(T|H_0(T))$, and the last subspace is $\{0\}$ since the restriction of T on the Banach space $H_0(T)$ is a quasi-nilpotent operator (this will be proved in the next Corollary 2.71). Hence $H_0(T) \cap K(T) = \{0\}$. From this it then follows that

$$K(T) \subseteq K(T) \cap X = K(T) \cap [\ker P_0 \oplus P_0(X)]$$

$$= \ker P_0 + K(T) \cap H_0(T) = \ker P_0,$$

so the desired inclusion is proved.   ∎

We now consider the case where $0 \in \operatorname{iso} \sigma(T)$.

Theorem 2.46 *If $T \in L(X)$ the following statements are equivalent:*

(i) $0 \in \operatorname{iso} \sigma(T)$.
(ii) *Both $H_0(T)$ and $K(T)$ are closed and $X = H_0(T) \oplus K(T)$, $T|K(T)$ is invertible, $T|H_0(T)$ quasi-nilpotent.*
(iii) *There exist two closed T-invariant subspaces M and N such that $X = M \oplus N$, $T|M$ is invertible, $T|N$ is quasi-nilpotent.*
(iv) *There exists a projection $0 \neq P \in L(X)$ such that $PT = TP$, $T + P$ is invertible, TP is quasi-nilpotent.*

Proof The case where $0 \notin \sigma(T)$ is trivial, so we can consider only the case $0 \in \operatorname{iso} \sigma(T)$.

(i) $\Rightarrow$ (ii) Since $0 \in \operatorname{iso} \sigma(T)$, $H_0(T)$ and $K(T)$ are both closed and $X = H_0(T) \oplus K(T)$, by Theorem 2.45. Furthermore, if P_0 denotes the spectral projection associated with $\{0\}$, then $H_0(T) = \ker P_0$ and $K(T) = P_0(X)$, so, by the spectral decomposition Theorem (see Appendix A) we have $0 \notin \sigma(T|K(T))$ and $\{0\} = \sigma(T|H_0(T))$.

(ii) $\Rightarrow$ (iii) is obvious.

(iii) $\Rightarrow$ (i) We have $\sigma(T) = \sigma(T|M) \cup \sigma(T|N) = \sigma(T|M) \cup \{0\}$. Since $0 \notin \sigma(T|M)$ it then follows that 0 is an isolated point of $\sigma(T)$.

(ii) $\Rightarrow$ (iv) Let P be a projection of X onto $H_0(T)$ along $K(T)$. Then $P(X) = H_0(T)$ and $\ker P = K(T)$. Since the pair of subspaces $(H_0(T), K(T))$ reduces T, we have $PT = TP$. Let $x := x_1 + x_2$ be arbitrary in X, with $x_2 \in H_0(T)$ and $x_1 \in K(t)$. Then

$$\|(TP)^n x\|^{\frac{1}{n}} = \|T^n P^n x\|^{\frac{1}{n}} = \|T^n P x\|^{\frac{1}{n}} = \|T^n x_2\|^{\frac{1}{n}} \to 0,$$

as $n \to \infty$. Then $H_0(TP) = X$ and hence TP is quasi-nilpotent, by Theorem 2.35. Clearly, the restriction $(T + P)|K(T) = T|K(T)$ is invertible, and also $(T + P)|H_0(T)$ is invertible, since $T|H_0(T)$ is quasi nilpotent and

$$(T + P)|H_0(T) = T|H_0(T) + I_{H_0(T)},$$

where $I_{H_0(T)}$ denotes the identity on $H_0(T)$. Thus, $T + P = (T+P)|H_0(T) \oplus (T+P)|K(T)$ is invertible.

(iv) $\Rightarrow$ (iii) Since $X = \ker P \oplus P(X)$ and $TP = PT$, $\ker P$ and $P(X)$ are T-invariant. The restriction $T|\ker P = (T + P)|\ker P$ is invertible, since $T + P$ is invertible by assumption. Suppose now that $x \in P(X)$. Then

$$\|(T^n|P(X))x\|^{\frac{1}{n}} = \|T^n P^n x\|^{\frac{1}{n}} = \|(TP)^n x\|^{\frac{1}{n}} \to 0,$$

as $n \to \infty$. This means that $H_0(T|P(X)) = P(X)$ and hence $T|P(X)$ is quasi-nilpotent, by Theorem 2.35.

The last assertion is clear.   ■

If λ_0 is a pole of the resolvent we can say much more:

Corollary 2.47 *Let $T \in L(X)$ and suppose that λ_0 is a pole of the resolvent of T, or, equivalently, $p := p(\lambda I - T) = q(\lambda I - T) < \infty$. Then*

$$H_0(\lambda_0 I - T) = \ker(\lambda_0 I - T)^p,$$

and

$$K(\lambda_0 I - T) = (\lambda_0 I - T)^p(X).$$

Proof Combine Theorem 1.36 of Appendix A with Theorem 2.45.   ■

In the next result we consider isolated points of the spectrum.

Theorem 2.48 *Let $T \in L(X)$ and suppose that λ_0 is an isolated point of $\sigma(T)$. If $\lambda \neq \lambda_0$ then*

$$\{0\} \neq H_0(\lambda_0 I - T) \subseteq K(\lambda I - T).$$

Proof If P_0 denotes the spectral projection associated with $\{\lambda_0\}$ then every $\lambda \neq \lambda_0$ does not belong to the spectrum of $T|P_0(X)$, so $\lambda I - T|P_0(X)$ is invertible, and hence $(\lambda I - T)(P_0(X) = P_0(X)$. Since $P_0(X)$ is closed it then follows, by Theorem 1.39, that $P_0(X) \subseteq K(\lambda I - T)$, while, by Theorem 2.45, we have $\{0\} \neq P_0(X) = H_0(\lambda_0 I - T)$. ∎

Corollary 2.49 *If $T \in L(X)$ and $\lambda_0 \neq 0$ is an isolated point of $\sigma(T)$ then $T(H_0(\lambda_0 I - T)) = H_0(\lambda_0 I - T)$.*

Corollary 2.50 *If $T \in L(X)$ and $\lambda_0 \in \sigma(T)$ satisfies $K(\lambda_0 I - T) = \{0\}$ then λ_0 is the only possible isolated point of $\sigma(T)$.*

Proof Suppose that $\sigma(T)$ has an isolated point $\lambda \neq \lambda_0$. By Theorem 2.48 we have $\{0\} \neq H_0(\lambda I - T) \subseteq K(\lambda_0 I - T)$, a contradiction. ∎

Recall that an operator $R \in L(X)$ is said to be a *Riesz operator* if $\lambda I - R$ is a Fredholm operator for every $\lambda \in \mathbb{C} \setminus \{0\}$. Denote by $\mathcal{R}(X)$ the class of all Riesz operators. The spectrum $\sigma(R)$ of a Riesz operator is at most countable and has no nonzero cluster point. Furthermore, each nonzero element of the spectrum is an eigenvalue and the spectral projection associated with every $\lambda \neq 0$ is finite-dimensional (see the next Chap. 3 for more information on Riesz operators). Examples of Riesz operators are quasi-nilpotent operators and compact operators, see Heuser [179]. An example of a quasi-nilpotent operator which is neither nilpotent nor compact is the operator $T := T_1 \oplus T_2$, defined in $\ell^2(\mathbb{N}) \oplus \ell^2(\mathbb{N})$, where

$$T_1(x_1, x_2, \ldots) := (0, x_1, 0, , x_3 \ldots) \quad \text{for all } x = (x_n) \in \ell^2(\mathbb{N}),$$

and

$$T_2(x_1, x_2, \ldots) := \left(0, x_1, \frac{x_2}{2}, \frac{x_3}{3}, \ldots\right) \quad \text{for all } x = (x_n) \in \ell^2(\mathbb{N}).$$

In Chap. 3 we shall see that the restriction of a Riesz operator to a closed invariant subspace is still a Riesz operator. The class of Riesz operators is very large, for instance it contains, properly, the two-sided ideal of all *strictly singular operators* $\mathcal{S}(X)$, introduced by Kato [195] and defined as the operators $T \in L(X)$ such that no restriction T_M to an infinite-dimensional closed subspace M of X is an isomorphism. The class $\mathcal{R}(X)$ also contains the two-sided ideal $\mathcal{C}(X)$ of all *strictly cosingular operators* introduced by Pełczyński [256] as the class of all operators for which there is no infinite-codimensional closed subspace N of X such that $Q_N T$ is surjective, where Q_N denotes the canonical quotient homomorphism of X onto

$X|N$. More generally, if we consider the two-sided ideals of $\Phi_+(X)$-perturbations and $\Phi_-(X)$-perturbations, defined by

$$P\Phi + (X) := \{T \in L(X) : T + \Phi_+(X) \subseteq \Phi_+(X)\},$$

and

$$P\Phi_-(X) := \{T \in L(X) : T + \Phi_-(X) \subseteq \Phi_-(X)\}$$

respectively, see [155], these ideals are contained in the two-sided ideal

$$\mathcal{I}(X) := \{I - TS \in \Phi(X) \text{ for all } S \in L(X)\}$$

known in the literature as the ideal of *inessential operators*, introduced by Klenecke. We have $\mathcal{I}(X) \subseteq \mathcal{R}(X)$, more precisely, $\mathcal{I}(X)$ is the uniquely determined largest ideal consisting of Riesz operators (for details, see Chapter 7 of [1]).

In the next results we characterize the case where 0 is an isolated point of $\sigma(R)$ with the help of the analytic core and the quasi-nilpotent part.

Theorem 2.51 *Let R be a Riesz operator on an infinite-dimensional Banach space. Then the following statements are equivalent.*

(i) *0 is an isolated point of $\sigma(R)$;*
(ii) *$K(R)$ is closed;*
(iii) *$K(R)$ has finite dimension;*
(iv) *$K(R^*)$ is closed;*
(v) *$K(R^*)$ has finite dimension.*

Proof Since $\sigma(R) = \sigma(R^*)$ and R^* is also a Riesz operator (this will be proved in Chap. 3) it suffices to prove only the equivalence of (i), (ii) and (iii). The implication (i) $\Rightarrow$ (ii) is clear by Theorem 2.45. To show the implication (ii) $\Rightarrow$ (iii), assume that $K(R)$ is closed. Then the restriction $R|K(R)$ is a Riesz operator and since $R(K(R)) = K(R)$ we also have that $R|K(R)$ is onto, in particular lower semi-Fredholm. Therefore, $(\lambda I - R)|K(R)$ is lower semi-Fredholm for all $\lambda \in \mathbb{C}$. This implies that $K(R)$ is finite-dimensional. To prove (iii) $\Rightarrow$ (i) observe first that if $K(R)$ is finite-dimensional then the surjective operator $R|K(R)$ is invertible, so there exists a $\delta > 0$ such that $(\lambda I - R)|K(R)$ is invertible for all $|\lambda| < \delta$. Since for $\lambda \neq 0$ we have $\ker(\lambda I - R) \subseteq K(R)$, we have

$$\ker(\lambda I - R) = \ker((\lambda I - R)|K(R)) = \{0\} \quad \text{for all } 0 < |\lambda| < \delta.$$

But since the index of $(\lambda I - R)$ is 0 for every $0 < |\lambda| < \delta$ it then follows that $\beta(\lambda I - R) = 0$, i.e., $(\lambda I - R)$ is onto and hence invertible for all $0 < |\lambda| < \delta$. Therefore, 0 is an isolated point of $\sigma(R)$. ∎

We have seen before that $T \in L(X)$ is quasi-nilpotent if and only if $H_0(T) = X$. A quasi-nilpotent operator may also be characterized in terms of the analytic core:

Corollary 2.52 *Let $R \in L(X)$ be a Riesz operator. Then R is quasi-nilpotent if and only if $K(R) = \{0\}$. The spectrum $\sigma(R)$ is a finite set which contains 0 precisely when $K(R)$ is a closed set and $K(R) \neq 0$.*

Proof If R is quasi-nilpotent then, by Theorem 2.35, $H_0(R) = X$ and since 0 is an isolated point of $\sigma(T)$ we have, by Theorem 2.45, $\{0\} = H_0(R) \cap K(R) = K(R)$. On the other hand, if $K(R) = \{0\}$ then, by Corollary 2.50, 0 is the only isolated point of $\sigma(R)$, hence $\sigma(R) = \{0\}$, since R is Riesz. The second assertion easily follows from Theorem 2.51. ∎

Theorem 2.53 *Suppose that $R \in L(X)$ is a Riesz operator on an infinite-dimensional Banach space X. If for some $\lambda_0 \neq 0$ we have $H_0(\lambda_0 I - R) + H_0(R) = X$ then 0 is an isolated point of $\sigma(R)$.*

Proof If $\lambda_0 \notin \sigma(R)$ then $H_0(\lambda_0 I - R) = \{0\}$, so $H_0(R) = X$ and hence $\sigma(R) = \{0\}$, by Theorem 2.35. Suppose that $\lambda_0 \in \sigma(R)$. Then λ_0 is a pole of the resolvent, so, by Corollary 2.47, we have $K(\lambda_0 I - R) = (\lambda_0 I - R)^p(X)$ and $H_0(\lambda_0 I - R) = \ker(\lambda_0 I - R)^p$ for some $p \in \mathbb{N}$. Since $\alpha(\lambda_0 I - R) < \infty$, $H_0(\lambda_0 I - R) = \ker(\lambda_0 I - R)^p$ is finite-dimensional. Observe that $0 \in \sigma(R)$, otherwise $H_0(R) = \{0\}$ and from the assumption we would have $X = H_0(\lambda_0 I - R)$ finite-dimensional. Now, by Corollary 2.49, we have $R(H_0(\lambda_0 I - R)) = H_0(\lambda_0 I - R)$, so the restriction of R to $H_0(\lambda_0 I - R)$ is onto and hence invertible. This implies that there exists a $\delta > 0$ such that

$$(\lambda I - R)(H_0(\lambda_0 I - R) = H_0(\lambda_0 I - R) \quad \text{for all } |\lambda| < \delta.$$

Since for $\lambda \neq 0$ we have $H_0(R) \subseteq (\lambda I - R)$, see Lemma 2.34, we then obtain

$$X = H_0(\lambda_0 I - R) + H_0(R) \subseteq (\lambda I - R)(X) \quad \text{for all } 0 < |\lambda| < \delta.$$

Hence $X = (\lambda I - R)(X)$, i.e. $\lambda I - R$ is onto. Since R is a Riesz operator we then conclude that $\lambda I - R$ is injective, for all $0 < |\lambda| < \delta$, i.e., 0 is isolated in $\sigma(R)$. ∎

Corollary 2.54 *Let $R \in L(X)$ be a Riesz operator. If there exists a $\lambda_0 \neq 0$ for which $H_0(R) = K(\lambda_0 I - R)$ then 0 is an isolated point of $\sigma(R)$.*

Proof Since $\lambda_0 \neq 0$, $\lambda_0 I - R$ is either invertible (in this case $K(\lambda_0 I - R) = X$) or a pole of the resolvent. In both cases $X = H_0(\lambda I - R) \oplus K(\lambda_0 I - T)$. Hence $X = H_0(\lambda I - R) \oplus H_0(R)$, so, by Theorem 2.53, 0 is an isolated point of $\sigma(R)$. ∎

We now consider the isolated points of $\sigma_{\mathrm{ap}}(T)$.

Theorem 2.55 *Suppose that $T \in L(X)$ and $\lambda \in \mathbb{C}$ is an isolated point of $\sigma_{\mathrm{ap}}(T)$. Then*

(i) *Both $H_0(\lambda I - T)$ and $K(\lambda I - T)$ are closed subspaces.*
(ii) *$H_0(\lambda I - T) \cap K(\lambda I - T) = \{0\}$.*

(iii) *The direct sum $H_0(\lambda I - T) \oplus K(\lambda I - T)$ is closed and there exists a $\lambda_0 \neq 0$
such that*

$$H_0(\lambda I - T) \oplus K(\lambda I - T) = K(\lambda_0 I - T) = \bigcup_{n=0}^{\infty} T(\lambda_0 I - T)^n(X).$$

Proof We may assume $\lambda = 0$. Since 0 is an isolated point of $\sigma_{\mathrm{ap}}(T)$, there exists a
$\delta > 0$ such that $\lambda I - T$ is bounded below for all $0 < |\lambda| < \delta$. By Theorem 1.50,
the map $\lambda \to K(\lambda I - T)$ is constant on the punctured disc $\mathbb{D}(0, \delta) \setminus \{0\}$, and fixing
$\lambda_0 \in \mathbb{D}(0, \delta) \setminus \{0\}$ we have, by Theorem 1.44, that $K(\lambda_0 I - T) = (\lambda_0 I - T)^\infty(X)$.
Set $X_0 := (\lambda_0 I - T)^\infty(X)$, and denote by $T_0 : X_0 \to X_0$ the operator induced by
T on X_0. X_0 is a Banach space, by Theorem 1.44. Clearly, $\lambda I - T_0$ is bijective for
all $\lambda \in \mathbb{D}(0, \delta) \setminus \{0\}$. Since T is not surjective, $K(T) \neq X$, hence $H_0(T) \neq 0$ by
Theorem 2.41. From Theorem 2.23, part (ii), we know that

$$(\lambda I - T)(H_0(T)) = (\lambda I - T)(\mathcal{X}_T(\{0\})) = \mathcal{X}_T(\{0\} = H_0(T) \quad \text{for all } \lambda \neq 0,$$

from which we deduce that

$$(\lambda I - T)^n(H_0(T)) = H_0(T) \subseteq (\lambda I - T)^n(X) \quad \text{for all } n \in \mathbb{N},$$

so that $H_0(T) \subseteq X_0$, hence $0 \in \sigma(T_0)$, and since $\lambda I - T_0$ is bijective for all $\lambda \in
\mathbb{D}(0, \delta) \setminus \{0\}$, we then have that 0 is an isolated point of $\sigma(T_0)$. By Theorem 2.45
we then have $X_0 = H_0(T_0) \oplus K(T_0)$. Clearly, $H_0(T) = H_0(T_0)$, hence, to finish
the proof it suffices to prove $K(T) = K(T_0)$. Let $x_0 \in K(T)$, $T x_{n+1} = x_n$ and
$\|x_n\| \leq c^n \|x_0\|$ for all n. Then

$$\phi(\lambda) := \sum_{n=0}^{\infty} x_{n+1} \lambda^n$$

defines an analytic function on the open disc $\mathbb{D}(0, \frac{1}{c})$ that satisfies the equality

$$(\lambda I - T)(\phi(\lambda) = \sum_{n=0}^{\infty} x_n \lambda^n - \sum_{n=0}^{\infty} x_{n+1} \lambda^{n+1} \quad \text{for all } \lambda \in \mathbb{D}(0, \frac{1}{c}).$$

In particular, $x_0 \in (\lambda I - T)(X)$ for all $\lambda \in \mathbb{D}(0, \frac{1}{c})$. Therefore, by Theorem 1.52,
$x_0 \in K(\lambda_0 I - T)$, and hence $K(T) \subseteq X_0$. Note that

$$\phi(\lambda) = (\lambda I - T_0)^{-1} x_0 \in X_0 \quad \text{for all } \lambda \in \mathbb{D}(0, \frac{1}{c}).$$

By continuity, $x_1 = \phi(0) \in X_0$. A similar argument shows that $x_n \in X_0$ for $n \geq 1$,
thus $x_0 \in K(T_0)$, from which we conclude that $K(T) = K(T_0)$. $\blacksquare$

Corollary 2.56 *Let $T \in L(X)$ and suppose that* dim $K(T) < \infty$. *Then T has the SVEP. Moreover, if M is a closed invariant subspace for which $T(M) = M$ then M is finite-dimensional.*

Proof By Theorem 1.38 we know that for each $\lambda \neq 0$ we have ker $(\lambda I - T) \subseteq K(T)$, hence ker $(\lambda I - T)$ is finite-dimensional. Moreover, a set of eigenvectors, each of them corresponding to a different eigenvalue of T, is linearly independent, so our assumption dim $K(T) < \infty$ implies that the point spectrum $\sigma_{\mathrm{p}}(T)$ is finite, and consequently T has the SVEP. The second assertion is an obvious consequence of the inclusion $M \subseteq K(T)$, established in Theorem 1.39.   ∎

The condition dim $K(T) < \infty$ is clearly satisfied if $T^{\infty}(X) = \{0\}$, since $K(T) \subseteq T^{\infty}(X)$. The condition $T^{\infty}(X) = \{0\}$ may be thought of as an abstract shift condition since it is satisfied by every unilateral weighted right shift, see Chap. 4.

We show now that the subspaces $\overline{H_0(\lambda I - T)}$ are constant as λ ranges through a connected component of the semi-regular resolvent.

Theorem 2.57 *Let $T \in L(X)$ and let $\Omega \subset \mathbb{C}$ be a connected component of $\rho_{\mathrm{se}}(T)$. If $\lambda_0 \in \Omega$ then*

$$\overline{H_0(\lambda I - T)} = \overline{H_0(\lambda_0 I - T)} \quad \text{for all } \lambda \in \Omega.$$

Proof By Theorem 1.43 we know that $\rho_{\mathrm{se}}(T) = \rho_{\mathrm{se}}(T^*)$. Further, Theorem 1.50 shows that $K(\lambda I^* - T^*) = K(\lambda_0 I^* - T^*)$ for all $\lambda \in \Omega$. From Theorem 2.37 we then conclude that

$$\overline{H_0(\lambda I - T)} = {}^{\perp}K(\lambda I^* - T^*) = {}^{\perp}K(\lambda_0 I^* - T^*) = \overline{H_0(\lambda_0 I - T)},$$

for all $\lambda \in \Omega$.   ∎

In the sequel by ∂K we denote the boundary of $K \subseteq \mathbb{C}$.

Theorem 2.58 *Let $T \in L(X)$, $X \neq \{0\}$ a Banach space. Then the semi-regular spectrum $\sigma_{\mathrm{se}}(T)$ is a non-empty compact subset of $\mathbb{C}$ containing $\partial \sigma(T)$. In particular, $\partial \sigma(T)$ is contained in $\sigma_{\mathrm{ap}}(T) \cap \sigma_{\mathrm{s}}(T)$.*

Proof Let $\lambda_0 \in \partial \sigma(T)$ and suppose $\lambda_0 \in \rho_{\mathrm{se}}(T) := \mathbb{C} \setminus \sigma_{\mathrm{se}}(T)$. The set $\rho_{\mathrm{se}}(T)$ is open, by Theorem 1.44, so we can consider a connected component Ω of $\rho_{\mathrm{se}}(T)$ containing λ_0. The set Ω is open so there exists a neighborhood $\mathcal{U}$ of λ_0 contained in Ω, and since $\lambda_0 \in \partial \sigma(T)$, $\mathcal{U}$ also contains points of $\rho(T)$. Hence $\Omega \cap \rho(T) \neq \emptyset$.

Now, consider a point $\lambda_1 \in \Omega \cap \rho(T)$. Clearly, ker $(\lambda_1 I - T)^n = \{0\}$ for every $n \in \mathbb{N}$, thus $\mathcal{N}^{\infty}(\lambda_1 I - T) = \{0\}$. Combining Theorems 2.57 and 2.37 we then have

$$\overline{H_0(\lambda_0 I - T)} = \overline{H_0(\lambda_1 I - T)} = \mathcal{N}^{\infty}(\lambda_1 I - T) = \{0\}.$$

Hence $\ker(\lambda_0 I - T) = \{0\}$, so $\lambda_0 I - T$ is injective. On the other hand, $\lambda_1 \in \rho(T)$ and hence from Theorem 1.50 we infer that

$$K(\lambda_0 I - T) = K(\lambda_1 I - T) = X,$$

so $\lambda_0 I - T$ is surjective. Hence $\lambda_0 \in \rho(T)$ and this is a contradiction, since $\lambda_0 \in \sigma(T)$. Therefore $\lambda_0 \in \sigma_{se}(T)$ and $\partial \sigma(T) \subseteq \sigma_{se}(T)$, so the last set is a compact non-empty subset of $\mathbb{C}$. The last assertion is clear, since $\sigma_{se}(T) \subseteq \sigma_{ap}(T) \cap \sigma_s(T)$.   ∎

By Theorem 2.37, if $\lambda I - T$ is semi-regular then $\overline{\mathcal{N}^\infty(\lambda I - T)} = \overline{H_0(\lambda I - T)}$, hence the statement of Theorem 2.57 is equivalent to saying that $\overline{\mathcal{N}^\infty(\lambda I - T)}$ is constant as λ ranges through a connected component of the semi-regular resolvent.

For an operator $T \in L(X)$ we have $\mathcal{N}^\infty(T) \subseteq H_0(T)$ and $T^\infty(X) \supseteq K(T)$. In the special situation of operators of Kato-type, there is equality in the sum of these subspaces:

Theorem 2.59 *Suppose that $T \in L(X)$ is of Kato-type. Then:*

(i) $\mathcal{N}^\infty(T) + T^\infty(X) = H_0(T) + K(T)$;
(ii) $\overline{\mathcal{N}^\infty(T) \cap T^\infty(X)} = \overline{H_0(T) \cap K(T)}$.

Proof

(i) Let (M, N) be a GKD for T such that $(T|N)^d = 0$ for some integer $d \in \mathbb{N}$. We know, from part (i) of Theorem 1.63, that $K(T) = K(T|M) = K(T) \cap M$. Moreover, by part (iii) of Theorem 2.37, the semi-regularity of $T|M$ entails that $H_0(T|M) \subseteq K(T|M) = K(T)$. Consequently,

$$H_0(T) \cap K(T) = H_0(T) \cap (K(T) \cap M) = (H_0(T) \cap M) \cap K(T)$$

$$= H_0(T|M) \cap K(T) = H_0(T|M).$$

This shows that $H_0(T) \cap K(T) = H_0(T|M)$.
   We claim that

$$H_0(T) + K(T) = N \oplus K(T).$$

From $N \subseteq \ker T^d \subseteq H_0(T)$ we obtain that $N \oplus K(T) \subseteq H_0(T) + K(T)$. Conversely, from Lemma 2.36 we have

$$H_0(T) = N \oplus H_0(T|M) = N \oplus (H_0(T) \cap K(T)) \subseteq N \oplus K(T),$$

and hence

$$H_0(T) + K(T) \subseteq (N \oplus K(T)) + K(T) \subseteq N \oplus K(T),$$

so our claim is proved.

From the inclusion $N \subseteq \ker T^d \subseteq \mathcal{N}^\infty(T)$, and, since the equality $K(T) = T^\infty(X)$ holds for every operator of Kato-type, we then deduce that

$$H_0(T) + K(T) = N \oplus K(T) \subseteq \mathcal{N}^\infty(T) + T^\infty(X) \subseteq H_0(T) + K(T).$$

Hence the equality $\mathcal{N}^\infty(T) + T^\infty(X) = H_0(T) + K(T)$ is proved.

(ii) Let (M, N) be a GKD for T such that for some $d \in \mathbb{N}$ we have $(T|N)^d = 0$. Then $\ker T^n = \ker(T|M)^n$ for every natural $n \geq d$. Since $\ker T^n \subseteq \ker T^{n+1}$ for all $n \in \mathbb{N}$ we then have

$$\mathcal{N}^\infty(T) = \bigcup_{n \geq d}^\infty \ker T^n = \bigcup_{n \geq d}^\infty \ker(T|M)^n = \mathcal{N}^\infty(T|M).$$

From part (ii) of Theorem 2.37 the semi-regularity of $T|M$ entails that

$$\overline{\mathcal{N}^\infty(T)} = \overline{\mathcal{N}^\infty(T|M)} = \overline{H_0(T|M)} = \overline{H_0(T) \cap M}. \tag{2.17}$$

We show now the equality $\overline{H_0(T) \cap M} = \overline{H_0(T)} \cap M$. Clearly we have $\overline{H_0(T) \cap M} \subseteq \overline{H_0(T)} \cap M$. Conversely, suppose that $x \in \overline{H_0(T)} \cap M$. Then there is a sequence $(x_n) \subset H_0(T)$ such that $x_n \to x$ as $n \to \infty$. Let P denote the projection of X onto M along N. Then $Px_n \to Px = x$ and $Px_n \in H_0(T) \cap P(H_0(T))$. From Lemma 2.36 we know that $H_0(T) = (H_0(T) \cap M) \oplus N$, so

$$P(H_0(T)) = P(H_0(T) \cap M) = H_0(T) \cap M,$$

and hence $Px_n \in H_0(T) \cap M$, from which we deduce that $x \in \overline{H_0(T) \cap M}$. Consequently, $\overline{H_0(T) \cap M} = \overline{H_0(T)} \cap M$. Finally, from equality (2.17) and taking into account that, by Theorems 1.63 and 1.64, we have $T^\infty(X) = K(T) \subseteq M$, we conclude that

$$\overline{\mathcal{N}^\infty(T)} \cap T^\infty(X) = (\overline{H_0(T)} \cap M) \cap K(T) = (\overline{H_0(T)} \cap M) \cap K(T)$$
$$= \overline{H_0(T)} \cap (M \cap K(T)) = \overline{H_0(T)} \cap K(T),$$

so the proof is complete.   ∎

2.4 The Localized SVEP

We have seen in Theorem 2.14 that the SVEP for T holds precisely when for every element $0 \neq x \in X$ we have $\sigma_T(x) = \emptyset$. The next fundamental theorem, which establishes a localized version of this result, will be useful in the sequel.

Theorem 2.60 *If $T \in L(X)$ the following statements are equivalent:*

(i)   *T has the SVEP at λ_0;*
(ii) $\ker(\lambda_0 I - T) \cap X_T(\emptyset) = \{0\}$;
(iii) $\ker(\lambda_0 I - T) \cap K(\lambda_0 I - T) = \{0\}$;
(iv) *For each $0 \neq x \in \ker(\lambda_0 I - T)$ we have $\sigma_T(x) = \{\lambda_0\}$.*

Proof By replacing T with $\lambda_0 I - T$ we may assume without loss of generality that $\lambda_0 = 0$.

(i) $\Leftrightarrow$ (ii) Suppose that for $x \in \ker T$ we have $\sigma_T(x) = \emptyset$. Then $0 \in \rho_T(x)$, so there is an open disc $\mathbb{D}(0, \varepsilon)$ and an analytic function $f : \mathbb{D}(0, \varepsilon) \to X$ such that $(\lambda I - T)f(\lambda) = x$ for every $\lambda \in \mathbb{D}(0, \varepsilon)$. Then

$$T((\lambda I - T)f(\lambda)) = (\lambda I - T)T(f(\lambda)) = Tx = 0$$

for every $\lambda \in \mathbb{D}(0, \varepsilon)$. Since T has the SVEP at 0, $Tf(\lambda) = 0$, and consequently

$$T(f(0)) = x = 0.$$

Conversely, suppose that for every $0 \neq x \in \ker T$ we have $\sigma_T(x) \neq \emptyset$ and consider an analytic function $f : \mathbb{D}(0, \varepsilon) \to X$ for which $(\lambda I - T)f(\lambda) = 0$ holds for every $\lambda \in \mathbb{D}(0, \varepsilon)$. We can represent the function f as

$$f(\lambda) = \sum_{n=0}^{\infty} \lambda^n u_n,$$

for a suitable sequence $(u_n) \subset X$. Evidently, $Tu_0 = T(f(0)) = 0$, from which we obtain $u_0 \in \ker T$. Furthermore, the equalities

$$\sigma_T(f(\lambda)) = \sigma_T(0) = \emptyset \text{ for every } \lambda \in \mathbb{D}(0, \varepsilon)$$

imply that

$$\sigma_T(f(0)) = \sigma_T(u_0) = \emptyset,$$

and hence, $u_0 = 0$. For all $0 \neq \lambda \in \mathbb{D}(0, \varepsilon)$ we have

$$0 = (\lambda I - T)f(\lambda) = (\lambda I - T)\sum_{n=1}^{\infty}\lambda^n u_n = \lambda(\lambda I - T)\sum_{n=1}^{\infty}\lambda^n u_{n+1},$$

and hence

$$0 = (\lambda I - T)(\sum_{n=0}^{\infty}\lambda^n u_{n+1}) \quad \text{for every } 0 \neq \lambda \in \mathbb{D}(0, \varepsilon).$$

By continuity this is still true for every $\lambda \in \mathbb{D}(0, \varepsilon)$. At this point, by repeating the same argument as in the first part of the proof, it is possible to show that $u_1 = 0$, and by iterating this procedure we can easily conclude that $u_2 = u_3 = \cdots = 0$. This shows that $f \equiv 0$ on $\mathbb{D}(0, \varepsilon)$, and therefore T has the SVEP at 0.

(ii) $\Leftrightarrow$ (iii) It suffices to prove the equality

$$\ker\ T \cap K(T) = \ker\ T \cap X_T(\emptyset).$$

By Theorem 2.30 we have $\ker T \subseteq H_0(T) \subseteq X_T(\{0\})$, and hence, from Theorem 2.20, we obtain

$$\ker\ T \cap K(T) = \ker\ T \cap X_T(\mathbb{C} \setminus \{0\}) \subseteq X_T(\{0\}) \cap X_T(\mathbb{C} \setminus \{0\}) = X_T(\emptyset).$$

Since $X_T(\emptyset) \subseteq X_T(\mathbb{C} \setminus \{0\}) = K(T)$, we then conclude that

$$\ker\ T \cap K(T) = \ker\ T \cap K(T) \cap X_T(\emptyset) = \ker\ T \cap X_T(\emptyset).$$

(ii) $\Rightarrow$ (iv) Since $\ker\ T \subseteq H_0(T)$, from Theorem 2.30 it then follows that $\sigma_T(x) \subseteq \{0\}$ for every $0 \neq x \in \ker\ T$. But, by assumption $\sigma_T(x) \neq \emptyset$, so $\sigma_T(x) = \{0\}$.

(iv) $\Rightarrow$ (ii) Obvious. ∎

For an arbitrary operator $T \in L(X)$ on a Banach space X let

$$\Xi(T) := \{\lambda \in \mathbb{C} : T \text{ fails to have the SVEP at } \lambda\}.$$

Clearly, $\Xi(T)$ is contained in the interior of the spectrum $\sigma(T)$, and, from the identity theorem for analytic functions it readily follows that $\Xi(T)$ is open. This implies that if T has the SVEP at all $\lambda \in \mathbb{D}(\lambda_0, \varepsilon) \setminus \{\lambda_0\}$, where $\mathbb{D}(\lambda_0, \varepsilon)$ is an open disc centered at λ_0, then T also has the SVEP at λ_0. Clearly, $\Xi(T)$ is empty precisely when T has the SVEP.

Corollary 2.61 *If $T \in L(X)$ is surjective, then T has the SVEP at 0 if and only if T is injective. Consequently, the equality $\sigma(T) = \sigma_s(T) \cup \Xi(T)$ holds for every $T \in L(X)$. Furthermore, $\sigma_s(T)$ contains $\partial \Xi(T)$, the topological boundary of $\Xi(T)$.*

Proof If T is onto and has the SVEP at 0 then $K(T) = X$. By Theorem 2.60 we have $\ker\ T \cap X = \ker\ T = \{0\}$, hence T is injective. The converse is clear. To show the equality $\sigma(T) = \sigma_s(T) \cap \Xi(T)$ we have only to show the inclusion $\subseteq$. Suppose that $\lambda \notin \sigma_s(T) \cap \Xi(T)$. From the first part we obviously have $\lambda \notin \sigma(T)$ from which we obtain $\sigma(T) = \sigma_s(T) \cup \Xi(T)$. The last claim is immediate: since $\partial \Xi(T) \subseteq \sigma(T)$ and $\Xi(T)$ is open it then follows that $\partial \Xi(T) \cap \Xi(T) = \emptyset$. This obviously implies that $\partial \Xi(T) \subseteq \sigma_s(T)$. ∎

An immediate consequence of Corollary 2.61 is that every unilateral left shift on the Hilbert space $\ell_2(\mathbb{N})$ fails to have the SVEP at 0.

Remark 2.62 Evidently if Y is a closed subspace of the Banach space X such that $(\lambda_0 I - T)(Y) = Y$ and the restriction $(\lambda_0 I - T) |Y$ does not have the SVEP at λ_0 then T also does not have the same property at λ_0.

This property, together with Corollary 2.61, suggests how to obtain operators without the SVEP: if for an operator $T \in L(X)$ there exists a closed subspace Y such that

$$(\lambda_0 I - T)(Y) = Y \quad \text{and} \quad \ker (\lambda_0 I - T) \cap Y \neq \{0\}$$

then T does not have the SVEP at λ_0.

Theorem 2.63 *Let X and Y be Banach spaces and consider two operators $S \in L(X, Y)$ and $R \in L(Y, X)$. Suppose that either of the following cases hold:*

(i) *R and S are both injective.*
(ii) *R or S is injective with dense range.*
        Then $\sigma(RS) = \sigma(SR)$.

Proof

(i) Assume that S and R are injective. By Theorems 2.21 and 2.4 we then have

$$\sigma(RS) = \bigcup_{x \in X} \sigma_{RS}(x) \cup \Xi(RS) = \bigcup_{x \in X} \sigma_{SR}(Sx) \cup \Xi(SR)$$

$$\subseteq \bigcup_{y \in Y} \sigma_{SR}(y) \cup \Xi(SR) = \sigma(SR),$$

thus $\sigma(RS) \subseteq \sigma(SR)$. By symmetry then $\sigma(RS) = \sigma(SR)$.
(ii) The statement follows by duality.   ∎

Remark 2.64 Let L denote the unilateral left shift on the Hilbert space $\ell_2(\mathbb{N})$, defined as

$$L(x_1, x_2, x_3, \cdots) := (x_2, x_3, \cdots) \quad \text{for all } x = (x_n) \in \ell_2(\mathbb{N}).$$

Evidently, L is onto but not injective, since every vector $(x_1, 0, 0, \cdots)$, with $x_1 \neq 0$, belongs to ker L. Corollary 2.61 then shows that L fails to have the SVEP at 0. Later, we shall see that other examples of operators which do not have the SVEP at 0 are semi-Fredholm operators on a Banach space having index strictly greater than 0.

Theorem 2.65 *For a bounded operator T on a Banach space X and $\lambda_0 \in \mathbb{C}$, the following implications hold:*

$$p(\lambda_0 I - T) < \infty \Rightarrow N^\infty(\lambda_0 I - T) \cap (\lambda_0 I - T)^\infty(X) = \{0\}$$

$$\Rightarrow T \text{ has the SVEP at } \lambda_0,$$

and

$$q(\lambda_0 I - T) < \infty \Rightarrow X = \mathcal{N}^\infty(\lambda_0 I - T) + (\lambda_0 I - T)^\infty(X)$$
$$\Rightarrow T^\star \text{ has the SVEP at } \lambda_0.$$

Proof There is no loss of generality in assuming $\lambda_0 = 0$.

Assume that $p := p(T) < \infty$. Then $\mathcal{N}^\infty(T) = \ker T^p$, and therefore from Lemma 1.19 we obtain that

$$\mathcal{N}^\infty(T) \cap T^\infty(X) \subseteq \ker T^p \cap T^p(X) = \{0\}.$$

From Theorem 2.60 we then conclude that T has the SVEP at 0.

To show the second chain of implications suppose that $q := q(T) < \infty$. Then $T^\infty(X) = T^q(X)$ and

$$\mathcal{N}^\infty(T) + T^\infty(X) = \mathcal{N}^\infty(T) + T^q(X) \supseteq \ker T^q + T^q(X). \tag{2.18}$$

Now, the condition $q = q(T) < \infty$ yields that $T^{2q}(X) = T^q(X)$, so for every element $x \in X$ there exists a $y \in T^q(X)$ such that $T^q y = T^q(x)$. Obviously $x - y \in \ker T^q$, and therefore $X = \ker T^q + T^q(X)$. From the inclusion (2.18) we conclude that $X = \mathcal{N}^\infty(T) + T^\infty(X)$, and therefore, by Theorem 2.65, $T^\star$ has the SVEP at 0. ∎

In the remaining part of this section we want show that the relative positions of all the subspaces introduced in the previous chapter are intimately related to the localized SVEP.

To see this, let us consider for an arbitrary $\lambda_0 \in \mathbb{C}$ and an operator $T \in L(X)$ the following increasing chain of kernel-type spaces:

$$\ker(\lambda_0 I - T) \subseteq \mathcal{N}^\infty(\lambda_0 I - T) \subseteq H_0(\lambda_0 I - T) \subseteq X_T(\{\lambda_0\}),$$

and the decreasing chain of the range-type spaces:

$$X_T(\emptyset) \subseteq X_T(\mathbb{C} \setminus \{\lambda_0\}) = K(\lambda_0 I - T) \subseteq (\lambda_0 I - T)^\infty(X) \subseteq (\lambda_0 I - T)(X).$$

The next corollary is an immediate consequence of Theorem 2.60 and the inclusions considered above.

Corollary 2.66 *Suppose that $T \in L(X)$ satisfies one of the following conditions:*

(i) $\mathcal{N}^\infty(\lambda_0 I - T) \cap (\lambda_0 I - T)^\infty(X) = \{0\}$;
(ii) $\mathcal{N}^\infty(\lambda_0 I - T) \cap K(\lambda_0 I - T) = \{0\}$;
(iii) $\mathcal{N}^\infty(\lambda_0 I - T) \cap X_T(\emptyset) = \{0\}$;
(iv) $H_0(\lambda_0 I - T) \cap K(\lambda_0 I - T) = \{0\}$;
(v) $\ker(\lambda_0 I - T) \cap (\lambda_0 I - T)(X) = \{0\}$.
  Then T has the SVEP at λ_0. ∎

Corollary 2.67 *For a bounded operator $T \in L(X)$, X a Banach space, the following implications hold:*

(i) $H_0(\lambda_0 I - T)$ *closed* $\Rightarrow H_0(\lambda_0 I - T) \cap K(\lambda_0 I - T) = \{0\} \Rightarrow T$ *has the SVEP at* λ_0.

(ii) $X = H(\lambda_0 I - T) + K(\lambda_0 I - T) \Rightarrow T^*$ *has the SVEP at* λ_0.

Proof Without loss of generality, we may consider $\lambda_0 = 0$. For the first implication in (i) see the proof of Theorem 2.39. The second implication of (i) has been proved in Corollary 2.66.

(ii) This follows from Theorem 2.41, since the condition $X = H(\lambda_0 I - T) + K(\lambda_0 I - T)$ is equivalent to saying that λ is isolated in $\sigma_s(T)$. ∎

The operator in the Example 2.33 shows that, in general, the converse of Theorem 2.39 does not hold. Indeed, T has the SVEP, since the point spectrum $\sigma_p(T)$ is empty, while $H_0(T)$ is not closed.

If T or T^* have the SVEP then some spectra coincide:

Theorem 2.68 *For $T \in L(X)$, the following statements hold:*

(i) *If T has the SVEP then $\sigma_s(T) = \sigma(T)$ and $\sigma_{se}(T) = \sigma_{ap}(T)$.*

(ii) *If T^* has the SVEP then $\sigma_{ap}(T) = \sigma(T)$ and $\sigma_{se}(T) = \sigma_s(T)$.*

(iii) *If both T and T^* have the SVEP then*

$$\sigma(T) = \sigma_s(T) = \sigma_{ap}(T) = \sigma_{se}(T).$$

Proof The first equality (i) is an obvious consequence of Corollary 2.61, since $\Xi(T)$ is empty. To prove the second equality of (i) observe first that the inclusion $\sigma_{se}(T) \subseteq \sigma_{ap}(T)$ is trivial, since every bounded below operator is semi-regular. Conversely, let $\lambda \notin \sigma_{se}(T)$. From the definition of semi-regularity and Theorem 1.33 we know

$$\ker(\lambda I - T) \subseteq (\lambda I - T)^\infty(X) = K(\lambda I - T).$$

But $\ker(\lambda I - T) \subseteq H_0(\lambda I - T)$, for all $\lambda \in \mathbb{C}$, so we have

$$\ker(\lambda I - T) \subseteq K(\lambda I - T) \cap H_0(\lambda I - T).$$

From Corollary 2.66 we then obtain $\ker(\lambda I - T) = \{0\}$, i.e. $\lambda I - T$ is injective. This implies, since $\lambda I - T$ has closed range by assumption, that $\lambda \notin \sigma_{ap}(T)$.

The two equalities of part (ii) are easily obtained by duality, while (iii) follows from part (i) and part (ii). ∎

Then converse of Corollary 2.66 need not be true. The next bilateral weighted shift provides an example of an operator T which has the SVEP at 0 while $H_0(T) \cap K(T) \neq \{0\}$.

Example 2.69 Let $\beta := (\beta_n)_{n \in \mathbb{Z}}$ be the sequence of real numbers defined as follows:

$$\beta_n := \begin{cases} 1 + |n| & \text{if } n < 0, \\ e^{-n^2} & \text{if } n \geq 0. \end{cases}$$

Let $X := L_2(\beta)$ denote the Hilbert space of all formal Laurent series

$$\sum_{n=-\infty}^{\infty} a_n z^n \quad \text{for which} \quad \sum_{n=-\infty}^{\infty} |\alpha_n|^2 \beta_n^2 < \infty.$$

Let us consider the *bilateral weighted right shift* defined by

$$T(\sum_{n=-\infty}^{\infty} a_n z^n) := \sum_{n=-\infty}^{\infty} a_n z^{n+1},$$

or equivalently, $T z^n := z^{n+1}$ for every $n \in \mathbb{Z}$. The operator T is bounded on $L_2(\beta)$ and

$$\|T\| = \sup \left\{ \frac{\beta_{n+1}}{\beta_n} : n \in \mathbb{Z} \right\} = 1.$$

Clearly T is injective, so it has the SVEP at 0. We show now that $H_0(T) \cap K(T) \neq \{0\}$. From $\|z^n\|_\beta = \beta_n$ for all $n \in \mathbb{Z}$ we obtain that

$$\lim_{n \to \infty} \|z^{n-1}\|_\beta^{1/n} = 0$$

and

$$\lim_{n \to \infty} \|z^{-n-1}\|_\beta^{1/n} = 1.$$

By the formula for the radius of convergence of a power series we then conclude that the two series

$$f(\lambda) := \sum_{n=1}^{\infty} \lambda^{-n} z^{n-1} \quad \text{and} \quad g(\lambda) := -\sum_{n=1}^{\infty} \lambda^n z^{-n-1}$$

converge in $L_2(\beta)$ for all $|\lambda| > 0$ and $|\lambda| < 1$, respectively. Evidently, the function f is analytic on $\mathbb{C} \setminus \{0\}$, and

$$(\lambda I - T) f(\lambda) = -\sum_{n=1}^{\infty} \lambda^{-n} z^n - \sum_{n=1}^{\infty} \lambda^{1-n} z^{n-1} = 1 \quad \text{for all } \lambda \neq 0,$$

while the function g, which is analytic on the open unit disc $\mathbb{D}(0, 1)$, satisfies

$$(\lambda I - T)g(\lambda) = \sum_{n=0}^{\infty} \lambda^n z^{-n} - \sum_{n=0}^{\infty} \lambda^{1+n} z^{-n-1} = 1 \quad \text{for all } \lambda \in \mathbb{D}(0, 1).$$

This implies that

$$1 \in \mathcal{X}_T(\{0\}) \cap \mathcal{X}_T(\mathbb{C} \setminus \mathbb{D}(0, 1)) = H_0(T) \cap K(T),$$

where the last equality follows from Theorems 2.20 and 2.30.

The SVEP may be characterized as follows.

Theorem 2.70 *A bounded operator $T \in L(X)$ has the SVEP if and only if $H_0(\lambda I - T) \cap K(\lambda I - T) = \{0\}$ for every $\lambda \in \mathbb{C}$.*

Proof If T has the SVEP, from Theorem 2.20 we know that

$$K(\lambda I - T) = X_{\lambda I - T}(\mathbb{C} \setminus \{0\}) = X_T(\mathbb{C} \setminus \{\lambda\}) \quad \text{for every } \lambda \in \mathbb{C},$$

and hence, by Theorem 2.30,

$$H_0(\lambda I - T) = X_{\lambda I - T}(\{0\}) = X_T(\{\lambda\}) \quad \text{for every } \lambda \in \mathbb{C}.$$

Consequently, by Theorem 2.14 we conclude that

$$H_0(\lambda I - T) \cap K(\lambda I - T) = X_T(\{\lambda\}) \cap X_T(\mathbb{C} \setminus \{\lambda\}) = X_T(\emptyset) = \{0\}.$$

The converse implication is clear by Corollary 2.66.   ∎

For a quasi-nilpotent operator we have $H_0(T) = X$. The analytic core $K(T)$ is "near" to being the complement of $H_0(T)$. Indeed we have:

Corollary 2.71 *If $T \in L(X)$ is quasi-nilpotent then $K(T) = \{0\}$.*

Proof By Theorem 2.35 we have $H_0(T) = X$. On the other hand, since T has the SVEP, from Theorem 2.70 we have $\{0\} = K(T) \cap H_0(T) = K(T)$.   ∎

The next result shows that a quasi-nilpotent operator $T \in L(X)$ cannot be essentially semi-regular if X is infinite-dimensional.

Theorem 2.72 *Let $T \in L(X)$ be essentially semi-regular and quasi-nilpotent. Then X is finite-dimensional and T is nilpotent. In particular, this holds for semi-Fredholm operators.*

Proof Let (M, N) be a GKD for T such that $T|N$ is nilpotent and N is finite-dimensional. Since T is quasi-nilpotent, Corollary 2.36 entails that $X = H_0(T) = $

$H_0(T|M) \oplus N$. Moreover, $T|M$ is semi-regular, hence by Theorems 2.37 and 1.63 $H_0(T|M) \subseteq K(T|M) = K(T)$. By Corollary 2.71 we know that $K(T) = \{0\}$, so $H_0(T|M) = \{0\}$. This implies that $X = \{0\} \oplus N = N$, hence X is finite-dimensional and T is nilpotent. ∎

Theorem 2.73 *Suppose that for a bounded operator $T \in L(X)$, the sum $H_0(\lambda_0 I - T) + (\lambda_0 I - T)(X)$ is norm dense in X. Then T^* has the SVEP at λ_0.*

Proof Also here we assume that $\lambda_0 = 0$. From Theorem 2.37 we have that $K(T^*) \subseteq H_0(T)^\perp$. From a standard duality argument we now obtain

$$\ker(T^*) \cap K(T^*) \subseteq T(X)^\perp \cap H_0(T)^\perp = (T(X) \cap H_0(T))^\perp.$$

If the subspace $H(T) + T(X)$ is norm-dense in X, then the last annihilator is zero, hence $\ker T^* \cap K(T^*) = \{0\}$, and from Theorem 2.60 we conclude that T^* has the SVEP at 0. ∎

Corollary 2.74 *Suppose either that $H_0(\lambda_0 I - T) + K(\lambda_0 I - T)$ or $\mathcal{N}^\infty(\lambda_0 I - T) + (\lambda_0 I - T)^\infty(X)$ is norm dense in X. Then T^* has the SVEP at λ_0.* ∎

The next theorem is, in a certain sense, dual to Theorem 2.73.

Theorem 2.75 *Suppose that for a bounded operator $T \in L(X)$ the sum $H_0(\lambda_0 I^* - T^*) + (\lambda_0 I^* - T^*)(X^*)$ is weak*-dense in X^*. Then T has the SVEP at λ_0.*

Proof From Theorem 2.37 we know that $K(T) \subseteq {}^\perp H_0(T^*)$. Therefore

$$\ker T \cap K(T) \subseteq {}^\perp T^*(X^*) \cap {}^\perp H_0(T^*) = {}^\perp(T^*(X^*) + H_0(T^*)).$$

But the sum $H_0(T^*) + T^*)(X^*)$ is weak*-dense in X^*, so, by the Hahn–Banach theorem, the last annihilator is zero and therefore T has the SVEP at 0, again by Theorem 2.60. ∎

The following corollary is clear, since the analytic core of an operator is contained in the range, while the hyper-kernel is contained in the quasi-nilpotent part.

Corollary 2.76 *Suppose that for a bounded operator $T \in L(X)$, either $H_0(\lambda_0 I - T^*) + K(\lambda_0 I - T^*)$ or $\mathcal{N}^\infty(\lambda_0 - T^*) + (\lambda_0 - T^*)^\infty(X^*)$ is weak*-dense in X^*. Then T has the SVEP at λ_0.* ∎

The result of Corollary 2.74 cannot be reversed, as the following example shows:

Example 2.77 Let V denote the *Volterra operator* on the Banach space $X :=$ $C[0, 1]$, defined by

$$(Vf)(t) := \int_0^t f(s)ds \quad \text{for all } f \in C[0, 1] \quad \text{and } t \in [0, 1].$$

V is injective and quasi-nilpotent. Consequently $\mathcal{N}^\infty(V) = \{0\}$ and $K(V) = \{0\}$ by Corollary 2.71. It is easy to check that

$$V^\infty(X) = \{f \in C^\infty[0, 1] : f^{(n)}(0) = 0, \; n \in \mathbb{Z}_+\},$$

thus $V^\infty(X)$ is not closed and hence is strictly larger than $V(T) = \{0\}$. Clearly the sum $\mathcal{N}^\infty(V) + V^\infty(X)$ is not norm dense in X, while V^* has the SVEP, because it is quasi-nilpotent.

Lemma 2.78 *Suppose that $\lambda_0 I - T$ has a GKD (M, N). Then*

$$\lambda_0 I - T | M \text{ is surjective } \Leftrightarrow \lambda_0 I^\star - T^\star | N^\perp \text{is injective.}$$

Proof We can assume $\lambda_0 = 0$. Suppose first that $T(M) = M$ and consider an arbitrary element $x^* \in \ker T^* | N^\perp = \ker T^* \cap N^\perp$. For every $m \in M$ there exists an $m' \in M$ such that $T m' = m$. Then we have

$$x^*(m) = x^*(T m') = (T^\star x^*)(m') = 0,$$

and therefore $x^* \in M^\perp \cap N^\perp = \{0\}$.

Conversely, suppose that $T | M$ is not onto, i.e., $T(M) \subseteq M$ and $T(M) \neq M$. By assumption $T(M)$ is closed, since $T | M$ is semi-regular, and hence via the Hahn–Banach theorem there exists a $z^* \in X^*$ such that $z^* \in T(M)^\perp$ and $z^* \notin M^\perp$.

Now, from the decomposition $X^* = N^\perp \oplus M^\perp$ we have $z^* = n^* + m^*$ for some $n^* \in N^\perp$ and $m^* \in M^\perp$. For every $m \in M$ we obtain

$$T^* n^*(m) = n^*(T m) = z^*(T m) - m^*(T m) = 0.$$

Hence $T^* n^* \in N^\perp \cap M^\perp = \{0\}$, and therefore $0 \neq n^* \in \ker T^* \cap N^\perp$.   ∎

Theorem 2.79 *Suppose that $T \in L(X)$ admits a GKD (M, N). Then the following statements are equivalent:*

(i)   *T has the SVEP at 0;*
(ii) *$T | M$ has the SVEP at λ_0;*
(iii) *$(\lambda_0 I - T) | M$ is injective;*
(iv) *$H_0(\lambda_0 - T) = N$;*
(v)   *$H_0(\lambda_0 I - T)$ is closed;*
(vi) *$H_0(\lambda_0 I - T) \cap K(\lambda_0 I - T) = \{0\}$;*
(vii) *$H_0(\lambda_0 I - T) \cap K(\lambda_0 I - T)$ is closed.*

Proof Also here we shall consider the particular case $\lambda_0 = 0$.

The implication (i) $\Rightarrow$ (ii) is clear, since the SVEP at 0 of T is inherited by the restrictions on every closed invariant subspace.

(ii) $\Rightarrow$ (iii) $T | M$ is semi-regular, so $T | M$ has the SVEP at 0 if and only if $T | M$ is injective.

(iii) $\Rightarrow$ (iv) If $T|M$ is injective, from Theorem 2.37 the semi-regularity of $T|M$ implies that $\overline{H_0(T|M)} = \mathcal{N}^\infty(T|M) = \{0\}$, and hence

$$H_0(T) = H_0(T|M) \oplus H_0(T|N) = \{0\} \oplus N = N.$$

The implications (iv) $\Rightarrow$ (v) and (vi) $\Rightarrow$ (vii) are obvious, while the implications (v) $\Rightarrow$ (vi) and (vii) $\Rightarrow$ (i) have been proved in Theorem 2.39.

The last assertion is clear since the pair $M := X$ and $N := \{0\}$ is a GKD for every semi-regular operator. ∎

The next result shows that if the operator $\lambda_0 I - T$ admits a generalized Kato decomposition then all the implications of Theorem 2.73 are actually equivalences.

Theorem 2.80 *Suppose that $\lambda_0 I - T \in L(X)$ admits a GKD (M, N). Then the following assertions are equivalent:*

(i) T^* *has the SVEP at λ_0;*
(ii) $(\lambda_0 I - T)|M$ *is surjective;*
(iii) $K(\lambda_0 I - T) = M$;
(iv) $X = H_0(\lambda_0 I - T) + K(\lambda_0 I - T)$;
(v) $H_0(\lambda_0 I - T) + K(\lambda_0 I - T)$ *is norm dense in X.*
    In particular, if $\lambda_0 I - T$ is semi-regular then the conditions (i)–(v) are equivalent to the following statement:
(vi) $K(\lambda_0 I - T) = X$.

Proof We suppose $\lambda_0 = 0$ here.

(i) $\Leftrightarrow$ (ii) We know that the pair $(N^\perp, M^\perp)$ is a GKD for $T^\star$, and hence, by Theorem 2.79, $T^\star$ has the SVEP at 0 if and only if $T^\star |N^\perp$ is injective. By Lemma 2.78 T^* then has the SVEP at 0 if and only if $T|M$ is onto.

(ii) $\Rightarrow$ (iii) If $T|M$ is surjective then $M = K(T|M) = K(T)$, by Theorem 2.37.

(iii) $\Rightarrow$ (iv) By assumption $X = M \oplus N = K(T) \oplus N$, and therefore $X = H_0(T) + K(T)$, since $N = H_0(T|N) \subseteq H_0(T)$.

The implication (iv) $\Rightarrow$ (v) is obvious, while (v) $\Rightarrow$ (i) has been established in Theorem 2.73.

The last assertion is obvious since $M := X$ and $N := \{0\}$ provides a GKD for T. ∎

Lemma 2.81 *If $T \in L(X)$ admits a GKD (M, N) and $0 \in \sigma(T)$, then*

(i) T *has the SVEP at* $0 \Leftrightarrow 0 \in \text{iso } \sigma_{\text{ap}}(T)$.
(ii) T^* *has the SVEP at* $0 \Leftrightarrow 0 \in \text{iso } \sigma_s(T)$.

Proof

(i) The implication ($\Leftarrow$) has been observed before. To show the reverse implication, suppose that T has the SVEP at 0 and that T has a GKD (M, N). Then the restriction $T|M$ has the SVEP at 0, so $T|M$ is injective and $H_0(T) = N$, by Theorem 2.79. Hence $X = M \oplus H_0(T)$. But $T|M$ is semi-regular, so $T(M)$ is closed and hence $T|M$ is bounded below. Since $T|N$ is quasi-nilpotent we then have $\sigma_{\mathrm{ap}}(T|N) = \{0\}$. Therefore,

$$\sigma_{\mathrm{ap}}(T) = \sigma_{\mathrm{ap}}(T|M) \cup \sigma_{\mathrm{ap}}(T|N) = \sigma_{\mathrm{ap}}(T|M) \cup \{0\}.$$

But $0 \notin \sigma_{\mathrm{ap}}(T|M)$ and $\sigma_{\mathrm{ap}}(T|M)$ is closed, from which we conclude that 0 is an isolated point of $\sigma_{\mathrm{ap}}(T)$.

(ii) The implication ($\Leftarrow$) has been already observed. To show the reverse implication assume that T^* has the SVEP at 0. The pair $(N^\perp, M^\perp)$ is a GKD for T^* and the restriction $T^*|N^\perp$ has the SVEP at 0 and, as above, we deduce that $T^*|N^\perp$ is injective. By Lemma 2.78 we then have that $T|M$ is onto. Since $T|N$ is quasi-nilpotent then $\sigma_{\mathrm{s}}(T|N) = \{0\}$. Finally

$$\sigma_{\mathrm{s}}(T) = \sigma_{\mathrm{s}}(T|M) \cup \sigma_{\mathrm{s}}(T|N) = \sigma_{\mathrm{s}}(T|M) \cup \{0\}.$$

But $0 \notin \sigma_{\mathrm{s}}(T|M)$ and $\sigma_{\mathrm{s}}(T|M)$ is closed, from which we conclude that 0 is an isolated point of $\sigma_{\mathrm{s}}(T)$.   ■

The operators which admit a GKD may be characterized by means of commuting projections as follows:

Theorem 2.82 *If $T \in L(X)$ the following statements are equivalent:*

(i)   *T admits a GKD (M,N);*
(ii) *there exists a commuting projection P such that $T + P$ is semi-regular and $T P$ is quasi-nilpotent.*
      In this case $K(T)$ is closed. If $0 \in \sigma(T)$ then $0 \in \mathrm{iso}\,\sigma_{\mathrm{se}}(T)$.

Proof (i) $\Rightarrow$ (ii) Suppose that T has a GKD (M, N). Then $T|M$ is semi-regular and $T|N$ is quasi-nilpotent. Let P be the projection of X onto N along M. Then $M = \ker P$ and $N = P(X)$. The pair (M, N) reduces T, so $PT = TP$. Since $T + P$ and $T P$ are also reduced by the pair (M, N), by Theorem 1.46 the restriction $(T + P)|M = T|M$ is also semi-regular. Furthermore, $(T + P)|N = (T + P)|P(X) = T|N + I|N$ is invertible, since $T|N$ is quasi-nilpotent, hence $T + P = (T + P)|M \oplus (T + P)|N$ is semi-regular. We have, by Lemma 2.36 and since $T|N$ is quasi-nilpotent,

$$H_0(TP) = H_0((TP)|M) \oplus H_0((TP)|N) = H_0(0) \oplus H_0(T|N) = M \oplus N = X,$$

so, by Theorem 2.35, $T P$ is quasi-nilpotent.

(ii) $\Rightarrow$ (i) Suppose that P is a commuting projection for which $T + P$ is semi-regular and $T P$ is quasi-nilpotent. Then $X = \ker P \oplus P(X)$ and the pair $(\ker P, P(X))$ reduces T and hence also $T + P$ and $T P$. By Theorem 1.46 the restriction $(T + P)|\ker P = T|\ker P$ is semi-regular.

We show now that the restriction $T|P(X)$ is quasi-nilpotent. Let $x \in P(X)$ be arbitrarily chosen. Then

$$\|(T|P(X))^n\|^{\frac{1}{n}} = \|T^n P^n x\|^{\frac{1}{n}} = \|(T P)^n x\|^{\frac{1}{n}} \to 0 \text{ as } n \to \infty,$$

so $x \in H_0(T|P(X))$ and hence $P(X) \subseteq H_0(T|P(X))$. The reverse inclusion $H_0(T|P(X)) \subseteq P(X)$ is obvious, so $H_0(T|P(X)) = P(X)$, and this implies that $T|P(X)$ is quasi-nilpotent. Consequently, the pair $(\ker P, P(X))$ is a GKD for T, and hence $K(T) = K(T|\ker P)$ by Theorem 1.63. On the other hand, since the restriction $T|\ker P$ is semi-regular, again by Theorem 1.63 we then have that $K(T|\ker P)$ is closed.

Suppose that 0 belongs to $\sigma(T)$. By Theorem 1.46 we have

$$\sigma_{se}(T) = \sigma_{se}(T|\ker P) \cup \sigma_{se}(T|P(X)) = \sigma_{se}(T|\ker P) \cup \{0\}.$$

We know that $T|P(X)$ is quasi-nilpotent, and the Kato spectrum is non-empty, so $\sigma_{se}(T|P(x)) = \{0\}$. Since $0 \notin \sigma_{se}(T|\ker P)$, and $\sigma_{se}(T|\ker P)$ is closed, it then follows that 0 is an isolated point of $\sigma_{se}(T)$.   ∎

Corollary 2.83 *If $T \in L(X)$ then the following statements are equivalent:*

(i) *T is of Kato-type;*
(ii) *there exists a commuting projection P such that $T + P$ is semi-regular and TP is nilpotent.*

Proof (i) $\Rightarrow$ (ii) Let (M, N) be a GKD for T such that $T|N$ is nilpotent. If P is the projection of X onto N along M we have $(TP)|M = (TP)|\ker P = 0$. Let $\nu \in \mathbb{N}$ be such that $(T|N)^\nu = ((TP)|P(X))^\nu = 0$. Then

$$(TP)^\nu = ((TP)|M)^\nu \oplus (TP|N)^\nu = 0,$$

so TP is nilpotent.

(ii) $\Rightarrow$ (i) Suppose that P is a commuting projection such that $T + P$ is bounded below and TP is nilpotent. As in the proof of Theorem 2.82, the pair $(\ker P, P(X))$ is a GKD for T. Furthermore, $T|P(X) = (TP)|P(X)$ is nilpotent.   ∎

The following result may be considered as a refinement of Theorem 2.46.

Theorem 2.84 *Let $T \in L(X)$. Then the following statements are equivalent:*

(i) *$H_0(T)$ is complemented by a T-invariant subspace M such that $T(M)$ is closed;*
(ii) *there exists a pair of proper closed subspaces (M, N) which reduces T such that the restriction $T = T|M \oplus T|N$, $T|M$ is bounded below, while $T|N$ quasi-nilpotent;*
(iii) *there exists a commuting projection $P \neq 0$ such that $T + P$ is bounded below and TP is quasi-nilpotent.*

    In this case both subspaces $H_0(T)$ and $K(T)$ are closed. More precisely, for every projection P which satisfies (ii) we have $H_0(T) = P(X)$. Moreover,

(iv) *$H_0(T) \cap K(T) = \{0\}$.*
(v) *$0 \in \text{iso}\, \sigma_{ap}(T)$.*

Proof (i) ⇔ (ii) Let M be a closed T-invariant subspace such that $X = H_0(T) \oplus M$ and $T(M)$ is closed. If $N := H_0(T)$ then $T = T|M \oplus T|N$ and $T_2 := T|N$ is quasi-nilpotent. Moreover, ker $T|M = M \cap$ ker $T \subseteq M \cap H_0(T) = \{0\}$, so $T_1 := T|M$ is bounded below, since $T(M)$ is closed. Conversely, let $X = M \oplus N$ such that $T|M$ is bounded below and $T|N$ is quasi-nilpotent. Then $H_o(T|M) = \{0\}$ and $H_0(T|N) = N$, hence $H_0(T) = H_0(T|M) \oplus H_=(T|N) = N$, so $X = M \oplus H_0(T)$ and $T(M)$ is closed.

(ii) ⇒ (iii) Clearly, (M, N) is a GKD for T, since every bounded below operator is semi-regular. As in the proof of Theorem 2.82, if P is the projection of X onto N along M then $M = $ ker P and $N = P(X)$, and TP is quasi-nilpotent. Moreover, $T + P = (T + P)|M \oplus (T + P)|N$ and $(T + P)|M = T|M$ is bounded below, by Lemma 1.28, while $(T + P)|N = T|N + I_N$ is invertible. Therefore, again by Lemma 1.28, $T + P$ is bounded below.

(iii) ⇒ (ii) Take $M := $ ker P and $N := P(X)$. As in the proof of Theorem 2.82, $T|M = (T + P)|M$ is bounded below, while $T|N = T|P(X)$ is quasi-nilpotent.

Observe that $K(T)$ is closed by Theorem 2.82. To show that $H_0(T)$ is closed observe that, by Lemma 2.36, we have $H_0(T) = H_0(T|M) \oplus N$. Since $T|M$ is bounded below, $H_0(T|M) = \{0\}$, so $H_0(T) = \{0\} \oplus N = N$ is closed.

(iii) $H_0(T) \cap K(T)$ is closed, so, by Theorem 2.39, we have $H_0(T) \cap K(T) = \{0\}$.

(iv) Because $H_0(T)$ is closed, T has the SVEP at 0. By Lemma 2.81, it then follows that $0 \in $ iso $\sigma_{ap}(T)$. ∎

Theorem 2.85 *Let $T \in L(X)$. Then the following statements are equivalent:*

(i) *$K(T)$ is closed and complemented by a T-invariant subspace N such that $N \subseteq H_0(T)$;*

(ii) *there exists a pair of proper closed subspaces (M, N) which reduces T such that $T = T|M \oplus T|N$, $T|M$ is onto and $T|N$ is quasi-nilpotent;*

(iii) *there exists a commuting projection $P \neq 0$ such that $T + P$ is onto and TP is quasi-nilpotent.*

If the equivalent conditions (i) and (iii) are satisfied we have:

(iii) *$K(T)$ is closed and $X = H_0(T) + K(T)$;*

(iv) *$0 \in $ iso $\sigma_s(T)$.*

Proof (i) ⇔ (ii) Let $T_1 := T|K(T)$ and $T_2 := T|N$. Obviously, T_1 is onto and $H_0(T_2) = H_0(T) \cap N = N$, so T_2 is quasinilpotent. Conversely, suppose that $X = M \oplus N$ with $T_1 := T|M$ onto and $T_2 := T|N$ quasi-nilpotent. Then $K(T) = K(T_1)$ is closed. Since T_1 is onto then $M = K(T)$, so $X = K(T) \oplus N$ and from the inclusion $N \subseteq H_0(T)$ we obtain that $T|N$ is quasi-nilpotent.

(ii) ⇒ (iii) If T satisfies (ii) then (M, N) is a GKD for T since every onto operator is semi-regular. If P is the projection of X onto N along M then $(T+P)|M = T|M$ is onto, while $(T + P)|N$ is invertible. By Lemma 1.28 it then follows that $T + P$ is onto. As in the proof of Theorem 2.82, we know that TP is quasi-nilpotent.

(iii) ⇒ (ii) Also here, take $M := $ ker P and $N := P(X)$. As in the proof of Theorem 2.82, $T|M = (T+P)|M$ is onto, while $T|N = T|P(X)$ is quasi-nilpotent.

(iii) Clearly $K(T)$ is closed by Theorem 2.82. Now,

$$H_0(T) = H_0(T|M) \oplus H_0(T|N) = H_0(T|M) \oplus N \supseteq N,$$

while $K(T) = K(T|M) = M$, since $T|M$ is onto. Therefore, $X = M \oplus N \subseteq K(T) + H_0(T)$ and hence $X = K(T) + H_0(T)$.

(iv) The condition $X = H_0(T) + K(T)$ entails that T^* has the SVEP at 0, so, by Lemma 1.70, $0 \in \text{iso } \sigma_s(T)$. ∎

2.5 A Local Spectral Mapping Theorem

Given an operator $T \in L(X)$, X a Banach space, and an analytic function f defined on an open neighborhood $\mathcal{U}$ of $\sigma(T)$, let $f(T)$ denote the corresponding operator defined by the functional calculus. One may be tempted to conjecture that the spectral theorem holds for the local spectrum, i.e. $f(\sigma_T(x)) = \sigma_{f(T)}(x)$ for all $x \in X$. It can easily be seen that this equality is not true in general. Indeed, if we consider the constant function $f \equiv c$ on the neighborhood $\mathcal{U}$ and an operator T without the SVEP, then there exists, by Theorem 2.20, a vector $0 \neq x \in X$ such that $\sigma_T(x) = \emptyset$. Clearly $f(\sigma_T(x)) = \emptyset$, while

$$\sigma_{f(T)}(x) = \sigma(f(T)) = \{c\} \neq \emptyset.$$

Denote by $\mathcal{H}(\sigma(T))$ the set of all analytic functions, defined on an open neighborhood of $\sigma(T)$. In order to show that the spectral theorem for the local spectrum holds if T has the SVEP, we first need to prove that the SVEP is preserved under the functional calculus.

Theorem 2.86 *If $T \in L(X)$ has the SVEP then $f(T)$ has the SVEP for every $f \in \mathcal{H}(\sigma(T))$.*

Proof Suppose that T has the SVEP and $\sigma_{f(T)}(x) = \emptyset$ for some $x \in X$. By Theorem 2.14 it suffices to show that $x = 0$. For every $\lambda \in \mathbb{C}$ there exists an analytic function $g : \mathcal{U}_\lambda \to X$ defined in an open neighborhood $\mathcal{U}_\lambda$ of λ such that $(\mu I - f(T))g(\mu) = x$. Set $F_\lambda := \mathbb{C} \setminus \mathcal{U}_\lambda$. Trivially,

$$\bigcap_{\lambda \in \mathbb{C}} F_\lambda = \emptyset,$$

and

$$x \in \bigcap_{\lambda \in \mathbb{C}} \mathcal{X}_{f(T)}(F_\lambda) = \bigcap_{\lambda \in \mathbb{C}} \mathcal{X}_T(f^{-1}(F_\lambda))$$

$$\subseteq X_T(\bigcap_{\lambda \in \mathbb{C}} f^{-1}(F_\lambda)) = X_T(\emptyset).$$

Since T has the SVEP, $X_T(\emptyset) = \{0\}$, hence $x = 0$, so $f(T)$ has the SVEP. ∎

Denote by $\mathcal{H}_{nc}(\sigma(T))$ the set of all analytic functions, defined on an open neighborhood of $\sigma(T)$, such that f is non-constant on each of the components of its domain.

Theorem 2.87 *If $T \in L(X)$ the following statements hold:*

(i) $f(\sigma_T(x)) \subseteq \sigma_{f(T)}(x)$ *for all $x \in X$ and $f \in \mathcal{H}(\sigma(T))$.*
(ii) *If T has the SVEP, or if the function $f \in \mathcal{H}_{nc}(\sigma(T))$, then*

$$f(\sigma_T(x)) = \sigma_{f(T)}(x) \quad \text{for all } x \in X.$$

Proof

(i) Let $x \in X$, and for every $\lambda \notin \sigma_{f(T)}(x)$ let $\mathcal{U}_\lambda$ denote an open neighborhood of λ such that $x \in \mathcal{X}_{f(T)}(\mathbb{C} \setminus \mathcal{U}_\lambda)$. From Theorem 2.29 we have

$$x \in \mathcal{X}_T(f^{-1}(\mathbb{C} \setminus \mathcal{U}_\lambda)) \subseteq X_T(f^{-1}(\mathbb{C} \setminus \mathcal{U}_\lambda)),$$

so $f(\sigma_T(x)) \subseteq \mathbb{C} \setminus \mathcal{U}_\lambda$ for all $\lambda \notin \sigma_{f(T)}(x)$. From this it then follows that $f(\sigma_T(x)) \subseteq \sigma_{f(T)}(x)$.

(ii) If T has the SVEP then, by Theorem 2.86 $f(T)$ has the SVEP and hence, by Theorem 2.23, local spectral and glocal spectral subspaces corresponding to the same closed set coincide. By Theorem 2.29 we then have $X_f(T)(F) = \mathcal{X}_T(f^{-1}(F))$ for all closed sets $F \subseteq \mathbb{C}$, and applying this equality to the closed set $F_0 := f(\sigma_T(x))$ we then obtain the reverse inclusion $\sigma_{f(T)}(x) \subseteq f(\sigma_T(x))$.

The proof in the case when $f \in \mathcal{H}_{nc}(\sigma(T))$ is rather technical and may be found in [216, Theorem 3.3.8]. ∎

The next result shows that the localized SVEP is preserved under the functional calculus under appropriate conditions on the analytic function.

Theorem 2.88 *Let $T \in L(X)$ and $f \in \mathcal{H}_{nc}(\sigma(T))$. Then $f(T)$ has the SVEP at $\lambda \in \mathbb{C}$ if and only if T has the SVEP at every point $\mu \in \sigma(T)$ for which $f(\mu) = \lambda$.*

Proof Suppose first that $f(T)$ has the SVEP at $\lambda_0 \in \mathbb{C}$. By Theorem 2.60 then

$$\ker(\lambda_0 I - f(T)) \cap X_{f(T)}(\emptyset) = \{0\}.$$

Suppose now that for some $\mu_0 \in \sigma(T)$ we have $f(\mu_0) = \lambda_0$. To show the SVEP of T at μ_0 it suffices, again by Theorem 2.60, to show that $\ker(\mu_0 I - T) \cap X_{\mu_0 I - T}(\emptyset) = \{0\}$.

Let $x \in \ker(\mu_0 I - T) \cap X_T(\emptyset)$ be arbitrarily given and define $h(\mu) := \lambda_0 - f(\mu)$ for all $\mu \in \mathcal{U}$. Then $h(T) = \lambda_0 I - f(T)$ and, since $h(\mu_0) = 0$, we can write $h(\mu) = (\mu_0 - \mu)g(\mu)$, where g is analytic on $\mathcal{U}$. Clearly

$$h(T) = (\mu_0 I - T)g(T) = g(T)(\mu_0 I - T),$$

so that $x \in \ker h(T) = \ker (\lambda_0 I - f(T)$. On the other hand, from $x \in X_T(\emptyset)$ we obtain $\sigma_T(x) = \emptyset$, and hence

$$\sigma_{f(T)}(x) = f(\sigma_T(x)) = f(\emptyset) = \emptyset,$$

so $x \in X_{f(T)}(\emptyset)$. Therefore

$$\ker (\mu_0 I - T) \cap X_T(\emptyset) \subseteq \ker (\lambda_0 I - f(T)) \cap X_{f(T)}(\emptyset) = \{0\},$$

which shows, by Theorem 2.60, that T has the SVEP at λ_0.

Conversely, let $\lambda_0 \in \mathbb{C}$ and assume that T has the SVEP at every $\mu_0 \in \sigma(T)$ for which $f(\mu_0) = \lambda_0$. Write $h(\mu) := \lambda_0 - f(\mu)$, where $\mu \in \mathcal{U}$. By assumption f is non-constant on each connected component of $\mathcal{U}$, so, by the identity theorem for analytic functions, the function h has only finitely many zeros in $\sigma(T)$ and these zeros are of finite multiplicity. Hence there exists an analytic function g defined on $\mathcal{U}$ without zeros in $\sigma(T)$ and a polynomial p of the form

$$p(\mu) = (\mu_1 - \mu) \cdots (\mu_n - \mu),$$

with not necessarily distinct elements $\mu_1, \cdots, \mu_n \in \sigma(T)$ such that

$$h(\mu) = \lambda_0 - f(\mu) = p(\mu)g(\mu) \quad \text{for all } \mu \in \mathcal{U}.$$

Assume that $x \in \ker (\lambda_0 I - f(T)) \cap X_{f(T)}(\emptyset)$. In order to prove that $f(T)$ has the SVEP at λ_0 it suffices to show, again by Theorem 2.60, that $x = 0$. From the classical spectral mapping theorem we know that $g(T)$ is invertible, so the equality

$$\lambda_0 I - f(T) = p(T)g(T) = g(T)p(T)$$

implies that $p(T)x \in \ker g(T) = \{0\}$. If we put

$$q(\mu) := (\mu_2 - \mu) \cdots (\mu_n - \mu)$$

and $y = q(T)x$ then $(\mu_1 I - T)y = 0$.

On the other hand, $x \in X_{f(T)}(\emptyset)$ and f is non-constant on each of the connected components of $\mathcal{U}$. Part (ii) of Theorem 2.87 then ensures that

$$f(\sigma_T(x)) = \sigma_{f(T)}(x) = \emptyset$$

and therefore since T and $q(T)$ commute

$$\sigma_T(y) = \sigma_T(q(T)x) \subseteq \sigma_T(x) = \emptyset.$$

But T has the SVEP at μ_1, by assumption, so, again by Theorem 2.60, $y = 0$. A repetition of this argument for $\mu_2, \cdots, \mu_n$ then leads to the equality $x = 0$, thus $f(T)$ has the SVEP at λ_0.

The last claim is obvious, being nothing else than a reformulation of the equivalence proved above.   ∎

Combining Theorems 2.88 and 2.86 we then have:

Corollary 2.89 *Let $T \in L(X)$ and $f \in \mathcal{H}_{nc}(\sigma(T))$. Then T has the SVEP if and only if $f(T)$ has the SVEP.*

An immediate consequence of Theorem 2.88 is that, in the characterization of the SVEP at a point $\lambda_0 \in \mathbb{C}$ given in Theorem 2.60, the kernel ker $(\lambda_0 I - T)$ may be replaced by the hyper-kernel $\mathcal{N}^\infty(\lambda_0 I - T)$.

Corollary 2.90 *For every bounded operator on a Banach space X the following properties are equivalent:*

  (i)   *T has the SVEP at λ_0;*
 (ii)   *$(\lambda_0 I - T)^n$ has the SVEP at 0 for every $n \in \mathbb{N}$;*
(iii)   *$\mathcal{N}^\infty(\lambda_0 I - T) \cap X_T(\emptyset) = \{0\}$;*
 (iv)   *$\mathcal{N}^\infty(\lambda_0 I - T) \cap K(\lambda_0 I - T) = \{0\}$.*

Proof The equivalence (i) $\Leftrightarrow$ (ii) is obvious from Theorem 2.88. Combining this equivalence with Theorem 2.60 we then obtain that T has the SVEP at λ_0 if and only if ker $(\lambda_0 I - T)^n \cap X_T(\emptyset) = \{0\}$, for every $n \in \mathbb{N}$. Therefore the equivalence (i) $\Leftrightarrow$ (iii) is proved. The equivalence (i) $\Leftrightarrow$ (iv) follows from Theorem 2.60 in a similar way.   ∎

Note that in condition (ii) of Corollary 2.90 the power $(\lambda_0 I - T)^n$ may be replaced by $f(T)$, where f is any analytic function on some neighborhood $\mathcal{U}$ of $\sigma(T)$ such that f is non-constant on each of the connected components of $\mathcal{U}$ and such that 0 is the only zero of f in $\sigma(T)$.

It is easily seen that if $X_T(F) = \{0\}$ then $F \cap \sigma_p(T) = \emptyset$. In fact, suppose that $X_T(F) = \{0\}$ and assume that there is a $\lambda_0 \in F \cap \sigma_p(T)$. Then there is an $0 \neq x \in \ker(\lambda_0 I - T)$. Clearly $\sigma_T(x) \subseteq \{\lambda_0\}$, and since $\lambda_0 \in F$ this implies that $x \in X_T(F) = \{0\}$, a contradiction.

We also have that $X_T(F) = X$ precisely when $\sigma_s(T) \subseteq F$. In fact, if $X_T(F) = X$ and $\lambda \notin F$ then

$$K(\lambda I - T) = X_T(\mathbb{C} \setminus \{\lambda\}) \supseteq X_T(F \setminus \{\lambda\}) = X_T(F) = X,$$

so that $X = K(\lambda I - T)$ and hence $\lambda I - T$ is surjective, namely $\lambda \notin \sigma_s(T)$. Conversely, suppose that $\sigma_s(T) \subseteq F$. By Theorem 2.21 we obtain that $\sigma_T(x) \subseteq F$ for all $x \in X$ so that $X = X_T(F)$.

One of the deepest results of local spectral theory, due to Laursen and Neumann [215], shows that analogous results hold for the glocal subspaces in the case where Ω is closed subset of $\mathbb{C}$, see also [216, Theorem 3.3.12].

The following result generalizes Corollary 2.61 to semi-regular operators.

Theorem 2.91 *Suppose that $\lambda_0 I - T$ is a semi-regular operator on the Banach space X. Then the following equivalences hold:*

(i) *T has the SVEP at λ_0 precisely when $\lambda_0 I - T$ is injective or, equivalently, when $\lambda_0 I - T$ is bounded below;*
(ii) *T^* has the SVEP at λ_0 precisely when $\lambda_0 I - T$ is surjective.*

Proof

(i) We can assume that $\lambda_0 = 0$. We have only to prove that if T has the SVEP at 0 then T is injective. Suppose that T is not injective. Then, by Theorem 1.44, the semi-regularity of T entails $T^\infty(X) = K(T)$ and $\{0\} \neq \ker T \subseteq T^\infty(X) = K(T)$. Thus T does not have the SVEP at 0 by Theorem 2.60.
(ii) If $\lambda_0 I - T$ is semi-regular then $\lambda_0 I^* - T^*$ is also semi-regular and $\lambda_0 I - T$ is surjective if and only if $\lambda_0 I^* - T^*$ is bounded below.   ∎

Corollary 2.92 *Let X be a Banach space and $T \in L(X)$. The following assertions hold:*

(i) *If $\lambda_0 \in \sigma(T) \setminus \sigma_{ap}(T)$ then T has the SVEP at λ_0, but $T^\star$ fails to have the SVEP at λ_0.*
(ii) *If $\lambda_0 \in \sigma(T) \setminus \sigma_s(T)$ then $T^\star$ has the SVEP at λ_0, but T fails to have the SVEP at λ_0.*

Proof The condition $\lambda_0 \in \sigma(T) \setminus \sigma_{ap}(T)$ implies that $\lambda_0 I - T$ has closed range and is injective but not surjective, so we can apply Theorem 2.91. Analogously, if $\lambda_0 \in \sigma(T) \setminus \sigma_s(T)$ then $\lambda_0 I - T$ is surjective but not injective, so we can apply again Theorem 2.91.   ∎

As observed before, the semi-regular resolvent $\rho_{se}(T)$ is an open subset of $\mathbb{C}$, so it may be decomposed into connected disjoint open non-empty components.

Theorem 2.93 *Let $T \in L(X)$, X a Banach space, and Ω a component of $\rho_{se}(T)$. Then we have the following alternative:*

(i) *T has the SVEP at every point of Ω. In this case $\sigma_p(T) \cap \Omega = \emptyset$;*
(ii) *For every $\lambda \in \Omega$, T does not have the SVEP. In this case $\sigma_p(T) \supseteq \Omega$.*

Proof Suppose that T has the SVEP at a point $\lambda_0 \in \Omega$ and consider an arbitrary point λ of Ω. In order to show that T has the SVEP at λ it suffices to show, by Theorem 2.91, that $\lambda I - T$ is injective. By Theorem 2.91 $\lambda_0 I - T$ is injective, so $\mathcal{N}^\infty(\lambda_0 I - T) = \{0\}$ and therefore $H_0(\lambda_0 I - T) = \{0\}$, by part (ii) of Theorem 2.37. From Theorem 2.57 we know that the subspaces $\overline{H_0(\lambda I - T)}$ are constant for λ ranging through Ω, so that $H_0(\lambda I - T) = \{0\}$ for every $\lambda \in \Omega$. This shows that T has the SVEP at every $\lambda \in \Omega$.

The assertions on the point spectrum are clear from Theorem 2.91.   ∎

A very special situation is given when $\sigma_{ap}(T)$ and $\sigma_s(T)$ are contained in the boundary $\partial\sigma(T)$ of the spectrum, or, equivalently, are equal, since both contain

$\partial\sigma(T)$. Later we shall see that this situation is fulfilled by several classes of operators. Note first that both T and T^* have the SVEP at every point $\lambda \in \overline{\rho(T)}$.

Theorem 2.94 *Suppose that for a bounded operator $T \in L(X)$, X a Banach space, we have $\sigma_{\mathrm{ap}}(T) = \partial\sigma(T)$. Then T has the SVEP while $\Xi(T^*)$ coincides with the interior of $\sigma(T)$. Similarly, if $\sigma_{\mathrm{s}}(T) = \partial\sigma(T)$ then T^* has the SVEP, while $\Xi(T)$ coincides with the interior of $\sigma(T)$.*

Proof Suppose that $\sigma_{\mathrm{ap}}(T) = \partial\sigma(T)$. If λ belongs to the interior of $\sigma(T)$ then $\lambda \notin \sigma_{\mathrm{ap}}(T)$, hence T has the SVEP at λ while T^* does not have the SVEP at λ, by part (i) of Corollary 2.92. Similarly the last claim is a consequence of part (ii) of Corollary 2.92. ∎

Theorem 2.94 has a nice application to the so-called *Cèsaro operator C_p* defined on the classical Hardy space $H_p(\mathbb{D})$, $\mathbb{D}$ the open unit disc and $1 < p < \infty$. The operator C_p is defined by

$$(C_p f)(\lambda) := \frac{1}{\lambda} \int_0^\lambda \frac{f(\mu)}{1 - \mu} \, d\mu \quad \text{for all } f \in H_p(\mathbf{D}) \text{ and } \lambda \in \mathbf{D}.$$

As noted by Miller et al. [241], the spectrum of the operator C_p is the entire closed disc Γ_p, centered at $p/2$ with radius $p/2$, and $\sigma_{\mathrm{ap}}(C_p)$ is the boundary $\partial\Gamma_p$. Hence, the Cesàro operator has the SVEP, while its adjoint does not have the SVEP at any point of the interior of Γ_p.

2.6   The Localized SVEP and Topological Uniform Descent

In this section we characterize in several ways the localized SVEP for operators having uniform descent, and in particular for quasi-Fredholm operators. We begin first by extending to operators with topological descent the results of Theorems 1.64 and 2.37. Recall that if $T \in L(X)$, the operator range topology on $T(X)$ is the topology induced by the norm $\| \cdot \|_T$ defined by:

$$\|y\|_T := \inf_{x \in X} \{\|x\| : y = Tx\}.$$

Theorem 2.95 *If $T \in L(X)$ has topological uniform descent for $n \geq d$, then we have:*

(i) $\overline{H_0(T)} = \overline{\mathcal{N}^\infty(T)}$.
(ii) $T^\infty(X)$ *is closed in the range topology on $T^d(X)$.*
(iii) $K(T) = T^\infty(X)$.
(iv) $\mathcal{N}^\infty(T) \subseteq T^\infty(X) + \ker T^d$.

Proof

(i) The inclusion $\overline{\mathcal{N}^\infty(T)} \subseteq \overline{H_0(T)}$ is clear for every operator, since $\mathcal{N}^\infty(T) \subseteq H_0(T)$. To show the reverse inclusion, observe first that by part (iv) of Theorem 1.79, the map $\widehat{T}$ induced by T on $X/\overline{\mathcal{N}^\infty(T)}$ is bounded below, in particular upper semi-Fredholm. Hence, by Theorem 2.37, $\overline{H_0(\widehat{T})} = \mathcal{N}^\infty(\widehat{T}) = \{0\}$. Let $\pi : X \to X/\overline{\mathcal{N}^\infty(\widehat{T})}$ be the canonical surjection, defined by $\pi x := [x]$. Since $\|\pi x\| \le \|x\|$ and $\pi(Tx) = \widehat{T}(\pi x)$ we then have

$$\|\widehat{T}^n(\pi x)\|^{\frac{1}{n}} = \|\pi(T^n x)\|^{\frac{1}{n}} \le \|T^n x\|^{\frac{1}{n}}.$$

If $x \in H_0(T)$ we then have $\|\widehat{T}^n(\pi x)\|^{\frac{1}{n}} \to 0$, so $\pi x \in H_0(\widehat{T})$. Therefore, $\pi(H_0(T) \subseteq H_0(\widehat{T}) = \{0\}$, and hence $\pi(H_0(T) = \{0\}$. From the definition of π we easily see that $H_0(T) \subseteq \overline{\mathcal{N}^\infty(T)}$, and hence $\overline{H_0(T)} \subseteq \overline{\mathcal{N}^\infty(T)}$.

(ii) Since T has topological uniform descent for $n \ge d$, the restriction $T|T^\infty(X)$ is onto, by Theorem 1.79, and $K(T|T^\infty(X)) = T^\infty(X)$ is closed in the operator range topology of $T^d(X)$.

(iii) For all $x \in K(T|T^\infty(X))$ there exists, by definition, a $\delta > 0$ and a sequence $(x_n) \subset T^\infty(X)$ such that $x_0 = x$, $Tx_{n+1} = x_n$, and

$$\|Tx_{n+1}\|_d \le \delta^n \|x\|_d \text{ for all } n = 0, 1, \dots.$$

Let $y := y_0$, where $y_0 := x_d$, and $y_n := x_{n+d}$, for all $n \in \mathbb{N}$. Clearly,

$$Ty_{n+1} = Tx_{n+d+1} = x_{n+d} = y_n.$$

From the definition of the operator range topology we know that there exists a constant $C > 0$ for which

$$C\|Ty_{n+1}\| \le \|Ty_{n+1}\|_d = \|Tx_{n+d+1}\|_d \le \delta^{n+d}\|x\|_d$$
$$= \delta^{n+d}\|T^d x_d\|_d \le \delta^{n+d}\|x_d\| = \delta^n \delta^d \|y\|.$$

Consequently, we can obtain a $\delta_1 > 0$ for which $\|Ty_{n+1}\| \le \delta_1^n \|y\|$, so $y = x_d \in K(T)$ and $x = T^d x_d \in K(T)$. Thus,

$$T^\infty(X) = K(T|T^\infty(X)) \subseteq K(T) \subseteq T^\infty(X),$$

and hence the equality $K(T) = T^\infty(X)$ is proved.

(iv) By part (i) of Lemma 1.80 we have

$$T^\infty(X) + \overline{\mathcal{N}^\infty(T)} = T^\infty(X) + \ker T^d,$$

hence $\mathcal{N}^\infty(T) \subseteq T^\infty(X) + \ker T^d$.   ∎

Corollary 2.96 *If $T \in L(X)$ is quasi-Fredholm then $K(T) = T^{\infty}(X)$ is closed*

Proof Every quasi-Fredholm operator has topological uniform descent, so, by Theorem 2.95, we have $K(T) = T^{\infty}(X)$. If $T \in QF(d)$ then $T^n(X)$ is closed for all $n \geq d$, thus $T^{\infty}(X)$ is closed.   ∎

The SVEP at a point for operators having topological uniform descent may be characterized in several ways:

Theorem 2.97 *Suppose that $\lambda_0 I - T$ has topological uniform descent for $n \geq d$. Then the following conditions are equivalent:*

   (i) *T has the SVEP at λ_0;*
  (ii) *the restriction $T|(\lambda_0 I - T)^d(X)$ has the SVEP at λ_0, where the subspace $(\lambda_0 I - T)^d(X)$ is equipped with the operator range topology;*
 (iii) *the restriction $(\lambda_0 I - T)|(\lambda_0 I - T)^d(X)$ is bounded below, where $(\lambda_0 I - T)^d(X)$ is equipped with the operator range topology;*
  (iv) *$\lambda_0 I - T$ has finite ascent, or equivalently $\lambda_0 I - T$ is left Drazin invertible;*
   (v) *$\sigma_{\mathrm{ap}}(T)$ does not cluster at λ_0;*
  (vi) *$\lambda_0 \notin \mathrm{int}\,\sigma_{\mathrm{ap}}(T)$, where $\mathrm{int}\,\sigma(T)$ is the interior of $\sigma_{\mathrm{ap}}(T)$;*
 (vii) *there exists a $p \in \mathbb{N}$ such that $H_0(\lambda_0 I - T) = \ker(\lambda_0 I - T)^p$;*
(viii) *$H_0(\lambda_0 I - T)$ is closed;*
  (ix) *$H_0(\lambda_0 I - T) \cap K(\lambda_0 I - T) = \{0\}$;*
   (x) *$\mathcal{N}^{\infty}(\lambda_0 I - T) \cap (\lambda_0 I - T)^{\infty}(X) = \{0\}$.*
       In particular, these equivalences hold for semi B-Fredholm operators.

Proof We may assume $\lambda_0 = 0$. We show that the first six statements are equivalent.

(i) $\Rightarrow$ (ii) Let $\mathcal{U}$ be an open neighborhood of 0 and let $f : \mathcal{U} \rightarrow (T^d(X), \|\cdot\|_d)$ be an analytic function such that $(\lambda I - T)f(\lambda) = 0$ for all $\lambda \in \mathcal{U}$. Because $(T^d(X), \|\cdot\|_d)$ may be continuously imbedded in X and

$$(\lambda I - T)|T^d(X)f(\lambda) = (\lambda I - T)f(\lambda),$$

f may be viewed as an analytic function from $\mathcal{U}$ into X which satisfies $(\lambda I - T)f(\lambda) = 0$ for all $\lambda \in \mathcal{U}$. This implies $f = 0$ on $\mathcal{U}$, by the SVEP of T at 0, and hence $T|T^d(X)$ has the SVEP at 0.

(ii) $\Rightarrow$ (iii) Let $Y := T^d(X)$ and $S := T|Y$. Clearly, S has topological uniform descent for $n \geq d$. By Theorem 1.78 the subspace $T^{d+1}(X)$, which is the range of S, is closed with respect the operator range topology. Moreover, by Theorem 1.74 we have $\ker S = \ker T \cap T^d(X) \subseteq S^n(Y)$ for all $n \geq d$, i.e., S is semi-regular. The SVEP of S at 0 then implies that S is bounded below, by Theorem 2.91.

(iii) $\Rightarrow$ (iv) Since $T|T^d(X)$ is injective we have $\{0\} = \ker(T|T^d(X)) = \ker T \cap T^d(X)$, and this implies, by Lemma 1.19, that $p(T) \leq d < \infty$. This, by Theorem 1.142, is equivalent to saying that T is left Drazin invertible.

(iv) $\Rightarrow$ (v) This is a consequence of Theorem 1.92, putting $S = \lambda I - T$ with λ sufficiently small.

(v) $\Rightarrow$ (vi) Obvious.

(vi) $\Rightarrow$ (i) This is an immediate consequence of the identity theorem for analytic functions.

Therefore, the conditions (i)–(vi) are equivalent.

(vi) $\Rightarrow$ (vii) By Theorem 2.95 we have $\overline{H_0(T)} = \overline{\mathcal{N}^\infty(T)}$. If $p := p(T)$ then

$$H_0(T) \subseteq \overline{H_0(T)} = \overline{\mathcal{N}^\infty(T)} = \ker T^p \subseteq H_0(T),$$

so $H_0(T) = \ker T^p$.

(vii) $\Rightarrow$ (viii) Obvious.

(viii) $\Rightarrow$ (ix)$\Rightarrow$ (i) See part (i) of Theorem 2.39.

(ix) $\Rightarrow$ (x) By Theorem 2.95 we have $T^\infty(X) = K(T)$. Since $\mathcal{N}^\infty(T) \subseteq H_0(T)$ holds for every operator, we then have

$$\mathcal{N}^\infty \cap T^\infty(X) \subseteq H_0(T) \cap K(T) = \{0\}.$$

(x) $\Rightarrow$ (i) See Theorem 2.66. $\blacksquare$

Next, we will consider some characterizations of SVEP for T^* at λ_0 in the case when $\lambda_0 I - T$ has topological uniform descent. Recall that the property of having topological uniform descent is not transmitted by duality, so we cannot use the results of Theorem 2.97.

Theorem 2.98 *Suppose that $\lambda_0 I - T$ has topological uniform descent for $n \geq d$. Then the following conditions are equivalent:*

(i) *T^* has the SVEP at λ_0;*
(ii) *there exists an $n \in \mathbb{N}$ such that the restriction $(\lambda_0 I - T)|(\lambda_0 I - T)^n(X)$ is onto, where $(\lambda_0 I - T)^n(X)$ is equipped with the operator range topology;*
(iii) *$\lambda_0 \notin \operatorname{int} \sigma_s(T)$;*
(iv) *$\sigma_s(T)$ does not cluster at λ_0;*
(v) *$\lambda_0 I - T$ has finite descent, or equivalently $\lambda_0 I - T$ is right Drazin invertible;*
(vi) *there exists a $q \in \mathbb{N}$ such that $K(\lambda_0 I - T) = (\lambda_0 I - T)^q(X)$;*
(vii) *$X = H_0(\lambda I - T) + K(\lambda_0 I - T)$;*
(viii) *$H_0(\lambda I - T) + K(\lambda_0 I - T)$ is norm dense in X;*
(ix) *$X = \mathcal{N}^\infty(\lambda_0 I - T) + (\lambda_0 I - T)^\infty(X)$;*
(x) *$\mathcal{N}^\infty(\lambda_0 I - T) + (\lambda_0 I - T)^\infty(X)$ is norm dense in X.*

In particular, the equivalences hold for semi B-Fredholm operators.

Proof Here we can assume $\lambda_0 = 0$.

(i) $\Rightarrow$ (v) Since T has topological uniform descent, for $n \geq d$, the operator $\hat{T} : X/\overline{\mathcal{N}^\infty(T)} \to X/\overline{\mathcal{N}^\infty(T)}$, defined by $\hat{T}[x] = [Tx]$, is bounded below by Theorem 1.79. Therefore the dual $\hat{T}^* : (X/\overline{\mathcal{N}^\infty(T)})^* \to (X/\overline{\mathcal{N}^\infty(T)})^*$ of $\hat{T}$ is onto. Now, by the Annihilator theorem (see Appendix A) there is an linear isometry J of $(X/\overline{\mathcal{N}^\infty(T)})^*$ onto $\overline{\mathcal{N}^\infty(T)}^\perp$ such that

$$J\hat{T}^* J^{-1} = T^*|\overline{\mathcal{N}^\infty(T)}^\perp.$$

Hence, the restriction of T^* on $\overline{\mathcal{N}^\infty(T)}^\perp$ is onto. Note that the invariance of $\overline{\mathcal{N}^\infty(T)}$ under T implies that $\overline{\mathcal{N}^\infty(T)}^\perp$ is invariant under T^*. The restriction $T^*|\overline{\mathcal{N}^\infty(T)}^\perp$ has the SVEP at 0, so, by Corollary 2.61, $T^*|\overline{\mathcal{N}^\infty(T)}^\perp$ is also injective. From the Annihilator theorem we then deduce that $\widehat{T}^*$ is invertible and hence $\widehat{T}$ is also invertible. This implies that the ranges of $\widehat{T}^d$ and $\widehat{T}^{d+1}$ coincide, consequently

$$\frac{T^d(X) + \overline{\mathcal{N}^\infty(T)}}{\overline{\mathcal{N}^\infty(T)}} = \frac{T^{d+1}(X) + \overline{\mathcal{N}^\infty(T)}}{\overline{\mathcal{N}^\infty(T)}}.$$

By Lemma 1.80 we also have $T^d(X) \cap \overline{\mathcal{N}^\infty(T)} = T^{d+1}(X) \cap \overline{\mathcal{N}^\infty(T)}$ and, by using Lemma 1.68, a simple calculation gives

$$\frac{T^d(X)}{T^d(X) \cap \overline{\mathcal{N}^\infty(T)}} = \frac{T^{d+1}(X)}{(T^d(X) \cap \overline{\mathcal{N}^\infty(T)})}.$$

Since $T^{d+1}(X) \subseteq T^d(X)$ we then have that $T^{d+1}(X) = T^d(X)$, hence T has finite descent. By Theorem 1.142 this is equivalent to saying that T is right Drazin invertible.

(v) $\Rightarrow$ (iv) If T has finite descent, then, by Corollary 1.92, there exists an $\varepsilon > 0$ such that $\lambda I - T$ is onto for all $0 < |\lambda| < \varepsilon$, thus $\sigma_s(T)$ does not cluster at 0.

(iv) $\Rightarrow$ (iii) Obvious.

(iii) $\Rightarrow$ (i) This easily follows from the equality $\sigma_s(T) = \sigma_{ap}(T^*)$ and from the identity theorem for analytic functions.

(ii) $\Rightarrow$ (v) If for some natural n the restriction $T|T^n(X)$ is onto, then

$$T^{n+1}(X) = T(T^n(X) = (T|T^n(X))(T^n(X)) = T^n(X),$$

and hence T has finite descent.

(v) $\Rightarrow$ (ii) If $q := q(T) < \infty$, then

$$(T|T^q(X))(T^q(X)) = T(T^q(X)) = T^q(X),$$

thus $T|T^q(X)$ is onto.

(v) $\Rightarrow$ (vi) By Theorem 2.95 we have $K(T) = T^\infty(X)$. Clearly, if T has finite descent q then $T^\infty(X) = T^q(X)$.

(vi) $\Rightarrow$ (v) From Theorem 2.95 we obtain $K(T) = T^\infty(X) = T^q(X)$, so T has finite descent.

(v) $\Rightarrow$ (ix) If $q := q(T) < \infty$, then $T^q(X) = T^\infty(X)$ and hence, by Theorem 1.19,

$$X = \ker T^q + T^q(X) \subseteq \mathcal{N}^\infty(T) + T^\infty(TX).$$

Therefore, $X = \mathcal{N}^\infty(T) + T^\infty(TX)$.

(ix) $\Rightarrow$ (vii) We have $\mathcal{N}^\infty(T) \subseteq H_0(T)$ and, by Theorem 2.95, $K(T) = T^\infty(X)$. Hence, $X = H_0(T) + K(T)$.

Finally, the implications (vii) $\Rightarrow$ (viii) and (ix) $\Rightarrow$ (x) are obvious, while the implications (viii) $\Rightarrow$ (i) and (x) $\Rightarrow$ (i) have been shown in Theorem 2.73 and Corollary 2.74.   ∎

The next corollary is an obvious consequence of Theorems 2.97 and 2.98.

Corollary 2.99 *If $\lambda_0 I - T$ has uniform topological descent then the following statements are equivalent:*

(i) *Both T and T^* have the SVEP at λ_0;*
(ii) *λ_0 is a pole of the resolvent;*
(iii) *$X = H_0(\lambda_0 I - T) \oplus K(\lambda_0 I - T)$;*
(iv) *$X = \mathcal{N}^\infty(\lambda_0 I - T) + (\lambda_0 I - T)^\infty(X)$.*

We now consider the case where $\lambda \in \mathrm{iso}\, \sigma(T)$.

Theorem 2.100 *Suppose that $T \in L(X)$ and $\lambda \in \mathrm{iso}\, \sigma(T)$. Then the following statements are equivalent:*

(i) *$\lambda I - T \in \Phi(X)$;*
(ii) *$\lambda I - T \in \Phi_+(X)$;*
(iii) *$\lambda I - T \in \Phi_-(X)$;*
(iv) *$\lambda I - T$ is Browder;*
(v) *$\dim H_0(\lambda I - T) < \infty$;*
(vi) *$\mathrm{codim}\, K(\lambda I - T) < \infty$.*

Proof If $\lambda \in \mathrm{iso}\, \sigma(T)$ then both T and T^* have the SVEP at λ. The implication (i) $\Rightarrow$ (ii) is obvious. Suppose that $\lambda I - T \in \Phi_+(X)$. The SVEP of T^* at λ then implies, by Theorem 2.97, that $q(\lambda I - T) < \infty$, and hence by Theorem that $\beta(\lambda I - T) < \alpha(\lambda I - T) < \infty$. Thus, $\lambda I - T \in \Phi_-(X)$ and hence (ii) $\Rightarrow$ (iii).

(iii) $\Rightarrow$ (iv) If $\lambda I - T \in \Phi_-(X)$, then $p(\lambda I - T) = q(\lambda I - T) < \infty$, by Corollary 2.99, and hence, again by Theorem 1.22, $\alpha(\lambda I - T) = \beta(\lambda I - T) < \infty$, so $\lambda I - T$ is Browder.

(iv) $\Rightarrow$ (v) By Corollary 2.47 $H_0(\lambda I - T) = \ker (\lambda I - T)^p$, where $p; = p(\lambda I - T)$. Since $\alpha(\lambda I - T) < \infty$, $\alpha(\lambda I - T)^p < \infty$, so $H_0(\lambda I - T)$ is finite-dimensional.

(v) $\Rightarrow$ (vi) Clear, since by Theorem 2.45, or by Corollary 2.99, we have $X = H_0(\lambda I - T) \oplus K(\lambda I - T)$.

(vi) $\Rightarrow$ (i) From the inclusion $K(\lambda I - T) \subseteq (\lambda I - T)$ it then follows that $\beta(\lambda I - T) < \infty$. The SVEP of T and T^* at λ entails that $p(\lambda I - T) = q(\lambda I - T) < \infty$, so, by Theorem 1.22, $\alpha(\lambda I - T) = \beta(\lambda I - T) < \infty$. Hence, $\lambda I - T \in \Phi(X)$. ∎

Another consequence of Theorems 2.97 and 2.98 is the following theorem, whose proof is omitted since it is similar to that of Theorem 2.100.

Theorem 2.101 *Suppose that $T \in L(X)$ and $\lambda \in \operatorname{iso} \sigma_{\mathrm{ap}}(T)$. Then the following statements are equivalent:*

(i) $\lambda I - T \in \Phi(X)$;
(ii) $\lambda I - T \in \Phi_{+}(X)$;
(iii) $\lambda I - T$ *is upper semi-Browder.*
       Analogously, if $\lambda \in \operatorname{iso} \sigma_{\mathrm{s}}(T)$ the following statements are equivalent:
(iv) $\lambda I - T \in \Phi(X)$;
(v) $\lambda I - T \in \Phi_{-}(X)$;
(vi) $\lambda I - T$ *is lower semi-Browder.*

Theorems 2.97 and 2.98 has some other applications. If $T \in L(X)$, define, as in Chap. 2,

$$c_n(T) := \dim \frac{T^n(X)}{T^{n+1}(X)}$$

and

$$c_n'(T) := \dim \frac{\ker T^{n+1}}{\ker T^n}.$$

Corollary 2.102 *Suppose that $T \in L(X)$ has uniform topological descent for $n \geq d$.*

(i) *If T^* has the SVEP at 0 then $c_d'(T) \leq c_d(T)$.*
(ii) *If $S \in L(X)$ commutes with T and is sufficiently small and invertible, then T^* has the SVEP at 0 if and only if $T^* + S^*$ has the SVEP at 0.*

Proof

(i) Suppose that $c_d(T) > c_d'(T)$. Then $c_d(T) > 0$. Let $\lambda \neq 0$ be small enough. By Theorem 1.89 $c_n(\lambda I - T) = c_d(T) > 0$ for all $n \in \mathbb{N}$, so $\lambda I - T$ is not onto. By Corollary 1.92 T has infinite descent, so, by Theorem 2.98, T^* does not have the SVEP at 0.
(ii) From Theorem 1.89 and Corollary 1.92, $T + S$ has topological uniform descent for $n \geq d$, and descent $q(T) > \infty$ precisely when $q(T + S) < \infty$. ∎

In the case when $\lambda_0 I - T$ is quasi-Fredholm, the SVEP at λ_0 may be characterized in several other ways:

Theorem 2.103 *Let $T \in L(X)$ and suppose that $\lambda_0 I - T$ is quasi-Fredholm of degree d. Then the following assertions are equivalent:*

(i)   *T has the SVEP at λ_0;*
(ii) *$\mathcal{N}^{\infty}(\lambda_0 I - T) \cap (\lambda_0 I - T)^{\infty}(X) = \{0\}$;*
(iii) *$\mathcal{N}^{\infty}(\lambda_0 I - T)^{\perp} + (\lambda_0 I - T)^{\infty}(X)^{\perp} = X^*$;*
(iv) *$\mathcal{N}^{\infty}(\lambda_0 I^* - T^*) + (\lambda_0 I^* - T^*)^{\infty}(X)^{\perp}$ is weak*-dense in X^*;*
(v)   *$H_0(\lambda_0 I^* - T^*) + K(\lambda_0 I^* - T^*)$ is weak*-dense in X^*;*
(vi) *$H_0(\lambda_0 I^* - T^*) + (\lambda_0 I^* - T^*)(X^*)$ is weak*-dense in X^*.*

Proof We may assume that $\lambda_0 = 0$.

(i) $\Rightarrow$ (ii) Assume that T has the SVEP at 0, or equivalently, by part (ix) of Theorem 2.97, that $H_0(T) \cap K(T) = \{0\}$. Since T is quasi-Fredholm we have $T^\infty(X) = K(T)$, by Corollary 2.96. Therefore,

$$\mathcal{N}^\infty(T) \cap T^\infty(X) \subseteq H_0(T) \cap T^\infty(X) = H_0(T) \cap K(T) = \{0\},$$

thus the statement (i) holds.

(ii) $\Rightarrow$ (iii) T has topological uniform descent for $n \geq d$, so by using part (i) of Lemma 1.80 and part (v) of Theorem 1.78, we deduce that $\mathcal{N}^\infty(T) + T^\infty(X) =$ ker $T^d + T^\infty(X)$ is closed. Consequently,

$$\mathcal{N}^\infty(T)^\perp + T^\infty(X)^\perp = [\mathcal{N}^\infty(T) + T^\infty(X)]^\perp = X^*.$$

(iii) $\Rightarrow$ (iv) By Theorem 1.104, T^* is quasi-Fredholm and hence $T^{*n}(X^*)$ is closed for all $n \geq d$. Consequently,

$$\mathcal{N}^\infty(T)^\perp \subseteq (\ker T^n)^\perp = T^{*n}(X^*) \quad \text{for all } n \geq d.$$

From this it then follows that $\mathcal{N}^\infty(T)^\perp \subseteq T^{*\infty}(X^*)$. Similarly, since T is quasi-Fredholm we have $^\perp\mathcal{N}^\infty(T^*) \subseteq T^\infty(X)$, from which we obtain

$$T^\infty(X)^\perp \subseteq [^\perp\mathcal{N}^\infty(T^*)]^\perp = \overline{\mathcal{N}^\infty(T^*)}^{w*}.$$

From our assumption it then follows that

$$X^* = \mathcal{N}^\infty(T)^\perp + T^\infty(X)^\perp \subseteq T^{*\infty}(X^*) + \overline{\mathcal{N}^\infty(T^*)}^{w*}$$
$$\subseteq \overline{\mathcal{N}^\infty(T^*) + T^{*\infty}(X^*)}^{w*} \subseteq X^*,$$

so that $\mathcal{N}^\infty(T^*) + T^{*\infty}(X^*)$ is weak*-dense in X^*.

(iv) $\Rightarrow$ (v). This is clear, since T^* is quasi-Fredholm and hence $T^{*\infty}(X^*) = K(T^*)$, by Corollary 2.96.

(v) $\Rightarrow$ (vi) Obvious, since $K(T^*) \subseteq T^*(X^*)$.

The implication (vi) $\Rightarrow$ (i) has been proved in Theorem 2.75. ∎

The next result is dual to Theorem 2.103.

Theorem 2.104 *Let* $T \in L(X)$ *and suppose that* $\lambda_0 I - T$ *is quasi-Fredholm of degree* d. *Then the following assertions are equivalent:*

(i) T^* *has the SVEP at* λ_0;
(ii) $X = \mathcal{N}^\infty(\lambda_0 I - T) + (\lambda_0 I - T)^\infty(X)$;
(iii) $\mathcal{N}^\infty(\lambda_0 I - T)^\perp \cap (\lambda_0 I - T)^\infty(X)^\perp = \{0\}$;
(iv) $\mathcal{N}^\infty(\lambda_0 I^* - T^*) \cap (\lambda_0 I^* - T^*)^\infty(X^*) = \{0\}$;
(v) $\ker(\lambda_0 - T^*) \cap (\lambda_0 I^* - T^*)^\infty(X^*) = \{0\}$.

Proof Here we can suppose that $\lambda_0 = 0$.

(i) $\Rightarrow$ (ii) Suppose that T^* has the SVEP at 0, or equivalently, by Theorem 2.98, that the descent $q := q(T) < \infty$. Then $T^\infty(X) = T^q(X)$ and hence, by Lemma 1.19,

$$X = \ker T^q + T^q(X) \subseteq \mathcal{N}^\infty(T) + T^q(X) = \mathcal{N}^\infty(T) + T^\infty(X),$$

thus (ii) holds.

(ii) $\Rightarrow$ (iii) Clear.

(iii) $\Rightarrow$ (iv) Since T is quasi-Fredholm of degree d, $T^n(X)$ is closed for all $n \geq d$. Hence $\ker(T^n)^\perp = T^*(X^*)$ and consequently $T^{*\infty}(X^*) = \mathcal{N}^\infty(T)^\perp$. By assumption

$$\mathcal{N}^\infty(T)^\perp \cap T^\infty(X)^\perp = [\mathcal{N}^\infty(T) + T^\infty(X)]^\perp = \{0\},$$

thus $X = \mathcal{N}^\infty(T) + T^\infty(X)$. Since $K(T) = T^\infty(X)$ it then follows that $X = \mathcal{N}^\infty(T) + K(T)$, and hence $H_0(T) + K(T) = X$, or equivalently $q := q(T) < \infty$, by Theorem 2.98. This implies that $p(T^*) \leq \infty$, so

$$\mathcal{N}^\infty(T^*) = \ker T^{*q} = T^q(X)^\perp = T^\infty(X)^\perp.$$

Thus,

$$\mathcal{N}^\infty(T^*) \cap T^{*\infty}(X^*) = \mathcal{N}^\infty(T)^\perp \cap T^\infty(X)^\perp = \{0\}.$$

(iv) $\Rightarrow$ (v) Obvious, since $\ker T^* \subseteq \mathcal{N}^\infty(T^*)$.

(v) $\Rightarrow$ (i) This follows from Theorem 2.60.   ∎

From Corollary 1.83 we know that every essentially semi-regular operator has topological uniform descent, so the results established in Theorems 2.97 and 2.98 are valid for this class of operators.

Theorem 2.105 *Suppose that $\lambda_0 I - T$ is essentially semi-regular. Then we have*

(i) *T has the SVEP at λ_0 if and only if $H_0(\lambda_0 I - T)$ is finite-dimensional. In this case $\lambda_0 I - T \in \Phi_+(X)$.*

(ii) *T^* has the SVEP at λ_0 if and only if $K(\lambda_0 I - T)$ has finite codimension. In this case $\lambda_0 I - T \in \Phi_-(X)$.*

In particular, these equivalences hold if $\lambda_0 I - T$ is semi-Fredholm

Proof Suppose also here that $\lambda_0 = 0$.

(i) Since T is essentially semi-regular, in the corresponding GKD (M, N) for T the subspace N is finite-dimensional. Suppose that T has the SVEP at 0. Then $T|M$ has the SVEP at 0, since the SVEP at 0 of T is inherited by the restrictions on every closed invariant subspaces. Moreover, $T|M$ is semi-regular, so, by Theorem 2.91, $T|M$ has the SVEP at 0 if and only if

$T|M$ is injective. From Theorem 2.37 the semi-regularity of $T|M$ implies that $\overline{H_0(T|M)} = \mathcal{N}^\infty(T|M) = \{0\}$, and hence

$$H_0(T) = H_0(T|M) \oplus H_0(T|N) = \{0\} \oplus N = N.$$

Consequently, $H_0(T) = N$ is finite-dimensional. Conversely, if $H_0(T)$ is finite-dimensional then $H_0(T)$ is closed, which is equivalent to saying that T has the SVEP at 0, by Theorem 2.97.

Now, if $H_0(T)$ is finite-dimensional then ker T is also finite-dimensional, since ker $T \subseteq H_0(T)$. On the other hand, $T(X) = T(M) + T(N)$ and $T(M)$ is closed, since $T|M$ is semi-regular, so $T(X)$ is closed, because it is the sum of a closed subspace and a finite-dimensional subspace of X. Therefore, $T \in \Phi_+(X)$.

(ii) Since T is essentially semi-regular, in the GKD (M, N) for T the subspace M has finite codimension in X, since N is finite-dimensional. We also know that the pair $(N^\perp, M^\perp)$ is a GKD for T^*, so $T^*|N^\perp$ is semi-regular, and hence, by Theorem 2.79, T^* has the SVEP at 0 if and only if $T^* |N^\perp$ is injective. By Lemma 2.78, T^* has the SVEP at 0 if and only if $T |M$ is onto. If $T|M$ is surjective then $M = K(T|M) = K(T)$, by Theorem 2.37, so $K(T)$ has finite codimension. Conversely, suppose that $K(T)$ has finite codimension. By Theorem 1.64 we have $K(T) = T^\infty(X)$ and $T^\infty(X) \subseteq T^n(X)$ for all $n \in \mathbb{N}$, from which we obtain that the descent $q(T) < \infty$, and this is equivalent to saying that T^* has the SVEP at 0, by Theorem 2.98. Finally, from $K(T) = T^\infty(X) \subseteq T(X)$ we see that $T(X)$ has finite codimension, so $T \in \Phi_-(X)$.

The last assertion is clear: every semi-Fredholm operator is essentially semi-regular.   ∎

For semi-Fredholm operators we have the following important result.

Corollary 2.106 *Suppose that $\lambda_0 I - T$ is semi-B-Fredholm. Then the following statements hold:*

(i) *If T has the SVEP at λ_0 then* ind $(\lambda_0 I - T) \leq 0$.
(ii) *If T^* has the SVEP at λ_0 then* ind $(\lambda_0 I - T) \geq 0$.
     Consequently, if both T and T^* have the SVEP at λ_0 then $\lambda_0 I - T \in \Phi(X)$ and ind $(\lambda_0 I - T) = 0$.

Proof

(i) By Theorem 2.97 we know that if T has the SVEP at λ_0 then $p(\lambda_0 I - T) < \infty$, and this implies, by part (i) of Theorem 1.22, that $\alpha(\lambda_0 I - T) \leq \beta(\lambda_0 I - T)$, thus ind $(\lambda_0 I - T) \leq 0$.

(ii) By Theorem 2.98 we know that if T^* has the SVEP at λ_0 then $q(\lambda_0 I - T) < \infty$, and this implies, by part (i) of Theorem 1.22, that $\beta(\lambda_0 I - T) \leq \alpha(\lambda_0 I - T)$, hence ind $(\lambda_0 I - T) \geq 0$.

The last assertion is clear.   ∎

The converses of the results of Corollary 2.106 do not hold, i.e., a semi-Fredholm operator with index less than or equal to 0 may fail to have the SVEP at 0. For instance, if R and L denote the right shift and the left shift on the Hilbert space $\ell_2(\mathbb{N})$, then it is easy to see that R is injective, and $\alpha(R \oplus L) = \alpha(L) = 1$, while $\beta(R \oplus L) = \beta(R) = 1$, so $R \oplus L$ is a Fredholm operator having index 0. Since L fails SVEP at 0, $R \oplus L$ does not have the SVEP at 0, by Theorem 2.15.

Corollary 2.107 *If $\lambda I - T \in L(X)$ is B-Fredholm with* ind $(\lambda I - T) = 0$ *then the following statements are equivalent:*

  (i) *T has the SVEP at λ;*
 (ii) *$\lambda I - T$ is Drazin invertible;*
(iii) *T^* has the SVEP at λ.*

Proof (i) $\Leftrightarrow$ (ii) The SVEP of T at λ is equivalent to $p(\lambda I - T) < \infty$, by Theorem 2.97, and this is equivalent to saying that $\lambda I - T$ is Drazin invertible, by Theorem 1.143. (iii) $\Leftrightarrow$ (ii) The SVEP of T^* at λ is equivalent to $q(\lambda I - T) < \infty$, by Theorem 2.98, and this is equivalent to saying that $\lambda I - T$ is Drazin invertible, again by Theorem 1.143. ∎

It has been observed in Remark 2.62 that if an operator $T \in L(X)$ has the SVEP at λ_0 and if Y is a closed subspace of X such that $(\lambda_0 I - T)(Y) = Y$, then ker $(\lambda_0 I - T) \cap Y = \{0\}$.

The following useful result shows that this result is even true if we assume that Y is complete with respect to a new norm and Y is continuously embedded in X.

Lemma 2.108 *Suppose that X is a Banach space and that the operator $T \in L(X)$ has the SVEP at λ_0. Let Y be a Banach space which is continuously embedded in X and satisfies $(\lambda_0 I - T)(Y) = Y$. Then* ker $(\lambda_0 I - T) \cap Y = \{0\}$.

Proof By the closed graph theorem the restriction $T|Y$ is continuous with respect to the given norm $\| \cdot \|_1$ on Y. Moreover, since every analytic function $f : \mathcal{U} \to (Y, \| \cdot \|_1)$ on an open set $\mathcal{U} \subseteq \mathbb{C}$ remains analytic, when considered as a function from $\mathcal{U}$ to X, it is clear that $T|Y$ inherits the SVEP at λ_0 from T. Hence Corollary 2.61 applies to $T|Y$ with respect to the norm $\| \cdot \|_1$. ∎

By Theorem 2.97, T has the SVEP at λ_0 precisely when $p(\lambda_0 I - T) < \infty$. The next result shows that this equivalence also holds under the assumption that $q(\lambda_0 I - T) < \infty$.

Theorem 2.109 *Let $T \in L(X)$, X a Banach space, and suppose that $0 < q(\lambda_0 I - T) < \infty$. Then the following conditions are equivalent:*

  (i) *T has the SVEP at λ_0;*
 (ii) *$p(\lambda_0 I - T) < \infty$;*
(iii) *λ_0 is a pole of the resolvent;*
 (iv) *λ_0 is an isolated point of $\sigma(T)$.*

Proof There is no harm in assuming $\lambda_0 = 0$.

(i) $\Rightarrow$ (ii) Let $q := q(T)$ and $Y := T^q(X)$. Let us consider the map $\widehat{T}$: $X/\ker T^q \to Y$ defined by $\widehat{T}(\widehat{x}) := Tx$ where $x \in \widehat{x}$. Clearly, since $\widehat{T}$ is continuous and bijective we can define in Y a new norm

$$\|y\|_1 := \inf\{\|x\| : T^q(x) = y\},$$

for which $(Y, \|\cdot\|_1)$ becomes a Banach space. Moreover, if $y = T^q(x)$ from the estimate

$$\|y\| = \|T^q(x)\| \le \|T^q\|\|x\|$$

we deduce that Y can be continuously embedded in X. Since $T(T^q(X)) = T^{q+1}(X) = T^q(X)$, by Corollary 2.108 we conclude that $\ker T \cap T^q(X) = \{0\}$ and hence by Lemma 1.19 $p(T) < \infty$.

(ii) $\Rightarrow$ (iii) If $p := p(\lambda_0 I - T) = q(\lambda_0 I - T) < \infty$ then λ_0 is a pole of order p.

(iii) $\Rightarrow$ (iv) Obvious.

(iv) $\Rightarrow$ (i) This has been observed above.   ∎

The quasi-nilpotent operators may be characterized in the following way:

Theorem 2.110 *If $T \in L(X)$ then the following statements are equivalent:*

(i) *T is quasi-nilpotent;*
(ii) *$K(T) = \{0\}$ and $0 \in \mathrm{iso}\,\sigma(T)$;*
(iii) *$\overline{H_0(T)} = X$ and $0 \in \mathrm{iso}\,\sigma(T)$.*

Proof (i) $\Leftrightarrow$ (ii). The implication (i) $\Rightarrow$ (ii) is clear by Theorem 2.71. Conversely, suppose that $0 \in \mathrm{iso}\,\sigma(T)$ and $K(T) = \{0\}$. If P_0 is the spectral projection associated with $\{0\}$ then, by Theorem 2.45, $\ker P_0 = K(T) = \{0\}$, and $P_0(X) = H_0(T)$. This implies that P_0 is the identity and hence $H_0(T) = X$, so T is quasi-nilpotent by Theorem 2.35.

(i) $\Leftrightarrow$ (iii) The implication (i) $\Rightarrow$ (ii) is clear, since $H_0(T) = X$. The reverse is also clear, since $H_0(T) = P_0(X)$ is closed, so $\overline{H_0(T)} = H_0(T) = X$.   ∎

We know, from Theorem 2.45, that if λ_0 is isolated in $\sigma(T)$ then $X = K(\lambda_0 I - T) \oplus H_0(\lambda_0 I - T)$, where the direct sum is in the topological sense. We now show that the reverse implication holds if we assume that only $K(\lambda_0 I - T)$ is closed and the direct sum is in the algebraic sense.

Theorem 2.111 *For a bounded operator $T \in L(X)$, where X is a Banach space, the following assertions are equivalent:*

(i) *λ_0 is an isolated point of $\sigma(T)$;*
(ii) *$K(\lambda_0 I - T)$ is closed and $X = K(\lambda_0 I - T) \oplus H_0(\lambda_0 I - T)$, where $\oplus$ is the algebraic sum.*

Proof Here we assume $\lambda_0 = 0$. By assumption $K(T)$ is closed and by Theorem 1.39 we know $K(T)$ is T-invariant. Let $\overline{T} := T|K(T)$. Since $\ker T \subseteq H_0(T)$, $\overline{T}$ is invertible. Hence there exists an $\varepsilon > 0$ such that $\lambda I - \overline{T}$ is invertible for every $|\lambda| < \varepsilon$. Consequently,

$$(\lambda I - T)(K(T)) = K(T) \text{ for every } |\lambda| < \varepsilon. \tag{2.19}$$

Since $\ker (\lambda I - T) \subseteq K(T)$ for all $\lambda \neq 0$, we then have

$$\ker (\lambda I - T) = \{0\} \quad \text{for every } 0 < |\lambda| < \varepsilon. \tag{2.20}$$

From part (iv) of Lemma 2.34, we also have

$$H_0(T) \subseteq (\lambda I - T)(X) \quad \text{for every } \lambda \neq 0. \tag{2.21}$$

The equality (2.19) and the inclusion (2.20) then imply

$$X = K(T) \oplus H_0(T) \subseteq (\lambda I - T)(X) \quad \text{for every } 0 < |\lambda| < \varepsilon.$$

Consequently

$$\{\lambda \in \mathbb{C} : 0 < |\lambda| < \varepsilon\} \subseteq \rho(T),$$

and hence 0 is an isolated point of $\sigma(T)$. ∎

Theorem 2.112 *Suppose that either $H_0(T)$ or $H_0(T) \cap K(T)$ are closed. Then the following statements are equivalent:*

(i) *there exists a commuting projection P such that $T + P$ is onto and TP is quasi-nilpotent.*
(ii) $0 \in \operatorname{iso} \sigma(T)$.

Proof (ii) $\Rightarrow$ (i) There is nothing to prove, by Theorem 2.46. To show (i) $\Rightarrow$ (ii) observe that $X = H_0(T) + K(T)$ by Theorem 2.85. Our assumption that $H_0(T)$ is closed, or that $H_0(T) \cap K(T)$ is closed, entails by Theorem 2.39 that $H_0(T) \cap K(T) = \{0\}$, so X is the algebraic direct sum of $H_0(T)$ and $K(T)$. By Theorem 2.85 $K(T)$ is closed, and Theorem 2.111 then entails that $0 \in \operatorname{iso} \sigma(T)$. ∎

Remark 2.113 The assumption that $H_0(T)$ is closed is essential in Theorem 2.112. To see this, let R denote the *right shift* on the Hilbert space $\ell_2(\mathbb{N})$, defined as

$$R(x_1, x_2, \dots) := (0, x_1, x_2, \dots) \quad \text{for all } (x_n) \in \ell_2(\mathbb{N}).$$

The Hilbert adjoint of R is the *left shift* L defined as

$$L(x_1, x_2, \dots) := (x_2, , x_3, \dots) \quad \text{for all } (x_n) \in \ell_2(\mathbb{N}).$$

The operator L is onto, indeed we have $\sigma_{\mathrm{ap}}(R) = \sigma_s(L)) = \Gamma$, where Γ denotes the unit circle of $\mathbb{C}$, and $\sigma(L) = D(0, 1)$, the unit closed disc of $\mathbb{C}$. Define $T := L \oplus Q$, where Q is any quasi-nilpotent operator on $\ell_2(\mathbb{N})$. By Theorem 2.85 the condition (ii) of Theorem 2.112 is satisfied by T. Note that the quasi-nilpotent part $H_0(T)$ is not closed, otherwise T would have the SVEP at 0, and hence, by Theorem 2.15, L would also have the SVEP at 0, and we known that this is not true. Now, $0 \in \mathrm{iso}\, \sigma_s(T)$, by Theorem 2.85, while $\sigma(T) = \sigma(L) \cup \sigma(Q) = D(0, 1)$, so $0 \notin \mathrm{iso}\, \sigma(T)$.

2.7   Components of Semi B-Fredholm Regions

In this section we give a classification of the components of the semi B-Fredholm region of an operator, by using the localized SVEP.

Let us consider the *upper semi B-Fredholm region* of $T \in L(X)$, defined as

$$\Psi_u(T) := \{\lambda \in \mathbb{C} : \lambda I - T \text{ is upper semi B-Fredholm}\},$$

and analogously let us consider the *lower semi B-Fredholm region*, defined as

$$\Psi_l(T) := \{\lambda \in \mathbb{C} : \lambda I - T \text{ is lower semi B-Fredholm}\}.$$

According to Theorem 1.117, if $\lambda_0 I - T \in L(X)$ is upper semi B-Fredholm (respectively, lower semi B-Fredholm), then there exists an open disc $\mathbb{D}(\lambda_0, \varepsilon)$ centered at λ_0 such that $\lambda I - T$ is upper semi-Fredholm (respectively, lower semi-Fredholm) for all $\lambda \in \mathbb{D}(\lambda_0, \varepsilon) \setminus \{\lambda_0\}$. Moreover, $\alpha(\lambda I - T)$ (respectively, $\beta(\lambda I - T)$) is constant as λ ranges over $\mathbb{D}(\lambda_0, \varepsilon) \setminus \{\lambda_0\}$ and

$$\mathrm{ind}\,(\lambda I - T) = \mathrm{ind}\,(\lambda_0 - T) \quad \text{for all } \lambda \in \mathbb{D}(\lambda_0, \varepsilon).$$

From this it then follows that both $\Psi_u(T)$ and $\Psi_l(T)$ are open subsets of $\mathbb{C}$, so they can be decomposed into components, i.e. maximal open, connected, pairwise disjoint non-empty subsets of $\mathbb{C}$. Note that $\Psi_u(T)$, as well as $\Psi_l(T)$, may coincide with all of $\mathbb{C}$ (this is the case for an algebraic operator, whose spectrum is a finite set of poles, see the next chapter). Recall that $\lambda \in \mathbb{C}$ is said to be a *deficiency value* of T if $\lambda I - T$ is not onto.

In the sequel, by acc K we denote the set of all accumulation points of $K \subseteq \mathbb{C}$.

Theorem 2.114 *If Ω is a component of $\Psi_u(T)$, or a component of $\Psi_l(T)$, then the index $\mathrm{ind}\,(\lambda I - T)$ is constant as λ ranges over Ω.*

Proof Join a fixed point λ_0 of a component Ω of $\Psi_u(T)$ to an arbitrary point $\lambda_1 \in \Omega$ by a polygonal line Γ contained in Ω. Associate with each $\mu \in \Gamma$ an open disc in which $\mathrm{ind}\,(\lambda I - T) = \mathrm{ind}\,(\mu I - T)$. By the Heine–Borel theorem already finitely many of these discs cover Γ. Therefore, we have ind $(\lambda_1 I - T) = $ ind $(\lambda_0 I - T)$. The proof for a component of $\Psi_l(T)$ is the same.   ∎

Theorem 2.115 *Suppose that $\lambda_0 \in \Psi_u(T)$. Then the following alternative holds:*

(i) $p(\lambda_0 I - T) < \infty$. *In this case $\lambda_0 \notin \mathrm{acc}\, \sigma_{\mathrm{ap}}(T)$.*

(ii) $p(\lambda_0 I - T) = \infty$. *In this case there exists an open disc $\mathbb{D}_0$ centered at λ_0 such that all points $\lambda \in \mathbb{D}_0$ are eigenvalues of T.*

  Analogously, if $\lambda_0 \in \Psi_l(T)$ then we have the following alternative:

(iii) $q(\lambda_0 I - T) < \infty$. *In this case $\lambda_0 \notin \mathrm{acc}\, \sigma_s(T)$.*

(iv) $q(\lambda_0 I - T) = \infty$. *In this case there exists an open disc $\mathbb{D}_0$ centered at λ_0 such that all points $\lambda \in \mathbb{D}_0$ are deficiency values of T.*

Proof

(i) Every semi B-Fredholm operator is quasi-Fredholm, hence has uniform topological descent, so, by Theorem 2.97, $p(\lambda_0 I - T) < \infty$ precisely when $\sigma_{\mathrm{ap}}(T)$ does not cluster at λ_0.

(ii) Suppose that $p(\lambda_0 I - T) = \infty$. Clearly, $\alpha(\lambda_0 I - T) > 0$. Again by Theorem 2.97, $\lambda_0 \in \mathrm{acc}\, \sigma_{\mathrm{ap}}(T)$, so there exists a sequence $\{\lambda_n\} \subseteq \sigma_{\mathrm{ap}}(T)$, $\lambda_n \neq \lambda_0$, such that $\lambda_n \to \lambda_0$ as $n \to \infty$. By Theorem 1.117, there exists an $n_0 \in \mathbb{N}$ such that the operators $\lambda_n I - T$ are upper semi-Fredholm for $n \geq n_0$, and hence have closed range. Since $\lambda_n \in \sigma_{\mathrm{ap}}(T)$ it then follows that $\alpha(\lambda_n I - T) > 0$ for $n \geq n_0$.

  On the other hand, by Theorem 1.117 there exists an open disc $\mathbb{D}(\lambda_0, \varepsilon)$, centered at λ_0, such that $\alpha(\lambda I - T)$ is constant as λ ranges over $\mathbb{D}(\lambda_0, \varepsilon) \setminus \{\lambda_0\}$. Therefore, for $n \in \mathbb{N}$ sufficiently large and $\lambda \in \mathbb{D}(\lambda_0, \varepsilon) \setminus \{\lambda_0\}$ we have

$$\alpha(\lambda I - T) = \alpha(\lambda_n I - T) > 0,$$

so statement (ii) is proved.

(iii) Since $\lambda_0 I - T$ is lower semi B-Fredholm then, by Theorem 2.98, $q(\lambda I - T) < \infty$ if and only if $\sigma_s(T)$ does not cluster at λ_0.

(iv) Argue as in the proof of part (ii): just use the constancy of $\beta(\lambda I - T)$ as λ ranges over a suitable punctured disc $\mathbb{D}(\lambda_0, \varepsilon)$ centered at λ_0.   ∎

Now we give a complete classification of the components of $\Psi_u(T)$, or $\Psi_l(T)$.

Theorem 2.116 *Let $T \in L(X)$. Then the following statements hold:*

(i) *If Ω_1 is a component of $\Psi_u(T)$ then $\lambda - T$ is left Drazin invertible either for every point of Ω_1 or for no point of Ω_1, or equivalently, T has the SVEP either at every point of Ω_1 or at no point of Ω_1. In the first case $\mathrm{ind}\,(\lambda I - T) \leq 0$ for all $\lambda \in \Omega_1$.*

(ii) *If Ω_2 is a component of $\Psi_l(T)$ then $\lambda - T$ is right Drazin invertible either for every point of Ω_2 or for no point of Ω_2, or equivalently, T^* has the SVEP either at every point of Ω_2 or at no point of Ω_2. In the first case $\mathrm{ind}\,(\lambda I - T) \leq 0$ for all $\lambda \in \Omega_2$.*

Proof (i) Suppose that $\lambda I - T$ is upper semi B-Fredholm for all $\lambda \in \Omega_1$. It suffices to prove by Theorem 2.97 that $p(\lambda I - T)$ is finite either for every point or for no point of Ω_1. Define

$$\Lambda_1 := \{\lambda \in \Omega_1 : p(\lambda I - T) < \infty\}.$$

If $\lambda_0 \in \Lambda_1$ then, by part (i) of Theorem 2.115, there exists an open disc $\mathbb{D}_0$ centered at λ_0 such that $\lambda I - T$ is bounded below for all $\lambda \in \mathbb{D}_0 \setminus \{\lambda_0\}$, in particular $p(\lambda I - T) = 0$ for all $\lambda \in \mathbb{D}_0 \setminus \{\lambda_0\}$. Therefore, Λ_1 is an open subset of Ω_1.

Next, we want show that Λ_1 is also a closed subset of Ω_1. To see this, let $\mu \in \mathrm{acc}\, \Lambda_1$ and $\{\lambda_n\}$ be a sequence from Λ_1, $\lambda_n \neq \mu$, such that $\lambda_n \to \mu$. Suppose that $p(\mu I - T) = \infty$. By Theorem 2.115, part (ii), there exists an open disc $\mathbb{D}$ centered at μ consisting of eigenvalues. Therefore, for n sufficiently large, a neighborhood of λ_n would consist of eigenvalues of T, in contradiction with the case (i) of Theorem 2.115. Thus $\mu \in \Lambda_1$, Λ_1 is a closed subset of Ω_1 and since Ω_1 is connected it then follows that $\Lambda_1 = \Omega_1$. Therefore, $p := p(\lambda I - T) < \infty$ for all $\lambda \in \Omega_1$. By Theorem 1.23 we have $p(\lambda I - T_k) = 0$ for all $k \geq p$. But for k sufficiently large $\lambda I - T_k$ has closed range, so $\lambda I - T$ is left Drazin invertible, and $\mathrm{ind}\,(\lambda I - T) \leq 0$, by Theorem 1.141.

The proof of part (ii) is similar. If

$$\Lambda_2 := \{\lambda \in \Omega_2 : q(\lambda I - T) < \infty\},$$

by using part (iii) and part (iv) of Theorem 1.117 and by arguing as in part (i), just use Theorem 2.98, it then easily follows that $\Lambda_2 = \Omega_2$. Therefore, $q(\lambda I - T) < \infty$ for all $\lambda \in \Omega_2$, and hence, by Theorem 1.24, $q(\lambda I - T_k) = 0$ for all $k \geq q$. Again, since for k sufficiently large $\lambda I - T_k$ has closed range, $\lambda I - T$ is right Drazin invertible, and hence $\mathrm{ind}\,(\lambda I - T) \geq 0$, by Theorem 1.141. ∎

Theorem 2.117 *Suppose that $T \in L(X)$ and Ω is a component of $\Psi_u(T)$. If T has the SVEP then only the following cases are possible:*

(i) *T^* has the SVEP at every $\lambda \in \Omega$. In this case $p(\lambda I - T) = q(\lambda I - T) < \infty$ and $\mathrm{ind}\,(\lambda I - T) = 0$ for all $\lambda \in \Omega$. Every $\lambda \in \sigma(T) \cap \Omega$ is a pole. The eigenvalues and the deficiency values do not cluster in Ω. This case occurs exactly when Ω intersects the resolvent $\rho(T) := \mathbb{C} \setminus \sigma(T)$.*

(ii) *T^* fails the SVEP at some points $\lambda \in \Omega$. In this case T^* fails the SVEP at every point $\lambda \in \Omega$. Moreover, $p(\lambda I - T) < \infty$, $q(\lambda I - T) = \infty$ and $\mathrm{ind}\,(\lambda I - T) < 0$ for all $\lambda \in \Omega$. Every $\lambda \in \sigma(T) \cap \Omega$ is a left pole. The eigenvalues do not cluster in Ω, while every point of Ω is a deficiency value.*

Proof

(i) By Theorem 2.98, if $\lambda I - T$ is upper semi B-Fredholm, the SVEP for T^* at λ is equivalent to saying that $q(\lambda I - T) < \infty$, and, again by Theorem 2.97, the SVEP for T implies $p(\lambda I - T) < \infty$ for all $\lambda \in \Omega$. Therefore, $p(\lambda I - T) < \infty = q(\lambda I - T)$, so every $\lambda \in \sigma(T) \cap \Omega$ is a pole. Therefore $\lambda I - T$ is Drazin invertible, and hence, by Theorem 1.141, ind $(\lambda I - T) = 0$. The remaining assertion follows from Theorem 2.116.

(ii) If T^* fails the SVEP at a point $\lambda_0 \in \Omega$ then, by Theorem 2.98, $q(\lambda_0 I - T) = \infty$. The SVEP for T implies that $p(\lambda_0 I - T) < \infty$, and hence, by Theorem 1.22, we have ind $(\lambda_0 I - T) \leq 0$. On the other hand, if ind $(\lambda_0 I - T) = 0$, then the SVEP of T at λ_0 is equivalent to the SVEP of T^* at λ_0, by Theorem 2.107, and this is impossible. Therefore, ind $(\lambda_0 I - T) < 0$. By Theorem 2.114 then ind $(\lambda_0 I - T) < 0$ for all $\lambda \in \Omega$ and this implies that $q(\lambda I - T) = \infty$ for all $\lambda \in \Omega$, or equivalently T^* fails the SVEP at every point $\lambda \in \Omega$.   ∎

In a very similar way we can prove:

Corollary 2.118 *Suppose that $T \in L(X)$ and Ω is a component of $\Psi_l(T)$. If T^* has the SVEP then only the following cases are possible:*

(i) *T has the SVEP at every $\lambda \in \Omega$. In this case $p(\lambda I - T) = q(\lambda I - T) < \infty$ and ind $(\lambda I - T) = 0$ for all $\lambda \in \Omega$. Every $\lambda \in \sigma(T) \cap \Omega$ is a pole. The eigenvalues and the deficiency values do not cluster in Ω. This case occurs exactly when Ω intersects the resolvent $\rho(T) := \mathbb{C} \setminus \sigma(T)$.*

(ii) *T fails the SVEP at some point $\lambda \in \Omega$. In this case, T fails the SVEP at every point $\lambda \in \Omega$, $q(\lambda I - T) < \infty$, $p(\lambda I - T) = \infty$ and ind $(\lambda I - T) > 0$ for all $\lambda \in \Omega$. Every $\lambda \in \sigma(T) \cap \Omega$ is a right pole. The deficiency values do not cluster in Ω, while every point of Ω is an eigenvalue.*

Remark 2.119 By using analogous arguments the results of Theorem 2.117 and Corollary 2.118 are still valid for the components of the semi-Fredholm region $\rho_{sf}(T)$, $\rho_{usf}(T)$, or $\rho_{lsf}(T)$.

Let us denote by $\sigma_{tud}(T)$ the topological uniform descent spectrum, i.e., the set of all $\lambda \in \mathbb{C}$ such that $\lambda i - T$ does not have topological uniform descent. By Corollary 1.90 this spectrum is closed, and may be empty, since, by Theorem 1.142, it is contained in the Drazin spectrum $\sigma_d(T)$, and it will be shown in Chap. 3 that this spectrum is empty whenever T is algebraic.

Let $\rho_{tud}(T)$ be the topological uniform descent resolvent, i.e. $\rho_{tud}(T) := \mathbb{C} \setminus \sigma_{tud}(T)$. Clearly, $\rho_{tud}(T)$ is an open subset of $\mathbb{C}$, and hence can be decomposed into components, i.e. maximal open, connected, pairwise disjoint non-empty subsets of $\mathbb{C}$. We want show that the previous results on the components of semi B-Fredholm regions may be extended to the components of $\rho_{tud}(T)$. We first need a preliminary result.

Theorem 2.120 *If $T \in L(X)$ has topological uniform descent then there exists an $\varepsilon > 0$ such that;*

(i) $\overline{K(\lambda I - T) + H_0(\lambda I - T)} = \overline{K(T) + H_0(T)}$ *for all $0 < |\lambda| < \varepsilon$.*

(ii) $\overline{K(\lambda I - T) + H_0(\lambda I - T)} = \overline{K(T) + H_0(T)}$ *for all $0 < |\lambda| < \varepsilon$.*

Proof If T has topological uniform descent then, by Corollary 1.90, there exists an $\varepsilon > 0$ such that $\lambda I - T$ is semi-regular and

$$(\lambda I - T)^{\infty}(X) = T^{\infty}(X) + \mathcal{N}^{\infty}(T) \quad \text{for all } 0 < |\lambda| < \varepsilon.$$

From Theorem 1.44 we have $K(\lambda I - T) = (\lambda I - T)^{\infty}(X)$, so, from Theorem 2.37, we obtain

$$K(\lambda I - T) + H_0(\lambda I - T) = K(\lambda I - T) = (\lambda I - T)^{\infty}(X),$$

and hence

$$K(\lambda I - T) + H_0(\lambda I - T) = T^{\infty}(X) + \mathcal{N}^{\infty}(T). \tag{2.22}$$

From Theorem 2.95, together with Lemma 1.80, part (i), it then follows that

$$K(T) + H_0(T) \subseteq K(T) + \overline{H_0(T)} = T^{\infty}(X) + \overline{\mathcal{N}^{\infty}(T)}$$
$$= T^{\infty}(X) + \mathcal{N}^{\infty}(T) \subseteq K(T) + H_0(T).$$

Hence

$$K(T) + H_0(T) = T^{\infty}(X) + \mathcal{N}^{\infty}(T). \tag{2.23}$$

From (2.22) and (2.23) it then follows that the statement (i) holds.

(ii) By Corollary 1.90, there exists an $\varepsilon > 0$ such that $\lambda I - T$ is semi-regular and

$$\mathcal{N}^{\infty}(\lambda I - T) = \overline{T^{\infty}(X) \cap \mathcal{N}^{\infty}(T)} \quad \text{for all } 0 < |\lambda| < \varepsilon.$$

Again by Theorem 2.37, we have

$$\overline{K(\lambda I - T) \cap H_0(\lambda I - T)} = \overline{H_0(\lambda I - T)} = \overline{\mathcal{N}^{\infty}(\lambda I - T)},$$

and hence

$$\overline{K(\lambda I - T) \cap H_0(\lambda I - T)} = \overline{T^{\infty}(X) \cap \mathcal{N}^{\infty}(T)}. \tag{2.24}$$

From Lemma 1.80 and Theorem 2.95 we also deduce that

$$\overline{K(T) \cap H_0(T)} \subseteq \overline{K(T) \cap \overline{H_0(T)}} = \overline{T^{\infty}(X) \cap \overline{\mathcal{N}^{\infty}(T)}}$$
$$= \overline{T^{\infty}(X) \cap \mathcal{N}^{\infty}(T)} \subseteq \overline{K(T) \cap H_0(T)},$$

thus,

$$\overline{K(T) \cap H_0(T)} = \overline{T^\infty(X) \cap \mathcal{N}^\infty(T)}. \tag{2.25}$$

Combining (2.24) and (2.25) we then conclude that (ii) holds. ∎

By using the Heine–Borel theorem, as in the proof of Theorem 2.114, we easily obtain the following corollary.

Corollary 2.121 *Let $T \in L(X)$ and let Ω be a component of $\rho_{\mathrm{tud}}(T)$. If $\lambda_0 \in \Omega$ is arbitrarily given, then for all $\lambda \in \Omega$ we have*

$$K(\lambda I - T) + H_0(\lambda I - T) = K(\lambda_0 I - T) + H_0(\lambda_0 I - T),$$

and

$$\overline{K(\lambda I - T) \cap H_0(\lambda I - T)} = \overline{K(\lambda_0 I - T) \cap H_0(\lambda_0 I - T)}.$$

Consequently, the mappings

$$\lambda \to K(\lambda I - T) + H_0(\lambda I - T) \tag{2.26}$$

and

$$\lambda \to \overline{K(\lambda I - T) \cap H_0(\lambda I - T)} \tag{2.27}$$

are constant on the components of $\rho_{\mathrm{tud}}(T)$.

From the proof of Theorem 2.120, the two mappings

$$\lambda \to (\lambda I - T)^\infty(X) + \mathcal{N}^\infty(\lambda I - T)$$

and

$$\lambda \to \overline{(\lambda I - T)^\infty(X) \cap \mathcal{N}^\infty(\lambda I - T)}$$

coincide with the mappings (2.26) and (2.27), respectively, as λ ranges over a component of $\rho_{\mathrm{tud}}(T)$, so we have:

Corollary 2.122 *Suppose that $T \in L(X)$ has topological uniform descent. Then*

$$K(T) + H_0(T) = T^\infty(X) + \mathcal{N}^\infty(T),$$

and

$$\overline{K(T) \cap H_0(T)} = \overline{T^\infty(X) \cap \mathcal{N}^\infty(T)}.$$

The mappings

$$\lambda \to (\lambda I - T)^{\infty} + \mathcal{N}^{\infty}(\lambda I - T),$$

and

$$\lambda \to \overline{(\lambda I - T)^{\infty}(X) \cap \mathcal{N}^{\infty}(\lambda I - T)}$$

are constant on the components of $\rho_{\mathrm{tud}}(T)$.

By using the previous results we obtain the following classification of the components of $\rho_{\mathrm{tud}}(T)$, similar to the classification of the semi B-Fredholm regions.

Theorem 2.123 *Let Ω be a component of $\rho_{\mathrm{tud}}(T)$. Then the following alternative holds:*

(i) *T has the SVEP at every $\lambda \in \Omega$. In this case $\lambda I - T$ is left Drazin invertible for all $\lambda \in \Omega$. Furthermore, $\sigma_{\mathrm{ap}}(T)$ does not cluster in Ω, and no point of Ω is an eigenvalue of T, except for a subset of Ω which consists of at most countably many isolated points.*

(ii) *T does not have the SVEP at any point $\lambda \in \Omega$. In this case the ascent $p(\lambda I - T) = \infty$ for all $\lambda \in \Omega$. Furthermore, every point of Ω is an eigenvalue of T.*

Proof

(i) Suppose that T has the SVEP at some $\lambda_0 \in \Omega$. Then, by Theorem 2.97, $K(\lambda_0 I - T) \cap H_0(\lambda_0 I - T) = \{0\}$ and hence, by Corollary 2.121, $K(\lambda I - T) \cap H_0(\lambda I - T) = \{0\}$, so T has the SVEP at λ, and, again by Theorem 2.97, $\lambda I - T$ is left Drazin invertible and $\sigma_{\mathrm{ap}}(T)$ does not cluster at any point $\lambda \in \Omega$. Consequently, no point of Ω is an eigenvalue of T except for a subset of Ω which consists of at most countably many isolated points.

(ii) Suppose that T does not have the SVEP at any point of Ω. Then, again by Theorem 2.97, $p(\lambda I - T) = \infty$ for all $\lambda \in \Omega$. By Theorem 2.60 it then follows that $\ker(\lambda I - T) \neq \{0\}$ for every $\lambda \in \Omega$, hence every point of Ω is an eigenvalue of T.   ∎

With respect to the dual T^* we have the following classification:

Theorem 2.124 *Let Ω be a component of $\rho_{\mathrm{tud}}(T)$. Then the following alternative holds:*

(i) *T^* has the SVEP at every $\lambda \in \Omega$. In this case $\lambda I - T$ is right Drazin invertible for all $\lambda \in \Omega$. Furthermore, $\sigma_s(T)$ does not cluster in Ω, and no point of Ω is a deficiency value of T, except for a subset of Ω which consists of at most countable many isolated points.*

(ii) *T^* does not have the SVEP at any point $\lambda \in \Omega$. In this case the descent $q(\lambda I - T) = \infty$ for all $\lambda \in \Omega$. Furthermore, every point of Ω is a deficiency value of T.*

Proof

(i) If T^* has the SVEP at some $\lambda_0 \in \Omega$, then, by Theorem 2.98, $X = K(\lambda_0 I - T) + H_0(\lambda_0 I - T)$ and hence, by Corollary 2.121, $X = K(\lambda I - T) + H_0(\lambda I - T)$ for all $\lambda \in \Omega$, so T^* has the SVEP at λ, and, again by Theorem 2.98, $\lambda I - T$ is right Drazin invertible and $\sigma_s(T)$ does not cluster at any point $\lambda \in \Omega$. Consequently, no point of Ω is a deficiency value of T except for a subset of Ω which consists of at most countably many isolated points.

(ii) Suppose that T^* does not have the SVEP at any point of Ω. Then, again by Theorem 2.98, $q(\lambda I - T) = \infty$ for all $\lambda \in \Omega$. If there exists a $\lambda_0 \in \Omega$ such that $\lambda_0 I - T$ is surjective, then $\lambda I - T^*$ is injective, and hence T^* has the SVEP at λ_0, a contradiction.   ∎

For the SVEP we have the following classification:

Corollary 2.125 *Let $T \in L(X)$ and let Ω be a component of $\rho_{\mathrm{tud}}(T)$. For the SVEP then only the following cases are possible:*

(i) *Both T and T^* have the SVEP at every point of Ω. In this case $p(\lambda I - T) = q(\lambda I - T) < \infty$ and $\lambda I - T$ is Drazin invertible, for all $\lambda \in \Omega$. The spectra $\sigma_{\mathrm{ap}}(T)$ and $\sigma_s(T)$ do not have a limit point in Ω, and the same holds for the spectrum $\sigma(T)$. This case occurs exactly when $\Omega \cap \rho(T) \neq \emptyset$.*

(ii) *T has the SVEP at every point of Ω, while T^* fails to have the SVEP at every $\lambda \in \Omega$. In this case $\lambda I - T$ is left Drazin invertible, while $q(\lambda I - T) = \infty$, for all $\lambda \in \Omega$. The spectrum $\sigma_{\mathrm{ap}}(T)$ does not have a limit point in Ω, while $\Omega \subseteq \sigma_s(T)$.*

(iii) *T^* has the SVEP at every point of Ω, while T fails to have the SVEP at every $\lambda \in \Omega$. In this case $\lambda I - T$ is right Drazin invertible, while $p(\lambda I - T) = \infty$, for all $\lambda \in \Omega$. The spectrum $\sigma_s(T)$ does not have a limit point in Ω, while every $\lambda \in \Omega$ is an eigenvalue*

(iv) *Both T and T^* fail to have the SVEP at the points $\lambda \in \Omega$. In this case $p(\lambda I - T) = q(\lambda I - T) = \infty$, for all $\lambda \in \Omega$. Every $\lambda \in \Omega$ is an eigenvalue and $\Omega \subseteq \sigma_s(T)$.*

Theorem 2.126 *If X is an infinite-dimensional Banach space then the semi-Fredholm spectra $\sigma_{\mathrm{usf}}(T)$ and $\sigma_{\mathrm{lsf}}(T)$ are non-empty.*

Proof Suppose that $\sigma_{\mathrm{usf}}(T) = \emptyset$. Then $\lambda I - T \in \Phi_+(X)$ for all $\lambda \in \mathbb{C}$, so $\lambda I - T$ has topological uniform descent for all $\lambda \in \mathbb{C}$. Therefore, $\Omega = \mathbb{C}$ is a component of $\rho_{\mathrm{tud}}(T)$, so by part (i) of Corollary 2.125, $p(\lambda I - T) = q(\lambda I - T) < \infty$ for all $\lambda \in \mathbb{C}$. By Theorem 1.22, $\alpha(\lambda I - T) = \beta(\lambda I - T) < \infty$, and hence the essential spectrum $\sigma_e(T) = \emptyset$, so X is finite-dimensional by Remark 1.56.   ∎

2.8   The SVEP Under Commuting Riesz Perturbations

We first mention that SVEP is not preserved under non-commuting perturbations. In fact, by [292, Example 5.6.29], the sum of a decomposable operator and a rank-one operator may fail to have the SVEP, although decomposable operators and finite rank operators have the SVEP.

In general the SVEP is also not stable under arbitrary sums and products of commuting operators. A specific example based on the theory of weighted shifts may be found in [81], but here we present a general principle that shows that such examples exist in abundance.

Theorem 2.127 *Let $S \in L(X)$ and suppose that there exist $\alpha \neq \beta \in \mathbb{C}$ such that*

$$K(\alpha I - S) = K(\beta I - S) = \{0\}. \tag{2.28}$$

If $T \in L(X)$ commutes with S, then T is the sum of two commuting operators with SVEP, while $\exp(T)$ is the product of two commuting operators with SVEP.

Proof Since all quasi-nilpotent operators share the SVEP, we may assume that the spectral radius $r(T) > 0$. To verify that $T(S - \alpha I)$ has the SVEP, we consider an arbitrary open set $U \subseteq \mathbb{C}$ and an analytic function $f : U \to X$ for which $(\mu I - T(S - \alpha I))f(\mu) = 0$ for all $\mu \in U$. For fixed non-zero $\mu \in U$ and arbitrary $\lambda \in \mathbb{C}$ with $\lambda < |\mu|/r(T)$, the operator $\lambda T - \mu I$ is invertible and its inverse commutes with both S and T. Moreover,

$$(\lambda I - (S - \alpha I))T(\lambda T - \mu I)^{-1}f(\mu)$$
$$= (\lambda T - \mu I)^{-1}[(\mu I - T(S - \alpha I)) + (\lambda T - \mu I)]f(\mu) = f(\mu).$$

This shows that $0 \in \rho_{S-\alpha I}(f(\mu))$ and therefore, by Theorem 2.20, $f(\mu) \in K(S - \alpha I) = \{0\}$ for all non-zero $\mu \in U$. Thus $f \equiv 0$ on U, which establishes SVEP for $T(S - \alpha I)$ and, of course, similarly also for $T(S - \beta I)$. Because

$$(\beta - \alpha)T = T(S - \alpha I) + T(\beta I - S),$$

the first assertion is now immediate, and the last claim, concerning the product, follows from the fact that SVEP is preserved under the analytical calculus.   ∎

Note that in Theorem 2.127 the operators T and $\exp(T)$ may fail to have the SVEP, while the condition on S entails that $X_T(\emptyset) = \{0\}$ and hence, by Theorem 2.23, the SVEP for S. To provide concrete examples of operators that satisfy condition (2.28) on $K(\lambda I - S)$ of the preceding result, we now introduce the concept of a semi-shift.

A bounded operator $S \in L(X)$ is said to be a *semi-shift* if S is an isometry for which $\mathcal{S}^{\infty}(X) = \{0\}$. Examples of semi-shifts T are the unilateral right shift operators of arbitrary multiplicity on the sequence spaces $\ell^p(\mathbb{N})$, with $1 \leq p < \infty$, defined as

$$Tx := (0, x_1, x_2, \dots) \quad \text{for all } x = (x_n) \in \ell_p(\mathbb{N}).$$

Other important examples of semi-shifts are the *right translation* operators on the Lebesgue spaces $L^p([0, +\infty])$, $1 \leq p < \infty$. Note that if T is a semi-shift then $\sigma_T(x) = \sigma(T)$ coincides with the closed unit disc $\mathbf{D}(0, 1)$ of $\mathbb{C}$ for all non-zero $x \in X$, see [216, Proposition 1.6.8].

Now, if $x \neq 0$ and $\alpha \in \mathbf{D}(0, 1)$ then $\alpha \in \sigma_T(x)$ and hence $0 \in \sigma_{\alpha I - T}(x)$, so $x \notin K(\alpha I - T)$, by Theorem 2.20. Therefore, $K(\alpha I - T) = \{0\}$ for all $\alpha \in \mathbf{D}(0, 1)$.

To find an operator without SVEP that commutes with a semi-shift is perhaps not completely obvious, but this task can easily be accomplished when X is a separable Hilbert space. Indeed, in this case, for arbitrary $S, T \in L(X)$ the operators $T \otimes I$ and $I \otimes S$ on the Hilbert tensor product $X \otimes X$ commute, since

$$(T \otimes I)(I \otimes S) = T \otimes S = (I \otimes S)(T \otimes I);$$

see Kadison and Ringrose [197, Section 2.6] for a nice exposition of the theory of the Hilbert tensor product. Moreover, since $T \otimes I$ is unitarily equivalent to the Hilbert direct sum $\sum_{n=1}^{\infty} \oplus T$, it is easily seen that the failure of SVEP at a point λ extends from T to $T \otimes I$. In the same vein, it follows that $I \otimes S$ is a semi-shift whenever S is, since $I \otimes S$ is unitarily equivalent to $\sum_{n=1}^{\infty} \oplus S$. Note that, in the Hilbert space case, the semi-shifts are precisely the pure isometries.

Thus neither the SVEP nor the localized SVEP is, in general, preserved under sums and products of commuting perturbations. Next we will show that the SVEP is preserved under Riesz commuting perturbations. In the sequel we need a preliminary elementary result.

Lemma 2.128 *Let $R \in L(X)$ be a Riesz operator and Ω a spectral subset of $\sigma(R)$ such that $0 \notin \Omega$. Then the spectral projection P associated with Ω is finite-dimensional.*

Proof We know that every spectral point $\lambda \neq 0$ of the spectrum of a Riesz operator is an isolated point of $\sigma(R)$. Consequently, Ω is a finite subset of $\mathbb{C}$. Set $\Omega := \{\lambda_1, \dots, \lambda_k\}$. Every spectral projection P_i associated with $\{\lambda_i\}$ is finite dimensional, so $P = \sum_{i=1}^{n} P_i$ is finite-dimensional operator. ∎

We now show that the localized SVEP from an operator T is preserved under Riesz commuting perturbations.

Theorem 2.129 *Let X be a Banach space, $T, R \in L(X)$, where R is a Riesz operator such that $TR = RT$. If $\lambda \in \mathbf{C}$, then T has the SVEP at λ if and only if $T - R$ has the SVEP at λ. In particular, the SVEP is stable under Riesz commuting perturbations.*

Proof Without loss of generality we may assume that $\lambda = 0$. Suppose T does not have the SVEP at 0. We show that $T - R$ does not have the SVEP at 0. Since T does not have the SVEP at 0, $\ker T \cap K(T) \neq \{0\}$, by Theorem 2.60, so there exists a sequence of vectors $(x_i)_{i=0,1,...}$ of X such that $x_0 \neq 0$, $Tx_0 = 0$, $Tx_i = x_{i-1}$ $(i \geq 1)$ and $\sup_{i \geq 1} \|x_i\|^{1/i} < \infty$. Let $C := \sup_{i \geq 1} \|x_i\|^{1/i}$. Fix an ε, $0 < \varepsilon < \frac{1}{2C}$. Let

$$\Omega := \{\lambda \in \sigma(R) : |\lambda| \geq \varepsilon\}$$

and denote by P the spectral projection associated with Ω. Then P is finite-dimensional, by Lemma 2.128, and if $X_2 := P(X)$ and $X_1 := \ker P$, then we can write $X = X_1 \oplus X_2$. According the spectral decomposition theorem, we have $R(X_j) \subset X_j$ $(j = 1, 2)$,

$$\sigma(R|X_1) \subset \{\lambda : |\lambda| < \varepsilon\} \quad \text{and} \quad \sigma(R|X_2) \subset \{\lambda : |\lambda| \geq \varepsilon\}.$$

Since $TR = RT$, we also have $T(X_j) \subset X_j$ $(j = 1, 2)$. Clearly,

$$TPx_0 = PTx_0 = 0,$$

and

$$TPx_i = PTx_i = Px_{i-1} \quad (i \geq 1).$$

We claim that $Px_i = 0$ for all i. To see this, suppose that $Px_i \neq 0$ for some $i \geq 0$. From $TPx_{i+1} = Px_i \neq 0$ we then deduce that $Px_{i+1} \neq 0$, and by induction it then follows that $Px_n \neq 0$ for all $n \geq i$.

Let $k \geq 1$ be the smallest integer for which $Px_k \neq 0$. Then

$$TPx_k = Px_{k-1} = 0.$$

For all $n \geq k$ we have

$$T^{n-k} Px_n = T^{n-k-1}(TPx_n) = T^{n-k-1} Px_{n-1} = \ldots.$$
$$= TPx_{k+1} = Px_k \neq 0,$$

so $Px_n \notin \ker(T|X_2)^{n-k}$, for all $n \geq k$. Furthermore,

$$T^{n-k+1} Px_n = TT^{n-k} Px_n = TPx_k = Px_{k-1} = 0,$$

so $Px_n \in \ker(T|X_2)^{n-k+1}$. This implies that $T|X_2$ has infinite ascent, which is impossible, since $\dim X_2 < \infty$. Therefore, $Px_i = 0$, and hence $x_i \in \ker P = X_1$, for all $i \geq 0$.

Let us consider the restriction $R_1 = R|X_1$. We have $r(R_1) < \varepsilon$, so there exists a j_0 such that $\|R_1^j\| \leq \varepsilon^j$ for all $j \geq j_0$.

Set

$$y_0 := \sum_{i=0}^{\infty} R^i x_i,$$

and similarly, for every $k \geq 1$ let

$$y_k := \sum_{i=k}^{\infty} \binom{i}{k} R^{i-k} x_i.$$

This definition is correct, since

$$\sum_{i=k}^{\infty} \binom{i}{k} \|R^{i-k} x_i\| \leq \sum_{i=k}^{\infty} 2^i \|R_1^{i-k}\| C^i$$

$$\leq \sum_{i=k}^{j_0+k} 2^i C^i \|R_1^{i-k}\| + \sum_{i=j_0+k+1}^{\infty} 2^i C^i \varepsilon^{i-k} < \infty.$$

Moreover, for $k \geq 2 j_0$ we have

$$\|y_k\| \leq \sum_{i=k}^{2k-1} 2^i C^i \|R_1^{i-k}\| + \sum_{i=2k}^{\infty} (2C)^i \varepsilon^{i-k}$$

$$\leq k \max\{(2C)^k, (2C)^{2k-1} \|R_1\|^{k-1}\} + \frac{(2C)^{2k} \varepsilon^k}{1 - 2C\varepsilon}.$$

Thus,

$$\|y_k\|^{1/k} \leq k^{1/k} \left(\max\{(2K)^k, (2C)^{2k-1} \|R_1\|^{k-1}\} \right)^{1/k} + \left(\frac{(2C)^{2k} \varepsilon^k}{1 - 2C\varepsilon} \right)^{1/k}$$

$$\leq k^{1/k} \max\{2C, (2C)^{\frac{2k-1}{k}} \|R_1\|^{\frac{k-1}{k}}\} + \frac{4C^2 \varepsilon}{1 - 2C\varepsilon},$$

from which we obtain $\limsup_{k \to \infty} \|y_k\|^{1/k} < \infty$.

We also have

$$(T - R)y_0 = \sum_{i=1}^{\infty} R^i x_{i-1} - \sum_{i=0}^{\infty} R^{i+1} x_i = 0.$$

Now, for $k \geq 1$ we have

$$(T - R)y_k = \sum_{i=k}^{\infty} \binom{i}{k} R^{i-k} x_{i-1} - \sum_{i=k}^{\infty} \binom{i}{k} R^{i-k+1} x_i$$

$$= x_{k-1} + \sum_{i=k}^{\infty} R^{i-k+1} x_i \left(\binom{i+1}{k} - \binom{i}{k} \right) = y_{k-1}.$$

It remains to show that not all of the y_k's are equal to zero. Suppose on the contrary that $y_k = 0$ $(k \geq 0)$ and let $j_1 \geq j_0$. Then we have

$$\sum_{k=0}^{j_1} (-1)^k R^k y_k = \sum_{i=0}^{\infty} \alpha_i R^i x_i,$$

where, if we let $\nu := \min\{i, j_1\}$, we have

$$\alpha_i = \sum_{k=0}^{\nu} (-1)^k \binom{i}{k} \quad \text{for every } i = 0, 1, \ldots.$$

Clearly, $\alpha_0 = 1$. For $1 \leq i \leq j_1$ we obtain

$$\alpha_i = \sum_{k=0}^{i} (-1)^k \binom{i}{k} = 0.$$

For $i > j_1$ we have $|\alpha_i| \leq 2^i$, so

$$0 = \sum_{k=0}^{j_1} (-1)^k R^k y_k = x_0 + \sum_{i=j_1+1}^{\infty} \alpha_i R^i x_i$$

and

$$\|x_0\| \leq \sum_{i=j_1+1}^{\infty} 2^i \|R_1^i\| \|x_i\| \leq \sum_{i=j_1+1}^{\infty} 2^i \varepsilon^i C^i = \frac{(2C\varepsilon)^{j_1+1}}{1 - 2C\varepsilon}.$$

Letting $j_1 \to \infty$ yields $\|x_0\| = 0$, a contradiction. Therefore, $\ker(T - R) \cap K(T - R) \neq \{0\}$, and this implies, again by Theorem 2.60, that $T - R$ does not have the SVEP at 0.

By symmetry we then conclude that T has the SVEP at 0 if and only if $T - R$ has the SVEP at 0. ∎

Remark 2.130 Every Riesz operator is *meromorphic*, i.e., every nonzero $\lambda \in \sigma(T)$ is a pole of the resolvent of T. Meromorphic operators have the same spectral structure as Riesz operators, i.e., every $0 \neq \lambda \in \sigma(T)$ is an eigenvalue, and the spectrum is at most countable and has no nonzero cluster point. A simple example shows that the result of Theorem 2.129 cannot be extended to meromorphic operators. Denote by L the left shift on $\ell_2(\mathbb{N})$ and let $\lambda_0 \notin \sigma(L) = \mathbf{D}(0, 1)$, $\mathbf{D}(0, 1)$ the closed unit disc. Then L does not have the SVEP at 0. Since L has the SVEP at λ_0, $T := \lambda_0 I - L$ has the SVEP at 0, while $T - \lambda_0 I = -L$, does not have the SVEP at 0, and, obviously, $\lambda_0 I$ is meromorphic.

Remark 2.131 If $\sigma_e(T)$ is the essential Fredholm spectrum of T, and $r_e(T)$ denotes the essential spectral radius of T, i.e.,

$$r_e(T) := \sup\{|\lambda| : \lambda \in \sigma_e(T)\}.$$

Obviously, in the case of a Riesz operator K we have $r_e(K) = 0$. A closer look at the proof of Theorem 2.129 reveals that the stability of the localized SVEP also holds if we assume that $r_e(K)$ is small enough.

The assumption of commutativity is essential in Theorem 2.129. To see this, define

$$\rho_{\mathrm{sf}}^+(T) := \{\lambda \in \rho_{\mathrm{sf}}(T) : \mathrm{ind}\,(\lambda I - T) > 0\}.$$

Theorem 2.132 *Let* $T \in L(X)$ *and* $K \in K(X)$. *If* $T + K$ *has the SVEP then* $\rho_{\mathrm{sf}}^+(T) = \emptyset$ *and* $\sigma_{\mathrm{uw}}(T) = \sigma_{\mathrm{sf}}(T)$.

Proof Suppose that $\rho_{\mathrm{sf}}^+(T) = \rho_{\mathrm{sf}}^+(T + K) \neq \emptyset$. Then there exists a $\lambda \in \Phi_\pm(T + K)$ such that $\mathrm{ind}\,(\lambda I - T) > 0$. But this is impossible, since the SVEP of $T + K$ entails that $\mathrm{ind}\,(\lambda I - (T + K)) \leq 0$, by Corollary 2.106. The last assertion follows from Lemma 3.57. ∎

Let $\rho_{\mathrm{w}}(T)$ denote the Weyl region of T, i.e., the set of all $\lambda \in \mathbb{C} : \lambda I - T \in \Phi(X)$, and $\mathrm{ind}\,(\lambda I - T) = 0$, and denote by $\sigma_{\mathrm{w}}(T) := \mathbb{C} \backslash \rho_{\mathrm{w}}(T)$ the Weyl spectrum. In Chap. 5 we shall prove that if $\rho_{\mathrm{w}}(T)$ is connected and $\mathrm{int}\,\sigma_{\mathrm{w}}(T) = \emptyset$. Then both $T + K$ and $T^* + K^*$ have the SVEP. The stability of SVEP under (not necessarily commuting) compact perturbations for Hilbert space operators has been studied by Zhu and Li [307], which showed that the reverse of Theorem 5.6 holds for Hilbert space operators. The proof is omitted, since it involves rather technical results on Hilbert spaces operators, due to Herrero [177] and Ji [187].

Theorem 2.133 *Let* $T \in L(H)$, H *a Hilbert space. Then* $T + K$ *has the SVEP for all* $K \in K(H)$ *if and only if* $\rho_{\mathrm{w}}(T)$ *is connected and* $\mathrm{int}\,\sigma_{\mathrm{w}}(T) = \emptyset$.

The next example shows that for a (non commuting) compact perturbation $T + K$ of an operator T which has the SVEP, the SVEP may fail.

Example 2.134 Let $S \in L(X)$ be the bilateral shift on $\ell^2(\mathbb{Z})$). It is easily seen that $\sigma(S) = \sigma(S^*)$ is the unit circle $\mathbf{T}$. We also have $\sigma_{\mathrm{w}}(S) = \mathbf{T}$. The inclusion $\sigma_{\mathrm{w}}(S) \subseteq \mathbf{T}$ is obvious. Observe that both S and S^* have the SVEP, since every spectral point of S belongs to the boundary of $\sigma(S)$. Suppose now that $\lambda \in \mathbf{T}$ and $\lambda \notin \sigma_w(S)$. Since $\lambda I - S \in \Phi(X)$, the SVEP for S and S^* implies that $0 < p(\lambda I - S) = q(\lambda I - S) < \infty$. But this is impossible, since $\mathrm{iso}\,\sigma(S) = \emptyset$. Therefore $\mathbf{T} = \sigma_w(S)$, so $\rho_w(S)$ is not connected. Consequently, by Theorem 2.133, there exists a compact operator K such that $T + K$ does not satisfy the SVEP.

We now consider the question of whether the SVEP is preserved under small perturbations. By $\rho_{sf}(T) = \mathbb{C} \setminus \sigma_{sf}(T)$ we denote the semi-Fredholm domain of T. The following results are due to Zhu and Li [307], again we omit the proof.

Theorem 2.135 *Let $T \in L(H)$. Then the following statements are equivalent:*

(i) *Given $\varepsilon > 0$, there exists a compact operator $K \in K(H)$ with $\|K\| < \varepsilon$ such that $T + K$ has the SVEP.*
(ii) *Given $\varepsilon > 0$, there exists a $K \in L(H)$ with $\|K\| < \varepsilon$ such that $T + K$ has the SVEP.*
(iii) *There exists a $K \in K(H)$ such that $T + K$ has the SVEP.*
(iv) *$\rho_{sf}^+(T)$ is empty, where $\rho_{sf}^+(T) := \{\lambda \in \rho_{sf}(T) : \mathrm{ind}\,(\lambda I - T) > 0\}$.*

The following result characterizes those operators for which SVEP is stable under small compact perturbations.

Theorem 2.136 *Let $T \in L(H)$. Then there exists a $\delta > 0$ such that $T + K$ has the SVEP for all compact operators $K \in K(H)$ if and only if*

(i) *the interior of the set $\rho_{sf}(T) \cap \sigma_p(T)$ is empty, where $\rho_{sf}(T) = \mathbb{C} \setminus \sigma_{sf}(T)$,*
(ii) *the interior of the semi-Fredholm spectrum $\sigma_{sf}(T)$ is empty,*
(iii) *$\rho_{sf}(T)$ consists of finitely many connected components.*

In Chap. 4 we shall see that the class of operators which have the SVEP is very rich, and includes several important classes of operators. However, we conclude this section by showing that the class of operators which do not have the SVEP is very large. The following concept is due to Herrero [178].

Definition 2.137 A certain property (P) concerning Hilbert space operators is said to be a *bad property* (P) if the following conditions are fulfilled:

(a) If T has property (P) then $\mu I + \eta T$ has property (P) for all $\mu, \eta \in \mathbb{C}$ with $0 \neq \eta$.
(b) If T has property (P) and S is similar to T, then T has property (P).
(c) If T has property (P), and S is another operator for which $\sigma(T) \cap \sigma(S) = \emptyset$, then the orthogonal direct sum $T \oplus S$ has property (P).

A natural question is whether the SVEP is stable under small perturbations. The answer to this question is negative. As noted in [178, Theorem 3.51], if there exists an operator $T \in L(H)$ which has a bad property (P), then the set of all operators which has property (P) is dense in $L(H)$. The property of being an operator for which the SVEP fails is a bad property, so we have

Theorem 2.138 *If $T \in L(H)$ and $\varepsilon > 0$, then there exists an operator $S \in L(H)$ with $\|S\| < \varepsilon$ such that $T + S$ does not have the SVEP.*

2.9 Stability of the Localized SVEP Under Quasi-Nilpotent Equivalence

Local spectral theory is a powerful tool when the issue is that of relating the spectral properties of two operators $T \in L(X)$ and $S \in L(Y)$, X and Y Banach spaces, that are linked in some way by an operator $T A \in L(X, Y)$. We have seen that a natural link is provided by the intertwining condition $SA = AT$, by some non-zero operator $A \in L(X, Y)$. If A is bijective then the condition $SA = AT$ means that T and S are similar. If T and S are similar, it is easily seen that $\sigma(T) = \sigma(S)$, and that the various distinguished parts of the spectrum coalesce. In this section the condition of invertibility of A will be replaced by weaker conditions on the intertwiner A. We begin with some definitions.

Definition 2.139 An operator $A \in L(X, Y)$ between Banach spaces X and Y is a *quasi-affinity* if it has a trivial kernel and dense range. We say that $T \in L(X)$ is a *quasi-affine transform* of $S \in L(Y)$, and we write $T \prec S$, if there is a quasi-affinity $A \in L(X, Y)$ that intertwines T and S, i.e. $SA = AT$. If there exists two quasi-affinities $A \in L(X, Y)$, $B \in L(X, Y)$ for which $SA = AT$ and $BS = TB$ then we say that S and T are *quasi-similar*. If A is invertible and $SA = AT$ then S and T are said to be *similar*.

The *commutator* of two operators $S, T \in L(X)$ is the operator $C(S, T)$ on $L(X)$ defined by

$$C(S, T)(A) := SA - AT \quad \text{for all } A \in L(X).$$

By induction it is easy to show the binomial identity

$$C(S, T)^n(A) = \sum_{k=0}^{n} \binom{n}{k} (-1)^k S^{n-k} A T^k. \tag{2.29}$$

Obviously, $C(\lambda I - S, \lambda I - T)^n(A) = (-1)^n C(S, T)^n(A)$ for all $\lambda \in \mathbb{C}$, from which we obtain

$$C(S, T)^n(A) = (-1)^n C(\lambda I - S, \lambda I - T)^n(A)$$

$$= \sum_{k=0}^{n} \binom{n}{k} (-1)^{n-k} (\lambda I - S)^{n-k} A (\lambda I - T)^k$$

for all $A \in L(X)$, $n \in \mathbb{N}$. The equality (2.29) also entails that

$$C(S, T)^n(A)x = S^n A x \quad \text{for all } x \in \ker T. \tag{2.30}$$

Let us consider a very weak notion of intertwining which dates back to I. Colojoară and C. Foiaş, see [98, Chapter 4] or [216, Chapter 3].

Definition 2.140 Given the operators $T \in L(X)$ and $S \in L(Y)$, we say that the pair (S, T) is *asymptotically intertwined* by the operator $A \in L(X, Y)$ if $\|C(T, S)(A)\|^{1/n} \to 0$ as $n \to \infty$. The operators $S \in L(X)$ and $T \in L(X)$ are said to be *quasi-nilpotent equivalent* if (S, T) and (T, S) are asymptotically intertwined by the identity operator I on X.

Evidently, the notion of asymptotically intertwined pairs is a generalization of the intertwining condition $C(S, T)(A) = 0$ which appears in the definition of $T \prec S$. This notion is also a generalization of the higher order intertwining condition:

$$C(S, T)^n(A) = 0 \quad \text{for some } n \in \mathbb{N}.$$

Theorem 2.141 *Let $T \in L(X)$, $S \in L(Y)$ and suppose that for some injective map $A \in L(X, Y)$ there exists an integer $n \in \mathbb{N}$ for which $C(S, T)^n(A) = 0$. If S has the SVEP at λ_0 then T has the SVEP at λ_0. In particular, if $T \in L(X)$ and $S \in L(Y)$ are intertwined by an injective map $A \in L(X, Y)$ then the localized SVEP carries over from S to T.*

Proof Let $\mathcal{U} \subseteq \mathbb{C}$ be an open neighborhood of λ_0 and $f : \mathcal{U} \to X$ be an analytic function such that $(\lambda I - T)f(\lambda) = 0$, for all $\lambda \in \mathcal{U}$. Since $f(\lambda) \in \ker(\lambda I - T)$, taking into account (2.30) we then obtain

$$
\begin{aligned}
0 &= (\lambda I - S)[C(S, T)^n(A)f(\lambda)] = (\lambda I - S)[C(\lambda I - S, \lambda I - T)^n(A)f(\lambda)] \\
&= (\lambda I - S)^{n+1} Af(\lambda).
\end{aligned}
$$

Now,

$$(\lambda I - S)^{n+1} Af(\lambda) = (\lambda I - S)[(\lambda I - S)^n Af(\lambda)] \quad \text{on } \mathcal{U},$$

and the SVEP of S at λ_0 implies $(\lambda I - S)^n Af(\lambda) = 0$. Repeating this argument we easily deduce that $(\lambda I - S)(A(f\lambda)) = 0$. Since S has the SVEP at λ_0 it then follows that $Af(\lambda) = 0$ for all $\lambda \in \mathcal{U}$ and the injectivity of A entails $f(\lambda) = 0$ for all $\lambda \in \mathcal{U}$. Therefore T has the SVEP at λ_0. The last assertion is clear. ∎

The following example shows that the converse of Theorem 2.141 does not hold, i.e. if $T \prec S$ the SVEP from T may not be transmitted to S.

Example 2.142 Let C denote the Cesàro matrix. C is a lower triangular matrix such that the nonzero entries of the n-th row are n^{-1} ($n \in \mathbf{N}$)

$$
\begin{pmatrix}
1 & 0 & 0 & 0 & \cdots \\
1/2 & 1/2 & 0 & 0 & \cdots \\
1/3 & 1/3 & 1/3 & 0 & \cdots \\
1/4 & 1/4 & 1/4 & 1/4 & \cdots \\
\vdots & \vdots & \vdots & \vdots & \vdots
\end{pmatrix}.
$$

Let $1 < p < \infty$ and consider the matrix C as an operator C_p acting on ℓ_p. Let q be such that $1/p + 1/q = 1$. In [266] Rhoades proved that $\sigma(C_p)$ is the closed disc Γ_q, where

$$\Gamma_q := \{\lambda \in \mathbb{C} : |\lambda - q/2| \le q/2\}.$$

Moreover, it has been proved by González [157] that for each $\mu \in \operatorname{int}\Gamma_q$ the operator $\mu I - C_p$ is an injective Fredholm operator with $\beta(C_p) = 1$. Consequently, every $\mu \in \operatorname{int}\Gamma_q$ belongs to the surjectivity spectrum $\sigma_s(C_p)$.

Let $C_p{}^* \in L(\ell_q)$ denote the conjugate operator of C_p. Obviously, $\sigma_s(C_p)$ clusters at every $\mu \in \operatorname{int}\Gamma_q$ and since $\mu I - C_p$ is Fredholm it then follows that $C_p{}^*$ does not have the SVEP at these points μ, by Theorem 2.98. Every operator has the SVEP at the boundary of the spectrum, and since $\sigma(C_p^*) = \sigma(C_p) = \Gamma_q$ it then follows that $C_p{}^*$ has the SVEP at λ precisely when $\lambda \notin \operatorname{int}\Gamma_q$. Choose $1 < p' < p < \infty$ and let q' be such that $1/p' + 1/q' = 1$. Then $1 < q < q' < \infty$. If we denote by $A : \ell_q \to \ell_{q'}$ the natural inclusion then we have $C_p^* A = A C_p^*$ and clearly A is an injective operator with dense range, i.e., $C_p^* \prec C_{p'}^*$. As noted before the operator C_p^* has the SVEP at every point outside of Γ_q, in particular at the points $\lambda \in \Gamma_{q'} \setminus \Gamma_q$, while $C_{p'}^*$ fails SVEP at the points $\lambda \in \Gamma_{q'} \setminus \Gamma_q$ which do not belong to the boundary of $\Gamma_{q'}$.

The work required for the following permanence results is rather technical. The reader can be find the proofs in Laursen and Neumann [216, Chapter 3].

Theorem 2.143 *Quasi-nilpotent equivalence preserves SVEP. Moreover, quasi-nilpotent equivalent operators have the same local spectra, the same surjectivity spectrum, the same approximate point spectrum, and the same spectrum. Furthermore, if T and S are quasi-nilpotent equivalent then the identity $\mathcal{X}_T(\Omega) = \mathcal{X}_S(\Omega)$ holds for every closed subset Ω of $\mathbb{C}$.*

Theorem 2.143 then implies that the identity $X_T(\Omega) = X_S(\Omega)$ holds for every closed subset Ω of $\mathbb{C}$. Moreover, since by Theorem 2.14 an operator $T \in L(X)$ has the SVEP precisely when $X_T(\emptyset) = \{0\}$, and since quasi-nilpotent equivalence preserves the analytic spectral subspaces, it is clear that the SVEP is stable under quasi-nilpotent equivalence. If there exists an integer $n \in \mathbb{N}$ for which $C(S,T)^n(I) = C(T,S)^n(I) = 0$, then the operators S and T are said to be *nilpotent equivalent*. For $S, T \in L(X)$ with $ST = TS$, it is easily seen that

$$C(S,T)^n(I) = (S - T)^n \quad \text{for all } n \in \mathbb{N}.$$

Thus, in this case, S and T are quasi-nilpotent equivalent precisely when $S - T$ is quasi-nilpotent, while S and T are nilpotent equivalent if and only if $S - T$ is nilpotent.

Theorem 2.144 *Suppose that the operators $S, T \in L(X)$ are nilpotent equivalent, and let $\lambda \in \mathbb{C}$. Then T has the SVEP at λ precisely when S does. In particular, if*

T has the SVEP at λ, *and if* $N \in L(X)$ *is nilpotent and satisfies* $TN = NT$, *then* $T + N$ *also has the SVEP at* λ.

Proof By symmetry, it suffices to show that the SVEP at λ is transferred from S to T. By Theorem 2.66, the condition on S entails that

$$\mathcal{N}^\infty(\lambda I - S) \cap X_S(\emptyset) = \{0\},$$

while the nilpotent equivalence of S and T ensures that $X_S(\emptyset) = X_T(\emptyset)$. We now choose an $n \in \mathbb{N}$ for which $C(S, T)^n(I) = 0$, and consider an arbitrary $x \in \ker(\lambda I - T)$. Then $(\lambda I - T)^k x = 0$ for $k = 1, \ldots, n$, so that the identities (2.30) imply that $(\lambda I - S)^n x = 0$. Consequently, we obtain

$$\ker(\lambda I - T) \subseteq \ker(\lambda I - S)^n \subseteq \mathcal{N}^\infty(\lambda I - S)$$

and therefore

$$\ker(\lambda I - T) \cap X_T(\emptyset) \subseteq \mathcal{N}^\infty(\lambda I - S) \cap X_S(\emptyset) = \{0\}.$$

Hence Theorem 2.66 guarantees that T has the SVEP at λ.   ∎

The result of Theorem 2.129 implies that the localized SVEP is stable under quasi-nilpotent commuting perturbations. A natural question is if the SVEP at a point is preserved under quasi-nilpotent equivalence. Although we do not know the answer to this question in general, we can handle certain important special cases.

Nilpotent operators are special cases of algebraic operators. Recall that an operator $K \in L(X)$ is said to be *algebraic* if there exists a non-trivial complex polynomial h such that $h(K) = 0$. In addition to nilpotent operators, examples of algebraic operators are idempotent operators and operators for which some power has finite-dimensional range. Note that if K is algebraic, by the classical spectral mapping theorem we have $h(\sigma(K)) = \sigma(h(K)) = \{0\}$, so the spectrum $\sigma(K)$ is finite.

If $T \in L(X)$ has the SVEP at a point λ, then it may be tempting to conjecture that $T + K$ has the SVEP at λ for every algebraic operator $K \in L(X)$ that commutes with T. However, this cannot be true in general. Indeed, in the example given in Remark 2.130, the operator $K := -\lambda_0 I$ is obviously algebraic, T has the SVEP at 0 while $T + K$ does not have the SVEP at 0. Nevertheless, we obtain the following result.

Theorem 2.145 *Let* $T, K \in L(X)$ *be commuting operators, suppose that* K *is algebraic, and let* h *be a non-zero polynomial for which* $h(K) = 0$. *If* T *has the SVEP at each of the zeros of* h, *then* $T - K$ *has the SVEP at 0. In particular, if* T *has SVEP, then so does* $T + K$.

Proof We know that K has a finite spectrum, say $\sigma(K) = \{\mu_1, \ldots, \mu_n\}$. For $i = 1, \ldots, n$, let $P_i \in L(X)$ denote the spectral projection associated with K and with the spectral set $\{\mu_i\}$, and let $Y_i := P_i(X)$ be the range of P_i. From standard spectral theory it is known that $P_1 + \cdots + P_n = I$, that $Y_1, \ldots, Y_n$ are closed linear subspaces of X which are each invariant under both K and T, and that $X = Y_1 \oplus \cdots \oplus Y_n$. Moreover, for arbitrary $i = 1, \ldots, n$, the two restrictions $K_i := K \mid Y_i$ and $T_i := T \mid Y_i$ commute, and we have $\sigma(K_i) = \{\mu_i\}$. Because $h(K_i) = h(K) \mid Y_i = 0$, we obtain

$$h(\{\mu_i\}) = h(\sigma(K_i)) = \sigma(h(K_i)) = \{0\}.$$

Hence we may factor h in the form

$$h(\mu) = (\mu - \mu_i)^{n_i} q_i(\mu) \qquad \text{for all } \mu \in \mathbb{C},$$

where $n_i \in \mathbb{N}$ and q_i is a complex polynomial for which $q_i(\mu_i) \neq 0$. We conclude that

$$0 = h(K_i) = (K_i - \mu_i I)^{n_i} q_i(K_i),$$

where $q_i(K_i) \in L(Y_i)$ is invertible in light of $\sigma(q_i(K_i)) = q_i(\sigma(K_i)) = \{q_i(\mu_i)\}$ and $q_i(\mu_i) \neq 0$. Therefore $(K_i - \mu_i I)^{n_i} = 0$, which shows that the operator $N_i := K_i - \mu_i I$ is nilpotent. Now observe that

$$T_i - K_i = (T_i - \mu_i I) - (K_i - \mu_i I) = T_i - \mu_i I - N_i.$$

Because T has the SVEP at μ_i, we know that $T - \mu_i I$ has the SVEP at 0. Since this condition is inherited by restrictions to closed invariant subspaces, we conclude that $T_i - \mu_i I$ has the SVEP at 0, and hence, by Theorem 2.129, $T_i - K_i = T_i - \mu_i I - N_i$ also has the SVEP at 0 for all $i = 1, 2, \ldots, n$. From Theorem 2.15, it then follows that

$$T - K = (T_1 - K_1) \oplus \cdots \oplus (T_n - K_n)$$

has the SVEP at 0, as desired. An application of the main result to the operators $-K$ and $T - \lambda I$, for arbitrary $\lambda \in \mathbb{C}$, then establishes the final claim.   ∎

From a closer look at the proof of Theorem 2.145 it is easy to deduce that the SVEP is stable under commuting perturbations K which have finite spectrum.

The case of commuting quasi-nilpotent equivalence seems to be more complicated. In the next theorem we assume that $H_0(\lambda I - T) \cap X_T(\emptyset) = \{0\}$. This condition, as it has been shown in Theorem 2.39, is stronger than the SVEP for T at λ.

Theorem 2.146 *Suppose that $T \in L(X)$ satisfies $H_0(\lambda I - T) \cap X_T(\emptyset) = \{0\}$ for some $\lambda \in \mathbb{C}$, and let $S \in L(X)$ be quasi-nilpotent equivalent to T. Then S has the SVEP at λ.*

Proof Let $x \in \ker(\lambda I - S)$. Then $(\lambda I - S)^k x = 0$ for all $k \in \mathbb{N}$. Moreover, for arbitrary $n \in \mathbb{N}$, we know that

$$C(T, S)^n(I) = \sum_{k=0}^{n} \binom{n}{k} (-1)^{n-k} (\lambda I - T)^{n-k} (\lambda I - S)^k.$$

Consequently, we obtain that

$$\|(\lambda I - T)^n x\|^{1/n} = \|C(T, S)^n(I)x\|^{1/n} \leq \|C(T, S)^n(I)\|^{1/n} \|x\|^{1/n} \to 0$$

as $n \to \infty$. Thus $\ker(\lambda I - S) \subseteq H_0(\lambda I - T)$, while $X_S(\emptyset) = X_T(\emptyset)$, by quasi-nilpotent equivalence. We conclude that

$$ker\,(\lambda I - S) \cap X_S(\emptyset) \subseteq H_0(\lambda I - T) \cap X_T(\emptyset) = \{0\},$$

so that Theorem 2.60 ensures that S has the SVEP at λ. ∎

The SVEP at a point is preserved under quasi-nilpotent equivalence, if we assume that $\lambda I - T$ either admits a generalized Kato decomposition or is quasi-Fredholm.

Corollary 2.147 *Let $S, T \in L(X)$ be quasi-nilpotent equivalent operators, let $\lambda \in \mathbb{C}$, and suppose that $\lambda I - T$ either admits a generalized Kato decomposition or is quasi-Fredholm. If T satisfies SVEP at λ, then so does S.*

Proof Under either of the two conditions on $\lambda I - T$, it is known that SVEP for T at λ is equivalent to the condition that $H_0(\lambda I - T) \cap K(\lambda I - T) = \{0\}$; see Theorem 2.97. Consequently, the assertion is clear from Theorem 2.39. ∎

We finally address the permanence of the localized SVEP for the adjoint T^* of an operator $T \in L(X)$. The condition $H_0(\lambda I - T) + K(\lambda I - T) = X$ may be thought of as being dual to the condition $H_0(\lambda I - T) \cap K(\lambda I - T) = \{0\}$, and entails the SVEP for T^* at λ, by Theorem 2.39. These observations concerning the localized SVEP are improved in the following result.

Theorem 2.148 *For every pair of quasi-nilpotent equivalent operators $S, T \in L(X)$ and arbitrary $\lambda \in \mathbb{C}$, the following assertions hold:*

(i) *if $K(\lambda I - T) + H_0(\lambda I - T)$ is norm dense in X, then S^* has the SVEP at λ;*
(ii) *if $H_0(\lambda I - T^*) + K(\lambda I - T^*)$ is weak-*-dense in X^*, then S has the SVEP at λ.*

Proof

(i) By Theorem 2.37 we have the inclusions

$$H_0(\lambda I - T) \subseteq {}^\perp K(\lambda I - T^*) \quad \text{and} \quad K(\lambda I - T) \subseteq {}^\perp H_0(\lambda I - T^*),$$

and therefore, by duality,

$$K(\lambda I - T^*) \subseteq H_0(\lambda I - T^*)^\perp \quad \text{and} \quad H_0(\lambda I - T^*) \subseteq K(\lambda I - T)^\perp.$$

We conclude that

$$\begin{aligned} K(\lambda I - T^*) \cap H_0(\lambda I - T^*) &\subseteq H_0(\lambda I - T)^\perp \cap K(\lambda I - T)^\perp \\ &= [H_0(\lambda I - T) + K(\lambda I - T)]^\perp = \{0\}, \end{aligned}$$

where the last equality follows from the condition that $H_0(\lambda I - T) + K(\lambda I - T)$ is norm dense in X. Moreover, since

$$[C(S, T)(A)]^* = (-1)^n C(T^*, S^*)^n (A^*)$$

for all $A \in L(X)$ and $n \in \mathbb{N}$, it is clear that the pair (S^*, T^*) inherits quasi-nilpotent equivalence from the pair (S, T). The assertion is now immediate from Theorem 2.146.

(ii) Similarly, we obtain

$$\begin{aligned} H_0(\lambda I - T) \cap K(\lambda I - T) &\subseteq {}^\perp K(\lambda I - T^*) \cap {}^\perp H_0(\lambda I - T^*) \\ &= {}^\perp[K(\lambda I - T^*) + H_0(\lambda I - T^*)] = \{0\}, \end{aligned}$$

where the last identity follows from the Hahn–Banach theorem and the weak*-density of $H_0(\lambda I - T^*) + K(\lambda I - T^*)$ in X^*. Another application of Theorem 2.146 now ensures that S has the SVEP at λ. ∎

2.10   Spectral Properties of Products of Operators

Let X and Y be Banach spaces and consider two operators $S \in L(X, Y)$ and $R \in L(Y, X)$. We begin this section by proving that the non-zero points of the spectra $\sigma(RS)$ and $\sigma(SR)$ are the same, and the same holds for a number of distinguished parts of the spectrum.

Theorem 2.149 *Let $S \in L(X, Y)$, $R \in L(Y, X)$ and $\lambda \neq 0$. Then we have:*

(i) $\alpha(\lambda I - SR) = \alpha(\lambda I - RS)$.
(ii) $\beta(\lambda I - SR) = \beta(\lambda I - RS)$.
(iii) $\lambda I - SR$ *has closed range if and only if $\lambda I - RS$ has closed range.*

Proof

(i) If $x \in X$ is an eigenvector of $\lambda I - RS$ then

$$(\lambda I - RS)x = 0 = T(\lambda I - RS)x = (\lambda I - SR)Sx.$$

Since $\lambda \neq 0$ implies $Sx \neq 0$, Sx is an eigenvector of $(\lambda I - SR)$. Let (x_k) $k = 1, 2, \ldots n$ be a set of linearly independent eigenvectors for $\lambda I - RS$. We show that the set $\{Sx_1, Sx_2, \ldots Sx_n\}$ is linearly independent. Assume not, so there exist non-zero scalars $\alpha_1, \alpha_2, \ldots \alpha_n$ such that $\sum_{k=1}^{n} \alpha_k Sx_k = 0$. Then

$$S(\sum_{k=1}^{n} \alpha_k x_k) = 0 = RS(\sum_{k=1}^{n} \alpha_k x_k) = \lambda(\sum_{k=1}^{n} \alpha_k x_k).$$

Since $\lambda \neq 0$, this contradicts the linearly independence of the x_k's. Therefore, $\{Sx_1, Sx_2, \ldots Sx_n\}$ is a linearly independent set and, consequently, $\alpha(\lambda I - SR) \leq \alpha(\lambda I - RS)$. The reverse inequality follows by symmetry.

(ii) Let $\hat{X} := X/(\lambda I - SR)(X)$ and $y \in X$ such that $\hat{y} \neq 0$. Then $\hat{R}y \neq 0$. Assume $Ry \in \overline{(\lambda I - RS)(X)}$. Then there exists a sequence $(z_n) \subseteq (\lambda I - RS)(X)$ such that $(\lambda I - RS)z_n = Ry$, as $n \to \infty$. Therefore,

$$(\lambda I - SR)Sz_n = SRy,$$

and $SRy \in \overline{(\lambda I - SR)(X)}$. Since

$$\lambda y = (\lambda I - SR)y + SRy \in \overline{(\lambda I - SR)(X)}$$

contradicts our assumption $\hat{y} \neq 0$, it then follows that $Ry \notin \overline{(\lambda I - RS)(X)}$.

Let $(\hat{y}_k)$ be a set of linearly independent vectors in $\hat{X}$, and set $\tilde{X} := X/\overline{(\lambda I - RS)(X)}$. We claim that $\widetilde{(Ry_k)}$ are linearly independent vectors in $\tilde{X}$. Indeed, assume that there are scalars (α_k) such that $\sum_{k=1}^{n} \alpha_k \widetilde{Ry_k} = 0$. Then there exists a sequence (z_k) in X such that $(\lambda I - RS)z_n \to \sum_{k=1}^{n} \alpha_k Ry_k$, which implies that

$$\sum_{k=1}^{n} \alpha_k SRy_k = \lim_{k \to \infty} (\lambda I - SR)Sz_n.$$

Now,

$$-\lambda \sum_{k=1}^{n} \alpha_k SRy_k = \sum_{k=1}^{n} \alpha_k[(\lambda I - SR)y_k - SRy_k],$$

hence

$$\lambda(\sum_{k=1}^{n} \alpha_k y_k) \in \overline{(\lambda I - SR)(X)}$$

and

$$\alpha_1 = \alpha_2 = \ldots \alpha_n = 0.$$

Thus, the vectors $\widetilde{(Ry_k)}$ are linearly independent and consequently, we have $\beta(\lambda I - SR) \leq \beta(\lambda I - RS)$. Similarly, $\beta(\lambda I - RS) \leq \beta(\lambda I - SR)$ and the equality then follows.

(iii) Suppose that $(\lambda I - RS)(X)$ is closed and $y \in X$ is such that $(\lambda I - RS)x_n \to y$, with $x_n \in X$. Then $(\lambda I - RS)Rx_n \to Ry$ and since $(\lambda I - RS)(X)$ is closed, $Ry \in (\lambda I - RS)(X)$. That is, there exists an $x \in X$ such that $Ry = (\lambda I - RS)x$. But

$$\lambda y = SRy - (SR - \lambda I)y = S(\lambda I - RS)x - (SR - \lambda I)y$$
$$= (\lambda I - SR)Sx + (\lambda I - SR)y$$

and since $\lambda \neq 0$ we have $y \in (\lambda I - SR)(X)$. In a similar fashion the reverse implication follows. ∎

Corollary 2.150 *If $R, S \in L(X)$ then the nonzero points of the spectrum, or of the approximate point spectrum, of RS and SR are the same. The same happens for the upper semi-Fredholm spectra and the essential spectrum of RS and SR.*

It is easy to find examples of operators for which the product SR is invertible, while RS is not invertible. For instance, if R is the right shift on $\ell_2(\mathbb{N})$ and L is the left shift, then $LR = I$ is invertible, while RL is not injective and hence not invertible.

Corollary 2.151 *If $S, R \in L(X)$ then RS is Riesz if and only if SR is Riesz.*

Note that by Theorem 2.4 the local spectrum of RS at x and the local spectrum of SR at Sx have the same non-zero points. Further results concerning other spectra of the products SR and RS will be given in the next chapter. We next want show that SR and RS also share some other local spectral properties.

It is not surprising that the property of having closed local spectral subspaces $X_T(F)$ for every closed set $F \subseteq \mathbb{C}$ is an important property. For instance, for every spectral operator $T \in L(X)$ with spectral measure $E(\cdot)$, the subspace $X_T(F)$ is closed, for every closed set F, since it coincides with the range of the projection $E(F)$, see [216, Corollary 1.2.25]. To label this situation we introduce the following definition.

Definition 2.152 A bounded operator $T \in L(X)$, X a Banach space, is said to have *Dunford's property (C)*, shortly property (C), if the analytic subspace $X_T(F)$ is closed for every closed subset $F \subseteq \mathbb{C}$.

Property (C) dates back to the earliest days of local spectral theory. It was first introduced by Dunford (see [143]) and plays an important role in the development of the theory of spectral operators (this condition is one of the three basic conditions that are used in the abstract characterization of spectral operators), and more generally in the development of the theory of decomposable operators. The monograph by Dunford and Schwartz [143] and the book by Laursen and Neumann [216] contain a number of pertinent results.

Trivially, by Theorem 2.14, we have the following relevant fact:

Theorem 2.153 *If $T \in L(X)$, X a Banach space, has property (C) then T has the SVEP.*

Note that if an operator T has property (C), and hence the SVEP, then the quasi-nilpotent part $H_0(T)$ is closed since $H_0(T) = X_T(\{0\})$, see Theorem 2.30. The operator T considered in Example 2.33 shows that the implication of Theorem 2.153 cannot be reversed in general. Further examples of operators with the SVEP but without property (C) may be found among the class of all multipliers of semi-simple commutative Banach algebras, which will be introduced in more detail in Chap. 3. In fact, these operators have the SVEP, since the quasi-nilpotent part of $\lambda I - T$ coincides with the kernel $\ker(\lambda I - T)$ for all $\lambda \in \mathbb{C}$, see the next Theorem 4.48, while property (C) plays a distinctive role in this context, see [216, Chapter 4].

A first example of operators which have property (C) is given by quasi-nilpotent operators.

Theorem 2.154 *Let $T \in L(X)$ be a quasi-nilpotent operator on a Banach space X. Then T has property (C).*

Proof Consider any closed subset of $F \subseteq \mathbb{C}$. Consider first the case $0 \notin F$. Then since T has the SVEP,

$$X_T(F) = X_T(F \cap \sigma(T)) = X_T(\emptyset) = \{0\}$$

is trivially closed. On the other hand, if $0 \in F$ then by Theorem 2.35 and Theorem 2.30

$$X_T(F) = X_T(F \cap \sigma(T)) = X_T(\{0\}) = H_0(T) = X.$$

Hence, also in this case $X_T(F)$ is closed.   ∎

Lemma 2.155 *Suppose that $T \in L(X)$ has the SVEP, and that $F \subseteq \mathbb{C}$ is a closed set for which $X_T(F)$ is closed. Then $\sigma(T|X_T(F)) \subseteq F \cap \sigma(T)$.*

Proof Set $A := T|X_T(F)$. Clearly, $\lambda I - A$ has the SVEP and part (ii) of Theorem 2.13 ensures that $\lambda I - S$ is onto for all $\lambda \in \mathbb{C} \setminus F$. By Corollary 2.61 then $\lambda I - S$ is invertible for all $\lambda \in \mathbb{C} \setminus F$. On the other hand, part (iii) of Theorem 2.13 shows that $\lambda I - S$ is invertible for all $\lambda \in F$ which belong to the resolvent, so $\sigma(S) \subseteq (\mathbb{C} \setminus F) \cup \sigma(T)$, and hence $\sigma(S) \subseteq F \cap \sigma(T)$.   ∎

Property (C) is inherited by restrictions to closed invariant subspaces.

Theorem 2.156 *Suppose that $T \in L(X)$, where X is a Banach space, has property (C). If Y is a T-invariant closed subspace of X then the restriction $T|Y$ has property (C).*

Proof Set $S := T \mid Y$ and let F be a closed subset of $\mathbb{C}$. Suppose that the sequence $(x_n) \subset Y_S(F)$ converges at $x \in X$. We have to show that $x \in Y_S(F)$. Evidently, $Y_S(F) \subseteq X_T(F) \cap Y$, so that $x \in Y_S(\Omega) \subseteq X_T(\Omega)$. By Theorem 2.153 we know that T has the SVEP, so there exists an analytic function $f : \mathbb{C} \setminus \Omega \to X$ such that $(\lambda I - T)f(\lambda) = x$ for all $\lambda \in \mathbb{C} \setminus \Omega$.

To show that $x \in Y_S(F)$ it suffices to prove that $f(\lambda)$ belongs to Y for all $\lambda \in \mathbb{C} \setminus F$. Since T has the SVEP, for every $n \in \mathbb{N}$ there exists an analytic function $f_n : \mathbb{C} \setminus F \to Y$ such that $(\lambda I - T)f_n(\lambda) = x_n$ for all $\lambda \in \mathbb{C} \setminus F$. The elements x and x_n belong to $X_T(F)$, so Theorem 2.9 implies that $f(\lambda)$ and $f_n(\lambda)$ belong to $X_T(F)$ for all $\lambda \in \mathbb{C} \setminus F$ and $n \in \mathbb{N}$. Since T has the SVEP, and $X_T(F)$ is closed by assumption, from Lemma 2.155 we know that $\sigma(T \mid X_T(F)) \subseteq F$, and therefore the bounded operator $\lambda I - T \mid X_T(F)$ on $X_T(F)$ has an inverse $(\lambda I - T \mid X_T(F))^{-1}$ for every $\lambda \in \mathbb{C} \setminus F$.

From this we then obtain that $f_n(\lambda) = (\lambda I - T \mid X_T(F))^{-1} x_n$ converges to the element $(\lambda I - T \mid X_T(F))^{-1} x$, as $n \to \infty$. Therefore $f(\lambda) \in Y$, so the proof is complete.   ∎

The next result shows that property (C) is preserved by the Riesz functional calculus. For a proof, see Theorem 3.3.6 of Laursen and Neumann [216].

Theorem 2.157 *If $T \in L(X)$ has property (C) and f is an analytic function on an open neighborhood U of $\sigma(T)$, then $f(T)$ has property (C). Similar statements hold for property (β).*

It could be reasonable to expect that the converse of Theorem 2.157 is true if we assume that f is non-constant on each connected component of U, as is the case, by Theorem 2.89, for the SVEP; but this is not known.

Lemma 2.158 *Let $S \in L(X, Y)$ and $R \in L(Y, X)$ and $\mu \in \mathbb{C}$. Then RS has the SVEP at λ if and only if SR has the SVEP at λ.*

Proof Suppose that RS has the SVEP at λ and let $f : \mathbb{D}_\lambda \to Y$ be an analytic function defined in an open disc centered at λ such that

$$(\mu I - SR)f(\mu) = 0 \quad \text{for all } \mu \in \mathbb{D}_\lambda. \tag{2.31}$$

Then $SRf(\mu) = \mu f(\mu)$ for all $\mu \in \mathbb{D}_\lambda$. From (2.31) we have

$$R(\mu I - SR)f(\mu) = (\mu I - RS)Rf(\mu) = 0 \quad \text{for all } \mu \in \mathbb{D}_\lambda,$$

and hence, since RS has the SVEP at λ, $Rf(\mu) = 0$. Then $0 = SRf(\mu) = \mu f(\mu)$ for all $\mu \in \mathbb{D}_\lambda$, from which we obtain $f(\mu) = 0$ for all $\mu \in \mathbb{D}_\lambda$. Therefore, SR has the SVEP at λ. The converse may be proved in a similar way.   ∎

Theorem 2.159 *Let F be a closed subset of $\mathbb{C}$ such that $0 \in F$. If $S \in L(X, Y)$ and $R \in L(Y, X)$ then $Y_{SR}(F)$ is closed if and only if $X_{RS}(F)$ is closed.*

Proof Suppose that $Y_{SR}(F)$ is closed and let (x_n) be a sequence in $X_{RS}(F)$ which converges to $x \in X$. Since $x_n \in X_{RS}(F)$, $\sigma_{RS}(x_n) \subseteq F$ for all $n \in \mathbb{N}$. Since $0 \in F$, $\sigma_{RS}(x_n) \cup \{0\} \subseteq F$. By Theorem 2.4, part (i), we have

$$\sigma_{RS}(x_n) \cup \{0\} = \sigma_{SR}(Sx_n) \cup \{0\},$$

so $\sigma_{SR}(Sx_n) \subseteq F$ and hence $Sx_n \in Y_{SR}(F)$. But $Sx_n \to Sx$ and $Y_{SR}(F)$ is closed, thus $Sx \in Y_{SR}(F)$, that is, $\sigma_{SR}(Sx) \subseteq F$. Again by Theorem 2.4 we obtain

$$\sigma_{RS}(x) \subseteq \sigma_{SR}(Sx) \cup \{0\} \subseteq F,$$

thus $x \in X_{RS}(F)$.

The converse implication follows in a similar way, just use part (ii) of Theorem 2.4. ∎

In order to study the case when $0 \notin F$ we need a preliminary result:

Lemma 2.160 *Suppose that $T \in L(X)$ has the SVEP and let F be a closed subset of $\mathbb{C}$ such that $Z := X_T(F)$ is closed. If $A := T | X_T(F)$ then $X_T(K) = Z_A(K)$ for all closed $K \subseteq F$.*

Proof Note first that A has the SVEP, so every glocal spectral subspace $\mathcal{Z}_A(K)$ coincides with the local spectral subspace $Z_A(K)$, and $X_T(K) \subseteq X_T(F) = Z$. The inclusion $\mathcal{Z}_A(K) \subseteq \mathcal{X}_T(K)$ is immediate. In order to prove the opposite inclusion, suppose that $x \in X_T(K) = \mathcal{X}_T(K)$. Then $\sigma_T(x) \subseteq K$ and there is an analytic function $f : \mathbb{C} \setminus K \to X$ such that $(\mu I - T)f(\mu) = x$ for all $\mu \in \mathbb{C} \setminus K$. By Theorem 2.9 we have

$$\sigma_T(f(\mu)) = \sigma_T(x) \subseteq K \quad \text{for all } \mu \in \mathbb{C} \setminus K,$$

thus $f(\mu) \in X_T(K) \subseteq Z$. Therefore, f is a Z-valued function and hence

$$(\mu I - T)f(\mu) = (\mu I - A)f(\mu) = x \quad \text{for all } \mu \in \mathbb{C} \setminus K,$$

i.e. $x \in \mathcal{Z}_A(K) = Z_A(K)$. ∎

Theorem 2.161 *Let F be a closed subset of $\mathbb{C}$ such that $\lambda \notin F$. If $T \in L(X)$ has the SVEP and $X_T(F \cup \{\lambda\})$ is closed then $X_T(F)$ is closed.*

Proof Let $Z := X_T(F \cup \{\lambda\})$ and $S := T | X_T(F \cup \{\lambda\})$. From Lemma 2.155 we know that $\sigma(S) \subseteq F \cup \{\lambda\}$. We consider two cases: Suppose first that $\lambda \notin \sigma(S)$. Then $\sigma(S) \subseteq F$ and hence $Z = Z_S(F)$. By Lemma 2.160 we then have $Z_S(F) = X_T(F)$, so $X_T(F)$ is closed. Suppose the other case that $\lambda \in \sigma(S)$ and set $F_0 := \sigma(S) \cap F$. Then $\sigma(S) = F_0 \cup \{\lambda\}$. Since $\lambda \in \sigma(S)$, by Theorem 2.26 we have $Z = Z_S(F_0) \oplus Z_S(\{\lambda\})$. From Lemma 2.160 it then follows that

$$Z_S(F_0) = Z_F(\sigma(S) \cap F) = Z_S(F) = X_T(F),$$

and hence $X_T(F)$ is closed.   ∎

Corollary 2.162 *Let $S \in L(X, Y)$ and $R \in L(Y, X)$ be such that RS has the SVEP, and denote by F a closed subset of $\mathbb{C}$ such that $0 \notin F$. Then we have:*

(i) *If $Y_{SR}(F \cup \{0\})$ is closed then $X_{RS}(F)$ is closed.*
(ii) *If $X_{RS}(F \cup \{0\})$ is closed then $Y_{SR}(F)$ is closed.*

Proof Theorem 2.159 ensures that $Z := X_{RS}(F \cup \{0\})$ is closed, since $0 \in F \cup \{0\}$. The SVEP for RS entails the SVEP for SR, by Lemma 2.158, thus Theorem 2.161 entails that $X_{RS}(F)$ is closed. An analogous argument proves (ii).   ∎

A remarkable consequence of the previous results is that property (C) for RS is equivalent to property (C) for SR.

Corollary 2.163 *If $S \in L(X, Y)$ and $R \in L(Y, X)$ then RS has Dunford's property (C) if and only if SR has Dunford's property (C).*

Proof Suppose that RS has Dunford's property (C), i.e. $X_{RS}(F)$ is closed for every closed subset $F \subseteq \mathbb{C}$. If $0 \in F$ then $Y_{SR}(F)$ is closed, by Theorem 2.159. Obviously, if $0 \notin F$ then $F \cup \{0\}$ is closed, so $X_{RS}(F \cup \{0\})$ is closed and hence $Y_{SR}(F)$ is closed, by Corollary 2.162. Therefore SR has Dunford's property (C). The proof of the opposite implication is similar.   ∎

Let us consider the particular case when F is a singleton set, say $F := \{\lambda\}$. Recall $H_0(\lambda I - T) = \mathcal{X}_T(\{\lambda\})$, where $\mathcal{X}_T(\{\lambda\})$ is the glocal spectral subspace associated with the closed set $\{\lambda\}$.

Definition 2.164 An operator $T \in L(X)$ is said to have *property (Q)* if $H_0(\lambda I - T)$ is closed for every $\lambda \in \mathbb{C}$.

Evidently, by Corollary 2.67,

$$\text{property } (C) \Rightarrow \text{property } (Q) \Rightarrow \text{SVEP}.$$

Consequently, by part (v) of Theorem 2.23, for operators T having property (Q) we have $H_0(\lambda I - T) = X_T(\{\lambda\})$. Property (Q) is strictly weaker than property (C), indeed every multiplier of a semi-simple commutative Banach algebra has property (Q), see the next Theorem 4.48, in particular every convolution operator T_μ, $\mu \in M(G)$, on the group algebra $L_1(G)$ has property (Q), but there are convolution operators which do not enjoy property (C), see [216, Chapter 4].

Lemma 2.165 *Let F be a closed subset of $\mathbb{C}$. If $S \in L(X, Y)$ and $R \in L(Y, X)$ are both injective, then $Y_{SR}(F)$ is closed if and only if $X_{RS}(F)$ is closed.*

Proof Suppose that $Y_{SR}(F)$ is closed and let (x_n) be a sequence in $\mathcal{X}_{RS}(F)$ which converges to $x \in X$. Since $x_n \in X_{RS}(F)$, $\sigma_{RS}(x_n) \subseteq F$ for all $n \in \mathbb{N}$. Because S is injective we have $\sigma_{RS}(x_n) = \sigma_{SR}(Sx_n)$, by Theorem 2.4, hence $Sx_n \in Y_{SR}(F)$, since $\sigma_{SR}(Sx_n) \subseteq F$. But $Sx_n \to Sx$, and $Y_{SR}(F)$ is closed, so that $Sx \in Y_{SR}(F)$ and hence $\sigma_{SR}(Sx) \subseteq F$. Again, by Theorem 2.4, we have $\sigma_{RS}(x) = \sigma_{SR}(Sx) \subseteq F$, hence $x \in X_{RS}(F)$. By using the same argument and, since R is injective, we easily obtain the reverse implication.   ∎

Theorem 2.166 *If $S \in L(X, Y)$ and $R \in L(Y, X)$ are both injective, then RS has property (Q) if and only if SR has property (Q).*

Proof Suppose that RS has property (Q). Then RS has the SVEP and hence SR also has the SVEP, by Lemma 2.158. By Theorem 2.23, part (iv), for every $\lambda \in \mathbb{C}$ we have $X_{RS}(\{\lambda\}) = \mathcal{X}_{RS}(\{\lambda\}) = H_0(\lambda I - RS)$, and analogously $Y_{SR}(\{\lambda\}) = \mathcal{Y}_{SR}(\{\lambda\}) = H_0(\lambda I - SR)$. Now, if $\lambda = 0$ then, by Theorem 2.159 $H_0(RS) = X_{RS}(\{0\})$ is closed if and only if $H_0(SR) = Y_{SR}(\{0\})$ is closed. If $\lambda \neq 0$ from Lemma 2.165 we see that $H_0(\lambda I - SR) = Y_{SR}(\lambda\}$ is closed if and only if $H_0(\lambda I - RS) = X_{RS}(\lambda\}$ is closed.   ∎

An interesting question is if the result of Theorem 2.166 is still valid without assuming that R, S are injective.

In order to establish some results concerning the analytic core of RS and SR, recall that, by Theorem 2.20, for every $T \in L(X)$ we have $K(\lambda I - T) = X_T(\mathbb{C} \setminus \{\lambda\}) = \{x \in X : \lambda \notin \sigma_T(x)\}$.

Theorem 2.167 *Let $S \in L(X, Y)$ and $R \in L(Y, X)$. We have:*

(i) *If $\lambda \neq 0$, then $K(\lambda I - SR)$ is closed if and only if $K(\lambda I - RS)$ is closed.*
(ii) *If R and S are injective then $K(SR)$ is closed if and only if $K(RS)$ is closed.*

Proof

(i) Suppose that $K(\lambda I - SR)$ is closed and let (x_n) be a sequence of elements of $K(\lambda I - RS)$ which converges to $x \in X$. Then $\lambda \notin \sigma_{RS}(x_n)$ for every $n \in \mathbb{N}$, and hence, by Theorem 2.4, $\lambda \notin \sigma_{SR}(Sx_n)$, that is, $Sx_n \in K(\lambda I - SR)$. Since $Sx_n \to Sx$ and $K(\lambda I - SR)$ is closed, $Sx \in K(\lambda I - SR)$, i.e. $\lambda \notin \sigma_{SR}(Sx)$. Again from Theorem 2.4 we then have $\lambda \notin \sigma_{RS}(x)$, hence $x \in K(\lambda I - RS)$.
(ii) The proof is similar to that of part (i), just recalling the equalities between the local spectra of SR and RS, established in Theorem 2.4 in the case where R and S are injective.   ∎

Another important property which plays a central role in local spectral theory is the property (β) introduced by Bishop [79]. Let $H(U, X)$ denote the space of all analytic functions from U into X. With respect to pointwise vector space operations and the topology of locally uniform convergence, $H(U, X)$ is a Fréchet space. Denote, as usual, by $\mathbb{D}(\lambda, r)$ an open disc centered at λ with radius $r < 0$.

Definition 2.168 An operator $T \in L(X)$ is said to have Bishop's property (β) at $\lambda \in \mathbb{C}$ if there exists an $r > 0$ such that for every open subset $U \subseteq \mathbb{D}(\lambda, r)$, and for every sequence of analytic functions $(f_n) \subset H(U, X)$ for which $(\mu I - T) f_n(\mu) \to$ 0 in $H(U, X)$, we have $f_n(\mu) \to 0$ in $H(U, X)$. $T \in L(X)$ is said to have property (β) if T has property (β) at every $\lambda \in \mathbb{C}$.

The relevance of property (β) in local spectral theory is not immediately evident at first glance. It should be noted that the following implication holds:

$$\text{property } (\beta) \Rightarrow \text{property } (C) \Rightarrow \text{SVEP}.$$

For a proof, see Laursen and Neumann's book [216, Proposition 1.2.19]. A dual property of property (β) is given by the following decomposition property (δ).

Definition 2.169 An operator $T \in L(X)$ is said to have the *decomposition property* (δ) if the decomposition

$$X = \mathcal{X}_T(\overline{U}) + \mathcal{X}_T(\overline{V})$$

holds for every open cover $\{U, V\}$ of $\mathbb{C}$.

A remarkable result of local spectral theory is that the there is a complete duality between property (δ) and property (β). Indeed, $T \in L(X)$ has property (δ) (respectively, property (β)) if and only if T^* has property (β) (respectively, property (δ)), see [216, Theorem 2.5.5].

We now introduce an important class of operators on Banach spaces which admits a rich spectral theory and contains many important classes of operators.

Definition 2.170 Given a Banach space X, an operator $T \in L(X)$ is said to be *decomposable* if, for any open covering $\{\mathcal{U}_1, \mathcal{U}_2\}$ of the complex plane $\mathbb{C}$, there are two closed T-invariant subspaces Y_1 and Y_2 of X such that $Y_1 + Y_2 = X$ and $\sigma(T|Y_k) \subseteq \mathcal{U}_k$ for $k = 1, 2$.

A very deep result in local spectral theory is that decomposability may be described as the union of the two weaker properties (β) and (δ), more precisely:

$$T \text{ is decomposable} \Leftrightarrow T \text{ has both properties } (\beta) \text{ and } (\delta),$$

or,

$$T \text{ is decomposable} \Leftrightarrow T \text{ has both properties } (C) \text{ and } (\delta),$$

see [216, Theorem 1.2.29].

Theorem 2.171 *Let $S \in L(X, Y)$ and $R \in L(Y, X)$. Then we have*

(i) *SR has property (β) if and only if RS has property (β).*
(ii) *SR has property (δ) if and only if RS has property (δ).*
(iii) *SR is decomposable if and only if RS is decomposable.*

Proof

(i) Suppose that SR has property (β) at $\lambda \in \mathbb{C}$. Then there exists an $r > 0$ such that for every open subset $U \subseteq \mathbb{D}(\lambda, r)$ and for every sequence $(g_n) \subset H(U, X)$ for which

$$\lim_{n \to \infty} (\mu I - SR)g_n(\mu) = 0 \quad \text{in } H(U, Y)$$

we have

$$\lim_{n \to \infty} g_n(\mu) = 0 \quad \text{in } H(U, Y).$$

Let $W \subseteq \mathbb{D}(\lambda, r)$ be open and suppose that for $(f_n) \subset H(W, X)$ we have

$$\lim_{n \to \infty} (\mu I - RS) f_n(\mu) = 0 \quad \text{in } H(W, X).$$

From this we obtain

$$\lim_{n \to \infty} (\mu I - SR)Sf_n(\mu) = \lim_{n \to \infty} S(\mu I - RS)f_n(\mu) = 0,$$

in $H(W, X)$, and hence $\lim_{n \to \infty} Sf_n(\mu) = 0$ in $H(W, X)$. This implies that $\lim_{n \to \infty} SRSf_n(\mu) = 0$ in $H(W, X)$, and consequently,

$$\lim_{n \to \infty} \mu f_n(\mu) = 0 \quad \text{in } H(W, X). \tag{2.32}$$

Obviously, if we set $T = 0$, the identity (2.32) may be rewritten as

$$\lim_{n \to \infty} (\mu I - T) f_n(\mu) = 0 \quad \text{in } H(W, X).$$

Therefore, since T has property (β), we have $\lim_{n \to \infty} f_n(\mu) = 0$ in $H(W, X)$, and hence RS has property (β) at λ. The reverse implication easily follows by interchanging S and R.

(ii) This follows from the duality between properties (β) and (δ).

(iii) As observed above, an operator is decomposable if and only if has both properties (β) and (δ).   ∎

2.11   Operators Which Satisfy $RSR = R^2$ and $SRS = S^2$

In this section we consider some local spectral properties of operators $S, R \in L(X)$ for which the operator equations

$$RSR = R^2 \quad \text{and} \quad SRS = S^2 \tag{2.33}$$

hold. This is a rather special case, but an easy example of operators which satisfy these equations can be obtained by putting $R := PQ$ and $S := QP$, where P and Q are idempotent operators. A remarkable result of Vidav [293, Theorem 2] shows that if R, S are self-adjoint operators on a Hilbert space then the Eq. (2.33) hold if and only if there exists a (uniquely determined) idempotent P such that $R = PP^*$ and $S = P^*P$, where P^* is the adjoint of P.

It is easily seen that if $0 \notin \sigma(R) \cap \sigma(S)$ then $R = S = I$, so this case is without interest. For this reason we shall assume that $0 \in \sigma(R) \cap \sigma(S)$. Evidently, the operator equations $RSR = R^2$ implies $(SR)^2 = SR^2$, while $SRS = S^2$ implies $(RS)^2 = RS^2$.

Theorem 2.172 *Let $S, R \in L(X)$ be such that $RSR = R^2$. Suppose that F is a closed subset of $\mathbb{C}$ and $0 \in F$. Then $X_R(F)$ is closed if and only if $X_{SR}(F)$ is closed.*

Proof Suppose that $X_R(F)$ is closed and let (x_n) be a sequence in $X_{SR}(F)$ which converges to $x \in X$. We need to show that $x \in X_{SR}(F)$. For every $n \in \mathbb{N}$ we have $\sigma_{SR}(x_n) \subseteq F$ and hence, by the first inclusion of Lemma 2.7, we have $\sigma_R(Rx_n) \subseteq F$, i.e., $Rx_n \in X_R(F)$. Since $0 \in F$ we then have, by part (iii) of Theorem 2.13, $x_n \in X_R(F)$, and since by assumption $X_R(F)$ is closed it then follows that $x \in X_R(F)$, hence $\sigma_R(x) \subseteq F$. From the second inclusion of Lemma 2.7 we then have $\sigma_{SR}(SRx) \subseteq F$, and this implies that $SRx \in X_{SR}(F)$. Again by part (iii) of Theorem 2.13, we conclude that $x \in X_{SR}(F)$, thus $X_{SR}(F)$ is closed.

Conversely, suppose that $X_{SR}(F)$ is closed and let (x_n) be a sequence of $X_R(F)$ which converges to $x \in X$. Then $\sigma_R(x_n) \subseteq F$ for every $n \in \mathbb{N}$ and hence, from the second inclusion of Lemma 2.7, we have $\sigma_{SR}(SRx_n) \subseteq F$, so $SRx_n \in X_{SR}(F)$. But $0 \in F$, so, by part (iii) of Theorem 2.13, $x_n \in X_{SR}(F)$. Since, by assumption, $X_{SR}(F)$ is closed, $x \in X_{SR}(F)$, and hence $\sigma_{SR}(x) \subseteq F$. From the first inclusion of Lemma 2.7 we then obtain $\sigma_R(Rx) \subseteq F$, so $Rx \in X_R(F)$, and the condition $0 \in F$ then implies $x \in X_R(F)$. ∎

Lemma 2.173 *Let $S, R \in L(X)$ be such that $RSR = R^2$. Then if one of the operators R, SR, RS has the SVEP, all of them have the SVEP. Additionally, if $SRS = S^2$, and one of R, S, SR, RS has the SVEP then all of them have the SVEP.*

Proof By Lemma 2.158, SR has the SVEP if and only if RS has the SVEP. So it suffices only to prove that R has the SVEP at λ_0 if and only if so has RS.

Suppose that R has the SVEP at λ_0 and let $f : \mathcal{U}_0 \to X$ be an analytic function on an open neighborhood $\mathcal{U}_0$ of λ_0 for which $(\lambda I - RS)f(\lambda) \equiv 0$ on U_0. Then $RSf(\lambda) = \lambda f(\lambda)$ and

$$0 = RS(\lambda I - RS)f(\lambda) = (\lambda RS - (RS)^2)f(\lambda) = (\lambda RS - (R^2 S)f(\lambda)$$

$$= (\lambda I - R)RSf(\lambda).$$

The SVEP of R at λ_0 implies that

$$RSf(\lambda) = \lambda f(\lambda) = 0 \quad \text{for all } \lambda \in \mathcal{U}_0.$$

Therefore, $f \equiv 0$ on U_0, and hence RS has the SVEP at λ_0.

Conversely, suppose that RS has the SVEP at λ_0 and let $f : \mathcal{U}_0 \to X$ be an analytic function on an open neighborhood $\mathcal{U}_0$ of λ_0 such that $(\lambda I - R)f(\lambda) \equiv 0$ on U_0. Clearly, $Rf(\lambda) = \lambda f(\lambda)$, and hence

$$R^2 f(\lambda) = \lambda R f(\lambda) = \lambda^2 f(\lambda) \quad \text{for all } \lambda \in \mathcal{U}_0.$$

Furthermore,

$$0 = RS(\lambda I - R)f(\lambda) = \lambda RSf(\lambda) - R^2 f(\lambda) = \lambda RSf(\lambda) - \lambda^2 f(\lambda)$$
$$= (\lambda I - RS)(-\lambda f(\lambda)),$$

and since RS has the SVEP at λ_0 we then have $\lambda f(\lambda) \equiv 0$, hence $f(\lambda) \equiv 0$, thus R has the SVEP at λ_0.

The second assertion is clear, if $SRS = S^2$, just interchanging R and S in the argument above, the SVEP for S holds if and only if SR, or equivalently RS, has the SVEP. ∎

We now consider the case where $0 \notin F$.

Theorem 2.174 *Let F be a closed subset of $\mathbb{C}$ such that $0 \notin F$. Suppose that $R, S \in L(X)$ satisfy $RSR = R^2$ and R has the SVEP. Then we have*

(i) *If $X_R(F \cup \{0\})$ is closed then $X_{SR}(F)$ is closed.*
(ii) *If $X_{SR}(F \cup \{0\})$ is closed then $X_R(F)$ is closed.*

Proof

(i) Let $F_1 := F \cup \{0\}$. Clearly, F_1 is closed and by assumption $X_R(F_1)$ is closed. Since $0 \in F_1$, $X_{SR}(F_1)$ is closed, by Theorem 2.172. Moreover, the SVEP for R is equivalent to the SVEP for SR by Lemma 2.173. By Theorem 2.161 then $X_{SR}(F)$ is closed.
(ii) The argument is similar to that of part (i): if $X_{SR}(F \cup \{0\})$ is closed then $X_R(F \cup \{0\})$ by Theorem 2.172, and since R has the SVEP then $X_R(F)$ is closed, by Theorem 2.161. ∎

Theorem 2.175 *Suppose that $S, R \in L(X)$ satisfy the operator equation $RSR = R^2$ and one of the operators R, SR, RS has property (C). Then all of them have property (C). If, additionally, $SRS = S^2$ and one of the operators R, S, RS, SR has property (C), then all of them have property (C).*

Proof Since property (C) entails the SVEP, all the operators have the SVEP, by Lemma 2.173. Moreover the equivalence of property (C) for SR and RS has been proved in Corollary 2.163, so it is enough to show that R has property (C) if and only if so has RS.

Suppose that R has property (C) and let F be a closed set. If $0 \in F$ then $X_{SR}(F)$ is closed, by Theorem 2.172, while in the case where $0 \notin F$ we have that $X_R(F \cup \{0\})$ is closed, and hence, by part (i) of Theorem 2.174, the SVEP for R ensures that also in this case $X_{SR}(F)$ is closed. Therefore, SR has property (C).

Conversely, suppose that SR has property (C). For every closed subset F which contains 0 then $X_R(F)$ is closed, again by Theorem 2.172. If $0 \notin F$ then $X_{SR}(F \cup \{0\})$ is closed and hence, by part (ii) of Theorem 2.174, $X_R(F)$ is closed. Hence R has property (C).

If additionally $SRS = S^2$ then, by interchanging S with R, the same argument above proves the second, so the proof is complete.   ∎

Lemma 2.176 *Let $S, R \in L(X)$ be such that $RSR = R^2$, $SRS = S^2$, and $\lambda \in \mathbb{C}$. Then the following statements are equivalent:*

 (i) $\lambda I - R$ *is injective;*
 (ii) $\lambda I - SR$ *is injective;*
 (iii) $\lambda I - RS$ *is injective;*
 (iv) $\lambda I - S$ *is injective.*

Proof We consider first the case $\lambda \neq 0$. For $\lambda \neq 0$, the equivalence (ii) ⇔ (iii) follows from Theorem 2.149, without any assumption on R and S. We show that if $\lambda \neq 0$, then (i) ⇒ (ii). Suppose that $\ker(\lambda I - R) = \{0\}$ and $(\lambda I - SR)x = 0$ for some $x \neq 0$. Then

$$0 = R(\lambda I - SR)x = (\lambda I - R)Rx,$$

so $Rx = 0$ and hence $SRx = \lambda x = 0$, from which we conclude that $x = 0$, a contradiction. Therefore, (i) ⇒ (ii).

(iii) ⇒ (iv). Suppose that $\ker(\lambda I - RS) = \{0\}$ and $(\lambda I - S)z = 0$ for some $z \neq 0$, i.e., $Sz = \lambda z$. Then

$$0 = RS(\lambda I - S)z = R(\lambda S - S^2) = R(\lambda S - SRS)z = (\lambda I - RS)RSz,$$

so $RSz = 0$, and hence

$$0 = SRSz = S^2 z = \lambda^2 z,$$

i.e., $z = 0$, a contradiction.

(iv) ⇒ (i) Observe first that $SR^2 = S^2 R$ and hence $SR(\lambda I - R) = (\lambda I - S)SR$. Now, suppose that $\ker(\lambda I - S) = \{0\}$ and $(\lambda I - R)u = 0$ for some $u \neq 0$, i.e., $Ru = \lambda u$. Then

$$0 = SR(\lambda I - R)u = (\lambda I - S)SRu,$$

from which we obtain $SRu = 0$. Consequently,

$$0 = RSRu = R^2u = \lambda^2 u,$$

i.e., $u = 0$, a contradiction. Hence (iv) $\Rightarrow$ (i) for $\lambda \neq 0$.

Therefore, the equivalence of the statements (i)–(iv) are proved for $\lambda \neq 0$.

Now, consider the case $\lambda = 0$. Evidently, ker $SR = \{0\}$ (respectively, ker $SR = \{0\}$) implies that ker $R = \{0\}$ (respectively, ker $S = \{0\}$). If ker $R = \{0\}$ (respectively, ker $S = \{0\}$), then $RSR = R^2$ (respectively. $SRS = S^2$) implies ker $SR = \{0\}$ (respectively, ker $RS = \{0\}$). Hence, (i) $\Leftrightarrow$ (ii) and (iii) $\Leftrightarrow$ (iv) for $\lambda = 0$. Suppose that ker $RS = \{0\}$ and $Ry = 0$. Then $RSRy = R^2y = 0$, so $SR^2y = RS^2y = 0$, which implies $Sy \in$ ker RS and hence $Sy = 0$. Then $RSy = 0$, i.e. $y = 0$. Therefore, (iii) $\Rightarrow$ (i). In a similar way we can prove that (ii) $\Rightarrow$ (iv). $\blacksquare$

The following construction, known in the literature as the *Sadovskii/Buoni, Harte, Wickstead* construction, leads to a representation of the Calkin algebra $L(X)/K(X)$ as an algebra of operators on a suitable Banach space.

Let us consider the Banach space $\ell^\infty(X)$ of all bounded sequences $\tilde{x} := (x_n)$ of elements of X, endowed with the norm $\|\tilde{x}\|_\infty := \sup_{n \in \mathbb{N}} \|x_n\|$, and if $T \in L(X)$ define the induced operator on $\ell^\infty(X)$, as

$$T_\infty \tilde{x} := (Tx_n) \quad \text{for all } \tilde{x} := (x_n).$$

Let $m(X)$ denote the set of all precompact sequences (x_n) of elements of X, (i.e. the closure of $\{x_n : n \in \mathbb{N}\}$ is compact in X). The set $m(X)$ is a closed subspace of $\ell^\infty(X)$ invariant under T_∞. Let $\widetilde{X} := \ell^\infty(X)/m(X)$, and let $\widetilde{T} : \widetilde{X} \to \widetilde{X}$ be defined by

$$\widetilde{T}(\tilde{x} + m(X)) := T^\infty \tilde{x} + m(X) \quad \text{for all } \tilde{x} \in \widetilde{X}.$$

The mapping $T \in L(X) \to \widetilde{T} \in L(\widetilde{X})$ is a unital homomorphism from $L(X)$ to $L(\widetilde{X})$ with kernel $K(X)$, which induces a norm decreasing monomorphism from $L(X)/K(X)$ to $L(\widetilde{X})$ with the following properties (see [245, §17, Chap. 3] for details):

(a) $T \in \Phi_+(X) \Leftrightarrow \widetilde{T}$ is injective $\Leftrightarrow \widetilde{T}$ is bounded below;
(b) $T \in \Phi_-(X) \Leftrightarrow \widetilde{T}$ is onto;
(c) $T \in \Phi(X) \Leftrightarrow \widetilde{T}$ is invertible.

These properties easily imply that the upper semi-Fredholm spectrum $\sigma_{\mathrm{usf}}(T)$ coincides with $\sigma_{\mathrm{ap}}(\widetilde{T})$, the lower semi-Fredholm spectrum $\sigma_{\mathrm{lsf}}(T)$ coincides with $\sigma_s(\widetilde{T})$, while the essential spectrum $\sigma_e(T)$ coincides with $\sigma(\widetilde{T})$.

Lemma 2.177 *Let $R, S \in L(X)$ be such that $RSR = R^2$ and $SRS = S^2$. If one of the operators R, RS, SR or S is upper semi-Fredholm (respectively, lower semi-Fredholm, Fredholm) then all are upper semi-Fredholm (respectively, lower semi-Fredholm, Fredholm).*

Proof Evidently,

$$\widetilde{R}\widetilde{S}\widetilde{R} = \widetilde{R}^2 \quad \text{and} \quad \widetilde{S}\widetilde{R}\widetilde{S} = \widetilde{R}S^2.$$

Suppose that $\lambda I - T \in \Phi_+(X)$, where U stands for one of the operators R, RS, SR and S. Then $\lambda \widetilde{I} - \widetilde{T}$ is injective, by part (a) above. This implies, by Lemma 2.176, that the operators $\widetilde{I} - \widetilde{R}$, $\widetilde{I} - \widetilde{RS}$, $\widetilde{I} - \widetilde{SR}$ and $\widetilde{I} - \widetilde{S}$ are all injective. This shows that the upper semi-Fredholm spectra of R, RS, SR and S coincide. The assertions concerning the other spectra may be proved in a similar way.   ∎

Similar results to those of Lemma 2.177 hold for the upper semi-Browder operators, lower semi-Browder operators, Browder operators, upper semi-Weyl operators, lower semi-Weyl operators, Weyl operators, and Drazin invertible operators, see Duggal [126, 282] and Schmoeger [283].

We consider now property (Q).

Theorem 2.178 *Let* $R, S \in L(X)$ *satisfy* $RSR = R^2$, *and* $R, S \in \Phi_+(X)$ *or* $R, S \in \Phi_-(X)$. *Then* R *has property* (Q) *if and only if* SR *has property* (Q).

Proof Suppose that $R, S \in \Phi_+(X)$ and R has property (Q). Then R has the SVEP and, by Lemma 2.173, SR also has the SVEP. Consequently, by part (iii) of Theorem 2.23, the local and glocal spectral subspaces relative to a closed set coincide for R and SR. By assumption $H_0(\lambda I - R) = X_R(\{\lambda\})$ is closed for every $\lambda \in \mathbb{C}$, and $H_0(SR) = X_{SR}(\{0\})$ is closed by Theorem 2.172. Let $0 \neq \lambda \in \mathbb{C}$. From part (iv) of Theorem 2.23 we have

$$X_R(\{\lambda\} \cup \{0\}) = X_R(\{\lambda\}) + X_R(\{0\}) = H_0(\lambda I - R) + H_0(R).$$

Since $R \in \Phi_+(X)$ has the SVEP at 0, Theorem 2.105 implies that $H_0(R)$ is finite-dimensional, so $X_R(\{\lambda\} \cup \{0\})$ is closed. Then, by Theorem 2.174, we conclude that $H_0(\lambda I - SR) = X_{SR}(\{\lambda\})$ is closed, hence SR has property (Q).

Conversely, suppose that SR has property (Q). If $\lambda = 0$ then $H_0(SR) = X_{RS}(\{0\})$ is closed by assumption, and $H_0(R) = X_R(\{0\})$ is closed by Theorem 2.172. In the case $\lambda \neq 0$ we have

$$X_{SR}(\{\lambda\} \cup \{0\}) = X_{SR}(\{\lambda\}) + X_{SR}(\{0\}) = H_0(\lambda I - SR) + H_0(SR).$$

Since SR has the SVEP and $SR \in \Phi_+(X)$ then $H_0(SR)$ is finite-dimensional, again by Theorem 2.105. So $X_{SR}(\{\lambda\} \cup \{0\})$ is closed. By Theorem 2.174 then $X_R(\{\lambda\}) = H_0(\lambda I - R)$ is closed. Therefore R has property (Q).

The proof in the case where $R, S \in \Phi_-(X)$ is analogous.   ∎

Corollary 2.179 *Let* $R, S \in L(X)$ *be such that* $RSR = R^2$. *If both* R, S *are bounded below, then* R *has property* (Q) *if and only if* SR *has property* (Q).

Proof By Lemma 2.176, the operators R, S, RS, and SR are all injective when one of them is injective, and the same is true for being upper semi-Fredholm, by Lemma 2.177. Hence, if one of the operators is bounded below, then all of them are bounded below. By Theorem 2.179 property (Q) for R and for SR are equivalent. So the same is true for S and RS, and also for RS and SR since R and S are injective. ∎

Corollary 2.180 *Let $S, R \in L(X)$ satisfy the operator equations $RSR = R^2$ and $SRS = S^2$, and suppose that R and S are bounded below. If any one of the operators $R, S, RS,$ and SR has property (Q), then all have property (Q).*

Proof Also RS, and SR are bounded below, as observed in the proof of Theorem 2.178, and property (Q) for R and SR are equivalent. By interchanging R and S we deduce that property (Q) is equivalent for S and RS. Finally, since R and S are injective, as noted before, property (Q) for RS and SR are equivalent. ∎

We conclude this section by giving some results on the analytic core of RS and SR.

Theorem 2.181 *Suppose that $R, S \in L(X)$ satisfy $RSR = R^2$.*

(i) *If $0 \neq \lambda \in \mathbb{C}$, then $K(\lambda I - R)$ is closed if and only $K(\lambda I - SR)$ is closed, or equivalently $K(\lambda I - RS)$ is closed.*
(ii) *If R is injective then $K(\lambda I - R)$ is closed if and only $K(\lambda I - SR)$ is closed, or equivalently $K(\lambda I - RS)$ is closed, for all $\lambda \in \mathbb{C}$.*

Proof

(i) Suppose $\lambda \neq 0$ and $K(\lambda I - R)$ is closed. Let (x_n) be a sequence of $K(\lambda I - SR)$ which converges to $x \in X$. Then $\lambda \notin \sigma_{SR}(x_n)$ and hence, by Lemma 2.7, $\lambda \notin \sigma_R(Rx_n)$, thus $Rx_n \in K(\lambda I - R)$. Since $Rx_n \to Rx$ and $K(\lambda I - R)$ is closed, it then follows that $Rx \in K(\lambda I - R)$, i.e., $\lambda \notin \sigma_R(Rx)$. Since $\lambda \neq 0$, by part (i) of Corollary 2.5 we have $\lambda \notin \sigma_R(x)$ and hence $\lambda \notin \sigma_{SR}(SRx)$, again by Lemma 2.7. This implies, again by part (i) of Corollary 2.5, that $\lambda \notin \sigma_{SR}(x)$. Therefore $x \in K(\lambda I - SR)$, and consequently, $K(\lambda I - SR)$ is closed.

Conversely, suppose that $\lambda \neq 0$ and $K(\lambda I - SR)$ is closed. Let (x_n) be a sequence in $K(\lambda I - R)$ which converges to $x \in X$. Then $\lambda \notin \sigma_R(x_n)$ and, by Lemma 2.7, we have $\lambda \notin \sigma_{SR}(SRx_n)$. From part (i) of Corollary 2.5 we have $\lambda \notin \sigma_{SR}(x_n)$, so $x_n \in K(\lambda I - SR)$, and hence $x \in K(\lambda I - SR)$, since the last set is closed. This implies that $\lambda \notin \sigma_{SR}(x)$, and hence $\lambda \notin \sigma_R(Rx)$, again by Lemma 2.7. Again by part (i) of Corollary 2.5 we then have $\lambda \notin \sigma_R(x)$, so that $x \in K(\lambda I - R)$. Therefore, $K(\lambda I - R)$ is closed. The equivalence, $K(\lambda I - SR)$ is closed if and only if $K(\lambda I - RS)$ is closed, has already been proved.

(ii) The proof is analogous to that of part (i), just use part (ii) of Corollary 2.4. ∎

Corollary 2.182 *If $RSR = R^2$ and $SRS = S^2$ and $\lambda \neq 0$ then the following statements are equivalent:*

(i) $K(\lambda I - R)$ *is closed;*
(ii) $K(\lambda I - SR)$ *is closed;*
(iii) $K(\lambda I - RS)$ *is closed;*
(iv) $K(\lambda I - S)$ *is closed.*

If R is injective then these equivalences also hold for $\lambda = 0$.

Proof The equivalence (iii)–(iv) follows from Theorem 2.181 by interchanging R and S. Because the injectivity of R is equivalent to the injectivity of S, the equivalences (i)–(iv) also hold for $\lambda = 0$.   ∎

2.12   Local Spectral Properties for Drazin Invertible Operators

We have seen in Theorem 1.135 that a generalization of the notion of invertibility, which satisfies the relationships of "reciprocity" observed between the nonzero parts of the spectrum is provided by the concept of Drazin invertibility. In this section we see that many local spectral properties are transmitted from a Drazin invertible operator T to its Drazin inverse S. Recall from Chap. 1 that if $T \in L(X)$ is Drazin invertible then there exist two closed invariant subspaces Y and Z such that $X = Y \oplus Z$ and, with respect to this decomposition, we can write $T = T_1 \oplus T_2$ with $T_1 := T|Y$ nilpotent and $T_2 := T|Z$ invertible. The Drazin inverse S of T, if it exists, is uniquely determined, and in the proof of Theorem 1.132, it has been observed that S may be represented, with respect to the decomposition $X = Y \oplus Z$, as the direct sum $S := 0 \oplus S_2$ with $S_2 := T_2^{-1}$. It has already been observed in Sect. 2.10 of this chapter that if T has property (C) then so does $f(T)$ for every function f analytic on an open neighborhood $\mathcal{U}$ of $\sigma(T)$. The same happens for the SVEP, by Theorem 2.86. An easy consequence is that the SVEP and property (C) are transmitted from $T \in L(X)$ to T^{-1} in the case when T is invertible:

Lemma 2.183 *Suppose that $T \in L(X)$ is invertible and let $\lambda_0 \neq 0$.*

(i) *If T has the SVEP at λ_0 then T^{-1} has the SVEP at $1/\lambda_0$.*
(ii) *If T has property (C) then T^{-1} has property (C).*

Proof

(i) Let $g(\lambda) = \frac{1}{\lambda}$. Since $0 \notin \sigma(T)$, there is an open neighborhood U containing the spectrum such that $0 \notin U$ and obviously g is analytic on D. Since T has the SVEP at λ_0 then, by Theorem 2.88, $g(T) = T^{-1}$ has the SVEP at $\frac{1}{\lambda_0}$.
(ii) This is proved by using a similar argument and Theorem 2.157.   ∎

A natural question, suggested by Lemma 2.183, is if property (C) or the other local spectral properties are transmitted to the Drazin inverse, if the operator is Drazin invertible. In this section we show that the answer to this question is positive. We begin with the SVEP.

Theorem 2.184 *Suppose that $R \in L(X)$ is Drazin invertible with Drazin inverse S.*

(i) *R has the SVEP at $\lambda_0 \neq 0$ if and only if S has the SVEP at $\frac{1}{\lambda_0}$.*
(ii) *R has the SVEP if and only if S has the SVEP.*

Proof

(i) Let $X = Y \oplus Z$, $R = R_1 \oplus R_2$ and $S = 0 \oplus S_2$, where $S_2 = R_2^{-1}$. Suppose that R has the SVEP at λ_0. Then $R_2 = R|Z$ has the SVEP at λ_0, since the localized SVEP is inherited by the restriction on invariant closed subspaces. By Lemma 2.183 then S_2 has the SVEP at $\frac{1}{\lambda_0}$. Since the null operator has the SVEP at every point, and by Theorem 2.15, S has the SVEP at $\frac{1}{\lambda_0}$. The reverse is proved similarly, since every nilpotent operator has the SVEP.

(ii) Suppose that R has the SVEP and that it is Drazin invertible with Drazin inverse S. We can assume that $0 \in \sigma(R)$. Then there exist two closed invariant subspaces Y and Z of X such that $X = Y \oplus Z$, $R_1 := R|Y$ is nilpotent and $R_2 := R|Z$ is invertible. The operator R_1 has the SVEP, since it is nilpotent, and also the restriction $R_2 := R|Z$ has the SVEP. By Lemma 2.183, $S_2 := R_2^{-1}$ also has the SVEP, so the Drazin inverse $S = 0 \oplus S_2$ has the SVEP.

Conversely, if $S = 0 \oplus S_2$ has the SVEP then the restriction $S_2 = S|Z$ has the SVEP, and hence its inverse $R_2 = R|Z$ has the SVEP. Consequently, $R = R_1 \oplus R_2$ has the SVEP, since R_1 is nilpotent and hence has the SVEP. ∎

Recall that $T \in L(X)$ is *relatively regular* if there exists an $S \in L(X)$ such that $TST = T$. S is called a *pseudo-inverse* of T. The reciprocal relationship between the nonzero part of the local spectrum of a relatively regular operator T and the nonzero part of the local spectrum of any of its pseudo-inverses is not satisfied. For instance, if T is the unilateral right shift in $\ell_2(\mathbb{N})$ then, as observed in Example 1.137, since

$$\sigma_T(x) = \sigma(T) = \mathbf{D}(0, 1),$$

where $\mathbf{D}(0, 1)$ is the closed unit disc of $\mathbb{C}$, $\sigma_T(x) \setminus \{0\}$ is the punctured disc $\mathbf{D}(0, 1) \setminus \{0\}$. Consequently, the points of $\sigma_S(x) \setminus \{0\}$, for any pseudo-inverse S, cannot be the reciprocals of $\sigma_T(x) \setminus \{0\}$, otherwise $\sigma_S(x)$ would be unbounded.

It should be noted that if S is a pseudo-inverse of $T \in L(X)$ the SVEP for T does not entail, in general, the SVEP for S. Indeed, the unilateral left shift L is a pseudo-inverse of the right shift T, since trivially $T = TLT$, but T has the SVEP, while L does not have the SVEP.

Theorem 2.185 *Suppose that $T \in L(X)$ is Drazin invertible with Drazin inverse S. If T has the SVEP then for every $x \in X$ we have:*

$$\sigma_S(x) \setminus \{0\} = \left\{ \frac{1}{\lambda} : \lambda \in \sigma_T(x) \setminus \{0\} \right\}. \tag{2.34}$$

Proof Suppose that T has the SVEP. If $0 \notin \sigma(T)$ then $S = T^{-1}$ and the equality (2.34) follows from part (ii) of Theorem 2.87, applied to the function $f(\lambda) := \frac{1}{\lambda}$. Suppose that $0 \in \sigma(T)$. According to the decomposition $X = Y \oplus Z$, $T_1 := T|Y$ nilpotent and $T_2 := RT|Z$ invertible, the restrictions T_1 and T_2 have the

SVEP. If $x = y + z$, $y \in Y$, $z \in Z$, we then have

$$\sigma_T(x) = \sigma_{T_1}(y) \cup \sigma_{T_2}(z),$$

by Theorem 2.15. The Drazin inverse $S := 0 \oplus S_2$, with $S_2 := T_2^{-1}$, has the SVEP, by Theorem 2.184, so, again by Theorem 2.15, we have

$$\sigma_S(x) = \sigma_0(y) \cup \sigma_{T_2}(z).$$

Since S_2 is the inverse of T_2, from the spectral mapping theorem of the local spectrum applied to the function $f(\lambda) := \frac{1}{\lambda}$, we have

$$\sigma_{S_2}(z) = \left\{ \frac{1}{\lambda} : \lambda \in \sigma_{T_2}(z) \right\} \quad \text{for all } z \in Z.$$

Now, in the decomposition $x = y + z$, $y \in Y$, $z \in Z$, consider first the case $y = 0$. Then $\sigma_{T_1}(y) = \emptyset$ and hence $\sigma_T(x) = \sigma_{T_2}(z)$ and, analogously, $\sigma_S(x) = \sigma_{S_2}(z)$. Therefore,

$$\sigma_S(x) = \left\{ \frac{1}{\lambda} : \lambda \in \sigma_T(x) \right\}.$$

Suppose that $x = y + z$, with $y \neq 0$. Since T_1 is nilpotent (and hence has the SVEP), $\sigma_{T_1}(y) \neq \emptyset$, hence $\sigma_{T_1}(y) = \{0\}$ and analogously, $\sigma_0(y) = \{0\}$. Therefore, $\sigma_T(x) \setminus \{0\} = \sigma_{T_2}(z)$ and $\sigma_S(x) \setminus \{0\} = \sigma_{S_2}(z)$, since T_2 and S_2 are invertible, from which we obtain:

$$\sigma_S(x) \setminus \{0\} = \sigma_{S_2}(z) = \left\{ \frac{1}{\lambda} : \lambda \in \sigma_{T_2}(z) \right\}$$

$$= \left\{ \frac{1}{\lambda} : \lambda \in \sigma_T(x) \setminus \{0\} \right\},$$

so the proof is complete.   ∎

Note that the right shift considered above has the SVEP, so the SVEP for a relatively regular operator does not ensure the reciprocal relationship noted for Drazin invertible operators with SVEP.

The following elementary result will be needed in the sequel.

Lemma 2.186 *Suppose that $T \in L(X)$ is quasi-nilpotent and F is a closed subset of $\mathbb{C}$. If $0 \in F$ then $X_T(F) = X$, while $X_T(F) = \{0\}$ if $0 \notin F$.*

Proof If T is quasi-nilpotent we have

$$X_T(F) = X_T(F \cap \sigma(T)) = X_T(F \cap \{0\}),$$

so $X_T(F) = X_T(\emptyset) = \{0\}$ if $0 \notin F$, while $X_T(F) = X_T(\{0\}) = H_0(T) = X$, if $0 \in F$, and T is quasi-nilpotent. ∎

Property $C)$ is transmitted from a Drazin invertible operator T to its Drazin inverse S:

Theorem 2.187 *Let $T \in L(X)$, X a Banach space, be Drazin invertible with Drazin inverse S. Then T has property (C) if and only if S has property (C).*

Proof Suppose that T has property (C) and that it is Drazin invertible. Also here we can assume, by Lemma 2.183, that $0 \in \sigma(T)$. Let $X = Y \oplus Z$ such that $T_1 := T|Y$ is nilpotent and $T_2 := T|Z$ is invertible. For the Drazin inverse $S := 0 \oplus S_2$ of T, with $S_2 := T_2^{-1}$, we have, by Corollary 2.16,

$$X_S(\mathcal{F}) = Y_0(F) \oplus Z_{S_2}(F), \quad \text{for every closed } F \subseteq \mathbb{C}.$$

Since T has property (C), the restriction $T_2 = T|Z$ also has property (C). But T_2 is invertible, so $S_2 = T_2^{-1}$ has property (C), by Lemma 2.183, and hence $Z_{S_2}(F)$ is closed for all closed $F \subseteq \mathbb{C}$. From Lemma 2.186 we know that if $0 \notin F$ then the spectral subspace $Y_0(F)$ of the null operator 0 is $\{0\}$, while if $0 \in F$ then $Y_0(F) = X$. Hence $X_S(\mathcal{F})$ coincides with $\{0\} \oplus Z_{S_2}(F)$, if $0 \notin F$, or coincides with X, if $0 \in F$. In both cases $X_S(F)$ is closed, and consequently S has property (C).

Conversely, suppose that S has property (C) and F is a closed subset of $\mathbb{C}$. Then $X_S(F)$ is closed, and as above $X_S(F) = \{0\} \oplus Z_{S_2}(F)$ if $0 \notin F$, or $X_S(F) = Y \oplus Z_{S_2}(F)$ if $0 \in F$. This implies that $Z_{S_2}(F)$ is closed, and hence S_2 has property (C). Consequently its inverse T_2 has property (C), by Lemma 2.183. Since $X_T(F) = Y_{T_1}(F) \oplus Z_{T_2}(F)$, by Theorem 2.19, and T_1 is nilpotent it then follows, again by Lemma 2.186, that $X_T(F)$ is either $\{0\} \oplus Z_{T_2}(F)$ (if $0 \notin F$), or is $Y \oplus Z_{T_2}(F)$ (if $0 \in F$), so $X_T(F)$ is closed for every closed $F \subseteq \mathbb{C}$, thus T has property (C). ∎

Consider now the case where the singleton set $F := \{\lambda\}$. Recall that since property (Q) for T entails SVEP then $H_0(\lambda I - T) = X_T(\{\lambda\}) = \mathcal{X}_T(\{\lambda\})$.

Lemma 2.188 *If $T \in L(X)$ has property (Q) and f is an injective analytic function defined on an open neighbourhood U of $\sigma(T)$, then $f(T)$ also has property (Q).*

Proof Recall that, by Theorem 2.29, the equality $\mathcal{X}_{f(T)}(F) = \mathcal{X}_T(f^{-1}(F))$ holds for every closed subset F of $\mathbb{C}$ and every analytic function f on an open neighbourhood U of $\sigma(T)$. Now, to show that $f(T)$ has property (Q), f injective, we have to prove that $H_0(\lambda I - f(T))$ is closed for every $\lambda \in \mathbb{C}$. If $\lambda \notin \sigma(f(T))$ then $H_0(\lambda I - f(T)) = \{0\}$, while if $\lambda \in \sigma(f(T)) = f(\sigma(T))$, then

$$H_0(\lambda I - f(T)) = X_{f(T)}(\{\lambda\}) = \mathcal{X}_T(f^{-1}\{\lambda\}) = H_0(\mu I - T),$$

where $f(\lambda) = \mu$, and, consequently, $H_0(\lambda I - f(T))$ is closed. In particular, considering the function $f(\lambda) := \frac{1}{\lambda}$, we see that if T is invertible and has property (Q) then its inverse has property (Q). ∎

Property (Q) is transmitted to the Drazin inverse:

Theorem 2.189 *Let $T \in L(X)$, X a Banach space, be Drazin invertible with Drazin inverse S. Then T has property (Q) if and only if S has property (Q).*

Proof Suppose that T is Drazin invertible and has property (Q). We need only consider the case $0 \in \sigma(T)$. Then there exist two closed invariant subspaces Y and Z of X such that $X = Y \oplus Z$, $T_1 := T|Y$ is nilpotent, $T_2 := T|Z$ is invertible and the Drazin inverse of T is given by $S := 0 \oplus S_2$, with $S_2 := T_2^{-1}$. Since T has the SVEP, S has the SVEP, by Theorem 2.184, so we have

$$H_0(\lambda I - S) = X_S(\{\lambda\}) = Y_0(\{\lambda\}) \oplus Z_{S_2}(\{\lambda\}), \quad \text{for all } \lambda \in \mathbb{C}.$$

Moreover, property (Q) for T implies that the restriction $T_2 = T|Z$ has property (Q). Since $0 \notin \sigma(T_2)$ then $S_2 := T_2^{-1}$ has property (Q), and hence $Z_{S_2}(\{\lambda\}) = H_0(\lambda I - S_2)$ is closed. By Lemma 2.186, the spectral subspace $H_0(\lambda I) = Y_0(\{\lambda\})$ of the null operator 0 is $\{0\}$ if $0 \neq \lambda$, while $Y_0(\{\lambda\}) = Y$ if $0 = \lambda$. Therefore, $X_S(\{\lambda\})$ is either $\{0\} \oplus Z_{S_2}(\{\lambda\})$ or $Y \oplus Z_{S_2}(\{\lambda\})$, so it is closed, and consequently S has property (Q).

Conversely, suppose that $S = 0 \oplus S_2$ has property (Q). Then $S_2 = S|Z$ has property (Q) and hence $T_2 = S_2^{-1}$ has property (Q). From

$$H_0(\lambda I - T) = H_0(\lambda I - T_1) \oplus H_0(\lambda I - T_2) = X_{T_1}(\{\lambda\}) \oplus H_0(\lambda I - T_2),$$

we obtain that $H_0(\lambda I - T)$ is either $Y \oplus H_0(\lambda I - T_2)$ or $\{0\} \oplus H_0(\lambda I - T_2)$, so $H_0(\lambda I - T)$ is closed for all $\lambda \in \mathbb{C}$.   ∎

Property (β) is transmitted from a Drazin invertible operator to its Drazin inverse. Recall that property (β) is inherited by the restriction to closed invariant subspaces, see [216, Theorem 3.3.6]. Property (β) is also preserved by the functional calculus, i.e. if $T \in L(X)$ has property (β) and f is an analytic function on an open neighbourhood U of $\sigma(T)$, then $f(T)$ has property (β), see [216, Theorem 3.3.6]. Consequently, if $T \in L(X)$ is invertible and has property (β) then T^{-1} has property (β). Let $H(U, X)$ denote the space of all analytic functions from U into X. With respect to pointwise vector space operations and the topology of locally uniform convergence, $H(U, X)$ is a Fréchet space. For every $T \in L(X)$ and every open set $U \subseteq \mathbb{C}$, define $T_U : H(U, X) \to H(U, X)$ by

$$(T_U f)(\lambda) := (\lambda I - T) f(\lambda) \quad \text{for all } f \in H(U, X) \text{ and } \lambda \in U.$$

Property (β) for T holds precisely when, for every open set $U \subseteq \mathbb{C}$, the operator T_U has closed range in $H(U, X)$, see Laursen and Neumann [216, Proposition 3.3.5]. Evidently, the restriction of an operator with property (β) to a closed invariant subspace inherits this property.

Theorem 2.190 *Let* $T \in L(X)$, X *a Banach space, be Drazin invertible with Drazin inverse* S. *Then* T *has property* (β) *if and only if* S *has property* (β).

Proof Suppose that T has property (β). Also here we may assume $0 \in \sigma(T)$. Let $X = Y \oplus Z$, where $T_1 := T|Y$ is nilpotent and $T_2 := T|Z$ is invertible. Consider the Drazin inverse of R given by $S := 0 \oplus S_2$, with $S_2 := T_2^{-1}$. As noted above, to show property (β) for S it suffices to prove that the operator $(S_U f)(\lambda) := (\lambda I - S)f(\lambda)$ defined on $H(U, X)$, U an open subset of $\mathbb{C}$, has closed range. By a result of Gleason [153] we can identify $H(U, X)$ with the direct sum $H(U, Y) \oplus H(U, Z)$ (for a proof, see also [216, Proposition 1.2.2] in the case when U is an open disc, and [216, Proposition 2.1.4] for the more general case of arbitrary open subsets of $\mathbb{C}$). The restriction R_2 has property (β) and hence $S_2 := R_2^{-1}$ has property (β). Consequently, the operator S_{2U} defined as

$$(S_{2U}g)(\lambda) := (\lambda I - S_2)g(\lambda) \quad \text{for all } g \in H(U, Z), \ \lambda \in U,$$

has closed range in $H(U, Z)$. Now,

$$\begin{aligned} S_U[H(U, X)] &= (0 \oplus S_2)_U[H(U, Y) \oplus H(U, Z)] \\ &= 0_U[H(U, Y)] \oplus S_{2U}[H(U, Z)]. \end{aligned}$$

Clearly, the operator 0_U has closed range in $H(U, Y)$, since, trivially, the null operator has property (β). Consequently, S_U has closed range in $H(U, X)$ and hence S has property (β).

Conversely, suppose that S has property (β). Then $S_2 := S|Z$ has property (β) and, consequently, its inverse R_2 has property (β) and hence R_{2U} has closed range in $H(U, Z)$. Since the nilpotent operator R_1 has property (β), R_{1U} also has closed range in $H(U, Y)$. From the decomposition

$$R_U[H(U, X)] = R_{1U}[H(U, Y)] \oplus R_{2U}[H(U, Z)],$$

we then conclude that $R_U[H(U, X)]$ is closed, so property (β) holds for R. ∎

Corollary 2.191 *Suppose that* T *is Drazin invertible with Drazin inverse* S. *If* T *has property* (δ) *then* S *has property* (δ), *and analogously, if* T *is decomposable then* S *is decomposable.*

Proof Clearly, from the definition of Drazin invertibility it follows that if T is Drazin invertible then its dual T^* is Drazin invertible, with Drazin inverse S^*. If T has property (δ) then T^* has property (β) and hence, by Theorem 2.190, S^* also has property (β). By duality this implies that S has property (δ). The second assertion is clear: if T is decomposable then T has both properties (δ) and (β) and the same holds for S, by Theorem 2.190 and the first part of the proof. Thus, S is decomposable. ∎

2.13   Comments

An extensive treatment of the role of the local spectral subspaces in the theory of spectral decomposition may be found in the book of Laursen and Neumann [216]. This book also provides a large variety of examples and applications to several concrete cases. Lemma 2.4 is due to Benhida and Zerouali [66]. The characterization of the analytic core of an operator given in Theorem 2.20 is due to Vrbová [293] and Mbekhta [230].

The concept of a glocal spectral subspace dates back to the early days of local spectral theory and may be found, for instance, in Bishop [79]. However, the precise relationship between local spectral subspaces and glocal spectral subspaces has been established, together with some other basic properties, by Laursen and Neumann [214]. The result that the glocal spectral subspaces behave canonically under the functional calculus, established in Theorem 2.29, is due to Bartle and Kariotis [59], which also showed Theorem 2.30, see also Laursen and Neumann [215]. Theorem 2.30 is due to Vrbová [294]. The equality $H_0(T) = X_T(\{0\})$ for an operator having the SVEP may also be found in Mbekhta [230].

The localized SVEP at a point was introduced by Finch [148], and the characterization of the SVEP at a single point λ_0 given in Theorem 2.60 is taken from Aiena and Monsalve [19], while the classical result of Corollary 2.61 is owed to Finch [148]. Except for Theorem 2.70 and Corollary 2.71, owed to Mbekhta [232], the source of the results of the second section is essentially that of Aiena et al. [33, 34]. The relations between the local spectrum and the surjectivity spectrum established in Theorem 2.21 are taken from Laursen and Vrbová [218] and Vrbová [293]. The section concerning the localized SVEP for operators having topological uniform descent is modeled after Jiang and Zhong [191], which extended previous results concerning the localized SVEP for quasi-Fredholm operators established in [3]. Theorem 2.104 is taken from [305], while Theorems 2.118 and 2.117 is taken from [24]. The SVEP on the components of the semi-Fredholm regions was first studied in Aiena and Villafãne [31]. Successively, this study has been extended to the components of a Kato-type resolvent in [192], and extended to the components of a quasi-Fredholm region in [305]. Finally, the SVEP on the components of the topological uniform descent resolvent, established here in Theorem 2.124 and Corollary 2.125, has been studied in Jiang et al. [193]. Theorem 2.111 is due to Schmoeger [273].

The stability of the localized SVEP under commuting Riesz perturbations is due to Aiena and Muller [20], while the subsequent material concerning the localized SVEP and quasi-nilpotent equivalence is modeled after Aiena and Neumann [21]. Property (C) was introduced by Dunford and plays a large role in the development of the theory of spectral operators. In the book by Dunford and Schwartz [143] property (C) was one of the three basic conditions used in the abstract characterization of spectral operators, and another one was the SVEP. The SVEP as a consequence of property (C) was observed by Laursen and Neumann [214].

The section concerning property (C) for RS and SR is inspired by the work of Aiena and Gonzalez [14, 15], and the work of Zeng and Zhong [302]. The spectral properties and the local spectral properties of operators which satisfy $RSR = R^2$ and $SRS = S^2$ have been investigated by some other authors, see Schmoeger [283], Duggal [126] and Aiena and Gonzalez [15]. The section concerning the local spectral theory of Drazin invertible operators is modelled after Aiena and Triolo [27]. Some of these results have been extended to generalized Drazin invertible operators, i.e. operators which are either invertible or such that $0 \in \text{iso}\,\sigma(T)$, by Duggal in [128].

Chapter 3
Essential Spectra Under Perturbations

The spectrum of a bounded linear operator on a Banach space X can be sectioned into subsets in many different ways, depending on the purpose of the inquiry. This chapter plays a central role in this book, since we establish the relationships between the various parts of the spectrum. More precisely, we look more closely at some parts of the spectrum of many bounded linear operators from the viewpoint of Fredholm theory, and we shall study in detail, by using the localized SVEP, some of the spectra generated by the classes of operators introduced in the first chapter. Moreover, we shall give further results concerning the stability of these essential spectra under commuting perturbations. In this chapter the interaction between the localized SVEP and Fredholm theory appears in full strength and elegancy: indeed, many classical results from Fredholm theory may be deduced by using the localized SVEP.

The first section of this chapter concerns the class of Riesz operators, and in particular we shall see, by using the stability of the localized SVEP under Riesz commuting perturbations, that the Browder spectra are invariant under Riesz commuting perturbations. The second section regards some representation theorems for Weyl and Browder operators, while the third section is focused on Drazin invertible operators. The fourth section addresses the class of meromorphic operators, a class of operators which contains the class of Riesz operators, while the fifth section concerns the subclass of all algebraic operators. The sixth section is mainly devoted to the study of Drazin spectra and the relationship of these spectra with the spectra generated by the B-Fredholm theory.

We also give in this section some other results concerning the spectra of the products of operators TR and RT of two operators R and T. In the following section we introduce the concept of regularity in order to establish the spectral mapping theorem for several classes of operators. The eighth section addresses the concept of the pole of the resolvent, and some generalizations of it, such as the concept of the left pole or right pole of the resolvent.

© Springer Nature Switzerland AG 2018   213
P. Aiena, *Fredholm and Local Spectral Theory II*, Lecture Notes in Mathematics
2235, https://doi.org/10.1007/978-3-030-02266-2_3

3.1   Weyl, Browder and Riesz Operators

We begin this section by proving several perturbation results concerning the spectra relative to some important classes of operators in Fredholm theory that have already been introduced in the first chapter. Recall that a bounded operator $T \in L(X)$ is said to be a *Weyl operator*, $T \in W(X)$, if T is a Fredholm operator having index 0. The classes of *upper semi-Weyl* and *lower semi-Weyl* operators are defined, respectively, as:

$$W_+(X) := \{T \in \Phi_+(X) : \operatorname{ind} T \le 0\},$$

and

$$W_-(X) := \{T \in \Phi_-(X) : \operatorname{ind} T \ge 0\}.$$

Clearly, $W(X) = W_+(X) \cap W_-(X)$. The *Weyl spectrum* is defined as

$$\sigma_w(T) := \{\lambda \in \mathbb{C} : \lambda I - T \notin W(X)\},$$

the *upper semi-Weyl spectrum* is defined as

$$\sigma_{uw}(T) := \{\lambda \in \mathbb{C} : \lambda I - T \notin W_+(X)\},$$

and the *lower semi-Weyl spectrum* is defined as

$$\sigma_{lw}(T) := \{\lambda \in \mathbb{C} : \lambda I - T \notin W_-(X)\}.$$

By duality we have $\sigma_w(T) = \sigma_w(T^*)$,

$$\sigma_{uw}(T) = \sigma_{lw}(T^*) \quad \text{and} \quad \sigma_{lw}(T) = \sigma_{uw}(T^*).$$

Let us now consider the following spectra associated with the Browder operators defined in Chap. 1. The *Browder spectrum*, defined by

$$\sigma_b(T) := \{\lambda \in \mathbb{C} : \lambda I - T \notin B(X)\},$$

the *upper semi-Browder spectrum* of T, defined as

$$\sigma_{ub}(T) := \{\lambda \in \mathbb{C} : \lambda I - T \notin B_+(X)\},$$

and the *lower semi-Browder spectrum* of T defined as

$$\sigma_{lb}(T) := \{\lambda \in \mathbb{C} : \lambda I - T \notin B_-(X)\}.$$

Clearly, every Browder (respectively, upper semi-Browder, lower semi-Browder) operator $T \in L(X)$ is Weyl (respectively, upper semi-Weyl, lower semi-Weyl), by Theorem 1.22, so

$$\sigma_w(T) \subseteq \sigma_b(T), \quad \sigma_{uw}(T) \subseteq \sigma_{ub}(T), \quad \sigma_{lw}(T) \subseteq \sigma_{lb}(T).$$

The operator $T := L \oplus R$, where R and L are the right unilateral shift and the left unilateral shift in $\ell^2(\mathbb{N})$, respectively, provides an example of a Weyl operator which is not Browder. Indeed, T is Fredholm with index ind $T = $ ind $L + $ ind $R = 0$, while $0 \in \sigma(T) = \mathbf{D}(0, 1)$ is not isolated in $\sigma(T)$, hence T cannot be Browder.

By duality, we have $\sigma_b(T) = \sigma_b(T^*)$,

$$\sigma_{ub}(T) = \sigma_{lb}(T^*) \quad \text{and} \quad \sigma_{lb}(T) = \sigma_{ub}(T^*).$$

Since $\sigma_{usf}(T) \subseteq \sigma_{ub}(T)$ and $\sigma_{lsf}(T) \subseteq \sigma_{lb}(T)$, Theorem 2.126 also entails that the semi-Browder spectra are non-empty. These spectra are closed subsets of $\mathbb{C}$, see the next Corollary 3.42.

We have already observed, in Chap. 2, in the case of a Hilbert space operator T, that the conjugate linear isometry defined as $U : y \in H \to f_y \in H^*$, where $f_y(x) := (x, y)$ for all $x \in H$, satisfies the identity

$$U(\bar{\lambda}I - T') = (\lambda I - T^*)U \quad \text{for all } \lambda \in \mathbb{C},$$

where as usual T' denotes the Hilbert adjoint of T. From this it then easily follows that

$$\lambda I - T^* \in \Phi_+(H^*) \Leftrightarrow \bar{\lambda}I - T' \in \Phi_+(H). \tag{3.1}$$

Theorem 3.1 *Let $T \in L(H)$, H a Hilbert space. Then $\sigma_b(T^*) = \overline{\sigma_b(T')}$ and*

$$\sigma_{ub}(T^*) = \overline{\sigma_{ub}(T')} \quad \text{and} \quad \sigma_{lb}(T^*) = \overline{\sigma_{lb}(T')}.$$

Analogously, $\sigma_w(T^*) = \overline{\sigma_w(T')}$ and

$$\sigma_{uw}(T^*) = \overline{\sigma_{uw}(T')} \quad \text{and} \quad \sigma_{lw}(T^*) = \overline{\sigma_{lw}(T')}.$$

Similar equalities hold for the approximate point spectrum and the surjective spectrum.

Proof We only prove that $\lambda I - T^* \in B_+(H^*)$ if and only if $\bar{\lambda}I - T' \in B_+(H)$. Suppose that $p := p(\lambda I - T^*) < \infty$ and let $x \in \ker(\bar{\lambda}I - T')^{p+1}$ be arbitrary. Then

$$U(\bar{\lambda}I - T')^{p+1}x = (\lambda I - T^*)^{p+1}x = 0, \tag{3.2}$$

so $Ux \in \ker(\lambda I - T^*)^{p+1} = \ker(\lambda I - T^*)^p$, from which we obtain

$$(\lambda I - T^*)^p Ux = U(\overline{\lambda} I - T')^p x = 0.$$

Since U is injective we then have $(\overline{\lambda} I - T')^p x = 0$, so $\ker(\overline{\lambda} I - T')^{p+1} \subseteq \ker(\overline{\lambda} I - T')^p$. Since the opposite inclusion always holds, we then conclude $p(\overline{\lambda} I - T') \leq p$. A similar argument shows that if $p(\overline{\lambda} I - T') < \infty$ then $p(\lambda I - T^*) \leq p(\overline{\lambda} I - T')$. Therefore, $p(\lambda I - T^*) = p(\overline{\lambda} I - T')$. Taking into account (3.1) we then conclude that $\lambda I - T^* \in B_+(H^*)$ if and only if $\overline{\lambda} I - T' \in B_+(H)$. Hence the first equality in (3.2) is proved. The other equalities may be shown in a similar way. ∎

We have already introduced the class of Riesz operators as those operators $R \in L(X)$ such that $\lambda I - R \in \Phi(X)$ for every $\lambda \in \mathbb{C} \setminus \{0\}$. Riesz operators may be characterized in several ways.

Theorem 3.2 *For a bounded operator T on a Banach space the following statements are equivalent:*

   (i) *T is a Riesz operator;*
   (ii) *$\lambda I - T \in \mathcal{B}(X)$ for all $\lambda \in \mathbb{C} \setminus \{0\}$;*
  (iii) *$\lambda I - T \in \mathcal{W}(X)$ for all $\lambda \in \mathbb{C} \setminus \{0\}$;*
  (iv) *$\lambda I - T \in \mathcal{B}_+(X)$ for all $\lambda \in \mathbb{C} \setminus \{0\}$;*
   (v) *$\lambda I - T \in \mathcal{B}_-(X)$ for all $\lambda \in \mathbb{C} \setminus \{0\}$;*
  (vi) *$\lambda I - T \in \Phi_+(X)$ for all $\lambda \in \mathbb{C} \setminus \{0\}$;*
 (vii) *$\lambda I - T \in \Phi_-(X)$ for all $\lambda \in \mathbb{C} \setminus \{0\}$;*
(viii) *$\lambda I - T$ is essentially semi-regular for all $\lambda \in \mathbb{C} \setminus \{0\}$;*
  (ix) *Each spectral point $\lambda \neq 0$ is isolated and the spectral projection associated with $\{\lambda\}$ is finite-dimensional.*

Proof (i) $\Rightarrow$ (ii) If T is a Riesz operator the topological uniform descent resolvent has a unique component $\mathbb{C} \setminus \{0\}$. By Theorems 2.123 and 2.124 it then follows that both T and T^* have the SVEP at every $\lambda \neq 0$. Therefore, again by Theorems 2.97 and 2.98, $\lambda I - T \in \mathcal{B}(X)$ for all $\lambda \neq 0$.

The implications (ii) $\Rightarrow$ (iii) $\Rightarrow$ (i) are clear, so (i), (ii), and (iii) are equivalent. The implications (ii) $\Rightarrow$ (iv) $\Rightarrow$ (vi) $\Rightarrow$ (viii) and (ii) $\Rightarrow$ (v) $\Rightarrow$ (viii) are evident, so in order to show that all these implications are equivalences we need only to show that (viii) $\Rightarrow$ (ii).

(viii) $\Rightarrow$ (ii) Suppose that (viii) holds. Then the topological uniform descent resolvent $\rho_k(T)$ contains $\mathbb{C} \setminus \{0\}$, and hence, by Theorem 2.125, both T and T^* have the SVEP at every $\lambda \neq 0$. By Theorems 2.97 and 2.98 we then conclude that $\lambda I - T \in \mathcal{B}(X)$ for all $\lambda \neq 0$.

(i) $\Rightarrow$ (ix) As above, both T and T^* have the SVEP at every $\lambda \neq 0$, so every non-zero spectral point λ is an isolated point of $\sigma(T)$. From Corollary 2.47 it then follows that the spectral projection associated with $\{\lambda\}$ is finite-dimensional.

(ix) $\Rightarrow$ (ii) If the spectral projection associated with the spectral set $\{\lambda\}$ is finite-dimensional then $H_0(\lambda I - T)$ is finite-dimensional and $K(\lambda I - T)$ has finite codimension, by Theorem 2.45. Hence, by Theorem 2.100, $\lambda I - T$ is Browder.   ∎

Since every non-zero spectral point of a Riesz operator T is isolated, the spectrum $\sigma(T)$ of a Riesz operator $T \in L(X)$ is a finite set or a sequence of eigenvalues which converges to 0. Moreover, since $\lambda I - T \in \mathcal{B}(X)$ for all $\lambda \in \mathbb{C} \setminus \{0\}$, every spectral point $\lambda \neq 0$ is a pole of $R(\lambda, T)$. Clearly, if X is an infinite-dimensional complex space the spectrum of a Riesz operator T contains at least the point 0. In this case $T \in L(X)$ is a Riesz operator if and only if $\widehat{T} := T + K(X)$ is a quasi-nilpotent element in the Calkin algebra $\widehat{L} := L(X)/K(X)$. This result is an easy consequence of the Atkinson characterization of Fredholm operators (see the Appendix A) and in the literature this result is known as the *Ruston characterization* of Riesz operators.

Further information on Riesz operators may be found in Aiena [1, Chapter 3 and Chapter 7]. A good treatment of Riesz operator theory may also be found in Heuser's book [179] and some generalization of Riesz operators are considered in the recent book by Jeribi [186].

In the sequel we collect some other basic facts about Riesz operators that will be used in the rest of the chapter.

Generally, the sum and the product of Riesz operators $T, S \in L(X)$ need not be Riesz. However, the next result shows this is true if we assume T and S commute.

Theorem 3.3 *If $T, S \in L(X)$ on a Banach space X the following statements hold:*

(i) *If T and S are commuting Riesz operators then $T + S$ is a Riesz operator.*
(ii) *If S commutes with the Riesz operator T then the products TS and ST are Riesz operators.*
(iii) *The limit of a uniformly convergent sequence of commuting Riesz operators is a Riesz operator.*
(iv) *If T is a Riesz operator and $K \in K(X)$ then $T + K$ is a Riesz operator.*

Proof If T and S commute, the equivalences classes $\widehat{T}$ and $\widehat{S}$ commute in $\widehat{L}$, so (i), (ii), and (iii) easily follow from the Ruston characterization of Riesz operators and from the well-known spectral radius formulas

$$r(\widehat{T} + \widehat{S}) \leq r(\widehat{T}) + r(\widehat{S}) \quad \text{and} \quad r(\widehat{TS}) \leq r(\widehat{T})r(\widehat{S}).$$

The assertion (iv) is obvious, again by the Ruston characterization of Riesz operators.   ∎

It should be noted that in part (i) and part (ii) of Theorem 3.3 the assumption that T and S commute may be relaxed into the weaker assumption that T and S commute modulo $K(X)$, i.e., $TS - ST \in K(X)$.

Theorem 3.4 *Let* $T \in L(X)$ *and suppose* $f \in \mathcal{H}(\sigma(T))$ *does not vanish on* $\sigma(T) \setminus \{0\}$. *Then we have:*

(i) *If* T *is a Riesz operator and* $f(0) = 0$ *then* $f(T)$ *is a Riesz operator.*

(ii) *If* $f(T)$ *is a Riesz operator and* $f \in \mathcal{H}(\sigma(T)$ *does not vanish on* $\sigma(T) \setminus \{0\}$ *then* T *is a Riesz operator. In particular, if* T^n *is a Riesz operator for some* $n \in \mathbb{N}$ *then* T *is a Riesz operator.*

Proof

(i) Suppose that T is a Riesz operator. Since $f(0) = 0$ there exists an analytic function g on a neighborhood of $\sigma(T)$ such that $f(\lambda) = \lambda g(\lambda)$. Hence $f(T) = Tg(T)$ and since T and $g(T)$ commute it then follows that $f(T)$ is a Riesz operator.

(ii) Assume that $f(T)$ is a Riesz operator and f vanishes only at 0. Then there exist an analytic function g on a neighborhood of $\sigma(T)$ and $n \in \mathbb{N}$ such that $f(\lambda) = \lambda^n g(\lambda)$ holds on the set of definition of f and $g(\lambda) \neq 0$. Hence $f(T) = T^n g(T)$ and $g(T)$ is invertible. The operators $f(T)$ and $g(T)^{-1}$ commute, so, by part (ii) of Theorem 3.3, $T^n = f(T)g(T)^{-1}$ is a Riesz operator. Hence T^n is quasi-nilpotent modulo $K(X)$ and from this it easily follows that T is quasi-nilpotent modulo $K(X)$. By the Ruston characterization we then conclude that T is a Riesz operator. ∎

Let $T \in L(X)$ and M a closed T-invariant subspace of X. Denote by $T|M$ the restriction of T to M and by $\widehat{T}_M : X/M \to X/M$ the operator induced by T, defined as $\widehat{T}_M \hat{x} := \widehat{Tx}$, for every $\hat{x} := x + M$.

Lemma 3.5 *Let* $T \in L(X)$ *and suppose that* M *is a closed* T-*invariant subspace of* X.

(i) *If* T *is invertible then* $T|M$ *is bounded below and* $\widehat{T}_M$ *is onto.*

(ii) *If* $T \in \Phi(X)$ *then* $T|M \in \Phi_+(M)$ *and* $\widehat{T}_M \in \Phi_-(X/M)$.

Proof The assertion (i) is easy to see. Suppose that $T \in \Phi(x)$. Since $\ker T|M = \ker T \cap M$, we have $\alpha(T|M) < \infty$. Define $\tilde{T} : x|\ker T \to T(X)$ by $\tilde{F}(x + \ker T) = Tx$. Since $T(X)$ is closed and $\tilde{T}$ is injective, $\tilde{T}$ is an open map, by the open mapping theorem. Hence $\tilde{T}(N + \ker T) = T(M)$ is closed, thus $T|M$ is upper semi-Fredholm.

To show that $\widehat{T}_M \in \Phi_-(X/M)$, observe first that since $T(X)$ has finite codimension, there exists a finite-dimensional subspace W of X such that $X = T(X) \oplus W$. Then

$$\widehat{T}_M(X/M) + (M + W)/M = (T(X) + M + W)/M = X/M.$$

Therefore, $\beta(\widehat{T}_M) < \infty$ and hence $\widehat{T}_M$ is lower semi-Fredholm. ∎

Remark 3.6 The result of Lemma 3.5 cannot, in general, be improved, in the sense that T is Fredholm does not imply that $T|M \in \Phi(M)$ and $\widehat{T}_M \in \Phi(X/M)$. For instance, suppose that H is an infinite-dimensional Hilbert space, and let R and L

denote the right shift and the left shift on the Hilbert space $Y := \ell_2(H)$. Denote by P the projection

$$P(x_1, x_2, \dots) := (x_1, 0, 0, \dots) \quad \text{for all } (x_n) \in \ell_2(H).$$

It is straightforward to check that

$$LR = I, \quad RL = I - P, \quad LP = 0, \quad \text{and } PR = 0.$$

Let $X = Y \oplus X$ and define the operator matrix $T := \begin{pmatrix} R & P \\ 0 & L \end{pmatrix}$. Then T is invertible with inverse $T^{-1} := \begin{pmatrix} L & 0 \\ P & R \end{pmatrix}$. Thus, $T \in \Phi(X)$ but neither R and L are Fredholm. Taking $M := Y \oplus 0$, we have, as is easy to check, that $T|M$ is similar to R, while $\widehat{T}_M$ is similar to L, from which we conclude that neither $T|M$ nor $\widehat{T}_M$ are Fredholm.

The property of being a Riesz operator is inherited by the restrictions on closed invariant subspaces:

Theorem 3.7 *Let* $R \in L(X)$ *be a Riesz operator. Then we have*

(i) *If* M *is a closed* R*-invariant subspace of* R *then* $R|M$ *and* $\widehat{R}_M$ *are Riesz.*
(ii) *The dual* R^* *is a Riesz operator. Conversely, if* R^* *is Riesz then* R *is Riesz.*

Proof

(i) If R is Riesz then $\lambda I - R \in \Phi(X)$ for all $\lambda \neq 0$. By Lemma 3.5, $(\lambda I - R)|M \in \Phi_+(M)$ for all $\lambda \neq 0$. By Theorem 3.2 then $T|M$ is Riesz. Analogously, by Lemma 3.5, $\lambda \widehat{I}_M - \widehat{R}_M \in \Phi_-(X/M)$ for all $\lambda \neq 0$, so, by Theorem 3.2, $\widehat{R}_M$ is Riesz.

(ii) If R is a Riesz operator then $\lambda I - R \in \Phi(X)$ for all $\lambda \neq 0$. Therefore $\lambda I^* - R^* \in \Phi(X^*)$ for all $\lambda \neq 0$, so R^* is Riesz. Conversely, if R^* is a Riesz operator, by what we have just proved the bi-dual R^{**} is also a Riesz operator. Since the restriction of R^{**} to the closed subspace X of X^{**} is R, it follows from part (i) that R itself must be a Riesz operator. ∎

We now turn to the stability of semi-Browder spectra under commuting Riesz perturbations.

Theorem 3.8 *Let* $T \in L(X)$ *and* R *be a Riesz operator such that* $TR = RT$*. Then we have:*

(i) $T \in B_+(X) \Leftrightarrow T + R \in B_+(X)$.
(ii) $T \in B_-(X) \Leftrightarrow T + R \in B_+(X)$.
(iii) $T \in B(X) \Leftrightarrow T + R \in B(X)$.

Proof

(i) If T is upper semi-Browder then $p(T) < \infty$ and this is equivalent to saying that T has the SVEP at 0, by Theorem 2.97. By Theorem 2.129 it then follows that $T + R$ has the SVEP at 0, and since $T + R$ is upper semi-Fredholm, $p(T + R) < \infty$, again by Theorem 2.97, so $T + R$ is upper semi-Browder. The converse follows by symmetry.

(ii) The proof is analogous to that of part (i). Let T be lower semi-Browder, so $q(T) < \infty$ and this is equivalent to saying that T^* has the SVEP at 0, by Theorem 2.98. The dual of a Riesz operator is also Riesz. By Theorem 2.129 it then follows that $T^* + R^*$ has the SVEP at 0, and since $T + R$ is lower semi-Fredholm it then follows that $q(T + R) < \infty$, so $T + R$ is lower semi-Browder.

(iii) Clear.

Corollary 3.9 *The Browder spectra $\sigma_{ub}(T)$, $\sigma_{lb}(T)$, and $\sigma_b(T)$ are stable under commuting Riesz perturbations.*

The following example shows that the assumption that the perturbation R commutes with T in Theorem 3.8, and Corollary 3.9, cannot be dropped, even in the case when R is finite-dimensional.

Example 3.10 Let H be a Hilbert space with an orthonormal basis $(e_k)_{k=-\infty}^{\infty}$ and consider the *bilateral shift* defined by $Te_k = e_{k+1}$. Let K be the one-dimensional operator defined by $Kx = (x, e_0)e_1$. Then $T - K$ has infinite descent, so $0 \in \sigma_{lb}(T - K)$, while $0 \notin \sigma_{lb}(T)$.

In the particular case of bounded below, or surjective, operators we can say something more:

Theorem 3.11 *Suppose that $T \in L(X)$ and $R \in L(X)$ is a Riesz operator commuting with T. Then*

(i) *If T is bounded below then $T + R \in B_+(X)$. Moreover,*

$$T(\ker (T + R)^n) = \ker (T + R)^n \quad \text{for all } n \in \mathbb{N}.$$

(ii) *If T is onto then $T + R \in B_-(X)$. Moreover,*

$$T^{-1}((T + R)^n(X)) = (T + R)^n(X) \quad \text{for all } n \in \mathbb{N}.$$

Proof

(i) The first assertion is clear by Theorem 3.8, since $T \in B_+(X)$.

   Define $S := T + R$. We have $S^n \in \Phi_+(X)$ for every $n \in \mathbb{N}$, so $\ker S^n$ is finite-dimensional. If $x \in \ker S^n$ then $Tx \in \ker S^n$, since

$$(T + R)^n Tx = T(T + R)^n x = 0,$$

hence ker S^n is T-invariant. Furthermore, the restriction of T to ker S^n is injective, since T is bounded below, so T maps ker S^n onto itself.

(ii) The first assertion is clear by Theorem 3.8, since $T \in B_-(X)$.

Let $S := T + R$. Then $S^n \in \Phi_-(X)$ for every $n \in \mathbb{N}$, so codim $S^n(X) = $ dim $X/S^n(X) < \infty$. Consider the map $\widehat{T} : X/S^n(X) \to X/S^n(X)$ induced by T, defined by $\widehat{T}\hat{x} := \widehat{Tx}$ for all $\hat{x} := x + S^n(X)$. Since T is onto, for every $y \in X$ there exists an element $z \in X$ such that $y = Tz$, and therefore, $\hat{y} = \widehat{T}\hat{z}$. Hence $\widehat{T}$ is onto. Since $X/S^n(X)$ is finite-dimensional it then follows that $\widehat{T}$ is also injective, and this implies that $Tx \in S^n(X)$ if and only if $x \in S^n(X)$. Consequently, $T^{-1}(S^n(X)) = S^n(X)$. $\blacksquare$

The class of essentially semi-regular operators, which properly contains the class of semi-Browder operators, is also stable under Riesz commuting perturbations:

Theorem 3.12 *Suppose that $T, R \in L(X)$ commutes. If T is essentially semi-regular and R is a Riesz operator, then $T + R$ is essentially semi-regular.*

Proof We know, by Theorems 1.63 and 1.64, that the hyper-range $T^\infty(X)$ is closed. Set $\widehat{T} := T|T^\infty(X)$ and denote by $\widetilde{T} : X/T^\infty(X) \to X/T^\infty(X)$ the operator induced by T. Observe that $TR = RT$ entails that $R(T^\infty(X)) \subseteq T^\infty(X)$, so $T^\infty(X)$ is both T-invariant and R-invariant. Since T has topological uniform descent, by Corollary 1.83, it then follows, by Theorem 1.79, that $\widehat{T}$ is onto. The restriction $\widehat{R} := R|T^\infty(X)$ is Riesz, and since $\widehat{T}\widehat{R} = \widehat{R}\widehat{T}$, from Theorem 3.11 we deduce that $\widehat{T} + \widehat{R}$ is lower semi-Browder. Now, let $\widetilde{T}$ and $\widetilde{R}$ denote the induced mappings on $X/T^\infty(X)$, by T and R, respectively. From Lemma 1.66 we know that $\widetilde{T}$ is upper semi-Browder, $\widetilde{R}$ is Riesz and $\widetilde{R}\widetilde{T} = \widetilde{T}\widetilde{R}$, thus, by Theorem 3.8, $\widetilde{T} + \widetilde{R}$ is upper semi-Browder. By Theorem 1.67, applied to $T + R$, we then conclude that $T + R$ is essentially semi-regular. $\blacksquare$

Theorem 3.13 *Let $T \in L(X)$ be essentially semi-regular. Then there exists an $\varepsilon > 0$ such that $T + S$ is essentially semi-regular for every $S \in L(X)$ which commutes with T having norm $\|S\| < \varepsilon$.*

Proof If $M := T^\infty(X)$, then M is closed and $T(M) = M$. By Lemma 1.66, the induced operator $\widetilde{T} : X/M \to X/M$ is upper semi-Browder. Suppose that S commutes with T and S has norm small enough. Then M is invariant under S, $(T + S)(M) = M$ and the operator induced $\widetilde{(T + S)} = \widetilde{T} + \widetilde{S}$ is upper semi-Browder, hence $T + S$ is essentially semi-regular, by Theorem 1.67. $\blacksquare$

Denote by

$$\sigma_{es}(T) := \{\lambda \in \mathbb{C} : \lambda I - T \text{ is not essentially semi-regular}\}$$

the *essentially semi-regular spectrum*. Obviously, $\sigma_{es}(T)$ is a subset of the semi-regular spectrum $\sigma_{se}(T)$. Further, since every Fredholm operator is essentially semi-regular, $\sigma_{es}(T) \subseteq \sigma_e(T)$.

Denote by $\mathcal{K}(X)$ and $\mathcal{F}(X)$ the two-sided ideals in $L(X)$ of all compact operators and all finite-dimensional continuous operators on the Banach space X, respectively. If $\mathcal{R}(X)$ denotes the set of all Riesz operators we have

$$\mathcal{F}(X) \subseteq \mathcal{K}(X) \subseteq \mathcal{R}(X).$$

It is well-known that $\mathcal{K}(X)$ is closed in $L(X)$.

Theorem 3.14 *Let $T \in L(X)$.*

(i) *$\sigma_{es}(T)$ is a non-empty compact subset. In particular, $\sigma_{es}(T)$ contains the boundary of the essential spectrum $\sigma_e(T)$.*

(ii) *If $R \in L(X)$ is a commuting Riesz operator then $\sigma_{es}(T)$ is invariant under R, i.e., $\sigma_{es}(T) = \sigma_{es}(T + R)$.*

(iii) *We have*

$$\sigma_{es}(T) = \bigcap_{F \in \mathcal{F}(X), FT = TF} \sigma_{se}(T + F)$$

$$= \bigcap_{R \in \mathcal{R}(X), RT = TR} \sigma_{se}(T + R).$$

Proof

(i) $\sigma_{es}(T)$ is closed, by Theorem 3.13. We show now that the boundary $\partial\sigma_e(T)$ is contained in $\sigma_{es}(T)$, from which it follows that $\sigma_{es}(T)$ is non-empty. Let $\lambda \in \partial\sigma_e(T)$ and suppose that $\lambda \notin \sigma_{es}(T)$. Then $\lambda I - T$ has closed range, since $\lambda I - T$ is essentially semi-regular, and hence there exist two invariant closed subspace M, N such that $X = M \oplus N$, $\dim N < \infty$, $(\lambda I - T)|M$ is semi-regular and $(\lambda I - T)|N$ is nilpotent. Choose a sequence (λ_n) which converges at λ and such that $\lambda_n \notin \sigma_e(T)$ for all $n \in \mathbb{N}$. Then $\lambda_n I - T$ is Fredholm and hence $\ker(\lambda_n I - T)|M$ is finite-dimensional, because it is contained in $\ker(\lambda_n I - T)$. Since $(\lambda I - T)|M$ is semi-regular we then have that $\dim \ker((\lambda I - T)|M) < \infty$, from which we conclude that $\dim \ker(\lambda I - T) < \infty$. In a similar way it is possible to prove that $\mathrm{codim}(\lambda I - T) < \infty$, so $\lambda I - T \in \Phi(X)$, and hence $\lambda \notin \sigma_e(T)$, a contradiction.

(ii) By Theorem 3.12 we have $\sigma_{es}(T + R) \subseteq \sigma_{es}(T)$ and by symmetry $\sigma_{es}(T) = \sigma_{es}((T + R) - R) \subseteq \sigma_{es}(T + R)$.

(iii) We show first the inclusion

$$\bigcap_{F \in \mathcal{F}(X), FT = TF} \sigma_{se}(T + F) \subseteq \sigma_{es}(T).$$

Let $\lambda \notin \sigma_{es}(T)$. There is no harm if we suppose $\lambda = 0$. Then T is essentially semi-regular, so there exists a decomposition $X = M \oplus N$ such that $\dim N < \infty$, $T|M$ is semi-regular and $T|N$ is nilpotent. Let $F := 0 \oplus I_N$, I_N the operator identity on N. Clearly, $F \in \mathcal{F}(X)$, $TF = FT$, and $T + F(X)$ is closed. Since $T|N$ is nilpotent we have

$$\ker(T + F) = \ker T|M \subseteq (T|M)^\infty(M) \subseteq (T|M)^\infty(M) \oplus N$$
$$= (T + F)^\infty(X),$$

thus $T + F$ is semi-regular and hence

$$0 \notin \bigcap_{F \in \mathcal{F}(X), FT=TF} \sigma_{se}(T + F),$$

and consequently

$$\bigcap_{F \in \mathcal{F}(X), FT=TF} \sigma_{se}(T + F) \subseteq \sigma_{es}(T).$$

The inclusion

$$\bigcap_{R \in \mathcal{R}(X), RT=TR} \sigma_{se}(T + R) \subseteq \bigcap_{F \in \mathcal{F}(X), FT=TF} \sigma_{se}(T + R)$$

is evident. To conclude the proof, let

$$\lambda \notin \bigcap_{R \in \mathcal{R}(X), RT=TR} \sigma_{se}(T + R).$$

Then there exists an $R \in \mathcal{R}(X)$, which commutes with T, such that $\lambda I - (T + R)$ is semi-regular. Adding R, by Theorem 3.12 we then deduce that $\lambda I - T = \lambda I - (T + R) + R$ is essentially semi-regular, so $\lambda \notin \sigma_{es}(T)$, and the proof is complete. ∎

The stability result observed in Theorem 3.14 for $\sigma_{es}(T)$ does not hold for the semi-regular spectrum. For instance, consider the identity I in a Hilbert space and let K denote a one-dimensional orthogonal projection. Then $0 \in \sigma_{se}(I - K)$, while, obviously, $0 \notin \sigma_{se}(I)$.

As a simple consequence of Theorem 3.14 we obtain the following characterization of Riesz operators:

Corollary 3.15 *Let $S \in L(X)$. Then the following statements are equivalent:*

(i) *S is a Riesz operator,*
(ii) *$\sigma_{es}(T + S) = \sigma_{es}(T)$ for all $T \in L(X)$ such that $TS = ST$.*

Proof (i) $\Rightarrow$ (ii) has been proved in Theorem 3.14. To show (ii) $\Rightarrow$ (i), take $T = 0$. Then $\sigma_{es}(S) = \sigma_{es}(0) = \{0\}$. From part (i) of Theorem 3.14 we then have $\sigma_e(S) = \{0\}$, thus S is a Riesz operator. ∎

Essentially semi-regular operators having finite ascent, or finite descent, are also stable under Riesz commuting perturbations:

Theorem 3.16 *Let $T \in L(X)$ be essentially semi-regular and $R \in L(X)$ a Riesz operator commuting with T. Then*

(i) *T has finite ascent if and only if $T + R$ has finite ascent.*
(ii) *T has finite descent if and only if $T + R$ has finite descent.*

Proof (i) Suppose that T has finite ascent and R is a Riesz operator commuting with T. We know, by Theorem 2.97, that the condition $p(T) < \infty$ entails that T has the SVEP at 0. Hence $T + R$ has the SVEP at 0, by Theorem 2.129. But $T + R$ is essentially semi-regular, by Theorem 3.12, and in particular has topological uniform descent. The SVEP of $T + R$ at 0 is then equivalent to saying that $p(T + R) < \infty$, by Theorem 2.97. The converse may be obtained by symmetry from the equality $p(T) = p((T + R) - R) = p(T + R)$, since $T + R$ commutes with R.

The proof of part (ii) is analogous: if T has finite descent and R is Riesz, then T^* has the SVEP at 0, hence, by Theorem 2.129, $T^* + R^*$ has the SVEP at 0, since R^* is a Riesz operator, and R^* commutes with T^*. Now, $T + R$ is essentially semi-regular, again by Theorem 3.12, and hence $T^* + R^*$ is essentially semi-regular, in particular it has topological uniform descent. The SVEP of $T^* + R^*$ at 0 then implies, by Theorem 2.97, that $q(T + R) < \infty$. The converse may be obtained again by symmetry.   ∎

To show the invariance of the semi-Fredholm spectra and Weyl spectra under Riesz commuting perturbations we shall use the *Sadovskii/Buoni, Harte, Wickstead* construction, already introduced in Chap. 2.

Recall that if $\widetilde{T} : \widetilde{X} \to \widetilde{X}$ is defined by

$$\widetilde{T}(\tilde{x} + m(X)) := T^\infty \tilde{x} + m(X) \quad \text{for all } \tilde{x} \in \widetilde{X},$$

then

$$T \in \Phi_+(X) \ \Leftrightarrow \ \widetilde{T} \text{ is injective} \ \Leftrightarrow \ \widetilde{T} \text{ is bounded below;}$$

while

$$T \in \Phi_-(X) \ \Leftrightarrow \ \widetilde{T} \text{ is onto;}$$

and

$$T \in \Phi(X) \ \Leftrightarrow \ \widetilde{T} \text{ is invertible.}$$

Theorem 3.17 *Let $T \in L(X)$ and R be a Riesz operator such that $TR = RT$. Then we have:*

(i) *$T \in W(X) \Leftrightarrow T + R \in W_+(X)$.*

(ii) $T \in W_+(X) \Leftrightarrow T + R \in W_+(X)$.
(iii) $T \in W_-(X) \Leftrightarrow T + R \in W(X)$.

Analogous statements hold for the classes $\Phi_+(X)$, $\Phi_-(X)$ *and* $\Phi(X)$.

Proof (i) If R is Riesz then $\lambda I - R \in B(X)$ for all $\lambda \neq 0$, hence $\widetilde{R}$ is quasi-nilpotent, from (c) above. Let $\mu \in [0, 1]$. Since T and μR commutes, we have $\widetilde{(T + \mu R)} = \widetilde{T} + \mu \widetilde{R}$. It then follows that $\widetilde{T}$ and $\widetilde{T} + \mu \widetilde{R}$ are quasi-nilpotent equivalent for all $\mu \in [0, 1]$, so, by Theorem 2.143, $\widetilde{(T + \mu R)}$ is invertible if and only if $\widetilde{T}$ is invertible. Hence, $T \in \Phi(X)$ if and only if $T + R \in \Phi(X)$. The Fredholm index being a continuous function, it then follows that $T \in W(X)$ if and only if $T + R \in W(X)$. The other statements may be proved in a similar way. ∎

Corollary 3.18 *The Weyl spectra* $\sigma_{uw}(T)$, $\sigma_{lw}(T)$, $\sigma_w(T)$, *the semi-Fredholm spectra* $\sigma_{usf}(T)$, $\sigma_{lsf}(T)$, *and the essential spectrum* $\sigma_e(T)$ *are stable under Riesz commuting perturbations.*

We now consider some other perturbation results concerning the whole spectrum. A well-known consequence of the Gelfand theory for commutative Banach algebras is that:

$$\sigma(T + S) \subseteq \sigma(T) + \sigma(S) \quad \text{for all } T, S \in L(X) \quad \text{with } TS = ST,$$

see for instance Theorem 11.23 of Rudin [269]. The following simple example shows that the spectrum in general is not stable under a commuting Riesz perturbation K, even in the case when K is finite-dimensional.

Example 3.19 Let $T := P \in L(X)$, X an infinite-dimensional Banach space and P a non-zero projection with finite-dimensional range. Set $K := 2P$. Then both P and $I - P$ are not injective, so $\sigma(T) = \{0, 1\}$. On the other hand, $\sigma(T + K) = \sigma(3P) = \{0, 3\} \neq \sigma(T)$.

The previous example shows that the isolated points of the two sets $\sigma(T)$ and $\sigma(T + K)$ may be different. This is not true for the sets of accumulation points.

Theorem 3.20 *Suppose that* $T, K \in L(X)$ *commutes. If* K^n *is a finite rank operator for some natural* $n \in \mathbb{N}$ *then* $\operatorname{acc} \sigma(T + K) = \operatorname{acc} \sigma(T)$.

Proof Suppose that $\lambda_0 \notin \operatorname{acc} \sigma(T)$. Then there exists an $\varepsilon > 0$ such that $\mu I - T$ is invertible for all $|\mu - \lambda_0| < \varepsilon$. Denote by $\mathbb{D}(\lambda_0, \varepsilon)$ the open disc centered at λ_0 with radius ε. Let $\mu_0 \in \mathbb{D}(\lambda_0, \varepsilon) \setminus \{\lambda_0\}$ and assume that $\mu_0 \in \sigma(T + K)$. Since $\mu_0 I - T$ is invertible, there exists an operator $S \in L(X)$ such that $S(\mu_0 I - T) = (\mu_0 I - T) = I$. It is easily seen that μ_0 is an eigenvalue of $T + K$. Indeed, if $\alpha(\mu_0 I - (T + K)) = 0$ then $\mu_0 I - (T + K)$ is Weyl, since $\mu_0 I - T$ is invertible, and hence

$$\alpha(\mu_0 I - (T + K)) = \beta(\mu_0 I - (T + K)) = 0,$$

thus $\mu_0 \notin \sigma(T + K)$, a contradiction.

Therefore, $\ker(\mu_0 I - (T + K)) \neq \{0\}$. Let $x \neq 0$ be an arbitrary eigenvector relative a μ_0, i.e., $(\mu_0 I - (T + K))x = 0$. Then

$$0 = [S(\mu_0 I - (T + K))]x = S(\mu_0 I - T)x - SKx = x - SKx,$$

thus $x = SKx$. Since $SK = KS$ it then follows that

$$x = SKSK = S^2 K^2 x = \cdots = S^n K^n x.$$

If $Z := S^n K^n(X)$ then $x \in Z$. It is well known that eigenvectors relative to distinct eigenvalues are linearly independent. But Z is a finite-dimensional subspace, since $K^n(X)$ is finite-dimensional, and consequently only finite many points of $\sigma(T+K)$ may be contained in $D(\lambda_0, \varepsilon) \setminus \{\lambda_0\}$, so $\lambda_0 \notin \mathrm{acc}\,\sigma(T+K)$. This prove the inclusion $\mathrm{acc}\,\sigma(T + K) \subseteq \mathrm{acc}\,\sigma(T)$. Since $T + K$ commutes with K, by symmetry, we have

$$\mathrm{acc}\,\sigma(T) = \mathrm{acc}\,\sigma[(T + K) - K] \subseteq \mathrm{acc}\,\sigma(T),$$

and hence the proof is complete.   ∎

In Example 3.19 we have $3 \in \mathrm{iso}\,\sigma(T + K)$ while $3 \notin \mathrm{iso}\,\sigma(T)$. The next corollary shows that isolated points of $\sigma(T + K)$ which belong to $\sigma(T)$ are isolated points of $\sigma(T + K)$.

Corollary 3.21 *Suppose that $T, K \in L(X)$ commutes, and K^n is a finite rank operator for some $n \in \mathbb{N}$. Then $\mathrm{iso}\,\sigma(T + K) \cap \sigma(T) \subseteq \mathrm{iso}\,\sigma(T)$.*

Proof Let $\lambda \in \mathrm{iso}\,\sigma(T + K) \cap \sigma(T)$ and suppose that $\lambda \notin \mathrm{iso}\,\sigma(T)$. Then $\lambda \in \mathrm{acc}\,\sigma(T) = \mathrm{acc}\,\sigma(T + K)$, a contradiction. Hence $\lambda \in \mathrm{iso}\,\sigma(T)$.   ∎

Since for every operator $T \in L(X)$ we have, trivially, $\sigma(T) = \mathrm{iso}\,\sigma(T) \cup \mathrm{acc}\,\sigma(T)$, from Theorem 3.20 we immediately have:

Corollary 3.22 *Suppose that $T, K \in L(X)$ commutes, and K^n is a finite rank operator for some $n \in \mathbb{N}$. Then $\sigma(T + K) = \sigma(T)$ if and only if $\mathrm{iso}\,\sigma(T + K) = \mathrm{iso}\,\sigma(T)$.*

In order to give further information about the approximate point spectrum of sums of operators we need to introduce the *Berberian–Quisley extension* for operators on Banach spaces. Given a non-trivial complex Banach space X, denote by $c_0(X)$ the subspace of all sequences of X which converge to 0. Denote by **X** the quotient $\ell^\infty(X)/c_0(X)$, endowed with the quotient canonical norm. Then **X** is a Banach space and X may be isometrically embedded into **X**. Every operator $T \in L(X)$ defines, by componentwise action, an operator on $\ell^\infty(X)$ which has as invariant subspace $c_0(X)$, and consequently T induces an operator $\mathbf{T} \in L(\mathbf{X})$. It is evident that **T** is an extension of T, when X is regarded as a subspace of **X**. Moreover, the mapping $T \in L(X) \rightarrow \mathbf{T} \in L(\mathbf{X})$ is an isometric algebra isomorphism. The Berberian–Quisley extension has an important property:

it converts points of $\sigma_{\text{ap}}(T)$ into eigenvalues of $\mathbf{T}$, i.e.

$$\sigma_{\text{ap}}(T) = \sigma_{\text{ap}}(\mathbf{T}) = \sigma_{\text{p}}(T) \quad \text{for all } T \in L(X),$$

see Choi and Davis [93].

Theorem 3.23 *Suppose that $T, S \in L(X)$ commutes. Then*

$$\sigma_{\text{ap}}(T + S) \subseteq \sigma_{\text{ap}}(T) + \sigma_{\text{ap}}(S) \quad \text{and} \quad \sigma_{\text{s}}(T + S) \subseteq \sigma_{\text{s}}(T) + \sigma_{\text{s}}(S).$$

Proof Let $\lambda \in \sigma_{\text{ap}}(T + S)$ and set $\mathbf{Z} := \ker(\lambda \mathbf{I} - (\mathbf{T} + \mathbf{S}))$. Evidently, $\mathbf{Z}$ is non-zero and $\mathbf{TS} = \mathbf{ST}$, so that $\mathbf{Z}$ is invariant under $\mathbf{T}$ and $\sigma_{\text{ap}}(\mathbf{T}|\mathbf{Z})$ is non-empty and $\mathbf{S}|\mathbf{Z} = \lambda(\mathbf{I} - \mathbf{T})|\mathbf{Z}$. Choosing $\mu \in \sigma_{\text{ap}}(\mathbf{T}|\mathbf{Z})$, we then obtain $\lambda - \mu \in \sigma_{\text{ap}}(\mathbf{S}|\mathbf{Z})$, and hence

$$\lambda = \mu + (\lambda - \mu) \in \sigma_{\text{ap}}(\mathbf{T}) + \sigma_{\text{ap}}(\mathbf{S}) = \sigma_{\text{ap}}(T) + \sigma_{\text{ap}}(S).$$

This proves the first inclusion, while the second one may be obtained by duality. ■

Corollary 3.24 *Suppose that $T \in L(X)$ and Q is a quasi-nilpotent operator commuting with T. Then $\sigma_{\text{ap}}(T) = \sigma_{\text{ap}}(T + Q)$ and $\sigma_{\text{s}}(T) = \sigma_{\text{s}}(T + Q)$.*

Proof From Theorem 3.23 we know that

$$\sigma_{\text{ap}}(T + Q) \subseteq \sigma_{\text{ap}}(T) + \{0\} = \sigma_{\text{ap}}(T).$$

The opposite inclusion is obtained by symmetry:

$$\sigma_{\text{ap}}(T) = \sigma_{\text{ap}}(T + Q - Q) \subseteq \sigma_{\text{ap}}(T + Q).$$

The equality $\sigma_{\text{s}}(T) = \sigma_{\text{s}}(T + Q)$ follows by duality. ■

We now show that the semi-regular spectrum $\sigma_{\text{se}}(T)$ and the essentially semi-regular spectrum $\sigma_{\text{es}}(T)$ are invariant under commuting quasi-nilpotent perturbations. Recall that for every $T \in L(X)$ the *spectral radius formula* $r(T) := \lim_{n \to \infty} \|T^n\|^{\frac{1}{n}}$ holds.

Theorem 3.25 *Suppose that $T \in L(X)$ and Q is a quasi-nilpotent operator commuting with T. Then*

$$\sigma_{\text{se}}(T) = \sigma_{\text{se}}(T + Q) \quad \text{and} \quad \sigma_{\text{es}}(T) = \sigma_{\text{es}}(T + Q).$$

Proof Suppose that $\lambda \notin \sigma_{\text{se}}(T)$. We may assume that $\lambda = 0$. Set $M := T^\infty(X)$. Since T and Q commute, $Q(M) \subseteq M$. Denote by $\widetilde{T} : X|M \to X/M$ and $\widetilde{Q} : X/M \to X/M$ the operators induced by T and Q, respectively. From the spectral radius formula we see that the restriction $Q|M$ is a quasi-nilpotent operator commuting with $T|M$, while $\tilde{Q}$ is a quasi-nilpotent operator which commutes with $\tilde{T}$. Now, by Theorem 1.45, $T|M$ is onto and $\tilde{T}$ is bounded below. By Corollary 3.24, then $\lambda \notin \sigma_s(T|M) = \sigma_s(T|M + Q|M)$, thus $(T + Q)|M = T|M + Q|M$ is onto. Again by Corollary 3.24, we have $\lambda \notin \sigma_{\mathrm{ap}}(\tilde{T}) = \sigma_{\mathrm{ap}}(\tilde{T} + \tilde{Q})$, so $\tilde{T} + \tilde{Q}$ is bounded below. Theorem 1.45 then yields that $T+Q$ is semi-regular, i.e., $\lambda \notin \sigma_{\mathrm{se}}(T+Q)$, and hence $\sigma_{\mathrm{se}}(T + Q) \subseteq \sigma_{\mathrm{se}}(T)$. A symmetric argument shows the reverse inclusion, thus the first equality is proved.

The second equality easily follows from Theorem 3.12.

∎

The approximate point spectrum in general is not stable under finite-rank perturbations commuting with T. The operator T considered in Example 3.19 shows that the isolated points of $\sigma_{\mathrm{ap}}(T)$ and $\sigma_{\mathrm{ap}}(T + K)$ can be different. Indeed, in this case we have $\sigma_{ap}(T) = \sigma(P) = \{0, 1\}$, while $\sigma_{\mathrm{ap}}(T + K) = \{0, 3\}$. By duality, it then follows that $\sigma_s(T)$ is also not stable under finite-rank perturbations commuting with T. However, we have:

Theorem 3.26 *Suppose that* $T, K \in L(X)$ *commute, and* K^n *is a finite rank operator for some* $n \in \mathbb{N}$. *Then we have*

(i) $\mathrm{acc}\, \sigma_{\mathrm{ap}}(T) = \mathrm{acc}\, \sigma_{\mathrm{ap}}(T + K)$.
(ii) $\mathrm{acc}\, \sigma_s(T) = \mathrm{acc}\, \sigma_s(T + K)$.

Proof Let $\lambda_0 \notin \mathrm{acc}\, \sigma_{\mathrm{ap}}(T + K)$ and assume that $\lambda_0 \in \mathrm{acc}\, \sigma_{\mathrm{ap}}(T)$. Then there exists a sequence $(\lambda_j) \subseteq \sigma_{\mathrm{ap}}(T)$ which converges to λ_0. We may assume that $\lambda_i \neq \lambda_j$ for $i \neq j$, and since $\lambda_0 \notin \mathrm{acc}\, \sigma_{\mathrm{ap}}(T + K)$ we may also assume that $\lambda_j I - (T + K)$ is bounded below for all $j \in \mathbb{N}$. Consequently, $\lambda_j I - (T + K) - K = \lambda_j I - T$ is upper semi-Browder, by Theorem 3.11, and hence has closed range. Note that $0 < \alpha(\lambda_j I - T) < \infty$, otherwise if $\alpha(\lambda_j I - T) = 0$ we would have that $\lambda_j I - T$ is bounded below, hence $\lambda_j \notin \sigma_{\mathrm{ap}}(T)$. Denote now by K_j the restriction $K|\ker(\lambda_j I - T)$. The operator K_j is injective. Indeed, if $K_j x = 0$ with $x \in \ker(\lambda_j I - T)$, then $(\lambda_j I - (T + K))x = 0$ and this implies that $x = 0$, because $\lambda_j I - (T + K)$ is injective. Since $\ker(\lambda_j I - T)$ is finite-dimensional it then follows that K_j is also onto, i.e.,

$$\ker(\lambda_j I - T) = K_j(\ker(\lambda_j I - T)) \quad \text{for every } j \in \mathbb{N}.$$

From this it easily follows that

$$\ker(\lambda_j I - T) = K_j{}^n[\ker(\lambda_j I - T)] \quad \text{for every } j \in \mathbb{N}.$$

In particular, $\ker(\lambda_j I - T)$ is contained in the range of $K_j{}^n$, and hence

$$\ker(\lambda_j I - T) \subseteq K^n(X) \quad \text{for all } j \in \mathbb{N}. \tag{3.3}$$

Let $0 \neq x_j \in \ker(\lambda_j I - T)$ with $j \in \mathbb{N}$. From the inclusion (3.3) we see that all the vectors x_j belong to $K^n(X)$ for every $j \in \mathbb{N}$. Since eigenvectors relative to distinct eigenvalues are linearly independent we then conclude that $K^n(X)$ has infinite dimension, a contradiction. Hence, $\lambda_0 \notin \mathrm{acc}\,\sigma_{\mathrm{ap}}(T)$. This proves the inclusion $\mathrm{acc}\,\sigma_{\mathrm{ap}}(T) \subseteq \mathrm{acc}\,\sigma_{\mathrm{ap}}(T + K)$. By symmetry we then obtain

$$\mathrm{acc}\,\sigma_{\mathrm{ap}}(T + K) = \mathrm{acc}\,\sigma_{\mathrm{ap}}[(T + K) - K] = \mathrm{acc}\,\sigma_{\mathrm{ap}}(T),$$

and hence $\mathrm{acc}\,\sigma_{\mathrm{ap}}(T + K) = \mathrm{acc}\,\sigma_{\mathrm{ap}}(T)$.

(ii) Since K^{*n} is finite-dimensional, by duality and part (i) we obtain

$$\mathrm{acc}\,\sigma_{\mathrm{s}}(T) = \mathrm{acc}\,\sigma_{\mathrm{ap}}(T^*) = \mathrm{acc}\,\sigma_{\mathrm{ap}}(T^* + K^*) = \mathrm{acc}\,\sigma_{\mathrm{s}}(T + K),$$

so the proof is complete. ∎

Theorem 3.27 *Suppose that* $T, K \in L(X)$ *commute, and* K^n *is a finite rank operator for some* $n \in \mathbb{N}$. *Then* $\sigma_{\mathrm{ap}}(T+K) = \sigma_{\mathrm{ap}}(T)$ *if and only if* $\mathrm{iso}\,\sigma_{\mathrm{ap}}(T+K) = \mathrm{iso}\,\sigma_{\mathrm{ap}}(T)$. *In this case we also have* $\sigma(T + K) = \sigma(T)$.

Proof Trivially, $\sigma_{\mathrm{ap}}(T + K) = \sigma_{\mathrm{ap}}(T)$ implies $\mathrm{iso}\,\sigma_{\mathrm{ap}}(T + K) = \mathrm{iso}\,\sigma_{\mathrm{ap}}(T)$. Conversely, assume that $\mathrm{iso}\,\sigma_{\mathrm{ap}}(T + K) = \mathrm{iso}\,\sigma_{\mathrm{ap}}(T)$. Then, by Theorem 3.26, we have

$$\sigma_{\mathrm{ap}}(T + K) = \mathrm{iso}\,\sigma_{\mathrm{ap}}(T + K) \cup \mathrm{acc}\,\sigma_{\mathrm{ap}}(T + K) = \mathrm{iso}\,\sigma_{\mathrm{ap}}(T) \cup \mathrm{acc}\,\sigma_{\mathrm{ap}}(T)$$
$$= \sigma_{\mathrm{ap}}(T).$$

To show that $\sigma(T) = \sigma(T + K)$, observe first that if $\lambda \in \mathrm{iso}\,\sigma(T)$, then $\lambda \in \sigma_{\mathrm{ap}}(T)$ and hence, $\lambda \in \mathrm{iso}\,\sigma_{\mathrm{ap}}(T) = \mathrm{iso}\,\sigma_{\mathrm{ap}}(T + K)$. Therefore, taking into account Theorem 3.20, we have

$$\sigma(T) = \mathrm{iso}\,\sigma(T) \cup \mathrm{acc}\,\sigma(T) \subseteq \mathrm{iso}\,\sigma_{\mathrm{ap}}(T) \cup \mathrm{acc}\,\sigma(T)$$
$$= \mathrm{iso}\,\sigma_{\mathrm{ap}}(T + K) \cup \mathrm{acc}\,\sigma(T + K) \subseteq \sigma_{\mathrm{ap}}(T + K) \cup \mathrm{acc}\,\sigma(T + K)$$
$$\subseteq \sigma(T + K).$$

Since K commutes with $T + K$, a symmetric argument shows that

$$\sigma(T + K) \subseteq \sigma((T + K) - K) = \sigma(T).$$

Therefore, $\sigma(T) = \sigma(T + K)$.

∎

Every isolated point of $\sigma_{ap}(T + K)$ which belongs to $\sigma_{ap}(T)$ is necessarily an isolated point of $\sigma_{ap}(T)$:

Corollary 3.28 *Suppose that $T, K \in L(X)$ commute, and K^n is a finite rank operator for some $n \in \mathbb{N}$. Then* iso $\sigma_{ap}(T + K) \cap \sigma_{ap}(T) \subseteq$ iso $\sigma_{ap}(T)$.

Proof Let $\lambda \in$ iso $\sigma_{ap}(T + K) \cap \sigma_{ap}(T)$ and suppose that $\lambda \notin$ iso $\sigma_{ap}(T)$. Since $\lambda \in \sigma_{ap}(T)$ it then follows that $\lambda \in$ acc $\sigma_{ap}(T) =$ acc $\sigma_{ap}(T + K)$, a contradiction. Hence $\lambda \in$ iso $\sigma_{ap}(T)$. ∎

The equality $\sigma_{ap}(T) = \sigma_{ap}(T + K)$ for a commuting finite-dimensional operator K is satisfied in a very particular case:

Theorem 3.29 *Let $T, K \in L(X)$ be commuting operators and suppose that* iso $\sigma_{ap}(T) = \emptyset$. *If K^n is a finite rank operator for some $n \in \mathbb{N}$, then $\sigma_{ap}(T) = \sigma_{ap}(T + K)$.*

Proof We have

$$\sigma_{ap}(T) = \text{iso } \sigma_{ap}(T) \cup \text{acc } \sigma_{ap}(T) = \text{acc } \sigma_{ap}(T)$$
$$= \text{acc } \sigma_{ap}(T + K) \subseteq \sigma_{ap}(T + K).$$

Now $\sigma(K)$, and hence also $\sigma_{ap}(K)$, is a finite set, for instance $\sigma_{ap}(K) = \{\lambda_1, \lambda_2, \ldots \lambda_n\}$. Then we have, by using Theorem 3.23,

$$\text{iso } \sigma_{ap}(T + K) \subseteq \text{iso } (\sigma_{ap}(T) + \sigma_{ap}(K)) = \text{iso } (\bigcup_{k=1}^{n} (\lambda_k + \sigma_{ap}(T)) = \emptyset,$$

thus, since iso $\sigma_{ap}(T) = \emptyset$,

$$\sigma_{ap}(T + K) = \text{iso } \sigma_{ap}(T + K) \cup \text{acc } \sigma_{ap}(T + K) = \text{acc } \sigma_{ap}(T + K)$$
$$= \text{acc } \sigma_{ap}(T) \cup \text{iso } \sigma_{ap}(T) = \sigma_{ap}(T).$$

∎

The condition iso $\sigma_{ap}(T) = \emptyset$ is satisfied by every non-quasi-nilpotent unilateral right shift T on $\ell^p(\mathbb{N})$, with $1 \le p < \infty$, see [216, Proposition 1.6.15].

We conclude this section by establishing some relationships between the cluster points of $\sigma_{se}(T)$, and the essential spectrum $\sigma_e(T)$. We start with a preliminary lemma.

Theorem 3.30 *Let $T \in L(X)$. Then we have:*

$$\sigma_{se}(T) \cup \rho_e(T) \subseteq \sigma_p(T) \cup \sigma_p(T^*), \tag{3.4}$$

and

$$\text{acc } \sigma_{se}(T) \subseteq \sigma_e(T), \tag{3.5}$$

and

$$\partial \sigma(T) \setminus \sigma_e(T) \subseteq \text{iso } \sigma(T). \tag{3.6}$$

Furthermore,

(i) *If T has the SVEP then* acc $\sigma_{ap}(T) \subseteq \sigma_e(T)$.
(ii) *If T^* has the SVEP then* acc $\sigma_s(T) \subseteq \sigma_e(T)$.

Proof Evidently, $\sigma_{ap}(T) \cup \rho_e(T) \subseteq \sigma_p(T)$, and hence

$$\sigma_s(T) \cup \rho_e(T) = \sigma_{ap}(T^*) \cup \rho_e(T^*) \subseteq \sigma_p(T^*).$$

Since $\sigma_{se}(T) \subseteq \sigma_{ap}(T) \cap \sigma_s(T)$, the inclusion (3.4) follows.

To show (3.4), let $\lambda \in \rho_e(T)$. Since every Fredholm operator is of Kato-type, by Theorem 1.65 there exists an open disc $\mathbb{D}(\lambda, \varepsilon)$ for which $\sigma_{se}(T) \cap \mathbb{D}(\lambda, \varepsilon) \subseteq \{\lambda\}$. This shows that every point of $\sigma_{se}(T) \cap \rho_e(T)$ is isolated in $\sigma_{se}(T)$, thus the inclusion (3.5) holds. The assertions (i) and (ii) are now immediate from Theorem 2.68, since the SVEP for T entails $\sigma_{ap}(T) = \sigma_{se}(T)$, while the SVEP for T^* entails that $\sigma_s(T) = \sigma_{se}(T)$. The inclusion (3.6) is clear: both T and T^* have the SVEP at the points $\lambda \in \partial \sigma(T)$, so, if $\lambda I - T \in \Phi(X)$, then, by Corollary 2.99, 0 is a pole and hence an isolated point of $\sigma(T)$. ∎

A simple consequence of Theorem 3.30 is the following result.

Corollary 3.31 *If $T \in L(X)$ has no eigenvalues then*

$$\sigma_{se}(T) = \sigma \sigma_e(T) = \sigma_{ap}(T), \tag{3.7}$$

while

$$\sigma(T) = \sigma_w(T). \tag{3.8}$$

Proof Immediate, since from Theorems 3.30 and 2.68, $\sigma_p(T) = \emptyset$ and T has the SVEP. ∎

3.2   Representation Theorems for Weyl and Browder Operators

This section addresses some characterizations of Weyl and Browder operators. The first simple characterization describes the semi-Browder operators in terms of the hyper-kernels or hyper-ranges.

Theorem 3.32 *If $T \in L(X)$ the following equivalences hold:*

(i) $T \in B_+(X)$ *if and only if $T(X)$ is closed and $\dim \mathcal{N}^\infty(T) < \infty$.*
(ii) $T \in B_-(X)$ *if and only if $\operatorname{codim} T^\infty(X) < \infty$.*
(i) $T \in B(X)$ *if and only if $\dim \mathcal{N}^\infty(T) < \infty$ and $\operatorname{codim} T^\infty(X) < \infty$.*

Proof

(i) If $T \in B_+(X)$ then the ascent $p := p(T) < \infty$ and $T(X)$ is closed. Clearly, $\mathcal{N}^\infty(T) = \ker T^p$, and since $T^n \in \Phi_+(X)$ for every $n \in \mathbb{N}$, $\ker T^p$ is finite-dimensional. Conversely, if $T(X)$ is closed and $\dim \mathcal{N}^\infty(T) < \infty$ then $\ker T$ is finite-dimensional, since $\ker T \subseteq \mathcal{N}^\infty(T)$, so $T \in \Phi_+(X)$. From the inclusions $\ker T^n \subseteq \ker T^{n+1} \subseteq \mathcal{N}^\infty(T)$, it then follows that $p(T) < \infty$.

(ii) If $T \in B_-(X)$ then the descent $q := q(T) < \infty$ and $T(X)$ is closed. Clearly, $T^\infty(X) = T^q(X)$, and since $T^n \in \Phi_-(X)$ for every $n \in \mathbb{N}$ we then have $\operatorname{codim} T^q(X) < \infty$. From the inclusion $T^\infty(X) \subseteq T^q(X)$ we conclude that $\operatorname{codim} T^\infty(X) < \infty$. Conversely, if $\operatorname{codim} T^\infty(X) < \infty$ then $T(X)$ has finite codimension, since $T^\infty(X) \subseteq T(X)$, so $T \in \Phi_-(X)$ and trivially $q(T) < \infty$.

(iii) Clear.   ∎

Lemma 3.33 *If $T \in B_+(X)$ is semi-regular then T is bounded below. If $T \in B_-(X)$ is semi-regular then T is onto.*

Proof To show the first assertion, assume that $0 \neq x_0 \in \ker T$. Since $\ker T \subseteq T(X)$ there exists an $x_1 \in X$ such that $x_0 = Tx_1$. Obviously, $x_1 \in \ker T^2 \subseteq T(X)$, so we can construct a sequence (x_n) such that $Tx_n = x_{n-1}$ for all $n \in \mathbb{N}$. Such vectors x_n are linearly independent, and they all belong to $\mathcal{N}^\infty(X)$, so $\mathcal{N}^\infty(X)$ is infinite-dimensional, contradicting Theorem 3.32. The second assertion follows by duality: if $T \in B_-(X)$ is semi-regular then $T^* \in B_+(X^*)$. By the first part it then follows that T^* is bounded below and hence T is onto.   ∎

Theorem 3.34 *Let $T \in L(X)$. Then we have:*

(i) $T \in B_+(X)$ *if and only if there exist two closed invariant subspaces M, N of X such that $X = M \oplus N$, $\dim N < \infty$, $T|M$ is bounded below, and $T|N$ is nilpotent. In this case $N = \mathcal{N}^\infty(T)$.*

(ii) $T \in B_-(X)$ *if and only if there exist two closed invariant subspaces M, N of X such that $X = M \oplus N$, $\dim N < \infty$, $T|M$ is onto, and $T|N$ is nilpotent. In this case $M = T^\infty(X)$.*

(i) $T \in B(X)$ *if and only if there exist two closed invariant subspaces M, N of X such that $X = M \oplus N$, $\dim N < \infty$, $T|M$ is invertible, and $T|N$ is nilpotent. In this case $N = \mathcal{N}^{\infty}(T) < \infty$ and $M = T^{\infty}(X)$.*

Proof (i) If $T \in B_{+}(X)$ then T is essentially semi-regular, so there exists a Kato decomposition $X = M \oplus N$, with $\dim N < \infty$, $T|M$ is semi-regular, and $T|N$ is nilpotent. Evidently, $\alpha(T|M) < \infty$ and since $T|M$ has closed range, so $T|M$ is upper semi-Fredholm. It is easily seen that $p(T|M) < \infty$, $T|M$ is upper semi-Browder. By Lemma 3.33 it then follows that $T|M$ is bounded below. Moreover, $N \subseteq \mathcal{N}^{\infty}(T)$. Suppose that $x = x_1 \oplus x_2 \in \ker T^n$ for some n. Then $T^n x_2 = 0$, so $x_2 = 0$. Therefore, $\ker T^n \subseteq N$ for all $n \in \mathbb{N}$, hence $N = \mathcal{N}^{\infty}(T)$. On the other hand, the direct sum $T_1 \oplus T_2$ of a finite-dimensional nilpotent operator T_1 and a bounded below operator T_2 is evidently upper semi-Browder.

Part (ii) and part (iii) can be obtained in a similar way. ∎

By a basic result of operator theory, every finite-dimensional operator $T \in \mathcal{F}(X)$ may be represented in the form

$$Tx = \sum_{k=1}^{n} f_k(x)x_k,$$

where the vectors $x_1, \ldots, x_n$ from X and the vectors $f_1, \ldots, f_n$ from X^* are linearly independent, see Heuser [179, p. 81]. Clearly, $T(X) \subseteq Y$, where Y is the subspace generated by the vectors $x_1, \ldots, x_n$.

Conversely, if $y := \lambda_1 x_1 + \ldots + \lambda_n x_n$ is an arbitrary element of Y we can choose $z_1, \ldots, z_n$ in X such that $f_i(z_j) = \delta_{i,j}$, where $\delta_{i,j}$ denotes the Kronecker delta (such a choice is always possible, see Heuser [179, Proposition 15.1]). Define $z := \sum_{k=1}^{n} \lambda_k z_k$, then

$$Tz = \sum_{k=1}^{n} f_k(z)x_k = \sum_{k=1}^{n} f_k\left(\sum_{k=1}^{n} \lambda_k z_k\right) x_k = \sum_{k=1}^{n} \lambda_k x_k = y,$$

thus the set $\{x_1, \ldots, x_n\}$ forms a basis for the subspace $T(X)$.

Theorem 3.35 *For a bounded operator T on a Banach space X, the following assertions are equivalent:*

(i) *T is a Weyl operator;*
(ii) *There exist a $K \in \mathcal{F}(X)$ and an invertible operator $S \in L(X)$ such that $T = S + K$ is invertible;*
(iii) *There exist a Riesz operator $R \in \mathcal{R}(X)$ and an invertible operator $S \in L(X)$ such that $T = S + R$ is invertible.*

Proof (i) $\Rightarrow$ (ii) Assume that $T \in W(X)$ and let $m := \alpha(T) = \beta(T)$. Let $P \in L(X)$ denote the projection of X onto the finite-dimensional space $\ker T$. Obviously, $\ker T \cap \ker P = \{0\}$ and we can represent the finite-dimensional operator P

in the form

$$Px = \sum_{i=1}^{m} f_i(x)x_i,$$

where the vectors $x_1, \ldots, x_m$ from X and the vectors $f_1, \cdots, f_m$ from X^*, are linearly independent. As observed before, the set $\{x_1, \ldots, x_m\}$ forms a basis of $P(X)$ and therefore $Px_i = x_i$ for every $i = 1, \ldots, m$, from which we obtain that $f_i(x_k) = \delta_{i,k}$.

Denote by Y the topological complement of the finite-codimensional subspace $T(X)$. Then $\dim Y = m$, so we can choose a basis $\{y_1, \ldots, y_m\}$ of Y. Define

$$Kx := \sum_{i=1}^{m} f_i(x)y_i.$$

Clearly, $K \in \mathcal{F}(X)$, so from classical Fredholm theory we obtain that $S := T + K \in \Phi(X)$ and $K(X) = Y$.

Finally, consider an element $x \in \ker S$. Then $Tx = Kx = 0$, and this easily implies that $f_i(x) = 0$ for all $i = 1, \ldots, m$. Consequently, $Px = 0$ and therefore $x \in \ker T \cap \ker P = \{0\}$, so S is injective.

In order to show that S is onto, observe first that

$$f_i(Px) = f_i\left(\sum_{k=1}^{m} f_k(x)x_k\right) = f_i(x).$$

From this we obtain that

$$KPx = \sum_{i=1}^{m} f_i(Px)y_i = \sum_{i=1}^{m} f_i(x)y_i = Kx. \tag{3.9}$$

Since $X = T(X) \oplus Y = T(X) \oplus K(X)$, every $z \in X$ may be represented in the form $z = Tu + Kv$, with $u, v \in X$. Set $u_1 := u - Pu$ and $v_1 := Pv$. From (3.9), and from the equality $P(X) = \ker T$, it then follows that

$$Ku_1 = 0, \quad Tv_1 = 0, \quad Kv_1 = Kv \quad \text{and} \quad Tu_1 = Tu.$$

Hence

$$S(u_1 + v_1) = (T + K)(u_1 + v_1) = Tu + Kv = z,$$

so S is surjective. Therefore $S = T + K$ is invertible.

(ii)$\Rightarrow$(iii) Clear.

(iii)$\Rightarrow$(i) Suppose $T + R = U$, where U is invertible and R is Riesz. Then U is Weyl, and hence $T = U - R$ is also Weyl, by Theorem 3.17.   ∎

A simple modification of the proof of Theorem 3.35 leads to the following characterizations of upper and lower semi-Weyl operators:

Theorem 3.36 *Let $T \in L(X)$. Then we have*

(i) $T \in W_+(X)$ *if and only if there exist a $K \in \mathcal{F}(X)$ (or a Riesz operator) and a bounded below operator $S \in L(X)$ such that $T = S + K$.*

(ii) $T \in W_-(X)$ *if and only if there exist a $K \in \mathcal{F}(X)$ (or a Riesz operator) and a surjective operator S such that $T = S + K$.*

Proof To show part (i), take $m := \alpha(T)$ and proceed as in the proof of Theorem 3.35. The operator $S = T + K$ is then injective and has closed range, since $T + K \in \Phi_+(X)$. To show part (ii), take $m := \beta(T)$ and proceed as in the proof of Theorem 3.35. The operator $S = T + K$ is then onto, since $T + K \in \Phi_-(X)$. $\blacksquare$

An immediate consequence of Theorems 3.35 and 3.36 is that the Weyl spectra may be characterized as follows.

Corollary 3.37 *Let $T \in L(X)$, X a Banach space. Then the Weyl spectra $\sigma_w(T)$, $\sigma_{uw}(T)$ and $\sigma_{lw}(T)$ are closed. Moreover,*

$$\sigma_w(T) = \bigcap_{K \in \mathcal{F}(X)} \sigma(T + K) = \bigcap_{R \in \mathcal{R}(X)} \sigma(T + R), \tag{3.10}$$

$$\sigma_{uw}(T) = \bigcap_{K \in \mathcal{F}(X)} \sigma_{ap}(T + K) = \bigcap_{R \in \mathcal{R}(X)} \sigma_{ap}(T + R), \tag{3.11}$$

and

$$\sigma_{lw}(T) = \bigcap_{K \in \mathcal{F}(X)} \sigma_s(T + K) = \bigcap_{R \in \mathcal{R}(X)} \sigma_s(T + R). \tag{3.12}$$

Proof Let $\rho_w(T) := \mathbb{C} \setminus \sigma_w(T)$. The equality (3.10) may be restated, taking complements, as follows

$$\rho_w(T) = \bigcup_{K \in \mathcal{F}(X)} \rho(T + K) = \bigcup_{R \in \mathcal{R}(X)} \rho(T + R). \tag{3.13}$$

The equalities (3.13) are now immediate from Theorem 3.35. The equalities (3.11) and (3.12) are proved in a similar way, by using Theorem 3.36. Clearly, all Weyl spectra are closed, since they are intersections of closed sets. $\blacksquare$

Semi-Browder operators admit similar characterizations to those given above for semi-Weyl operators. In this case the perturbations need to commute with T:

Theorem 3.38 *For an operator $T \in L(X)$, X a Banach space, the following statements are equivalent:*

(i) *T is essentially semi-regular and T has the SVEP at 0;*
(ii) *There exist an idempotent $P \in \mathcal{F}(X)$ and a bounded below operator $S \in L(X)$ such that $TP = PT$ and $T = S + P$;*
(iii) *There exist a Riesz operator $R \in \mathcal{R}(X)$ and a bounded below operator $S \in L(X)$ such that $TR = RT$ and $T = S + R$;*
(iv) *$T \in B_+(X)$.*

Proof (i) $\Rightarrow$ (ii) Suppose that T is essentially semi-regular and that T has the SVEP at 0. Let (M, N) be a GKD for T, where $T|N$ is nilpotent and N is finite-dimensional. Let P denote the finite-dimensional projection of X onto N along M. Clearly P commutes with T, because N and M reduce T. Since T has the SVEP at 0 it follows that $T|M$ is injective, by Theorem 2.91. Furthermore, the restriction $(I - T)|N$ is bijective, since $T|N$ is nilpotent and hence $1 \notin \sigma(T|N)$. Therefore $(I - T)(N) = N$ and $\ker(I - T)|N = \{0\}$. From this it follows that

$$\ker(T - P) = \ker(T - P)|M \oplus \ker(T - P)|N$$

$$= \ker T|M \oplus \ker(I - T)|N = \{0\},$$

thus the operator $T - P$ is injective. On the other hand, the equalities

$$(T - P)(X) = (T - P)(M) \oplus (T - P)(N)$$

$$= T(M) \oplus (T - I)(N) = T(M) \oplus N$$

show that the subspace $(T - P)(X)$ is closed, since it is the sum of the subspace $T(M)$, which is closed by semi-regularity, and the finite-dimensional subspace N. Therefore, the operator $T - P$ is bounded below.

(ii) $\Rightarrow$ (iii) Clear.

(iii) $\Rightarrow$ (iv) Suppose that there exists a commuting Riesz operator R such that $T + R$ is bounded below. By Theorem 3.11 it then follows that $T = (T + R) - R \in B_+(X)$.

The implication (iv) $\Rightarrow$ (i) is clear, since every upper semi-Browder operator is essentially semi-regular and T has the SVEP at 0, since $p(T) < \infty$. ∎

The next result is dual to that given in Theorem 3.38.

Theorem 3.39 *Let $T \in L(X)$, X a Banach space. Then the following properties are equivalent:*

(i) *T is essentially semi-regular and T^* has the SVEP at 0;*
(ii) *There exist an idempotent $P \in \mathcal{F}(X)$ and a surjective operator S such that $TP = PT$ and $T = S + P$;*

(iii) *There exist a Riesz operator $R \in \mathcal{R}(X)$ and a surjective operator S such that $TR = RT$ and $T = S + R$;*
(iv) $T \in B_-(X)$.

Proof (i) $\Rightarrow$ (ii) Let T be essentially semi-regular and suppose that T^* has the SVEP at 0. Let (M, N) be a GKD for T, where $T|N$ is nilpotent and N is finite-dimensional. Then $(N^\perp, M^\perp)$ is a GKD for T^*. In particular, $T^*|N^\perp$ is semi-regular.

Let P denote the finite rank projection of X onto N along M. Then P commutes with T, since N and M reduce T. Moreover, since $T^*|N^\perp$ has the SVEP at 0, $T^*|N^\perp$ is injective and this implies that $T|M$ is surjective, see Lemma 2.78. Since $T|N$ is nilpotent the restriction $(T - I)|N$ is bijective, so we have

$$(T - P)(X) = (T - P)(M) \oplus (T - P)(N) = T(M) \oplus (T - I)(N) = M \oplus N = X.$$

This shows that $T + P$ is onto.

(ii) $\Rightarrow$ (iii) Obvious.

(iii) $\Rightarrow$ (iv) Suppose that there exists a commuting Riesz operator $R \in L(X)$ such that $T + R$ is onto. By Theorem 3.11 it then follows that $T = (T + R) - R \in B_-(X)$.

The implication (iv) $\Rightarrow$ (i) is clear, since every lower semi-Bowder operator is essentially semi-regular and the condition $q(T) < \infty$ entails the SVEP for T^* at 0. ∎

Combining Theorems 3.38 and 3.39 we readily obtain the following characterizations of Browder operators.

Theorem 3.40 *Let $T \in L(X)$, X a Banach space. Then the following properties are equivalent:*

(i) *T is essentially semi-regular, both T and T^* have the SVEP at 0;*
(ii) *There exist an idempotent $P \in \mathcal{F}(X)$ and an invertible operator S such that $TP = PT$ and $T = S + P$;*
(iii) *There exist a Riesz operator $R \in \mathcal{R}(X)$ and an invertible operator S such that $TR = RT$ and $T = S + R$;*
(iv) $T \in B(X)$.

The following corollary is an immediate consequence of Theorem 3.40, once we observe that both the operators T and T^* have the SVEP at every $\lambda \in \partial \sigma(T)$, $\partial \sigma(T)$ the boundary of $\sigma(T)$.

Corollary 3.41 *Let $T \in L(X)$, X a Banach space, and suppose that $\lambda_0 \in \partial \sigma(T)$. Then $\lambda_0 I - T$ is essentially semi-regular if and only if $\lambda_0 I - T$ is semi-Fredholm, and this is the case if and only if $\lambda_0 I - T$ is Browder.* ∎

Corollary 3.42 *Let $T \in L(X)$. Then we have*

$$\sigma_{\mathrm{ub}}(T) = \bigcap_{K \in \mathcal{F}(X), KT=TK} \sigma_{\mathrm{ap}}(T + K) = \bigcap_{R \in \mathcal{R}(X), RT=TR} \sigma_{\mathrm{ap}}(T + R), \qquad (3.14)$$

$$\sigma_{\mathrm{lb}}(T) = \bigcap_{K \in \mathcal{F}(X), KT=TK} \sigma_{\mathrm{s}}(T + K) = \bigcap_{R \in \mathcal{R}(X), RT=TR} \sigma_{\mathrm{s}}(T + R), \qquad (3.15)$$

and

$$\sigma_{\mathrm{b}}(T) = \bigcap_{K \in \mathcal{F}(X), KT=TK} \sigma(T + K) = \bigcap_{R \in \mathcal{R}(X), RT=TR} \sigma(T + R). \qquad (3.16)$$

We now show that the Browder spectra may be obtained by adding to Weyl spectra the cluster points of some parts of the spectrum.

Theorem 3.43 *For a bounded operator $T \in L(X)$ the following statements hold:*

(i) $\sigma_{\mathrm{ub}}(T) = \sigma_{\mathrm{uw}}(T) \cup \operatorname{acc} \sigma_{\mathrm{ap}}(T)$.
(ii) $\sigma_{\mathrm{lb}}(T) = \sigma_{\mathrm{lw}}(T) \cup \operatorname{acc} \sigma_{\mathrm{s}}(T)$.
(iii) $\sigma_{\mathrm{b}}(T) = \sigma_{\mathrm{w}}(T) \cup \operatorname{acc} \sigma(T)$.

Proof (i) If $\lambda \notin \sigma_{\mathrm{uw}}(T) \cup \operatorname{acc} \sigma_{\mathrm{ap}}(T)$ then $\lambda I - T \in \Phi_+(X)$ and $\sigma_{\mathrm{a}}(T)$ does not cluster at λ. Then T has the SVEP at λ, and hence, by Theorem 2.97, $p(\lambda I - T) < \infty$ from which we conclude that $\lambda \notin \sigma_{\mathrm{ub}}(T)$. This shows the inclusion $\sigma_{\mathrm{ub}}(T) \subseteq \sigma_{\mathrm{uw}}(T) \cup \operatorname{acc} \sigma_{\mathrm{ap}}(T)$.

Conversely, suppose that $\lambda \in \sigma_{\mathrm{uw}}(T) \cup \operatorname{acc} \sigma_{\mathrm{ap}}(T)$. If $\lambda \in \sigma_{\mathrm{uw}}(T)$ then $\lambda \in \sigma_{\mathrm{ub}}(T)$, since $\sigma_{\mathrm{uw}}(T) \subseteq \sigma_{\mathrm{ub}}(T)$. If $\lambda \in \operatorname{acc} \sigma_{\mathrm{ap}}(T)$ then either $\lambda \in \sigma_{\mathrm{uw}}(T)$ or $\lambda \notin \sigma_{\mathrm{uw}}(T)$. In the first case $\lambda \in \sigma_{\mathrm{ub}}(T)$. In the second case, since $\lambda I - T \in W_+(X)$, the condition $\lambda \in \operatorname{acc} \sigma_{\mathrm{ap}}(T)$ entails, by Theorem 2.97, that $p(\lambda I - T) = \infty$. From this we conclude that $\lambda \in \sigma_{\mathrm{ub}}(T)$. Therefore the equality (i) is proved.

The proof of equality (ii) is similar. Equality (iii) follows combining (i) with (ii) and taking into account the equality $\sigma(T) = \sigma_{\mathrm{ap}}(T) \cup \sigma_{\mathrm{s}}(T)$. ∎

If either T or T^* has the SVEP we can say more:

Theorem 3.44 *Suppose that $T \in L(X)$.*

(i) *If T has the SVEP then $\sigma_{\mathrm{lw}}(T) = \sigma_{\mathrm{w}}(T) = \sigma_{\mathrm{b}}(T) = \sigma_{\mathrm{lb}}(T)$.*
(ii) *If T^* has the SVEP then $\sigma_{\mathrm{uw}}(T) = \sigma_{\mathrm{w}}(T) = \sigma_{\mathrm{ub}}(T) = \sigma_{\mathrm{b}}(T)$.*

Proof

(i) We have $\sigma_{\mathrm{lw}}(T) \subseteq \sigma_{\mathrm{w}}(T) \subseteq \sigma_{\mathrm{b}}(T)$ and $\sigma_{\mathrm{lw}}(T) \subseteq \sigma_{\mathrm{lb}}(T) \subseteq \sigma_{\mathrm{b}}(T)$, so it suffices to prove the inclusion $\sigma_{\mathrm{b}}(T) \subseteq \sigma_{\mathrm{lw}}(T)$. Suppose that $\lambda \notin \sigma_{\mathrm{lw}}(T)$ then $\lambda I - T \in W_-(X)$, and the SVEP of T at λ ensures that $p(\lambda I - T) < \infty$, by Theorem 2.97. By Theorem 1.22, part (i), we then have $\operatorname{ind}(\lambda I - T) \leq 0$ and since $\lambda I - T \in W_-(X)$ we also have $\operatorname{ind}(\lambda I - T) \geq 0$, hence $\operatorname{ind}(\lambda I - T) = 0$.

From part (iv) of Theorem 1.22 we then conclude that $\lambda I - T$ is Browder, hence $\lambda \notin \sigma_b(T)$.

(ii) The equalities may be shown from part (i) by duality. ∎

3.3 Semi B-Browder Spectra

It is natural to extend the concepts of Weyl and Browder operators in the context of B-Fredholm theory. Recall that by T_n we denote the restriction $T|T^n(X)$.

Definition 3.45 A bounded operator $T \in L(X)$ is said to be *B-Weyl* (respectively, *upper semi B-Weyl, lower semi B-Weyl*) if for some integer $n \geq 0$ $T^n(X)$ is closed and T_n is Weyl (respectively, upper semi-Weyl, lower semi-Weyl). Analogously, $T \in L(X)$ is said to be *B-Browder* (respectively, *upper semi B-Browder, lower semi B-Browder*) if for some integer $n \geq 0$ $T^n(X)$ is closed and T_n is Browder (respectively, upper semi-Browder, lower semi-Weyl).

The classes of operators previously defined generate the following spectra: the *B-Weyl spectrum*, defined as

$$\sigma_{bw}(T) := \{\lambda \in \mathbb{C} : \lambda I - T \text{ is not B-Weyl}\},$$

the *upper semi B-Weyl spectrum*, defined as

$$\sigma_{ubw}(T) := \{\lambda \in \mathbb{C} : \lambda I - T \text{ is not upper semi B-Weyl}\},$$

and the *lower semi B-Weyl spectrum*, defined as

$$\sigma_{lbw}(T) := \{\lambda \in \mathbb{C} : \lambda I - T \text{ is not lower semi B-Weyl}\}.$$

The *B-Browder spectrum* is defined as

$$\sigma_{bb}(T) := \{\lambda \in \mathbb{C} : \lambda I - T \text{ is not B-Browder}\},$$

the *upper semi B-Browder spectrum* is defined as

$$\sigma_{ubb}(T) := \{\lambda \in \mathbb{C} : \lambda I - T \text{ is not upper semi B-Browder}\},$$

and the *lower semi B-Browder spectrum* is defined as

$$\sigma_{lbb}(T) := \{\lambda \in \mathbb{C} : \lambda I - T \text{ is not lower semi B-Browder}\}.$$

Obviously,

$$\sigma_{bw}(T) \subseteq \sigma_w(T) \quad \text{and} \quad \sigma_{ubw}(T) \subseteq \sigma_{uw}(T),$$

and the *B-Fredholm spectrum*, defined as

$$\sigma_{\mathrm{bf}}(T) := \{\lambda \in \mathbb{C} : \lambda I - T \text{ is not B-Fredholm}\},$$

is a subset of the essential spectrum $\sigma_{\mathrm{e}}(T)$.

These inclusions in general are proper. For instance, if $X := \ell_2(\mathbb{N}) \oplus \ell_2(\mathbb{N})$ and $T := 0 \oplus R \in L(X)$, R the right shift, then $\sigma_{\mathrm{lbw}}(T) = \Gamma$, Γ the closed unit circle of $\mathbb{C}$, while $\sigma_{\mathrm{uw}}(T) = \Gamma \cup \{0\}$. If $V \in L(\ell_2(\mathbb{N}))$ is defined as

$$V(x_1, x_2, \ldots) := \left(0, \frac{x_1}{2}, 0, 0\right) \quad \text{for all } (x_1, x_2, \ldots) \in \ell_2(\mathbb{N}),$$

then we have $\sigma_{\mathrm{bf}}(V) = \sigma_{\mathrm{bw}}(V) = \emptyset$, while $\sigma_{\mathrm{e}}(V) = \sigma_{\mathrm{w}}(V) = \{0\}$.

Given $n \in \mathbb{N}$ let us denote by $\widehat{T}_n : X/\ker T^n \to X/\ker T^n$ the quotient map defined canonically by $\widehat{T}_n \hat{x} := \widehat{Tx}$ for each $\hat{x} \in \widehat{X} := X/\ker T^n$, where $x \in \hat{x}$.

Lemma 3.46 *Suppose that $T \in L(X)$ and $T^n(X)$ is closed for some $n \in \mathbb{N}$. If T_n is upper semi-Fredholm then $\widehat{T}_n$ is upper semi-Fredholm and* $\mathrm{ind}\ \widehat{T}_n = \mathrm{ind}\ T_n$. *Analogous statements hold if T_n is assumed to be lower semi-Fredholm, upper or lower semi-Weyl, upper or lower semi-Browder, respectively. Moreover, if T has the SVEP at 0 then $\widehat{T}_n$ also has the SVEP at 0.*

Proof It is easily seen that the operator $[T^n] : X/\ker T^n \to T^n(X)$ defined by

$$[T^n]\hat{x} = T^n x, \quad \text{where } x \in \hat{x}$$

is a bijection. Moreover,

$$[T^n]\widehat{T}_n = T_n[T^n] \quad \text{for all } n \in \mathbb{N}, \tag{3.17}$$

from which the statements easily follow.

If T has the SVEP at 0 then the restriction $T_n = T|T^n(X)$ has the SVEP at 0. By Lemma 2.141 the equality (3.17) entails that $\widehat{T}_n$ also has the SVEP at 0. ∎

Every bounded below operator $T \in L(X)$ is upper semi-Browder, while every surjective operator $T \in L(X)$ is lower semi-Browder, so, by Theorem 1.140, every left Drazin invertible operator is upper semi B-Browder, while every right Drazin invertible operator is lower semi B-Browder. Actually, we have the following equivalences:

Theorem 3.47 *If $T \in L(X)$ then the following equivalences hold:*

(i) *T is upper semi B-Browder $\Leftrightarrow$ T is left Drazin invertible.*
(ii) *T is lower semi B-Browder $\Leftrightarrow$ T if T is right Drazin invertible.*

   Consequently, T is B-Browder if and only if T is Drazin invertible.

Proof

(i) Suppose that T is upper semi B-Browder. By Lemma 3.46, $\widehat{T}_n$ is upper semi-Browder for some $n \in \mathbb{N}$ and hence has uniform topological descent. By Theorem 2.97 the condition $p(\widehat{T}_n) < \infty$ is equivalent to saying that $\sigma_{\mathrm{ap}}(\widehat{T}_n)$ does not cluster at 0. Let $\mathbb{D}(0, \varepsilon)$ be an open disc centered at 0 such that $\mathbb{D}(0, \varepsilon) \setminus \{0\}$ does not contain points of $\sigma_{\mathrm{ap}}(\widehat{T}_n)$, so

$$\ker(\lambda I - \widehat{T}_n) = \{0\} \quad \text{for all } 0 < |\lambda| < \varepsilon. \tag{3.18}$$

Since the restriction $T \,|\, \ker T^n$ is nilpotent we also have that $\mathbb{D}(0, \varepsilon) \setminus \{0\} \subseteq \rho(T \,|\, \ker T^n)$, where $\rho(T \,|\, \ker T^n)$ is the resolvent of $T \,|\, \ker T^n$, hence

$$(\lambda I - T)(\ker T^n) = \ker T^n \quad \text{for all } 0 < |\lambda| < \varepsilon. \tag{3.19}$$

Since for all $0 < |\lambda| < \varepsilon$ we also have $\ker(\lambda I - T) \,|\, \ker T^n = \{0\}$, it then easily follows that $\ker(\lambda I - T) = \{0\}$, so $\lambda I - T$ is injective for all $0 < |\lambda| < \varepsilon$.

We show now that $(\lambda I - T)(X)$ is closed for all $0 < |\lambda| < \varepsilon$. Set $\hat{X} := X/\ker T^n$ and let $w \in (\lambda I - T)(X)$ be arbitrary. Then there exists an $x \in X$ such that $w = (\lambda I - T)x$ and hence

$$\hat{w} = (\lambda I - \widehat{T}_n)\hat{x} \in (\lambda I - \widehat{T}_n)(\hat{X}).$$

Because $\lambda \notin \sigma_{\mathrm{ap}}(\widehat{T}_n)$, $(\lambda I - \widehat{T}_n)(\hat{X})$ is closed, and hence there exists a sequence $(w_n) \subset X$ such that $(\lambda I - \widehat{T}_n)\hat{w}_n \to \hat{w}$ as $n \to +\infty$, thus

$$(\lambda I - T)w_n - w \to z_n \in \ker T^n.$$

From (3.19) we know that there exists a $y_n \in \ker T^n$ such that $z_n = (\lambda I - T)y_n$, and hence

$$(\lambda I - T)w_n - (\lambda I - T)y_n = (\lambda I - T)(w_n - y_n) \to w,$$

so $(\lambda I - T)(X)$ is closed. We have shown that $\lambda I - T$ is bounded below for all $0 < |\lambda| < \varepsilon$ and hence that 0 is an isolated point of $\sigma_{\mathrm{ap}}(T)$. By assumption T is upper semi B-Fredholm, and hence has topological uniform descent. Moreover, $p(T) < \infty$, by Theorem 2.97. From Theorem 1.142 we then conclude that T is left Drazin invertible.

(ii) Let T be lower semi B-Browder and let $n \in \mathbb{N}$ such that $T^n(X)$ is closed and T_n is lower semi-Browder. By Lemma 3.46, $\widehat{T}_n$ is lower semi-Browder and hence, by Theorem 2.98, the condition $q(\widehat{T}_n) < \infty$ is equivalent to saying that $\sigma_{\mathrm{s}}(\widehat{T}_n)$ does not cluster at 0. Let $\mathbb{D}(0, \varepsilon)$ be an open ball centered at 0 such that $\mathbb{D}(0, \varepsilon) \setminus \{0\}$ does not contain points of $\sigma_{\mathrm{s}}(\widehat{T}_n)$. As in the proof of part (i) we have $(\lambda I - T)(\ker T^n) = \ker T^n$ for all $0 < |\lambda| < \varepsilon$. We show that $(\lambda I - T)(X) = X$ for all $0 < |\lambda| < \varepsilon$. Since $\lambda I - \widehat{T}_n$ is onto, for each $x \in X$

there exists a $y \in X$ such that $(\lambda I - \widehat{T_n})\hat{y} = \hat{x}$ and hence

$$x - (\lambda I - T)y \in \ker T^n = (\lambda I - T)(\ker T^n).$$

Consequently, there exists a $z \in \ker T^n$ such that

$$x - (\lambda I - T)y = (\lambda I - T)z,$$

from which it follows that

$$x = (\lambda I - T)(z + y) \in (\lambda I - T)(X).$$

We have proved that $\lambda I - T$ is onto for all $0 < |\lambda| < \varepsilon$, thus $\sigma_s(T)$ does not cluster at 0 and consequently T^* has the SVEP at 0. Every lower semi B-Browder operator has topological uniform descent, so, by Theorem 2.98, T has finite descent $q := q(T)$. Consequently, $T^q(X) = T^\infty(X)$ is closed, by Corollary 2.96, and hence T is right Drazin invertible.   ∎

The *left Drazin spectrum* is defined as

$$\sigma_{\mathrm{ld}}(T) := \{\lambda \in \mathbb{C} : \lambda I - T \text{ is not left Drazin invertible}\},$$

the *right Drazin spectrum* is defined as

$$\sigma_{\mathrm{rd}}(T) := \{\lambda \in \mathbb{C} : \lambda I - T \text{ is not right Drazin invertible}\}.$$

The *Drazin spectrum* is defined as

$$\sigma_{\mathrm{d}}(T) := \{\lambda \in \mathbb{C} : \lambda I - T \text{ is not Drazin invertible}\}.$$

Obviously, $\sigma_{\mathrm{d}}(T) = \sigma_{\mathrm{ld}}(T) \cup \sigma_{\mathrm{rd}}(T)$.

Remark 3.48 If $T \in L(X)$ is meromorphic and the spectrum $\sigma(T)$ is an infinite set, then $\sigma(T)$ clusters at 0, and T is not Drazin invertible. Therefore, $\sigma_{\mathrm{d}}(T) = \{0\}$. Note that for the topological uniform descent spectrum we have $\sigma_{\mathrm{utd}}(T) \subseteq \sigma_{\mathrm{d}}(T)$ for every operator $T \in L(X)$. On the other hand, if T is meromorphic with an infinite spectrum, then the opposite inclusion $\sigma_{\mathrm{d}}(T) \subseteq \sigma_{\mathrm{tud}}(T)$ is also true, so $\sigma_{\mathrm{tud}}(T) = \{0\}$, and hence T does not have topological uniform descent.

From Theorem 1.144 we obtain $\sigma_{\mathrm{d}}(T) = \sigma_{\mathrm{d}}(T^*)$ and

$$\sigma_{\mathrm{ld}}(T) = \sigma_{\mathrm{rd}}(T^*) \quad \text{and} \quad \sigma_{\mathrm{rd}}(T) = \sigma_{\mathrm{ld}}(T^*).$$

An easy consequence is that $T \in L(X)$ is meromorphic if and only if T^* is meromorphic. From Theorem 3.47 the next result immediately follows.

Corollary 3.49 *For every $T \in L(X)$ we have*

$$\sigma_{\text{ubb}}(T) = \sigma_{\text{ld}}(T), \quad \sigma_{\text{lbb}}(T) = \sigma_{\text{rd}}(T), \quad \sigma_{\text{bb}}(T) = \sigma_{\text{d}}(T).$$

The relationship between the B-Browder spectra and the B-Weyl spectra is similar to that observed for the Browder spectra and Weyl spectra, established in Theorem 3.43:

Theorem 3.50 *If $T \in L(X)$ then the following equalities hold:*

(i) $\sigma_{\text{ld}}(T) = \sigma_{\text{ubw}}(T) \cup \text{acc } \sigma_{\text{ap}}(T).$
(ii) $\sigma_{\text{rd}}(T) = \sigma_{\text{lbw}}(T) \cup \text{acc } \sigma_{\text{s}}(T).$
(iii) $\sigma_{\text{d}}(T) = \sigma_{\text{bw}}(T) \cup \text{acc } \sigma(T).$

Proof

(i) The inclusion $\sigma_{\text{ubw}}(T) \subseteq \sigma_{\text{ubb}}(T) = \sigma_{\text{ld}}(T)$ is clear from Corollary 3.49, so, in order to show that $\sigma_{\text{ubw}}(T) \cup \text{acc } \sigma_{\text{ap}}(T) \subseteq \sigma_{\text{ld}}(T)$ it suffices to prove that $\text{acc } \sigma_{\text{ap}}(T) \subseteq \sigma_{\text{ld}}(T)$. For this, let $\lambda \notin \sigma_{\text{ld}}(T)$. Then $\lambda I - T$ is left Drazin invertible, and hence has topological uniform descent. Since $p(\lambda_0 I - T) < \infty$ it then follows, from Corollary 1.92, that $\lambda I - T$ is bounded below in a punctured disc centered at λ_0, so $\lambda \notin \text{acc } \sigma_{\text{ap}}(T)$.

  To show the inclusion $\sigma_{\text{ubw}}(T) \cup \text{acc } \sigma_{\text{ap}}(T) \supseteq \sigma_{\text{ld}}(T)$, consider $\lambda \notin \sigma_{\text{ubw}}(T) \cup \text{acc } \sigma_{\text{ap}}(T)$. Since $\lambda \notin \text{acc } \sigma_{\text{ap}}(T)$ then T has the SVEP at λ. Moreover, since $\lambda I - T$ is upper semi B-Weyl, by Theorem 2.97, we have that $p(\lambda I - T) < \infty$, so $\lambda I - T$ is upper semi B-Browder, or equivalently, left Drazin invertible, and hence $\lambda \notin \sigma_{\text{ld}}(T)$. Hence the equality (i) is proved.

(ii) The proof is similar to part (i). Indeed, by Corollary 3.49, we have $\sigma_{\text{lbw}}(T) \subseteq \sigma_{\text{lbb}}(T) = \sigma_{\text{rd}}(T)$. In order show the inclusion $\sigma_{\text{rd}}(T) \supseteq \sigma_{\text{lbw}}(T) \cup \text{acc } \sigma_{\text{s}}(T)$ we need only to prove that $\text{acc } \sigma_{\text{s}}(T) \subseteq \sigma_{\text{rd}}(T)$. If $\lambda \notin \sigma_{\text{rd}}(T)$ then $\lambda I - T$ is right Drazin invertible, and hence $\lambda I - T$ is semi B-Fredholm with $q(\lambda I - T) < \infty$. By Corollary 1.92 it then follows that $\lambda I - T$ is onto in a punctured disc centered at λ, thus $\lambda \notin \text{acc } \sigma_{\text{s}}(T)$.

  To show the opposite inclusion $\sigma_{\text{rd}}(T) \subseteq \sigma_{\text{lbw}}(T) \cup \text{acc } \sigma_{\text{s}}(T)$, suppose that $\lambda \notin \sigma_{\text{lbw}}(T) \cup \text{acc } \sigma_{\text{s}}(T)$. Since $\lambda \notin \text{acc } \sigma_{\text{s}}(T)$, T^* has the SVEP at λ, and since $\lambda I - T$ is lower semi B-Fredholm, it then follows by Theorem 2.98 that $\lambda I - T$ is lower semi B-Browder, or equivalently, right Drazin invertible, i.e. $\lambda \notin \sigma_{\text{rd}}(T)$. Hence $\sigma_{\text{rd}}(T) \subseteq \sigma_{\text{lbw}}(T) \cup \text{acc } \sigma_{\text{s}}(T)$.

(iii) Clear.   ∎

Evidently, every upper semi-Browder operator is upper semi B-Browder, so, from Corollary 3.49 we obtain that

$$\sigma_{\text{ld}}(T) = \sigma_{\text{ubb}}(T) \subseteq \sigma_{\text{ub}}(T),$$

while

$$\sigma_d(T) = \sigma_{bb}(T) \subseteq \sigma_b(T).$$

The next result shows that $\sigma_{ub}(T)$ may be obtained by adding to $\sigma_{ld}(T)$ the isolated points of $\sigma_{ub}(T)$, and analogous results hold for the other Browder spectra.

Theorem 3.51 *For every $T \in L(X)$ we have:*

  (i) $\operatorname{acc} \sigma_{ub}(T) \subseteq \sigma_{ld}(T)$ *and* $\operatorname{acc} \sigma_b(T) \subseteq \sigma_d(T)$.
 (ii) $\operatorname{iso} \sigma_b(T) \subseteq \operatorname{iso} \sigma_{ub}(T)$ *and* $\operatorname{iso} \sigma_d(T) \subseteq \operatorname{iso} \sigma_{ld}(T)$.
(iii) $\sigma_{ub}(T) = \sigma_{ld}(T) \cup \operatorname{iso} \sigma_{ub}(T)$ *and* $\sigma_b(T) = \sigma_d(T) \cup \operatorname{iso} \sigma_b(T)$.

Proof

  (i) Let $\lambda_0 \notin \sigma_{ld}(T)$. Then $\lambda_0 I - T$ is left Drazin invertible, and hence upper semi B-Browder, by Theorem 3.47. By Theorem 1.117 we see that there exists an $\varepsilon > 0$ such that $\lambda I - T$ is upper semi-Browder in the open punctured disc $\mathbb{D}(\lambda_0, \varepsilon) \setminus \{\lambda_0\}$, thus $\lambda_0 \notin \operatorname{acc} \sigma_{ub}(T)$. The second equality follows by a similar argument.
 (ii) Since $\sigma_{ub}(T)$ is a subset of $\sigma_b(T)$ it suffices to prove that $\operatorname{iso} \sigma_b(T) \subseteq \sigma_{ub}(T)$. Let $\lambda_0 \in \operatorname{iso} \sigma_b(T)$ be arbitrary and suppose that $\lambda_0 \notin \sigma_{ub}(T)$. We can suppose that $\lambda_0 = 0$. Then $T \in B_+(X)$, thus $\alpha(T) < \infty$, and, by the punctured neighborhood theorem, there exists an $\varepsilon > 0$ such that $\lambda I - T$ is Browder for all $0 < |\lambda| < \varepsilon$. In particular, $q(\lambda I - T) < \infty$ for all $0 < |\lambda| < \varepsilon$, so T^* has the SVEP at every point of the punctured disc $\mathbb{D}(0, \varepsilon) \setminus \{0\}$. This implies that T^* has the SVEP at 0 and hence $q(T) < \infty$, by Theorem 2.98. Since $p(T) < \infty$ we then have $p(T) = q(T)$ and from Theorem 1.22 we obtain that $\alpha(T) = \beta(T) < \infty$. Hence $0 \notin \sigma_b(T)$, a contradiction. Therefore, $0 \in \sigma_{ub}(T)$. The second inclusion follows by a similar argument.
(iii) From part (i) we have

$$\sigma_{ub}(T) = \operatorname{acc} \sigma_{ub}(T) \cup \operatorname{iso} \sigma_{ub}(T) \subseteq \sigma_{ld}(T) \cup \operatorname{iso} \sigma_{ub}(T).$$

The opposite inclusion is always true, hence the first equality is proved. The second equality follows by a similar argument.   ∎

Corollary 3.52 *If* $\operatorname{iso} \sigma_{ub}(T) \subseteq \sigma_{ld}(T)$ *then* $\sigma_{ub}(T) = \sigma_{ld}(T)$ *and* $\sigma_b(T) = \sigma_d(T)$.

Proof The equality $\sigma_{ub}(T) = \sigma_{ld}(T)$ is clear from the first equality of (iii) of Theorem 3.51. Part (ii) of Theorem 3.51 entails that

$$\operatorname{iso} \sigma_b(T) \subseteq \operatorname{iso} \sigma_{ub}(T) \subseteq \sigma_{ld}(T) \subseteq \sigma_d(T),$$

thus $\sigma_b(T) = \sigma_d(T)$, again by part (iii) of Theorem 3.51.   ∎

The following result shows that many of the spectra considered before coincide whenever T or T^* has the SVEP.

Theorem 3.53 *Suppose that $T \in L(X)$. Then the following statements hold:*

(i) *If T has the SVEP then*

$$\sigma_{\mathrm{lbw}}(T) = \sigma_{\mathrm{rd}}(T) = \sigma_{\mathrm{d}}(T) = \sigma_{\mathrm{bw}}(T), \tag{3.20}$$

and

$$\sigma_{\mathrm{ubw}}(T) = \sigma_{\mathrm{ld}}(T). \tag{3.21}$$

(ii) *If T^* has the SVEP then*

$$\sigma_{\mathrm{ubw}}(T) = \sigma_{\mathrm{ld}}(T) = \sigma_{\mathrm{d}}(T) = \sigma_{\mathrm{bw}}(T), \tag{3.22}$$

and

$$\sigma_{\mathrm{lbw}}(T) = \sigma_{\mathrm{rd}}(T). \tag{3.23}$$

(iii) *If both T and T^* have the SVEP then*

$$\sigma_{\mathrm{ubw}}(T) = \sigma_{\mathrm{lbw}}(T) = \sigma_{\mathrm{bw}}(T) = \sigma_{\mathrm{ld}}(T) = \sigma_{\mathrm{rd}}(T) = \sigma_{\mathrm{d}}(T). \tag{3.24}$$

Proof

(i) By Corollary 3.49 and we have

$$\sigma_{\mathrm{lbw}}(T) \subseteq \sigma_{\mathrm{lbb}}(T) = \sigma_{\mathrm{rd}}(T) \subseteq \sigma_{\mathrm{d}}(T).$$

We show now that $\sigma_{\mathrm{d}}(T) \subseteq \sigma_{\mathrm{lbw}}(T)$. Assume that $\lambda \notin \sigma_{\mathrm{lbw}}(T)$. We may assume $\lambda = 0$. Then there exists an $n \in \mathbb{N}$ such that $T^n(X)$ is closed, T_n is lower semi-Fredholm and $\operatorname{ind} T_n \geq 0$. Since T has the SVEP at 0 then, by Theorem 2.97, we also have $p(T) < \infty$ and hence, by Lemma 1.23, there exists a $k \in \mathbb{N}$ such that T_k is bounded below, in particular $\operatorname{ind} T_k \leq 0$. Now, if $m = \max\{n, k\}$, then $\operatorname{ind} T_m = \operatorname{ind} T_n = \operatorname{ind} T_k = 0$, and since $\ker T_{j+1} \subseteq \ker T_j$ for all $j \in \mathbb{N}$, we also have $p(T_m) = 0$. By Theorem 1.22, it then follows that $q(T_m) = 0$, so T_m is Browder and hence T is B-Browder. By part (iii) of Theorem 3.47, T is Drazin invertible, so $0 \notin \sigma_{\mathrm{d}}(T)$, as desired. Thus, $\sigma_{\mathrm{lbw}}(T) = \sigma_{\mathrm{rd}}(T) = \sigma_{\mathrm{d}}(T)$. Clearly, $\sigma_{\mathrm{bw}}(T) \subseteq \sigma_{\mathrm{bb}}(T) = \sigma_{\mathrm{d}}(T)$, by Corollary 3.49. Suppose that $\lambda \notin \sigma_{\mathrm{bw}}(T)$, i.e., there exists a $k \in \mathbb{N}$ such that $(\lambda I - T)^n(X)$ is closed and $\alpha(\lambda I - T_n) = \beta(\lambda I - T_n) < \infty$ for all $n \geq k$. Since T has the SVEP at λ, $p := p(\lambda I - T) < \infty$, by Theorem 2.97, hence, see Remark 1.25, $\lambda I - T_n$ is injective for all $n \geq p$. Therefore, by Theorem 1.22, for n sufficiently large we have $q(\lambda I - T_n) = p(\lambda I - T_n) = 0$, so $\lambda I - T_n$ is invertible, hence Browder. Therefore, by Corollary 3.49, $\lambda \notin \sigma_d(T)$ and hence $\sigma_d(T) \subseteq \sigma_{lbw}(T)$.

To conclude the proof we need only to show that $\sigma_d(T) = \sigma_{bw}(T)$. The inclusion $\sigma_{bw}(T) \subseteq \sigma_d(T)$ follows from Theorem 1.141. Suppose that $\lambda \notin \sigma_{bw}(T)$. Then $\lambda I - T$ is B-Weyl and the SVEP for T at λ implies, by Theorem 2.97, that $p(\lambda I - T) < \infty$. From part (iii) of Theorem 1.143 we then conclude that $\lambda \notin \sigma_d(T)$, so $\sigma_d(T) \subseteq \sigma_{bw}(T)$ and hence $\sigma_d(T) = \sigma_{bw}(T)$.

To show the equalities (3.22), note first that the inclusion $\sigma_{ubw}(T) \subseteq \sigma_{ld}(T)$ holds for every operator. Let $\lambda \notin \sigma_{ubw}(T)$. Then $\lambda I - T$ is upper semi B-Weyl and from Theorem 2.97 the SVEP for T implies that $\lambda I - T$ is left Drazin invertible. Hence, $\lambda \notin \sigma_{ld}(T)$, thus $\sigma_{ld}(T) \subseteq \sigma_{ubw}(T)$.

(ii) The inclusion

$$\sigma_{lbw}(T) \subseteq \sigma_{lbb}(T) = \sigma_{rd}(T) \subseteq \sigma_d(T)$$

holds for every $T \in L(X)$, by Theorem 3.50 and Corollary 3.49.

We show that $\sigma_d(T) \subseteq \sigma_{ubw}(T)$. Suppose that $\lambda \notin \sigma_{ubw}(T)$ and assume that $\lambda = 0$. Then there exists an $n \in \mathbb{N}$ such that T_n is upper semi-Fredholm with ind $T_n \leq 0$. Since T^* has the SVEP at 0, by Theorem 2.98 we have $q(T) < \infty$ and hence T_k is onto for some $k \in \mathbb{N}$, by Lemma 1.24. Clearly, ind $T_k \geq 0$. For n sufficiently large we then have ind $T_n = 0$ and $q(T_n) = 0$. By Theorem 1.22, it then follows that $p(T_n) = 0$, so that T_n is Browder and hence T is B-Browder, or equivalently T is Drazin invertible. Therefore $0 \notin \sigma_d(T)$, as desired. Also here, to finish the proof, we have to prove that $\sigma_d(T) = \sigma_{bw}(T)$. The inclusion $\sigma_{bw}(T) \subseteq \sigma_d(T)$ follows from Theorem 1.141. Suppose that $\lambda \notin \sigma_{bw}(T)$. Then $\lambda I - T$ is B-Weyl and the SVEP for T^* at λ implies, by Theorem 2.98, that $q(\lambda I - T) < \infty$. From part (iii) of Theorem 1.143 we then conclude that $\lambda \notin \sigma_d(T)$, so $\sigma_d(T) = \sigma_{bw}(T)$.

To show (3.23), note that $\sigma_{lbw}(T) \subseteq \sigma_{rd}(T)$ holds for every operator. Let $\lambda \notin \sigma_{lbw}(T)$. Then $\lambda I - T$ is lower semi B-Weyl and from Theorem 2.98 the SVEP for T^* implies that $\lambda I - T$ is right Drazin invertible. Hence, $\lambda \notin \sigma_{rd}(T)$, so $\sigma_{rd}(T) \subseteq \sigma_{lbw}(T)$.

(iii) Clear from part (i) and part (ii).   ■

It makes sense to ask if the Drazin spectra, and the B-Weyl spectra are also stable under commuting Riesz perturbations. The following example shows that the answer in the case of Riesz operators is negative, even in the simple case of a commuting quasi-nilpotent perturbation.

Example 3.54 Let $X := \ell^2(\mathbb{N})$ and $\{e_i\}$ the canonical basis of X. Denote by P the orthogonal projection of X on the subspace generated by the set $\{e_1 : 1 \leq i \leq n\}$ and let S be the quasi-nilpotent operator defined by

$$S(x_1, x_2, \dots) := \left(\frac{x_2}{2}, \frac{x_3}{3}, \dots \right) \quad \text{for all } x = (x_1, x_2, \dots) \in \ell^2(\mathbb{N}).$$

If $T := 0 \oplus P \in L(X \oplus X)$ then T has the SVEP, since P is finite-dimensional, hence, from part (i) of Theorem 3.53 we have

$$\sigma_{\mathrm{ubw}}(T) = \sigma_{\mathrm{ld}}(T) = \sigma_{\mathrm{d}}(T) = \sigma_{\mathrm{bw}}(T) = \emptyset.$$

Consider the quasi-nilpotent operator $Q = S \oplus 0$. Clearly, Q has infinite ascent, and $TQ = QT = 0$. It is easily seen that

$$\sigma_{\mathrm{ubw}}(T + Q) = \sigma_{\mathrm{ld}}(T + Q) = \sigma_{\mathrm{d}}(T + Q) = \sigma_{\mathrm{bw}}(T + Q) = \{0\}.$$

We want show now that the stability of the Drazin spectra, as well as the B-Weyl spectra, under Riesz commuting perturbations hold in some special cases. First we need the following theorem.

Theorem 3.55 *If $T \in L(X)$ the following statements hold:*

(i) $\sigma_{\mathrm{uw}}(T) = \sigma_{\mathrm{ubw}}(T) \cup \mathrm{iso}\,\sigma_{\mathrm{uw}}(T)$.
(ii) $\sigma_{\mathrm{w}}(T) = \sigma_{\mathrm{bw}}(T) \cup \mathrm{iso}\,\sigma_{\mathrm{w}}(T)$.
(iii) $\sigma_{\mathrm{e}}(T) = \sigma_{\mathrm{bf}}(T) \cup \mathrm{iso}\,\sigma_{\mathrm{e}}(T)$.

Proof (i) Let $\lambda_0 \in \sigma_{\mathrm{uw}}(T) \setminus \sigma_{\mathrm{ubw}}(T)$. Then $\lambda_0 I - T$ is upper semi B-Weyl, and hence, by Theorem 1.117, there exists an $\varepsilon > 0$ such that $\lambda I - T$ is upper semi-Weyl for all $|\lambda - \lambda_0| < \varepsilon$. Therefore, $\lambda_0 \in \mathrm{iso}\,\sigma_{\mathrm{uw}}(T)$, so the inclusion $\subseteq$ holds in (i). The opposite inclusion is always true, thus (i) is proved.

(ii) and (iii) Use a similar argument. ∎

The following corollary is an easy consequence of Theorem 3.55.

Corollary 3.56 *If $\mathrm{iso}\,\sigma_{\mathrm{uw}}(T) \subseteq \sigma_{\mathrm{ubw}}(T)$ then $\sigma_{\mathrm{uw}}(T) = \sigma_{\mathrm{ubw}}(T)$. Analogously, if $\mathrm{iso}\,\sigma_{\mathrm{w}}(T) \subseteq \sigma_{\mathrm{bw}}(T)$ then $\sigma_{\mathrm{w}}(T) = \sigma_{\mathrm{bw}}(T)$.*

For an operator $T \in L(X)$ set

$$\rho_{\mathrm{sf}}^+(T) := \{\lambda \in \mathbb{C} : \lambda I - T \in \Phi_\pm(X),\ \mathrm{ind}\,(\lambda I - T) > 0\},$$

and

$$\rho_{\mathrm{sf}}^-(T) := \{\lambda \in \mathbb{C} : \lambda I - T \in \Phi_\pm(X),\ \mathrm{ind}\,(\lambda I - T) < 0\}.$$

If, as usual, $\sigma_{\mathrm{sf}}(T)$ denotes the semi-Fredholm spectrum of T we have:

Lemma 3.57 *If $T \in L(X)$ then*

(i) $\sigma_{\mathrm{w}}(T) = \sigma_{\mathrm{sf}}(T) \cup \rho_{\mathrm{sf}}^+(T) \cup \rho_{\mathrm{sf}}^-(T)$.
(ii) $\sigma_{\mathrm{uw}}(T) = \sigma_{\mathrm{sf}}(T) \cup \rho_{\mathrm{sf}}^+(T)$.
(iii) $\sigma_{\mathrm{lw}}(T) = \sigma_{\mathrm{sf}}(T) \cup \rho_{\mathrm{sf}}^-(T)$.

Proof

(i) The inclusion ($\subseteq$) is evident. Conversely, if $\lambda \notin \sigma_w(T)$, then $\lambda I - T \in W(X)$, so $\lambda \notin \sigma_{sf}(T) \cup \rho_{sf}^+(T) \cup \rho_{sf}^-(T)$.

(ii) The inclusion $\sigma_{sf}(T) \cup \rho_{sf}^+(T) \subseteq \sigma_{uw}(T)$ is clear. Conversely, suppose that $\lambda \notin \sigma_{sf}(T) \cup \rho_{sf}^+(T)$. Then $\lambda I - T \in \Phi_\pm(X)$ and ind $(\lambda I - T) \leq 0$, so $\alpha(\lambda I - T) \leq \beta(\lambda I - T)$, which obviously implies that $\lambda I - T \in \Phi_+(X)$ and hence $\lambda \notin \sigma_{uw}(T)$. Therefore, the equality (i) holds.

(iii) The proof is analogous to that of part (ii). ∎

The isolated points of the Weyl spectra are related as follows:

Theorem 3.58 *If $T \in L(X)$ we have*

(i) iso $\sigma_w(T) \subseteq$ iso $\sigma_{uw}(T) \subseteq$ iso $\sigma_{sf}(T)$.

(ii) iso $\sigma_w(T) \subseteq$ iso $\sigma_{lw}(T) \subseteq$ iso $\sigma_{sf}(T)$.

Proof We prove only the inclusions (i). Let $\lambda_0 \in$ iso $\sigma_w(T)$. Then there exists an $\varepsilon > 0$ such that $\lambda I - T \in W(X)$ for all $0 < |\lambda| < \varepsilon$. This easily implies that $\lambda_0 \in \sigma_{sf}(T)$. In fact, if not, then $\lambda_0 I - T \in \Phi_+(X)$. By the continuity of the index function we then obtain ind $(\lambda_0 I - T) = 0$, i.e. $\lambda_0 \notin \sigma_w(T)$, a contradiction. Since, by Lemma 3.57, we have $\sigma_{uw}(T) = \sigma_{sf}(T) \cup \rho_{sf}^+(T)$, then $\mathbb{D}(\lambda_0, \varepsilon) \cap \sigma_{uw}(T) = \{\lambda_0\}$, so $\lambda_0 \in$ iso $\sigma_{uw}(T)$.

Now, choose an arbitrary $\mu_0 \in$ iso $\sigma_{uw}(T)$. To show the second inclusion it suffices to prove that $\mu_0 \in \sigma_{sf}(T)$. Assume that $\mu_0 \notin \sigma_{sf}(T)$. Then $\mu_0 I - T \in \Phi_\pm(X)$, and, since $\Phi_\pm(X)$ is an open subset of $L(X)$, there exists a $\delta > 0$ such that $\mu I - T \in \Phi_\pm(X)$ for all $\mu \in \mathbb{D}(\mu_0, \delta)$. Again by the continuity of the index function, there exists an $n \in \mathbb{Z} \cup \{-\infty, +\infty\}$ such that ind $(\mu I - T) = n$ for all $\mu \in \mathbb{D}(\mu_0, \delta)$. Note that $\sigma_{uw}(T) = \sigma_{sf}(T) \cup \rho_{sf}^+(T)$, by Lemma 3.57. If $n \leq 0$ then $\mu_o \notin \sigma_{uw}(T)$, a contradiction. If $n > 0$ then $\mathbb{D}(\mu_0, \delta) \subseteq \rho_{uw}(T) = \mathbb{C} \setminus \sigma_{uw}(T)$, and hence μ_0 is an interior point of $\sigma_{uw}(T)$, again a contradiction. Therefore, $\mu_0 \in \sigma_{sf}(T)$. ∎

We now consider the case when the T is perturbed by a commuting Riesz operator.

Theorem 3.59 *Let $T \in L(X)$ and let $R \in L(X)$ be a Riesz operator which commutes with T. Then we have:*

(i) *If iso $\sigma_w(T) = \emptyset$ then*

$$\sigma_{bw}(T + R) = \sigma_{bw}(T).$$

(ii) *If iso $\sigma_{uw}(T) = \emptyset$ then*

$$\sigma_{ubw}(T + R) = \sigma_{ubw}(T) \text{ and } \sigma_{bw}(T + R) = \sigma_{bw}(T).$$

(iii) *If* iso $\sigma_e(T) = \emptyset$ *then*

$$\sigma_{bf}(T + R) = \sigma_{bf}(T).$$

(iv) *If* iso $\sigma_b(T) = \emptyset$ *then*

$$\sigma_d(T + R) = \sigma_d(T).$$

(v) *If* iso $\sigma_{ub}(T) = \emptyset$ *then*

$$\sigma_{ld}(T + R) = \sigma_{ld}(T).$$

Proof (i) By Corollary 3.18 we have $\sigma_w(T + R) = \sigma_w(T)$ and hence iso $\sigma_w(T) =$ iso $\sigma_w(T+R) = \emptyset$. From Corollary 3.56 we then obtain $\sigma_{bw}(T+R) = \sigma_w(T+R) = \sigma_w(T) = \sigma_{bw}(T)$.

The proofs of (ii), (iii), (iv) and (v) are similar to that of part (i). Clearly, the equality $\sigma_{bw}(T + R) = \sigma_{bw}(T)$ in (ii) is a consequence of Theorem 3.58, part (i). ∎

Corollary 3.60 *Suppose that* $T \in L(X)$ *has the SVEP,* $R \in L(X)$ *is a Riesz operator which commutes with* T *and* iso $\sigma_b(T) = \emptyset$. *Then*

$$\sigma_d(T) = \sigma_d(T + R) = \sigma_{ld}(T) = \sigma_{ld}(T + R) = \sigma_{ubw}(T)$$
$$= \sigma_{ubw}(T + R) = \sigma_{bw}(T) = \sigma_{bw}(T + R).$$

Proof By Corollary 3.9 we have iso $\sigma_b(T + R) = \sigma_b(T) = \emptyset$, and $T + R$ has the SVEP, by Theorem 2.129. The equalities then follow from Theorems 3.53 and 3.59. ∎

3.4 Meromorphic Operators

An operator $T \in L(X)$ is *meromorphic* if all the non-zero spectral points are poles of the resolvent. Evidently, an operator $T \in L(X)$ is meromorphic if and only if $\sigma_d(T) \subseteq \{0\}$. An obvious consequence is that every meromorphic operator possesses at most countably many spectral points, which can cluster only at 0. Obviously, Riesz operators, and in particular compact operators, are meromorphic. Moreover,

$$T \text{ is meromorphic} \Leftrightarrow T^* \text{ is meromorphic}.$$

Note that, in contrast with Riesz operators, the sum of commuting meromorphic operators may not be meromorphic. For instance, the identity I and a quasi-nilpotent operator are trivially meromorphic and commute, while $I + Q$ is not meromorphic. The product of two commuting operators, one of which is meromorphic, may not be meromorphic. For instance, the operators I and $I + Q$ above commute and $I + Q = I(I + Q)$ is not meromorphic.

Theorem 3.61 *If $T \in L(X)$ then the following statements are equivalent:*

  (i) *T is meromorphic;*
  (ii) *$\lambda I - T$ is left Drazin invertible for all $\lambda \neq 0$;*
 (iii) *$\lambda I - T$ is right Drazin invertible for all $\lambda \neq 0$;*
 (iv) *$\lambda I - T$ is upper semi B-Fredholm for all $\lambda \neq 0$;*
  (v) *$\lambda I - T$ is lower semi B-Fredholm for all $\lambda \neq 0$;*
 (vi) *$\lambda I - T$ is quasi-Fredholm for all $\lambda \neq 0$;*
(vii) *$\lambda I - T$ has topological uniform descent for all $\lambda \neq 0$.*

Proof If T is meromorphic then $\lambda I - T$ is Drazin invertible, and hence left Drazin invertible, for all $\lambda \neq 0$. Hence (i) $\Rightarrow$ (ii). Clearly, (ii) $\Rightarrow$ (iv) since every left Drazin invertible operator is upper semi B-Fredholm, and (iv) $\Rightarrow$ (vi) $\Rightarrow$ (vii), since every upper semi B-Fredholm is quasi-Fredholm and has topological uniform descent. Analogously, (i) $\Rightarrow$ (iii) $\Rightarrow$ (vi) $\Rightarrow$ (vii).

(vii) $\Rightarrow$ (i) If $\lambda I - T$ has topological uniform descent for all $\lambda \neq 0$ then $\mathbb{C} \setminus \{0\}$ is contained in a component of the topological uniform descent resolvent which intersects the resolvent. By Corollary 2.125, $\lambda I - T$ is Drazin invertible for all $\lambda \neq 0$. ∎

Corollary 3.62 *If $T \in L(X)$ is meromorphic and $K \in L(X)$ is a finite-dimensional operator then $T + K$ is meromorphic.*

Proof If T is meromorphic, $\lambda I - T$ is quasi-Fredholm for every $\lambda \neq 0$, hence $\lambda I - (T + K)$ is quasi-Fredholm, by Theorem 1.110, and this is equivalent to saying that $T + K$ is meromorphic, by Theorem 3.61. ∎

As before, if M is a closed subspace invariant under $T \in L(X)$, we denote by $\widehat{T}_M$ the quotient operator induced by T on X/M defined by $\widehat{T}_M \hat{x} = \widehat{Tx}$ for every $\hat{x} \in \widehat{X} := X/M$. The spectra of the three operators T, $T|M$ and $\widehat{T}_M$ are related as follows:

Theorem 3.63 *If $T \in L(X)$ then we have:*

  (i) $\sigma(T) \subseteq \sigma(T|M) \cup \sigma(\widehat{T}_M)$.
 (ii) $\sigma(T|M) \subseteq \sigma(T) \cup \sigma(\widehat{T}_M)$.
(iii) $\sigma(\widehat{T}_M) \subseteq \sigma(T) \cup \sigma(T|M)$.

Proof

  (i) Suppose that $\lambda \notin \sigma(T|M) \cup \sigma(\widehat{T}_M)$. We may assume $\lambda = 0$. Then $T|M$ and $\widehat{T}_M$ are invertible. First we show that T is onto. Let $y \in X$. Then there exists an $x \in X$ such that $\widehat{T}_M \hat{x} = \hat{y}$, thus $y - Tx \in M$. Since $T|M$ is surjective, there exists $w \in M$ such that $Tw = y - Tx$, hence $T(w + x) = y$, so T

is surjective. Next, to show that T is injective, assume that $Tx = 0$. Then $\widehat{T}_M \hat{x} = \hat{0}$, so $x \in M$. Since $T|M$ is invertible and $T|Mx = 0$, we then have $x = 0$. Hence $0 \notin \sigma(T)$, thus the inclusion (i) is proved.

(ii) Suppose that $\lambda \notin \sigma(T) \cup \sigma(\widehat{T}_M)$. We may suppose $\lambda = 0$. Then T is invertible, so $T|M$ is bounded below. Let $y \in M$ be arbitrary. Then there exists an $x \in X$ such that $y = Tx$. Clearly, $\widehat{T}_M \hat{x} = \widehat{Tx} = \hat{Y}$, and since $y \in M$ we then have $\hat{y} = \hat{0}$, i.e., $y \in M$. Hence $y = T|Mx$, so $T|M$ is onto and consequently, $0 \notin \sigma(T|M)$.

(iii) Let $\pi : X \rightarrow X/M$ denote the quotient homomorphism. Evidently, $\pi T = \widehat{T}_M \pi$. Let $\lambda \notin \sigma(T) \cup \sigma(T|M)$ be arbitrary. Also here we can assume $\lambda = 0$. Then $\widehat{T}_M$ is onto, since both π and T are onto. If $x \in X$ satisfies $\widehat{T}_M \hat{x} = \widehat{T}_M \pi x = 0$, then $Tx \in M$, hence $x \in M$, because $0 \notin \sigma(T|M)$. Therefore, $\widehat{T}_M$ is invertible, i.e. $0 \notin \sigma(\widehat{T}_M)$. ■

Corollary 3.64 *If* $T \in L(X)$ *then:*

(i) $\sigma(T|M) \cup \sigma(\widehat{T}_M) = \sigma(T) \cup \{\sigma(T|M) \cap \sigma(\widehat{T}_M)\}$.
(ii) $\sigma(T) \cup \sigma(\widehat{T}_M) = \sigma(T|M) \cup \{\sigma(T) \cap \sigma(\widehat{T}_M)\}$.
(iii) $\sigma(T) \cup \sigma(T|M) = \sigma(\widehat{T}_M) \cup \{\sigma(T) \cap \sigma(T|M)\}$.

Consequently, if λ is an isolated point in the spectra of any two of the operators T, $T|M$, $\widehat{T}_M$, then λ is an isolated point of the spectrum of the third one or is not in its spectrum.

It is meaningful to find necessary or sufficient conditions under which the spectrum of T coincides with the union of the spectra of $T|M$ and $\widehat{T}_M$. Clearly, the equality $\sigma(T) = \sigma(T|M) \cup \sigma(\widehat{T}_M)$ holds if $\sigma(T|M) \cap \sigma(\widehat{T}_M) = \emptyset$, or $\sigma(T|M) \subseteq \sigma(T)$, or $\sigma(\widehat{T}_M) \subseteq \sigma(T)$. Another important condition which entails this equality is given in the following theorem.

Theorem 3.65 *Let* $T \in L(X)$ *and M be a closed T-invariant subspace of X. Then we have:*

(i) *$T|M$ and $\widehat{T}_M$ are invertible if and only if T is invertible and M is invariant under T^{-1}.*
(ii) *If M is a closed subspace of X hyperinvariant under T then $\sigma(T) = \sigma(T|M) \cup \sigma(\widehat{T}_M)$.*

Proof

(i) By part (i) of Theorem 3.63, if $T|M$ and $\widehat{T}_M$ are invertible then T is invertible. Suppose that $w \in M$. Then there exists an $x \in M$ such that $T|M x = w$, thus $T^{-1}(M) \subseteq M$. Conversely, assume that T is invertible and $T^{-1}(M) \subseteq M$. Clearly the restriction $T^{-1}|M$ is the inverse of $T|M$ and the quotient operator $\widehat{T_M^{-1}}$, induced by T^{-1} on X/M, is the inverse of $\widehat{T}_M$.

(ii) Clear, from part (i). ■

In [115] S.V. Djordjević and B.P. Duggal showed that this equality $\sigma(T) = \sigma(T|M) \cup \sigma(\widehat{T}_M)$ also holds if we assume that $(T|M)^*$ or $\widehat{T}_M$ has the SVEP. We mention that similar results to those of Theorem 3.63 and Corollary 3.64 hold for the essential spectra and the Browder spectra of T, $T|M$ and $\widehat{T}_M$, see [115].

It is easily seen that the ascent and descent of the three operators are related as follows:

$$p(T|M) \leq p(T) \leq p(T|M) + p(\widehat{T}_M), \tag{3.25}$$

and

$$q(\widehat{T}_M) \leq q(T) \leq q(T|M) + q(\widehat{T}_M). \tag{3.26}$$

The next theorem shows that the property of being meromorphic for T, $T|M$ and $\widehat{T}_M$ are strongly related.

Theorem 3.66 *Let $T \in L(X)$. Then we have:*

(i) *If $T := \oplus_{k=1}^{n} T_k$. Then T is meromorphic if and only if T_k is meromorphic for all $1 \leq k \leq n$.*
(ii) *T is meromorphic if and only if the restriction $T|M$ and $\widehat{T}_M$ are both meromorphic.*

Proof

(i) We have $\sigma(T) = \cup_{k=1}^{n} \sigma(T_k)$. For all $1 \leq j \leq n$ and $\lambda \in \mathbb{C}$ we have

$$p(\lambda I_j - T_j) \leq p(\lambda I - T) \leq \sum_{k=1}^{n} p(\lambda_k I_k - T_k)$$

and

$$q(\lambda I_j - I_j) \leq q(\lambda I - T) \leq \sum_{k=1}^{n} q(\lambda_k I_k - T_k),$$

see [291, Exercise 7, p. 293], so the statement follows.
(ii) The implication ($\Leftarrow$) is clear from the inequalities (3.25) and (3.26). Suppose that T is meromorphic. We show first that $(\lambda I - T)(M) = M$ for all $\lambda \in \rho(T)$. The inclusion $(\lambda I - T)(M) \subseteq M$ is clear. Let $|\lambda| > r(T)$. If $R_\lambda := (\lambda I - T)^{-1}$, from the well-known representation

$$R_\lambda = \sum_{n=0}^{\infty} T^n / \lambda^{n+1}$$

it easily follows that $R_\lambda(M) \subseteq M$. For every $x' \in M^\perp$ and $x \in M$ let us consider the analytic function $\lambda \in \rho(T) \to x'(R_\lambda x)$. This function vanishes outside the spectral disc of T, so, since $\rho(T)$ is connected, we infer from the identity theorem for analytic functions that $x'(R_\lambda x) = 0$ for all $\lambda \in \rho(T)$. Therefore $R_\lambda x \in M^{\perp\perp} = M$ and consequently

$$x = (\lambda I - T)R_\lambda x \in (\lambda I - T)(M).$$

This shows that $M \subseteq (\lambda I - T)(M)$, and hence $(\lambda I - T)(M) = M$ for all $\lambda \in \rho(T)$.

Now, $\lambda I - T$ is injective for all $\lambda \in \rho(T)$, so $\rho(T) \subseteq \rho(T|M)$ and hence

$$\sigma_{ap}(T|M) = \sigma(T|M) \subseteq \sigma(T) = \sigma_{ap}(T).$$

Now, let $0 \neq \lambda \in \sigma(T|M)$. Then λ is isolated in $\sigma(T)$ and hence isolated in $\sigma(T|M)$ and $p(\lambda I - T|M) \leq p(\lambda I - T) < \infty$. Observe that λ is isolated in $\sigma(T)$ and $\sigma(T|M)$ forces λ to be isolated in $\sigma(\widehat{T}_M)$. Thus, since $q(\lambda I - \widehat{T}_M) \leq q(\lambda I - T) < \infty$, $\lambda I - \widehat{T}_M$ is Drazin invertible (trivially, if $\lambda \in \sigma(\widehat{T}_M)$, $\lambda I - \widehat{T}_M$ is invertible). This forces $\lambda I - T|M$ to be Drazin invertible. Consequently, $T|M$ is meromorphic.

To show that $\widehat{T}_M$ is meromorphic, observe that $\lambda I - T^*$ is Drazin invertible for every $\lambda \neq 0$, and an argument similar to the above implies that the restriction of T^* to the annihilator $M^\perp$ has spectrum $\sigma(T^*|M^\perp) \subseteq \sigma(T^*)$ and $\lambda I - T^*|M^\perp$ is Drazin invertible for all $0 \neq \lambda \in \sigma(T^*|M^\perp)$. Identifying the dual of $\widehat{T}_M$ with $T^*|M^\perp$ it follows that $\sigma(\widehat{T}_M) = \sigma(\widehat{T}_M)^* \subseteq \sigma(T^*)$ and $(\widehat{T}_M)^*$ is meromorphic, which implies that the bidual of $\widehat{T}_M$ is also meromorphic. Finally, identifying $\widehat{T}_M$ with the restriction of the bidual of $\widehat{T}_M$ on X/M (X/M is a closed subspace of $(X/M)^{**}$) under $(\widehat{T}_M)^{**}$ it then follows that $\widehat{T}_M$ is meromorphic. ∎

3.5 Algebraic Operators

In this section we study in more detail the class of algebraic operators, i.e., those operators $T \in L(X)$ for which there exists a nontrivial polynomial h such that $h(T) = 0$. The algebraic operators find many applications in applied linear algebra and they have been investigated by several mathematicians, see for instance Aupetit [53].

We shall need the following lemma.

Lemma 3.67 *Let $T \in L(X)$ and let $\{\lambda_1, \cdots, \lambda_k\}$ be a finite subset of $\mathbb{C}$ such that $\lambda_i \neq \lambda_j$ for $i \neq j$. Assume that $\{n_1, \cdots, n_k\} \subset \mathbb{N}$ and set $p(\lambda) := \prod_{i=1}^{k}(\lambda_i - \lambda)^{n_i}$. Then*

$$\ker p(T) = \bigoplus_{i=1}^{k} \ker(\lambda_i I - T)^{n_i} \tag{3.27}$$

and

$$p(T)(X) = \bigcap_{i=1}^{k} (\lambda_i I - T)^{n_i}(X). \tag{3.28}$$

Proof We shall show (3.27) for $k = 2$ and the general case then follows by induction. Clearly ker $(\lambda_i I - T)^{n_i} \subseteq$ ker $p(T)$ for $i = 1, 2$, so that if $p_i(T) := (\lambda_i I - T)^{n_i}$ then

$$\text{ker } p_1(T) + \text{ker } p_2(T) \subseteq \text{ker } p(T).$$

In order to show the converse inclusion, observe first that p_1, p_2 are relatively prime, hence, by Lemma 1.13, there exist two polynomials q_1, q_2 such that

$$q_1(T)p_1(T) + q_2(T)p_2(T) = I.$$

Consequently, every $x \in X$ admits the decomposition

$$x = q_1(T)p_1(T)x + q_2(T)p_2(T)x. \tag{3.29}$$

Now if $x \in$ ker $p(T)$ then

$$0 = p(T)x = p_1(T)p_2(T)x = p_2(T)p_1(T)x = p(T)x,$$

from which we deduce that $p_2(T)x \in$ ker $p_1(T)$ and $p_1(T)x \in$ ker $p_2(T)$. Moreover, since every polynomial in T maps the subspaces ker $(\lambda_i I - T)^{n_i}$ into themselves, we have

$$x_1 := q_2(T)p_2(T)x \in \text{ker } p_1(T)$$

while

$$x_2 := q_1(T)p_1(T)x \in \text{ker } p_2(T).$$

From the equality (3.29) we have $x = x_1 + x_2$, so

$$\text{ker } p(T) \subseteq \text{ker } p_1(T) + \text{ker } p_2(T).$$

Therefore ker $p(T) = $ ker $p_1(T) + $ ker $p_2(T)$. It only remains to prove that ker $p_1(T) \cap$ ker $p_2(T) = \{0\}$. This is an immediate consequence of the identity (3.29).

As before we prove (3.27) only for $n = 2$. Evidently,

$$p(T)(X) = p_1(T)p_2(T)(X) = p_2(T)p_1(T)(X)$$

is a subset of $p_1(T)(X)$, as well as a subset of $p_2(T)(X)$.

Conversely, suppose that $x \in p_1(T)(X) \cap p_2(T)(X)$ and let $y \in X$ such that $x = p_2(T)y$. Then

$$p_1(T)x = p(T)y \in p(T)(X).$$

Analogously $p_2(T)x \in p(T)(X)$. Let q_1, q_2 be two polynomials for which the equality (3.29) holds. Then we have

$$x = q_1(T)p_1(T)x + q_2(T)p_2(T)x \in q_1(T)p(T)(X) + q_2(T)p(T)(X)$$
$$= p(T)q_1(T)(X) + p(T)q_2(T)(X) \subseteq p(T)(X) + p(T)(X) \subseteq p(T)(X).$$

Hence

$$p_1(T)(X) \cap p_2(T)(X) \subseteq p(T)(X),$$

and this completes the proof.   ∎

Algebraic operators may be characterized in several ways:

Theorem 3.68 *If $T \in L(X)$ the following assertions are equivalent:*

(i) $\lambda I - T$ *is essentially right Drazin invertible for all $\lambda \in \mathbb{C}$;*
(ii) $\lambda I - T$ *has finite descent for all $\lambda \in \mathbb{C}$;*
(iii) $\lambda I - T$ *has finite descent for all λ which belong to the boundary of the spectrum $\partial\sigma(T)$;*
(iv) *The spectrum is a finite set of poles;*
(v) *T is algebraic.*

Proof (i) $\Rightarrow$ (ii) By duality we know that $\lambda I^* - T^*$ is essentially left Drazin invertible, hence upper semi B-Fredholm, for all $\lambda \in \mathbb{C}$. Therefore, in the denotation of Theorem 2.116, the set $\Psi_u(T^*)$ has a unique component and consequently T^* has the SVEP. Since $\lambda I - T$ is quasi-Fredholm, hence has topological uniform descent, it then follows, by Theorem 2.98 that the descent $q(\lambda I - T)$ is finite for all $\lambda \in \mathbb{C}$.

(ii)$\Rightarrow$ (iii) is obvious.

(iii) $\Rightarrow$ (iv) Observe first that T has the SVEP at every $\lambda \in \partial\sigma(T)$, so, by Theorem 2.109, $p(\lambda I - T) < \infty$ for every $\lambda \in \partial\sigma(T)$, so λ is a pole, hence an isolated point of $\sigma(T)$. This implies that $\sigma(T) = \partial\sigma(T)$, hence $\sigma(T)$ is a finite set of poles.

(iv) $\Rightarrow$ (i) It suffices to prove that $\lambda I - T$ is semi B-Fredholm for all $\lambda \in \sigma(T)$. Suppose that $\sigma(T)$ is a finite set of poles of $R(\lambda, T)$. If $\lambda \in \sigma(T)$, let P be the spectral projection associated with the singleton $\{\lambda\}$. Then $X = M \oplus N$, where $M := K(\lambda I - T) = \ker P$ and $N := H_0(\lambda I - T)$, by Theorem 2.45. Since $I - T$ has positive finite ascent and descent, if $p := p(\lambda_0 I - T) = q(\lambda I - T)$ then $N = \ker(\lambda I - T)^p$. From the classical Riesz functional calculus we know that $\sigma(T|M) = \sigma(T) \setminus \{\lambda\}$, so that $(\lambda I - T)|M$ is bijective, while $(\lambda I - T|N)^p = 0$.

Therefore, by Theorem 1.119 $\lambda I - T$ is B-Fredholm for every $\lambda \in \mathbb{C}$. In particular, $\lambda I - T$ is upper semi B-Fredholm for every $\lambda \in \mathbb{C}$, or equivalently, is essentially left Drazin invertible for every $\lambda \in \mathbb{C}$.

Therefore the statements (i)–(iv) are equivalent.

(iv) $\Rightarrow$ (v) Assume that $\sigma(T)$ is a finite set of poles $\{\lambda_1, \ldots, \lambda_n\}$, where for every $i = 1, \ldots, n$ with p_i we denote the order of λ_i. Let $h(\lambda) := (\lambda_1 - \lambda)^{p_1} \ldots (\lambda_n - \lambda)^{p_n}$. Then by Lemma 3.67

$$h(T)(X) = \bigcap_{i=1}^{n} (\lambda_i I - T)^{p_i}(X) = \bigcap_{i=1}^{n} K(\lambda_i I - T),$$

where the last equality follows from Theorem 2.98 since T has the SVEP. But the last intersection is $\{0\}$, because if $x \in K(\lambda_i I - T) \cap K(\lambda_j I - T)$, with $\lambda_i \neq \lambda_j$, then $\sigma_T(x) \subseteq \{\lambda_i\} \cap \{\lambda_j\} = \emptyset$, and hence $x = 0$, since T has the SVEP. Therefore $h(T) = 0$.

(v) $\Rightarrow$ (i) Let h be a polynomial such that $h(T) = 0$. From the spectral mapping theorem we easily deduce that $\sigma(T)$ is a finite set $\{\lambda_1, \ldots, \lambda_n\}$. The points $\lambda_1, \ldots, \lambda_n$ are zeros of finite multiplicities of h, say $k_1, \cdots, k_n$, respectively, so that

$$h(\lambda) = (\lambda_1 - \lambda)^{k_1} \ldots (\lambda_n - \lambda)^{k_n},$$

and hence, by Lemma 3.67,

$$X = \ker h(T) = \bigoplus_{i=1}^{n} \ker(\lambda_i I - T)^{k_i}.$$

Now suppose that $\lambda = \lambda_j$ for some j and define

$$h_0(\lambda) := \prod_{i \neq j} (\lambda_i - \lambda)^{k_i}.$$

We have

$$M := \ker h_0(T) = \bigoplus_{i \neq j} \ker(\lambda_i I - T)^{k_i},$$

and if $N := \ker(\lambda_j I - T)^{k_j}$ then $X = M \oplus N$ and M, N are invariant under $\lambda_j I - T$. From the inclusion $\ker(\lambda_j I - T) \subseteq \ker(\lambda_j I - T)^{k_j} = N$ we infer that the restriction of $\lambda_j I - T$ on M is injective. It is easily seen that

$$(\lambda_j I - T)(\ker(\lambda_i I - T)^{k_i}) = \ker(\lambda_i I - T)^{k_i}, \quad i \neq j,$$

so $(\lambda_j I - T)(M) = M$. Therefore, the restriction of $\lambda_j I - T$ on M is also surjective and hence bijective. Obviously $(\lambda_j I - T)|N)^{k_j} = 0$, so, by Theorem 1.119, $\lambda I - T$ is B-Fredholm when λ belongs to the spectrum, in particular is upper semi B-Fredholm, or equivalently essentially left Drazin invertible.   ∎

From Theorem 3.68 it immediately follows that:

Corollary 3.69 *Every algebraic operator is meromorphic.*

Evidently, every nilpotent operator is algebraic. A less trivial example of an algebraic operator is provided by an operator $K \in L(X)$ for which a power K^n is finite-dimensional:

Corollary 3.70 *Let $K \in L(X)$ be such that K^n is finite-dimensional for some $n \in \mathbb{N}$. Then K is algebraic.*

Proof K^n is a Riesz operator and hence K is also a Riesz operator, by Theorem 3.7, part (iv). Therefore, $q(\lambda I - T) < \infty$ for all $\lambda \neq 0$. On the other hand, $K^{n+1}(X) \subseteq K^n(X)$ for all $n \in \mathbb{N}$, and since $K^n(X)$ is finite-dimensional it follows that $q(K) < \infty$. Therefore, by Theorem 3.68, K is algebraic.   ∎

Now, we want to characterize the operators having a finite-dimensional power among the class of all algebraic operators. We first need a preliminary lemma.

Lemma 3.71 *Let $T \in L(X)$ and let p_1 and p_2 be two relatively prime polynomials. Then $\ker p_1(T) \subseteq p_2(T)(X)$.*

Proof Since p_1 and p_2 are two relatively prime polynomials there exist two polynomial q_1, q_2 such that $p_1(\lambda)q_1(\lambda) + p_2(\lambda)q_2(\lambda) = 1$, so $p_1(T)q_1(T) + p_2(T)q_2(T) = I$. Let $x \in \ker p_1(T)$. Then

$$x = q_2(T)p_2(T)x = p_2(T)q_2(T)x \in p_2(T)(X).$$

                                    ∎

Let $\mathbb{P}$ denote the set of all polynomials $h(\lambda) := a_0 + a_1\lambda + \cdots a_n\lambda^n$ with $a_n = 1$. Operators which have a finite-dimensional power may be characterized in the following way.

Theorem 3.72 *Let $T \in L(X)$. Then T has a finite-dimensional power if and only if T is algebraic and $\dim \ker h(T) < \infty$ for all $h \in \mathbb{P}$.*

Proof Suppose that $\dim T^n(X) < \infty$ and set $\widetilde{T} := T|T^n(X)$. Let ν be the characteristic polynomial of $\widetilde{T}$, so $\nu(\widetilde{T}) = 0$ and set $\nu_0(\lambda) = \lambda^n \nu_0(\lambda)$. Then

$$\nu_0(T)x = T^n \nu(T)x = \nu(\widetilde{T}) T^n x = 0 \quad \text{for all } x \in X.$$

From Lemma 3.71, we have $\ker h(T) \subseteq T^n(X)$ for all $h \in \mathbb{P}$. Moreover, T is algebraic, by Corollary 3.70.

Conversely, if dim ker $h(T) < \infty$ for all $h \in \mathbb{P}$ and T is algebraic, then ker $h(T)^n$ is finite-dimensional for all $n \in \mathbb{N}$. ∎

The operators K for which a power is finite-dimensional may be characterized among Riesz operators in the following way:

Theorem 3.73 *Let $K \in L(X)$. The following statements are equivalent:*

(i)   *K^n is finite-dimensional for some $n \in \mathbb{N}$;*
(ii) *K is Riesz and Drazin invertible;*
(iii) *K is Riesz and left Drazin invertible;*
(iv) *K is Riesz and right Drazin invertible;*
(v)   *K is Riesz and essentially left Drazin invertible;*
(vi) *K is Riesz and essentially right Drazin invertible;*
(vii) *K is Riesz and $q(K) < \infty$;*
(viii)*K is Riesz and has essential descent $q_e(K) < \infty$;*
(viii)*K is Riesz and has topological uniform descent.*

Proof (i) $\Rightarrow$ (ii) K is algebraic, by Corollary 3.70. The implications (ii) $\Rightarrow$ (iii) $\Rightarrow$ (v) are clear. We show (v) $\Rightarrow$ (i). Suppose that K is Riesz and essentially left Drazin invertible. As observed in Remark 1.146, then the restriction $K_n := K|K^n(X)$ is upper semi-Fredholm. But K is a Riesz operator, so the restriction K_n is also Riesz, by Theorem 3.7, so $\lambda I - K_n$ is upper semi-Fredholm for all $\lambda \in \mathbb{C}$. This implies, by Theorem 2.126, that $K^n(X)$ is finite-dimensional.

Clearly, (ii) $\Rightarrow$ (iv) $\Rightarrow$ (vi) $\Rightarrow$ (viii), and (ii) $\Rightarrow$ (iv) $\Rightarrow$ (vii) $\Rightarrow$ (viii).

We show now the implication (viii) $\Rightarrow$ (ix). Suppose that K is Riesz and $q_e(K) < \infty$ and set $q := q_e(K)$. Then $K(X) + \mathrm{ker}\ K^q$ has finite-codimension in X, and since the sequence $\{K(X) + \mathrm{ker}\ K^n\}$ is increasing, we infer that there exists a $d \geq q$ such that $K(X) + \mathrm{ker}\ K^{n+1} = K(X) + \mathrm{ker}\ K^n$ for all $n \geq d$. Note that the quantity

$$\dim \frac{K^d(X)}{K^n(X)} = \dim \frac{K^d(X)}{K^{n+1}(X)} + \dim \frac{K^{d+1}(X)}{K^{d+2}(X)} \cdots + \dim \frac{K^{n-1}(X)}{K^n(X)}$$

is finite for all n d. Note that $K^n(X)$ can be viewed as the operator range of the restriction $K^{n-d}|K^d(X) : K^d(X) \rightarrow K^d(X)$, where $K^d(X)$ is provided with the range topology. From Corollary 1.7 it then follows that $K^n(X)$ is closed in the operator range topology of $K^d(X)$ for all $n > d$. Therefore, K has topological uniform descent.

(ix) $\Rightarrow$ (i) Suppose that K is Riesz and has topological uniform descent. Then K^* is a Riesz operator, so K^* has the SVEP at every $\lambda \in \mathbb{C}$, in particular K^* has the SVEP at 0. By Theorem 2.98 then K has descent $q := q(T) < \infty$. Consequently, the dimension of the quotient space $\frac{X}{K(X) + \mathrm{ker}\ K^n}$ is zero and hence the induced operator $\widehat{K}$ on $X/\mathrm{ker}\ K^n$, defined by

$$\widehat{K}(x + \mathrm{ker}\ K^n) := Kx + \mathrm{ker}\ K^n,$$

is onto, and hence lower semi-Fredholm. Since K is Riesz, then $\widehat{K}$ is also Riesz, by Theorem 3.7, hence $\lambda I - \widehat{K}$ is lower semi-Fredholm for all $\lambda \in \mathbb{C}$. Since the upper semi-Fredholm spectrum of an operator acting on an infinite-dimensional space is non-empty, see Theorem 2.126, $X/K^n(X)$ is finite-dimensional, and, consequently, $K^n(X)$ has finite-dimension. ∎

Finite-dimensional Banach spaces may be characterized as follows:

Theorem 3.74 *Let X be a Banach space. The following assertions are equivalent:*

(i) $\dim X < \infty$;
(ii) *Every $T \in L(X)$ has finite ascent;*
(iii) *Every $T \in L(X)$ has finite essential descent.*

Proof The implications (i) $\Rightarrow$ (ii) $\Rightarrow$ (iii) are obvious. Suppose that X has infinite dimension and take an infinite sequence (x_n) of linearly independent vectors of X and suppose that (f_n) is a sequence in X^* such that $f_j(e_i) = \delta_{ij}$ for all $i, j \in \mathbb{N}$. Define

$$T := \sum_{k=1}^{\infty} \lambda_k f_{2k} \otimes e_k,$$

where (λ_k) is sequence of non-zero scalars such that the sum

$$\sum_{k=1}^{\infty} |\lambda_k| \|f_{2k}\| \|e_k\| < \infty.$$

The sequence $(e_{2^{m+1}k+2^k})$ consists of linearly independent vectors of the difference $\ker T^{m+1} \setminus \ker T^m$, hence T has infinite essential ascent. ∎

Lemma 3.75 *Let $T \in L(X)$ be essentially left Drazin invertible. Then the operator $\widetilde{T}$ induced by T on the quotient $X/\ker T^\nu$ is both semi-regular and upper semi-Fredholm.*

Proof Clearly, $\ker \widetilde{T}$ has finite dimension and

$$\widetilde{T}(X/\ker T^\nu) = (T(X) + \ker T^\nu)/\ker T^\nu = (T^{-\nu}(T^{\nu+1}(X)))/\ker T^\nu$$

is closed, so $\widetilde{T}$ is upper semi-Fredholm. Now, since $\ker T \cap T^\nu(X) = \ker T \cap T^{\nu+n}(X)$ for all $n \in \mathbb{N}$, it is easy to see that $\ker T^{\nu+1} \subseteq T^n(X) + \ker T^\nu$ for all $n \in \mathbb{N}$. Therefore $\ker \widetilde{T}$ is contained in the hyper-range of $\widetilde{T}$, thus $\widetilde{T}$ is semi-regular. ∎

Theorem 3.76 *Suppose that $T, K \in L(X)$ commutes and that K^n is finite-dimensional for some $n \in \mathbb{N}$. Then we have*

(i) *If T has finite essential ascent then $T + K$ also has finite essential ascent.*
(ii) *If T is essentially left Drazin invertible then $T + K$ is essentially left Drazin invertible.*

Proof

(i) Suppose that $p^e(T) < \infty$ and let $\nu := \nu(T)$. Given $k \geq n + \nu$, since T^{n+k} maps $\ker (T + K)^k$ into $K^n(X)$ it is clear that

$$\dim \ker(T + K)^k / (\ker(T + K)^k \cap \ker T^{n+k}) < \infty.$$

Moreover, from

$$\dim \ker(T + K)^k / (\ker T^\nu) < \infty,$$

we deduce that

$$\dim \ker(T + K)^k / (\ker(T + K)^k \cap \ker T^\nu) < \infty.$$

We also have

$$\ker T^n \cap \ker T^\nu \subseteq \ker(T + K)^k \cap \ker T^\nu \subseteq \ker T^\nu$$

and, since K^n is finite-dimensional, we conclude that

$$\dim \ker T^\nu / (\ker K^n \cap \ker T^\nu) < \infty.$$

From this we obtain

$$\dim \ker(T + K)^k / (\ker K^n \cap \ker T^\nu) < \infty,$$

which implies that $p^e(T + K) \leq n + \nu$.

(ii) By part (i) we have only to prove that if $T^{p_e(T)+1}(X)$ is closed then $(T + K)^{p_e(T+K)+1}(X)$ is closed. From the equality (3.31) it suffices to prove that $(T + K)^k(X)$ is closed for some $k > n + \nu$. Denote by $\widetilde{T}$ and $\widetilde{K}$ the operators induced by T and K on $X / T^\nu(X)$, respectively. By Lemma 3.75 we know that $\widetilde{T}$ is upper semi-Fredholm and since K is Riesz then $\widetilde{K}$ is Riesz, so $\widetilde{T} + \widetilde{K}$ is upper semi-Fredholm. Consequently, $(T + K)^k(X) + \ker T^\nu$ is closed. To conclude the proof, by Lemma 1.6 it suffices to prove that $(T+K)^k(X) \cap \ker T^\nu$ is closed. Note that since $(T + K)^k$ maps $\ker T^\nu$ into $K^n(X)$ we have that

$$\dim \ker T^\nu / (\ker (T + K)^k \cap \ker T^\nu) < \infty.$$

Therefore,

$$\dim (T + K)^k(X) \cap \ker T^\nu / ((T + K)^k(X) \cap \ker (T + K)^k \cap \ker T^\nu) < \infty.$$

Since $p^e(T + K) < \infty$ it then follows that $(T + K)^k(X) \cap \ker (T + K)^k$ is finite-dimensional, consequently $(T + K)^k(X) \cap \ker T^\nu$ is also finite-dimensional, in particular a closed subspace of X.   ∎

Corollary 3.77 *Suppose that* $T, K \in L(X)$ *commutes and that* K^n *is finite-dimensional for some* $n \in \mathbb{N}$.

(i) *If* T *is upper semi B-Fredholm then* $T + K$ *is also upper semi B-Fredholm.*
(ii) *If* T *is lower semi B-Fredholm then* $T + K$ *is also lower semi B-Fredholm.*

Proof

(i) As noted in Remark 1.146, T is upper semi B-Fredholm if and only if T is essentially left Drazin invertible, so this statement has already been proved in Theorem 3.76.
(ii) Observe that $(K^*)^n$ is finite-dimensional. Indeed, $K^n(X)$ is finite-dimensional and hence a closed subspace of X, so, by the annihilator theorem

$$\dim (K^*)^n(X^*) = \dim (\ker K^n)^\perp = \dim K^n(X).$$

Obviously $K^*T^* = T^*K^*$. Now, if T is lower semi B-Fredholm then T is essentially right Drazin invertible, hence T^* is essentially left Drazin invertible, so $T^* + K^*$ is essentially left Drazin invertible, by Theorem 3.76, and this implies that $T + K$ is lower semi B-Fredholm.   ∎

A consequence of Corollary 3.77 is that the essential left Drazin spectrum and the essential right Drazin spectrum are invariant under commuting perturbations K for which K^n is finite-dimensional for some $n \in \mathbb{N}$. We show now that this is also true for the Drazin spectra.

Theorem 3.78 *The Drazin spectra* $\sigma_{\mathrm{ld}}(T)$, $\sigma_{\mathrm{rd}}(T)$, *and* $\sigma_{\mathrm{d}}(T)$ *are stable under commuting perturbations* K *for which* K^n *is finite-dimensional for some* $n \in \mathbb{N}$.

Proof Suppose that $\lambda \notin \sigma_{\mathrm{ld}}(T)$. Then $\lambda I - T$ is left Drazin invertible, or equivalently, upper semi B-Browder, in particular $\lambda - T$ is upper semi B-Fredholm. By Corollary 3.77 it then follows that $\lambda I - T + K$ is also lower semi B-Fredholm, and hence quasi-Fredholm. Now, K is a Riesz operator, since K^n is Riesz. From $p(\lambda I - T) < \infty$ we know that T has the SVEP at λ, and hence $T + K$ has the SVEP at λ, by Theorem 2.129. Since every quasi-Fredholm operator has topological uniform descent, then, by Theorem 2.97, the SVEP of $T + K$ at λ entails that $p(\lambda I - (T + K)) < \infty$. By Theorem 1.142 we then conclude that $\lambda I - (T + K)$ is left Drazin invertible, i.e., $\lambda \notin \sigma_{\mathrm{ld}}(T + K)$. Hence, $\sigma_{\mathrm{ld}}(T + K) \subseteq \sigma_{\mathrm{ld}}(T)$.

By symmetry, the same argument shows that

$$\sigma_{\mathrm{ld}}(T) = \sigma_{\mathrm{ld}}((T + K) - K) \subseteq \sigma_{\mathrm{ld}}(T + K).$$

Therefore, $\sigma_{\mathrm{ld}}(T + K) = \sigma_{\mathrm{ld}}(T)$.

The stability for the other spectra $\sigma_{\mathrm{rd}}(T)$, and $\sigma_{\mathrm{d}}(T)$ may be proved in a similar way. ∎

Corollary 3.79 *If $T \in L(X)$ is meromorphic, $K \in L(X)$ commutes with T and K^n is finite-dimensional for some $n \in \mathbb{N}$, then $T + K$ is meromorphic.*

The *descent spectrum* $\sigma_{\mathrm{desc}}(T)$ (defined as the set of all $\lambda \in \mathbb{C}$ for which $q(\lambda I - T)$ is infinite) is also invariant under commuting perturbations K for which there exists a finite-dimensional power, see Kaashoek and Lay [194]. This property characterizes those operators, i.e., K^n is finite-dimensional for some natural n precisely when $\sigma_{\mathrm{desc}}(T) = \sigma_{\mathrm{desc}}(T + K)$ for all $T \in L(X)$ which commutes with K, see Burgos et al. [85]. Bel Hadj Fredj [149] has shown that this characterization may be extended to the essential descent spectrum, i.e. K^n is finite-dimensional for some natural n if and only if the essential descent spectra of T and of $T + K$ coincide for all $T \in L(X)$ which commutes with K. Analogously, K^n is finite-dimensional for some natural n if and only if the essential left Drazin spectra of T and of $T + K$ coincide for all $T \in L(X)$ which commute with K, or, equivalently, the left Drazin spectra of T and of $T + K$ coincide for all $T \in L(X)$ which commute with K, see [150]. By duality, the same statement holds for the right Drazin spectrum.

It should be noted that the spectra $\sigma_{\mathrm{bw}}(T)$, $\sigma_{\mathrm{bf}}(T)$, $\sigma_{\mathrm{ubw}}(T)$ and $\sigma_{\mathrm{lbw}}(T)$ are also stable under commuting perturbations K for which K^n is finite-dimensional for some $n \in \mathbb{N}$. The proof of this is omitted, see [304].

We know that if $\lambda I - T$ has finite ascent then T has the SVEP at λ. The following example reveals that this is not true for the essential ascent.

Example 3.80 It has been noted in Remark 2.64 that the left shift L on $\ell^2(\mathbb{N})$ fails the SVEP at 0. It is easily seen that $\lambda I - L$ has essential finite ascent for all $\lambda \in \mathbb{C}$, since $\ker(\lambda I - L) = \{0\}$ for $|\lambda| \geq 1$, and $\ker(\lambda I - T)$ has dimension 1 for all $|\lambda| < 1$.

We conclude this section by mentioning some recent results on the perturbation class of algebraic operators due to Oudghiri and Souilah [254]. Denote by $\mathcal{A}(X)$ the set of all algebraic operators on X.

Lemma 3.81 *If $K \in L(X)$ is algebraic and $N \in L(X)$ is nilpotent and commutes with K, then $T + N$ is algebraic.*

Proof Since K is algebraic we can write $X = X_1 \oplus \cdots \oplus X_n$, where $X_j = \ker(\lambda_j I - K)^{m_j}$, where the scalars λ_j are distinct. All the restrictions $K_j := (\lambda_j I - K)|X_j$ are nilpotent, and $K = (\lambda_1 + K_1) \oplus \cdots (\lambda_n + K_n)$. Moreover, every X_j is invariant under N. With respect the same decomposition of X we can write $N = N_1 \oplus \cdots \oplus N_n$. Since every N_j is nilpotent and $K_j N_j = N_j K_j$ it then follows that $T_j + N_j$ is nilpotent for every $1 \le j \le n$, so $K + N$ is algebraic. ∎

Let us consider the class of operators $P_c(\mathcal{A}(X))$ of all $T \in L(X)$ such that $T + S \in \mathcal{A}(X)$ for all $S \in \mathcal{A}(X)$ which commute with T.

Theorem 3.82 *We have*

$$P_c(\mathcal{A}(X)) = \mathcal{A}(X).$$

Proof Evidently, $P_c(\mathcal{A}(X)) \subseteq \mathcal{A}(X)$. To show the reverse inclusion suppose that $K, S \in \mathcal{A}(X)$ and $SK = KS$. Then $X = X_1 \oplus \cdots \oplus X_n$, where $X_j = \ker(\lambda_j I - K)^{m_j}$ and λ_j are distinct. The restriction $K_J := (\lambda_j I - K)|X_j$ are nilpotent for every $j = 1, \ldots n$, and $K = \bigoplus_{j=1}^{n}(\lambda_j I + K_j)$. Since $SK = KS$, X_j is invariant under S, and with respect to the decomposition above we can write $S = S_1 \oplus \cdots \oplus S_n$. From this one can easily see that every $K_j + S_j$ is algebraic, thus $K + S$ is algebraic. ∎

The class $\mathcal{A}(X)$ of nilpotent operators is not stable under small perturbations:

Lemma 3.83 *Let $N \in L(X)$ be nilpotent. Then, for every $\varepsilon > 0$ there exists a $T \in L(X)$ for which $\|T\| < \varepsilon$ and $T \notin \mathcal{A}(X)$.*

Proof Since N is nilpotent then $\ker N$ is infinite-dimensional. Let $0 \ne x_0 \in \ker N$ and write $X = \text{span}\{x_0\} \oplus X_0$. Evidently, X_0 is a closed subspace. Choose $f_0 \in X^*$ such that $f_0(x_0) = 1$ and $f_0 \equiv 0$ on X_0. Let $0 \ne x_1 \in \ker N \cap X_0$ and write $X_0 = \text{span}\{x_1\} \oplus X_1$. In particular, the vectors x_0 and x_1 are linearly independent, and $X = \text{span}\{x_0, x_1\} \oplus X_1$. Hence there exists a linear form $f_1 \in X^*$ which satisfies

$$f_1(x_i) = \delta_{i1} \quad \text{for } i = 0, 1 \quad \text{and} \quad f_1(x) = 0 \quad \text{for } x \in X_1.$$

Note that $f_0(x_1) = 0$. Repeating this argument we then obtain two linearly independent sets $\{x_n\}$ and $\{f_n\}$ such that $f_i(x_j) = \delta_{ij}$ for all $i, j \ge 0$. Let $\varepsilon > 0$ and define

$$T := \sum_{n=0}^{\infty} \mu_n x_{n+1} \otimes f_n,$$

where the scalars μ_n satisfy

$$\sum_{n=0}^{\infty} |\mu_n| \, \|x_{n+1} \otimes f_n\| < \varepsilon.$$

We have

$$(T + K)^n x_0 = \mu_0 \mu_1 \cdots \mu_{n-1} x_n \quad \text{for every } n \geq 1,$$

so the set $\{(T + K)^j x_0\}$ for $0 \leq j \leq n$ is a linearly independent set, and this implies that $T + K$ is non-algebraic.   ∎

The previous result has a remarkable consequence.

Corollary 3.84 *If X is a Banach space then X is infinite-dimensional if and only if $L(X)$ contains a non-algebraic operator.*

The class of all algebraic operators is not stable under small perturbations:

Theorem 3.85 *Let $K \in \mathcal{A}(X)$. Then, for every $\varepsilon > 0$ there exists a $T \in L(X)$ for which $\|T\| < \varepsilon$ and $T \notin \mathcal{A}(X)$. Consequently, the interior of $\mathcal{A}(X)$ is empty.*

Proof Let $K \in \mathcal{A}(X)$ and $\varepsilon > 0$. Write $X = X_1 \oplus \cdots X_n$, where $X_j = \ker(\lambda_j I - K)^{m_j}$, $j = 1, \ldots, n$ and $\lambda_j \neq \lambda_i$ for $j \neq i$. The operators $N_j = (\lambda_j I - K)|X_j$ are nilpotent and $K = \bigoplus_{j=1}^{n} (\lambda_j I + N_j)$. Without loss of generality we may assume $\dim X_1 = \infty$. Let $P \in L(X)$ be the idempotent operator defined as $P := I \oplus 0 \oplus 0 \oplus \cdots \oplus 0$, with respect to the above decomposition of X. By Lemma 3.83 there exists a bounded operator T_1 on X_1 such that $\|T_1\| < \varepsilon \|P\|^{-1}$ and $T_1 + N_1$ is non-algebraic. If we set $T := T_1 \oplus 0 \oplus \cdots 0$, we then have $\|T\| < \varepsilon$ and $T \notin \mathcal{A}(X)$.   ∎

Now, let us consider the perturbation class of $\mathcal{A}(X)$ defined as

$$P(\mathcal{S}(X)) := \{T \in L(X) : T + S \in \mathcal{A}(X) \text{ for all } S \in \mathcal{A}(X)\}.$$

In the case of Hilbert space operators we have:

Theorem 3.86 ([254, Theorem 2.1]) *If H is an infinite-dimensional complex Hilbert space, then*

$$P(\mathcal{A}(H)) = \mathbb{C}I + \mathcal{F}(H),$$

where $\mathcal{F}(H)$ is the ideal of finite-dimensional operators.

A natural question arises from the previous result: *Does the result of Theorem 3.86 hold for Banach spaces?*

Theorem 3.87 ([254, Theorem 2.1]) *If H is an infinite-dimensional complex Hilbert space, then*

$$P(\mathcal{A}(H)) = \mathbb{C}I + \mathcal{F}(H),$$

where $\mathcal{F}(H)$ is the ideal of finite-dimensional operators.

3.6 Essentially Left and Right Drazin Invertible Operators

In this section we will give further results concerning the Drazin spectra. To do this we first need some preliminary results on essentially left and right Drazin invertible operators, which have been defined in Chap. 1.

Define, as we have done in Chap. 1,

$$c_n(T) := \dim \frac{T^n(X)}{T^{n+1}(X)} \quad \text{and} \quad c'_n(T) := \dim \frac{\ker T^{n+1}}{\ker T^n}.$$

If the essential ascent $p_e(T)$ is finite let us denote by $v(T)$ the smallest positive integer k with $c'_n(T) = c'_k(T)$ for all $n \geq k$. Clearly, $p_e(T) \leq v(T)$. Analogously if the essential descent $q_e(T)$ is finite, let $\mu(T)$ denote the smallest positive integer k with $c_n(T) = c_k(T)$ for all $n \geq k$. Clearly, $q_e(T) \leq \mu(T)$.

Lemma 3.88 *Let $T \in L(X)$ be semi-regular.*

(i) *If $\ker T$ finite-dimensional then $\dim \ker T^n = n \dim \ker T$ for all $n \in \mathbb{N}$.*
(ii) *If $T(X)$ has finite codimension then $\operatorname{codim} T^n(X) = n \operatorname{codim} T(X)$ for all $n \in \mathbb{N}$.*

Proof

(i) Let $n \in \mathbb{N}$. Since $\ker T^{n-1} \subseteq T(X)$, T is a surjection from $\ker T^n$ to $\ker T^{n-1}$, hence $\dim \ker T^n = \dim \ker T + \dim \ker T^{n-1}$. The statement then follows by induction.
(ii) Let $n \geq 2$ and let $S : X \to X/T^n(X)$ be the operator defined by

$$Sx := T^{n-1}x + T^n(X).$$

Since T is semi-regular we have

$$\ker S = T(X) + \ker T^{n-1} = T(X),$$

consequently the quotient $X/T(X)$ is isomorphic to $T^{n-1}(X)/T^n(X)$. On the other hand, we know that $X/T^{n-1}(X) \times T^{n-1}(X)/T^n(X)$ is isomorphic to $X/T^n(X)$, so $X/T^{n-1}(X) \times X/T(X)$ is isomorphic to $X/T^n(X)$. Hence

$$\operatorname{codim} T^n(X) = \operatorname{codim} T^{n-1}(X) + \operatorname{codim} T(X),$$

and a successive repetition of this argument easily leads to the equality $\operatorname{codim} T^n(X) = n \operatorname{codim} T(X)$ for all $n \in \mathbb{N}$. ∎

Theorem 3.89 *Let $T \in L(X)$ be essentially left Drazin invertible. Then there exists a $\delta > 0$ such that $\lambda I - T$ is semi-regular for every $0 < |\lambda| < \delta$. Moreover,*

(i) *$\dim \ker (\lambda I - T)^n = n \dim \ker T^{v+1}/\ker T^v$ for all $n \in \mathbb{N}$.*

(ii) $\text{codim}(\lambda I - T)^n(X) = n \dim T^\nu(X)/T^{\nu+1}(X)$ *for all* $n \in \mathbb{N}$.

Analogously, if T has finite essential descent then there exists a $\delta > 0$ such that $\lambda I - T$ is semi-regular for every $0 < |\lambda| < \delta$. Moreover,

(iii) $\dim \ker (\lambda I - T)^n = n \dim \ker T^{\mu+1}/\ker T^\mu$ *for all* $n \in \mathbb{N}$.

(iv) $\text{codim}(\lambda I - T)^n(X) = n \dim T^\mu(X)/T^{\mu+1}(X)$ *for all* $n \in \mathbb{N}$.

Proof Suppose first that T is essentially left Drazin invertible. By Lemma 3.75 the operator $\tilde{T}$ induced by T on $Y := X/\ker T^\nu$ is both semi-regular and upper semi-Fredholm. From part (iii) of Theorem 1.44, and from the punctured neighborhood theorem for semi-Fredholm operators, we then have that there exists a $\delta > 0$ such that $\lambda \tilde{I} - \tilde{T}$ is semi-regular and upper semi-Fredholm with

$$\dim \ker (\lambda \tilde{I} - \tilde{T}) = \dim \ker \tilde{T} \quad \text{for all } |\lambda| < \delta.$$

Fix $\lambda \in \mathbb{C}$ such that $|\lambda| < \delta$. We have

$$\ker (\lambda \tilde{I} - \tilde{T})^n = \ker ((\lambda I - T)^n T^\nu)/\ker T^\nu = (\ker (\lambda I - T)^n \oplus \ker T^\nu)/\ker T^\nu, \tag{3.30}$$

and

$$(\lambda \tilde{I} - \tilde{T})(Y) = ((\lambda I - T)(X) + \ker T^\nu)/\ker T^\nu = (\lambda I - T)(X)/\ker T^\nu. \tag{3.31}$$

Consequently, $(\lambda I - T)(X)$ is closed and contains the finite-dimensional subspace $\ker (\lambda I - T)$ for all $n \in \mathbb{N}$. From this it then follows that $\lambda I - T$ is both semi-regular and upper semi-Fredholm. Moreover, by (3.30) and Lemma 3.88, we have

$$\dim (\lambda I - T)^n(X) = \dim (\lambda \tilde{I} - \tilde{T})^n(Y) = n \dim \ker (\lambda \tilde{I} - \tilde{T})$$
$$= n \dim \ker \tilde{T} = n \dim \ker T^{\nu+1}/\ker T^\nu.$$

From the continuity of the index we then deduce that

$$\text{codim} (\lambda I - T)^n(X) = \text{codim}(\lambda I - T)^n(X)/\ker T^\nu = \text{codim}(\lambda \tilde{I} - \tilde{T})^n(Y)$$
$$= \dim \ker (\lambda \tilde{I} - \tilde{T})^n - \text{ind} (\lambda \tilde{I} - \tilde{T})^n$$
$$= n \dim \ker \tilde{T} - \text{ind} (\lambda \tilde{I} - \tilde{T})$$
$$= n \dim \ker \tilde{T} - \text{ind} (\tilde{T}) = n \, \text{codim} \tilde{T}(Y)$$
$$= n \dim X/(T(X) + \ker T^\nu) = n \dim T^\nu(X)/T^{\nu+1}(X),$$

for all $n \in \mathbb{N}$.

Assume the other case that T has finite essential descent $q^e(T)$. Define on $T^\mu(X)$ a new norm

$$|y| := \|y\| + \inf \{\|x\| : y = T^n x\} \quad \text{for all } y \in T^\mu(X).$$

Then the space $Y := T^\mu(X)$ equipped with this norm is a Banach space and the restriction $T_0 := T(Y)$ is semi-Fredholm, since the range of T_0 is $T^{\mu+1}(X)$ which has finite codimension in $Y = T\mu(X)$. Moreover, since $q_e(T) < \infty$, by Theorem 1.74 we have

$$\ker T \cap T^\mu(X) = \ker T \cap T^{\mu+n}(X) \quad \text{for all } n \in \mathbb{N},$$

and so

$$\ker T_0 = \ker T \cap T^\mu(X) = \ker T \cap T^{\mu+n}(X) \subseteq T^{\mu+n}(X) = T_0^n(Y),$$

which shows that T_0 is also semi-regular. According to Theorem 1.44, let $\delta > 0$ be such that $\lambda I - T_0$ is both semi-Fredholm and semi-regular for $|\lambda| < \delta$. We show now that, without any restriction on T, we have

$$X = (\lambda I - T)^n + T^\mu(X) \quad \text{for all } n \in \mathbb{N} \, \lambda \neq 0. \tag{3.32}$$

Indeed, the two polynomials $h(z) := (z - \lambda)^n$ and $k(z) := z^\mu$ have no common divisors, so, by Lemma 1.13, there exist two polynomials u, v such that

$$I = h(T)u(T) + k(T)v(T),$$

from which we obtain the equality (3.32). Consequently, by Lemma 3.88, for each $0 < |\lambda| < \delta$ we have

$$\begin{aligned}
\operatorname{codim}(\lambda I - T)^n(X) &= \dim X/(\lambda I - T)^n(X) \\
&= \dim\left((T^\mu(X) + (\lambda I - T)^n(X))/(\lambda I - T)^n(X)\right) \\
&= \dim\left(T^\mu(X)/(T^\mu(X) \cap (\lambda I - T)^n(X))\right) \\
&= \operatorname{codim}(\lambda I - T_0)^n(Y) = n \operatorname{codim}(\lambda I - T_0)(Y) \\
&= n \dim T^\mu(X)/T^{\mu+1}(X).
\end{aligned}$$

In particular, $\lambda I - T$ is semi-Fredholm. Moreover, for all $k \in \mathbb{N}$ we have

$$\begin{aligned}
\ker(\lambda I - T) &= T^\mu(X) \cap \ker(\lambda I - T) = \ker(\lambda I - T_0) \\
&\subseteq (\lambda I - T_0)^k(Y) \subseteq (\lambda I - T)^k(X),
\end{aligned}$$

so $\lambda I - T$ is semi-regular. It remains to show part (iii). We have

$$\begin{aligned}
\dim \ker(\lambda I - T)^n &= \ker(\lambda I - T_0) = \operatorname{ind}(\lambda I - T_0)^n + \operatorname{codim}(\lambda I - T_0)^n(Y) \\
&= n\left[\operatorname{ind}(\lambda I - T_0) + \operatorname{codim}(\lambda I - T_0)(Y)\right] \\
&= n\left[\operatorname{ind} T_0 + \operatorname{codim} T_0(Y)\right] = n \dim \ker T_0 \\
&= n \dim(T^\mu(X) \cap \ker T).
\end{aligned}$$

Since T^μ induces an isomorphism from ker $T^{\mu+1}/$ ker T^μ onto $T^\mu(X) \cap \ker T$, it then follows that

$$\dim \ker(\lambda I - T)^n = n \dim (T^\mu(X) \cap \ker T) = n \dim (\ker T^{\mu+1}/ \ker T^\mu),$$

so the proof is complete.   ■

Taking into account the continuity of the index, from Theorem 3.89 we easily deduce that a semi-Fredholm operator T has finite essential ascent if and only if $T \in \Phi_+(X)$. Note that, if T has a finite essential descent, then there exists a finite-dimensional subspace M of X for which $X = (\lambda I - T)(X) \oplus M$ for every $\lambda \neq 0$ sufficiently small. Indeed, arguing as in the second part of proof of Theorem 3.89, since T_0 is semi-regular and its range has finite codimension, there exists a $\delta > 0$ and a finite-dimensional subspace M for which

$$T^\mu(X) = (\lambda - T_0)(Y) \oplus M \quad \text{for all } |\lambda| < \delta.$$

Hence,

$$X = (\lambda I - T)(X) + T^\mu(X) = (\lambda I - T)(X) \oplus M \quad \text{for } 0 < |\lambda| < \delta.$$

Corollary 3.90 *Suppose that $T \in L(X)$ has finite ascent $p := p(T)$ and that $T^{p+1}(X)$ is closed. Then there exists a $\delta > 0$ such that for every $0 < |\lambda| < \delta$ the following assertions hold:*

(i) $\lambda I - T$ *is bounded below.*
(ii) $\text{codim}(\lambda I - T)^n(X) = n \dim T^p(X)/T^{p+1}(X)$ *for all $n \in \mathbb{N}$.*

Analogously, if $T \in L(X)$ has finite descent $q := q(T)$ then there exists a $\delta > 0$ such that for every $0 < |\lambda| < \delta$ the following assertions hold:

(i) $\lambda I - T$ *is onto.*
(ii) $\dim \ker (\lambda I - T)^n = n \dim \ker T^{q+1}/ \ker T^p$ *for all $n \in \mathbb{N}$.*

The *essential left Drazin spectrum* $\sigma_{ld}^e(T)$ of $T \in L(X)$ is defined as the complement of the *essential resolvent ascent* $\rho_{ld}^e(T)$, defined as the set of all $\lambda \in \mathbb{C}$ such that $\lambda I - T$ is essentially left Drazin invertible. In a similar way the *essential right Drazin spectrum* $\sigma_{rd}^e(T)$ of $T \in L(X)$ is defined as the complement of the set $\rho_{rd}^e(T)$, defined as the set of all $\lambda \in \mathbb{C}$ such that $\lambda I - T$ is essentially right Drazin invertible. Clearly, $\sigma_{ld}^e(T)$ is a subset of $\sigma_{ld}(T)$, while $\sigma_{rd}^e(T)$ is a subset of $\sigma_{rd}(T)$.

Corollary 3.91 *The essential left Drazin spectrum, the essential right Drazin spectrum, the left Drazin spectrum and the right Drazin spectrum are compact subsets of $\mathbb{C}$.*

Denote by $\Pi(T)$ the set of poles of T and by $\rho_{ld}(T)$ the set of all $\lambda \in \mathbb{C}$ such that $\lambda I - T$ is left Drazin invertible.

Lemma 3.92 *If $T \in L(X)$ then*

$$\rho_{\mathrm{ld}}^e(T) \cap \partial\sigma(T) = \rho_{\mathrm{ld}}(T) \cap \partial\sigma(T) = \Pi(T),$$

and

$$\rho_{\mathrm{rd}}^e(T) \cap \partial\sigma(T) = \rho_{\mathrm{rd}}(T) \cap \partial\sigma(T) = \Pi(T).$$

Proof The inclusion

$$\Pi(T) \subseteq \rho_{\mathrm{ld}}(T) \cap \partial\sigma(T) = \rho_{\mathrm{ld}}^e(T) \cap \partial\sigma(T)$$

is clear. Now, let $\lambda_0 \in \rho_{\mathrm{ld}}^e(T) \cap \partial\sigma(T)$ be arbitrary and let $\nu := \nu(T)$ be defined as before. By Theorem 3.89 there exists a punctured neighborhood U of λ_0 such that

$$\dim \ker(\lambda I - T) = \dim \ker(\lambda_0 I - T)^{\nu+1} / \ker(\lambda_0 I - T)^\nu,$$

and

$$\mathrm{codim}\,(\lambda I - T)(X) = \dim\,(\lambda_0 I - T)^\nu(X)/(\lambda_0 I - T)^{\nu+1}(X)$$

for all $\lambda \in U$. Since $U \setminus \sigma(T)$ is non-empty, we then have

$$\dim \ker(\lambda_0 I - T)^{\nu+1}/\ker(\lambda_0 I - T)^\nu = \dim\,(\lambda_0 I - T)^\nu(X)/(\lambda_0 I - T)^{\nu+1}(X) = 0,$$

thus $\lambda_0 I - T$ has finite ascent and descent, i.e., λ_0 is a pole of the resolvent of T. The proof concerning essentially right Drazin invertible operators is analogous. ■

Algebraic operators may be characterized in the following way.

Theorem 3.93 *If $T \in L(X)$ then the following statements are equivalent:*

(i) *T is algebraic;*
(ii) *$\sigma_{\mathrm{ld}}(T) = \emptyset$;*
(iii) *$\sigma_{\mathrm{ld}}^e(T) = \emptyset$;*
(iv) *$\partial\sigma(T) \cap \sigma_{\mathrm{ld}}^e(T) = \emptyset$;*
(v) *$\sigma_{\mathrm{rd}}(T) = \emptyset$.*

Proof If T is algebraic then $\sigma(T)$ is a finite set of poles, so $\lambda I - T$ is left Drazin invertible for all $\lambda \in \mathbb{C}$. Hence the implications (i) $\Rightarrow$ (ii) $\Rightarrow$ (iii) $\Rightarrow$ (iv) hold. Assume (iv). By Lemma 3.92 we have $\partial\sigma(T) = \Pi(T)$, and this implies that $\sigma(T) = \partial\sigma(T)$, so $\sigma(T)$ is a finite set of poles and hence, by Theorem 3.68, T is algebraic. Therefore, the statements (i)–(iv) are equivalent.

(i) $\Leftrightarrow$ (v) T is algebraic if and only if T^* is algebraic. From the first part of the proof, T^* is algebraic if and only if $\sigma_{\mathrm{ld}}(T^*) = \emptyset$. The equivalence then follows from the equality $\sigma_{\mathrm{rd}}(T) = \sigma_{\mathrm{ld}}(T^*)$. ■

As in Chap. 2, set

$$\Xi(T) := \{\lambda \in \mathbb{C} : T \text{ fails to have the SVEP at } \lambda\} .$$

The next result shows that the left Drazin spectrum and the essential left Drazin spectrum coincide if T has the SVEP.

Theorem 3.94 *For every $T \in L(X)$ we have $\sigma_{\mathrm{ld}}(T) = \sigma_{\mathrm{ld}}^e(T) \cup \Xi(T)$. If T has the SVEP then $\sigma_{\mathrm{ld}}(T) = \sigma_{\mathrm{ld}}^e(T) \subseteq \sigma_{\mathrm{ld}}^e(T^*)$.*

Proof Clearly, $\sigma_{\mathrm{ld}}^e(T) \cup \Xi(T) \subseteq \sigma_{\mathrm{ld}}(T)$. The opposite inclusion easily follows from Corollary 2.99. Hence if T has the SVEP then $\sigma_{\mathrm{ld}}(T) = \sigma_{\mathrm{ld}}^e(T)$. Let $\lambda_0 \notin \sigma_{\mathrm{ld}}^e(T^*)$ be arbitrary, and suppose that T has the SVEP. Let $\nu := \nu(\lambda_0 I - T^*)$, according to the definition before Lemma 3.75. From Theorem 3.89 we know that for all λ which belong to a small punctured open disc centered at λ_0, $\lambda I - T^*$, and hence $\lambda I - T$, is semi-Fredholm with

$$\begin{aligned}
\dim \ker(\lambda I - T) &= \operatorname{codim}(\lambda I - T^*)(X^*) \\
&= \dim (\lambda_0 I - T^*)^\nu(X^*)/(\lambda_0 I - T^*)^{\nu+1}(X^*) \\
&= \dim \ker(\lambda_0 I - T) \cap (\lambda_0 I - T)^\nu(X) \\
&= \dim (\ker(\lambda_0 I - T)^{\nu+1}/\ker(\lambda_0 I - T)^\nu).
\end{aligned}$$

Since T has the SVEP the index of $\lambda I - T$ is less than or equal to 0, by Corollary 2.106. This implies that

$$\dim \ker(\lambda I - T) = \dim (\ker(\lambda_0 I - T)^{\nu+1}/\ker(\lambda_0 I - T)^\nu) < \infty,$$

thus $\lambda_0 \notin \sigma_{\mathrm{ld}}^e(T)$.   ∎

In the sequel we will give further results for the products RT and TR of operators $T, R \in L(X)$. First we need some preliminary results.

Theorem 3.95 *Let $T, R \in L(X)$ and $\lambda \neq 0$. Then we have:*

(i) $p(\lambda I - TR) = p(\lambda I - RT)$.
(ii) $q(\lambda I - TR) = q(\lambda I - RT)$.
(iii) *If $n \in \mathbb{N}$ then $(\lambda I - TR)^n(X)$ is closed if and only if $(\lambda I - RT)^n(X)$ is closed.*

Proof

(i) Let $x \in \ker(\lambda I - RT)^n$ but $x \notin \ker(\lambda I - RT)^{n-1}$. We show that $Tx \in \ker(\lambda I - TR)^n$, but $Tx \notin \ker(\lambda I - TR)^{n-1}$. Notice that

$$(\lambda I - RT)^n x = 0 = T(\lambda I - RT)^n x = (\lambda I - TR)^n Tx,$$

and consequently that $Tx \in \ker(\lambda I - TR)^n$.

Assume now that $Tx \in \ker(\lambda I - TR)^{n-1}$. We have

$$(\lambda I - TR)^{n-1} Tx = 0 = T(\lambda I - RT)^{n-1} x,$$

where $(\lambda I - RT)^{n-1} x \neq 0$. Hence,

$$(\lambda I - RT)(\lambda I - RT)^{n-1} x = 0 = RT(\lambda I - RT)^{n-1} x - \lambda(\lambda I - RT)^{n-1} x$$
$$= \lambda(\lambda I - RT)^{n-1} x.$$

Since $\lambda \neq 0$, this contradicts the fact that $(\lambda I - RT)^{n-1} x \neq 0$. Therefore, $Tx \notin \ker(\lambda I - TR)^{n-1}$ and hence, $p(\lambda I - TR) \leq p(\lambda I - RT)$. In a similar fashion we obtain the reverse inequality, so (i) is proved.

(ii) Suppose that $y \in (\lambda I - TR)^{n-1}(X)$ and $y \notin (\lambda I - TR)^n(X)$. We show that $Ry \in (\lambda I - RT)^{n-1}(X)$ and $Ry \notin (\lambda I - RT)^n(X)$. There exists an $x \in X$ such that $(\lambda I - TR)^{n-1} x = y$, so

$$(\lambda I - RT)^{n-1} Rx = Ry,$$

and

$$Ry \in (\lambda I - RT)^{n-1}(X).$$

Assume that $Ry \in (\lambda I - RT)^n(X)$. Then there exists a $z \in X$ with

$$(\lambda I - TR)y = T(\lambda I - RT)^n z - \lambda y.$$

Since $y \in (\lambda I - TR)^{n-1}(X)$ we have $(\lambda I - TR)y \in (\lambda I - TR)^n(X)$, hence

$$y = \frac{1}{\lambda}[(\lambda I - TR)Tz - (\lambda I - TR)y] \in (\lambda I - TR)^n(X),$$

and this is a contradiction. Thus $Ry \notin (\lambda I - RT)^n(X)$ and consequently, $q(\lambda I - TR) \leq q(\lambda I - RT)$. Similarly, $q(\lambda I - RT) \leq q(\lambda I - TR)$, so (ii) is proved.

(iii) Assume that $(\lambda I - TR)^n(X)$ is closed for some $n \in \mathbb{N}$. Let $y \in X$ such that $(\lambda I - RT)^n x_k \to y$ as $k \to \infty$. Since $(\lambda I - RT)^n Tx_k \to Ty$ and $(\lambda I - TR)^n(X)$ is closed, there exists a $z \in X$ such that $Ty = (\lambda I - TR)^n z$. Now,

$$\lambda y = (\lambda I - RT)y + RTy = (\lambda I - RT)y + R(\lambda I - TR)^n z$$
$$= (\lambda I - RT)y + (\lambda I - RT)^n Rz,$$

thus, $y \in (\lambda I - RT)(X)$ and there exists a $u \in X$ such that $y = (\lambda I - RT)u$. From this we obtain

$$\lambda y = (\lambda I - RT)y + RTy = (\lambda I - RT)^2 u + (\lambda I - RT)^n Rz,$$

and $y \in (\lambda I - RT)^2(X)$. By continuing this argument, it then follows that $y \in (\lambda I - RT)^n(X)$, so $(\lambda I - RT)^n(X)$ is closed. The converse follows by symmetry. ∎

We now consider the case where $\lambda = 0$.

Lemma 3.96 *If $T, R \in L(X)$ then:*

(i) $p(TR) - 1 \le p(RT) \le p(TR) + 1$.
(ii) $q(TR) - 1 \le q(RT) \le q(TR) + 1$.

Proof

(i) Suppose that $p(RT) := p < \infty$. We show first that $p(TR) - 1 \le p$. Suppose that this is not true, i.e., $p(TR) > p + 1$. Then there exists a $y \in \ker (TR)^{p+2}$ such that $y \notin \ker (TR)^{p+1}$. We have

$$(RT)^{p+2} Ry = R(TR)^{p+1} TRy = R(TR)^{p+2} y = 0,$$

and from

$$T(RT)^p Ry = (TR)^{p+1} \ne 0$$

we obtain that $(RT)^p Ry \ne 0$. Thus, $Ry \in \ker (RT)^{p+2}$ but $Ry \notin \ker (R)S^p$, a contradiction.

To show that $p(RT) \le p(TR) + 1$, let $p(TR) := p < \infty$ and suppose that $p(RT) > p + 1$. Then there exists an $x \in \ker (RT)^{p+2}$ such that $x \notin \ker (RT)^{p+1}$. We have

$$(TR)^{p+2} Tx = T(RT)^{p+2} x = 0,$$

and from

$$R(TR)^p Tx = R(TR)^p Tx = (RT)^{p+1} x \ne 0$$

we deduce that $(TR)^p Tx \ne 0$. Thus $Tx \in \ker (TR)^{p+2}$ and $Tx \notin \ker (TR)^p$, which is impossible. Therefore, $p(RT) \le p(TR) + 1$.

(ii) Let $q := q(RT) < \infty$. To show $q(RT) \le q + 1$, suppose that $q(RT) > q + 1$. Then there exists a $y \in (TR)^{q+1}(X)$ such that $y \notin (TR)^{q+2}(X)$. Now, let $z \in X$ be such that

$$y = (TR)^{q+1} z = T(RT)^q Rz,$$

and consider the element

$$u := (RT)^q Rz \in (RT)^q(X) = (RT)^{q+2}(X).$$

Then $y = Tu$, so there exists a $w \in X$ such that

$$y = T(RT)^{q+2}w = (TR)^{q+2}Tw \in (TR)^{q+2}(X)$$

and this is impossible. Therefore, $p(RT) \leq p(TR) + 1$ is proved.

To complete the proof, suppose that $q := q(TR) < \infty$ and that $q(RT) > q + 1$. Then there exists an $x \in (RT)^{q+1}(X)$ such that $x \notin (RT)^{q+2}(X)$. Let $v \in X$ such that

$$x = (RT)^{q+1}v = R(TR)^q Tv,$$

and set

$$t := (TR)^q Tv \in (TR)^q(X) = (TR)^{q+2}(X).$$

Then there exists an $s \in X$ such that

$$x = (R(TR)^{q+2}s = (RT)^{q+2}Rs \in (RT)^{q+2}(X),$$

and this is impossible. Therefore, $q(RT) \leq q + 1$. ∎

Example 3.97 An elementary example shows that we may have $p(TR) \neq p(RT)$ and $q(TR) \neq q(RT)$. Let us consider the 2×2 matrices

$$T := \begin{pmatrix} 0 & 1 \\ 0 & 0 \end{pmatrix} \quad \text{and} \quad R := \begin{pmatrix} 0 & 0 \\ 0 & 1 \end{pmatrix}.$$

Then

$$TR = \begin{pmatrix} 0 & 1 \\ 0 & 0 \end{pmatrix} \quad \text{and} \quad RT = \begin{pmatrix} 0 & 0 \\ 0 & 0 \end{pmatrix},$$

and it is easy to check that $p(TR) = q(TR) = 2$, while $p(RT) = q(RT) = 1$.

Theorem 3.98 *Let $T, R \in L(X)$. Then the nonzero points of the Weyl spectra, or the nonzero points of the Browder spectra, of TR and RT are the same. Furthermore,*

$$\sigma_d(TR) = \sigma_d(RT), \quad \sigma_{ld}(TR) = \sigma_{ld}(RT), \quad \sigma_{ld}(TR) = \sigma_{ld}(RT).$$

Proof The assertion concerning Weyl spectra is a consequence of Theorem 2.149. The assertion concerning Browder spectra is a consequence of Theorem 3.95. The assertion concerning the Drazin spectra is a consequence of Theorem 3.95 and Lemma 3.96. ∎

From Theorem 3.68 we know that $T \in L(X)$ is algebraic if and only if $\sigma_d(T) = \emptyset$, while T is meromorphic if $\sigma_d(T) \subseteq \{0\}$. An easy consequence of Theorem 3.98 is the following corollary:

Corollary 3.99 *Let $T, R \in L(X)$. Then we have:*

(i) *TR is meromorphic if and only if RT is meromorphic.*
(ii) *TR is algebraic if and only if RT is algebraic.*

Lemma 3.100 *Let $T, R \in L(X)$ and $\lambda \neq 0$. Then $T(\ker(I - RT)) = \ker(I - TR)$.*

Proof Let $x \in \ker(I - RT)$, so $RTx = x$. Then $TRTx = Tx$, and thus, $T(\ker(I - RT)) \subseteq \ker(I - TR)$. To verify the opposite inclusion, suppose that $y \in \ker(I - TR)$. Arguing as above then

$$R(\ker(I - TR)) \subseteq \ker(I - RT).$$

Therefore, $Ry \in \ker(I - RT)$, and hence $y = TRy \in T(\ker(I - RT))$. ∎

Also the non-zero points of the essential Drazin spectra of RS and SR, respectively, are the same:

Theorem 3.101 *If $T, R \in L(X)$ then*

$$\sigma_{ld}^e(TR) \setminus \{0\} = \sigma_{ld}^e(RT) \setminus \{0\} \quad and \quad \sigma_{rd}^e(TR) \setminus \{0\} = \sigma_{rd}^e(RT) \setminus \{0\}.$$

Proof By part (iii) of Theorem 3.95 we have only to show that the essential ascent, or the essential descent, of $\lambda I - RT$ and $\lambda I - TR$, with $\lambda \neq 0$, are equal. We can take $\lambda = 1$. Observe that if for every integer $n \geq 0$ we set

$$U_n := \sum_{k=1}^{n+1} (-1)^{k-1} \binom{n+1}{k} R(TR)^{k-1},$$

then by direct computation we obtain.

$$(I - TR)^{n+1} = I - TU_n \quad \text{and} \quad (I - RT)^{n+1} = I - U_nT.$$

Now, by Lemma 3.100 we have

$$T(\ker(I - U_nT)) = \ker(I - TU_n) \quad \text{for all } n \in \mathbb{N},$$

so the operator $\widetilde{S}$ induced by T from $\frac{\ker(I-U_{n+1}T)}{\ker(I-U_nT)}$ to $\frac{\ker(I-TU_{n+1})}{\ker(I-TU_n)}$ is onto. Because $\ker T \cap \ker(I - U_{n+1}) = \{0\}$, it then follows that $\widetilde{S}$ is an isomorphism, so the essential ascent of $p_e(I - RT)$ and $p_e(I - TR)$ coincide.

The assertion concerning the essential right Drazin spectrum is obtained by duality. ∎

Recall that, by Theorem 1.125, if $\mathcal{F}(X)$ is the ideal of all finite-dimensional operators in $L(X)$ and $\pi : L(X) \to \mathcal{L} = L(X)/\mathcal{F}(X)$ is the quotient mapping, then $T \in L(X)$ is B-Fredholm if and only if $\pi(T)$ is a Drazin invertible element of the algebra $\mathcal{L}$.

Theorem 3.102 *Let* $T, R \in L(X)$. *Then* TR *is B-Fredholm if and only if* RT *is B-Fredholm. In this case,* $\mathrm{ind}\,(TR) = \mathrm{ind}\,(RT)$. *Moreover,* TR *is B-Weyl if and only if* RT *is B-Weyl.*

Proof If TR is B-Fredholm then $\pi(TR) = \pi(T)\pi(R)$ is Drazin invertible in $\mathcal{L}$. By Theorem 1.124 then $\pi(RT) = \pi(R)\pi(T)$ is Drazin invertible in $\mathcal{L}$, or equivalently, by Theorem 1.125, TR is B-Fredholm. To show that the indexes of TR and RT are the same, recall that, since TR and RT are B-Fredholm, by Theorem 1.117 we have that $\frac{1}{n}I - TR$ and $\frac{1}{n}I - RT$ are both Fredholm, for n sufficiently large. Moreover,

$$\mathrm{ind}\,(TR) = \mathrm{ind}\,\left(\frac{1}{n}I - TR\right) \quad \text{and} \quad \mathrm{ind}\,(RT) = \mathrm{ind}\,\left(\frac{1}{n}I - RT\right).$$

An obvious consequence of Theorem 2.150 is that

$$\mathrm{ind}\,(TR) = \mathrm{ind}\,\left(\frac{1}{n}I - TR\right) = \mathrm{ind}\,\left(\frac{1}{n}I - RT\right) = \mathrm{ind}\,(RT).$$

∎

3.7 Regularities

It is well known that the spectral mapping theorem holds for T, see [179, Theorem 48.2], i.e.,

$$\sigma(f(T)) = f(\sigma(T)) \quad \text{for all } f \in \mathcal{H}(\sigma(T)).$$

It is natural to ask whether the spectral theorem holds for all, or for some, of the spectra previously defined. A straightforward consequence of the Atkinson characterization of Fredholm operators shows that the spectral mapping theorem holds for the essential spectrum $\sigma_e(T)$, i.e.

$$f(\sigma_e(T)) = \sigma_e(f(T)) \quad \text{for all } f \in \mathcal{H}(\sigma(T)).$$

We have

$$\sigma_b(T)) = \sigma_e(T) \cup \text{acc}\,\sigma(T).$$

Indeed, $\sigma_e(T) \subseteq \sigma_b(T)$, and $\text{acc}\,\sigma(T) \subseteq \sigma_b(T)$, since every $\lambda \notin \sigma_b(T)$ is an isolated point of the spectrum. Conversely, if $\lambda \in \sigma_e(T) \cup \text{acc}\,\sigma(T)$, then either $\lambda I - T \notin \Phi(X)$, or $\lambda \notin \text{iso}\,\sigma(T)$, and in both cases $\lambda I - T$ cannot be Browder. An analogous argument shows the following two equalities, just take into account Theorem 2.101:

$$\sigma_{ub}(T)) = \sigma_{usf}(T) \cup \text{acc}\,\sigma_{ap}(T) \quad \text{and} \quad \sigma_{lb}(T)) = \sigma_{lsf}(T) \cup \text{acc}\,\sigma_s(T).$$

From the analyticity of the function $f \in \mathcal{H}(\sigma(T))$, it is easily seen that $f(\text{acc}\,\sigma(T)) = \text{acc}\,(f(\sigma(T))$. Therefore,

$$f(\sigma_b(T)) = f(\sigma_e(T) \cup \text{acc}\,\sigma(T)) = f(\sigma_e(T)) \cup f(\text{acc}\,\sigma(T))$$
$$= \sigma_e(f(T)) \cup \text{acc}\,(f(\sigma(T)) = \sigma_b(f(T)),$$

so the spectral mapping theorem also holds for the Browder spectrum.

The following example shows that the spectral theorem in general does not hold for the Weyl spectrum $\sigma_w(T)$.

Example 3.103 Let $T := R \oplus (R^* + 2I)$, where R is the right unilateral shift in $\ell^2(\mathbb{N})$. Then R^* is the left unilateral shift in $\ell^2(\mathbb{N})$. Set $f(\lambda) := (\lambda(\lambda - 2)$. Then

$$f(T) = T(T - 2I) = [R \oplus (R^* + 2I)][R - 2I] \oplus R^*.$$

Since $\text{ind}\,(R) = -1$, and both $R^* + 2I$, $R - 2I)$ are invertible, it then follows that T as well as $T - 2I$ are Fredholm, $\text{ind}(T) = -1$ and $\text{ind}\,(T - 2I) = 1$. We have

$$\text{ind}\,f(T) = \text{ind}(T) + \text{ind}\,(T - 2I) = 0,$$

so $f(T)$ is Weyl, hence $0 \notin \sigma_w(f(T))$, whereas $0 = f(0) \in f(\sigma_w(T))$.

The last example shows that the notion of Weyl spectrum does not derive from the notion of invertibility in some algebras, since in this case the spectral mapping theorem holds for all polynomials, see Dieudonné [110].

However, we show now that the spectral mapping theorem holds for many other kinds of spectra which originate from Fredholm theory. To do this, we briefly outline an axiomatic theory of the spectrum, in particular we introduce the concept of regularity. This concept may be introduced in the more general context of Banach algebras.

Let $\mathcal{A}$ be a unital Banach algebra with unit u, and let us denote by $\text{inv}\,\mathcal{A}$ the set of all invertible elements.

Definition 3.104 A non-empty subset $\mathbf{R}$ of $\mathcal{A}$ is said to be a *regularity* if the following conditions are satisfied:

(i) $a \in \mathbf{R} \Leftrightarrow a^n \in \mathbf{R}$ for all $n \in \mathbb{N}$.
(ii) If a, b, c, d are mutually commuting elements of $\mathcal{A}$ and $ac + bd = u$ then

$$ab \in \mathbf{R} \Leftrightarrow a \in \mathbf{R} \text{ and } b \in \mathbf{R}.$$

Denote by $\operatorname{inv} \mathcal{A}$ the set of all invertible elements of $\mathcal{A}$. It is easily seen that if $\mathbf{R}$ is a regularity then $u \in \mathbf{R}$ and $\operatorname{inv} \mathcal{A} \subseteq \mathbf{R}$. Moreover, if $a, b \in \mathcal{A}$, $ab = ba$, and $a \in \operatorname{inv} \mathcal{A}$, then

$$ab \in \mathbf{R} \Leftrightarrow b \in \mathbf{R}. \tag{3.33}$$

In fact $a\, a^{-1} + b\, 0 = u$, so property (ii) above applies.

It is easy to verify the following criterion.

Theorem 3.105 *Let $\mathbf{R} \neq \emptyset$ be a subset of $\mathcal{A}$. Suppose that for all commuting elements $a, b \in \mathcal{A}$ we have*

$$ab \in \mathbf{R} \Leftrightarrow a \in \mathbf{R} \text{ and } b \in \mathbf{R}. \tag{3.34}$$

Then $\mathbf{R}$ is a regularity.

Denote by

$$\sigma_{\mathbf{R}}(a) := \{\lambda \in \mathbb{C} : \lambda u - a \notin \mathbf{R}\}$$

the *spectrum corresponding to the regularity* $\mathbf{R}$. Obviously, $\operatorname{inv} \mathcal{A}$ is a regularity by Theorem 3.105, and the corresponding spectrum is the ordinary spectrum. Note that $\sigma_{\mathbf{R}}(a)$ may be empty. For instance, if $\mathcal{A} := L(X)$ and $\mathbf{R} = L(X)$, or if $T \in L(X)$ is algebraic and $\mathbf{R}$ is the set of all Drazin invertible operators (see below). Indeed, $\sigma_d(T)$ is empty since the spectrum is a finite set of poles, by Theorem 3.68. In particular, $\sigma_d(N) = \emptyset$ for every nilpotent operator N, since N is algebraic. In general, $\sigma_{\mathbf{R}}(a)$ is not compact, and $\sigma_{\mathbf{R}}(a) \subseteq \sigma(a)$ for every regularity $\mathbf{R}$ and $a \in \mathcal{A}$. The proof of the following result is immediate.

Theorem 3.106 *The intersection $\mathbf{R}$ of a family $(\mathbf{R}_\alpha)_\alpha$ of regularities is again a regularity. Moreover,*

$$\sigma_{\mathbf{R}}(a) = \bigcup_\alpha \sigma_{\mathbf{R}_\alpha}(a), \quad a \in \mathcal{A}.$$

The union $\mathbf{R}$ of a directed system of regularities $(\mathbf{R}_\alpha)_\alpha$ is again a regularity. Moreover,

$$\sigma_{\mathbf{R}}(a) = \bigcap_\alpha \sigma_{\mathbf{R}_\alpha}(a), \quad a \in \mathcal{A}.$$

Let $\mathcal{H}_{nc}(\sigma(a))$ denote the set of all analytic functions, defined on an open neighborhood of $\sigma(a)$, such that f is non-constant on each of the components of its domain.

Theorem 3.107 *Suppose that* $\mathbf{R}$ *is a regularity in a Banach algebra* $\mathcal{A}$ *with unit* u. *Then* $\sigma_{\mathbf{R}}(f(a)) = f(\sigma_{\mathbf{R}}(a))$ *for every* $a \in \mathcal{A}$ *and every* $f \in \mathcal{H}_{nc}(\sigma(a))$.

Proof It is sufficient to prove that

$$\mu \notin \sigma_{\mathbf{R}}(f(a)) \Leftrightarrow \mu \notin f(\sigma_{\mathbf{R}}(a)). \tag{3.35}$$

Since $f(\lambda) - \mu$ has only a finite number of zeros $\lambda_1, \ldots, \lambda_n$ in the compact set $\sigma(a)$, we can write

$$f(\lambda) - \mu = (\lambda - \lambda_1)^{\nu_1} \cdots (\lambda - \lambda_n)^{\nu_n} \cdot g(\lambda),$$

where g is an analytic function defined on an open set containing $\sigma(a)$ and $g(\lambda) \neq 0$ for $\lambda \in \sigma(a)$. Then

$$f(a) - \mu u = (a - \lambda_1 u)^{\nu_1} \cdots (a - \lambda_n u)^{\nu_n} \cdot g(a),$$

with $g(a)$ invertible by the spectral mapping theorem for the ordinary spectrum. Therefore, (3.35) is equivalent to

$$f(a) - \mu u \in \mathbf{R} \Leftrightarrow a - \lambda_k u \in \mathbf{R} \quad \text{for all } k = 1, \ldots n. \tag{3.36}$$

But $g(a)$ is invertible, so, by (3.33) and by the definition of a regularity, the equivalence (3.36) holds if and only if

$$(a - \lambda_1 u)^{\nu_1} \cdots (a - \lambda_n u)^{\nu_n} \in \mathbf{R} \Leftrightarrow (a - \lambda_k)^{\nu_k} \in \mathbf{R} \quad \text{for all } k = 1, \ldots n. \tag{3.37}$$

To show (3.37) observe first that for all relatively prime polynomials p, q there exist polynomials p_1, q_1 such that $pp_1 + qq_1 = 1$ and we have $p(a)p_1(a) + q(a)q_1(a) = u$. Applying property (ii) of Definition 3.104 we then obtain, by induction, that the equivalence (3.37) holds. ∎

In the assumptions of Theorem 3.107 the condition that f is non-constant on each component cannot dropped. In fact, the spectral mapping theorem for constant functions cannot be true if $\sigma_{\mathbf{R}}(a) = \emptyset$ for some $a \in \mathcal{A}$ and $0 \notin \mathbf{R}$.

Now we give, in the case of $\mathcal{A} = L(X)$, a criterion which ensures that the spectral mapping theorem holds for *all* analytic functions $f \in \mathcal{H}(\sigma(T))$, defined on an open neighborhood of $\sigma(a)$. Let us consider a regularity $\mathbf{R}(X) \subseteq L(X)$ and let X_1, X_2 be a pair of closed subspaces of X for which $X = X_1 \oplus X_2$. Define

$$\mathbf{R}^1 := \{T_1 \in L(X_1) : T_1 \oplus I_{X_2} \in \mathbf{R}\}$$

and

$$\mathbf{R}^2 := \{T_2 \in L(X_2) : I_{X_1} \oplus T_2 \in \mathbf{R}\}.$$

It is easy to see that if $X_1 \neq 0$, both $\mathbf{R}^1$ and $\mathbf{R}^2$ are regularities in $L(X_1)$ and $L(X_2)$, respectively. Now, assume that a regularity $\mathbf{R}$ satisfies the following condition:

$$\sigma_{\mathbf{R}^1}(T_1) \neq \emptyset \text{ for all } T_1 \in L(X_1) \text{ and } \mathbf{R}_1 \neq L(X_1). \tag{3.38}$$

It is easily seen that if $\mathbf{R}$ satisfies the condition (3.38) then we have

$$\sigma_{\mathbf{R}}(T) = \sigma_{\mathbf{R}^1}(T_1) \cup \sigma_{\mathbf{R}^2}(T_2). \tag{3.39}$$

The proof of the following result may be found in [243, Chapter 6].

Theorem 3.108 *Let $\mathbf{R}$ be a regularity in $L(X)$, and suppose that for all closed subspaces X_1 and X_2, $X = X_1 \oplus X_2$, such that the regularity $\mathbf{R}^1 \neq L(X_1)$ and $\sigma_{\mathbf{R}^1}(T_1) \neq \emptyset$ for all $T_1 \in L(X_1)$. Then*

$$\sigma_{\mathbf{R}}(f(T)) = f(\sigma_{\mathbf{R}}(T))$$

for every $T \in L(X)$ and every $f \in \mathcal{H}(\sigma(T))$.

In many situations a regularity decomposes as required in Theorem 3.108. For instance, if $\mathbf{R} := \{T \in L(X) : T \text{ is onto}\}$ and $X = X_1 \oplus X_2$ then $\mathbf{R}_i = \{T_i \in L(X_i) : T_i \text{ is onto}\}$, $i = 1, 2$, and $T_1 \oplus T_2$ is onto if and only if T_1, T_2 are onto. Thus the spectral mapping theorem for all $f \in \mathcal{H}(\sigma(T))$ is reduced, by Theorem 3.108, to the question of the non-emptiness of the spectrum.

The axioms of regularity are usually rather easy to verify and there are many classes of operators in Fredholm theory satisfying them. An excellent survey concerning the regularity of various classes of bounded linear operators in Banach spaces may be found in Mbekhta and Müller [235].

In the sequel we just list some of these classes. Let us consider the following sets:

(1) $\mathbf{R}_1 := \{T \in L(X) : T \text{ is bounded below}\}$. In this case $\sigma_{\mathbf{R}_1}(T) = \sigma_{\mathrm{ap}}(T)$.
(2) $\mathbf{R}_2 := \{T \in L(X) : T \text{ is onto}\}$. In this case $\sigma_{\mathbf{R}_2}(T) = \sigma_{\mathrm{s}}(T)$.
(3) $\mathbf{R}_3 := \{T \in L(X) : T \in \Phi(X)\}$. In this case $\sigma_{\mathbf{R}_3}(T) = \sigma_{\mathrm{e}}(T)$ is the essential spectrum.
(4) $\mathbf{R}_4 := \Phi_+(X)$. The corresponding spectrum is the *upper semi-Fredholm spectrum* $\sigma_{\mathrm{usf}}(T)$, also known in the literature as *the essential approximate point spectrum*.
(5) $\mathbf{R}_5 := \Phi_-(X)$. The corresponding spectrum is the *lower semi-Fredholm spectrum* $\sigma_{\mathrm{lsf}}(T)$, also known in the literature as *the essential surjective spectrum*.

(6) $\mathbf{R_6} := B_+(X)$. The corresponding spectrum is the upper semi-Browder spectrum $\sigma_{ub}(T)$.

(7) $\mathbf{R_7} := B_-(X)$. The corresponding spectrum is the lower semi-Browder spectrum $\sigma_{lb}(T)$.

(8) $\mathbf{R_8} := B(X)$. The corresponding spectrum is the Browder spectrum $\sigma_b(T)$.

(9) $\mathbf{R_9} := \{T \in L(X) : T \text{ is semi-regular}\}$. In this case $\sigma_{\mathbf{R_9}}(T)$ is the semi-regular spectrum, also known as the *Kato spectrum*.

(10) $\mathbf{R_{10}} := \{T \in L(X) : T \text{ is essentially semi-regular}\}$. In this case $\sigma_{\mathbf{R_9}}(T)$ is the essentially semi-regular spectrum.

(11) The complemented version of semi-regular operators is given by *Saphar operators*. A bounded operator $T \in L(X)$ is said to be *Saphar* if T is both semi-regular and relatively regular (i.e. there exists an $S \in L(X)$ such that $TST = T$), see for details [274] or [245, Chapter II]. Obviously, in a Hilbert space, since every closed subspace is complemented, $T \in L(H)$ is Saphar if and only if T is semi-regular. If

$$\mathbf{R_{11}} := \{T \in L(X) : T \text{ is Saphar}\},$$

the corresponding spectrum $\sigma_{\mathbf{R_{11}}}(T)$ is called the *Saphar spectrum*.

(12) $\mathbf{R_{12}} := \{T \in L(X) : T \text{ is Drazin invertible}\}$. In this case $\sigma_{\mathbf{R_{12}}}(T)$ is the Drazin spectrum.

(13) $\mathbf{R_{13}} := \{T \in L(X) : T \text{ is left Drazin invertible}\}$. In this case $\sigma_{\mathbf{R_{13}}}(T)$ is the left Drazin spectrum.

(13) $\mathbf{R_{14}} := \{T \in L(X) : T \text{ is right Drazin invertible}\}$. In this case $\sigma_{\mathbf{R_{14}}}(T)$ is the right Drazin spectrum.

(14) $\mathbf{R_{15}} := \{T \in L(X) : T \text{ is essentially right Drazin invertible}\}$. In this case $\sigma_{\mathbf{R_{15}}}$ is the essential right Drazin spectrum.

(15) $\mathbf{R_{16}} := \{T \in L(X) : T \text{ is essentially left Drazin invertible}\}$. In this case $\sigma_{\mathbf{R_{16}}}$ is the essential left Drazin spectrum.

(16) $\mathbf{R_{17}} := \{T \in L(X) : T \text{ is quasi-Fredholm}\}$. In this case the spectrum $\sigma_{\mathbf{R_{17}}}$ is the quasi-Fredholm spectrum $\sigma_{qf}(T)$.

(17) $\mathbf{R_{18}} := \{T \in L(X) : T \text{ has topological uniform descent}\}$.

All the sets $\mathbf{R_i}$, $i = 1, 2, \ldots, 18$ are regularities.

Theorem 3.109 *If $T \in L(X)$ and $f \in \mathcal{H}(\sigma(T))$ then*

$$\sigma_{\mathbf{R_i}}(f(T)) = f(\sigma_{\mathbf{R_i}}(T)) \quad \text{for all } i = 1, 2, \ldots, 10.$$

Moreover, if the function f is non-constant on each component of its domain of definition then

$$\sigma_{\mathbf{R_i}}(f(T)) = f(\sigma_{\mathbf{R_i}}(T)) \quad \text{for all } i = 11, \ldots, 18.$$

Proof All the regularities $\mathbf{R_i}$ satisfy the conditions of Theorem 3.108, see Chapter III of [243]. ∎

The spectral mapping theorem for the Drazin spectrum entails the following result:

Corollary 3.110 *If* $T \in L(X)$ *is meromorphic and* $f \in \mathcal{H}_{nc}(\sigma(T))$ *such that* $f(0) = 0$ *then* $f(T)$ *is meromorphic. The converse is true whenever* f *vanishes only at 0.*

Proof If T is meromorphic then $\sigma_d(T) \subseteq \{0\}$, so $\sigma_d(f(T)) = f(\sigma_d(T)) \subseteq f(\{0\}) \subseteq \{0\}$, hence $f(T)$ is meromorphic. Conversely, assume that $f(T)$ is meromorphic and that f vanishes only at 0. Then $f(\sigma_d(T)) = \sigma_d(f(T)) \subseteq \{0\}$, from which we obtain $\sigma_d(T) \subseteq \{0\}$. ∎

We have already seen that, for some i, for the products TR and RT we have $\sigma_{\mathbf{R_i}}(TR) \setminus \{0\} = \sigma_{\mathbf{R_i}}(TR)$. This may be extended to every regularity $\mathbf{R}_i$ for $1 \leq i \leq 16$. More precisely, we have;

Theorem 3.111 *Let* $T, R \in L(X)$. *Then* $\sigma_{\mathbf{R_i}}(TR) \setminus \{0\} = \sigma_{\mathbf{R_i}}(TR) \setminus \{0\}$ *for every* $1 \leq i \leq 18$. *Furthermore, for the B-Fredholm spectrum and the B-Weyl spectrum we have*

$$\sigma_{bf}(TR) = \sigma_{bf}(RT) \quad and \quad \sigma_{bw}(TR) = \sigma_{bw}(RT). \tag{3.40}$$

For a proof, see Zeng and Zhong [303] and [301]. The equalities (3.40) are clear from Theorem 3.102.

We have already observed that the spectral mapping theorem does not hold for the Weyl spectrum. In order to give some sufficient conditions for which the spectral mapping theorem holds for the Weyl spectra we introduce the abstract concept of a Φ-semigroup.

Let $\mathcal{A}$ be a complex Banach algebra with identity u and $\mathcal{J}$ a closed two-sided ideal of $\mathcal{A}$. Let ϕ be the canonical homomorphism of $\mathcal{A}$ onto $\widehat{\mathcal{A}} := \mathcal{A}/\mathcal{J}$. Let us denote by $\widehat{\mathcal{G}}$ the group of all invertible elements in $\widehat{\mathcal{A}}$.

Definition 3.112 An open semigroup $\mathcal{S}$ of $\mathcal{A}$ is said to be a Φ-semigroup if the following properties hold:

(i) If $a, b \in \mathcal{A}$ and $ab = ba \in \mathcal{S}$ then $a \in \mathcal{S}$ and $b \in \mathcal{S}$;
(ii) There exists a closed two-sided ideal $\mathcal{J}$ and an open semi-group $\widehat{\mathcal{R}}$ in $\widehat{\mathcal{A}} = \mathcal{A}/\mathcal{J}$ such that $\widehat{\mathcal{G}} \subseteq \widehat{\mathcal{R}}$, $\widehat{\mathcal{R}} \setminus \widehat{\mathcal{G}}$ is open and $\mathcal{S} = \phi^{-1}(\widehat{\mathcal{R}})$.

Evidently, any Φ-semi-group $\mathcal{S}$ contains all invertible elements of $\mathcal{A}$ and $\mathcal{S} + \mathcal{J} \subseteq \mathcal{S}$.

For every $a \in \mathcal{S}$ let us denote by $\mathcal{S}_a$ the component of $\mathcal{S}$ containing a. If $b \in \mathcal{J}$ and

$$\Gamma := \{a + tb : 0 \leq t \leq 1\}$$

is a path joining the two elements a and $a + b$ then the inclusion $S + J \subseteq S$ implies that $\Gamma \subseteq S$. From this it follows that $a + J \subseteq S_a$. This also implies that $S = S_1 \cup S_2$, where S_1 and S_2 are open disjoint subsets of S, so $S_i + J \subseteq S_i$ for $i = 1, 2$.

The *index* on a Φ-semigroup S is defined as a locally constant homomorphism of J into $\mathbb{N}$. Evidently, if $i : S \to \mathbb{N}$ is an index, then

$$i(a + b) = i(a) \quad \text{for all } a \in S, \ b \in J.$$

The sets $\Phi(X)$, $\Phi_+(X)$ and $\Phi_-(X)$ satisfy condition (i) of Definition 3.112. The Atkinson characterization of Fredholm operators establishes that $\Phi(X) = \phi^{-1}(\widehat{\mathcal{G}})$, where $\widehat{\mathcal{G}}$ is the set of all invertible elements of $\widehat{\mathcal{A}} := L(X)/K(X)$, so $\Phi(X)$ also satisfies condition (ii) and hence is a Φ-semigroup of $\mathcal{A} = L(X)$. Since the canonical homomorphism ϕ is an open mapping, it follows that $\Phi_+(X)$ and $\Phi_-(X)$ are also Φ-semigroups of $\mathcal{A} := L(X)/K(X)$.

For every $a \in \mathcal{A}$ and a Φ-semigroup S of $\mathcal{A}$ let us consider the spectrum generated by S:

$$\sigma_S(a) := \{\lambda \in \mathbb{C} : \lambda u - a \notin S\}.$$

The following result establishes an abstract spectral mapping theorem for spectra generated by Φ-semigroups. A proof of it may be found in [1, Theorem 3.60].

Theorem 3.113 *Let $\mathcal{A}$ be a Banach algebra with identity u and S any Φ-semigroup. Suppose that $i : S \to \mathbb{N}$ is an index such that $i(b) = 0$ for all $b \in \mathrm{inv}\,\mathcal{A}$. If f is an analytic function on an open domain $\mathbb{D}$ containing $\sigma(a)$, then the following statements hold:*

(i) *$f(a) \in S$ if and only if $f(\lambda) \neq 0$ for all $\lambda \in \sigma_S(a)$;*
(ii) *$\sigma_S(f(a)) = f(\sigma_S(a))$.*

An immediate consequence of Theorem 3.113 is that the spectral mapping theorem holds for the Fredholm spectrum $\sigma_f(T)$, for the upper semi-Fredholm spectrum $\sigma_{uf}(T)$, and for the lower semi-Fredholm spectrum $\sigma_{uf}(T)$, since $\Phi(X)$, $\Phi_+(X)$ and $\Phi_-(X)$ are Φ-semigroups.

Let $a \in \mathcal{A}$ and S be a Φ-semigroup of $\mathcal{A}$ with an index i. For every $n \in \mathbb{N}$ let us define

$$\Omega_n := \{\lambda \in \sigma(T) : i(\lambda u - a) = n\}.$$

A proof of the following may be found in [1, Lemma 3.62].

Lemma 3.114 *Suppose that $f(a) \in S$ and let α_n be the number of zeros of f on Ω_n, counted according to their multiplicities, ignoring components of Ω_n where f is identically 0. Then*

$$i(f(a)) = \sum_n n\alpha_n. \tag{3.41}$$

For the Weyl spectra we have, in general, only the following inclusion:

Theorem 3.115 *Let $T \in L(X)$ be a bounded operator on the Banach space X. If $f \in \mathcal{H}(\sigma(T))$, then the following inclusions hold:*

(i) $f(\sigma_{\mathrm{sf}}(T)) \subseteq \sigma_{\mathrm{sf}}(f(T))$;
(ii) $\sigma_{\mathrm{w}}(f(T)) \subseteq f(\sigma_{\mathrm{w}}(T))$;
(iii) $\sigma_{\mathrm{uw}}(f(T)) \subseteq f(\sigma_{\mathrm{uw}}(T))$;
(iv) $\sigma_{\mathrm{lw}}(f(T)) \subseteq f(\sigma_{\mathrm{lw}}(T))$.

Proof

(i) We have

$$\sigma_{\mathrm{sf}}(f(T)) = \sigma_{\mathrm{uf}}(f(T)) \cap \sigma_{\mathrm{uf}}(f(T)) = gf(\sigma_{\mathrm{uf}}(T)) \cap f(\sigma_{\mathrm{lf}}(T))$$
$$\supseteq f[\sigma_{\mathrm{uf}}(T) \cap (\sigma_{\mathrm{lf}}(T)] = f(\sigma_{\mathrm{sf}}(T)).$$

(ii) For every $n \in \mathbb{N}$ define

$$\Phi_n(X) := \{T \in \Phi(X) : \mathrm{ind}\, T = n\},$$

and

$$\Omega_n := \{\lambda \in \mathbb{C} : \lambda I - T \in \Phi_n(X)\}.$$

Evidently

$$\sigma_{\mathrm{w}}(T) = \sigma_{\mathrm{f}}(T) \cup \left(\bigcup_{n \neq 0} \Omega_n \right). \tag{3.42}$$

Now, let $\mu \notin f(\sigma_{\mathrm{w}}(T))$ be arbitrary given. Then $\mu - f(\lambda)$ has no zeros on $\sigma_{\mathrm{w}}(T)$, and in particular has no zero on $\sigma_{\mathrm{f}}(T)$. From part (i) of Theorem 3.113, applied to the Φ-semigroup $\Phi(X)$, we then conclude that $\mu - f(T) \in \Phi(X)$ and

$$\mathrm{ind}\,(\mu I - f(T)) = \sum_{n \neq 0} n\alpha_n,$$

where α_n is the number of isolated zeros of $\mu - f(\lambda)$ on Ω_n, counted according to their multiplicities. From the equality (3.42) we deduce that $\alpha_n = 0$ for every $n \neq 0$. Therefore, $\mathrm{ind}\,(\mu I - f(T)) = 0$ and hence $\mu \notin \sigma_{\mathrm{w}}(f(T))$.

(iii) We consider first the case when f is non-constant on each of the components of the domain of f. Let $\lambda \in \sigma_{\mathrm{uw}}(f(T))$. Write

$$\lambda I - f(T) = \Pi_{i=1}^{n}(\mu_i I - T)^{k_i} h(T),$$

where the scalars μ_i are different and $h(T)$ is invertible. Then for some j, $1 \le j \le n$, we have $\mu_j \in \sigma_{\mathrm{uw}}(T)$. Hence $\lambda = f(\mu_j) \in f(\sigma_{\mathrm{uw}}(T))$, so (iii) is proved in this case.

Suppose now that $f \in \mathcal{H}(\sigma(T))$ is arbitrarily given, and let $\lambda \in \sigma_{\mathrm{uw}} f((T))$. If $g(z) := \lambda - f(z)$ then g is defined on an open set $\mathcal{U} = \mathcal{U}_1 \cup \mathcal{U}_2$, with $\mathcal{U}_1$ and $\mathcal{U}_2$ open, $\mathcal{U}_1 \cap \mathcal{U}_2 = \emptyset$, with

$$\sigma_1 := \sigma(T) \cap \mathcal{U}_1 \neq \emptyset, \quad \sigma_1 := \sigma(T) \cap \mathcal{U}_1 \neq \emptyset,$$

and $g \equiv 0$ on $\mathcal{U}_1$, $g|\mathcal{U}_2$ non-constant on an open set containing σ_2. Let P be the spectral projection associated with σ_2, and set $T_1 := T|\ker P$ and $T_2 := T|P(X)$. From the spectral canonical decomposition we then have $X = \ker P \oplus P(X)$ and $\sigma(T_i) = \sigma_i$, $i = 1, 2$.

Assume that $\lambda \notin f(\sigma_{\mathrm{uw}}(T))$. Then $\lambda \notin f(\sigma_{\mathrm{uf}}(T)) = \sigma_{\mathrm{uf}}(f(T))$, since the spectral mapping theorem holds for the upper semi-Fredholm spectrum and $\sigma_{\mathrm{uf}}(T) \subseteq \sigma_{\mathrm{uw}}(T)$. Hence $\lambda I - f(T) \in \Phi_+(X)$, and according to Lemma 3.114 we have

$$\operatorname{ind} g(T) = \sum_{n \neq 0} n \alpha_n,$$

where α_n is the number of zeros of g on Ω_n. Since

$$\sigma_{\mathrm{uw}}(T) = \sigma_{\mathrm{uf}}(T) \cup \{\bigcup_{n>0} \Omega_n\},$$

and $\lambda \notin f(\sigma_{\mathrm{uw}}(T))$ it then follows that

$$\operatorname{ind}(g(T) = \sum_{n<0} n \alpha_n < 0.$$

Hence $\lambda \notin \sigma_{\mathrm{uw}}(f(T))$, so (iii) is proved.
(iv) Proceed by duality.   ∎

It is already noted that the equality $\sigma_{\mathrm{w}}(f(T)) = f(\sigma_{\mathrm{w}}(T))$ in general does not hold. The following example shows that also in (i) of Theorem 3.115, the equality, in general, is not satisfied.

Example 3.116 Let us consider an operator $T \in L(X)$, X a Banach space, for which $(I + T)(X)$ is closed and that is such that

$$\alpha(I + T) < \infty, \quad \beta(I + T) = \infty, \quad \alpha(I - T) = \infty, \quad \beta(I - T) < \infty.$$

Clearly $I + T \in \Phi_+(X)$ and $I - T \in \Phi_-(X)$, so $\{1, -1\} \subseteq \sigma_{\mathrm{sf}}(T)$. Define

$$g(\lambda) := (1 + \lambda)(1 - \lambda).$$

Then $\alpha(g(T)) = \beta(g(T)) = \infty$, thus $0 \in \sigma_{sf}(g(T))$. On the other hand, it is clear that $0 \notin g(\sigma_{sf}(T))$. This shows that the equality (i) of Theorem 3.115 does not hold.

Let $\rho_*(T)$ denote one of the following resolvent sets $\rho_f(T) := \mathbb{C} \setminus \sigma_f(T)$, $\rho_{uf}(T) := \mathbb{C} \setminus \sigma_{uf}(T)$, $\rho_{bf}(T) := \mathbb{C} \setminus \sigma_{bf}(T)$, and $\rho_{ubf}(T) := \mathbb{C} \setminus \sigma_{ubf}(T)$.

Definition 3.117 Let $T \in L(X)$. We say that T has stable sign index on $\rho_*(T)$ if for every $\lambda, \mu \in \rho_*(T)$ the sign of $\mathrm{ind}(\lambda I - T)$ and the sign of $\mathrm{ind}(\mu I - T)$ are the same.

Remark 3.118 Recall that, by the index theorem (see Appendix A) if $T, S \in \Phi(X)$ then $TS \in \Phi(X)$ with $\mathrm{ind}\, TS = \mathrm{ind}\, T + \mathrm{ind}\, S$. Hence if T, S are Weyl operators then TS is Weyl. Moreover, if $TS = ST$ then $TS \in \Phi(X)$ if and only if T and S are Fredholm. Analogously, if $T, S \in \Phi_+(X)$ then $ST \in \Phi_+(X)$. If $TS = ST$ and $ST \in \Phi_+(X)$ then both T and S belong to $\Phi_+(X)$.

Theorem 3.119 *Let $T \in L(X)$. Then the spectral mapping theorem holds for $\sigma_w(T)$ if and only if T is of stable sign index on the Fredholm region $\rho_f(T)$. Analogously, the spectral mapping theorem holds for $\sigma_{uw}(T)$ if and only if T is of stable sign index on the upper semi-Fredholm region $\rho_{uf}(T)$.*

Proof To prove the first assertion, suppose that T is of stable sign on $\rho_f(T)$. To show that the spectral mapping theorem holds for $\sigma_w(T)$ we have only to prove, by part (i) of Theorem 3.115, the inclusion

$$f(\sigma_w(T)) \subseteq \sigma_w(f(T)). \tag{3.43}$$

Assume first that f is not identically zero in any component of its domain. Let $g(z) := \lambda - f(z)$ and write $g(z) = u(z)h(z)$, where h has no zeros in $\sigma(T)$ and $p(z) := \Pi_{i=1}^n (\lambda_i - \lambda)^{k_i}$, k_i the multiplicity of λ_i. Then $g(T) = \lambda I - f(T) = p(T)h(T)$, where $h(T)$ invertible. Suppose that $\lambda \notin \sigma_w(f(T))$. Then $p(T)$ is Weyl and hence all $\lambda_i I - T$ are Fredholm operators for all $i = 1, \ldots, n$. Clearly, $\mathrm{ind}\, h(T) = 0$, so we have

$$0 = \mathrm{ind}\, g(T) = \mathrm{ind}\, p(T) + \mathrm{ind}\, h(T) = \sum_{i=1}^n k_i \mathrm{ind}\, (\lambda_i I - T).$$

Since T is of stable sign on $\rho_f(T)$, $\mathrm{ind}\, (\lambda_i I - T) = 0$ for every $i = 1, \ldots, n$, so $\lambda_i \notin \sigma_w(T)$, and consequently $\lambda \notin f(\sigma_{uw}(T))$. Therefore the inclusion (3.43) is shown in this case.

In the general case, g is defined on an open set $\mathcal{U} = \mathcal{U}_1 \cup \mathcal{U}_2$, where $\mathcal{U}_1$ and $\mathcal{U}_2$ are two disjoint open sets, $g \equiv 0$ on $\mathcal{U}_1$ and g is not identically zero in any component of $\mathcal{U}_2$. Thus $\sigma(T)$ is the union of two disjoint compact sets $\sigma_i \subseteq \mathcal{U}_i$, $i = 1, 2$. If P is the spectral projection associated with σ_2 and we set $X_1 := \ker P$, $X_2 := P(X)$, $T_i := T|X_i$, by the spectral decomposition theorem we have $X = X_1 \oplus X_2$ and $\sigma(T_i) = \sigma_i$, $i = 1, 2$. But $g \equiv 0$ on σ_1, so $g(T_1) = 0$, thus $g(T) = 0 \oplus g(T_2)$ and $g(T) = g(T)P = Pg(T)$. Further, P is a Weyl operator if and only if dim $X_1 < \infty$ and this is equivalent to saying that σ_1 is a finite set consisting of eigenvalues of T of finite multiplicity. Hence, P is Weyl if and only if $\sigma_w(T) \cap \mathcal{U}_1 = \emptyset$. Since codim $P(X) = \dim \ker P$, and taking into account Remark 3.118, we then conclude that $g(T)$ is a Weyl operator if and only if both P and $g(T_2)$ are Weyl operators. These arguments then show that $\lambda \in f(\sigma_w(T))$ precisely when $\lambda \in \sigma_w(f(T))$. Therefore, the inclusion (3.43) is proved for every $f \in \mathcal{H}(\sigma(T))$.

Assume now that the spectral mapping theorem holds for $\sigma_w(T)$ and that T is not of stable sign on $\rho_f(T)$, i.e., there are $\lambda_1, \lambda_2 \in \rho_f(T)$ such that $\lambda_1 I - T$ and $\lambda_2 I - T$ are Fredholm operators with ind $(\lambda_1 I - T) > 0$ and ind $(\lambda_2 I - T) < 0$. Put

$$m_1 := \text{ind}\,(\lambda_1 I - T),$$

and

$$m_2 := -\text{ind}\,(\lambda_2 I - T).$$

Let us consider the polynomial $p(\lambda) := (\lambda_1 - \lambda)^{m_2}(\lambda_2 - \lambda)^{m_1}$. Then $u(T)$ is Fredholm, with

$$\text{ind}\, p(T) = m_1 m_2 + m_2(-m_1) = 0,$$

thus $0 \notin \sigma_w(p(T)$. Because $\lambda_1 \in \sigma_w(T)$ we have $0 = p(\lambda_1) \in p(\sigma_w(T)) = \sigma_w(p(T))$ since, by assumption, the spectral mapping theorem holds in the particular case of polynomials. This is a contradiction, so the proof of the first assertion is complete.

The proof of the second assertion is similar and it is omitted.   ∎

A closer look at the proof of Theorem 3.119 shows that T is of stable sign index in $\rho_f(T)$ if and only if $\sigma_w(f(T)) \supseteq f(\sigma_w(T))$, and, similarly, T is of stable sign index in $\rho_{uf}(T)$ if and only if $\sigma_{uw}(f(T)) \supseteq f(\sigma_{uw}(T))$.

By Corollary 2.106, if T or T^* has the SVEP then T is of stable sign index on $\rho_f(T)$ and $\rho_{uf}(T)$.

Corollary 3.120 *Let $T \in L(X)$ be such that either T or T^* has the SVEP. Then the spectral mapping theorem holds for the Weyl spectrum $\sigma_w(T)$ and the semi-Weyl spectra $\sigma_{uw}(T)$, $\sigma_{lw}(T)$.*

The spectral mapping theorem for the Weyl spectra also holds if the function $f \in \mathcal{H}(\sigma(T))$ is injective:

Theorem 3.121 *Let $T \in L(X)$ and suppose that $f \in \mathcal{H}(\sigma(T))$ is injective on $\sigma_w(T)$. Then $\sigma_w(f(T)) = f(\sigma_w(T))$.*   ∎

Proof By part (ii) of Theorem 3.115 we need only to prove the inclusion $\sigma_w(f(T)) \subseteq f(\sigma_w(T))$. Suppose that $\mu_0 \in f(\sigma_w(T))$ and let $\lambda_0 \in \sigma_w(T)$

such that $\mu_0 = f(\lambda_0)$. Define $g \in \mathcal{H}(\sigma(T))$ in the same domain of F as

$$g(\lambda) := \begin{cases} \dfrac{f(\lambda) - f(\lambda_0)}{\lambda - \lambda_0} & \text{if } \lambda \neq \lambda_0, \\ f'(\lambda_0) & \text{if } \lambda = \lambda_0. \end{cases}$$

Since f is injective on $\sigma_w(T)$, the function g does not vanish on $\sigma_w(T)$, hence $0 \notin g(\sigma_w(T))$. By part (ii) of Theorem 3.115 we then have that $0 \notin \sigma_w(g(T))$ so $g(T)$ is a Weyl operator. Since $g(T)(\lambda_0 I - T) = \mu_0 I - f(T)$ and $\lambda_0 I - T$ is not a Weyl operator, it then follows, see Remark 3.118, that $\mu_0 I - f(T)$ is not Weyl. Thus $\mu_0 \in f(\sigma_w(T))$. ∎

Analogous results to those established in Theorems 3.115 and 3.119 may be established for B-Weyl spectra. The arguments for proving these results are rather similar to those used in Theorems 3.115 and 3.119. In the following theorem we only enunciate these results.

Theorem 3.122 *Let $T \in L(X)$ and $f \in \mathcal{H}(\sigma(T))$. Then we have*

(i) $\sigma_{bw}(f(T)) \subseteq f(\sigma_{bw}(T))$.
(ii) $\sigma_{ubw}(f(T)) \subseteq f(\sigma_{ubw}(T))$.

Moreover, T is of stable sign index on $\rho_{bf}(T)$ if and only if the spectral mapping theorem holds for $\sigma_{bw}(T)$ for every $f \in \mathcal{H}_{nc}(\sigma(T))$. Analogously, T is of stable sign index on $\rho_{ubf}(T)$ if and only if the spectral mapping theorem holds for $\sigma_{ubw}(T)$ for every $f \in \mathcal{H}_{nc}(\sigma(T))$.

Again by Corollary 2.106, if T or T^* has the SVEP then T is of stable sign index on $\rho_{bf}(T)$ and $\rho_{ubf}(T)$, so we have:

Corollary 3.123 *Let $T \in L(X)$ be such that either T or T^* has the SVEP. Then the spectral mapping theorem holds for the Weyl spectrum $\sigma_{bw}(T)$ and the semi-Weyl spectrum $\sigma_{ubw}(T)$ for every $f \in \mathcal{H}_{nc}(\sigma(T))$.*

The following examples of operators show that the condition $f \in \mathcal{H}_{nc}(\sigma(T))$ cannot be dropped in Corollary 3.123.

Example 3.124 Let R denote the unilateral right shift on $\ell_2(\mathbb{N})$ and $f \equiv 0$. Since R has the SVEP, R has stable sign index on $\rho_{bw}(R)$. We have $\sigma_{bw}(R) = \sigma(R) = \mathbf{D}(0,1)$, $\sigma(f(R)) = \{0\}$, $\sigma_{bw}(f(R)) = \emptyset$ while $f(\sigma_{bw}(R)) = \{0\}$. Hence, $\sigma_{bw}(f(R))$ does not contain $f(\sigma_{bw}(R))$.

Let L be the unilateral left shift on $\ell_2(\mathbb{N})$ and $f \equiv 0$. Then $L = R'$ and R has stable sign index on $\rho_{bw}(L)$. We have $\sigma_{bw}(L) = \sigma_{ubw}(L) = \sigma(L) = \mathbf{D}(0,1)$, $\sigma(f(L)) = \{0\}$ while $\sigma_{ubw}(f(L)) \subseteq \sigma_{bw}(f(L)) = \emptyset$. Since $f(\sigma_{ubw}(L)) = \{0\}$, $\sigma_{ubw}(f(L))$ does not contain $f(\sigma_{ubw}(L))$.

3.8   Spectral Properties of the Drazin Inverse

An obvious consequence of Theorem 3.109 is that if $T \in L(X)$ is invertible and $\mathbf{R}$ is a regularity then

$$\sigma_{\mathbf{R}}(T^{-1}) = \left\{ \frac{1}{\lambda} : \lambda \in \sigma_{\mathbf{R}}(T) \right\}. \tag{3.44}$$

Indeed, if $T \in L(X)$ is invertible then $0 \notin \sigma(T)$. Consider the function $f(\lambda) = \frac{1}{\lambda}$ defined on an open neighborhood U of $\sigma(T)$ which does not contain 0. Then $f(T) = T^{-1}$, so, by Theorem 3.109, the equality (3.44) follows.

A similar relation of reciprocity holds for the Drazin inverse.

Theorem 3.125 *Let* $\mathbf{R}$ *be a regularity in* $L(X)$ *which satisfies the condition (3.38). Let* $R \in L(X)$ *be a Drazin invertible operator with Drazin inverse S. If R is not invertible or nilpotent, then*

$$\sigma_{\mathbf{R}}(S) \setminus 0 = \{1/\lambda : \lambda \in \sigma_{\mathbf{R}}(R) \setminus \{0\}\}.$$

Proof Let R be Drazin invertible, with Drazin inverse S. Suppose that $0 \in \sigma(R)$ (and hence $0 \in \sigma(S)$) and that R is not nilpotent. Then in the decomposition $X = Y \oplus Z$, $R_1 = R|Y$, $R_2 = R|Z$, with R_1 nilpotent and R_2 invertible, we have $Y \neq \{0\}$ and $Z \neq \{0\}$. If $\mathbf{R}$ is a regularity in $L(X)$, let $\mathbf{R}^1$ and $\mathbf{R}^2$ be as in (3.38). Since R_1 is nilpotent and, by assumption, $\sigma_{\mathbf{R}^1}(R_1) \neq \emptyset$, $\sigma_{\mathbf{R}^1}(R_1) = \{0\}$, while $0 \notin \sigma_{\mathbf{R}^2}(R_2)$, since $0 \notin \sigma(R_2)$. Therefore, from the equality (3.39) we have

$$\sigma_{\mathbf{R}}(R) = \sigma_{\mathbf{R}^1}(R_1) \cup \sigma_{\mathbf{R}^2}(R_2) = \{0\} \cup \sigma_{\mathbf{R}^2}(R_2),$$

and hence $\sigma_{\mathbf{R}}(R) \setminus \{0\} = \sigma_{\mathbf{R}^2}(R_2)$. Analogously, $\sigma_{\mathbf{R}}(S) \setminus \{0\} = \sigma_{\mathbf{R}^2}(S_2)$. In view of the equality (3.44), we then have

$$\sigma_{\mathbf{R}}(S) \setminus \{0\} = \sigma_{\mathbf{R}^2}(S_2) = \left\{ \frac{1}{\lambda} : \lambda \in \sigma_{\mathbf{R}^2}(R_2) \right\}$$

$$= \left\{ \frac{1}{\lambda} : \lambda \in \sigma_{\mathbf{R}}(R) \setminus \{0\} \right\},$$

as desired.   ∎

Theorem 3.126 *Suppose that* $R \in L(X)$ *is Drazin invertible with Drazin inverse S. Then we have*

(i) R *is Browder if and only if S is Browder.*
(ii) $\sigma_{\mathrm{b}}(S) \setminus \{0\} = \{\frac{1}{\lambda} : \lambda \in \sigma_{\mathrm{b}}(R) \setminus \{0\}\}.$
(iii) $\sigma_{\mathrm{ub}}(S) \setminus \{0\} = \{\frac{1}{\lambda} : \lambda \in \sigma_{\mathrm{ub}}(R) \setminus \{0\}\}.$
(iv) $\sigma_{\mathrm{lb}}(S) \setminus \{0\} = \{\frac{1}{\lambda} : \lambda \in \sigma_{\mathrm{lb}}(R) \setminus \{0\}\}.$

Proof

(i) If $0 \notin \sigma(R)$ then R is invertible and the Drazin inverse is $S = R^{-1}$ so the assertion is trivial in this case. Suppose that $0 \in \sigma(R)$ and that R is Browder. Then 0 is a pole of the resolvent of R and is also a pole (of the first order) of the resolvent of S. Let $X = Y \oplus Z$ such that $R = R_1 \oplus R_2$, $R_1 = R|Y$ is nilpotent and $R_2 = R|Z$ is invertible. Observe that

$$\ker R = \ker R_1 \oplus \ker R_2 = \ker R_1 \oplus \{0\}, \qquad (3.45)$$

and, analogously, since $S = 0 \oplus S_2$ with $S_2 = R_2^{-1}$, we have

$$\ker S = \ker 0 \oplus \ker S_2 = Y \oplus \{0\}. \qquad (3.46)$$

Since R is Browder we have $\alpha(R) = \dim \ker R < \infty$, and since $\ker R_1 \subseteq \ker R$ it then follows that $\alpha(R_1) < \infty$. Consequently, $\alpha(R_1^n) < \infty$ for all $n \in \mathbb{N}$. Let $R_1^\nu = 0$. Since $Y = \ker R_1^\nu$ we then conclude that the subspace Y is finite-dimensional and hence $\ker S = Y \oplus \{0\}$ is finite-dimensional, i.e. $\alpha(S) < \infty$. Now, S is Drazin invertible, so $p(S) = q(S) < \infty$ and hence, by Theorem 1.22, $\alpha(S) = \beta(S) < \infty$. Hence S is Browder.

Conversely, suppose that S is Browder. Then $\alpha(S) < \infty$ and hence by (3.46) the subspace Y is finite-dimensional from which it follows that $\ker R_1 = \ker R|Y$ is also finite-dimensional. From (3.45) we then have that $\alpha(R) < \infty$ and since $p(R) = q(R) < \infty$ we then conclude that $\alpha(R) = \beta(R)$, again by Theorem 1.22. Hence R is a Browder operator.

(ii) The class of Browder operators is a regularity and the spectrum $\sigma_b(T)$ is non-empty for all $T \in L(X)$. Hence from Theorem 3.125 we have:

$$\sigma_b(S) \setminus \{0\} = \left\{ \frac{1}{\lambda} : \lambda \in \sigma_b(S) \setminus \{0\} \right\}.$$

(iii) Also the class of upper semi-Browder operators is a regularity and the spectrum $\sigma_{ub}(T)$ is non-empty for all $T \in L(X)$. Again from Theorem 3.125 we have

$$\sigma_{ub}(S) \setminus \{0\} = \left\{ \frac{1}{\lambda} : \lambda \in \sigma_{ub}(S) \setminus \{0\} \right\}.$$

(iv) Proceed by duality.   ■

Corollary 3.127 *If a Drazin invertible operator $R \in L(X)$ is a Riesz operator then its Drazin inverse is also Riesz.*

Proof Since X is infinite-dimensional and R is Riesz, $\sigma_b(R) = \{0\}$. Suppose that the Drazin inverse S is not Riesz. Then there exists a $0 \neq \lambda$ such that $\lambda \in \sigma_b(S)$. From part (ii) of Theorem 3.126 then $\frac{1}{\lambda} \in \sigma_b(R)$, a contradiction.   ■

Theorem 3.128 *Let $\lambda \neq 0$.*

(i) *If $T \in L(X)$ is invertible then*

$$(\lambda I - T)^k(X) = \left(\frac{1}{\lambda}I - T^{-1}\right)^k (X) \text{ for all } k \in \mathbb{N}.$$

(ii) *If R is Drazin invertible with Drazin inverse S then*

$$\ker (\lambda I - S)^k = \ker \left(\frac{1}{\lambda}I - R\right)^k \text{ for all } k \in \mathbb{N}.$$

(iii) *If R is Drazin invertible with Drazin inverse S then*

$$(\lambda I - S)^k(X) = \left(\frac{1}{\lambda}I - R\right)^k (X) \text{ for all } k \in \mathbb{N}.$$

(iv) *If R is Drazin invertible with Drazin inverse S, then λ is a pole of the resolvent of R if and only if $\frac{1}{\lambda}$ is a pole of the resolvent of S.*

(v) *If R is Drazin invertible with Drazin inverse S, then*

$$H_0(\lambda I - S) = H_0\left(\frac{1}{\lambda}I - R\right).$$

Proof

(i) Let $y = (\lambda I - T)^k x$. Then

$$\left(\frac{1}{\lambda}I - T^{-1}\right)^k T^k x = \left(\frac{1}{\lambda}T^k - I\right)^k x = \left(-\frac{1}{\lambda}\right)^k y,$$

so $(\lambda I - T)^k(X) \subseteq (\frac{1}{\lambda}I - T^{-1})^k(X)$. The reverse inclusion follows by symmetry.

(ii) It suffices to show that $\ker (\lambda I - T)^k \subseteq \ker(\frac{1}{\lambda}I - T^{-1})^k$. Let $x \in \ker (\lambda I - T)^k$. Then $(\lambda I - T)^k x = 0$, so

$$0 = (T^{-1})^k(\lambda I - T)^k x = (\lambda T^{-1} - I)^k x,$$

and hence $x \in \ker(\frac{1}{\lambda}I - T^{-1})^k$.

(iii) Let $X = Y \oplus Z$, $R = R_1 \oplus R_2$ and $S = 0 \oplus S_2$ with $S_2 = R_2^{-1}$. Since R_1 is nilpotent then $\frac{1}{\lambda}I - R_1$ is invertible, and hence $(\frac{1}{\lambda}I - R_1)^k(Y) = Y$. Hence

$$\left(\frac{1}{\lambda}I - R\right)^k (X) = \left(\frac{1}{\lambda}I - R_1\right)^k (Y) \oplus \left(\frac{1}{\lambda}I - R_2\right)^k (Z) = Y \oplus \left(\frac{1}{\lambda}I - R_2\right)^k (Z),$$

and analogously

$$(\lambda I - S)^k(X) = Y \oplus (\lambda I - S_2)^k(Z).$$

From part (i) we have $(\lambda I - S_2)^k(Z) = (\frac{1}{\lambda}I - R_2)^k(Z)$, so

$$\left(\frac{1}{\lambda}I - R\right)^k (X) = Y \oplus (\lambda I - S_2)^k(Z) = (\lambda I - S)^k(X).$$

(iv) From part (ii) and part (iii) we have $p(\lambda I - R) = p(\frac{1}{\lambda}I - S)$ and $q(\lambda I - R) = q(\frac{1}{\lambda}I - S)$.

(v) We have, for every $n \in \mathbb{N}$ and $\lambda \neq 0$,

$$\|(\lambda I - R_1)^n x\| \leq |\lambda|^n \|R_1\|^n \left\|\left(\frac{1}{\lambda}I - S_1\right)^n x\right\|,$$

and

$$\left\|\left(\frac{1}{\lambda}I - S_1\right)^n x\right\| \leq |\lambda|^{-n} \|S_1\|^n \|(\lambda I - R_1)^n x\|,$$

from which the equality $H_0(\lambda I - S) = H_0(\frac{1}{\lambda}I - R)$ follows. ∎

Recall that if T is algebraic then, by Theorem 3.68, the spectrum of T is a finite set of poles of the resolvent. Obviously, every algebraic operator T is Drazin invertible.

Corollary 3.129 *If $T \in L(X)$ is algebraic then its Drazin inverse is also algebraic.*

Proof Let S be the Drazin inverse of T. Since $\sigma(T)$ is a finite set it then follows that $\sigma(S)$ is also a finite set. We show that every point of $\sigma(S)$ is a pole of the resolvent. If $0 \in \sigma(S)$ then, since S is Drazin invertible, 0 is a pole (of the first order) of the resolvent of S. Let $0 \neq \lambda \in \sigma(S)$. Then $\frac{1}{\lambda} \in \sigma(T)$ and hence $\frac{1}{\lambda}$ is a pole of the resolvent of T. From part (iv) of Theorem 3.128 it then follows that λ is a pole of the resolvent of S. ∎

The class of Weyl operators is not a regularity, and as observed before, the spectral theorem may fail for the Weyl spectrum $\sigma_w(T)$. Note that if $T \in L(X)$ is invertible then $0 \notin \sigma_w(T)$ and $0 \notin \sigma_w(T^{-1})$. Although the spectral mapping theorem does not hold for $\sigma_w(T)$, we show that for a Drazin invertible operator R the relationship of reciprocity between the nonzero parts of the $\sigma_w(R)$, and the Weyl spectrum of its Drazin inverse $\sigma_w(S)$, remains true. We first need the following lemma.

Lemma 3.130 *Suppose that* $T \in L(X)$ *is invertible. Then*

$$\sigma_w(T^{-1}) = \left\{ \frac{1}{\lambda} : \lambda \in \sigma_w(T) \right\},$$

and

$$\sigma_{uw}(T^{-1}) = \left\{ \frac{1}{\lambda} : \lambda \in \sigma_{uw}(T) \right\}.$$

Proof Consider the analytic function $f(\lambda) := \frac{1}{\lambda}$ defined on a open neighborhood U containing the spectrum of T and such that $0 \notin U$. Then $T^{-1} = f(T)$, and since f is injective the statements follow from Theorem 3.121. ∎

Theorem 3.131 *Let* $R \in L(X)$ *be Drazin invertible with Drazin inverse S. Then we have*

(i) $\sigma_w(S) \setminus \{0\} = \{\frac{1}{\lambda} : \lambda \in \sigma_w(R) \setminus \{0\}\}.$
(ii) $\sigma_{uw}(S) \setminus \{0\} = \{\frac{1}{\lambda} : \lambda \in \sigma_{uw}(R) \setminus \{0\}\},$ *and*

$$\sigma_{lw}(S) \setminus \{0\} = \left\{ \frac{1}{\lambda} : \lambda \in \sigma_{lw}(R) \setminus \{0\} \right\}.$$

Proof With respect to the decomposition $R = R_1 \oplus R_2$ and $S = 0 \oplus S_2$, with $S_2 = R_2^{-1}$, we have

$$\sigma_w(R) = \sigma_w(R_1) \cup \sigma_w(R_2) = \{0\} \cup \sigma_w(R_2)$$

and

$$\sigma_w(S) = \sigma_w(0) \cup \sigma_w(R_2) = \{0\} \cup \sigma_w(S_2).$$

Observe that R_2 and S_2 are invertible, so $0 \notin \sigma_w(R_2)$ and $0 \notin \sigma_w(S_2)$. Hence, $\sigma_w(R) \setminus \{0\} = \sigma_w(R_2)$ and $\sigma_w(S) \setminus \{0\} = \sigma_w(S_2)$. By Lemma 3.130, the points of $\sigma_w(R_2)$ and $\sigma_w(S_2)$ are reciprocal, hence

$$\sigma_w(S) \setminus \{0\} = \sigma_w(S_2) = \left\{ \frac{1}{\lambda} : \lambda \in \sigma_w(R_2) \right\} = \left\{ \frac{1}{\lambda} : \lambda \in \sigma_w(R) \setminus \{0\} \right\},$$

so the first equality is proved.

(ii) Let $0 \neq \lambda$ and suppose that $\frac{1}{\lambda} \notin \sigma_{uw}(R)$, i.e., $\frac{1}{\lambda}I - R$ is upper semi-Weyl. Then $\frac{1}{\lambda}I - R \in \Phi_+(X)$ and ind $(\frac{1}{\lambda}I - R) \leq 0$. By Theorem 3.128 we have ker $(\frac{1}{\lambda}I - R) = \ker(\lambda I - S)$, so $\alpha(\lambda I - S) < \infty$. Moreover,

$$(\lambda I - S)(X) = (\lambda I - 0)(Y) \oplus (\lambda I - S_2)(Z) = Y \oplus (\lambda I - S_2)(Z).$$

Now, R_1 is nilpotent so $\frac{1}{\lambda}I - R_1$ is invertible, and hence $(\frac{1}{\lambda}I - R_1)(Y) = Y$, while $(\frac{1}{\lambda}I - R_2)(Z) = (\lambda I - S_2)(Z)$, by part (i) of Theorem 3.128. Therefore,

$$\left(\frac{1}{\lambda}I - R\right)(X) = \left(\frac{1}{\lambda}I - R_1\right)(Y) \oplus \left(\frac{1}{\lambda}I - R_2\right)(Z) = Y \oplus (\lambda I - S_2)(Z)$$
$$= (\lambda I - S)(X),$$

so $(\lambda I - S)(X)$ is closed, because $(\frac{1}{\lambda}I - R)(X)$ is closed by assumption, and this shows that $\lambda I - S \in \Phi_+(X)$. It remains only to prove that $\text{ind}(\lambda I - S) \leq 0$. Clearly, $\beta(\frac{1}{\lambda}I - R) = \beta(\lambda I - S)$ and $\alpha(\frac{1}{\lambda}I - R) = \alpha(\lambda I - S)$, by Theorem 3.128, so $\text{ind}(\lambda I - S) = \text{ind}(\frac{1}{\lambda}I - R) \leq 0$. Therefore, $\lambda I - S$ is upper semi-Weyl and hence $\lambda \notin \sigma_{\text{uw}}(S)$.

Conversely, suppose that $\lambda \notin \sigma_{\text{uw}}(S)$, i.e. $\lambda I - S$ is upper semi-Weyl. From the equalities $\ker(\frac{1}{\lambda}I - R) = \ker(\lambda I - S)$ and $(\frac{1}{\lambda} - R)(X) = (\lambda I - S)(X)$ we then obtain that $\frac{1}{\lambda}I - R \in \Phi_+(X)$. As above, $\text{ind}(\frac{1}{\lambda}I - R) = \text{ind}(\lambda I - S) \leq 0$, so $\frac{1}{\lambda}I - R$ is upper semi-Weyl, and hence $\frac{1}{\lambda} \notin \sigma_{\text{uw}}(R)$.

The equality for the lower semi-Weyl spectrum may be obtained by duality. ∎

3.9 Comments

The invariance of the Browder spectra under commuting Riesz perturbations is a classical result of Rakočević [265], but the proof given of Theorem 3.8, based on the stability of the localized SVEP under commuting Riesz perturbations, is taken from Aiena and Müller [20]. The proof of Theorem 3.11 is taken from Aiena and Triolo [26] and extends a previous result obtained by Grabiner [160] in the case of semi-Fredholm operators and compact commuting perturbations.

The stability of the essentially semi-regular spectrum under Riesz commuting perturbations established in Corollary 3.15 was observed by Kordula and Müller [204], while Theorem 3.16 is modeled after [26] and extends a previous result, proved by Grabiner [160], in the case of semi-Fredholm operators and compact commuting perturbations. The stability of Weyl operators under Riesz commuting perturbations given in Theorem 3.17 is a classical result due to Schechter and Whitley [271], but the proof given here is modeled after Duggal et al. [136].

Theorems 3.20 and 3.26 extend results of Djordjević [111], which proved that the approximate point spectra of T and $T + K$ have the same accumulation points, where K is a commuting finite rank operator.

Semi-Browder operators have been treated extensively in the books by Harte [170], Müller [245], and Aiena [1]. These classes of operators have also been investigated by several other authors, see Rakočević [265], Kordula et al. [205], and Laursen [213]. The characterizations of semi-Browder operators given in Theorems 3.38, 3.39 and 3.40 is modeled after Aiena and Carpintero [12], while the material of Theorems 3.43 and 3.44 is modeled after Aiena and Biondi [10]. The fact that the classes of upper and lower semi-B-Browder operators coincide with the classes of left Drazin invertible operators and right Drazin invertible operators, respectively, was observed by Berkani [63, Theorem 3.6], but the proof given here is from [36].

Theorems 3.55 and 3.59 is modeled after Berkani and Zariouh [77], while Theorem 3.53 is taken from [36]. The stability of essentially Drazin spectra under a commuting perturbation K having a finite-dimensional power is taken from the works by Burgos et al. [85], Bel Hadj Fredj [149], and [150], see also [304], but the proof of Theorem 3.78, using the localized SVEP, is modeled after Aiena and Triolo [26]. The section concerning meromorphic operators is inspired to the work of Barnes [58], Djordjević and Duggal [115, 116]. Theorems 3.83 and 3.85 have been taken from Oudghiri and Souilah [254]. The notion of regularity and the corresponding axiomatic spectral theory was studied by Kordula and Müller [203]. Further developments are given in Kordula, Mbekhta and Müller [206, 244], Mbektha and Müller [235]. A good treatment of the concept of regularity may be found in the monograph by Müller [245]. The material concerning spectra generated by Φ-semigroups is modeled after Gramsch and Lay [164], while the results concerning the spectral theorem for Weyl spectra is taken from Schmoeger [276]. The relationship of reciprocity between the Browder spectra and the Weyl spectra of the Drazin inverse R and its Drazin inverse S was observed in Aiena and Triolo [28].

Chapter 4
Polaroid-Type Operators

In this chapter we introduce the classes of polaroid-type operators, i.e., those operators $T \in L(X)$ for which the isolated points of the spectrum $\sigma(T)$ are poles of the resolvent, or the isolated points of the approximate point spectrum $\sigma_{ap}(T)$ are left poles of the resolvent. We also consider the class of all hereditarily polaroid operators, i.e., those operators $T \in L(X)$ for which all the restrictions to closed invariant subspaces are polaroid. The class of polaroid operators, as well as the class of hereditarily polaroid operators, is very large. We shall see that every generalized scalar operator is hereditarily polaroid, and this implies that many classes of operators acting on Hilbert spaces, obtained by relaxing the condition of normality, are hereditarily polaroid. Multipliers of commutative semi-simple Banach algebras, and in particular every convolution operators T_μ, defined in the group algebras $L^1(G)$, where G is a locally compact abelian group, are also hereditarily polaroid.

We also introduce some other classes of operators: the class of paranormal operators on Banach spaces, and more generally, the larger class of quasi totally hereditarily normaloid operators. These operators are also hereditarily polaroid.

The remaining part of the chapter is devoted to several other examples of polaroid operators. In particular, the fifth section is devoted to the spectral properties of isometries, invertible or non-invertible, while the sixth section regards spectral theory and the local spectral theory of weighted unilateral shifts, as well as bilateral weighted shifts. The seventh section is devoted to the important class of Toeplitz operators on Hardy spaces $H^2(\mathbf{T})$, where $\mathbf{T}$ denotes the unit circle in $\mathbb{C}$.

The last section is devoted to the topic of spectral inclusions. We are mainly interested in some spectral consequences of the intertwining condition $SA = AT$ considered in Chap. 2, in particular we study the preservation of the polaroid properties under quasi-affinities.

© Springer Nature Switzerland AG 2018
P. Aiena, *Fredholm and Local Spectral Theory II*, Lecture Notes in Mathematics
2235, https://doi.org/10.1007/978-3-030-02266-2_4

4.1 Left and Right Polaroid Operators

The concept of pole may be sectioned as follows:

Definition 4.1 Let $T \in L(X)$, X a Banach space. If $\lambda I - T$ is left Drazin invertible and $\lambda \in \sigma_{\mathrm{ap}}(T)$ then λ is said to be a *left pole*. A left pole λ is said to have finite rank if $\alpha(\lambda I - T) < \infty$. If $\lambda I - T$ is right Drazin invertible and $\lambda \in \sigma_{\mathrm{s}}(T)$ then λ is said to be a *right pole*. A right pole λ is said to have finite rank if $\beta(\lambda I - T) < \infty$.

Clearly, λ is pole of the resolvent if and only if λ is both a right pole and a left pole.

Theorem 4.2 *For every $T \in L(X)$ the following equivalences hold:*

  (i) λ *is a left pole of the resolvent of T if and only if λ is a right pole of the resolvent of T^*.*
 (ii) λ *is a right pole of the resolvent of T if and only if λ is a left pole of the resolvent of T^*.*
(iii) λ *is a pole of the resolvent of T if and only if λ is a pole of the resolvent of T^*.*

Proof The proof for the dual T^* is immediate from Theorem 6.4, taking into account that $\sigma_{\mathrm{ap}}(T) = \sigma_{\mathrm{s}}(T^*)$ and $\sigma_{\mathrm{s}}(T) = \sigma_{\mathrm{ap}}(T^*)$. ∎

In the sequel $\mathcal{H}(\Omega, Y)$, Y any Banach space, denotes the Fréchet space of all analytic functions from the open set $\Omega \subseteq \mathbb{C}$ to Y. We have proved in Theorem 2.55 that if $\lambda \in \mathrm{iso}\, \sigma_{\mathrm{ap}}(T)$ then $H_0(\lambda I - T)$ is closed. If λ is a left pole we can say much more:

Theorem 4.3 *Let $T \in L(X)$, X a Banach space.*

  (i) *If λ is a left pole of $T \in L(X)$ then λ is an isolated point of $\sigma_{\mathrm{ap}}(T)$ and there exists a $\nu \in \mathbb{N}$ such that*

$$H_0(\lambda I - T) = \ker(\lambda I - T)^{\nu}.$$

Moreover, λ is a left pole of finite rank then $H_0(\lambda I - T)$ is finite-dimensional.
 (ii) *If λ is a right pole of $T \in L(X)$ then λ is an isolated point of $\sigma_{\mathrm{s}}(T)$, and there exists a $\nu \in \mathbb{N}$ such that*

$$K(\lambda I - T) = (\lambda I - T)^{\nu}(X).$$

Moreover, if λ is a right pole of finite rank then $K(\lambda I - T)$ has finite codimension in X.

Proof There is no loss of generality if we assume $\lambda = 0$.

  (i) If 0 is a left pole then T is left Drazin invertible, or equivalently, by Theorem 3.47, T is upper semi B-Browder. The condition $p(T) < \infty$ also entails that T has the SVEP at 0, and this is equivalent, by Theorem 2.97, to saying that $\sigma_{ap}(T)$ does not cluster at 0.

To show the equality $H_0(T) = \ker T^\nu$ for some $\nu \in \mathbb{N}$, observe first since T_n is upper semi-Browder for some $n \in \mathbb{N}$, by Lemma 3.46, the canonical map $\widehat{T}_n : X/\ker T^n \to X/\ker T^n$ is upper semi-Browder, and hence has the SVEP at 0. Since every semi-Fredholm operator has topological uniform descent then, by Theorem 2.97, $H_0(\widehat{T}_n) = \ker(\widehat{T}_n)^k$, where k is the ascent of $\widehat{T}_n$.

Let $x \in H_0(T)$. We show that $\hat{x} \in H_0(\widehat{T}_n)$. We know, since $H_0(T) = \mathcal{X}_T(\{0\})$ by definition, that there exists a $g \in \mathcal{H}(\mathbb{C} \setminus \{0\}, X)$ such that

$$x = (\mu I - T)g(\mu) \quad \text{for all } \mu \in \mathbb{C} \setminus \{0\}.$$

If $\gamma : X \to \widehat{X} := X/\ker T^n$ denotes the canonical quotient map, then $\hat{g} := \gamma \circ g \in \mathcal{H}(\mathbb{C} \setminus \{0\}, \widehat{X})$, and for all $\mu \in \mathbb{C} \setminus \{0\}$ we have

$$\hat{x} = (\mu I - \widehat{T}_n)\widehat{g(\mu)} = (\mu I - \widehat{T}_n)\hat{g}(\mu).$$

Therefore,

$$\hat{x} \in \mathcal{X}_{\widehat{T}_n}(\{0\}) = H_0(\widehat{T}_n) = \ker(\widehat{T}_n)^k,$$

and hence

$$\widehat{T}_n^k \hat{x} = \widehat{T^k x} = \hat{0}.$$

Consequently, $T^k x \in \ker T^n$ and this implies that $H_0(T) \subseteq \ker T^{k+n}$. The opposite inclusion is true for every operator, hence we have $H_0(T) = \ker T^\nu$, where $\nu := k + n$.

Finally, suppose that 0 is a left pole of finite rank. Then $\ker T$ is finite-dimensional and this implies that $H_0(T) = \ker T^\nu$ is finite-dimensional.

(ii) If 0 is a right pole then T is right Drazin invertible and hence lower semi B-Browder, by Theorem 3.47. The condition $q(T) < \infty$ also entails that T^* has the SVEP at 0, and this is equivalent, by Theorem 2.98, to saying that $\sigma_s(T)$ does not cluster at 0. Since every right Drazin invertible is quasi-Fredholm, we have, by Corollary 2.96, $K(T) = T^\infty(X) = T^q(X)$.

Finally, if 0 is a right pole of finite rank then $K(T)$ has finite codimension. ∎

Remark 4.4 It should be noted that a left pole, as well as a right pole, need not be an isolated point of $\sigma(T)$. For instance, let $R \in \ell^2(\mathbb{N})$ be the classical unilateral right shift and

$$U(x_1, x_2, \dots) := (0, x_2, x_3, \cdots) \quad \text{for all } (x_n) \in \ell^2(\mathbb{N}).$$

Define $T := R \oplus U$. Then $\sigma(T) = \mathbf{D}(0, 1)$, $\mathbf{D}(0, 1)$ the closed unit disc of $\mathbb{C}$. Moreover, $\sigma_{ap}(T) = \Gamma \cup \{0\}$, Γ the unit circle, and T is upper semi-Browder, in particular left Drazin invertible. Hence 0 is a left pole (of finite rank, since $\alpha(T) = 1$) but $0 \notin \mathrm{iso}\, \sigma(T) = \emptyset$. Note that 0 is a right pole of the dual T^*, but is not an isolated point of $\sigma(T^*) = \sigma(T) = \mathbf{D}(0, 1)$.

In the case of Hilbert space operators we have much more.

Theorem 4.5 *Let $T \in L(H)$, H a Hilbert space.*

(i) $\lambda \in \sigma_{\mathrm{ap}}(T)$ *is a left pole if and only if there exist two T-invariant closed subspaces M, N such that $H = M \oplus N$, $\lambda I - T|M$ is bounded below, and $\lambda I - T|N$ is nilpotent. In this case $N = H_0(\lambda I - T)$.*

(ii) *If $\lambda \in \sigma_s(T)$ then λ is a right pole if and only if there exist two T-invariant closed subspaces M, N such that $H = M \oplus N$, $\lambda I - T|M$ is onto, and $\lambda I - T|N$ is nilpotent. In this case, $M = K(\lambda I - T)$.*

Proof We may assume that $\lambda = 0$.

(i) If 0 is a left pole then T is left Drazin invertible and hence quasi-Fredholm. By Theorem 1.107, T is of Kato-type, so there exist two T-invariant subspaces M, N such that $H = M \oplus N$, $T|M$ is semi-regular, and $T|N$ is nilpotent. Since $p(T) < \infty$, T has the SVEP at 0, and hence $T|M$ has the SVEP at 0. By Theorem 2.91 it then follows that $T|M$ is bounded below.

Conversely, suppose that there exist two T-invariant subspaces M, N such that $H = M \oplus N$, $T|M$ is bounded below, $T|N$ is nilpotent. Then $T^n(H)$ is closed and T_n is bounded below for some $n \in \mathbb{N}$, so T is left Drazin invertible by Theorem 1.140. By assumption $0 \in \sigma_{\mathrm{ap}}(T)$, thus 0 is a left pole.

To complete the proof, observe first that since the restriction $T|N$ is quasi-nilpotent then $N = H_0(T|N)$. Moreover, since $T|M$ is bounded below, hence semi-regular, by Theorem 2.37

$$H_0(T|M) = \bigcup_{n=1}^{\infty} \ker (T|M)^n = \{0\}.$$

Therefore,

$$H_0(T) = H_0(T|M) \oplus H_0(T)|N = \{0\} \oplus N = N.$$

(ii) If 0 is a right pole of T then 0 is a left pole of the adjoint T', so T' is left Drazin invertible. By part (i) there exist two closed subspaces U, V of H such that $H = U \oplus V$, $T'|U$ is bounded below, $T'|V$ is nilpotent. Consider the orthogonal sets $M := U^{\perp}$ and $N := V^{\perp}$. Proceeding as in the proof of Lemma 2.78, just adapting the arguments to the Hilbert adjoint, it is easy to see that $T|M$ is bounded below, while $T|N$ is nilpotent.

Conversely, suppose that there exist two T-invariant subspaces M, N such that $H = M \oplus N$, $T|M$ is onto, and $T|N$ is nilpotent. Then $T^n(H) = M$ is closed and T_n is onto for some $n \in \mathbb{N}$, hence T is right Drazin invertible by Theorem 1.140. By assumption $\lambda \in \sigma_s(T)$, thus λ is a right pole.

To prove the last assertion, observe that for $n \in \mathbb{N}$ sufficiently large we have

$$T^n(H) = (T|M)^n(M) \oplus (T|N)^n(N) = M \oplus \{0\} = M.$$

Therefore,

$$T^\infty(H) = \bigcap_{n=1}^{\infty} T^n(H) = M,$$

and, by Theorem 2.95, we conclude that $M = T^\infty(H) = K(T)$. ∎

Corollary 4.6 *If $T \in L(H)$, H a Hilbert space, then λ is a left pole of finite rank if and only if there exist two closed T-invariant subspaces M, N such that $X = M \oplus N$, N is finite-dimensional, $\lambda I - T|M$ is bounded below and $\lambda I - T|N$ is nilpotent. Analogously, λ is a right pole of finite rank if and only if there exist two closed T-invariant subspaces M, N such that $X = M \oplus N$, M is finite-codimensional, $\lambda I - T|M$ is onto and $\lambda I - T|N$ is nilpotent*

Definition 4.7 A bounded operator $T \in L(X)$ is said to be *left polaroid* if every $\lambda \in \mathrm{iso}\,\sigma_{\mathrm{ap}}(T)$ is a left pole of the resolvent of T. $T \in L(X)$ is said to be *right polaroid* if every $\lambda \in \mathrm{iso}\,\sigma_s(T)$ is a right pole of the resolvent of T. $T \in L(X)$ is said to be *polaroid* if every $\lambda \in \mathrm{iso}\,\sigma(T)$ is a pole of the resolvent of T.

An immediate consequence of Theorem 4.2 is that the concepts of left and right polaroid are dual each other:

Theorem 4.8 *$T \in L(X)$ is left polaroid (respectively, right polaroid, polaroid) if and only if T^* is right polaroid (respectively, left polaroid, polaroid). Analogously, a Hilbert space operator $T \in L(H)$ is left polaroid (respectively, right polaroid, polaroid) if and only if its adjoint T' is right polaroid (respectively, left polaroid, polaroid).*

Proof The assertions concerning the dual T^* follow from Theorem 4.2.

To show the statement for Hilbert space operators, suppose that T is a left polaroid and $\lambda \in \mathrm{iso}\,\sigma_s(T')$. Then $\bar{\lambda} \in \mathrm{iso}\,\sigma_s(T^*) = \mathrm{iso}\,\sigma_{\mathrm{ap}}(T)$, so $\bar{\lambda}$ is a left pole of T, hence $p := p(\bar{\lambda}I - T) < \infty$ and $(\bar{\lambda}I - T)^{p+1}(H)$ is closed. From the closed range theorem, see Appendix A, we have $\ker(\bar{\lambda}I - T)^p = \ker(\bar{\lambda}I - T)^{p+1}$, thus

$$(\lambda I - T')^p(H) = [\ker(\bar{\lambda}I - T)^p]^\perp = [\ker(\bar{\lambda}I - T)^{p+1}]^\perp = (\lambda I - T')^{p+1}(H),$$

where, as usual, by $N^\perp$ we denote the orthogonal of $N \subseteq H$. Therefore $q := q(\lambda I - T') \le p < \infty$ and consequently $(\lambda I - T')^q(H) = (\lambda I - T')^p(H)$ is closed, by Lemma 1.100. Thus λ is a right pole of T'.

Conversely, suppose that T' is right polaroid and let $\lambda \in$ iso $\sigma_{\mathrm{ap}}(T)$. Then $\lambda \in$ iso $\sigma_{\mathrm{s}}(T^*) = $ iso $\overline{\sigma_{\mathrm{s}}(T')}$, thus $\overline{\lambda}$ is a right pole of T'. Consequently, $q := q(\overline{\lambda}I - T') < \infty$ and $(\overline{\lambda}I - T')^q(H) = (\overline{\lambda}I - T')^{q+1}(H)$ is closed. Since

$$\ker(\lambda I - T)^{q+1} = [(\overline{\lambda}I - T')^{q+1}(H)]^{\perp} = [(\overline{\lambda}I - T')^q(H)]^{\perp} = \ker(\lambda I - T)^q$$

it then follows that $p := p(\lambda I - T) \le q < \infty$. By Lemma 1.100, we have that $(\lambda I - T)^{p+1}(H)$ is closed, so λ is a left pole of T.

Part (ii) and part (iii) for Hilbert space operators may be proved in a similar way. ∎

Definition 4.9 A bounded operator $T \in L(X)$ is said to be *a-polaroid* if every $\lambda \in$ iso $\sigma_{\mathrm{ap}}(T)$ is a pole of the resolvent of T.

Note that if T is a-polaroid then iso $\sigma_{\mathrm{ap}}(T) =$ iso $\sigma(T)$. Indeed, if $\lambda \in$ iso $\sigma_{\mathrm{ap}}(T)$ then λ is a pole of the resolvent, hence an isolated point of $\sigma(T)$. Conversely, if $\lambda \in$ iso $\sigma(T)$ then $\lambda \in \sigma_{\mathrm{ap}}(T)$, by part (ii) of Theorem 1.12, so λ is an isolated point of $\sigma_{\mathrm{ap}}(T)$.

Trivially,

$$T \ a\text{-polaroid} \Rightarrow T \text{ left polaroid.} \tag{4.1}$$

The following example provides an operator that is left polaroid but not a-polaroid.

Example 4.10 Let $R \in \ell^2(\mathbb{N})$ be the unilateral right shift defined as

$$R(x_1, x_2, \dots) := (0, x_1, x_2, \cdots) \quad \text{for all } (x_n) \in \ell^2(\mathbb{N}),$$

and

$$U(x_1, x_2, \dots) := (0, x_2, x_3, \cdots) \quad \text{for all } (x_n) \in \ell^2(\mathbb{N}).$$

If $T := R \oplus U$ then $\sigma(T) = \mathbf{D}(0, 1)$, so iso $\sigma(T) = \emptyset$. Moreover, $\sigma_{\mathrm{ap}}(T) = \Gamma \cup \{0\}$, where Γ is the unit circle, so iso $\sigma_{\mathrm{ap}}(T) = \{0\}$. Since R is injective and $p(U) = 1$ it then follows that $p(T) = p(R) + p(U) = 1$. Furthermore, $T \in \Phi_+(X)$ and hence $T^2 \in \Phi_+(X)$, so that $T^2(X)$ is closed. Therefore 0 is a left pole and hence T is left polaroid. On the other hand $q(R) = \infty$, so that $q(T) = q(R) + q(U) = \infty$, so T is not a-polaroid. Note that T is also polaroid.

Theorem 4.11 *Let $T, R \in L(X)$. Then TR is polaroid (respectively, left polaroid, right polaroid, a-polaroid) if and only if RT is polaroid (respectively, left polaroid, right polaroid, a-polaroid)*

Proof Suppose that TR is polaroid and $\lambda \in$ iso $\sigma(RT)$. Suppose first that $\lambda = 0$. Since 0 is a pole of the resolvent of TR then, by Lemma 3.96, 0 is a pole of the resolvent of RT. Suppose the other case, $\lambda \ne 0$. Then $\lambda \in$ iso $\sigma(TR)$, by

Corollary 2.150, and hence λ is a pole of the resolvent of TR. By Theorem 3.95 it then follows that λ is a pole of the resolvent of RT. The other assertions may be proved in a similar way. ∎

The condition of being polaroid may be characterized by means of the quasi-nilpotent part as follows:

Theorem 4.12 *If $T \in L(X)$ the following statements hold:*

(i) *T is polaroid if and only if there exists a $p := p(\lambda I - T) \in \mathbb{N}$ such that*

$$H_0(\lambda I - T) = \ker(\lambda I - T)^p \quad \text{for all } \lambda \in \text{iso}\,\sigma(T). \tag{4.2}$$

(ii) *If T is left polaroid then there exists a $p := p(\lambda I - T) \in \mathbb{N}$ such that*

$$H_0(\lambda I - T) = \ker(\lambda I - T)^p \quad \text{for all } \lambda \in \text{iso}\,\sigma_{\text{ap}}(T). \tag{4.3}$$

Proof Suppose T satisfies (4.2) and that λ is an isolated point of $\sigma(T)$. Since λ is isolated in $\sigma(T)$ then, by Lemma 2.47,

$$X = H_0(\lambda I - T) \oplus K(\lambda I - T) = \ker(\lambda I - T)^p \oplus K(\lambda I - T),$$

from which we obtain

$$(\lambda I - T)^p(X) = (\lambda I - T)^p(K(\lambda I - T)) = K(\lambda I - T).$$

So $X = \ker(\lambda I - T)^p \oplus (\lambda I - T)^p(X)$, which implies, by Theorem 1.35, that $p(\lambda I - T) = q(\lambda I - T) \leq p$, hence λ is a pole of the resolvent, so that T is polaroid. Conversely, suppose that T is polaroid and λ is an isolated point of $\sigma(T)$. Then λ is a pole, and if p is its order then $H_0(\lambda I - T) = \ker(\lambda I - T)^p$, again by Lemma 2.47.

(ii) This follows from Theorem 4.3. ∎

Corollary 4.13 *If $T \in L(X)$ is either left or right polaroid then T is polaroid.*

Proof Assume that T is left polaroid and let $\lambda \in \text{iso}\,\sigma(T)$. The boundary of the spectrum is contained in $\sigma_{\text{ap}}(T)$, in particular every isolated point of $\sigma(T)$, see Theorem 2.58, thus $\lambda \in \text{iso}\,\sigma_{\text{ap}}(T)$ and hence λ is a left pole of the resolvent of T. By Theorem 4.12, there exists a $\nu := \nu(\lambda I - T) \in \mathbb{N}$ such that $H_0(\lambda I - T) = \ker(\lambda I - T)^\nu$. But λ is isolated in $\sigma(T)$, so λ is a pole of the resolvent, i.e. T is polaroid.

To show the last assertion suppose that T is right polaroid. By Theorem 4.8, T^* is left polaroid and hence, by the first part, T^* is polaroid, or equivalently T is polaroid. ∎

The following example shows that the converse of Corollary 4.13, in general, does not hold.

Example 4.14 Let R denote the right shift on $\ell^2(\mathbb{N})$ defined by

$$R(x_1, x_2, \dots) := (0, x_1, x_2, \dots) \quad (x_n) \in \ell^2(\mathbb{N}),$$

and let Q be the weighted left shift defined by

$$Q(x_1, x_2, \dots) := (x_2/2, x_3/3, \dots) \quad (x_n) \in \ell^2(\mathbb{N}).$$

Q is a quasi-nilpotent operator, $\sigma(R) = \mathbf{D}(0, 1)$, where $\mathbf{D}(0, 1)$ denotes the closed unit disc of $\mathbb{C}$, and $\sigma_{ap}(R) = \Gamma$, where Γ is the unit circle of $\mathbb{C}$. Moreover, if $e_n := (0, \dots, 0, 1, 0 \dots)$, where 1 is the n-th term, then $e_{n+1} \in \ker Q^{n+1}$ while $e_{n+1} \notin \ker Q^n$ for every $n \in \mathbb{N}$, so $p(Q) = \infty$.

Define $T := R \oplus Q$ on $X := \ell^2(\mathbb{N}) \oplus \ell^2(\mathbb{N})$. Clearly, $\sigma(T) = \mathbf{D}(0, 1)$, and $\sigma_{ap}(T) = \Gamma \cup \{0\}$. We have $p(T) = p(R) + p(Q) = \infty$, so 0 is not a left pole. Therefore, T is polaroid, since iso $\sigma(T) = \emptyset$, but not left polaroid. It is easily seen that the dual T^* is polaroid but not right polaroid, since $q(T^*) = \infty$.

Theorem 4.15 *Let $T \in L(X)$. Then we have:*

(i) *If T^* has the SVEP at every $\lambda \notin \sigma_{uw}(T)$ then $\sigma_{ap}(T) = \sigma(T)$. Furthermore, the properties of being polaroid, a-polaroid and left polaroid for T are all equivalent.*

(ii) *If T has the SVEP at every $\lambda \notin \sigma_{lw}(T)$ then $\sigma_s(T) = \sigma(T)$. Furthermore, the properties of being polaroid, a-polaroid and left polaroid for T^* are all equivalent.*

Proof

(i) Suppose that $\lambda \notin \sigma_{ap}(T)$. Then $p(\lambda I - T) = 0$ and $\lambda I - T \in W_+(X)$, so $\lambda \notin \sigma_{uw}(T)$ and hence, by assumption, T^* has the SVEP at λ. Since $\lambda I - T$ is upper semi-Weyl, by Theorem 2.98 it then follows that $q(\lambda I - T) < \infty$ and hence, by Theorem 1.22, $p(\lambda I - T) = q(\lambda I - T) = 0$, i.e., $\lambda \notin \sigma(T)$. This proves the equality $\sigma_{ap}(T) = \sigma(T)$. The equivalence of the polaroid conditions is now clear: if T is polaroid then T is a-polaroid, since iso $\sigma_{ap}(T) = $ iso $\sigma(T)$. Thus, by Corollary 4.13, the equivalence is proved.

(ii) Using dual arguments to those of the proof of part (i) we have $\sigma_s(T) = \sigma(T)$ and hence, by duality $\sigma_{ap}(T^*) = \sigma(T^*)$. Therefore, if T^* is polaroid then T^* is a-polaroid, so the equivalence follows from Corollary 4.13. ∎

Define

$$\Pi(T) := \{\lambda \in \mathbb{C} : \lambda \text{ is a pole}\}.$$

Analogously, set

$$\Pi_l(T) := \{\lambda \in \mathbb{C} : \lambda \text{ is a left pole}\}$$

and

$$\Pi_r(T) := \{\lambda \in \mathbb{C} : \lambda \text{ is a right pole}\}.$$

Clearly, $\sigma(T) = \Pi(T) \cup \sigma_d(T)$, and

$$\sigma_{ap}(T) = \Pi_l(T) \cup \sigma_{ld}(T) \quad \text{and} \quad \sigma_s(T) = \Pi_r(T) \cup \sigma_{rd}(T).$$

The sets $\Pi(T)$, $\Pi_l(T)$ and $\Pi_r(T)$ may be empty. This, trivially, happens if iso $\sigma_{ap}(T) = \emptyset$, or iso $\sigma_s(T) = \emptyset$, for instance if T is a non-quasi-nilpotent weighted right shift, as we shall see in Chap. 4.

Theorem 4.16 *If $T \in L(X)$ and $f \in \mathcal{H}_{nc}(\sigma(T))$, then λ is a left pole (respectively, right pole, pole) of $f(T)$ if and only if there exists a left pole (respectively, right pole, pole) μ of T such that $f(\mu) = \lambda$.*

Proof We show that $f(\Pi_l(T)) = \Pi_l(f(T))$. We know that the spectral mapping theorem holds for $\sigma_{ap}(T)$. Moreover, $\Pi_l(T)$ and $\sigma_{ld}(T)$ are disjoint. We have

$$f(\sigma_{ap}(T)) = f(\Pi_l(T) \cup \sigma_{ld}(T)) = f(\Pi_l(T)) \cup f(\sigma_{ld}(T)).$$

On the other hand,

$$\sigma_{ap}(f(T)) = \Pi_l(f(T)) \cup \sigma_{ld}(f(T)) = \Pi_l(f(T)) \cup f(\sigma_{ld}(T)),$$

and since $f(\sigma_{ap}(T)) = \sigma_{ap}(f(T))$ it then follows that $f(\Pi_l(T)) = \Pi_l(f(T))$.

The case of right poles may be proved in a similar way: to prove that $f(\Pi_r(T)) = \Pi_r(f(T))$, just replace $\Pi_l(T)$ by $\Pi_r(T)$ and $\sigma_{ap}(T)$ by $\sigma_s(T)$. Proceed similarly for the set of poles. ∎

To obtain further insight into the properties of polaroid operators, we need some preliminary results concerning the kernel and the quasi-nilpotent part of $p(T)$ where p is a polynomial.

Lemma 4.17 *Let $T \in L(X)$ and let p be a complex polynomial. If $p(\lambda_0) \neq 0$ then $H(\lambda_0 I - T) \cap \ker p(T)) = \{0\}$. If T has the SVEP then*

$$H_0(p(T)) = H_0(\lambda_1 I - T) \oplus H_0(\lambda_2 I - T) \cdots \oplus H_0(\lambda_n I - T),$$

where $\lambda_1, \lambda_2, \ldots, \lambda_n$ are the distinct roots of p.

Proof Suppose that there is a non-zero element $x \in H(\lambda_0 I - T) \cap \ker p(T)$ and set

$$p(\lambda_0)I - p(T) := q(T)(\lambda_0 I - T),$$

where q denotes a polynomial. Then $q(T)(\lambda_0 I - T)x = p(\lambda_0)x$, and hence

$$q(T)(\lambda_0 I - T)^n x = p(\lambda_0)^n x.$$

Therefore,

$$|p(\lambda_0)| \, \|x\|^{1/n}\| \le \|q(T)^n\|^{1/n} \|(\lambda_0 I - T)x\|^{1/n} \quad \text{for all } n \in \mathbb{N}.$$

Since x is a non-zero element of $H(\lambda_0 I - T)$ we then have $p(\lambda_0) = 0$, which is a contradiction.

To prove the second assertion, consider an element $x \in H_0(p(T))$. Since $H_0(p(T)) = \mathcal{X}_{p(T)}(\{0\})$, the glocal spectral space of $p(T)$ relative to the set $\{0\}$, there exists an analytic function f for which

$$x = (\mu I - p(T))f(\mu) \quad \text{for all } \mu \in \mathbb{C} \setminus \{0\}.$$

Thus, for $\lambda \in \mathbb{C} \setminus \{\lambda_1, \ldots, \lambda_n\}$ we have

$$x = (p(\lambda)I - p(T))f(p(\lambda)) = (\lambda I - T)q(T, \lambda)f(p(\lambda)),$$

where q is a polynomial of T and λ. Consequently

$$\sigma_T(x) \subseteq \{\lambda_1, \ldots, \lambda_n\},$$

and hence

$$x \in X_T(\{\lambda_1, \ldots, \lambda_n\}) = \bigoplus_{i=1}^{n} X_T(\{\lambda_i\}).$$

Since T has the SVEP, by Theorem 2.23 we then have that

$$X_T(\{\lambda_i\}) = \mathcal{X}_T(\{\lambda_i\}) = H_0(\lambda_i I - T) \quad \text{for all } i = 1, \ldots, n,$$

and hence

$$H_0(p(T)) \subseteq \bigoplus_{i=1}^{n} H_0(\lambda_i I - T).$$

The opposite inclusion is clear, since each λ_i is a root of the polynomial p. ∎

Remark 4.18 It is easy to check, from the definition of a quasi-nilpotent part, the following properties:

(i) $H_0(T) \subseteq H_0(T^k)$, for all $k \in \mathbb{N}$.
(ii) If $T, U \in L(X)$ commute and $S = TU$ then $H_0(T) \subseteq H_0(S)$.

Theorem 4.19 *For an operator $T \in L(X)$ the following statements are equivalent.*

(i) T is polaroid;
(ii) $f(T)$ is polaroid for every $f \in \mathcal{H}_{nc}(\sigma(T))$;
(iii) there exists a non-trivial polynomial p such that $p(T)$ is polaroid;
(iv) there exists an $f \in \mathcal{H}_c(\sigma(T))$ such that $f(T)$ is polaroid.

Proof (i) $\Rightarrow$ (ii) Let $\lambda_0 \in \text{iso } \sigma(f(T))$. The spectral mapping theorem implies $\lambda_0 \in \text{iso } f(\sigma(T))$. Let us show that $\lambda_0 \in f(\text{iso } \sigma(T))$.

Select $\mu_0 \in \sigma(T)$ such that $f(\mu_0) = \lambda_0$. Denote by Ω the connected component of the domain of f which contains μ_0 and suppose that μ_0 is not isolated in $\sigma(T)$. Then there exists a sequence $(\mu_n) \subset \sigma(T) \cap \Omega$ of distinct scalars such that $\mu_n \to \mu_0$. Since $K := \{\mu_0, \mu_1, \mu_2, \dots\}$ is a compact subset of Ω, the principle of isolated zeros of analytic functions tells us that the function f may assume the value $\lambda_0 = f(\mu_0)$ at only a finite number of points of K; so for n sufficiently large $f(\mu_n) \neq f(\mu_0) = \lambda_0$, and since $f(\mu_n) \to f(\mu_0) = \lambda_0$ it then follows that λ_0 is not an isolated point of $f(\sigma(T))$, a contradiction. Hence $\lambda_0 = f(\mu_0)$, with $\mu_0 \in \text{iso } \sigma(T)$. Since T is polaroid, μ_0 is a pole of T and by Theorem 4.16, λ_0 is a pole for $f(T)$, which proves that $f(T)$ is polaroid.

The implications (ii) $\Rightarrow$ (iii) $\Rightarrow$ (iv) are obvious.

(iv) $\Rightarrow$ (i) Suppose $f(T)$ is polaroid for some $f \in \mathcal{H}_c(\sigma(T))$ and let $\lambda_0 \in \text{iso } \sigma(T)$ be arbitrary. Then $\mu_0 := f(\lambda_0) \in f(\text{iso } \sigma(T))$. We show now that $\mu_0 \in \text{iso } f(\sigma(T))$. Indeed, suppose that μ_0 is not isolated in $f(\sigma(T))$. Then there exists a sequence $(\mu_n) \subset f(\sigma(T))$ of distinct scalars such that $\mu_n \to \mu_0$ as $n \to +\infty$. Let $\lambda_n \in \sigma(T)$ such that $\mu_n = f(\lambda_n)$ for all n. Clearly, $\lambda_n \neq \lambda_m$ for $n \neq m$, and since

$$\mu_n = f(\lambda_n) \to \mu_0 = p(\lambda_0),$$

we then have $\lambda_n \to \lambda_0$, and this is impossible since, by assumption, $\lambda_0 \in \text{iso } \sigma(T)$. By the spectral mapping theorem, $\mu_0 \in \text{iso } f(\sigma(T)) = \text{iso } \sigma(f(T))$. Now, since $f(T)$ is polaroid, part (i) of Theorem 4.12 entails that there exists a natural ν such that

$$H_0(\mu I - f(T)) = \ker (\mu I - f(T))^\nu. \tag{4.4}$$

Let $g(\lambda) := \mu_0 - f(\lambda)$. Trivially, λ_0 is a zero of g, and g may have only a finite number of zeros. Let $\{\lambda_0, \lambda_1, \dots, \lambda_n\}$ be the set of all zeros of g, with $\lambda_i \neq \lambda_j$, for all $i \neq j$. Define

$$p(\lambda) := \Pi_{i=1}^n (\lambda_i - \lambda)^{\nu_i},$$

where ν_i is the multiplicity of λ_i. Then we can write, for some $k \in \mathbb{N}$,

$$g(\lambda) = (\lambda_0 - \lambda)^k \, p(\lambda) \, h(\lambda),$$

where $h(\lambda)$ is an analytic function which does not vanish in $\sigma(T)$. Consequently,

$$g(T) = \mu_0 I - f(T) = (\lambda_0 I - T)^k p(T) h(T),$$

where $h(T)$ is invertible, and hence

$$H_0(\mu_0 I - f(T)) = H_0((\lambda_0 I - T)^k p(T) h(T)) = H_0((\lambda_0 I - T)^k p(T)).$$

According to Remark 4.18 we then have

$$H_0(\lambda_0 I - T) \subseteq H_0((\lambda_0 I - T)^k) \subseteq H_0((\lambda_0 I - T)^k p(T))$$
$$= H_0(\mu_0 I - f(T)),$$

and, evidently,

$$\ker g(T) = \ker [(\lambda_0 I - T)^k p(T)].$$

By Lemma 3.67, we also have

$$\ker g(T) = \ker (\mu_0 I - f(T)) = \ker [(\lambda_0 I - T)^k \oplus \ker \, p(T)].$$

and hence, from the equality (4.4),

$$H_0(\mu_0 I - f(T)) = \ker (\lambda_0 I - T)^{k\nu} \oplus \ker \, p(T)^k.$$

Therefore,

$$H_0(\lambda_0 I - T) \subseteq \ker (\lambda_0 I - T)^{k\nu} \oplus \ker \, p(T)^k.$$

From Lemma 4.17, we obtain

$$H_0(\lambda_0 I - T) \cap \ker \, p(T)^k = \{0\},$$

and hence $H_0(\lambda_0 I - T) \subseteq \ker (\lambda_0 I - T)^{k\nu}$. The opposite of the latter inclusion also holds, so we have $H_0(\lambda_0 I - T) = \ker (\lambda_0 I - T)^{k\nu}$. Theorem 4.12 then entails that T is polaroid. ∎

Remark 4.20 A natural question is if an analogue of Theorem 4.19 holds for left polaroid operators. By using the same arguments as the proof of Theorem 4.19

(just use the spectral mapping theorem for $\sigma_{ap}(T)$ and $\sigma_s(T)$) we easily obtain that the implication

$$T \text{ left polaroid} \Rightarrow f(T) \text{ left polaroid}$$

holds for every $f \in \mathcal{H}_{nc}(\sigma(T))$, and a similar implication also holds for right polaroid operators.

Denote by $\mathcal{H}_c^i(\sigma(T))$ the set of all $f \in \mathcal{H}_{nc}(\sigma(T))$ such that f is injective.

Theorem 4.21 *For an operator $T \in L(X)$ the following statements are equivalent.*

(i) T is left polaroid;
(ii) $f(T)$ is left polaroid for every $f \in \mathcal{H}_c^i(\sigma(T))$;
(iii) there exists an $f \in \mathcal{H}_c^i(\sigma(T))$ such that $f(T)$ is left polaroid.

Proof We have only to show that (iii) $\Rightarrow$ (i). Let λ_0 be an isolated point of $\sigma_{ap}(T)$ and let $\mu_0 := f(\lambda_0)$. As in the proof of Theorem 4.19 it then follows that $\mu_0 \in$ iso $\sigma_{ap}(f(T))$, so μ_0 is a left pole of $f(T)$. Now, by Theorem 4.16 there exists a left pole η of T such that $f(\eta) = \mu_0$ and since f is injective then $\eta = \lambda_0$. Therefore, T is left polaroid. ∎

The polaroid properties are transmitted from a Drazin invertible operator to its Drazin inverse:

Theorem 4.22 *Suppose that $T \in L(X)$ is Drazin invertible with Drazin inverse S. If T is polaroid then S is polaroid. Analogously, if T is a-polaroid then S is a-polaroid.*

Proof Suppose that T is polaroid and suppose that $\lambda \in$ iso $\sigma(S)$. If $\lambda = 0$, then 0 is a pole, since S is Drazin invertible. Hence we can suppose that $0 \neq \lambda$. We know that we can write, by Theorem 1.132, $T = T_1 \oplus T_2$ with T_1 nilpotent and T_2 invertible. Write $S = 0 \oplus S_2$ where S_2 is the inverse of T_2. From Theorem 1.135 we then have $\frac{1}{\lambda} \in$ iso $\sigma(T)$, so, T being polaroid, by Theorem 4.12 we have

$$H_0\left(\frac{1}{\lambda}I - T\right) = \ker\left(\frac{1}{\lambda}I - T\right)^p \text{ for some } p \in \mathbb{N}.$$

Since $\frac{1}{\lambda}I - T_1$ is invertible, we have

$$H_0\left(\frac{1}{\lambda}I - T\right) = H_0\left(\frac{1}{\lambda}I - T_1\right) \oplus H_0\left(\frac{1}{\lambda}I - T_2\right) = \{0\} \oplus H_0\left(\frac{1}{\lambda}I - T_2\right),$$

and analogously

$$H_0(\lambda I - S) = H_0(\lambda I - 0) \oplus H_0(\lambda I - S_2) = \{0\} \oplus H_0(\lambda I - S_2).$$

From Theorem 2.29, applied to the function $f(\lambda) := \frac{1}{\lambda}$, we know that

$$H_0(\lambda I - S_2) = \mathcal{Y}_{S_2}(\{\lambda\}) = \mathcal{Y}_{T_2}\left(\left\{\frac{1}{\lambda}\right\}\right) = H_0\left(\frac{1}{\lambda}I - T_2\right).$$

Hence, $H_0(\lambda I - S) = H_0(\frac{1}{\lambda}I - RT)$. By Lemma 3.128 we also have

$$\ker(\lambda I - S)^p = \ker\left(\frac{1}{\lambda}I - T\right)^p,$$

so $H_0(\lambda I - S) = \ker(\lambda I - S)^p$ for all $\lambda \in \mathrm{iso}\,\sigma(S)$. Hence S is polaroid.

Observe that if T is a-polaroid then $\mathrm{iso}\,\sigma(T) = \mathrm{iso}\,\sigma_{\mathrm{ap}}(T)$. Suppose now that T is a-polaroid. Then $\mathrm{iso}\,\sigma(T) = \mathrm{iso}\,\sigma_{\mathrm{ap}}(T)$. Let $\lambda \in \mathrm{iso}\,\sigma_{\mathrm{ap}}(S)$. If $\lambda = 0$ then $0 \in \sigma(S)$ and since S is Drazin invertible it then follows that 0 is a pole (of first order) of the resolvent of S. If $\lambda \neq 0$ then $\frac{1}{\lambda} \in \mathrm{iso}\,\sigma_{\mathrm{ap}}(T) = \mathrm{iso}\,\sigma(T)$, so $\lambda \in \sigma(S)$. Since S is polaroid by the first part of the proof, it then follows that λ is a pole of the resolvent of S.   ∎

Next we shall consider the preservation of the polaroid condition under suitable commuting perturbations. We start by considering nilpotent commuting perturbations.

Lemma 4.23 *If $T \in L(X)$ and N is a nilpotent operator commuting with T then $H_0(T + N) = H_0(T)$. Consequently, T is polaroid if and only if $T + N$ is polaroid.*

Proof It is enough to prove $H_0(T) \subseteq H_0(T + N)$, since the opposite inclusion may be obtained by symmetry. Let $x \in H_0(T)$ and suppose $N^\nu = 0$. Then we have $(T + N)^\nu = TS$, where

$$S := \sum_{j=0}^{\nu-1} c_{\nu,j} T^{\nu-1-j} N^j,$$

with suitable binomial coefficients $c_{\nu,j}$. We have

$$\|(T + N)^{\nu n}\|^{\frac{1}{n}} \leq \|T^n x\|^{\frac{1}{n}} \|S^n x\|^{\frac{1}{n}}.$$

From this estimate we then obtain $\lim_{n \to +\infty} \|(T + N)^{\nu n}\|^{\frac{1}{n}} = 0$ and hence $x \in H_0(T + N)$.

By Theorem 3.78, λ is a pole of the resolvent of T if and only if λ is a pole of the resolvent of $T + N$. Moreover, $\mathrm{iso}\,\sigma(T) = \mathrm{iso}\,\sigma(T + N)$.   ∎

The result above cannot be extended to non-nilpotent quasi-nilpotent operators. To see this, consider the case $T = 0$, and a non-nilpotent quasi-nilpotent operator Q. The perturbation of a polaroid operator by a compact operator may or may not affect the polaroid property of the operator. For instance, if R is the right shift operator on

$\ell^2(\mathbb{N})$, let $T := R \oplus R^*$ and

$$K = \begin{pmatrix} T0 & I - RR^* \\ 0 & 0 \end{pmatrix}.$$

Then both T and $T + K$ are polaroid, since iso $\sigma(T) = \emptyset$ and $T + K$ is unitary. Trivially, the identity I is polaroid, but its perturbation $I + Q$ by a compact quasi-nilpotent is not polaroid. The polaroid conditions are also preserved if K is a commuting operator for which K^n is a finite rank operator for some $n \in \mathbb{N}$:

Theorem 4.24 *Let $T \in L(X)$ and let $K \in L(X)$ be a commuting operator for which K^n is finite-dimensional for some $n \in \mathbb{N}$. Then:*

(i) *If T is polaroid then $T + K$ is polaroid.*
(ii) *If T is left polaroid (respectively, right polaroid) then $T + K$ is left polaroid (respectively, right polaroid).*
(iii) *If T is a-polaroid then $T + K$ is a-polaroid.*

Proof

(i) Suppose T is polaroid. If $\lambda \in$ iso $\sigma(T + K)$ then there are two possibilities: $\lambda \notin \sigma(T)$ or $\lambda \in \sigma(T)$. If $\lambda \notin \sigma(T)$ then $\lambda I - T$ is invertible and hence $\lambda I - (T + K)$ is Browder, by Theorem 3.11. Therefore, λ is a pole of $T + K$. If $\lambda \in \sigma(T)$ then $\lambda \in$ iso $\sigma(T)$, by Corollary 3.21. Since T is polaroid then λ is a pole of T, or equivalently $\lambda I - T$ is Drazin invertible. By Theorem 3.78 it then follows that $\lambda I + (T + K)$ is Drazin invertible. Since $\lambda \in$ iso $\sigma(T + K)$, λ is a pole of the resolvent of T.

(ii) Suppose T is left polaroid. If $\lambda \in$ iso $\sigma_{ap}(T + K)$ then there are two possibilities: $\lambda \notin \sigma_{ap}(T)$ or $\lambda \in \sigma_{ap}(T)$. If $\lambda \notin \sigma_{ap}(T)$ then $\lambda I - T$ is bounded below and hence $\lambda I - (T + K)$ is upper semi-Browder, by Theorem 3.11, in particular it is left Drazin invertible. Therefore, λ is a left pole of $T + K$. If $\lambda \in \sigma_{ap}(T)$ then $\lambda \in$ iso $\sigma_{ap}(T)$, by Corollary 3.28. Since T is left polaroid then λ is a left pole of T, and hence left Drazin invertible. By Theorem 3.78 it then follows that $\lambda I + (T + K)$ is left Drazin invertible. Since $\lambda \in$ iso $\sigma_{ap}(T + K)$ then λ is a left pole of $T + K$. The proof in the case of right polaroid operators is similar.

(iii) Suppose T is a-polaroid. If $\lambda \in$ iso $\sigma_{ap}(T + K)$ then there are two possibilities: $\lambda \notin \sigma(T)$ or $\lambda \in \sigma(T)$. If $\lambda \notin \sigma(T)$ then $\lambda I - T$ is invertible and hence $\lambda I - (T + K)$ is Browder, by Theorem 3.11, in particular it is Drazin invertible. If $\lambda \in \sigma(T)$ then $\lambda \in$ iso $\sigma(T)$, by Corollary 3.21. Since T is a- polaroid then λ is a pole of T, and hence $\lambda I - T$ is Drazin invertible. By Theorem 3.78 it then follows that $\lambda I - (T + K)$ is Drazin invertible. Since $\lambda \in \sigma(T + K)$ we then conclude that λ is a pole of the resolvent of $T + K$. ∎

Obviously Theorem 4.24 applies to nilpotent commuting perturbations. The next example shows that this result cannot be extended to quasi-nilpotent operators Q commuting with T.

Example 4.25 Let $Q \in L(\ell^2(\mathbb{N}))$ be defined by

$$Q(x_1, x_2, \ldots) = \left(\frac{x_2}{2}, \frac{x_3}{3}, \ldots\right) \quad \text{for all } (x_n) \in \ell^2(\mathbb{N}),$$

Then Q is quasi-nilpotent and if $e_n := (0, \ldots, 1, 0, \ldots)$, where 1 is the n-th term and all others are 0, then $e_{n+1} \in \ker Q^{n+1}$ while $e_{n+1} \notin \ker Q^n$, so that $p(Q) = \infty$. If we take $T = 0$, the null operator, then T is both left and a-polaroid, while $T + Q = Q$ is not left polaroid, as well as not a-polaroid or polaroid.

However, the following theorem shows that $T + Q$ is polaroid in a very special case. Recall first that if $\alpha(T) < \infty$ then $\alpha(T^n) < \infty$ for all $n \in \mathbb{N}$.

Theorem 4.26 *Suppose that* $Q \in L(X)$ *is a quasi-nilpotent operator which commutes with* $T \in L(X)$ *and suppose that all eigenvalues of* T *have finite multiplicity.*

(i) *If* T *is a polaroid operator then* $T + Q$ *is polaroid.*
(ii) *If* T *is a left polaroid operator then* $T + Q$ *is left polaroid.*
(iii) *If* T *is an a-polaroid operator then* $T + Q$ *is a-polaroid.*

Proof

(i) Let $\lambda \in \operatorname{iso}\sigma(T + Q)$. It is well-known that the spectrum is invariant under commuting quasi-nilpotent perturbations, thus $\lambda \in \operatorname{iso}\sigma(T)$ and hence is a pole of the resolvent of T (consequently, an eigenvalue of T). Therefore, $p := p(\lambda I - T) = q(\lambda I - T) < \infty$ and since by assumption $\alpha(\lambda I - T) < \infty$ we then have $\alpha(\lambda I - T) = \beta(\lambda I - T)$, by Theorem 1.22, so $\lambda I - T$ is Browder. By Theorem 3.8 we then obtain that $\lambda I - (T + Q)$ is Browder, hence λ is a pole of $T + Q$, thus $T + Q$ is polaroid.

(ii) Let $\lambda \in \operatorname{iso}\sigma_{\mathrm{ap}}(T + Q)$. We know that $\sigma_{\mathrm{ap}}(T)$ is invariant under commuting quasi-nilpotent perturbations, so $\lambda \in \operatorname{iso}\sigma_{\mathrm{ap}}(T)$ and hence, since T is left-polaroid, λ is a left pole of the resolvent of T. Therefore, $p := p(\lambda I - T) < \infty$ and $(\lambda I - T)^{p+1}(X)$ is closed. Now, $\lambda I - T$ is injective or λ is an eigenvalue of T. In both cases we have $\alpha(\lambda I - T) < \infty$ and hence $\alpha(\lambda I - T)^{p+1} < \infty$. Thus, $(\lambda I - T)^{p+1} \in \Phi_+(X)$ and this implies that $\lambda I - T \in \Phi_+(X)$. Consequently, $\lambda I - T \in B_+(X)$ and hence, by Theorem 3.8, $\lambda I - (T + Q)$ is upper-Browder. This implies that $p' := p(\lambda I - (T + Q)) < \infty$ and since $(\lambda I - (T+Q))^{p'+1}$ is still upper semi-Browder, $\lambda I - (T+Q))^{p'+1}(X)$ is closed and hence $(\lambda I - (T + Q))$ is left Drazin invertible. Since $\lambda \in \operatorname{iso}\sigma_{\mathrm{ap}}(T + Q)$ it then follows that λ is a left pole of $T + Q$ and hence $T + Q$ is left polaroid.

(iii) The proof is analogous to that of part (i). In fact, if $\lambda \in \operatorname{iso}\sigma_{\mathrm{ap}}(T + Q)$ then $\lambda \in \operatorname{iso}\sigma_{\mathrm{ap}}(T)$ and hence, since T is a-polaroid, λ is a pole of the resolvent of T. By assumption, $\alpha(\lambda I - T) < \infty$. Proceeding as in part (i) we then have that λ is a pole of $T + Q$, thus $T + Q$ is a-polaroid.   ∎

The argument of the proof of part (i) of Theorem 4.26 also works if we assume that every isolated point of $\sigma(T)$ is a finite rank pole (in this case T is said to be *finitely polaroid*). This is the case, for instance, for Riesz operators having infinite spectrum. Evidently, $T+Q$ is also finitely polaroid, since for every $\lambda \in \text{iso}\,\sigma\,(T+Q)$ we have $\alpha(\lambda I - (T + Q)) < \infty$.

Theorem 4.27 *Let* $T \in L(X)$ *and* Q *be a quasi-nilpotent operator which commutes with* T. *If* $\text{iso}\,\sigma_b(T) = \emptyset$ *then* T *is polaroid if and only if* $T + Q$ *is polaroid. Analogously, if* $\text{iso}\,\sigma_{ub}(T) = \emptyset$ *then* T *is a-polaroid if and only if* $T + Q$ *is a-polaroid*

Proof We know that T and $T + Q$ have the same spectrum. By Theorem 3.59 T and $T + Q$ have the same set of poles. The proof of the second assertion is similar. ∎

4.2 Hereditarily Polaroid Operators

Every operator K for which K^n is finite-dimensional is algebraic, so it makes sense to find conditions for which the polaroid condition is preserved under algebraic commuting perturbations. By a *part of an operator* T we mean the restriction of T to a closed T-invariant subspace.

Definition 4.28 An operator $T \in L(X)$ is said to be *hereditarily polaroid* if every part of T is polaroid.

A simple example shows that a polaroid operator need not be necessarily hereditarily polaroid. Let $T := R \oplus Q$ on $H \oplus H$, where $H := \ell^2(\mathbb{N})$, R is the right shift and Q is quasi-nilpotent. Then $\sigma(T)$ is the unit disc, so iso $\sigma(T)$ is empty and hence T is polaroid. On the other hand, if $M := \{0\} \oplus \ell^2(\mathbb{N})$, then $T|M$ is not polaroid, since Q is not polaroid.

It is easily seen that the property of being hereditarily polaroid is similarity invariant, but is not preserved by a quasi-affinity. We now want to show that every hereditarily polaroid operator has the SVEP. First we need to introduce two concepts of orthogonality on Banach spaces.

Definition 4.29 A closed subspace M of a Banach space X is said to be *orthogonal* to a closed subspace N of X in the sense of Birkhoff and James, in symbols $M \perp N$ if $\|x\| \leq \|x + y\|$ for all $x \in M$ and $y \in N$.

A study of this concept of orthogonality may be found in [143]. Note that this concept of orthogonality is asymmetric and reduces to the usual definition of orthogonality in the case of Hilbert spaces. This concept of orthogonality may be weakened as follows:

Definition 4.30 A closed subspace M of a Banach space X is said to be *approximately orthogonal* to a closed subspace N of X, in symbols $M \perp_a N$, if there exists a scalar $\alpha \geq 1$ such that $\|x\| \leq \alpha\|x + y\|$ for all $x \in M$ and $y \in N$.

What $M \perp_a N$ means is that M meets N at an angle θ, $0 \leq \theta \leq \frac{\pi}{2}$, where by definition

$$\sin \theta = \inf\{\|x - y\|, \|y\| = 1\} \quad \text{for all } x \in M, y \in N.$$

If $\theta = \frac{\pi}{2}$, then M is orthogonal in the Birkhoff–James sense. If M meets N at an angle $\theta > 0$ then N meets M at an angle $\phi > 0$, where in general $\theta \neq \phi$.

Theorem 4.31 *Every hereditarily polaroid operator $T \in L(X)$ has the SVEP.*

Proof Let T be hereditarily polaroid. For distinct eigenvalues λ and μ of T, let M denote the subspace generated by $\ker(\lambda I - T)$ and $\ker(\mu I - T)$. Set $S := T|M$. Then S is polaroid and $\sigma(S) = \{\lambda, \mu\}$. Denote by P_μ the spectral projection corresponding to the spectral set $\{\mu\}$. Then

$$P_\mu(M) = \ker(\mu I - S) = \ker(\mu I - T),$$

while

$$\ker P_\mu = (I - P_\mu)(M) = \ker(\lambda I - S) = \ker(\lambda I - T).$$

Set $\alpha := \|P_\mu\|$. Then $\alpha \geq 1$, and

$$\|x\| = \|p_\mu x\| = \|P_\mu(x.y)\| \leq \alpha \|x - y\|$$

for all $x \in P_\mu(M) = \ker(\mu I - T)$ and $y \in (I - P_\mu)(M) = \ker(\lambda I - T)$.

Now, suppose that T does not have the SVEP at a point $\delta_0 \in \mathbb{C}$. Then there exists an open disc $\mathbb{D}_0$ centered at δ_0 and a non-trivial analytic function $f : \mathbb{D}_0 \to X$ such that

$$f(\delta) \in \ker(\delta I - T) \text{ for all } \delta \in \mathbb{D}_0.$$

Let $\lambda \in \mathbb{D}_0$ and $\mu \in \mathbb{D}_0$ be two distinct complex numbers such that $f(\lambda)$ and $f(\mu)$ are non-zero. Since $\ker(\mu I - T) \perp_a \ker(\lambda I - T)$,

$$0 < \|f(\mu)\| \leq \alpha \|f(\mu) - f(\lambda)\|.$$

But then f is not continuous at μ, a contradiction. Hence T has the SVEP.   ∎

We have seen that T^* is polaroid if and only if T is polaroid. An immediate consequence of Theorem 4.31 is that this equivalence in general is not true for hereditarily polaroid operators. Indeed, the right shift R is trivially hereditarily polaroid while its dual, the left shift L, cannot be hereditarily polaroid, since it does not have the SVEP. Note that, by Theorem 3.44, for a hereditarily polaroid operator

T we have

$$\sigma_{\mathrm{lw}}(T) = \sigma_{\mathrm{w}}(T) = \sigma_{\mathrm{b}}(T) = \sigma_{\mathrm{lb}}(T),$$

since T has the SVEP.

Theorem 4.32 *Suppose that $T \in L(X)$ and $K \in L(X)$ is an algebraic operator which commutes with T.*

(i) *Suppose that T is hereditarily polaroid. Then $T + K$ is polaroid and $T^* + K^*$ is a-polaroid. If T^* has the SVEP then $T + K$ is a-polaroid.*

(ii) *Suppose that T^* is hereditarily polaroid. Then $T^* + K^*$ is polaroid and $T + K$ is a-polaroid. If T has the SVEP then $T^* + K^*$ is a-polaroid.*

Proof

(i) An easy consequence of the spectral mapping theorem is that an algebraic operator has a finite spectrum. Let $\sigma(K) = \{\lambda_1, \lambda_2, \dots, \lambda_n\}$ and denote by P_j the spectral projection associated with K and the spectral sets $\{\lambda_j\}$. Set $Y_j := P_j(X)$ and $Z_j := \ker P_j$. From the classical spectral decomposition we know that $X = Y_j \oplus Z_j$, Y_j and Z_j are invariant closed subspaces under T and K. Moreover, if we let $K_j := K|Y_j$ and $T_j := T|Y_j$, then K_j and T_j commutes, $\sigma(K_j) = \{\lambda_j\}$ and

$$\sigma(T + K) = \bigcup_{j=1}^{n} \sigma(T_j + K_j).$$

We claim that $N_j := \lambda_j I - K_j$ is nilpotent for every $j = 1, 2, \dots, n$. To prove this, denote by h a non-trivial polynomial for which $h(K) = 0$. Then

$$h(K_j) = h(K|Y_j) = 0 \quad \text{for all } j = 1, 2, \dots, n,$$

and since

$$h(\{\lambda_j\}) = h(\sigma(K_j)) = \sigma(h(K_j)) = \{0\}$$

it then follows that $h(\lambda_j) = 0$. Set

$$h(\mu) := (\lambda_j - \mu)^{\nu} q(\mu) \text{ with } q(\lambda_j) \neq 0.$$

Then

$$0 = h(K_j) = (\lambda_j - K_j)^{\nu} q(K_j),$$

where all $q(K_j)$ are invertible. Therefore, $(\lambda_j - K_j)^{\nu} = 0$ and hence $N_j := \lambda_j - K_j$ is nilpotent for every $j = 1, 2, \dots, n$, as desired.

We show now that $T + K$ is polaroid. Let $\lambda \in \text{iso}\,\sigma(T + K)$. Then $\lambda \in \text{iso}\,\sigma(T_j + K_j)$ for some $j = 1, 2, \ldots, n$, hence $\lambda - \lambda_j \in \text{iso}\,\sigma(T_j + K_j - \lambda_j I)$. The restriction T_j is polaroid, by assumption, and as proved before $\lambda_j I - K_j$ is nilpotent. By Lemma 4.23 then $T_j + K_j - \lambda_j I$ is polaroid. Therefore, $\lambda - \lambda_j$ is a pole of the resolvent of $T_j + K_j - \lambda_j I$ and hence, by Corollary 2.47, there exists a $\nu_j \in \mathbb{N}$ such that

$$H_0((\lambda - \lambda_j)I - (T_j + K_j + \lambda_j I)) = H_0(\lambda I - (T_j + K_j))$$
$$= \ker(\lambda I - (T_j + K_j))^{\nu_j}.$$

Taking into account that $H_0(\lambda I - (T_j + K_j)) = \{0\}$ if $\lambda \notin \sigma(T_j + K_j)$, it then follows that

$$H_0(\lambda I - (T + K)) = \bigoplus_{j=1}^{n} H_0((\lambda I - (T_j + K_j))$$
$$= \bigoplus_{j=1}^{n} \ker(\lambda I - (T_j + K_j))^{\nu_j}.$$

Clearly, if we put $\nu := \max\{\nu_1, \nu_2, \cdots, \nu_n\}$ we then obtain

$$H_0(\lambda I - (T + K)) = \ker(\lambda I - (T + K))^{\nu}.$$

As λ is an arbitrary isolated point of $\sigma(T + K)$, it then follows, by Theorem 4.12, that λ is a pole of the resolvent of $T + K$. Hence $T + K$ is polaroid.

To show that $T^* + K^*$ is a-polaroid observe that by duality $T^* + K^*$ is polaroid. Since T has the SVEP, by Theorem 4.31, $T + K$ also has the SVEP, by Theorem 2.145. By Theorem 4.15 it then follows that $T^* + K^*$ is a-polaroid.

Suppose now that T^* has the SVEP. Obviously, K^* is algebraic and commutes with T^*. Therefore, $T^* + K^*$ has the SVEP, again by Theorem 2.145, and hence $T + K$ is a-polaroid, by Theorem 4.15.

(ii) By part (i) we know that $T^* + K^*$ polaroid or, equivalently, $T + K$ is polaroid. Since T^* has the SVEP, again by Theorem 4.31, $T^* + K^*$ also has the SVEP, so, by Theorem 4.15, $T + K$ is a-polaroid. If we suppose that T has the SVEP, then $T + K$ has the SVEP, hence $T^* + K^*$ is a-polaroid, by Theorem 4.15. ∎

The result of Theorem 4.32 may be extended to $f(T)$ for every function $f \in \mathcal{H}_{nc}(\sigma(T + K))$.

Theorem 4.33 *Suppose $K \in L(X)$ is an algebraic operator commuting with $T \in L(X)$ and let $f \in \mathcal{H}_{nc}(\sigma(T + K))$. Then we have*

(i) *If T is hereditarily polaroid then $f(T + K)$ is polaroid, while $f(T^* + K^*)$ is a-polaroid.*

(ii) *If T^* is hereditarily polaroid then $f(T + K)$ is a-polaroid, while $f(T^* + K^*)$ is polaroid.*

Proof

(i) Let T be hereditarily polaroid. Then $T + K$ is polaroid, by Theorem 4.32, and hence $f(T + K)$ is polaroid, by Theorem 4.19. We also know that the SVEP holds for T, and this entails that $T + K$ also has the SVEP, by Theorem 2.145. From Theorem 2.86 it then follows that $f(T + K)$ has the SVEP. Since $f(T + K)^* = f(T^* + K^*)$ is also polaroid, by Theorem 4.15, we then conclude that $f(T^* + K^*)$ is a-polaroid.

(ii) The proof is analogous.   ∎

Remark 4.34 In the case of Hilbert space operators, the assertions of Theorem 4.32 are still valid if T^* is replaced by the Hilbert adjoint T'.

A natural question is whether the polaroid property, or the hereditarily polaroid property, for an operator is preserved under compact perturbations. The answer to these questions is negative, see Zhu and Li [227], or Duggal [127].

4.3   Examples of Polaroid Operators

The class of hereditarily polaroid operators is substantial; it contains several important classes of operators. The first class that we consider is the following one introduced by Oudghiri [251].

Definition 4.35 A bounded operator $T \in L(X)$ is said to *belong to the class $H(p)$* if there exists a natural $p := p(\lambda)$ such that:

$$H_0(\lambda I - T) = \ker(\lambda I - T)^p \quad \text{for all } \lambda \in \mathbb{C}. \tag{4.5}$$

It should be noted that the integer $p := p(\lambda)$ may assume different values. We shortly say that T *belongs to the class $H(1)$* if $p(\lambda) = 1$ for all $\lambda \in \mathbb{C}$. The class $H(1)$ has been studied by Aiena and Villafãne in [32].

Evidently, since $H_0(\lambda I - T)$ is closed for each $\lambda \in \mathbb{C}$, every $H(p)$-operator has the SVEP, by Theorem 2.39. Moreover, every $H(p)$-operator is polaroid, by Theorem 4.12.

The property $H(p)$ is inherited by restrictions to closed invariant subspaces:

Theorem 4.36 *Let $T \in L(X)$ be a bounded operator on a Banach space X. If T has the property $H(p)$ and Y is a closed T-invariant subspace of X then $T|Y$ has the property $H(p)$.*

Proof If $H_0(\lambda I - T) = \ker(\lambda I - T)^p$ then

$$H_0((\lambda I - T)|Y) \subseteq \ker(\lambda I - T)^p \cap Y = \ker((\lambda I - T)|Y)^p,$$

from which we obtain $H_0((\lambda I - T)|Y) = \ker((\lambda I - T)|Y)^p$.   ∎

The following result is an easy consequence of Theorems 4.36 and 4.33.

Corollary 4.37 *Every $H(p)$-operator T is hereditarily polaroid. Moreover, if K is algebraic and commutes with T, then $f(T + K)$ is a-polaroid, while $f(T^* + K^*)$ is polaroid, for every $f \in \mathcal{H}_{nc}(\sigma(T + K))$.*

The next result shows that property $H(p)$ is preserved by quasi-affine transforms.

Theorem 4.38 *If $S \in L(Y)$ has property $H(p)$ and $T \prec S$, then T has property $H(p)$.*

Proof We consider the case when $p := (\lambda I - T) = 1$ for all $\lambda \in \mathbb{C}$. Suppose S has property $H(1)$, i.e. $SA = AT$, with A injective. If $\lambda \in \mathbb{C}$ and $x \in H_0(\lambda I - T)$ then

$$\|(\lambda I - S)^n Ax\|^{1/n} = \|A(\lambda I - T)^n x\|^{1/n} \leq \|A\|^{1/n} \|(\lambda I - T)^n x\|^{1/n},$$

from which it follows that $Ax \in H_0(\lambda I - S) = \ker(\lambda I - S)$. Hence,

$$A(\lambda I - T)x = (\lambda I - S)Ax = 0$$

and, since A is injective, we then conclude that $(\lambda I - T)x = 0$, i.e., $x \in \ker(\lambda I - T)$. Therefore $H_0(\lambda I - T) = \ker(\lambda I - T)$ for all $\lambda \in \mathbb{C}$.

The more general case of $H(p)$-operators is proved by a similar argument. ∎

The class of $H(p)$-operators is very large. To see this, we first introduce a special class of operators which plays an important role in local spectral theory. Let $\mathcal{C}^\infty(\mathbb{C})$ denote the Fréchet algebra of all infinitely differentiable complex-valued functions on $\mathbb{C}$.

Definition 4.39 An operator $T \in L(X)$, X a Banach space, is said to be *generalized scalar* if there exists a continuous algebra homomorphism $\Psi : \mathcal{C}^\infty(\mathbb{C}) \to L(X)$ such that

$$\Psi(1) = I \quad \text{and} \quad \Psi(Z) = T,$$

where Z denotes the identity function on $\mathbb{C}$.

The interested reader can find a well-organized treatment of generalized scalar operators in Laursen and Neumann [216, Section 1.5]. It should be noted that every quasi-nilpotent generalized scalar operator is nilpotent [216, Proposition 1.5.10]. Moreover, if T is generalized scalar then T has the Dunford property (C), i.e. $\mathcal{X}_T(\Omega)$ is closed for all closed subset $\Omega \subseteq \mathbb{C}$, see [216, Theorem 1.5.4 and Proposition 1.4.3]. In particular, $H_0(\lambda I - T) = \mathcal{X}_T(\{\lambda\})$ is closed for each $\lambda \in \mathbb{C}$, so every generalized scalar operator has the SVEP, by Theorem 2.39.

An operator similar to a restriction of a generalized scalar operator to one of its closed invariant subspaces is called *subscalar*.

Theorem 4.40 *Every generalized scalar, as well as every subscalar operator* $T \in L(X)$ *is* $H(p)$. *Consequently, every generalized scalar and every subscalar operator is hereditarily polaroid.*

Proof By Lemma 4.36 and Theorem 4.38 we may assume that T is generalized scalar. Consider a continuous algebra homomorphism $\Psi : C^\infty(\mathbb{C}) \to L(X)$ such that $\Psi(1) = I$ and $\Psi(Z) = T$. Let $\lambda \in \mathbb{C}$. Since every generalized scalar operator has property (C), $H_0(\lambda I - T) = \mathcal{X}_T(\{\lambda\})$ is closed. On the other hand, if $f \in C^\infty(\mathbb{C})$ then

$$\Psi(f)(H_0(\lambda I - T)) \subseteq H_0(\lambda I - T),$$

because $T = \Psi(Z)$ commutes with $\Psi(f)$. Define

$$\tilde{\Psi} : C^\infty(\mathbb{C}) \to L(H_0(\lambda I - T))$$

by

$$\tilde{\Psi}(f) = \Psi(f)|H_0(\lambda I - T) \quad \text{for every } f \in C^\infty(\mathbb{C}).$$

Clearly, $T|H_0(\lambda I - T)$ is generalized scalar and quasi-nilpotent, so it is nilpotent. Thus there exists a $p \geq 1$ for which $H_0(\lambda I - T) = \ker(\lambda I - T)^p$. ∎

Definition 4.41 An operator $T \in L(X)$ is said to be *paranormal* if

$$\|Tx\| \leq \|T^2 x\| \quad \text{for all unit vectors } x \in X. \tag{4.6}$$

The restriction $T|M$ of a paranormal operator $T \in L(X)$ to a closed subspace M is evidently paranormal. The property of being paranormal is not translation-invariant, see Chō and Lee [91]. An operator $T \in L(X)$ is called *totally paranormal* if $\lambda I - T$ is paranormal for all $\lambda \in \mathbb{C}$. Note that every isometry is paranormal.

Theorem 4.42 *Every totally paranormal operator has property* $H(1)$.

Proof In fact, if $x \in H_0(\lambda I - T)$ then $\|(\lambda I - T)^n x\|^{1/n} \to 0$ and since T is totally paranormal then $(\lambda I - T)^n x\|^{1/n} \geq \|(\lambda I - T)x\|$. Therefore, $H_0(\lambda I - T) \subseteq \ker(\lambda I - T)$, and since the reverse inclusion holds for every operator, we have $H_0(\lambda I - T) = \ker(\lambda I - T)$. ∎

Theorem 4.40 implies that some important classes of operators are $H(p)$. In the sequel we list some of these classes acting on Hilbert spaces. Let H be a Hilbert space, with inner product $(\cdot, \cdot)$ and, as usual, denote by T' the adjoint of T.

(a) *Hyponormal operators.* A bounded operator $T \in L(H)$ is said to be *hyponormal* if

$$T'T \geq TT'.$$

It is easily seen that T is hyponormal if and only if

$$\|T'x\| \le \|Tx\| \quad \text{for all } x \in H.$$

Indeed, $T'T \ge TT'$ means that $(T'Tx, x) \ge (TT'x, x)$ for all $x \in H$, or equivalently

$$\|T'x\|^2 = (T'x, T'x) = (TT'x, x) \le (T'Tx, x) = (Tx, Tx) = \|Tx\|^2.$$

Clearly, $\|T'x\|^2 \le \|Tx\|^2$ if and only if $\|T'x\| \le \|Tx\|$.

By an important result due to Putinar [258], every hyponormal operator is similar to a subscalar operator, see also [216, section 2.4], so, by Theorem 4.40, hyponormal operators are $H(p)$. A routine computation shows that that a weighted right shift, see later for the definition, on the Hilbert space $\ell^2(\mathbb{N})$ is hyponormal if and only if the corresponding weight sequence is increasing. Since every generalized scalar operator is decomposable, see [216, Theorem 1.5.4], and hence has property (β), for every increasing weight sequence the corresponding weighted right shift has property (β). Examples of hyponormal operators are the quasi-normal operators, see Conway [99] or Furuta [151], where $T \in L(H)$ is said to be *quasi-normal* if

$$T(T'T) = (T'T)T.$$

A very easy example of a quasi-normal operator is given by the right shift R on $\ell_2(\mathbb{N})$. Indeed, the adjoint of R is the left shift L and obviously $R(R'R) = (RR')R$. Note that R is not normal, since $RR' = RL \ne R'R = LR = I$. An operator $T \in L(H)$ is said to be *subnormal* if there exists a normal extension N, i.e. there exists a Hilbert space K such that $H \subseteq K$ and a normal operator $N \in L(K)$ such that $N|H = T$. We have

$$T \text{ quasi-normal} \Rightarrow T \text{ subnormal} \Rightarrow T \text{ hyponormal}.$$

For details, see Furuta [151, p. 105]. We give in the sequel some relevant properties of hyponormal operators.

Lemma 4.43 *Let $T \in L(H)$ be hyponormal. Then we have:*

(i) $\lambda I - T$ *is hyponormal for every $\lambda \in \mathbb{C}$.*
(ii) *If M is a closed invariant subspace of H then $T|M$ is hyponormal.*

Proof

(i) We have

$$(\lambda I - T)'(\lambda I - T) - (\lambda I - T)(\lambda I - T)'$$
$$= (\bar{\lambda}I - T)(\lambda I - T) - (\lambda I - T)(\bar{\lambda}I - T) = T'T - TT' \le 0,$$

thus, $\lambda I - T$ is hyponormal.

(ii) Observe first that if P_M is the projection of T onto M, then $(T|M)' = (P_M T')|M$. For every $x \in M$ we then have

$$\|(T|M)'x\| = \|(P_M T')x\| \leq \|P_M\| \|T'x\|$$
$$= \|T'x\| \leq \|Tx\| = \|T|Mx\|,$$

so $T|M$ is hyponormal. ∎

Lemma 4.44 *Let $T \in L(H)$ be a self-adjoint operator such that $\lambda I \leq T$ for some $\lambda \geq 0$. Then T is invertible. In particular, if $I \leq T$ then $0 \leq T^{-1} \leq I$.*

Proof To show the first assertion, observe that by the Schwarz inequality we have

$$\|Tx\| \|x\| \geq (Tx, x) \geq c\|x\|^2,$$

so $\|Tx\| \geq c\|x\|$, and hence T is bounded below by Lemma 1.9. Let y be orthogonal to $T(H)$, that is

$$0 = (y, Tx) = (Ty, x) \text{ for all } x \in H.$$

Then $Ty = 0$ and since T is injective we then have $y = 0$. Therefore, $T(H)^\perp = \overline{T(H)}^\perp = \{0\}$, and hence T is surjective, thus T is invertible.

To show the second assertion, note that if $I \leq T$ then T is invertible and T^{-1} is also positive. Since the product of two commuting positive operators is also positive, it then follows that

$$T^{-1}(T - I) = I - T^{-1} \geq 0,$$

thus $T^{-1} \leq I$. ∎

It is easily seen that if T is self-adjoint then STS' is also self-adjoint for every $S \in L(H)$. Moreover, if T is positive then $STS' \geq 0$ for all $S \in L(H)$.

Theorem 4.45 *If $T \in L(H)$ is an invertible hyponormal operator then its inverse T^{-1} is also hyponormal.*

Proof Suppose that T is hyponormal. Then $T'T - TT' \geq 0$ and hence, as noted above, the product $T^{-1}(T'T - TT')(T^{-1})'$ is still positive. From this we obtain $T^{-1}(T'T)(T^{-1})' - I \geq 0$, and hence

$$T^{-1}(T'T)(T^{-1})' \geq I,$$

thus, by Lemma 4.44, the product $T^{-1}(T'T)(T^{-1})'$ is invertible with

$$0 \leq [T^{-1}(T'T)(T^{-1})']^{-1} \leq I.$$

From the last inequality we then obtain that

$$S := I - T'(T^{-1}(T')^{-1})T$$

is positive, so $T^{-1}ST^{-1} \geq 0$, from which we easily obtain that

$$(T^{-1})'T^{-1} - T^{-1}(T^{-1})' \geq 0.$$

Therefore, T^{-1} is hyponormal. ■

Theorem 4.46 *Every hyponormal operator $T \in L(H)$ is totally paranormal. Consequently, every hyponormal operator is $H(1)$.*

Proof To show that T is totally paranormal it suffices to prove, by Lemma 4.43, that every hyponormal operator is paranormal. Since T is hyponormal we have, for every $x \in H$,

$$\|Tx\|^2 = (Tx, Tx) = (T'Tx, x) \leq \|T'(Tx)\| \|x\|$$
$$\leq \|T(Tx)\| \|x\| = \|T^2x\| \|x\|.$$

Taking $\|x\| = 1$ we then have $\|Tx\|^2 \leq \|T^2x\|$, so T is paranormal.
By Theorem 4.42 T is $H(1)$. ■

For $T \in L(H)$ let $T = W|T|$ be the polar decomposition of T. Then

$$R := |T|^{1/2}W|T|^{1/2}$$

is said to be the *Aluthge transform* of T, see [47]. If $R = V|R|$ is the polar decomposition of R (see Appendix A) let us define

$$\widetilde{T} := |R|^{1/2}V|R|^{1/2}.$$

(b) *Log-hyponormal operators.* An operator $T \in L(H)$ is said to be log-hyponormal if it is invertible and satisfies

$$\log(T^*T) \geq \log(TT^*).$$

If T is log-hyponormal then $\widetilde{T}$ is hyponormal and $T = K\widetilde{T}K^{-1}$, where $K := |R|^{1/2}|T|^{1/2}$, see Tanahashi [290], and Chō et al. [92]. Hence T is similar to a hyponormal operator and therefore, by Theorem 4.38, has property $H(1)$.

(c) *p-hyponormal operators.* An operator $T \in L(H)$ is said to be *p*-hyponormal, with $0 < p \leq 1$, if

$$(T'T)^p \geq (TT')^p.$$

If $p = \frac{1}{2}$, T is said to be semi-hyponormal. The class of p-hyponormal operators has been studied by Aluthge [47], while semi-hyponormal operators were introduced by Xia [298]. Any p-hyponormal operator is q-hyponormal if $q < p$, but there are examples to show that the converse is not true, see [47]. Every invertible p-hyponormal is subscalar [200], and is quasi-similar to a log-hyponormal operator. Consequently, by Theorem 4.38, every invertible p-hyponormal is operator has property $H(1)$, see Aiena and Miller [17], and Duggal and Djordjević [129]. This is also true for p-hyponormal operators which are not invertible, see Duggal and Jeon [134]. Every p-hyponormal operator is paranormal, see [49] or [90].

(d) *M-hyponormal operators.* Recall that $T \in L(H)$ is said to be *M-hyponormal* if there exists an $M > 0$ such that

$$TT^* \leq MT^*T.$$

Every M-hyponormal operator is subscalar [216, Proposition 2.4.9] and hence $H(p)$.

(e) *w-hyponormal operators.* If $T \in L(H)$ and $T = U|T|$ is the polar decomposition, define

$$\hat{T} := |T|^{\frac{1}{2}} U |T|^{\frac{1}{2}}.$$

$T \in L(H)$ is said to be w-hyponormal if

$$|\hat{T}| \geq |T| \geq |\hat{T}^*|.$$

Examples of w-hyponormal operators are p-hyponormal operators and log-hyponormal operators. Each w-hyponormal operator is subscalar, together with its Aluthge transformation, see Chō et al. [228], and hence $H(p)$. In [168, Theorem 2.5] it is shown that for every isolated point λ of the spectrum of a w-hyponormal operator T we have $H_0(\lambda I - T) = \ker(\lambda I - T)$ and hence λ is a simple pole of the resolvent.

(f) *Multipliers of semi-simple Banach algebras.* Let A denote a complex Banach algebra (not necessarily commutative) with or without a unit.

Definition 4.47 The mapping $T : A \rightarrow A$ is said to be a *multiplier* of A if

$$x(Ty) = (Tx)y \quad \text{for all } x, y \in A. \tag{4.7}$$

The set of all multipliers of A is denoted by $M(A)$.

An immediate example of a multiplier of a Banach algebra A is given by the multiplication operator $L_a : x \in A \rightarrow ax \in A$ by an element a which commutes with every $x \in A$. In the case in which A is a commutative Banach algebra with unit u the concept of multiplier reduces, trivially, to the multiplication operator by an element of A. To see this, given a multiplier $T \in M(A)$, let us consider the multiplication operator L_{Tu} by the element Tu. For each $x \in A$ we have

$$L_{Tu}x = (Tu)x = u(Tx) = Tx,$$

thus $T = L_{Tu}$. In this case we can identify A with $M(A)$. A very important example of a multiplier is given in the case where A is the semi-simple commutative Banach algebra $L^1(G)$, the group algebra of a locally compact abelian group G with convolution as multiplication. Indeed, in this case to any complex Borel measure μ on G there corresponds a multiplier T_μ defined by

$$T_\mu(f) := \mu \star f \ \text{ for all } f \in L^1(G),$$

where

$$(\mu \star f)(t) := \int_G f(t - s)d\mu(s).$$

The classical Helson–Wendel Theorem shows that each multiplier is a convolution operator and the multiplier algebra of $A := L^1(G)$ may be identified with the measure algebra $M(G)$, see Larsen [210, Chapter 0].

We recall that an algebra A is said to be *semi-prime* if $\{0\}$ is the only two-sided ideal J for which $J^2 = \{0\}$. A left ideal J of a Banach algebra $\mathcal{A}$ is said to be *regular* (or also *modular*) if there exists an element $v \in \mathcal{A}$ such that $\mathcal{A}(1 - v) \subseteq J$, where

$$\mathcal{A}(1 - v) := \{x - xv : x \in \mathcal{A}\}.$$

Similar definitions apply to right regular ideals and regular ideals. It is clear that if $\mathcal{A}$ has a unit u then every ideal, left, right, or two-sided, is regular. A two-sided ideal J of $\mathcal{A}$ is called *primitive* if there exists a maximal regular left ideal L of $\mathcal{A}$ such that

$$J = \{x \in \mathcal{A} : x\mathcal{A} \subseteq L\}.$$

It is well known that J is a primitive ideal of $\mathcal{A}$ if and only if J is the kernel of an irreducible representation of $\mathcal{A}$, see Bonsall and Duncan [80, Proposition 24.12].

The (Jacobson) *radical* of an algebra is the intersection of the primitive ideals of A, or, equivalently, the intersection of the maximal regular left (right) ideals of A, see Bonsall and Duncan [80, Proposition 24.14]. An algebra A is said to be *semi-simple* if its radical rad A is equal to $\{0\}$. If $A = \text{rad} A$ then A is said to be a *radical algebra*. Each semi-simple Banach algebra is semi-prime. Note that in a commutative Banach algebra $\mathcal{A}$ the radical is the set of all quasi-nilpotent elements of $\mathcal{A}$, see [80, Corollary 17.7], and consequently $\mathcal{A}$ is semi-simple precisely when it contains non-zero quasi-nilpotent elements, while a commutative Banach algebra $\mathcal{A}$ is semi-prime if and only if it contains non-zero nilpotent elements.

The *weighted convolution* algebra $L_1(\mathbb{R}_+, \omega)$, where the weight ω is chosen so that $\omega^{1/t} \to 0$ as $t \to 0$, is an example of a semi-prime Banach algebra which is not semi-simple [80], so these two classes of Banach algebras are distinct. For an extensive treatment of multiplier theory we refer to the books by Laursen and Neumann [216] and Aiena [1].

Theorem 4.48 *Let A be a semi-simple Banach algebra. Every $T \in M(A)$ has property $H(1)$, i.e.,*

$$H_0(\lambda I - T) = \ker(\lambda I - T) \quad \text{for all } \lambda \in \mathbb{C}. \tag{4.8}$$

Consequently, every $T \in M(A)$ is hereditarily polaroid.

Proof Since $\lambda I - T$ is a multiplier, it suffices to show (4.8) for $\lambda = 0$. We know that $\ker T \subseteq H_0(T)$, so it remains to prove the inverse inclusion.

Suppose that $x \in H_0(T)$. By an easy inductive argument we have

$$(Ty)^n = (T^n y)y^{n-1} \quad \text{for every } y \in A \text{ and } n \in \mathbb{N}.$$

From this it follows that

$$\|(aTx)^n\| = \|(Tax)^n\| = \|T^n(ax)(ax)^{n-1}\|$$
$$\leq \|a\|\|T^n x\|\|(ax)^{n-1}\|$$

for every $a \in A$, so the spectral radius of the element aTx is

$$r(aTx) = \lim_{n \to \infty} \|(aTx)^n\|^{1/n} = 0$$

for every $a \in A$. This implies that $Tx \in \operatorname{rad} A$, see [80, Proposition 1, p. 126]. Since A is semi-simple $Tx = 0$, hence $x \in \ker T$, and consequently $H_0(T) \subseteq \ker T$, which concludes the proof. ∎

Corollary 4.49 *Let A be a semi-simple Banach algebra and $T \in M(A)$. Then T is quasi-nilpotent if and only if $T = 0$.*

Proof Suppose $T \in M(A)$ is quasi-nilpotent. Combining Theorems 4.48 and 2.35 we have $A = H_0(T) = \ker T$ and hence $T = 0$. ∎

Remark 4.50 Note that the assumption of semi-simplicity in Theorem 4.48 is crucial, since, in general, a multiplier of a non-semi-simple Banach algebra A, also semi-prime, does not satisfy property $H(1)$. To see this, let $\omega := (\omega_n)_{n \in \mathbb{N}}$ be a sequence with the property that

$$0 < \omega_{m+n} \leq \omega_m \omega_n \text{ for all } m, n \in \mathbb{N},$$

and let $\ell^1(\omega)$ denote the space of all complex sequences $x := (x_n)_{n\in\mathbb{N}}$ for which

$$\|x\|_\omega := \sum_{n=0}^{\infty} \omega_n |x_n| < \infty.$$

The Banach space $\ell^1(\omega)$ equipped with convolution as multiplication

$$(x \star y)_n := \sum_{j=0}^{n} x_{n-j} y_j \text{ for all } n \in \mathbb{N},$$

is a commutative Banach algebra with unit. Denote by A_ω the maximal ideal of $\ell^1(\omega)$ defined by

$$A_\omega := \{(x_n)_{n\in\mathbb{N}} \in \ell^1(\omega) : x_0 = 0\}.$$

The Banach algebra A_ω is an integral domain (in the sense that the product of two non-zero elements of A_ω is always non-zero) and hence semi-prime. Suppose now that the weight sequence ω satisfies the condition

$$\rho_\omega := \lim_{n\to\infty} \omega_n^{\frac{1}{n}} = 0.$$

Then A_ω is a radical algebra (see Laursen and Neumann [216, Example 4.1.9]), i.e., A_ω coincides with its radical, and hence is not semi-simple.

For every $0 \neq a \in A_\omega$, let $T_a(x) := a \star x$, $x \in A_\omega$, denote the multiplication operator by the element a. It is easily seen that T_a is quasi-nilpotent, thus $H_o(T_a) = A_\omega$. On the other hand, A_ω is an integral domain so that $\ker T_a = \{0\}$. Hence, the operator T_a does not satisfy property $H(1)$.

4.4   Paranormal Operators

The paranormal operators on Banach spaces provide important examples of operators which are not $H(p)$. An operator $T \in L(X)$, X a Banach space, is said to be *normaloid* if its spectral radius $r(T)$ is equal to the norm $\|T\|$, or equivalently, $\|T^n\| = \|T\|^n$ for every $n = 1, 2, \ldots$.

Theorem 4.51 *If $T \in L(X)$ is paranormal then we have:*

(i) *Any scalar multiple, and the inverse (if it exists), of a paranormal operator, is paranormal.*
(ii) *Every power T^n is paranormal.*
(iii) *T is normaloid.*

Proof (i) Obvious. To show (ii) observe that from the definition (4.6) we have

$$\frac{\|T^{k+1}x\|}{\|T^kx\|} \le \frac{\|T^{k+2}x\|}{\|T^{k+1}x\|}$$

from which we obtain

$$\frac{\|T^nx\|}{\|x\|} = \frac{\|Tx\|}{\|x\|}\frac{\|T^2x\|}{\|Tx\|}\cdots\frac{\|T^nx\|}{\|T^{n-1}x\|}$$

$$\le \frac{\|T^{n+1}x\|}{\|T^nx\|}\frac{\|T^{n+2}x\|}{\|T^{n+1}x\|}\cdots\frac{\|T^{2n}x\|}{\|T^{2n-1}x\|} = \frac{\|T^{2n}x\|}{\|T^nx\|}.$$

Consequently, $\|T^nx\|^2 \le \|(T^n)^2x\|\|x\|$.

(iii) For every paranormal operator we have

$$\|Tx\|^2 \le \|T^2x\|\|x\| \le \|T^2\|\|x\|^2,$$

thus $\|T^2\| = \|T\|^2$. Since T^n is paranormal, $\|T^{2n}\| = \|T\|^{2n}$ for every $n \in \mathbb{N}$. Hence

$$r(T) = \lim_{n\to\infty}\|T^{2^n}\|^{\frac{1}{2^n}} = \|T\|. \qquad\blacksquare$$

Remark 4.52 In [151, p. 113] it is shown that there exists a hyponormal operator T such that T^2 is not hyponormal. Since every hyponormal operator is paranormal, T is paranormal and hence, by Theorem 4.51, T^2 is paranormal. Therefore T^2 provides an example of an operator which is paranormal, but not hyponormal.

Corollary 4.53 *If $T \in L(X)$ is quasi-nilpotent and paranormal, then $T = 0$.*

Proof T is normaloid, so $r(T) = \|T\| = 0$. $\qquad\blacksquare$

Definition 4.54 Recall that an invertible operator $T \in L(X)$ is said to be *doubly power-bounded* if $\sup\{\|T^n\| : n \in \mathbb{Z}\} < \infty$.

The following theorem is due to Gelfand, see [216, Theorem 1.5.14] for an elegant proof.

Theorem 4.55 *If T is doubly power-bounded then $T = I$.*

Evidently, every isometry is paranormal. Note that if $T \in L(X)$ is paranormal and $\sigma(T) \subseteq \Lambda$, Λ the unit circle in $\mathbb{C}$, then T is an invertible isometry. Indeed, T and its inverse T^{-1} are paranormal, and hence normaloid. Hence $\|T\| = \|T^{-1}\| = 1$ and

$$\|x\| = \|T^{-1}Tx\| \le \|Tx\| \le \|x\|$$

for all $x \in X$, thus $\|Tx\| = \|x\|$.

We know that every totally paranormal operator is $H(1)$, by Theorem 4.42, and hence hereditarily polaroid. In the next theorem we show that this is true for every paranormal operator.

Theorem 4.56 *If $T \in L(X)$ is paranormal then every $\lambda \in \operatorname{iso} \sigma(T)$ is a pole of the resolvent of order 1. Moreover, T is hereditarily polaroid.*

Proof Let $\lambda \in \operatorname{iso} \sigma(T)$ and denote by P_λ the spectral projection associated with $\{\lambda\}$. If $\lambda = 0$ then the paranormal operator $T \mid P_0(X)$ has spectrum $\{0\}$, i.e, is quasi-nilpotent. By Corollary 4.53, $T \mid P_0(X) = 0$. By Theorem 2.45 we have $P_0(X) = H_0(T)$, so $H_0(T) \subseteq \ker T$. Since the opposite inclusion holds we then conclude that $H_0(T) = \ker T$. Since $X = H_0(T) \oplus K(T)$, we then have $T(X) = T(K(T)) = K(T)$, hence $X = \ker T \oplus T(X)$, which is equivalent to saying, by Theorem 1.35, that 0 is a pole of the first order. Suppose that $\lambda \neq 0$. Then $T_\lambda := \frac{1}{\lambda}(T \mid P_\lambda)$ is paranormal with spectrum equal to $\{1\}$. Therefore, T_λ and its inverse T_λ^{-1} are both isometries, and hence $\|T_\lambda^n\| = 1$ for all $n \in \mathbb{Z}$. By Lemma 4.55 we then deduce that $T_\lambda = I$. So again, $(\lambda I - T_\lambda)(P_\lambda(X)) = \{0\}$ and proceeding as in the case $\lambda = 0$ we obtain that λ is a pole of the first order. This shows that T is polaroid. Since the restriction of T to a closed invariant subspace is paranormal, we then conclude that T is hereditarily polaroid. ∎

We say that $T \in L(X)$ is *analytically paranormal* if there exists a function $f \in \mathcal{H}_{nc}(\sigma(T))$ such that $f(T)$ is paranormal.

Corollary 4.57 *Analytically paranormal operators on Banach spaces are hereditarily polaroid.*

Proof Let $T \in L(X)$ be analytically paranormal and M a closed T-invariant subspace of X. By assumption there exists an analytic function h such that $h(T)$ is paranormal. The restriction of any paranormal operator to an invariant closed subspace is also paranormal, so $h(T \mid M) = h(T) \mid M$ is paranormal and hence polaroid, by Theorem 4.56. From Theorem 4.19 we then conclude that $T \mid M$ is polaroid. ∎

Let $\mathcal{C}$ be any class of operators. We say that T is an *analytically $\mathcal{C}$-operator* if there exists some analytic function $f \in \mathcal{H}_{nc}(\sigma(T))$ such that $f(T) \in \mathcal{C}$.

Lemma 4.58 *The property of being analytically $\mathcal{C}$ is translation invariant.*

Proof We have to show that

$$T \text{ analytically } \mathcal{C} \text{ and } \lambda_0 \in \mathbb{C} \Rightarrow \lambda_0 I - T \text{ analytically } \mathcal{C}.$$

Suppose that $f(T) \in \mathcal{C}$ for some $f \in \mathcal{H}_{nc}(\sigma(T))$. Let $\lambda_0 \in \mathbb{C}$ be arbitrary and set $g(\mu) := f(\lambda_0 - \mu)$. Then g is analytic and

$$g(\lambda_0 I - T) = f(\lambda_0 I - (\lambda_0 I - T)) = f(T),$$

thus $\lambda_0 I - T$ is analytically $\mathcal{C}$. ∎

Definition 4.59 An operator $T \in L(X)$ is said to be *hereditarily normaloid*, $T \in \mathcal{HN}$, if the restriction $T|M$ of T to any closed T-invariant subspace M is normaloid. Finally, $T \in L(X)$ is said to be *totally hereditarily normaloid*, $T \in \mathcal{THN}$, if $T \in \mathcal{HN}$ and every invertible restriction $T|M$ has a normaloid inverse.

Evidently, every paranormal operator, and in particular every hyponormal operator, is totally hereditarily normaloid.

Theorem 4.60 *Suppose that $T \in L(X)$ is quasi-nilpotent. If T is an analytically $\mathcal{THN}$ operator, then T is nilpotent.*

Proof Let $T \in L(X)$ and suppose that $f(T)$ is a $\mathcal{THN}$-operator for some $f \in \mathcal{H}_{nc}(\sigma(T))$. From the spectral mapping theorem we have

$$\sigma(f(T)) = f(\sigma(T)) = \{f(0)\}.$$

We claim that $f(T) = f(0)I$. To see this, let us consider the two possibilities: $f(0) = 0$ or $f(0) \neq 0$.

If $f(0) = 0$ then $f(T)$ is quasi-nilpotent and $f(T)$ is normaloid, and hence $f(T) = 0$. The equality $f(T) = f(0)I$ then trivially holds.

Suppose the other case $f(0) \neq 0$, and set $f_1(T) := \frac{1}{f(0)} f(T)$. Clearly, $\sigma(f_1(T)) = \{1\}$ and $\|f_1(T)\| = 1$. Further, $f_1(T)$ is invertible and is $\mathcal{THN}$. This easily implies that its inverse $f_1(T)^{-1}$ has norm 1. The operator $f_1(T)$ is then doubly power-bounded and hence, by Theorem 4.55, $f_1(T) = I$, and consequently $f(T) = f(0)I$, as claimed.

Now, let $g(\lambda) := f(0) - f(\lambda)$. Clearly, $g(0) = 0$, and g may have only a finite number of zeros in $\sigma(T)$. Let $\{0, \lambda_1, \ldots, \lambda_n\}$ be the set of all zeros of g, where $\lambda_i \neq \lambda_j$, for all $i \neq j$, and λ_i has multiplicity $n_i \in \mathbb{N}$. We have

$$g(\lambda) = \mu \lambda^m \prod_{i=1}^{n} (\lambda_i I - T)^{n_i} h(\lambda),$$

where $h(\lambda)$ has no zeros in $\sigma(T)$. From the equality $g(T) = f(0)I - f(T) = 0$ it then follows that

$$0 = g(T) = \mu T^m \prod_{i=1}^{n} (\lambda_i I - T)^{n_i} h(T) \quad \text{with } \lambda_i \neq 0,$$

where all the operators $\lambda_i I - T$ and $h(T)$ are invertible. This, obviously, implies that $T^m = 0$, i.e., T is nilpotent. ∎

If $T \in L(X)$ the *numerical range* of T is defined as

$$W(T) := \{f(T) : f \in L(X)^*, \|f\| = f(I) = 1\},$$

while the *numerical radius* of T is defined by

$$w(T) := \sup\{|\lambda| : \lambda \in W(T)\}.$$

In the case of Hilbert space operators the numerical range may be described as the set

$$W(T) = \{(Tx, x) : \|x\| = 1\},$$

and the well-known *Toeplitz–Hausdorff theorem* establishes that $W(T)$ is a convex set in the complex plane (for a proof, see Furuta [151, p. 91]). It is known that

$$r(T) \le w(T) \le \|T\|.$$

The next non-trivial result was proved in Sinclair [285]. We omit the difficult proof.

Theorem 4.61 *Let $T \in L(X)$ and suppose that 0 is in the boundary of the numerical range of T. Then the kernel of T is orthogonal to the range of T.*

In the case of paranormal operators we have:

Theorem 4.62 *Suppose that $T \in L(X)$ is totally hereditarily normaloid and $\lambda, \mu \in \mathbb{C}$, with $\lambda \ne 0$ and $\lambda \ne \mu$. Then $\ker(\lambda I - T) \perp \ker(\mu I - T)$, i.e. $\ker(\lambda I - T)$ is orthogonal to $\ker(\mu I - T)$ in the Birkhoff and James sense.*

Proof Suppose first that $|\lambda| \ge |\mu|$, let $x \in \ker(\lambda I - T)$ and $y \in \ker(\mu I - T)$. Then $Tx = \lambda x$ and $Ty = \mu y$. Denote by M the subspace generated by x and y and set $T_M := T|M$. Clearly, $\sigma(T|M) = \{\lambda, \mu\}$ and since $T|M$ is normaloid,

$$\|T|M\| = r(T|M) = |\lambda, |$$

so that $\nu(T|M) = |\lambda|$. Consequently, λ belongs to the boundary of the numerical range of $T|M$ and hence, by Theorem 4.61, $\ker(\lambda I - T|M) \perp (\lambda I - T|M)(M)$. Evidently, λ and μ are poles of the resolvent of $T|M$ having order 1. Denoting by P_λ and P_μ the spectral projections for $T|M$ associated with $\{\lambda\}$ and $\{\mu\}$, respectively, we then have

$$(\lambda I - T|M)(M) = (I - P_\lambda)(M) = P_\mu(M) = \ker(\beta I - T|M).$$

Now, $x \in \ker(\lambda I - T|M)$ and $y \in \ker(\mu I - T|M)$, hence $\|x + y\| \ge \|x\|$.

Consider now the case where $|\lambda| < |\mu|$. Then $|\mu| > 0$, so $T|M$ is invertible and

$$\sigma(T|M)^{-1} = \left\{\frac{1}{\lambda}, \frac{1}{\mu}\right\},$$

with $|\frac{1}{\lambda}| > |\frac{1}{\mu}|$. Since $T|M$ is normaloid then $(T|M)^{-1}$ is also normaloid. As in the first case we then see that the kernels $\ker(\frac{1}{\lambda}I - (T|M)^{-1})$ and $\ker(\frac{1}{\mu}I - (T|M)^{-1})$ are orthogonal. Obviously, $x \in \ker(\frac{1}{\lambda}I - (T|M)^{-1})$ and $y \in \ker(\frac{1}{\mu}I - (T|M)^{-1})$, so the proof is complete. ∎

Theorem 4.63 *Every totally hereditarily normaloid operator T on a separable Banach space has the SVEP.*

Proof To prove the first assertion, we show that the point spectrum $\sigma_p(T)$ is countable, hence its interior part is empty. If $\sigma_p(T)$ were not countable we would have an uncountable set of unit vectors such that $\|x_i - x_j\| \geq 1$. Since X is separable this is not possible. ∎

Every normal operator on a Hilbert space is paranormal. Indeed, if $T \in L(H)$ is normal then $\|Tx\| = \|T'x\|$ for every $x \in H$. Consequently,

$$\|Tx\|^2 = (Tx, Tx) = (T'Tx, x) \leq \|T'Tx\|\|x\| = \|T^2x\|\|x\|.$$

Theorem 4.64 *If $T \in L(H)$ is paranormal and has finite spectrum then H is the direct sum of eigenspaces of T.*

Proof Let $\sigma(T) = \{\lambda_1, \ldots, \lambda_n\}$ and denote by P_k the spectral projection associated with $\{\lambda_k\}$. Then

$$H = \bigoplus \sum_{k=1}^{k=n} P_k(X).$$

By Theorem 4.56 we have $P_k(X) = \ker(\lambda_k I - T)$. ∎

Corollary 4.65 *If $T \in L(H)$ is paranormal and has finite spectrum then T is normal. In particular, any algebraic paranormal operator is normal.*

Proof We have

$$H = \bigoplus \sum_{k=1}^{k=n} \ker(\lambda_k I - T),$$

and $\ker(\lambda_k I - T) \perp \ker(\lambda_j I - T)$ for $k \leq j$, and this entails that T is normal. The last assertion is clear, since algebraic operators have finite spectrum. ∎

The class of paranormal operators includes some other classes of operators defined on Hilbert spaces:

(g) *p-quasihyponormal operators.* It has been observed before that every *p*-hyponormal operator is paranormal. A Hilbert space operator $T \in L(H)$ is said to be p-quasihyponormal for some $0 < p \leq 1$ if

$$T'|T'|^{2p}T \leq T'|T|^{2p}T.$$

Every p-quasi-hyponormal is paranormal, see Lee and Lee [225].

(h) *Class A operators.* An operator $T \in L(H)$ is said to be a class A operator if $|T^2| \geq |T|^2$. Every log-hyponormal operator is a class A operator, see Furuta et al. [152], but the converse is not true, see Furuta [151, p. 176]. Every class A operator is paranormal (an example of a paranormal operator which is not a class A operator can be found in [151, p. 177]).

4.5 Isometries

Let us consider, for an arbitrary operator $T \in L(X)$ on a Banach space X, the so-called *lower bound* of T defined by

$$k(T) := \inf\{\|Tx\| : x \in X, \|x\| = 1\}.$$

It is obvious that if T is invertible then $k(T) = \|T^{-1}\|$. Clearly

$$k(T^n)k(T^m) \leq k(T^{n+m}) \quad \text{for all } n, m \in \mathbb{N} \tag{4.9}$$

and consequently $k(T) = 0$ whenever $k(T^n) = 0$ for some $n \in \mathbb{N}$. The converse is also true: if $k(T) = 0$ then $0 \in \sigma_{\mathrm{ap}}(T)$ and therefore $k(T^n) = 0$ for all $n \in \mathbb{N}$.

Theorem 4.66 *If $T \in L(X)$ then*

$$\lim_{n \to \infty} k(T^n)^{1/n} = \sup_{n \in \mathbb{N}} k(T^n)^{1/n}. \tag{4.10}$$

Proof Fix $m \in \mathbb{N}$ and write for all $n \in \mathbb{N}, n = mq + r, 0 \leq r \leq m$, where $q := q(n)$ and $r := r(n)$ are functions of n. Note that

$$\lim_{n \to \infty} \frac{q(n)}{n} = \frac{1}{m} \quad \text{and} \quad \lim_{n \to \infty} \frac{r(n)}{n} = 0.$$

From (4.9) we obtain that $k(T^n) \geq k(T^m)^q k(T)^r$ and hence

$$\lim_{n \to \infty} \inf(k(T^n))^{1/n} \geq k(T^m)^{1/m} \quad \text{for all } m \in \mathbb{N}.$$

Therefore

$$\lim_{n \to \infty} \inf(k(T^n))^{1/n} \geq \sup_{n \in \mathbb{N}} k(T^n)^{1/n} \geq \lim_{n \to \infty} \sup(k(T^n))^{1/n},$$

from which the equality (4.10) follows.   ∎

Put

$$i(T) := \lim_{n \to \infty} k(T^n)^{1/n}.$$

If $r(T)$ denotes the spectral radius of T it is obvious that $i(T) \leq r(T)$. For every bounded operator $T \in L(X)$, X a Banach space, let us consider the (possible degenerate) closed annulus

$$\Lambda(T) := \{\lambda \in \mathbb{C} : i(T) \leq |\lambda| \leq r(T)\}.$$

The next result shows that the approximate point spectrum is located in $\Lambda(T)$.

Theorem 4.67 *For every bounded operator $T \in L(X)$, X a Banach space, we have $\sigma_{ap}(T) \subseteq \Lambda(T)$.*

Proof If $\lambda \in \sigma_{ap}(T)$ then $|\lambda| \leq r(T)$. Assume $|\lambda| < i(T)$ and let $c > 0$ be such that $|\lambda| < c < i(T)$. Take $n \in \mathbb{N}$ such that $c^n \leq k(T^n)$. For every $x \in X$ we have $c^n \|x\| \leq \|T^n x\|$ and hence

$$\|(\lambda^n I - T^n)x\| \geq \|T^n x\| - |\lambda^n| \|x\| \geq (c^n - |\lambda^n|)\|x\|,$$

thus $\lambda^n I - T^n$ is bounded below, so $\lambda^n \notin \sigma_{ap}(T)$. Writing

$$\lambda^n I - T^n = (\lambda I - T)(T^{n-1} + \lambda T^{n-2} + \cdots + \lambda^n I),$$

we then conclude that $\lambda \notin \sigma_{ap}(T)$. Therefore, $\sigma_{ap}(T) \subseteq \Lambda(T)$. ∎

As usual by $\mathbb{D}(0, \varepsilon)$ and $\mathbf{D}(0, \varepsilon)$ we shall denote the open disc and the closed disc centered at 0 with radius ε, respectively.

Theorem 4.68 *For a bounded operator $T \in L(X)$, X a Banach space, the following properties hold:*

(i) *If T is invertible then $\mathbb{D}(0, i(T)) \subseteq \rho(T)$, and consequently $\sigma(T) \subseteq \Lambda(T)$. If T is non-invertible then $\mathbf{D}(0, i(T)) \subseteq \sigma(T)$;*

(ii) *Suppose that $i(T) = r(T)$. If T is invertible then $\sigma(T) \subseteq \partial\mathbb{D}(0, r(T))$, while if T is non-invertible then*

$$\sigma(T) = \mathbf{D}(0, r(T)) \quad and \quad \sigma_{ap}(T) = \partial\sigma(T).$$

Proof

(i) Let T be invertible and suppose that there is some $\lambda \in \sigma(T)$ for which $|\lambda| < i(T)$. We have $0 \in \rho(T)$, so there is some μ in the boundary of $\sigma(T)$ such that $|\mu| \leq |\lambda| < i(T)$. But this is impossible since, by Theorem 1.12, we have $\mu \in \sigma_{ap}(T)$. Hence, from Theorem 4.67 we deduce that $|\mu| \geq i(T)$, and this shows the first assertion of (i).

Suppose now that T is non-invertible and that there is an element $\lambda \in \rho(T)$ for which $|\lambda| \leq i(T)$. By assumption $0 \in \sigma(T)$ and $\rho(T)$ is open, so there exists a $0 \leq c < 1$ such that $c\lambda$ belongs to the boundary of $\sigma(T)$. From Theorem 1.12 it then follows that $c\lambda \in \sigma_{ap}(T)$. On the other hand, $|c\lambda| < i(T)$, so, by Theorem 4.67, $c\lambda \notin \sigma_{ap}(T)$, and this is a contradiction. This shows the second assertion of part (i).

(ii) The inclusion $\sigma(T) \subseteq \partial\mathbb{D}(0, r(T))$, if T is invertible, and the equality $\sigma(T) = \mathbf{D}(0, r(T))$, if T is non-invertible, are simple consequences of part (i).

Suppose now that if T is not invertible and that there exists some $\lambda \in \sigma(T)$ such that $|\lambda| = 1$ and $\lambda \notin \sigma_{ap}(T)$. By Corollary 2.92 T^* then fails the SVEP at λ and this contradicts the fact that λ belongs to the boundary of the spectrum. Therefore $\partial\sigma(T) \subseteq \sigma_{ap}(T)$ and from Theorem 4.67 it then follows that $\partial\sigma(T) = \sigma_{ap}(T)$. ∎

Remark 4.69 Part (ii) of Theorem 4.68 shows that if $i(T) = r(T)$ and T is not invertible then T is *a*-polaroid, since iso $\sigma_{ap}(T) = \emptyset$.

Theorem 4.70 *Let $T \in L(X)$, X a Banach space, and suppose that $\lambda \in \mathbb{C}$ is a point for which $|\lambda| < i(T)$. Then T has the SVEP at λ, while T^* has the SVEP at λ if and only if T is invertible.*

Proof From Theorem 4.67 we know that if $|\lambda| < i(T)$ then $\lambda \notin \sigma_{ap}(T)$. Hence the assertions easily follow from Corollary 2.92. ∎

The following corollary describes the SVEP in the special case $i(T) = r(T)$.

Corollary 4.71 *Let $T \in L(X)$, X a Banach space, and suppose that $i(T) = r(T)$. Then the following dichotomy holds:*

(i) *If T is invertible then both T and T^* have the SVEP;*
(ii) *If T is non-invertible then T has the SVEP, while T^* has the SVEP at a point λ precisely when $|\lambda| \geq r(T)$.*

We now describe the SVEP for T or T^* for isometries:

Theorem 4.72 *Every isometry $T \in L(X)$ has the SVEP, while the adjoint T^* of a non-invertible isometry has the SVEP at a point $\lambda \in \mathbb{C}$ if and only if $|\lambda| \geq 1$. Every non-invertible isometry is a-polaroid. Every invertible isometry is hereditarily polaroid and also T^* has the SVEP.*

Proof The first assertion is clear from Corollary 4.71 and every non-invertible isometry is *a*-polaroid by Remark 4.69. An isometry T is invertible if and only if it is generalized scalar, or equivalently is decomposable, see [216, Theorem 1.6.7]. Consequently, T^* has the SVEP. Every invertible isometry is $H(p)$, by Theorem 4.40 and hence, by Theorem 4.37 is hereditarily polaroid. ∎

Actually, Douglas in [120] has shown that every isometry has property (β), see for a proof also [216, Theorem 1.6.7].

The next result on non-invertible isometries will be useful to settle this question in the case of certain operators.

Theorem 4.73 *Let $T \in L(X)$ be a non-invertible isometry and suppose that $f : \mathcal{U} \to \mathbb{C}$ is a non-constant analytic function on some connected open neighborhood of the closed unit disc. Then the following assertions hold:*

(i) $\sigma(f(T)) = f(\overline{\mathbb{D}})$ *and* $\sigma_{\mathrm{ap}}(f(T)) = f(\partial\mathbb{D})$, *where $\mathbb{D}$ denotes the open unit disc of $\mathbb{C}$.*

(ii) $f(T)$ *has the SVEP.*

(iii) $f(T)^*$ *has the SVEP at a point λ if and only if $\lambda \notin f(\mathbb{D})$.*

(iv) $f(\partial\mathbb{D}) \cap f(\mathbb{D}) = \{\lambda \in \mathbb{C} : f(T)^* \text{ does not have the SVEP at } \lambda\}$.

Proof Since $\sigma(f(T)) = \overline{\mathbb{D}}$ and, by Theorem 4.67, $\sigma_{\mathrm{ap}}(T) = \partial\mathbb{D}$, the equalities (i) follow from the spectral mapping theorems of $\sigma(T)$ and $\sigma_{\mathrm{ap}}(T)$. Assertion (ii) is a consequence of Corollary 4.71 and the spectral mapping theorem.

(iii) Since $f(T)^* = f(T^*)$, from Theorem 2.88 it follows that $f(T)^*$ has the SVEP at the point $\lambda \in \mathbb{C}$ if and only if T^* has the SVEP at each point $\mu \in \mathcal{U}$ for which $f(\mu) = \lambda$. Corollary 4.71 then ensures that the latter condition holds precisely when $\lambda \notin f(\mathbb{D})$.

The assertion (iv) easily follows from part (i) and part (iii). ∎

Part (iv) of Theorem 4.73 leads to many examples in which the SVEP for the adjoint fails to hold at the points which belong to the approximate point spectrum of T. In fact, if f is a non-constant analytic function on some connected open neighborhood $\mathcal{U}$ of the closed unit disc and $\Lambda := f(\partial\mathbb{D}) \cap f(\mathbb{D})$ is non-empty then for every $\lambda \in \Lambda$ the adjoint of $f(T)$ does not have the SVEP at λ. This situation is, for instance, fulfilled for every function of the form

$$f(\lambda) := (\lambda - \gamma)(\lambda - \omega)g(\lambda) \quad \text{for } \lambda \in \mathcal{U},$$

where g is an arbitrary analytic function on $\mathcal{U}$, $|\gamma| = 1$ and $|\omega| < 1$.

We conclude this section by mentioning two applications of Theorem 4.73 to operators defined on Hardy spaces. In the sequel by $H^p(\mathbb{D})$, $1 \le p < \infty$, we denote the *Hardy space* of all analytic functions $f : \mathbb{D} \to \mathbb{C}$ for which

$$\sup\left\{ \int_{-\pi}^{\pi} |f(re^{i\theta}|d\theta : 0 \le r < 1 \right\} < \infty.$$

By $H^\infty(\mathbb{D})$ we denote the Banach algebra of all bounded analytic functions on the open disc $\mathbb{D}$.

Example 4.74 If $f \in H^\infty(\mathbb{D})$, the operator T_f on $H^p(\mathbb{D})$ defined by the assignment

$$T_f g := fg \quad \text{for every } g \in H^p(\mathbb{D})$$

is called the *multiplication analytic Toeplitz operator with symbol f*.

Theorem 4.75 *Let f be a non-constant analytic function on some connected open neighborhood of the closed unit disc. The multiplication Toeplitz operator T_f on $H^2(\mathbb{D})$ is polaroid and has the SVEP. The adjoint T_f^* is a-polaroid and has the SVEP at λ if and only if $\lambda \notin f(\mathbb{D})$.*

Proof If T denotes the operator of multiplication by the independent variable, defined by

$$(Tg)(\lambda) := \lambda g(\lambda) \quad \text{for all } g \in H^2(\mathbb{D}), \ \lambda \in \mathbb{D},$$

then $T_f = f(T)$. The operator T is unitary equivalent to the unilateral right shift on $\ell^2(\mathbb{N})$, and hence is a non-invertible isometry. By part (i) of Theorem 4.73 we have

$$\sigma(T_f) = f(\sigma(T)) = f(\overline{\mathbb{D}}) \quad \text{and} \quad \sigma_{\mathrm{ap}}(T_f) = f(\partial\mathbb{D}).$$

From part (ii) of Theorem 4.73 we see that T_f has the SVEP, while from part (iv) of Theorem 4.73, we conclude that the adjoint T_f^* has the SVEP at $\lambda \in \mathbb{C}$ if and only if $\lambda \notin f(\mathbb{D})$. Since $\sigma(T) = \mathbf{D}(0, 1)$ we then have that there are no isolated spectral points, so T is polaroid. The SVEP for T also entails that T^* is a-polaroid, by Theorem 4.15.   ∎

Note that similar results hold for Toeplitz operators with arbitrary bounded analytic symbols. In fact, if $f \in H^\infty(\mathbb{D})$ the approximate point spectrum $\sigma_{\mathrm{ap}}(T_f)$ coincides with the essential range of the boundary function, which is obtained by taking non-tangential limits of f almost everywhere on the unit circle, and the operator T_f does not have the SVEP at any $\lambda \notin f(\mathbb{D})$. These results may be established using standard tools from the theory of Hardy spaces, see Porcelli [257].

Example 4.76 Let $C(\Omega)$ denote the Banach algebra of all continuous complex-valued functions on a compact Hausdorff space Ω and $\gamma : \Omega \to \Omega$ a homomorphism. Then we can define a *composition operator* $T_\gamma : C(\Omega) \to C(\Omega)$ by the assignment

$$T_\gamma(f) := f \circ \gamma \quad \text{for all } f \in C(\Omega).$$

The operator T_γ is a surjective isometry, and therefore is generalized scalar (see [216]). Thus T_γ is $H(p)$, by Theorem 4.40, and hence hereditarily polaroid, by Corollary 4.37.

In the same vein, every analytic function $\varphi : \mathbb{D} \to \mathbb{D}$ on the open unit disc $\mathbb{D}$ induces a composition operator on $H^p(\mathbb{D})$ defined by

$$T_\varphi(f) := f \circ \varphi \quad \text{for all } f \in H^p(\mathbb{D}).$$

The operator T_φ is an isometry which is invertible if and only if φ is an automorphism of $\mathbb{D}$, i.e., a mapping of the form

$$\varphi(\lambda) = \frac{a\lambda + \overline{b}}{b\lambda + \overline{a}} \quad \text{for all } \lambda \in \mathbb{D},$$

where a and b are complex numbers for which $|a|^2 - |b|^2 = 1$. These automorphisms φ are classified as follows:

- φ is *elliptic* if $|\mathrm{Im}\,a| > |b|$;
- φ is *parabolic* if $|\mathrm{Im}\,a| = |b|$;
- φ is *hyperbolic* if $|\mathrm{Im}\,a| < |b|$.

If φ is either elliptic or parabolic then a result of Smith [287] shows that the corresponding composition operator T_φ on $H^p(\mathbb{D})$ and its adjoint has the SVEP (actually we have much more, T_φ is generalized scalar and therefore decomposable, see §1.5 of Laursen and Neumann [216]). Therefore, if φ is either elliptic or parabolic, the operator T_φ is $H(p)$, by Theorem 4.40, and hence hereditarily polaroid.

On the other hand, from an inspection of the proof of Theorem 6 of Nordgreen [249] and Theorems 1.4 and 2.3 of Smith [287] it easily follows that if φ is hyperbolic then

$$\sigma(T_\varphi) = \left\{ \lambda \in \mathbb{C} : \frac{1}{r} \le |\lambda| \le r \right\} \text{ for some } r > 1.$$

Moreover, T_φ does not have the SVEP at λ if and only if $\frac{1}{r} < |\lambda| < r$. We mention that the adjoint T_φ^* is subnormal, see [216], hence hyponormal, by Conway [101, Proposition 2.4.2], in particular T_φ^* is $H(p)$ and hence hereditarily polaroid.

4.6 Weighted Shift Operators

We consider first the operators $T \in L(X)$ for which the condition $T^\infty(X) = \{0\}$ holds. This condition may be viewed, in a certain sense, as an abstract shift condition, since it is satisfied by every weighted right shift operator T on $\ell^p(\mathbb{N})$. Clearly the condition $T^\infty(X) = \{0\}$ entails that T is non-surjective and hence non-invertible. This condition also implies that $K(T) = \{0\}$, since $K(T)$ is a subset of $T^\infty(X)$, but the quasi-nilpotent Volterra operator defined in Example 2.77 shows that in general the converse is not true. In fact, for this operator we have, by Corollary 2.71, $K(T) = \{0\}$, while $T^\infty(X) \ne \{0\}$, see Example 2.77.

The proof of the following result may be found in [216, Theorem 3.1.12]. This may be viewed as a local analogue of the inclusion $\partial\sigma(T) \subseteq \sigma_{se}(T)$ proved in Theorem 2.58.

Theorem 4.77 *If $T \in L(X)$, then $\partial \sigma_T(x) \subseteq \sigma_{se}(T)$ for all $x \in X$.*

In the sequel we shall denote by $\mathbf{D}(0, i(T))$ the closed disc centered at 0 with radius $i(T)$.

Theorem 4.78 *Suppose that for $T \in L(X)$ we have $T^{\infty}(X) = \{0\}$. Then:*

(i) $\ker(\lambda I - T) = \{0\}$ *for all $0 \neq \lambda \in \mathbb{C}$;*

(ii) *T has the SVEP;*

(iii) *The local spectra $\sigma_T(x)$ and $\sigma(T)$ are connected, and the closed disc $\mathbf{D}(0, i(T))$ is contained in $\sigma_T(x)$ for all $x \neq 0$;*

(iv) $H_0(\lambda I - T) = \{0\}$ *for all $0 \neq \lambda \in \mathbb{C}$.*

Proof

(i) For every $\lambda \neq 0$ we have $\ker(\lambda I - T) \subseteq T^{\infty}(X)$.

(ii) This may be seen in several ways, for instance from Theorem 2.60, since $\ker(\lambda I - T) \cap K(\lambda I - T) = \{0\}$ for every $\lambda \in \mathbb{C}$.

(iii) It is easy to see that $0 \in \sigma_T(x)$ for every non-zero $x \in X$. Indeed, from Theorem 2.20 we have

$$\{0\} = K(T) = \{x \in X : 0 \in \rho_T(x)\},$$

and hence $0 \in \sigma_T(x)$ for every $x \neq 0$. Now, suppose that $\sigma_T(x)$ is non-connected for some element $x \neq 0$. Then there exist two non-empty closed subsets Ω_1, Ω_2 of $\mathbb{C}$ such that:

$$\sigma_T(x) = \Omega_1 \cup \Omega_2, \quad \text{and} \quad \Omega_1 \cap \Omega_2 = \varnothing.$$

From the local decomposition property established in Theorem 2.19, there exist two elements $x_1, x_2 \in X$ such that

$$x = x_1 + x_2 \quad \text{with } \sigma_T(x_i) \subseteq \Omega_i \ (i = 1, 2).$$

Now, from Theorem 2.19 we have $x_1 \neq 0$ and $x_2 \neq 0$, and hence

$$0 \in \sigma_T(x_1) \cap \sigma_T(x_2) \subseteq \Omega_1 \cap \Omega_2 = \varnothing,$$

a contradiction. Hence $\sigma_T(x)$ is connected.

To prove that $\sigma(T)$ is connected observe that since T has the SVEP we have by Theorem 2.21 and Corollary 2.68

$$\sigma(T) = \sigma_{su}(T) = \bigcup_{x \in X} \sigma_T(x).$$

Since the local spectra $\sigma_T(x)$ are connected for every non-zero $x \in X$, and $\sigma(0) = \varnothing$, then $\sigma(T)$ is connected.

It remains to prove the inclusion $\mathbf{D}(0, i(T)) \subseteq \sigma_T(x)$ for all $x \neq 0$. We know by Theorem 4.77 that $\partial \sigma_T(x) \subseteq \sigma_{se}(T) \subseteq \sigma_{ap}(T)$ for all $x \in X$. Since $i(T) \leq |\lambda|$ for all $\lambda \in \sigma_{ap}(T)$, it follows easily that $\mathbf{D}(0, i(T)) \subseteq \sigma_T(x)$, as desired.

(iv) Since T has the SVEP, $H_0(\lambda I - T) = \{x \in X : \sigma_T(x) \subseteq \{\lambda\}\}$ for every $\lambda \in \mathbb{C}$, see Theorem 2.30. Now, if $x \neq 0$ and $x \in H_0(\lambda I - T)$ the SVEP ensures that $\sigma_T(x) \neq \varnothing$, so $\sigma_T(x) = \{\lambda\}$. On the other hand, from part (iii) we have $0 \in \{\lambda\}$, a contradiction. ∎

Theorem 4.79 *Let $T \in L(X)$, where X is an infinite-dimensional Banach space, and suppose that $T^\infty(X) = \{0\}$. Then we have:*

(i) $\sigma(T) = \sigma_w(T) = \sigma_b(T)$;
(ii) $q(\lambda I - T) = \infty$ *for every* $\lambda \in \sigma(T) \setminus \{0\}$;
(iii) T *is nilpotent* $\Leftrightarrow q(T) < \infty$.

Proof

(i) By Theorem 4.78 T has the SVEP and hence, by Theorem 3.44, $\sigma_w(T) = \sigma_b(T)$. We show that $\sigma_b(T) = \sigma(T)$. The inclusion $\sigma_b(T) \subseteq \sigma(T)$ is obvious, so it remains to establish that $\sigma(T) \subseteq \sigma_b(T)$. Observe that if the spectral point $\lambda \in \mathbb{C}$ is not isolated in $\sigma(T)$ then $\lambda \in \sigma_b(T)$.

Suppose first that T is quasi-nilpotent. Then $\sigma_b(T) = \sigma(T) = \{0\}$ since $\sigma_b(T)$ is non-empty whenever X is infinite-dimensional. Suppose that T is not quasi-nilpotent and let $0 \neq \lambda \in \sigma(T)$. Since $\sigma(T)$ is connected, by Theorem 4.78, and $0 \in \sigma(T)$, it follows that λ is not an isolated point in $\sigma(T)$. Hence $\sigma(T) \subseteq \sigma_b(T)$.

(ii) Let $\lambda \in \sigma(T) \setminus \{0\}$ and suppose that $q(\lambda I - T) < \infty$. By Theorem 4.78 we have $p(\lambda I - T) = 0$ for every $0 \neq \lambda$, and hence by Theorem 1.20 $q(\lambda I - T) = p(\lambda I - T) = 0$, which implies $\lambda \in \rho(T)$, a contradiction.

(iii) Clearly, because T is nilpotent we have $q(T) < \infty$. Conversely, if $q := q(T) < \infty$ then $T^q(X) = T^\infty(X) = \{0\}$. ∎

It is evident that the proof of Theorem 4.78 also works if we assume $K(T) = \{0\}$, a condition which is less restrictive with respect to the condition $T^\infty(X) = \{0\}$. However, the next result shows that these two conditions are equivalent if $i(T) > 0$.

Corollary 4.80 *Suppose that for a bounded operator $T \in L(X)$, X a Banach space, we have $i(T) > 0$. Then the following statements are equivalent:*

(i) $T^\infty(X) = \{0\}$;
(ii) $\mathbf{D}(0, i(T)) \subseteq \sigma_T(x)$ *for all* $x \neq 0$;
(iii) $K(T) = \{0\}$.

Proof The implication (i) $\Rightarrow$ (ii) has been proved in Theorem 4.78, while (ii) $\Rightarrow$ (iii) is obvious. It remains only to prove the implication (iii) $\Rightarrow$ (i). From Theorem 4.67 the condition $i(T) > 0$ implies that $0 \notin \sigma_{ap}(T)$, T is bounded below and therefore semi-regular. By Theorem 1.44 it then follows that $T^\infty(X) = K(T) = \{0\}$. ∎

Corollary 4.81 *Suppose that for a bounded operator* $T \in L(X)$, X *a Banach space, we have* $T^{\infty}(X) = \{0\}$ *and* $i(T) = r(T)$. *Then we have*

$$\sigma_T(x) = \sigma(T) = \mathbf{D}(0, r(T)), \tag{4.11}$$

for every $x \neq 0$. *Furthermore, if* $i(T) = r(T) > 0$ *then the equalities* (4.11) *hold for every* $x \neq 0$ *if and only if* $T^{\infty}(X) = \{0\}$.

Proof If $T^{\infty}(X) = \{0\}$ then T is non-invertible, so by Theorem 4.68 the condition $i(T) = r(T)$ entails that $\sigma(T) = \mathbf{D}(0, r(T))$, and therefore $\sigma_T(x) \subseteq \sigma(T) = \mathbf{D}(0, r(T))$. The opposite inclusion is true by part (iii) of Theorem 4.78, so (4.11) is satisfied. The equivalence in the last assertion is clear from Corollary 4.80. ∎

It should be noted that if $T \in L(X)$ satisfies the conditions of the preceding corollary then T has property (C). In fact, for every closed subset F of $\mathbb{C}$ we have:

$$X_T(F) = \begin{cases} X & \text{if } F \supseteq \mathbf{D}(0, r(T)), \\ \{0\} & \text{otherwise,} \end{cases}$$

and hence all $X_T(F)$ are closed.

Let $\ell^p(\mathbb{N})$, where $1 \leq p < \infty$, denote the space of all p-summable sequences of complex numbers. Denote by $\omega := \{\omega_n\}_{n \in \mathbb{N}}$ any bounded sequence of strictly positive real numbers. The corresponding *unilateral weighted right shift* operator on the Banach space $\ell^p(\mathbb{N})$ is the operator defined by:

$$Tx := \sum_{n=1}^{\infty} \omega_n x_n e_{n+1} \quad \text{for all } x := (x_n)_{n \in \mathbb{N}} \in \ell^p(\mathbb{N}).$$

It is easily seen that T does not admit eigenvalues, thus T has the SVEP. Furthermore, the lower bound and the norms of the iterates T^n may be easily computed as follows:

$$k(T^n) = \inf_{k \in \mathbb{N}} \{\omega_k \cdots \omega_{k+n-1}\} \quad \text{for all } n \in \mathbb{N},$$

and

$$\|T^n\| = \sup_{k \in \mathbb{N}} \{\omega_k \cdots \omega_{k+n-1}\} \quad \text{for all } n \in \mathbb{N}.$$

Moreover, a routine calculation shows that the numbers $i(T)$ and $r(T)$ of a unilateral weighted right shift may be computed as follows:

$$i(T) = \lim_{n \to \infty} \inf_{k \in \mathbb{N}} (\omega_k \cdots \omega_{k+n-1})^{1/n}$$

and

$$r(T) = \lim_{n \to \infty} \sup_{k \in \mathbb{N}} (\omega_k \cdots \omega_{k+n-1})^{1/n}.$$

To determine further properties of the spectrum of an unilateral weighted right shift we recall two simple facts which will be used in the sequel.

Remark 4.82

(i) Let $\alpha \in \mathbb{C}$, with $|\alpha| = 1$ and define, on $\ell^p(\mathbb{N})$, the linear operator $U_\alpha x := (\alpha^n x_n)_{n \in \mathbb{N}}$ for all $x = (x_n)_{n \in \mathbb{N}} \in \ell^p(\mathbb{N})$. Evidently, $\lambda T U_\alpha = U_\alpha T$ and

$$U_\alpha U_{\overline{\alpha}} = U_{\overline{\alpha}} U_\alpha = I.$$

From this it follows that the operators αT and T are similar, and consequently have the same spectrum. This also shows that $\sigma(T)$ is circularly symmetric about the origin.

(ii) Let K be a non-empty compact subset of $\mathbb{C}$. If K is connected and invariant under circular symmetry about the origin, then there are two real numbers a and b, with $0 \le a \le b$, such that $K = \{\lambda \in \mathbb{C} : a \le |\lambda| \le b\}$.

Theorem 4.83 *For an arbitrary unilateral weighted right shift T on $\ell_p(\mathbb{N})$ we have $\sigma(T) = \mathbf{D}(0, r(T))$ and*

$$\sigma_{\mathrm{ap}}(T) = \{\lambda \in \mathbb{C} : i(T) \le |\lambda| \le r(T)\}.$$

Proof We know by Theorem 4.78 that $\sigma(T)$ is connected and contains the closed disc $\mathbf{D}(0, i(T))$. Since, by part (i) of Remark 4.82, $\sigma(T)$ is circularly symmetric about the origin, from part (ii) of the same Remark we deduce that $\sigma(T)$ is the whole closed disc $\mathbf{D}(0, r(T))$. For the description of $\sigma_{\mathrm{ap}}(T)$, see Proposition 1.6.15 of [216]. ∎

It is easily seen that the adjoint of a unilateral weighted right shift T is the *unilateral weighted left shift* on $\ell^q(\mathbb{N})$ defined by:

$$T^* x := \sum_{n=1}^{\infty} \omega_n x_{n+1} e_n \quad \text{for all } x := (x_n)_{n \in \mathbb{N}} \in \ell^q(\mathbb{N}),$$

where, as usual, $\dfrac{1}{p} + \dfrac{1}{q} = 1$, and $\ell^q(\mathbb{N})$ is canonically identified with the dual $(\ell^p(\mathbb{N}))^*$ of $\ell^p(\mathbb{N})$. Finally, from Corollary 4.71 we deduce that T^* does not have the SVEP whenever $i(T) > 0$.

To investigate more precisely the question of the SVEP for T^* we introduce the following quantity:

$$c(T) := \lim_{n \to \infty} \inf(\omega_1 \cdots \omega_n)^{1/n}.$$

It is clear that $i(T) \le c(T) \le r(T)$.

Theorem 4.84 *Let T be a unilateral weighted right shift on $\ell^p(\mathbb{N})$ for some $1 \le p < \infty$. Then T^* has the SVEP at a point $\lambda \in \mathbb{C}$ precisely when $|\lambda| \ge c(T)$. In particular, T^* has the SVEP if and only if $c(T) = 0$.*

Proof By the classical formula for the radius of convergence of a vector-valued power series we see that the series

$$f(\lambda) := \sum_{n=1}^{\infty} \frac{e_n \, \lambda^{n-1}}{\omega_1 \cdots \omega_{n-1}}$$

converges in $\ell^q(\mathbb{N})$ for every $|\lambda| < c(T)$. Moreover, this series defines an analytic function f on the open disc $\mathbb{D}(0, c(T))$. Clearly

$$(\lambda I - T^*)f(\lambda) = 0 \quad \text{for all } \lambda \in \mathbb{D}(0, c(T)),$$

and hence the set of all points where T^* does not have the SVEP is a subset of $\mathbb{D}(0, c(T))$.

On the other hand, it is not difficult to check that T^* has no eigenvalues outside the closed disc $\mathbf{D}(0, c(T))$. This implies that T^* has the SVEP at every point λ for which $|\lambda| \ge c(T)$, so the proof is complete. ∎

The result of Theorem 4.84 has a certain interest, since for every triple of real numbers i, c, and r for which $0 \le i \le c \le r$ it is possible to find a weighted right shift T on $\ell^p(\mathbb{N})$ for which $i(T) = i$, $c(T) = c$ and $r(T) = r$. The details of the construction of the sequences $\{\omega_n\}_{n \in \mathbb{N}}$ for which the corresponding weighted right shift T has these properties are outlined in Shields [284].

It is clear that for every weighted right shift operator T on $\ell^p(\mathbb{N})$ we have $e_1 \in \ker T^* \cap T^{*\infty}(X)$, so $\mathcal{N}^\infty(T^*) \cap T^{*\infty}(X)$ is non-trivial. On the other hand, if we consider a weighted right shift T such that $c(T) = 0$ then, by Theorem 4.84, T^* has the SVEP at 0 while $\mathcal{N}^\infty(T^*) \cap T^{*\infty}(X) \ne \{0\}$. This observation illustrates that the implication established in Corollary 2.66 cannot be reversed in general.

The next result shows that the converse of the implications provided in Theorems 2.73 and 2.75 also fails to be true in general.

Theorem 4.85 *Let $1 \le p < \infty$ be arbitrarily given and let T be the weighted right shift operator on $\ell^p(\mathbb{N})$ with weight sequence $\omega := (\omega_n)_{n \in \mathbb{N}}$. Then:*

(i) *$H_0(T) + T(X)$ is norm dense in $\ell^p(\mathbb{N})$ if and only if*

$$\lim_{n \to \infty} \sup(\omega_1 \cdots \omega_n)^{1/n} = 0; \qquad (4.12)$$

(ii) *T* has the SVEP at 0 if and only if*

$$\lim_{n \to \infty} \inf(\omega_1 \cdots \omega_n)^{1/n} = 0.$$

Proof By Theorem 4.84 we need only to prove the equivalence (i). Since

$$\|T^n e_1\| = \omega_1 \cdots \omega_n \quad \text{for all } n \in \mathbb{N},$$

the equality (4.12) holds precisely when $e_1 \in H_0(T)$. From this it follows that (4.12) implies that the sum $H_0(T) + T(X)$ is norm dense in $\ell^p(\mathbb{N})$ because $e_n \in T(X)$ for all $n \geq 2$.

Conversely, suppose that $H_0(T) + T(X)$ is norm dense in $\ell^p(\mathbb{N})$, and for every $k \in \mathbb{N}$ choose $u_k \in H_0(T)$ and $v_k \in T(X)$ such that $u_k + v_k \to e_1$ as $k \to \infty$. Let P denote the projection on $\ell^p(\mathbb{N})$ defined by

$$Px := x_1 e_1 \quad \text{for every } x := (x_n)_{n \in \mathbb{N}} \in \ell^p(\mathbb{N}).$$

It is clear that P vanishes on $T(X)$ and leaves $H_0(T)$ invariant. Moreover, the subspace $H_0(T) \cap T(X)$ is closed, since its dimension is at most 1. Finally,

$$P(u_k + v_k) \to Pe_1 = e_1 \quad \text{as } k \to \infty,$$

so that $e_1 \in H_0(T)$, which concludes the proof. ∎

Every weighted right shift operator T on $\ell^p(\mathbb{N})$ is injective, thus $\mathcal{N}^\infty(T) = \{0\}$. Moreover, $T^\infty(\ell^p(\mathbb{N})) = \{0\}$, and consequently $K(T) = \{0\}$. From this it follows that for these operators the implications provided in Corollary 2.74 and the implications provided in Corollary 2.76 are considerably weaker than those provided in Theorems 2.73 and 2.75.

We now give some information on the SVEP for the bilateral case of shift operators. Let $\ell^2(\mathbb{Z})$ denote the space of all two-sided 2-summable sequences of complex numbers. For a two-sided bounded sequence $\omega = (\omega_n)_{n \in \mathbb{Z}}$ of strictly positive real numbers, the corresponding *bilateral weighted right shift* on $\ell^p(\mathbb{Z})$, $1 \leq p \leq \infty$, is defined by

$$Tx := (\omega_{n-1} x_{n-1})_{n \in \mathbb{Z}} \quad \text{for all } x = (x_n)_{n \in \mathbb{Z}} \in \ell^p(\mathbb{Z}).$$

The dual of T is the bilateral weighted left shift on $\ell^q(\mathbb{Z})$, defined by

$$T^* x := (\omega_n x_{n+1})_{n \in \mathbb{Z}} \quad \text{for all } x = (x_n)_{n \in \mathbb{Z}} \in \ell^q(\mathbb{Z}),$$

where $\frac{1}{p} + \frac{1}{q} = 1$.

In contrast to the unilateral case, T may well have eigenvalues. In fact, we shall see that T need not have the SVEP. Define $\alpha_0 := 1$, $\alpha_n := \omega_0 \cdots \omega_{n-1}$ and $\alpha_{-n} := \omega_{-n} \cdots \omega_{-1}$, and let

$$c^{\pm}(T) := \liminf_{n\to\infty} \alpha_{\pm}^{1/n} \quad \text{and} \quad d^{\pm}(T) := \limsup_{n\to\infty} \alpha_{\pm}^{1/n}.$$

Theorem 4.86 *Let T be bilateral weighted right shift on $\ell^p(\mathbb{Z})$, $1 \le p \le \infty$. Then T does not have the SVEP at λ precisely when $d^+(T) < |\lambda| < c^-(T)$. In particular, T has the SVEP if and only if $c^-(T) \le d^+(T)$.*

Proof Suppose that λ is an eigenvalue of T and consider a corresponding non-zero eigenvector $x \in \ell^p(\mathbb{Z})$. A simple computation shows that $\lambda \ne 0$ and

$$x_n = x_0 \alpha_n / \lambda^n, \quad x_{-n} = x_0 \lambda^n / \alpha_{-n} \quad \text{for all } n \in \mathbb{N}.$$

Because $x \in \ell^p(\mathbb{Z})$, it then follows that $d^+(T) \le |\lambda| \le c^-(T)$. On the other hand, if $d^+(T) < c^-(T)$, then as in the proof of Theorem 4.84, the classical formula for the radius of convergence guarantees that the definition

$$f(\lambda) := \sum_{n=0}^{\infty} e_n \alpha_n / \lambda^n + \sum_{n=1}^{\infty} e_{-n} \lambda^n / \alpha_{-n},$$

for all $\lambda \in \mathbb{C}$ with $d^+(T) < |\lambda| < c^-(T)$, is an analytic solution of the equation $(\lambda I - T)f(\lambda) = 0$ on the annulus $\{\lambda \in \mathbb{C} : d^+(T) < |\lambda| < c^-(T)\}$. ∎

Corollary 4.87 *Let T be a bilateral weighted right shift on $\ell^p(\mathbb{Z})$, $1 \le p \le \infty$. Then the following assertions hold:*

(i) *T^* does not have the SVEP at λ if and only if $d(T) < |\lambda| < c(T)$.*
(ii) *T^* has the SVEP if and only if $c^-(T) \le d^+(T)$.*
(iii) *At least one of the operators T or T^* has the SVEP.*

Proof We have already observed that the dual of T is the bilateral weighted left shift on $\ell^q(\mathbb{Z})$, defined as $T^*x := (\omega_n x_{n+1})_{n\in\mathbb{Z}}$ for all $x = (x_n)_{n\in\mathbb{Z}} \in \ell^q(\mathbb{Z})$. Choose $\hat{\omega} := (\omega_{-n-1})_{n\in\mathbb{Z}}$ and set

$$Sx := (x_{-n})_{n\in\mathbb{Z}} \quad \text{for all } x = (x_n)_{n\in\mathbb{Z}} \in \ell^q(\mathbb{Z}).$$

It is easily seen that

$$(ST^*S)x = (\hat{\omega}_{n-1} x_{n-1})_{n\in\mathbb{Z}} \quad \text{for all } x = (x_n)_{n\in\mathbb{Z}} \in \ell^q(\mathbb{Z}).$$

This shows that T^* is similar to the bilateral weighted right shift on $\ell^q(\mathbb{Z})$ with weight sequence $\hat{\omega}$. In the sense of the right shift representation of T^*, we then obtain the identities $c^{\pm}(T^*) = c^{\mp}(T)$ and $d^{\pm}(T^*) = d^{\mp}(T)$, because $\hat{\alpha}_n = \alpha_{-n}$ for all $n \in \mathbb{Z}$. Hence the assertion (i) is clear from Theorem 4.86, and (ii) is immediate from (i). Finally, to prove (iii), assume that both T and T^* fail to have the SVEP. Then the preceding results entail that $d^+(T) < c^-(T)$ and $d^-(T) < c^+(T)$. But this leads to an obvious contradiction, since $c^-(T) \le d^-(T)$ and $c^+(T) \le d^+(T)$. ∎

Note that in part (c) of the preceding result, it is possible that both T and T^* have the SVEP. For instance, the classical bilateral shift has spectrum $\sigma(T) = \sigma(T^*) = \Gamma$, Γ the unit circle, and hence both T and T^* have the SVEP. There are many examples of decomposable bilateral weighted shifts beyond the quasi-nilpotent one, however, the precise characterization of those weight sequences for which the corresponding bilateral shift is decomposable remains an open problem.

Let us consider a bounded operator T on a Banach space X which satisfies the abstract shift condition $T^\infty(X) = \{0\}$. This condition entails that $0 \in \sigma(T)$ since T is not surjective.

Theorem 4.88 *Suppose that $T \in L(X)$, X an infinite-dimensional Banach space, is non-invertible and $i(T) = r(T)$. Then*

$$\sigma_{\mathrm{w}}(T) = \sigma_{\mathrm{b}}(T) = \sigma_{\mathrm{su}}(T) = \sigma(T) = \mathbf{D}(0, r(T)), \tag{4.13}$$

while, in particular, these equalities hold if $T^\infty(X) = \{0\}$ and $i(T) = r(T)$.

Proof If T is a non-invertible and $i(T) = r(T)$ then, by Theorem 4.68, $\sigma(T)$ is the whole closed disc $\mathbf{D}(0, r(T))$ and $\sigma_{\mathrm{ap}}(T)$ is the circle $\partial\mathbf{D}(0, r(T))$. Since, by Theorem 4.71, T has the SVEP then $\sigma_{\mathrm{su}}(T) = \sigma(T)$, by Theorem 2.68, and $\sigma_{\mathrm{w}}(T) = \sigma_{\mathrm{b}}(T)$, by Theorem 3.44.

Suppose first that $i(T) = r(T) = 0$. Then T is quasi-nilpotent. The equalities (4.13) are then trivially satisfied (note that since X is infinite-dimensional, $\sigma_{\mathrm{b}}(T)$ is non-empty and hence is $\{0\}$). Suppose then that $i(T) = r(T) > 0$. Also in this case $\sigma(T) = \sigma_{\mathrm{b}}(T)$, since every non-isolated point of the spectrum lies on $\sigma_{\mathrm{b}}(T)$. Therefore the equalities (4.13) are proved. ∎

Theorem 4.88 also applies to every non-invertible isometry T on a Banach space X since $i(T) = r(T) = 1$.

Definition 4.89 An operator $T \in L(X)$ is said to be a *semi-shift* if T is an isometry and $T^\infty(X) = \{0\}$.

Every semi-shift is non-invertible isometry, since the condition $T^\infty(X) = \{0\}$ entails that T is not surjective. It should be noted that for Hilbert space operators the semi-shifts coincide with the isometries for which none of the restrictions to a non-trivial reducing subspace is unitary, see Chapter I of Conway [101].

An operator $T \in L(X)$ for which the equality $\sigma_T(x) = \sigma(T)$ holds for every $x \ne 0$ is said to have *fat local spectra*, see Neumann [247]. Clearly, by Corollary 4.81 an isometry T is a semi-shift if and only if T has fat local spectra.

Examples of semi-shifts are the unilateral right shift operators of arbitrary multiplicity on $\ell^p(\mathbb{N})$, as well as every right translation operator on $L^p([0, \infty))$. In Laursen and Neumann [216, Proposition 1.6.9] it is shown that if X is the

Banach space of all analytic functions on a connected open subset $\mathcal{U}$ of $\mathbb{C}$, f is a non-constant analytic function on $\mathcal{U}$, and if $T_f \in L(X)$ denotes the point-wise multiplication operator by f, then the condition $\sigma(T_f) \subseteq \overline{f(\mathcal{U})}$ implies that T_f has local fat spectra. In particular, these conditions are verified by every multiplication operator T_f on the disc algebra $\mathcal{A}(\mathbb{D})$ of all complex-valued functions continuous on the closed unit disc of $\mathbb{C}$ and analytic on the open unit disc $\mathbb{D}$, where $f \in \mathcal{A}(\mathbb{D})$, and the same result holds for the Hardy algebra $H^\infty(\mathbb{D})$. If $f \in H^\infty(\mathbb{D})$ and $1 \leq p < \infty$ the operator on $H^p(\mathbb{D})$ defined by the multiplication by f also has a local fat spectra.

4.7 Toeplitz Operators on Hardy Spaces

An important class of polaroid operators is provided by the Toeplitz operators on the classical Hardy spaces $H^2(\mathbf{T})$, where $\mathbf{T}$ denotes the unit circle of $\mathbb{C}$. To define the Hardy space $H^2(\mathbf{T})$, for $n \in \mathbb{Z}$, let χ_n be the function on $\mathbf{T}$ defined by

$$\chi_n(e^{it}) := e^{int} \quad \text{for all } n \in \mathbb{N}.$$

Let μ be the normalized Lebesgue measure on $\mathbf{T}$, and $L^2(\mathbf{T})$ the classical Hilbert space defined with respect to μ. The set $\{\chi_n\}_{n \in \mathbb{Z}}$ is an orthogonal basis of $L^2(\mathbf{T})$. If $f \in L^2(\mathbf{T})$, then the *Fourier transform* of f is the map $\widehat{f} : \mathbb{Z} \to \mathbb{C}$, defined by

$$\widehat{f}(n) = \langle f, \chi_n \rangle = \int_0^{2\pi} f(t)e^{-int}\,\mathrm{dt}.$$

$\widehat{f}(n)$ is called the n-th Fourier coefficient of f, and by the classical Parseval's identity we have

$$f = \sum_{-\infty}^{\infty} \widehat{f}(n)z^n.$$

Note that if $f \in L^2(\mathbf{T})$ then $\widehat{f} \in \ell^2(\mathbb{Z})$ and the mapping $\Psi : L^2(\mathbf{T}) \to \ell^2(\mathbb{Z})$, defined by $\Psi(f) := \widehat{f}$, is an isomorphism.

The *Hardy space* $H^2(\mathbf{T})$ is defined as the closed subspace of all $f \in L^2(\mathbf{T})$ for which

$$\frac{1}{2\pi}\int_0^{2\pi} f\chi_n\,\mathrm{dt} = 0 \quad \text{for } n = 1, 2, \ldots.$$

The Hilbert space $H^2(\mathbf{T})$ is the closed linear span of the set $\{\chi_n\}_{n=0,1,\ldots}$. Moreover, $H^2(\mathbf{T})$ is a closed subspace of $L^\infty(\mathbf{T})$. We summarize in the sequel some basic results concerning the Hardy spaces $H^2(\mathbf{T})$. For further details the reader is invited to consult Douglas' book [121].

If $\phi \in L^\infty(\mathbf{T})$ and $f \in L^2(\mathbf{T})$ then $\phi f \in L^2(\mathbf{T})$, so we may define an operator $M_\phi : L^2(\mathbf{T}) \to L^2(\mathbf{T})$ by

$$M_\phi f = \phi f \quad \text{for all } f \in L^2(\mathbf{T}),$$

where ϕf is the pointwise product. Let P denote the projection of $L^2(\mathbf{T})$ onto $H^2(\mathbf{T})$.

Definition 4.90 If $\phi \in L^\infty(\mathbf{T})$, the *Toeplitz operator with symbol ϕ T_ϕ* on $H^2(\mathbf{T})$ is defined by

$$T_\phi f := P(\phi f) \quad \text{for } f \in H^2(\mathbf{T}).$$

Since the set $\Gamma := \{z^n : n = 0, 1, 2, \ldots\}$ is an orthonormal basis for $H^2(\mathbf{T})$, if for every $\phi \in L^\infty(\mathbf{T})$ we set

$$\hat{\phi}(n) := \frac{1}{2\pi} \int_0^{2\pi} \phi \bar{z} dt,$$

then, with respect to the basis Γ, T_ϕ may be represented by a matrix a_{ij}, where

$$a_{ij} = \langle T_\phi z^j, z^i \rangle = \frac{1}{2\pi} \int_0^{2\pi} \phi \bar{z} i - jdt = \hat{\phi}(i - j).$$

Thus the matrix for T_ϕ is constant on diagonals:

$$(a_{ij}) = \begin{pmatrix} c_0 & c_{-1} & c_{-2} & c_{-3} & \cdots \\ c_1 & c_0 & c_{-1} & c_{-2} & \cdots \\ c_2 & c_1 & c_0 & c_{-1} & \cdots \\ c_3 & c_2 & c_1 & c_0 & \cdots \\ \vdots & \vdots & \vdots & \vdots & \vdots \end{pmatrix}, \quad \text{where } c_j = \hat{\phi}(j).$$

Such a matrix is called a *Toeplitz matrix*. The Toeplitz operators with analytic symbols are particulary amenable to study. The adjoint of the Hilbert space operator M_ϕ on $L^2(\mathbf{T})$ is $M_\phi' = M_{\bar{\phi}}$ and obviously, $M_\phi M_\phi' = M_\phi' M_\phi$, so M_ϕ is a normal operator. Let $H^\infty(\mathbf{T})$ denote the Banach space of all $\phi \in L^\infty(\mathbf{T})$ such that

$$\frac{1}{2\pi} \int_0^{2\pi} \phi \chi_n dt = 0 \quad \text{for all } n = 1, 2, \ldots.$$

$H^\infty(\mathbf{T})$ is a closed subalgebra of $L^\infty(\mathbf{T})$ and $H^\infty(\mathbf{T}) = L^\infty(\mathbf{T}) \cap H^2(\mathbf{T})$. If $\phi \in H^\infty(\mathbf{T})$, the operator T_ϕ is the restriction of M_ϕ to the closed invariant subspace $H^2(\mathbf{T})$, so T_ϕ is subnormal. Note that every Toeplitz operator is normaloid, i.e. $\|T_\phi\| = r(T_\phi)$, see [82].

Theorem 4.91 *For* $\phi \in H^\infty(\mathbf{T})$, *the Toeplitz operator* T_ϕ *is hyponormal. In particular,* T_ϕ *is* $H(1)$ *and hence is hereditarily polaroid and has the SVEP.*

Proof T_ϕ, $\phi \in H^\infty(\mathbf{T})$ is subnormal and hence hyponormal, see Conway [101, Proposition 2.4.2]. By Theorem 4.46, every hyponormal operator is totally paranormal and hence $H(1)$. ∎

For each $\phi \in L^\infty(\mathbf{T})$, let

$$\epsilon(\phi) := \{k \in H^\infty(\mathbf{T}) : \|k\| \leq 1, \phi - k\phi \in H^\infty(\mathbf{T})\}.$$

An elegant theorem due to Cowen [103] characterizes the hyponormality of T_ϕ, where $\phi \in L^\infty(\mathbf{T})$, by means of some properties of the symbol ϕ. More precisely, the result of Cowen (in its Nakazi–Takahashi formulation) shows that T_ϕ is hyponormal if and only if $\epsilon(\phi) \neq \emptyset$. If $\epsilon(\phi) \neq \emptyset$ then T is $H(1)$, by Theorem 4.46. The reader may find further results on subnormality of Toeplitz operators in Lee [223].

Note that if $\phi \in L^\infty(\mathbf{T})$ then $\lambda I - T_\phi = T_{\lambda - \phi}$ is a Toeplitz operator and for the adjoint T_ϕ' we have $T_\phi' = T_{\bar{\phi}}$. The Fredholm theory of T_ϕ enjoys some important properties which we list in the sequel.

Theorem 4.92 *Suppose that* $\phi \in H^\infty(\mathbf{T})$. *We have:*

(i) *If* T_ϕ *is Fredholm then* ϕ *is invertible in* $L^\infty(\mathbf{T})$.
(ii) *If* T_ϕ *is quasi-nilpotent, or compact then* $\phi = 0$.

In particular, T_ϕ *is invertible if and only if* ϕ *is invertible in* $L^\infty(\mathbf{T})$.

Let $C(\mathbf{T})$ denote the Banach algebra of all complex-valued continuous functions on $\mathbf{T}$. By [121, Proposition 7.22] we also have

Theorem 4.93 *If* $\phi \in C(\mathbf{T})$ *and* $\psi \in L^\infty(\mathbf{T})$ *then* $T_\phi T_\psi - T_\psi T_\phi$ *and* $T_\psi T_\phi - T_\phi T_\psi$ *are both compact.*

An operator $T \in L(H)$, H a Hilbert space, is said to be *essentially normal* if $TT' - T'T$ is compact. Note that the operator $T := S + K$, with S normal and K compact, is evidently essentially normal, but the converse is not true, for instance, if T is the unilateral shift on $\ell^2(\mathbb{N})$ with basis $\{e_n\}, n = 0, 1, \ldots$ then $T'T - TT' = I - TT'$ is the rank one projection onto the 1-dimensional subspace $\mathbb{C}e_0$, thus T is essentially normal. However, T is a Fredholm operator having nozero index. The index is stable under compact perturbation, so the same persists for a normal operator on a Hilbert space H plus a compact operator. In [83] Brown et al. proved that T is a normal operator plus a compact operator precisely when T is essentially normal and $\sigma_e(T) = \sigma_w(T)$. Evidently, if $\phi \in C(\mathbf{T})$ then T_ϕ is essentially normal.

We now enunciate the classical Frechét and Riesz theorem, whose proof may be found in many standard books, see for instance Hofmann [180].

Theorem 4.94 (Frechét and Riesz Theorem) *If f is a nonzero function in $H^2(\mathbf{T})$ then the set $\{z \in \mathbf{T} : f(z) = 0\}$ has Lebesgue measure zero. In particular, if $f, g \in H^2(\mathbf{T})$ and if $fg = 0$ almost everywhere then $f = 0$ or $g = 0$, almost everywhere.*

The following result plays a crucial role in characterizing the Toeplitz operators which are Fredholm:

Theorem 4.95 (Coburn) *Suppose that $\phi \in L^\infty(\mathbf{T})$ is not almost everywhere 0. Then either $\alpha(T_\phi) = 0$ or $\beta(T_\phi) = \alpha(T'_\phi) = 0$.*

Proof Suppose that both $\alpha(T_\phi) \neq 0$ and $\alpha(T'_\phi) \neq 0$. Then there exist nonzero functions $f, g \in H^2(\mathbf{T})$ such that $T_\phi f = 0$ and $T'_\phi g = T_{\overline{\phi}} g = 0$. Then $P(\phi f) = P(\overline{\phi} g) = 0$, so that, by the standard properties of $H^2(\mathbf{T})$, there exist functions $h, k \in H^2(\mathbf{T})$ for which

$$\int_0^{2\pi} h\, dt = \int_0^{2\pi} k\, dt = 0 \quad \text{and} \quad \phi f = \overline{h}, \ \overline{\phi} g = \overline{h}.$$

From the Frechét and Riesz theorem, it then follows that ϕ, f, g, h, k are all nonzero except on a set of measure zero. Dividing the two sides of the equation $\overline{\phi} f = h$ by the corresponding sides of the equation $\overline{\phi} g = \overline{k}$, we see that

$$\frac{\overline{f}}{g} = \frac{h}{\overline{k}} \quad \text{pointwise},$$

so that $\overline{fk} = gh$ almost everywhere. By another standard property of $H^2(\mathbf{T})$, this is possible unless $gh = 0$ almost everywhere. Using again the theorem of Frechét and Riesz, we then conclude that either $f = 0$ almost everywhere or $g = 0$ almost everywhere, a contradiction. ∎

As a consequence of Theorem 4.95 we obtain:

Corollary 4.96 *Suppose that $\phi \in L^\infty(\mathbf{T})$ is not almost everywhere 0. Then T_ϕ is Weyl if and only if T_ϕ is invertible. Consequently, $\sigma(T_\phi) = \sigma_w(T_\phi)$.*

Proof Suppose that T_ϕ is a Weyl operator. Then $\alpha(T_\phi) = \beta(T_\phi)$ and from Theorem 4.95 we see that $\alpha(T_\phi) = \beta(T_\phi) = 0$, thus T_ϕ is invertible. ∎

It should be noted that for every compact operator $K \in K(L^2(\mathbf{T}))$ we have $\|T_\phi\| \leq \|T + K\|$. Indeed, the Fredholm spectrum $\sigma_f(T_\phi)$ coincides with the spectrum of the class $T_\phi + K(X)$ in the Calkin algebra $L(L^2(\mathbf{T})/K(L^2(\mathbf{T})))$, and hence $\sigma(T_\phi) \subseteq \sigma(T_\phi + K)$. Since T_ϕ is normaloid then $\|T_\phi\| \leq \|T + K\|$. The operators which satisfy the latter inequality are sometimes called *extremally noncompact*.

We next consider the *Toeplitz operators* T_ϕ *with continuous symbols*, i.e. $\phi \in C(\mathbf{T})$. First we recall that the *winding number* $wn(\phi, \lambda)$ of a closed curve ϕ in the plane around a given point λ is an integer representing the total number of times that curve travels counterclockwise around the point. The winding number depends on the orientation of the curve, and is negative if the curve travels around the point clockwise.

We recall the classical definition of homotopy. Suppose that γ_0 and γ_1 are closed curves in a topological space X, both with parameter interval $[0, 1]$. We say that γ_0 and γ_1 are *X-homotopic* if there is a continuous mapping $\Theta : [0, 1] \times [0, 1] \to X$ such that

$$\Theta(s, 0) = \gamma_0(s), \quad \Theta(s, 1) = \gamma_1(s), \quad \Theta(0, t) = \Psi(1, t)$$

for all $s, t \in [0, 1]$. Intuitively, this means that the curve γ_0 can be continuously deformed in γ_1, within X.

The following nice result is due to a number of authors (Krein [207], Widom [297], Devinatz [108]).

Theorem 4.97 *If $\phi \in C(\mathbf{T})$ then T_ϕ is a Fredholm operator if and only if ϕ does not vanish. In this case*

$$\mathrm{ind}\, T_\phi = -wn(\phi, 0),$$

where $wn(\phi, 0)$ is the winding number of the curve traced by ϕ with respect to the origin. In particular, T_ϕ is Weyl, or equivalently, invertible, if and only if $wn(\phi, 0) = 0$.

Proof The first assertion is clear. We show that if two functions ϕ and ψ determine homotopic curves in $\mathbb{C}\{0\}$ then $\mathrm{ind}\,(T_\phi) = \mathrm{ind}\,(T_\psi)$. To see this, let Φ be a constant map from $[0, 1] \times \mathbf{T}$ to $\mathbb{C} \setminus \{0\}$ such that

$$\Phi(0, e^{it}) = \phi(e^{it}), \quad \Phi(1, e^{it}) = \psi(e^{it}).$$

If we set $\Phi_\lambda := \Phi(\lambda, e^{it})$, then the mapping $\lambda \to T_{\Phi_\lambda}$ is norm continuous, and each T_{Φ_λ} is Fredholm. Since the map index is continuous, $\mathrm{ind}\,(T_\phi) = \mathrm{ind}\,(T_\psi)$, as claimed. Now, take $n := wn(\phi, 0)$. Then ϕ is homotopic in $\mathbb{C} \setminus \{0\}$ to $\psi(z) := z^n$, and since $\mathrm{ind}\, T_{z^n} = -n$, we then conclude that $\mathrm{ind}\, T_\phi = -wn(\phi, 0)$. The last assertion is evident.   ∎

Recall that given a compact set $\sigma \subset \mathbb{C}$, a *hole* of σ is a bounded component of the complement $\mathbb{C} \setminus \sigma$. Since $\mathbb{C} \setminus \sigma$ always has an unbounded component, $\mathbb{C} \setminus \sigma$ is connected precisely when σ has no holes. Now, set $\sigma = \phi(\mathbf{T})$, the range of ϕ. Then $\mathbb{C} \setminus \phi(\mathbf{T})$ has a unique unbounded component Ω and the winding number is 0 in Ω, while it is constant on each other component of $\mathbb{C} \setminus \phi(\mathbf{T})$.

The spectrum and the Weyl spectrum of a Toeplitz operator having a continuous symbol may be described in the following way.

Corollary 4.98 *If $\phi \in C(\mathbf{T})$ then*

$$\sigma(T_\phi) = \sigma_{\mathrm{w}}(T_\phi) = \sigma_{\mathrm{b}}(T_\phi) = \phi(\mathbf{T}) \cup \{\lambda \in \mathbb{C} : wn(\phi, \lambda) \neq 0\}$$

and

$$\sigma_{\mathrm{e}}(T_\phi) = \phi(\mathbf{T}).$$

In particular, $\sigma(T_\phi) = \sigma_{\mathrm{w}}(T)$ is connected.

Proof $\sigma(T_\phi)$ is connected since it is formed from the union of Γ and certain components of the resolvent of T_ϕ. Clearly, $\sigma(T_\phi) = \sigma_{\mathrm{w}}(T_\phi)$ by Corollary 4.96, while $\sigma_{\mathrm{b}}(T_\phi) = \sigma_{\mathrm{w}}(T_\phi)$ is clear, since $\sigma_{\mathrm{w}}(T_\phi) \subseteq \sigma_{\mathrm{b}}(T_\phi) \subseteq \sigma(T_\phi)$. For the equality $\sigma_{\mathrm{e}}(T_\phi) = \phi(\mathbf{T})$, see Douglas [121, Chapter 7]. ∎

The result of Corollary 4.98 may be improved. The spectra $\sigma(T_\phi)$ and the essential spectrum $\sigma_{\mathrm{e}}(T_\phi)$ are also connected if $\phi \in L^\infty(\mathbf{T})$, see [121, Corollary 7.47 and Theorem 7.45].

Theorem 4.99 *If $\phi \in C(\mathbf{T})$ then the following statements hold:*

  (i) *ϕ is non-constant.*
 (ii) *iso $\sigma_{\mathrm{w}}(T_\phi) = \emptyset$.*
(iii) *iso $\sigma_{\mathrm{uw}}(T_\phi) = \emptyset$.*

Moreover, T_ϕ is polaroid.

Proof If $\phi \in C(\mathbf{T})$ we have

$$\rho_{\mathrm{w}}(T_\phi) = \sigma(T_\phi) = \{\lambda \in \mathbb{C} : wn(\phi, \lambda) = 0\},$$

and

$$\rho_{\mathrm{sf}}^+(T_\phi) = \{\lambda \in \mathbb{C} : wn(\phi, \lambda) < 0\},$$

while

$$\rho_{\mathrm{sf}}^-(T_\phi) = \mathbb{C} \setminus \sigma_{\mathrm{lw}}(T_\phi) = \{\lambda \in \mathbb{C} : wn(\phi, \lambda) > 0\}.$$

From Lemma 3.57, we know that

$$\sigma_{\mathrm{w}}(T_\phi) = \sigma_{\mathrm{sf}}(T_\phi) \cup \rho_{\mathrm{sf}}^+(T_\phi) \cup \rho_{\mathrm{sf}}^-(T_\phi)$$

and

$$\sigma_{\mathrm{uw}}(T_\phi) = \sigma_{\mathrm{sf}}(T_\phi) \cup \rho_{\mathrm{sf}}^+(T_\phi),$$

so $\sigma_{\mathrm{w}}(T_\phi)$ consists of $\mathbf{\Gamma} = \phi(\mathbf{T})$ and those holes with respect to which the winding number of ϕ is nonzero, and analogously, $\sigma_{\mathrm{uw}}(T_\phi)$ consists of $\mathbf{\Gamma}$ and those holes with respect to which the winding number of ϕ is negative.

We see now that

$$\mathrm{iso}\,\sigma_{\mathrm{uw}}(T_\phi) = \emptyset \Leftrightarrow \mathrm{iso}\,\sigma_{\mathrm{w}}(T_\phi) = \emptyset \Leftrightarrow \phi \text{ is non-constant.}$$

Indeed, if $\mathrm{iso}\,\sigma_{\mathrm{uw}}(T_\phi) \neq \emptyset$, or $\mathrm{iso}\,\sigma_{\mathrm{w}}(T_\phi) \neq \emptyset$, then, by Theorem 3.58, we have $\mathrm{iso}\,\sigma_{\mathrm{sf}}(T_\phi) \neq \emptyset$. Because $\mathbf{\Gamma} = \sigma_{\mathrm{e}}(T_\phi)$ is connected, it then follows that $\mathbf{\Gamma}$ is a singleton and ϕ is constant. On the other hand, if ϕ is constant, for instance $\phi \equiv \lambda$, then it is obvious that $\sigma(T_\phi) = \sigma_{\mathrm{w}}(T_\phi) = \sigma_{\mathrm{uw}}(T_\phi) = \{\lambda\}$. Thus, $\mathrm{iso}\,\sigma_{\mathrm{w}}(T_\phi) = \mathrm{iso}\,\sigma_{\mathrm{uw}}(T_\phi) = \{\lambda\}$. These remarks also show that every Toeplitz operator with continuous symbol is polaroid. ∎

In the next result we show that if the orientation of the curve $\phi(\mathbf{T})$ does not change then either T_ϕ, or T_ϕ', has the SVEP.

Theorem 4.100 *Let $\phi \in C(T)$. Then we have:*

(i) *If the orientation of the curve $\phi(\mathbf{T})$ traced out by ϕ is counterclockwise then T_ϕ has the SVEP.*

(ii) *If the orientation of the curve $\phi(\mathbf{T})$ traced out by ϕ is clockwise then T_ϕ' has the SVEP.*

Proof

(i) Suppose first that the orientation of $\phi(\mathbf{T})$ is counterclockwise. Let Ω_1 be the bounded component of $\mathbb{C} \setminus \phi(\mathbf{T})$ and Ω_2 the unbounded component of $\mathbb{C} \setminus \phi(\mathbf{T})$. Then $wn(\phi, \lambda) > 0$ for every $\lambda \in \Omega_1$, while $wn(\phi, \lambda) = 0$ for every $\lambda \in \Omega_2$. Therefore, for every $\lambda \in \Omega_1$ we have $\mathrm{ind}\,(\lambda I - T_\phi) = -wn(\phi, \lambda) < 0$ and consequently

$$\sigma(T_\mu) = \sigma_{\mathrm{w}}(T_\mu) = \Omega_1 \cup \phi(\mathbf{T}).$$

Note that $\sigma_{\mathrm{uw}}(T_\phi) = \phi(\mathbf{T})$ is the boundary of the spectrum. Now, if $\lambda \in \Omega_1$ the condition $\mathrm{ind}\,(\lambda I - T_\phi) < 0$ entails that $\alpha(\lambda I - T_\phi) < \beta(\lambda I - T_\phi)$ and hence $\beta(\lambda I - T_\phi) > 0$. From Theorem 4.95 we have that $\alpha(\lambda I - T_\phi) = 0$, and $\lambda I - T_\phi$ having a closed range, since $\lambda I - T_\phi$ is upper semi-Weyl, we then deduce that $\lambda \notin \sigma_{\mathrm{ap}}(T_\phi)$. Therefore, $\sigma_{\mathrm{ap}}(T_\phi) \subseteq \phi(\mathbf{T})$, from which we obtain that

$$\phi(\mathbf{T}) = \sigma_{\mathrm{uw}}(T_\phi) \subseteq \sigma_{\mathrm{ap}}(T_\phi) \subseteq \phi(\mathbf{T}),$$

thus $\sigma_{\mathrm{ap}}(T_\phi) = \phi(\mathbf{T})$ is the boundary of the spectrum $\sigma(T_\phi)$. This entails that T_ϕ has the SVEP.

(ii) Suppose that the orientation of $\phi(\mathbf{T})$ is clockwise. Then $wn(\phi, \lambda) < 0$ for every $\lambda \in \Omega_1$, so, if $\lambda \in \Omega_1$ then $\mathrm{ind}\,(\lambda I - T_\phi) > 0$. Consequently, $\sigma_{\mathrm{lw}}(T_\phi) = \phi(\mathbf{T})$, and $\alpha(\lambda I - T_\phi) > \beta(\lambda I - T_\phi)$ for all $\lambda \in \Omega_1$, so $\alpha(\lambda I - T_\phi) > 0$.

From Theorem 4.95 we have that $\beta(\lambda I - T_\phi) = 0$, so $\lambda \notin \sigma_s(T_\phi)$ and hence $\sigma_s(T_\phi) \subseteq \phi(\mathbf{T})$. Again,

$$\phi(\mathbf{T}) = \sigma_{lw}(T_\phi) \subseteq \sigma_s(T_\phi) \subseteq \phi(\mathbf{T}),$$

from which we obtain that $\sigma_s(T_\phi) = \phi(\mathbf{T})$ is contained in the boundary of the spectrum. This entails that T'_ϕ has the SVEP. ∎

Example 4.101 The next example provides a Toeplitz operator T_ϕ having continuous symbol for which the SVEP for both T_ϕ and T_ϕ fails. Let ϕ be defined by

$$\phi(e^{i\theta}) := \begin{cases} -e^{2i\theta} + 1 & \text{if } 0 \le \theta \le \pi, \\ e^{-2i\theta} - 1 & \text{if } \pi \le \theta \le 2\pi. \end{cases}$$

The orientation of the graph of ϕ is shown in the following figure.

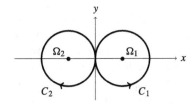

Let Ω_1 and Ω_2 be the interior of the circle C_1 and C_2, respectively. Since $wn(\phi, \lambda) = 1$ in Ω_1 and $wn(\phi, \lambda) = -1$ in Ω_2, we have ind $(\lambda I - T_\phi) < 0$ for $\lambda \in \Omega_1$, while ind $(\lambda I - T_\phi) > 0$ for $\lambda \in \Omega_2$. Since $\lambda I - T_\phi$ is Fredholm for every $\lambda \in \Omega_1 \cup \Omega_2$, the operator T_ϕ cannot have the SVEP, otherwise we would have ind $(\lambda I - T_\phi) \le 0$ for all $\lambda \in \Omega_2$, by Corollary 2.106, and, analogously, if T'_ϕ has the SVEP we would have ind $(\lambda I - T_\phi) \ge 0$ for all $\lambda \in \Omega_1$. A contradiction.

Example 4.101 provides an example of a Toeplitz operator T_ϕ which is not hereditarily polaroid, because T_ϕ does not have the SVEP.

In some sense an opposite result to that established in Theorem 4.100 holds: if ϕ has a unique orientation and T_ϕ has the SVEP then the orientation of ϕ is forced to be counterclockwise, and analogously if T'_ϕ has the SVEP then the orientation of ϕ is forced to be clockwise.

In general, for symbols $\phi \in L^\infty(\mathbf{T})$, the operators T_ϕ are not hyponormal, even if the symbol is continuous. The operator T_ϕ in Example 4.101 cannot be hyponormal since hyponormality entails SVEP. Toeplitz operators with continuous symbol which are hyponormal have also been studied by Farenick and Lee [145], in particular if ϕ is a trigonometric polynomial. A celebrated result of Brown and Halmos [82] shows that T_ϕ is normal if and only if $\phi = \alpha + \beta \psi$ where α and β are complex and ϕ is a real-valued function in $L^\infty(\mathbf{T})$. There are many results concerning hyponormality of Toeplitz operators in the literature and properties of hyponormal Toeplitz operators have played an important role in work on Halmos

Problem 5 [165]: "*Is every subnormal Toeplitz operator either normal or analytic?*"
but a characterization has been lacking. Here, the operator T_ϕ is said to be *analytic*
if $\phi \in H^2(\mathbf{T})$. This question appears natural since every normal operator is
obviously subnormal and, as has observed before, every analytic Toeplitz operator
is subnormal.

Theorem 4.102 *If $\phi \in C(\mathbf{T})$ is such that $\sigma(T_\phi)$ has planar Lebesgue measure zero
then both T_ϕ and T'_ϕ have the SVEP.*

Proof The planar measure of $\sigma(T_\phi)$ is zero, because $\sigma(T_\phi) = \sigma_e(T_\phi) = \phi(\mathbf{T})$ is a
compact set consisting of $\phi(\mathbf{T})$ and some of its holes, so $\partial\sigma(T_\phi) = \phi(\mathbf{T})$, which is
just a continuous curve. Therefore, both T_ϕ and T'_ϕ have the SVEP.   ∎

4.8 Polaroid Operators Under Affinities

Let $T \in L(X)$ and $\lambda \in \mathrm{iso}\,\sigma(T)$. Recall, according to Definition 2.139, that $A \in
L(X, Y)$ intertwines $T \in L(X)$ and $S \in L(Y)$ if $SA = AT$. If A is a quasi-
affinity then $T \prec S$. If A is bijective then T and S are similar, and we know that
similar operators have the same spectrum, as well as some distinguished parts of the
spectrum, for instance $\sigma_{ap}(T) = \sigma_{ap}(S)$ and $\sigma_s(T) = \sigma_s(S)$. The situation becomes
more complicated if the condition of similarity is replaced by the weaker condition
of quasi-similarity. For instance, if A is assumed to be a quasi-affinity, a result of
Fialkow [146] asserts that the essential spectra $\sigma_e(T)$ and $\sigma_e(S)$ always have non-
empty intersection. In [178] Herrero proved that if T and S are quasi-similar each
connected component of $\sigma_e(T)$ touches $\sigma_e(S)$, and vice versa. Herrero's proof is
rather long and complicated, to present it here would lead us too far afield.

Denote by $P_T(\lambda)$ the spectral projection associated to T and the spectral set $\{\lambda\}$.

Lemma 4.103 *Suppose that $T \in L(X)$ and $S \in (Y)$ are intertwined by $A \in
L(X, Y)$. If $\lambda \in \mathrm{iso}\,\sigma(T) \cap \mathrm{iso}\,\sigma(S)$ then $P_T(\lambda)$ and $P_S(\lambda)$ are also intertwined
by A, i.e. $P_S(\lambda)A = AP_T(\lambda)$.*

Proof If T and S are intertwined by $A \in L(X, Y)$ we have

$$(\mu I - S)A = A(\mu I - T) \quad \text{for all } \mu \in \mathbb{C}.$$

Suppose that μ belongs to the resolvent of T and to the resolvent of S. Then $A =
(\mu I - S)^{-1}A(\mu I - T)$ and hence $A(\mu I - T)^{-1} = (\mu I - S)^{-1}A$, from which it
easily follows that

$$P_S(\lambda)A = \left(\frac{1}{2\pi i}\int_\Gamma (\mu I - S)^{-1}d\mu\right)A = \frac{1}{2\pi i}\int_\Gamma (\mu I - S)^{-1}Ad\mu$$

$$= \frac{1}{2\pi i}\int_\Gamma A(\mu I - T)^{-1}d\mu = AP_T(\lambda).$$ ∎

If $T \prec S$ a classical result due to Rosenblum shows that $\sigma(S)$ and $\sigma(T)$ must overlap, see [268]. Quasi-similarity is, in general, not sufficient to preserve the spectrum. In fact, Hoover in [182] provides an example of a pair of quasi-similar operators T and S defined on a Hilbert space such that T is quasi-nilpotent and compact, hence $\sigma(T) = \{0\}$, while $\sigma(S)$ is the closed unit disc. The preservation of the spectrum happens only in some special cases, for instance, by a result of Clary [95], if T and S are quasi-similar hyponormal operators, or whenever T and S have totally disconnected spectra, see [161, Corollary 2.5]. Therefore, it is not quite surprising that, if $T \prec S$, the preservation of "certain" spectral properties from S to T requires that some spectral inclusions are satisfied.

Remark 4.104 Classical examples show the polaroid property is not preserved if two bounded operators are intertwined by an injective map. For instance, by Fialkow [146], or Hoover [182], there exist bounded linear operators U, V, B on a Hilbert space such that $BU = UV$, B and its Hilbert adjoint B' are injective, V is quasi-nilpotent and the spectrum of U the unit disc $\mathbf{D}(0, 1)$. Let $T := V'$, $S := U'$ and $A := B'$. Then $SA = AT$, so that T and S are intertwined by the injective operator A, S is polaroid, since $\sigma(S) = \overline{\sigma(U)} = \mathbf{D}(0, 1)$ has no isolated points, while T is also quasi-nilpotent and hence is not polaroid.

The next example shows that a polaroid operator may be the quasi-affine transform of an operator which is not polaroid.

Example 4.105 Let $S \in L(\ell^2(\mathbb{N}))$ be the weighted unilateral right shift defined as

$$S(x_1, x_2, \ldots) := \left(0, \frac{x_1}{2}, \frac{x_2}{3}, \ldots\right), \quad (x_n) \in \ell^2(\mathbb{N}),$$

and let $T \in L(\ell^2(\mathbb{N}))$ be the unilateral right shift defined by

$$T(x_1, x_2, \ldots) := (0, x_1, x_2, \ldots), \quad (x_n) \in \ell^2(\mathbb{N}).$$

If $A \in L(\ell^2(\mathbb{N}))$ is the operator defined by

$$A(x_1, x_2, \ldots) := \left(\frac{x_1}{1!}, \frac{x_2}{2!}, \ldots\right), \quad (x_n) \in \ell^2(\mathbb{N}),$$

then A is a quasi-affinity. Clearly, $SA = AT$, T is polaroid, since $\sigma(T)$ is the closed unit disc of $\mathbb{C}$, while S is quasi-nilpotent and hence not polaroid.

The polaroid condition is preserved if we assume some special conditions on the isolated points of the spectrum:

Theorem 4.106 *Let* $T \in L(X)$, $S \in L(Y)$ *be intertwined by an injective map* $A \in L(X, Y)$, *and suppose that*

(i) *If* S *is polaroid and* iso $\sigma(T) \subseteq$ iso $\sigma(S)$ *then* T *is polaroid.*
(ii) *If* S *is meromorphic and* $\sigma(T) \setminus \{0\} \subseteq \sigma(S) \setminus \{0\}$ *then* T *is meromorphic.*

Proof

(i) If $\sigma(T)$ has no isolated point then T is polaroid and hence there is nothing to
   prove. Suppose that iso $\sigma(T) \neq \emptyset$ and let $\lambda \in$ iso $\sigma(T)$. Then $\lambda \in$ iso $\sigma(S)$, so λ
   is a pole of the resolvent of S. Let $P_T(\lambda)$ and $P_S(\lambda)$ be the spectral projections
   associated to T and S with respect to $\{\lambda\}$, respectively. As we have seen in
   Lemma 4.103, $P_T(\lambda)$ and $P_S(\lambda)$ are intertwined by A, i.e. $P_S(\lambda)A = AP_T(\lambda)$.
   Since λ is a pole of the resolvent of S then $p := p(\lambda I - S) = q(\lambda I - S) < \infty$
   and ker$(\lambda I - S)^p$ coincides with the range of $P_S(\lambda)$, by Theorem 2.45 and
   Corollary 2.47. Therefore, $(\lambda I - S)^p P_S(\lambda) = 0$, and consequently

$$0 = (\lambda I - S)^p P_S(\lambda)A = (\lambda I - S)^p AP_T(\lambda) = A(\lambda I - T)^p P_T(\lambda).$$

   Since A is injective, $(\lambda I - T)^p P_T(\lambda) = 0$. Now, the range of $P_T(\lambda)$ coincides
   with the quasi-nilpotent part $H_0(\lambda I - T)$, by Theorem 2.45, hence

$$H_0(\lambda I - T) = P_T(\lambda)(X) \subseteq \ker(\lambda I - T)^p.$$

   The opposite inclusion also holds, since ker$(\lambda I - T)^n \subseteq H_0(\lambda I - T)$ for all
   natural $n \in \mathbb{N}$. Therefore, $H_0(\lambda I - T) = \ker(\lambda I - T)^p$ for all $\lambda \in$ iso $\sigma(T)$.
   By Theorem 4.12 we then conclude that T is polaroid.

(ii) The condition $\sigma(T) \backslash \{0\} \subseteq \sigma(S) \backslash \{0\}$ entails that iso $\sigma(T) \backslash \{0\} \subseteq$ iso $\sigma(S) \backslash \{0\}$.
   If $0 \neq \lambda$ is an isolated point of $\sigma(T)$ then $\lambda \in$ iso $\sigma(S)$ and proceeding as in
   the proof of part (i) we see that $H_0(\lambda I - T) = \ker(\lambda I - T)^p$. This shows that
   λ is a pole of the resolvent of T, see the proof of Theorem 4.12. ∎

The inclusion iso $\sigma(T) \subseteq$ iso $\sigma(S)$ has a crucial role in Theorem 4.106. If T,
S and A are as in Remark 4.104 we have iso $\sigma(S) = \emptyset$, iso $\sigma(T) = \{0\}$ and
the polaroid condition is not preserved by the quasi-affinity A. The example of
Remark 4.104 also shows that the condition iso $\sigma(T) \subseteq$ iso $\sigma(S)$ cannot be replaced
by the weaker condition iso $\sigma(T) \subseteq \sigma(S)$.

Corollary 4.107 *Suppose that $T \in L(X)$ and $S \in L(Y)$ are intertwined by a quasi-*
affinity $A \in L(X, Y)$ and iso $\sigma(T) =$ iso $\sigma(S)$. Then T is polaroid if and only if S
is polaroid.

Proof By Theorem 4.106 we need only to prove that if T is polaroid then S is
polaroid. Now, T^* is polaroid and $SA = AT$ implies $T^*A^* = A^*S^*$, where
$A^* \in L(Y^*, X^*)$ is injective, since A has a dense range. Moreover, iso $\sigma(T^*) =$
iso $\sigma(T) =$ iso $\sigma(S) =$ iso $\sigma(S^*)$. Since T^* is polaroid by Theorem 4.106 it then
follows that S^* is polaroid, or equivalently S is polaroid. ∎

The operator C_p^* considered in Example 2.142 shows that in general a polaroid
operator does not satisfy the SVEP. In fact, $\sigma(C_p^*)$ has no isolated points. A more
trivial example is given by the left shift T on $\ell^2(\mathbb{N})$. This operator is polaroid, since
$\sigma(T)$ is the unit disc of $\mathbb{C}$, and it is well known that T fails SVEP at 0.

The next result shows that hereditarily polaroid operators are transformed, always under the assumption $\operatorname{iso}\sigma(T) \subseteq \operatorname{iso}\sigma(S)$, by quasi-affinities into a-polaroid operators.

Theorem 4.108 *Suppose that $T \in L(X)$, $S \in L(Y)$ are intertwined by an injective map $A \in L(X, Y)$. If S is hereditarily polaroid and $\operatorname{iso}\sigma(T) \subseteq \operatorname{iso}\sigma(S)$ then T^* is a-polaroid.*

Proof By Theorem 4.106 T is polaroid, and hence T^* is also polaroid. As observed above, S has the SVEP, so T has the SVEP by Lemma 2.153. The SVEP for T by Theorem 2.68 entails that $\sigma(T^*) = \sigma(T) = \sigma_s(T) = \sigma_{\mathrm{ap}}(T^*)$, and this trivially implies that T^* is a-polaroid. ∎

Note that quasi-similar operators may have unequal approximate point spectra, for an example see Clary [95].

Theorem 4.109 *Let $T \in L(X)$, $S \in L(Y)$ be intertwined by an injective map $A \in L(X, Y)$ and suppose that $\operatorname{iso}\sigma_{\mathrm{ap}}(T) \subseteq \operatorname{iso}\sigma_{\mathrm{ap}}(S)$. If S is left polaroid then T is polaroid.*

Proof We first show that

$$A(H_0(\lambda I - T)) \subseteq H_0(\lambda I - S).$$

Let $x \in H_0(\lambda I - T)$. Then

$$\lim_{n \to \infty} \|(\lambda I - S)^n A x\|^{1/n} = \lim_{n \to \infty} \|A(\lambda I - T)^n x\|^{1/n}$$
$$\leq \lim_{n \to \infty} \|(\lambda I - T)^n x\|^{1/n} = 0,$$

thus $Ax \in H_0(\lambda I - S)$ and hence $A(H_0(\lambda I - T)) \subseteq H_0(\lambda I - S)$, as claimed.

Here we can also suppose that $\operatorname{iso}\sigma(T) \neq \emptyset$. Let $\lambda \in \operatorname{iso}\sigma(T)$. Since the approximate point spectrum of every operator contains the boundary of the spectrum, in particular every isolated point of the spectrum, $\lambda \in \operatorname{iso}\sigma_a(T) \subseteq \operatorname{iso}\sigma_a(S)$. Since S is left polaroid by part (ii) of Theorem 4.12 there exists a positive integer p such that $H_0(\lambda I - S) = \ker(\lambda I - S)^p$. Consequently,

$$A(H_0(\lambda I - T)) \subseteq H_0(\lambda I - S) = \ker(\lambda I - S)^p,$$

so, if $x \in H_0(\lambda I - T)$ then

$$A(\lambda I - T)^p x = (\lambda I - S)^p (Ax) = 0.$$

Since A is injective, $(\lambda I - T)^p x = 0$ and hence $H_0(\lambda I - T) \subseteq \ker(\lambda I - T)^p$. The opposite inclusion is still true, so that $H_0(\lambda I - T) = \ker(\lambda I - T)^p$ for every $\lambda \in \operatorname{iso}\sigma(T)$, and hence by Theorem 4.12 T is polaroid. ∎

In Theorem 4.109 the assumption that iso $\sigma_{ap}(T) \subseteq$ iso $\sigma_{ap}(S)$ is essential. For the operators S and T of Remark 4.104 we have $\sigma_{ap}(S) = \Gamma$, Γ the unit circle of $\mathbb{C}$, so iso $\sigma_{ap}(S) = \emptyset$, while $\{0\} = \sigma_{ap}(T) =$ iso $\sigma_{ap}(T)$. Evidently, S is both left and a-polaroid, while T is not polaroid.

Theorem 4.110 *Suppose that $S \in L(Y)$ and $T \in L(X)$ are intertwined by a map $A \in L(Y, X)$ which has dense range. If iso $\sigma_s(T) \subseteq$ iso $\sigma_s(S)$ and S is right polaroid then T is polaroid.*

Proof From $TA = AS$ we have $A^*T^* = S^*A^*$ with $A^* \in L(X^*, Y^*)$ injective. Since S is right-polaroid, S^* is left-polaroid and by duality we have $\sigma_s(T) = \sigma_{ap}(T^*)$ and $\sigma_s(S) = \sigma_{ap}(S^*)$. Therefore iso $\sigma_{ap}(T^*) \subseteq$ iso $\sigma_{ap}(S^*)$. By Theorem 4.106 it then follows that T^* is polaroid, or equivalently T is polaroid. ∎

Under the stronger conditions of quasi-similarity and property (β), the assumption on the isolated points of the spectra of T and S in Theorem 4.106 may be omitted:

Theorem 4.111 *Let $T \in L(X)$, $S \in L(Y)$ be quasi-similar.*

(i) *If both T and S have property (β) then T is polaroid if and only if S is polaroid. In this case, T^* is a-polaroid.*

(ii) *If both T and S are Hilbert spaces operators for which property (C) holds then T is polaroid if and only if S is polaroid. In this case, T^* is a-polaroid.* ∎

Proof

(i) By a result of Putinar [259] property (β) preserves the spectrum, i.e. $\sigma(S) = \sigma(T)$ and hence iso $\sigma(T) =$ iso $\sigma(S)$. By Corollary 4.107 we then obtain that T is polaroid exactly when S is polaroid. Evidently, in this case T^* is polaroid. Now, property (β) implies that S has the SVEP and hence, by Lemma 2.153, T also has the SVEP. The SVEP for T, again by Theorem 2.68, entails that $\sigma(T^*) = \sigma_{ap}(T^*)$, and hence T^* is a-polaroid.

(ii) Also in this case, by a result of Stampfli [289], we have $\sigma(S) = \sigma(T)$, and property (C) entails SVEP, so the assertion follows by using the same argument of part (i). ∎

Hyponormal operators on Hilbert spaces have property (β), see [216]. Theorem 4.111 then applies to these operators, since they are $H(1)$ and hence polaroid. Another class of polaroid operators to which Theorem 4.111 applies is the class of all $p_* - QH$ operators studied in [135]. In fact, these operators are $H(1)$ and have property (β), see Duggal and Jeon [135, Theorem 2.12 and Theorem 2.2].

Recall that if $T \in L(X)$ and $S \in L(Y)$, we say that the pair (S, T) is *asymptotically intertwined* by the operator $A \in L(X, Y)$ if $\|C(T, S)(A)\|^{1/n} \to 0$ as $n \to \infty$, where $C(T, S)$ is the commutator introduced in Chap. 2.

Definition 4.112 The pairs (S, T) and (T, S) are said to be *asymptotically quasi-similar* if both are asymptotically intertwined by some quasi-affinity.

The proof of the next theorem may be found in Laursen and Neumann [216, Corollary 3.4.5].

Theorem 4.113 *If a pair (S, T) is asymptotically intertwined by $A \in L(X, Y)$ then*

$$A\mathcal{X}_T(F) \subseteq \mathcal{Y}_S(F) \quad \text{for all closed sets } F \subseteq \mathbb{C}. \tag{4.14}$$

Example 4.114 The polaroid condition is not transmitted whenever S and T are asymptotically intertwined by a quasi-affinity, even in the case when the inclusion $\operatorname{iso}\sigma(T) \subseteq \operatorname{iso}\sigma(S)$ is satisfied. For instance, if $T \in L(\ell^2(\mathbb{N}))$ is defined by

$$T(x_1, x_2, \dots) = \left(\frac{x_2}{2}, \frac{x_3}{3}, \dots\right) \quad \text{for all } (x_n) \in \ell^2(\mathbb{N}).$$

If $S := 0$ then S is polaroid, while the quasi-nilpotent operator T is not polaroid. T and S are, as observed above, quasi-nilpotent equivalent.

Let us now consider the very particular case when $C(S, T)^n(I) = 0$ for some $n \in \mathbb{N}$. If T and S commute then $C(S, T)^n(I) = (S - T)^n = 0$. In this case T and S differ from a commuting nilpotent operator N and, without any condition, if S is polaroid then T is also polaroid, by Lemma 4.23.

Set

$$E_\infty(S) := \{\lambda \in \operatorname{iso}\sigma(S) : \alpha(\lambda I - S) < \infty\}.$$

Theorem 4.115 *Let $T \in L(X)$ and $S \in L(Y)$ be asymptotically intertwined by an injective map $A \in L(X, Y)$ and $\operatorname{iso}\sigma(T) \subseteq E_\infty(S)$. If S is polaroid then T is polaroid.*

Proof

(i) If $\lambda \in \operatorname{iso}(T)$ then $\lambda \in \operatorname{iso}\sigma(T)$. Since S is polaroid it then follows that $H_0(\lambda I - S) = \ker(\lambda I - S)^p$ for some positive integer p. Since $\lambda \in E_\infty(S)$ we have $\alpha(\lambda I - S) < \infty$, so $\alpha((\lambda I - S)^p) < \infty$, and hence $H_0(\lambda I - S)$ is finite-dimensional. From the inclusion (4.14) we have

$$A(H_0(\lambda I - T)) = A(\mathcal{X}_T(\{\lambda\}) \subseteq \mathcal{Y}_S(\{\lambda\}) = H_0(\lambda I - S),$$

and since A is injective it then follows that $H_0(\lambda I - T)$ is finite-dimensional. From the inclusion $\ker(\lambda I - T)^n \subseteq H_0(\lambda I - T)$ for all $n \in \mathbb{N}$ it then easily follows that $p(\lambda I - T) < \infty$. But λ is an isolated point of $\sigma(T)$, so the decomposition $X = H_0(\lambda I - T) \oplus K(\lambda I - T)$ holds, consequently $K(\lambda I - T)$ is finite co-dimensional, and since $K(\lambda I - T) \subseteq (\lambda I - T)(X)$ we then conclude that $\beta(\lambda I - T) < \infty$. Therefore, $\lambda I - T$ is Fredholm. But λ is an isolated point of $\sigma(T^*) = \sigma(T)$, so T^* has the SVEP at λ and, since $\lambda I - T$ is Fredholm, this implies that $q(\lambda I - T) < \infty$. Therefore, λ is a pole of the resolvent of T. ∎

Corollary 4.116 *Suppose that S and T are quasi-nilpotent equivalent. If S is polaroid and every eigenvalue of S has finite multiplicity then T is polaroid.*

Proof The quasi-nilpotent equivalence preserves the spectrum, see the book by Colojoară and Foiaş [98, Chapter 1,Theorem 2.2], hence iso $\sigma(T) = $ iso $\sigma(S)$. Now, if $\lambda \in$ iso $\sigma(S)$ then either $\lambda I - S$ is injective or λ is an eigenvalue of S. In both cases $\lambda \in E_\infty(S)$.   ∎

The Example 4.114 shows that the result of Corollary 4.116 fails if the eigenvalues of S do not have finite multiplicity. Define

$$E_\infty^a(S) := \{\lambda \in \text{iso}\,\sigma_{\text{ap}}(S) : \alpha(\lambda I - S) < \infty\}.$$

Clearly, $E_\infty(S) \subseteq E_\infty^a(S)$.

Theorem 4.117 *Let $T \in L(X)$ and $S \in L(Y)$ be asymptotically intertwined by an injective map $A \in L(X, Y)$ and iso $\sigma(T) \subseteq E_\infty^a(S)$. If S is left polaroid then T is polaroid.*

Proof

(i) If $\lambda \in$ iso $\sigma(T)$ then $\lambda \in$ iso $\sigma_{\text{ap}}(S)$. S is left polaroid so, by part (ii) of Theorem 4.12, there exists a positive integer p such that $H_0(\lambda I - S) = \ker(\lambda I - S)^p$. Since $\alpha(\lambda I - S) < \infty$ it then follows that $H_0(\lambda I - S)$ is finite-dimensional.   ∎

4.9   Comments

The class of polaroid operators was introduced in [138] by Duggal, Harte, and Jeon, while hereditarily polaroid operators were introduced by Duggal in [125], see also [127] and [7].

All the remaining material in the section concerning the perturbation $T + K$ of a hereditarily polaroid operators by an algebraic commuting operator is modeled after Aiena et al. [38], Aiena and Sanabria [24], Aiena et al. [42], and Aiena and Aponte [8]. The results of Theorems 4.19 and 4.21 were proved in Aiena [5]. The SVEP for hereditarily polaroid operators was first proved by Duggal and Djordjević [132, 133]. The class of $H(p)$ operators was introduced by Oudghiri [253], which showed the important fact that every subscalar operator is $H(p)$ for some $p \in \mathbb{N}$. The material concerning multipliers on semi-simple commutative Banach algebras is taken from Aiena and Villafãne [31]. Paranormal operators on Hilbert spaces were studied by Chourasia and Ramanujan [94], who first observed the SVEP for these operators.

All the material contained in the section on weighted shift operators is modeled after Aiena et al.[34]. The class of semi-shifts was introduced in Holub [181] and discussed in Laursen and Vrbová [217], and Laursen et al. [219]. The result that an isometry T is a semi-shift if and only if T has fat local spectra was first observed by Neumann, see [247].

The section concerning Toeplitz operators is modeled after Farenick and Lee [145], Jia and Feng [188], and Aiena and Triolo [29].

The section concerning polaroid operators under affinities is modeled after Aiena et al. [39].

Chapter 5
Browder-Type Theorems

This chapter may be viewed as the part of the book in which the interaction between local spectral theory and Fredholm theory comes into focus. The greater part of the chapter addresses some classes of operators on Banach spaces that have a very special spectral structure. We have seen that the Weyl spectrum $\sigma_w(T)$ is a subset of the Browder spectrum $\sigma_b(T)$ and this inclusion may be proper. In this chapter we investigate the class of operators on complex infinite-dimensional Banach spaces for which the Weyl spectrum and the Browder spectrum coincide. These operators are said to satisfy *Browder's theorem*. The operators which satisfy Browder's theorem have a very special spectral structure, indeed they may be characterized as those operators $T \in L(X)$ for which the spectral points λ that do not belong to the Weyl spectrum are all isolated points of the spectrum.

In some sense Browder's theorem is a local spectral property. Indeed, if T satisfies Browder's theorem then T has the SVEP at every $\lambda \notin \sigma_w(T)$, since λ belongs to the resolvent or is an isolated point of $\sigma(T)$. On the other hand, the converse is still true, by Theorem 2.97. Browder's theorem was introduced by Harte and Lee in [173], in an erroneous description of Weyl's theorem that will be studied in Chap. 6, and then developed a life of its own. We shall also consider classes of operators for which some other parts of the spectrum coincide, and these operators are said to satisfy a Browder-type theorem. Examples of Browder-type theorems are a-Browder's theorem, property (b) and property (ab). All these Browder-type theorems may also be characterized by means of the SVEP for T or T^* at the points of some suitable parts of the spectrum. Other variants of Browder-type theorems may be obtained by considering the B-Fredholm theory, treated in the previous chapters, instead of the classical Fredholm theory. These last variants are, rather improperly, called generalized Browder-type theorems, however it will be shown that Browder's theorem (respectively, a-Browder's theorem) in its generalized form or in its classical form, are equivalent.

© Springer Nature Switzerland AG 2018

P. Aiena, *Fredholm and Local Spectral Theory II*, Lecture Notes in Mathematics 2235, https://doi.org/10.1007/978-3-030-02266-2_5

Some other characterizations of Browder-type theorems, generalized or not, are obtained by showing that the quasi-nilpotent part $H_0(\lambda I - T)$ (respectively, the analytic core $K(\lambda I - T)$) has finite dimension (respectively, has finite codimension), as λ ranges over a certain subset of the spectrum. We also give some perturbation results concerning Browder-type theorems, by using the perturbation results of Chap. 3. In the last part of the chapter we show that Browder-type theorems are transferred from a Drazin invertible operator R to its Drazin inverse.

5.1 Browder's Theorem

In Theorem 3.44 we proved that if T or T^* has the SVEP then the Browder spectra $\sigma_b(T)$, $\sigma_{ub}(T)$ and $\sigma_{lb}(T)$ coincide with the corresponding Weyl spectra $\sigma_w(T)$, $\sigma_{uw}(T)$ and $\sigma_{lw}(T)$, respectively.

Definition 5.1 $T \in L(X)$ is said to satisfy *Browder's theorem* if

$$\sigma_w(T) = \sigma_b(T),$$

or equivalently, by Theorem 3.43, if

$$\mathrm{acc}\,\sigma(T) \subseteq \sigma_w(T). \tag{5.1}$$

Hence the SVEP for either T or T^* entails that both T and T^* satisfy Browder's theorem. However, the following example shows that SVEP for T or T^* is not a necessary condition for Browder's theorem.

Example 5.2 Let $T := L \oplus R \oplus Q$, where L and R are the unilateral left shift and the unilateral right shift on $\ell^2(\mathbb{N})$, respectively, while Q is any quasi-nilpotent operator on $\ell^2(\mathbb{N})$. Now, R is the adjoint L' of L, and analogously L is the adjoint R' of R. Moreover, L does not have the SVEP, see Remark 2.64, and consequently, by Theorem 2.15, both T and T' do not have the SVEP. On the other hand, it is easily seen that $\sigma_b(T) = \sigma_w(T) = \mathbf{D}(0, 1)$, thus Browder's theorem holds for T as well as for T'.

For a bounded operator $T \in L(X)$ let us denote by

$$p_{00}(T) := \sigma(T) \setminus \sigma_b(T) = \{\lambda \in \sigma(T) : \lambda I - T \in \mathcal{B}(X)\}$$

the set of all *Riesz points* in $\sigma(T)$. It is evident, from Theorem 3.7, that $p_{00}(T) = p_{00}(T^*)$. Let us consider the following set:

$$\Delta(T) := \sigma(T) \setminus \sigma_w(T).$$

We begin with an elementary lemma.

Lemma 5.3 *If $T \in L(X)$ then $p_{00}(T) \subseteq \Delta(T)$. Moreover,*

$$\Delta(T) = \{\lambda \in \mathbb{C} : \lambda I - T \in W(X), \ 0 < \alpha(\lambda I - T)\}. \tag{5.2}$$

Proof The inclusion $p_{00}(T) \subseteq \Delta(T)$ is obvious, since every Browder operator is Weyl. The equality (5.2) is clear: if $\lambda \in \Delta(T)$ then $\alpha(\lambda I - T) = \beta(\lambda I - T) > 0$, since $\lambda \in \sigma(T)$. ∎

The following result shows that Browder's theorem is equivalent to the localized SVEP at the points of the complement of $\sigma_w(T)$.

Theorem 5.4 *If $T \in L(X)$ then the following statements are equivalent:*

(i) $p_{00}(T) = \Delta(T)$;
(ii) *T satisfies Browder's theorem;*
(iii) *T^* satisfies Browder's theorem;*
(iv) *T has the SVEP at every $\lambda \notin \sigma_w(T)$;*
(v) *T^* has the SVEP at every $\lambda \notin \sigma_w(T)$.*

Consequently, if either T or T^* has SVEP then Browder's theorem holds for both T and T^*.

Proof (i) $\Rightarrow$ (ii) Suppose that $p_{00}(T) = \Delta(T)$. Let $\lambda \notin \sigma_w(T)$. We show that $\lambda \notin \sigma_b(T)$. Evidently, if $\lambda \notin \sigma(T)$ then $\lambda \notin \sigma_b(T)$. Consider the other case $\lambda \in \sigma(T)$. Then $\lambda \in \Delta(T) = p_{00}(T)$, thus $\lambda \notin \sigma_b(T)$. Hence $\sigma_b(T) \subseteq \sigma_w(T)$. The reverse inclusion is satisfied by every operator, so $\sigma_b(T) = \sigma_w(T)$.

(ii) $\Leftrightarrow$ (iii) Obvious, since $\sigma_b(T) = \sigma_b(T^*)$ and $\sigma_w(T) = \sigma_w(T^*)$.

(ii) $\Rightarrow$ (iv) Suppose that $\sigma_b(T) = \sigma_w(T)$. If $\lambda \notin \sigma_w(T)$ then $\lambda I - T \in B(X)$ so $p(\lambda I - T) < \infty$ and hence T has the SVEP at λ.

(iv) $\Rightarrow$ (v) Suppose that T has the SVEP at every point $\lambda \in \mathbb{C} \setminus \sigma_w(T)$. For every $\lambda \notin \sigma_w(T)$ then $\lambda I - T \in W(X)$, and the SVEP at λ implies that $p(\lambda I - T) < \infty$. Since $\alpha(\lambda I - T) = \beta(\lambda I - T) < \infty$ it then follows, by Theorem 1.22, that $q(\lambda I - T) < \infty$, and consequently T^* has the SVEP at λ.

(v) $\Rightarrow$ (i) Suppose that $\lambda \in \Delta(T)$. We have $\lambda I - T \in W(X)$ and hence $\mathrm{ind}(\lambda I - T) = 0$. By Theorem 2.98 the SVEP of T^* at λ implies that $q(\lambda I - T) < \infty$ and hence, again by Theorem 1.22, $p(\lambda I - T)$ is also finite. Therefore $\lambda \in \sigma(T) \setminus \sigma_b(T) = p_{00}(T)$. This shows that $\Delta(T) \subseteq p_{00}(T)$. By Lemma 5.3 we then conclude that the equality $p_{00}(T) = \Delta(T)$ holds. ∎

Corollary 5.5 *Let $T \in L(X)$. Then we have:*

(i) *If T satisfies Browder's theorem and $R \in L(X)$ is a Riesz operator which commutes with T, then $T + R$ satisfies Browder's theorem.*
(ii) *If T has the SVEP and $K \in L(X)$ is an algebraic operator which commutes with T, then $T + K$ satisfies Browder's theorem.*

(iii) *If T has the SVEP then both $f(T)$ and $f(T^*)$ satisfy Browder's theorem for every $f \in \mathcal{H}(\sigma(T))$.*
(iv) *If $T \in L(X)$ has the SVEP and T and $S \in L(Y)$ are intertwined by an injective map $A \in L(X, Y)$ then S satisfies Browder's theorem.*

Proof (i) follows from Theorems 2.129 and 5.4. (ii) follows from Theorem 2.145. (iii) By Theorem 2.86 $f(T)$ has the SVEP. (iv) S has the SVEP by Lemma 2.141. ∎

Theorem 5.6 *Let $T \in L(X)$ and suppose that $\operatorname{int}\sigma_w(T) = \emptyset$. Then the following statements are equivalent:*

 (i) *T satisfies Browder's theorem;*
 (ii) *T has SVEP;*
(iii) *T^* satisfies Browder's theorem;*
(iv) *T^* has SVEP.*

Proof (i) ⇔ (ii) Browder's theorem for T is equivalent to the SVEP of T at the points $\lambda \notin \sigma_w(T)$, by Theorem 5.4. Let $\lambda_0 \in \sigma_w(T)$. Then $\lambda_0 \notin \operatorname{int}\sigma_w(T)$, since the last set is empty. Hence $\lambda_0 \in \partial\sigma_w(T)$, the boundary of $\sigma_w(T)$. Let $(\lambda I - T)f(\lambda) = 0$ for all λ in an open disc $\mathbb{D}_{\lambda_0}$ centered at λ. Obviously, the disc $\mathbb{D}_\lambda$ contains a point μ of $\rho_w(T)$, and the SVEP of T at μ entails that $f \equiv 0$ in a suitable open disc $\mathbb{D}_\mu$ centered at μ, contained in $\mathbb{D}_{\lambda_0}$. From the identity theorem for analytic functions it then follows that $f \equiv 0$ in $\mathbb{D}_{\lambda_0}$, so T also has the SVEP at the points $\lambda \in \sigma_w(T)$. This shows the implication (i) ⇒ (ii). The reverse implication follows from Theorem 5.4. The statements (i) and (iii) are equivalent, by Theorem 5.4. Since $\operatorname{int}\sigma_w(T) = \operatorname{int}\sigma_w(T^*) = \emptyset$, the previous argument shows that (iii) ⇔ (iv). ∎

We have seen in Chap. 3 that if T or T^* has the SVEP then the spectral mapping theorem holds for the Weyl spectrum $\sigma_w(T)$. In general, Browder's theorem and the spectral mapping theorem are independent. In [173, Example 6] an example is given of an operator T for which the spectral mapping theorem holds for $\sigma_w(T)$ but Browder's theorem fails for T. Another example [173, Example 7] shows that there exist operators for which Browder's theorem holds while the spectral mapping theorem for the Weyl spectrum fails. However we have:

Theorem 5.7 *Let $T \in L(X)$ and $f \in \mathcal{H}(\sigma(T))$. If Browder's theorem holds for $f(T)$ then $f(\sigma_w(T)) = \sigma_w(f(T))$.*

Proof We have, since the spectral theorem holds for the Browder spectrum,

$$\sigma_w(f(T)) = \sigma_b(f(T)) = f(\sigma_b(T)) \supseteq f(\sigma_w(T)).$$

From part (ii) of Theorem 3.115 we then obtain $\sigma_w(f(T)) = f(\sigma_w(T))$. ∎

Part (iii) of Theorem 5.5 may be improved in the case when f is non-constant on each of the components of its domain:

Theorem 5.8 *Suppose that $T \in L(X)$ satisfies Browder's theorem and $f \in \mathcal{H}_c(\sigma(T))$. Then Browder's theorem holds for $f(T)$.*

Proof Suppose that $f \in \mathcal{H}_c(\sigma(T))$ and $f(\lambda_0) \in \sigma(f(T)) \setminus \sigma_w(f(T))$. Then there is a $\nu \in \mathbb{N}$ and two polynomials h and g in $H(\sigma(T))$ with no zero in $\sigma(T)$, such that

$$f(\lambda) - f(\lambda_0) = (\lambda_0 - \lambda)^\nu h(\lambda) g(\lambda),$$

with $h(\lambda_0) \neq 0$ and $h(\lambda_0) \notin g(\sigma(T))$. It then follows that

$$f(T) - f(\lambda_0 I) = (\lambda_0 I - T)^\nu h(T) g(T) \in W(X),$$

with $0 \notin \sigma(h(T)g(T))$ and, consequently, $\lambda_0 \notin \sigma_w(T)$. By Theorem 5.4 T has the SVEP at λ_0 and, by Theorem 2.88, $f(T)$ has the SVEP at $f(\lambda_0)$. This implies Browder's theorem for $f(T)$. ∎

Remark 5.9 It is easily seen that in the case of $f \in H(\sigma(T))$ the result of Theorem 5.8 remains valid if we assume that T has the SVEP at every $\lambda \in \sigma(T) \setminus \sigma_w(T)$, or f is injective.

Let us write iso K for the set of all isolated points of $K \subseteq \mathbb{C}$. A very clear spectral picture of operators for which Browder's theorem holds is given by the following theorem:

Theorem 5.10 *For an operator $T \in L(X)$ the following statements are equivalent:*

(i) *T satisfies Browder's theorem;*
(ii) *Every $\lambda \in \Delta(T)$ is an isolated point of $\sigma(T)$;*
(iii) *$\Delta(T) \subseteq \partial\sigma(T)$, $\partial\sigma(T)$ the topological boundary of $\sigma(T)$;*
(iv) *int $\Delta(T) = \emptyset$;*
(v) *$\sigma(T) = \sigma_w(T) \cup iso\,\sigma(T)$.*

Proof (i) $\Rightarrow$ (ii) If T satisfies Browder's theorem then $\Delta(T) = p_{00}(T)$, and in particular every $\lambda \in \Delta(T)$ is an isolated point of $\sigma(T)$.

(ii) $\Rightarrow$ (iii) Obvious.

(iii) $\Rightarrow$ (iv) Clear, since int $\partial\sigma(T) = \emptyset$.

(iv) $\Rightarrow$ (v) Suppose that int $\Delta(T) = \emptyset$. Let $\lambda_0 \in \Delta(T) = \sigma(T) \setminus \sigma_w(T)$. We show first that $\lambda_0 \in \partial\sigma(T)$. Suppose that $\lambda_0 \notin \partial\sigma(T)$. Then there exists an open disc centered at λ_0 contained in the spectrum. Since $\lambda_0 I - T \in W(X)$ by the classical punctured neighborhood theorem there exists another open disc $\mathbb{D}$ centered at λ_0 such that $\lambda I - T \in W(X)$ for all $\lambda \in \mathbb{D}$. Therefore $\lambda_0 \in$ int $\Delta(T)$, which is impossible. This argument shows that $\sigma(T) = \sigma_w(T) \cup \partial\sigma(T)$.

Now, if $\lambda \in \partial\sigma(T)$ and $\lambda \notin \sigma_w(T)$ then $\lambda I - T \in W(X)$ and, since both T and T^* have the SVEP at every point of $\partial\sigma(T) = \partial\sigma(T^*)$, by Theorems 2.97 and 2.98 we then have $p(\lambda I - T) = q(\lambda I - T) < \infty$. Therefore λ is an isolated point of $\sigma(T)$, and consequently $\sigma(T) = \sigma_w(T) \cup \text{iso}\,\sigma(T)$.

(v) $\Rightarrow$ (i) Suppose that $\sigma(T) = \sigma_w(T) \cup \text{iso}\,\sigma(T)$. Suppose that $\lambda \notin \sigma(T) \setminus \sigma_w(T)$. Then $\lambda \in \text{iso}\,\sigma(T)$ (otherwise, $\lambda \notin \sigma_w(T) \cup \text{iso}\,\sigma(T) = \sigma(T)$, a contradiction). Since T and T^* have the SVEP at every isolated point of $\sigma(T)$ and $\lambda I - T \in W(X)$ it then follows, by Theorems 2.97 and 2.98, that $p(\lambda I - T) = q(\lambda I - T) < \infty$, so $\lambda \notin \sigma_b(T)$. Therefore $\sigma_b(T) = \sigma_w(T)$. ∎

Let M, N denote two closed linear subspaces of a Banach space X. In Chap. 1 we have observed that the gap $\widehat{\delta}(M, N)$ is a metric on the set of all linear closed subspaces of X, and the convergence $M_n \to M$ is defined by $\widehat{\delta}(M_n, M) \to 0$ as $n \to \infty$. Browder's theorem may be characterized by means of the discontinuity of certain mappings:

Theorem 5.11 *For a bounded operator $T \in L(X)$ the following statements are equivalent:*

(i) *T satisfies Browder's theorem;*
(ii) *the mapping $\lambda \to \ker(\lambda I - T)$ is not continuous at every $\lambda \in \Delta(T)$ in the gap metric;*
(iii) *the mapping $\lambda \to \gamma(\lambda I - T)$ is not continuous at every $\lambda \in \Delta(T)$;*
(iv) *the mapping $\lambda \to (\lambda I - T)(X)$ is not continuous at every $\lambda \in \Delta(T)$ in the gap metric.*

Proof (i) $\Rightarrow$ (ii) By Theorem 5.10 if T satisfies Browder's theorem then $\Delta(T) \subseteq \text{iso}\,\sigma(T)$. For every $\lambda_0 \in \Delta(T)$ we have $\alpha(\lambda_0 I - T) > 0$ and since λ_0 is an isolated point of $\sigma(T)$ there exists an open disc $\mathbb{D}(\lambda_0, \varepsilon)$ such that $\alpha(\lambda I - T) = 0$ for all $\lambda \in \mathbb{D}(\lambda_0, \varepsilon) \setminus \{\lambda_0\}$. Therefore the mapping $\lambda \to \ker(\lambda I - T)$ is not continuous at λ_0 in the gap metric.

(ii) $\Rightarrow$ (i) Let $\lambda_0 \in \Delta(T)$ be arbitrary. By the punctured neighborhood theorem there exists an open disc $\mathbb{D}(\lambda_0, \varepsilon)$ such that $\lambda I - T \in \Phi(X)$ for all $\lambda \in \mathbb{D}(\lambda_0, \varepsilon)$, $\alpha(\lambda I - T)$ is constant as λ ranges on $\mathbb{D}(\lambda_0, \varepsilon) \setminus \{\lambda_0\}$,

$$\text{ind}(\lambda I - T) = \text{ind}(\lambda_0 I - T) \quad \text{for all} \quad \lambda \in \mathbb{D}(\lambda_0, \varepsilon),$$

and

$$0 \leq \alpha(\lambda I - T) \leq \alpha(\lambda_0 I - T) \quad \text{for all } \lambda \in \mathbb{D}(\lambda_0, \varepsilon).$$

The discontinuity of the mapping $\lambda \to \ker(\lambda I - T)$ at every $\lambda \in \Delta(T)$ implies that

$$0 \leq \alpha(\lambda I - T) < \alpha(\lambda_0 I - T) \quad \text{for all } \lambda \in \mathbb{D}(\lambda_0, \varepsilon) \setminus \{\lambda_0\}.$$

We claim that $\alpha(\lambda I - T) = 0$ for all $\lambda \in \mathbb{D}(\lambda_0, \varepsilon) \setminus \{\lambda_0\}$. To see this, suppose that there is a $\lambda_1 \in \mathbb{D}(\lambda_0, \varepsilon) \setminus \{\lambda_0\}$ such that $\alpha(\lambda_1 I - T) > 0$. Clearly, $\lambda_1 \in \Delta(T)$, so arguing as for λ_0 we obtain a $\lambda_2 \in \mathbb{D}(\lambda_0, \varepsilon) \setminus \{\lambda_0, \lambda_1\}$ such that

$$0 < \alpha(\lambda_2 I - T) < \alpha(\lambda_1 I - T),$$

and this is impossible since $\alpha(\lambda I - T)$ is constant for all $\lambda \in \mathbb{D}(\lambda_0, \varepsilon) \setminus \{\lambda_0\}$. Therefore $0 = \alpha(\lambda I - T)$ for $\lambda \in \mathbb{D}(\lambda_0, \varepsilon) \setminus \{\lambda_0\}$, and since $\lambda I - T \in W(X)$ for all $\lambda \in \mathbb{D}(\lambda_0, \varepsilon)$ we conclude that $\alpha(\lambda I - T) = \beta(\lambda I - T) = 0$ for all $\lambda \in \mathbb{D}(\lambda_0, \varepsilon) \setminus \{\lambda_0\}$. Hence $\lambda_0 \in \text{iso}\,\sigma(T)$, thus T satisfies Browder's theorem by Theorem 5.10.

To show the equivalences of the assertions (ii), (iii) and (iv) observe first that for every $\lambda_0 \in \Delta(T)$ we have $\lambda_0 I - T \in \Phi(X)$, and hence the range $(\lambda I - T)(X)$ is closed for all λ near to λ_0. The equivalences (ii) $\Leftrightarrow$ (iii) $\Leftrightarrow$ (iv) then follow from Theorem 1.51. ∎

Recall that a bounded operator $T \in L(X)$ is said be *relatively regular* if there exists an $S \in L(X)$ such that $TST = T$. It is well known that every Fredholm operator is relatively regular, see Appendix A. A "complemented" version of Kato operators is given by the *Saphar operators*, where $T \in L(X)$ is said to be *Saphar* if T is semi-regular and relatively regular. The *Saphar spectrum*, already introduced in Chap. 3, is denoted by $\sigma_{\text{sa}}(T)$. Clearly, $\sigma_{\text{se}}(T) \subseteq \sigma_{\text{sa}}(T)$.

Theorem 5.12 *For a bounded operator T each of the following statements is equivalent to Browder's theorem:*

(i) $\Delta(T) \subseteq \sigma_{\text{se}}(T)$;
(ii) $\Delta(T) \subseteq \text{iso}\,\sigma_{\text{se}}(T)$;
(iii) $\Delta(T) \subseteq \sigma_{\text{sa}}(T)$;
(iv) $\Delta(T) \subseteq \text{iso}\,\sigma_{\text{sa}}(T)$.

Proof By Theorem 1.51 the equivalent conditions of Theorem 5.11 are equivalent to saying that $\lambda I - T$ is not semi-regular for all $\lambda \in \Delta(T)$.

(i) $\Leftrightarrow$ (ii) The implication (ii) $\Rightarrow$ (i) is obvious. To show that (i) $\Rightarrow$ (ii) suppose that $\Delta_a(T) \subseteq \sigma_{\text{se}}(T)$. If $\lambda_0 \in \Delta_a(T)$ then $\lambda_0 I - T \in \Phi_+(X)$ so $\lambda_0 I - T$ is essentially semi-regular, in particular of Kato-type. By Theorem 1.65 then there exists an open disc $\mathbb{D}(\lambda_0, \varepsilon)$ such that $\lambda I - T$ is semi-regular for all $\lambda \in \mathbb{D}(\lambda_0, \varepsilon) \setminus \{\lambda_0\}$. But $\lambda_0 \in \sigma_{\text{se}}(T)$, so $\lambda_0 \in \text{iso}\,\sigma_{\text{se}}(T)$.

(i) $\Leftrightarrow$ (iii) The implication (i) $\Rightarrow$ (iii) is immediate, since $\sigma_{\text{se}}(T) \subseteq \sigma_{\text{sa}}(T)$.

To show the implication (iii) $\Rightarrow$ (i) suppose that $\Delta(T) \subseteq \sigma_{\text{sa}}(T)$. Let $\lambda \in \Delta(T)$. Then $\alpha(\lambda I - T) < \infty$ and since $\lambda I - T \in W(X)$ it follows that $\beta(\lambda I - T) < \infty$. Clearly, $\ker(\lambda I - T)$ is complemented, since it is finite-dimensional, and $(\lambda I - T)(X)$ is complemented, since it is closed and finite-codimensional. Therefore T is relatively regular, and from $\lambda \in \sigma_{\text{sa}}(T)$ it then follows that $\lambda I - T$ is not semi-regular. Thus $\Delta(T) \subseteq \sigma_{\text{se}}(T)$.

(iv) $\Rightarrow$ (iii) Obvious.

(ii) $\Rightarrow$ (iv) Let $\lambda_0 \in \Delta(T)$. Since $\lambda_0 I - T \in W(X)$, then there exists an open disc $\mathbb{D}(\lambda_0, \varepsilon)$ centered at λ_0 such that $\lambda I - T \in W(X)$ for all $\lambda \in (\lambda_0, \varepsilon)$, so $\lambda I - T$ is Fredholm, and hence is relatively regular for all $\lambda \in \mathbb{D}(\lambda_0, \varepsilon)$. On the other hand, λ_0 is isolated in $\sigma_{\text{se}}(T)$, so $\lambda_0 \in \text{iso}\,\sigma_{\text{sa}}(T)$. ∎

We now establish some characterizations of operators satisfying Browder's theorem in terms of the quasi-nilpotent parts $H_0(\lambda I - T)$.

Theorem 5.13 *For a bounded operator $T \in L(X)$ the following statements are equivalent:*

(i) *Browder's theorem holds for T;*

(ii) *For every $\lambda \in \Delta(T)$ there exists a $p := p(\lambda I - T)$ such that $H_0(\lambda I - T) = \ker(\lambda I - T)^p$;*

(iii) *$H_0(\lambda I - T)$ is finite-dimensional for every $\lambda \in \Delta(T)$;*

(iv) *$H_0(\lambda I - T)$ is closed for all $\lambda \in \Delta(T)$;*

(v) *$K(\lambda I - T)$ has finite-codimension for all $\lambda \in \Delta(T)$.*

Proof (i) $\Rightarrow$ (ii) Suppose that T satisfies Browder's theorem. By Theorem 5.4 then $\Delta(T) = p_{00}(T) = \sigma(T) \setminus \sigma_b(T)$. If $\lambda \in \Delta(T)$ then $\lambda I - T \in B(X)$, so λ is isolated in $\sigma(T)$ and hence T has the SVEP at λ. From Theorem 2.105 we then conclude that $H_0(\lambda I - T)$ is finite-dimensional.

(ii) $\Rightarrow$ (iii) If $\lambda \in \Delta(T) = \sigma(T) \setminus \sigma_w(T)$, then $\lambda I - T$ is upper semi-Fredholm and hence $(\lambda I - T)^p$ is upper semi-Fredholm, so $\alpha(\lambda I - T)^p < \infty$. Consequently, $H_0(\lambda I - T) = \ker(\lambda I - T)^p$ is finite-dimensional.

(iii) $\Rightarrow$ (iv) Clear.

(iv) $\Rightarrow$ (i) Suppose that $H_0(\lambda I - T)$ is closed for all $\lambda \in \Delta(T)$. Then T has the SVEP at λ, by Theorem 2.39, and hence, by Theorem 2.97, we have $p(\lambda I - T) < \infty$. Since $\lambda I - T$ is Weyl, we then have $\lambda I - T \in B(X)$, by Theorem 1.22, hence $\lambda \in p_{00}(T)$. This shows the inclusion $\Delta(T) \subseteq p_{00}(T)$ and since the reverse inclusion holds for every operator, we then have $\Delta(T) = p_{00}(T)$, so T satisfies Browder's theorem by Theorem 5.4.

(i) $\Rightarrow$ (v) By Theorem 5.10 every $\lambda \in \Delta(T)$ is an isolated point of $\sigma(T)$ and by the first part of the proof $H_0(\lambda I - T)$ is finite-dimensional. By Theorem 2.45, $X = H_0(\lambda I - T) \oplus K(\lambda I - T)$, so $K(\lambda I - T)$ has finite codimension.

(v) $\Rightarrow$ (i) Suppose that $K(\lambda I - T)$ has finite codimension for every $\lambda \in \Delta(T)$. Because $K(\lambda I - T) \subseteq (\lambda I - T)^n(X)$ for each $n \in \mathbb{N}$, it then follows that $q(\lambda I - T) < \infty$, so T^* has the SVEP at every $\lambda \in \Delta(T)$, and hence Browder's theorem holds for T, by Theorem 5.4.   ■

A natural generalization of Browder's theorem is suggested by considering B-Fredholm theory instead of Fredholm theory. In other words, by considering those operators $T \in L(X)$ for which the equality

$$\sigma_{bw}(T) = \sigma_d(T) \tag{5.3}$$

holds. Recall that by Corollary 3.49 we have $\sigma_d(T) = \sigma_{bb}(T)$. By Theorem 3.50 the equality (5.3) holds if and only if

$$\mathrm{acc}\,\sigma(T) \subseteq \sigma_{bw}(T). \tag{5.4}$$

The operators $T \in L(X)$ which satisfy the equality (5.3) have been investigated for some time by several authors, and these operators are said to satisfy the *generalized Browder's theorem*. Since $\sigma_{\text{bw}}(T) \subseteq \sigma_{\text{w}}(T)$ holds for every $T \in L(X)$, from the inclusion (5.4) we obtain acc $\sigma(T) \subseteq \sigma_{\text{w}}(T)$, which means that T satisfies Browder's theorem. Hence the generalized Browder's theorem apparently seems to be a stronger property than Browder's theorem. Next we show that this is not true. First we need to characterize the equality (5.3) by the localized SVEP at the points $\lambda \notin \sigma_{\text{bw}}(T)$.

Theorem 5.14 *Let $T \in L(X)$. Then the following statements are equivalent:*

(i) $\sigma_{\text{bw}}(T) = \sigma_{\text{d}}(T)$, *i.e. the generalized Browder's theorem holds for T ;*
(ii) *T has the SVEP at every $\lambda \notin \sigma_{\text{bw}}(T)$;*
(iii) *T^* has the SVEP at every $\lambda \notin \sigma_{\text{bw}}(T)$.*

Proof (i) $\Leftrightarrow$ (ii) If $\sigma_{\text{d}}(T) = \sigma_{\text{bw}}(T)$, then, by Theorem 2.97, T has the SVEP at the points $\lambda \notin \sigma_{\text{bw}}(T)$.

Conversely, assume that T has the SVEP at every point $\lambda \notin \sigma_{\text{bw}}(T)$. If $\lambda \notin \sigma_{\text{bw}}(T)$ then, by Corollary 2.107, $\lambda \notin \sigma_{\text{d}}(T)$. This shows that $\sigma_{\text{d}}(T) \subseteq \sigma_{\text{bw}}(T)$. On the other hand, by Theorem 1.141 we have $\sigma_{\text{bw}}(T) \subseteq \sigma_{\text{d}}(T)$, for all operators $T \in L(X)$, thus $\sigma_{\text{bw}}(T) = \sigma_{\text{d}}(T)$, and hence T satisfies the generalized Browder's theorem.

(i) $\Leftrightarrow$ (iii) If $\sigma_{\text{d}}(T) = \sigma_{\text{bw}}(T)$ then $q(\lambda I - T) < \infty$ for every $\lambda \notin \sigma_{\text{bw}}(T)$, so T^* has the SVEP at every $\lambda \notin \sigma_{\text{bw}}(T)$, by Theorem 2.98. Conversely, if $\lambda \notin \sigma_{\text{bw}}(T)$, the SVEP for T^* at λ entails, by Corollary 2.107, that $\lambda \notin \sigma_{\text{d}}(T)$, hence $\sigma_{\text{d}}(T) \subseteq \sigma_{\text{bw}}(T)$. The opposite inclusion is always true, thus $\sigma_{\text{d}}(T) = \sigma_{\text{d}}(T)$.

It is perhaps surprising, and somewhat unexpected, that the two concepts of Browder's theorem and the generalized Browder's theorem are equivalent:

Theorem 5.15 *If $T \in L(X)$ the following statements are equivalent:*

(i) $\sigma_{\text{w}}(T) = \sigma_{\text{b}}(T)$;
(ii) $\sigma_{\text{bw}}(T) = \sigma_{\text{d}}(T)$.

Consequently, for an operator $T \in L(X)$, Browder's theorem and the generalized Browder's theorem are equivalent.

Proof Suppose that $\sigma_{\text{w}}(T) = \sigma_{\text{b}}(T)$. Since, by Theorem 1.141, $\sigma_{\text{bw}}(T) \subseteq \sigma_{\text{d}}(T) =$ for all $T \in L(X)$, we need only show the opposite inclusion. Assume that $\lambda_0 \notin \sigma_{\text{bw}}(T)$, i.e. that $\lambda_0 I - T$ is B-Weyl. By Theorem 1.117, there exists an open disc $\mathbb{D}(\lambda_0, \varepsilon)$ such that $\lambda I - T$ is Weyl and hence Browder for all $\lambda \in \mathbb{D}(\lambda_0, \varepsilon) \setminus \{\lambda_0\}$. The condition $p(\lambda I - T) = q(\lambda I - T) < \infty$ implies that both T and T^* have the SVEP at every $\lambda \in \mathbb{D} \setminus \{\lambda_0\}$, so both T and T^* have the SVEP at λ_0. Since every B-Fredholm operator has topological uniform descent, by Theorem 2.97 and Theorem 2.98, we then have that $\lambda_0 I - T$ is Drazin invertible, i.e. $\lambda_0 \notin \sigma_d(T)$. Hence $\sigma_{bw}(T) = \sigma_d(T)$.

Conversely, suppose that $\sigma_{bw}(T) = \sigma_d(T)$. By Theorem 5.14, T has the SVEP at every $\lambda \notin \sigma_{bw}(T)$ and in particular T has the SVEP for every $\lambda \notin \sigma_w(T)$, since $\sigma_{bw}(T) \subseteq \sigma_w(T)$. Therefore, by Theorem 5.4, $\sigma_w(T) = \sigma_b(T)$. ∎

Define

$$\Delta^g(T) := \sigma(T) \setminus \sigma_{bw}(T),$$

and let

$$\Pi(T) := \sigma(T) \setminus \sigma_d(T)$$

be the set of all poles of the resolvent (no restriction on rank).

Lemma 5.16 *If* $T \in L(X)$ *then*

$$\Delta^g(T) = \{\lambda \in \mathbb{C} : \lambda I - T \text{ is B-Weyl and } 0 < \alpha(\lambda I - T)\}. \tag{5.5}$$

Furthermore, $\Pi(T) \subseteq \Delta^g(T)$ *and* $\Delta(T) \subseteq \Delta^g(T)$.

Proof The inclusion

$$\{\lambda \in \mathbb{C} : \lambda I - T \text{ is B-Weyl and } 0 < \alpha(\lambda I - T)\} \subseteq \Delta^g(T)$$

is obvious. To show the opposite inclusion, suppose that $\lambda \in \Delta^g(T)$. There is no harm in assuming $\lambda = 0$. Since $0 \in \sigma(T)$ and T is B-Weyl, hence, by Theorem 1.119, $T = T_1 \oplus T_2$, where T_1 is Weyl and T_2 is nilpotent. If $\alpha(T) = 0$ then $\alpha(T_1) = 0$ and since T_1 is Weyl then $\alpha(T_1) = \beta(T_1) = 0$ so T_1 is invertible and hence T is Drazin invertible, i.e., $p(T) = q(T) < \infty$. But this implies, by Theorem 1.22, that $0 = \alpha(T) = \beta(T)$, so $0 \notin \sigma(T)$, a contradiction. Therefore $\alpha(T) > 0$, so that $\Delta^g(T)$ is contained in the set on the right-hand side of (5.5). Therefore the equality (5.5) holds.

To show the inclusion $\Pi(T) \subseteq \Delta^g(T)$, let us assume that $\lambda \in \Pi(T) = \sigma(T) \setminus \sigma_d(T)$. Then $\lambda I - T$ is Drazin invertible, and since λ is a pole, λ is an isolated point of $\sigma(T)$. Moreover, by Theorem 1.141, $\lambda I - T$ is B-Weyl. We also have $0 < \alpha(\lambda I - T)$ (otherwise from $p(\lambda I - T) = q(\lambda I - T) < \infty$ we would obtain $\alpha(\lambda I - T) = \beta(\lambda I - T) = 0$, i.e. $\lambda \notin \sigma(T)$), hence $\Pi(T) \subseteq \Delta^g(T)$.

The inclusion $\Delta(T) \subseteq \Delta^g(T)$ is obvious, since $\sigma_{bw}(T) \subseteq \sigma_w(T)$. ∎

The following theorem is an improvement of Theorem 5.10.

Theorem 5.17 *For a bounded operator* $T \in L(X)$ *the following statements are equivalent:*

(i) *T satisfies Browder's theorem;*
(ii) $\Delta^g(T) = \Pi(T)$;
(iii) $\Delta^g(T) \subseteq \mathrm{iso}\,\sigma(T)$;
(iv) $\Delta^g(T) \subseteq \partial\sigma(T)$, $\partial\sigma(T)$ *the topological boundary of* $\sigma(T)$;
(v) $\mathrm{int}\,\Delta^g(T) = \emptyset$;
(vi) $\sigma(T) = \sigma_{bw}(T) \cup \partial\sigma(T)$;
(vii) $\sigma(T) = \sigma_{bw}(T) \cup \mathrm{iso}\,\sigma(T)$.

Proof (i) $\Rightarrow$ (ii) By Theorem 5.15 we have $\sigma_{bw}(T) = \sigma_d(T)$, and hence $\Delta^g(T) = \sigma(T) \setminus \sigma_{bw}(T) = \sigma(T) \setminus \sigma_d(T) = \Pi(T)$.

(ii) $\Rightarrow$ (iii) Clear, since $\Delta^g(T) = \Pi(T) \subseteq \mathrm{iso}\,\sigma(T)$.

(iii) $\Rightarrow$ (iv) Obvious.

(iv) $\Rightarrow$ (v) Clear, since $\mathrm{int}\,\partial\sigma(T) = \emptyset$.

(v) $\Rightarrow$ (vi) Suppose that $\mathrm{int}\,\Delta^g(T) = \emptyset$. Let $\lambda_0 \in \Delta^g(T) = \sigma(T) \setminus \sigma_{bw}(T)$ and suppose that $\lambda_0 \notin \partial\sigma(T)$. Then there exists an open disc $\mathbb{D}(\lambda_0, \varepsilon)$ centered at λ_0 contained in $\sigma(T)$. Since $\lambda_0 I - T$ is B-Weyl there exists, by Theorem 1.117, a punctured open disc $\mathbb{D}_1$ contained in $\mathbb{D}$ such that $\lambda I - T$ is Weyl for all $\lambda \in \mathbb{D}_1$. Clearly, $0 < \alpha(\lambda I - T)$ for all $\lambda \in \mathbb{D}_1$ (otherwise we would have $0 = \alpha(\lambda I - T) = \beta(\lambda I - T)$), hence by Lemma 5.16 λ_0 belongs to $\mathrm{int}\,\Delta^g(T)$, and this is a contradiction since $\mathrm{int}\,\Delta^g(T) = \emptyset$. This shows that $\sigma(T) = \sigma_{bw}(T) \cup \partial\sigma(T)$, as desired.

(vi) $\Rightarrow$ (vii) If $\lambda \in \partial\sigma(T)$ and $\lambda \notin \sigma_{bw}(T)$ then $\lambda I - T$ is B-Weyl and T has the SVEP at λ. By Theorem 2.107 it then follows that $\lambda I - T$ is Drazin invertible, i.e. $0 < p(\lambda I - T) = q(\lambda I - T) < \infty$ and hence λ is an isolated point of the spectrum. Therefore, $\sigma(T) = \sigma_{bw}(T) \cup \mathrm{iso}\,\sigma(T)$.

(vii) $\Rightarrow$ (i) Suppose that $\sigma(T) = \sigma_{bw}(T) \cup \mathrm{iso}\,\sigma(T)$. Let $\lambda \notin \sigma_{bw}(T)$. If $\lambda \notin \sigma(T)$ then $\lambda \notin \sigma_d(T)$. Suppose that $\lambda \in \sigma(T)$. Then $\lambda \in \sigma(T) \setminus \sigma_{bw}(T)$ and hence $\lambda \in \mathrm{iso}\,\sigma(T)$. This implies that T has the SVEP at λ. Since $\lambda I - T$ is B-Weyl it then follows, by Theorem 1.141, that $\lambda I - T$ is Drazin invertible and hence $\lambda \notin \sigma_d(T)$. This proves the inclusion $\sigma_d(T) \subseteq \sigma_{bw}(T)$. Since the opposite inclusion is satisfied by every operator we then conclude that $\sigma_d(T) = \sigma_{bw}(T)$, so that T satisfies the generalized Browder's theorem, or equivalently, Browder's theorem. ∎

Corollary 5.18 *If either T or T^* has the SVEP and* $\mathrm{iso}\,\sigma(T) = \emptyset$ *then* $\sigma(T) = \sigma_w(T) = \sigma_{bw}(T)$.

Proof By Theorem 5.17 we have $\emptyset = \Delta^g(T) = \sigma(T) \setminus \sigma_{bw}(T)$. Hence $\sigma(T) = \sigma_{bw}(T)$ and, obviously, these spectra coincide with $\sigma_w(T)$, since $\sigma_{bw}(T) \subseteq \sigma_w(T) \subseteq \sigma(T)$. ∎

Theorem 5.19 *Suppose that* $T^\infty(X) = 0$. *If T is not nilpotent then*

$$\sigma_{bw}(T) = \sigma_d(T) = \sigma(T). \tag{5.6}$$

Proof The condition $T^\infty(X) = 0$ entails that T satisfies SVEP. This may be proved in several ways, for instance, since $\ker(\lambda I - T) \subseteq T^\infty(X)$ for all $\lambda \neq 0$, the condition $T^\infty(X) = \{0\}$ entails that $\ker(\lambda I - T) = 0$ for every $\lambda \neq 0$. Therefore,

$$\ker(\lambda I - T)^n \cap (\lambda I - T)^\infty(X) = \{0\} \quad \text{for all } \lambda \in \mathbb{C}.$$

Since $K(\lambda I - T) \subseteq (\lambda I - T)^\infty(X)$ we then obtain $\mathcal{N}^\infty(\lambda_0 I - T) \cap K(\lambda_0 I - T) = \{0\}$, and hence T has the SVEP, by Corollary 2.66. Therefore, T satisfies the generalized Browder's theorem and hence $\sigma_{bw}(T) = \sigma_d(T)$. Suppose that T is quasi-nilpotent, but not nilpotent. Then $X = H_0(T) \neq \ker T^p$ for each $p \in \mathbb{N}$, so 0 cannot be a pole of the resolvent by Theorem 2.47. Hence $\sigma_{bw}(T) = \sigma_d(T) = \sigma(T) = \{0\}$.

Suppose that T is not quasi-nilpotent. Then $q(\lambda I - T) = \infty$ for all $\lambda \in \sigma(T) \setminus \{0\}$, otherwise, since $p(\lambda I - T) = 0$ for all $\lambda \neq 0$, we would have $q(\lambda I - T) = p(\lambda I - T) = 0$, hence $\lambda \notin \sigma(T)$. Therefore,

$$\sigma(T) \setminus \{0\} \subseteq \sigma_d(T) = \sigma_{bw}(T) \subseteq \sigma(T).$$

On the other hand, we also have $q(T) = \infty$, because if $q(T) < \infty$ we would have $T^q(X) = T^\infty(X) = \{0\}$ and hence T is nilpotent. Therefore $0 \in \sigma_d(T)$, from which we may conclude that the equalities (5.6) hold.   ∎

Remark 5.20 It should be noted that every nilpotent operator T on an infinite-dimensional complex Banach space satisfies the equality $\sigma_d(T) = \sigma_{bw}(T)$, since T has the SVEP, while $T^p(X) = \{0\}$ and $\ker T^p = X$ for some $p \in \mathbb{N}$ entail that $p(T) = q(T) < \infty$ and hence $\sigma_d(T) = \emptyset$. Therefore, for a nilpotent operator the equality (5.6) fails, since $\sigma(T) = \{0\}$.

The next two theorems improve the results of Theorems 5.11 and 5.35.

Theorem 5.21 *For a bounded operator $T \in L(X)$ the following statements are equivalent:*

(i) *T satisfies Browder's theorem;*
(ii) *The mapping $\lambda \to \ker(\lambda I - T)$ is discontinuous at every $\lambda \in \Delta^g(T)$ in the gap metric.*

Proof (i) $\Rightarrow$ (ii) By Theorem 5.17 we have $\Delta^g(T) \subseteq \text{iso}\,\sigma(T)$. If $\lambda_0 \in \Delta^g(T)$ then $\alpha(\lambda_0 I - T) > 0$ and there exists a punctured open disc $\mathbb{D}(\lambda_0)$ centered at λ_0 such that $\alpha(\lambda_0 I - T) = 0$ for all $\lambda \in \mathbb{D}(\lambda_0)$. Hence $\lambda \to \ker(\lambda I - T)$ is discontinuous at λ_0 in the gap metric.

(ii) $\Rightarrow$ (i) Let $\lambda_0 \in \Delta^g(T)$ be arbitrary. Then $\lambda_0 I - T$ is B-Weyl and by Theorem 1.117 we know that there exists an open disc $\mathbb{D}(\lambda_0, \varepsilon)$ such that $\lambda I - T \in \Phi(X)$ for all $\lambda \in \mathbb{D}(\lambda_0, \varepsilon) \setminus \{\lambda_0\}$, $\alpha(\lambda I - T)$ is constant as λ ranges on $\mathbb{D}(\lambda_0, \varepsilon) \setminus \{\lambda_0\}$,

$$\text{ind}(\lambda I - T) = \text{ind}(\lambda_0 I - T) \quad \text{for all} \quad \lambda \in \mathbb{D}(\lambda_0, \varepsilon),$$

and

$$0 \leq \alpha(\lambda I - T) \leq \alpha(\lambda_0 I - T) \quad \text{for all } \lambda \in \mathbb{D}(\lambda_0, \varepsilon).$$

The discontinuity of the mapping $\lambda \to \ker(\lambda I - T)$ at every $\lambda \in \Delta^g(T)$ implies that

$$0 \leq \alpha(\lambda I - T) < \alpha(\lambda_0 I - T) \quad \text{for all } \lambda \in \mathbb{D}(\lambda_0, \varepsilon) \setminus \{\lambda_0\}.$$

We claim that

$$\alpha(\lambda I - T) = 0 \quad \text{for all } \lambda \in \mathbb{D}(\lambda_0, \varepsilon) \setminus \{\lambda_0\}. \tag{5.7}$$

To see this, suppose that there exists a $\lambda_1 \in \mathbb{D}(\lambda_0, \varepsilon) \setminus \{\lambda_0\}$ such that $\alpha(\lambda_1 I - T) > 0$. Clearly, $\lambda_1 \in \Delta(T)$, so arguing as for λ_0 we obtain a $\lambda_2 \in \mathbb{D}(\lambda_0, \varepsilon) \setminus \{\lambda_0, \lambda_1\}$ such that

$$0 < \alpha(\lambda_2 I - T) < \alpha(\lambda_1 I - T),$$

and this is impossible since $\alpha(\lambda I - T)$ is constant for all $\lambda \in \mathbb{D}(\lambda_0, \varepsilon) \setminus \{\lambda_0\}$. Therefore (5.7) is satisfied and, since $\lambda I - T \in W(X)$ for all $\lambda \in \mathbb{D}(\lambda_0, \varepsilon) \setminus \{\lambda_0\}$ we then conclude that $\alpha(\lambda I - T) = \beta(\lambda I - T) = 0$ for all $\lambda \in \mathbb{D}(\lambda_0, \varepsilon) \setminus \{\lambda_0\}$. Hence $\lambda_0 \in \mathrm{iso}\, \sigma(T)$, thus T satisfies Browder's theorem by Theorem 5.17. $\blacksquare$

The generalized Browder's theorem, or equivalently Browder's theorem, may be characterized by means of the quasi-nilpotent parts $H_0(\lambda I - T)$ as λ ranges over $\Delta^g(T)$:

Theorem 5.22 *For a bounded operator $T \in L(X)$ the following statements are equivalent:*

(i) *T satisfies Browder's theorem;*
(ii) *For every $\lambda \in \Delta^g(T)$ there exists a $p := p(\lambda) \in \mathbb{N}$ such that $H_0(\lambda I - T) = \ker(\lambda I - T)^p$;*
(iii) *$H_0(\lambda I - T)$ is closed for all $\lambda \in \Delta^g(T)$;*
(iv) *$\mathcal{N}^\infty(\lambda I - T)$ is closed for all $\lambda \in \Delta^g(T)$.*

Proof (i) $\Rightarrow$ (ii) By Theorem 5.17 we have $\Delta^g(T) = \Pi(T)$. If $\lambda_0 \in \Delta^g(T)$ then λ_0 is a pole of the resolvent and $H_0(\lambda_0 I - T) = \ker(\lambda_0 I - T)^p$ where $p := p(\lambda_0 I - T) = q(\lambda_0 I - T)$, by Theorem 2.47.

(ii) $\Rightarrow$ (iii) Clear.

(iii) $\Rightarrow$ (i) Suppose that $H_0(\lambda_0 I - T)$ is closed for $\lambda_0 \in \Delta^g(T)$. Since $\lambda_0 I - T$ is B-Weyl there exists, by Theorem 1.119, two closed linear subspaces M, N such that $X = M \oplus N$, $\lambda_0 I - T|M$ is Weyl and $\lambda_0 I - T|N$ is nilpotent. Now, from Theorem 2.39 we know that T has the SVEP at λ_0 and hence $T|M$ also has the SVEP at λ_0, so, by Theorem 2.97, we have $p(\lambda_0 I - T|M) < \infty$. Since $\alpha(\lambda_0 I - T|M) = \beta(\lambda_0 I - T|M)$ it then follows by Theorem 1.22 that $q(\lambda_0 I - T|M) < \infty$.

Obviously, $p(\lambda_0 I - T) = p(\lambda_0 I - T|M) + p(\lambda_0 I - T|N) < \infty$, and a similar argument shows that also $q(\lambda_0 I - T) < \infty$. Therefore $\lambda_0 \in \Pi(T)$. This shows that $\Delta^g(T) \subseteq \mathrm{iso}\,\sigma(T)$, so, by Theorem 5.17, T satisfies Browder's theorem.

(ii) $\Rightarrow$ (iv) We have $\ker(\lambda I - T)^n \subseteq \mathcal{N}^\infty(\lambda I - T) \subseteq H_0(\lambda I - T)$ for all $n \in \mathbb{N}$, so $\mathcal{N}^\infty(\lambda I - T) = \ker(\lambda I - T)^p$ is closed for all $\lambda \in \Delta^g(T)$.

(iv) $\Rightarrow$ (i) We first prove that $p(\lambda I - T) < \infty$ for every $\lambda \in \Delta(T)$. We use a standard argument from the well-known Baire theorem. Suppose $p(\lambda I - T) = \infty$ for $\lambda \in \sigma(T) \setminus \sigma_{\mathrm{bw}}(T)$. By assumption $\mathcal{N}^\infty(\lambda I - T) = \bigcup_{n=1}^\infty \ker(\lambda I - T)^n$ is closed so it is of second category in itself. Moreover, $\ker(\lambda I - T)^n \neq \mathcal{N}^\infty(\lambda I - T)$ implies that $\ker(\lambda I - T)^n$ is of the first category as subset of $\mathcal{N}^\infty(\lambda I - T)$ and hence $\mathcal{N}^\infty(\lambda I - T)$ is also of the first category. From this it then follows that $\mathcal{N}^\infty(\lambda I - T)$ is not closed, a contradiction.

Therefore $p(\lambda I - T) < \infty$ for every $\lambda \in \sigma(T) \setminus \sigma_{\mathrm{bw}}(T)$ and consequently T has the SVEP at λ. Trivially, T also has the SVEP at every point $\lambda \notin \sigma(T)$, so T has the SVEP at every $\lambda \notin \sigma_{\mathrm{bw}}(T)$ and this is equivalent, by Theorem 5.14, to saying that T satisfies Browder's theorem. ∎

5.2 a-Browder's Theorem

An approximate point version of Browder's theorem is defined as follows:

Definition 5.23 A bounded operator $T \in L(X)$ is said to satisfy a-Browder's theorem if

$$\sigma_{\mathrm{uw}}(T) = \sigma_{\mathrm{ub}}(T),$$

or equivalently, by Theorem 3.43, if

$$\mathrm{acc}\,\sigma_{\mathrm{ap}}(T) \subseteq \sigma_{\mathrm{uw}}(T).$$

By Theorem 3.44 it then follows that if either T or T^* has the SVEP then a-Browder's theorem holds for both T and T^*.

Define

$$p_{00}^a(T) := \sigma_{\mathrm{ap}}(T) \setminus \sigma_{\mathrm{ub}}(T) = \{\lambda \in \sigma_{\mathrm{ap}}(T) : \lambda I - T \in \mathcal{B}_+(X)\}.$$

By Theorem 4.3 $p_{00}^a(T) \subseteq \mathrm{iso}\,\sigma_{\mathrm{ap}}(T)$.

Lemma 5.24 Let $T \in L(X)$. Then we have:

(i) $\lambda \in p_{00}^a(T)$ if and only if λ is a left pole of finite rank for T.

(ii) $\lambda \in p_{00}^a(T^*)$ if and only if λ is a right pole of finite rank for T.

Proof

(i) Suppose λ is a left pole of finite rank. We may assume $\lambda = 0$. Then, $0 \in \sigma_{\mathrm{ap}}(T)$, T is left Drazin invertible, so $p(T) < \infty$. By Corollary 3.49 the condition of left Drazin invertibility is equivalent to saying that T is upper semi B-Browder, i.e. there exists an $n \in \mathbb{N}$ such that $T^n(X)$ is closed and the restriction $T_n := T|T^n(X)$ is upper semi-Browder, in particular upper semi-Fredholm. Since λ is a left pole of finite rank we have $\alpha(T) < \infty$ and hence $\alpha(T^n) < \infty$, so $T^n \in \Phi_+(X)$ and from the classical Fredholm theory this implies that $T \in \Phi_+(X)$. Since $p(T) < \infty$ we then conclude that $T \in B_+(X)$, so $0 \notin \sigma_{\mathrm{ub}}(T)$, and consequently $0 \in \sigma_a(T) \setminus \sigma_{\mathrm{ub}}(T) = p_{00}^a(T)$.

Conversely, assume that $0 \in p_{00}^a(T)$. Then $0 \in \sigma_{\mathrm{ap}}(T) \setminus \sigma_{\mathrm{ub}}(T)$, hence $p := p(T) < \infty$ and $T \in \Phi_+(X)$. From Fredholm theory we know that $T^n \in \Phi_+(X)$ for all $n \in \mathbb{N}$, so $T^{p+1}(X)$ is closed. Thus T is left Drazin invertible. But $0 \in \sigma_{\mathrm{ap}}(T)$, thus 0 is a left pole having finite rank, since $\alpha(T) < \infty$.

(ii) Suppose λ is a right pole of finite rank. We may assume $\lambda = 0$. Then $0 \in \sigma_s(T) = \sigma_{\mathrm{ap}}(T^*)$, T is right Drazin invertible and $q(T) < \infty$. The condition of right Drazin invertibility is equivalent to saying that T is lower semi B-Browder, i.e. there exists an $n \in \mathbb{N}$ such that $T^n(X)$ is closed and the restriction $T_n := T|T^n(X)$ is lower semi-Browder, in particular lower semi-Fredholm. Since $\beta(T) < \infty$ then $\beta(T^n) < \infty$, hence $T^n \in \Phi_-(X)$, from which we obtain that $T \in \Phi_-(X)$. Since $q(T) < \infty$ we then conclude that $T \in B_-(X)$, or equivalently $T^* \in B_+(X^*)$, hence $0 \notin \sigma_{\mathrm{ub}}(T^*)$. Therefore, $0 \in \sigma_{\mathrm{ap}}(T^*) \setminus \sigma_{\mathrm{ub}}(T^*) = p_{00}^a(T^*)$.

Conversely, assume that $0 \in p_{00}^a(T^*)$. Then $0 \in \sigma_{\mathrm{ap}}(T^*) \setminus \sigma_{\mathrm{ub}}(T^*)$ and since, by duality $\sigma_{\mathrm{ub}}(T^*) = \sigma_{\mathrm{lb}}(T)$, it then follows that $0 \in \sigma_s(T) \setminus \sigma_{\mathrm{lb}}(T)$. Therefore, $q := q(T) < \infty$ and $T \in \Phi_-(X)$. From Fredholm theory we know that $T^n \in \Phi_-(X)$ for all $n \in \mathbb{N}$, in particular $T^q(X)$ is closed. Thus T is right Drazin invertible. But $0 \in \sigma_s(T)$, thus 0 is a right pole of T. Finally, since $T \in \Phi_-(X)$ we have $\beta(T) < \infty$ and consequently 0 is a right pole of finite rank for T. ∎

Define

$$\Delta_a(T) := \sigma_{\mathrm{ap}}(T) \setminus \sigma_{\mathrm{uw}}(T).$$

It should be noted that the set $\Delta_a(T)$ may be empty. This is, for instance, the case of a right shift on $\ell^2(\mathbb{N})$, see the next Corollary 5.34, since a right shift has the SVEP. If $\rho_{\mathrm{w}}(T) := \mathbb{C} \setminus \sigma_{\mathrm{w}}(T)$ and, as usual, $\sigma_{\mathrm{p}}(T)$ denotes the point spectrum, we have

Lemma 5.25 *For $T \in L(X)$ we have*

$$\Delta_a(T) = \{\lambda \in \mathbb{C} : \lambda I - T \in W_+(X),\ 0 < \alpha(\lambda I - T)\}$$
$$= [\rho_{\mathrm{sf}}^-(T) \cup \rho_{\mathrm{w}}(T)] \cap \sigma_{\mathrm{p}}(T).$$

Moreover, the following inclusions hold:

(i) $p_{00}(T) \subseteq p_{00}^a(T) \subseteq \Delta_a(T)$.

(ii) $p_{00}(T) \subseteq \Delta(T) \subseteq \Delta_a(T)$ and $p_{00}^a(T) \subseteq \Delta_a(T) \subseteq \sigma_{ap}(T)$.

Proof The first equality above is clear: if $\lambda \in \Delta_a(T)$ then $\lambda I - T$ is upper semi-Weyl and hence has closed range. Since $\lambda \in \sigma_{ap}(T)$ we then deduce that $0 < \alpha(\lambda I - T)$. The second equality is evident.

The inclusion $p_{00}(T) \subseteq p_{00}^a(T)$ is easy to see: every $\lambda \in p_{00}(T)$ is a pole and hence an eigenvalue of T, so $\lambda \in \sigma_{ap}(T)$. On the other hand, $\lambda \notin \sigma_{ub}(T)$ since $\lambda \notin \sigma_b(T)$ and $\sigma_{ub}(T) \subseteq \sigma_b(T)$.

The inclusion $p_{00}^a(T) \subseteq \Delta^a(T)$ is evident, since $\sigma_{uw}(T) \subseteq \sigma_{ub}(T)$. Clearly, $p_{00}(T) \subseteq \Delta(T)$, and $\Delta(T) \subseteq \Delta_a(T)$ follows from Lemma 5.3. ∎

Theorem 5.26 *For a bounded operator $T \in L(X)$, a-Browder's theorem holds for T if and only if $p_{00}^a(T) = \Delta_a(T)$. In particular, a-Browder's theorem holds whenever $\Delta_a(T) = \emptyset$.*

Proof Suppose that T satisfies a-Browder's theorem. Clearly, by Lemma 5.25, part (ii), the equality $p_{00}^a(T) = \Delta_a(T)$ holds whenever $\Delta_a(T) = \emptyset$. Suppose then $\Delta_a(T) \neq \emptyset$ and let $\lambda \in \Delta_a(T)$. Then $\lambda I - T \in W_+(X)$ and $\lambda \in \sigma_{ap}(T)$. From the equality $\sigma_{uw}(T) = \sigma_{ub}(T)$ it then follows that $\lambda I - T \in B_+(X)$, so $\lambda \in p_{00}^a(T)$. Hence $\Delta_a(T) \subseteq p_{00}^a(T)$, and by part (ii) of Lemma 5.25 we conclude that $p_{00}^a(T) = \Delta_a(T)$.

Conversely, suppose that $p_{00}^a(T) = \Delta_a(T)$. Let $\lambda \notin \sigma_{uw}(T)$. We show that $\lambda \notin \sigma_{ub}(T)$. If $\lambda \notin \sigma_{ap}(T)$ then $\lambda \notin \sigma_{ub}(T)$, since $\sigma_{ub}(T) \subseteq \sigma_{ap}(T)$. Consider the other case $\lambda \in \sigma_{ap}(T)$. Then $\lambda \in \Delta_a(T) = p_{00}^a(T)$, thus $\lambda \notin \sigma_{ub}(T)$. Therefore we have $\sigma_{ub}(T) \subseteq \sigma_{uw}(T)$ and, since the reverse inclusion is satisfied by every operator, then $\sigma_{ub}(T) = \sigma_{uw}(T)$, i.e. T satisfies a-Browder's theorem.

The last assertion is clear by Lemma 5.25, part (ii). ∎

a-Browder's theorem may also be described in terms of the localized SVEP at the points of a certain set:

Theorem 5.27 *If $T \in L(X)$ then the following statements hold:*

(i) *T satisfies a-Browder's theorem if and only if T has the SVEP at every $\lambda \notin \sigma_{uw}(T)$, or equivalently T has the SVEP at every $\lambda \in \Delta_a(T)$.*

(ii) *T^* satisfies a-Browder's theorem if and only if T^* has the SVEP at every $\lambda \notin \sigma_{lw}(T)$.*

(iii) *If T has the SVEP at every $\lambda \notin \sigma_{lw}(T)$ then a-Browder's theorem holds for T^*.*

(iv) *If T^* has the SVEP at every $\lambda \notin \sigma_{uw}(T)$ then a-Browder's theorem holds for T.*

Consequently, if either T or T^* has the SVEP then a-Browder's theorem holds for both T and T^*.

Proof

(i) Suppose that $\sigma_{ub}(T) = \sigma_{uw}(T)$. If $\lambda \notin \sigma_{uw}(T)$ then $\lambda I - T \in B_+(X)$ so $p(\lambda I - T) < \infty$ and hence T has the SVEP at λ. Conversely, if T has the SVEP at every point which is not in $\sigma_{uw}(T)$, then for every $\lambda \notin \sigma_{uw}(T)$, $\lambda I - T \in \Phi_+(X)$ and the SVEP at λ by Theorem 2.97 implies that $p(\lambda I - T) < \infty$, and consequently $\lambda \notin \sigma_{ub}(T)$. This shows that $\sigma_{ub}(T) \subseteq \sigma_{uw}(T)$. The opposite inclusion is clear, so $\sigma_{ub}(T) = \sigma_{uw}(T)$.

To show the last assertion, observe first that if T has the SVEP at every $\lambda \notin \sigma_{uw}(T)$ then T has the SVEP at every $\lambda \in \Delta_a(T)$, since $\Delta_a(T) \cap \sigma_{uw}(T) = \emptyset$. Conversely, suppose that T has the SVEP at every $\lambda \in \Delta_a(T)$. If $\lambda \notin \sigma_{uw}(T)$ then either $\lambda \notin \sigma_{ap}(T)$ or $\lambda \in \Delta_a(T) = \sigma_{ap}(T) \setminus \sigma_{uw}(T)$. In both cases T has the SVEP at λ, by Theorem 2.97.

(ii) Obvious, since $\sigma_{lw}(T) = \sigma_{uw}(T^*)$.

(iii) Suppose that T has the SVEP at every point which does not belong to $\sigma_{lw}(T)$. If $\lambda \notin \sigma_{uw}(T^*) = \sigma_{lw}(T)$ then $\lambda I - T \in \Phi_-(X)$ with $\mathrm{ind}(\lambda I - T) \geq 0$. By Theorem 2.97 the SVEP of T at λ entails that $p(\lambda I - T) < \infty$ and hence by Theorem 1.22 we have $\mathrm{ind}(\lambda I - T) \leq 0$. Therefore, $\mathrm{ind}(\lambda I - T) = 0$, and since $p(\lambda I - T) < \infty$ we conclude, from part (iv) of Theorem 1.22, that $q(\lambda I - T) < \infty$, and hence $\lambda \notin \sigma_{lb}(T) = \sigma_{ub}(T^*)$. Consequently, $\sigma_{ub}(T^*) \subseteq \sigma_{uw}(T^*)$, and since the reverse inclusion holds for every operator we then conclude that $\sigma_{ub}(T^*) = \sigma_{uw}(T^*)$.

(iv) This has been shown in Theorem 3.44. ∎

The next example shows that the reverse of the assertions (iii) and (iv) of Theorem 5.27 generally do not hold.

Example 5.28 Let $1 \leq p \leq \infty$ be given, and let $\omega := (\omega_n)$ be a bounded sequence of strictly positive real numbers. The corresponding *unilateral weighted right shift* on $\ell^p(\mathbb{N})$ is defined by

$$Tx := \sum_{n=1}^{\infty} \omega_n x_n e_{n+1} \quad \text{for all } x = (x_n) \in \ell^p(\mathbb{N}),$$

where (e_n) is the standard basis of $\ell^p(\mathbb{N})$. In this case the *spectral radius* of T is given by

$$r(T) = \lim_{n \to \infty} \sup_{k \in \mathbb{N}} (\omega_k \cdots \omega_{k+n-1})^{1/n}.$$

Define

$$i(T) := \lim_{n \to \infty} \inf_{k \in \mathbb{N}} (\omega_k \cdots \omega_{k+n-1})^{1/n},$$

and

$$c(T) := \lim_{n \to \infty} \inf_{k \in \mathbb{N}} (\omega_1 \cdots \omega_n)^{1/n}.$$

We have $i(T) \le c(T) \le r(T)$, and as observed by Shields [284], for every triple of real number $0 \le i \le c \le r$ it is possible to find a weighted right shift on $\ell^p(\mathbb{N})$ such that

$$i(T) = i, \quad c(T) = c, \quad r(T) = r.$$

Suppose now that $0 < i(T) \le c(T) \le r(T)$. We know that every unilateral weighted right shift has the SVEP, so T satisfies a-Browder's theorem. Moreover, by Theorem 1.6.15 of [216], we have that

$$\sigma_{\mathrm{ap}}(T) = \{\lambda \in \mathbb{C} : i(T) \le |\lambda| \le r(T)\},$$

and since $i(T)$ is strictly greater than 0 then $0 \notin \sigma_{\mathrm{ap}}(T)$. The dual T^* of T is the *unilateral weighted left shift* on $\ell^q(\mathbb{N})$ given by

$$T^* x = (\omega_n x_{n+1}) \quad \text{for all } x = (x_n) \in \ell^q(\mathbb{N}),$$

where, as usual, $1/p + 1/q = 1$ and $\ell^q(\mathbb{N})$ is canonically identified with the dual of $\ell^p(\mathbb{N})$. Since the inclusion $\sigma_{\mathrm{wa}}(T) \subseteq \sigma_{\mathrm{ap}}(T)$ holds for every operator, we conclude that $0 \notin \sigma_{\mathrm{wa}}(T)$. This example shows that the assertion (iv) of Theorem 5.27 cannot be reversed.

To show the converse of assertion (iii) of Theorem 5.27, let $S := T^*$. Then $S^* = T$ has the SVEP, so a-Browder's theorem holds for S, while S does not have the SVEP at 0. On the other hand, $0 \notin \sigma_{\mathrm{uw}}(T) = \sigma_{\mathrm{lw}}(T^*) = \sigma_{\mathrm{lw}}(S)$.

The SVEP is preserved under commuting Riesz or algebraic perturbations, and also by the functional calculus, so we have:

Corollary 5.29 *Let $T \in L(X)$. Then we have:*

(i) *If satisfies a-Browder's theorem and $R \in L(X)$ is a Riesz operator which commutes with T, then $T + R$ satisfies a-Browder's theorem.*

(ii) *If T has the SVEP and $K \in L(X)$ is an algebraic operator which commutes with T, then $T + K$ satisfies a-Browder's theorem.*

(iii) *If T has the SVEP then $f(T)$ satisfies a-Browder's theorem for every $f \in \mathcal{H}(\sigma(T))$.*

(iv) *If $T \in L(X)$ has the SVEP and T and $S \in L(Y)$ are intertwined by an injective map $A \in L(X, Y)$ then S satisfies a-Browder's theorem.*

Proof (i) follows from Theorems 2.129 and 5.27. (ii) follows from Theorem 2.145, while the assertion (iii) follows from Theorem 2.86. To show (iv), observe that S has the SVEP by Lemma 2.141.   ∎

Since $\sigma_{uw}(T) \subseteq \sigma_w(T)$, from Theorems 5.27 and 5.4 we readily obtain:

Corollary 5.30 *If $T \in L(X)$ then a-Browder's theorem for T implies Browder's theorem for T.*

The following results are analogous to the results of Theorem 5.10, and they give a precise spectral picture of *a*-Browder's theorem.

Theorem 5.31 *For a bounded operator $T \in L(X)$ the following statements are equivalent:*

  (i) *T satisfies a-Browder's theorem;*
 (ii) *$\Delta_a(T) \subseteq iso\,\sigma_{ap}(T)$;*
(iii) *$\Delta_a(T) \subseteq \partial\sigma_{ap}(T)$, $\partial\sigma_{ap}(T)$ the topological boundary of $\sigma_{ap}(T)$;*
 (iv) *the mapping $\lambda \to \ker(\lambda I - T)$ is not continuous at every $\lambda \in \Delta_a(T)$ in the gap metric;*
  (v) *the mapping $\lambda \to \gamma(\lambda I - T)$ is not continuous at every $\lambda \in \Delta_a(T)$;*
 (vi) *the mapping $\lambda \to (\lambda I - T)(X)$ is not continuous at every $\lambda \in \Delta_a(T)$ in the gap metric;*
(vii) *$\Delta_a(T) \subseteq \sigma_{se}(T)$;*
(viii) *$\Delta_a(T) \subseteq iso\,\sigma_{se}(T)$;*
 (ix) *$\sigma_{ap}(T) = \sigma_{uw}(T) \cup iso\,\sigma_{ap}(T)$.*

Proof The equivalences are obvious if $\Delta_a(T) = \emptyset$, so we may suppose that $\Delta_a(T)$ is non-empty.

(i) $\Leftrightarrow$ (ii) By Theorem 5.26 if T satisfies *a*-Browder's theorem then $\Delta_a(T) = p_{00}^a(T)$, so, by Lemma 5.25, every $\lambda \in \Delta_a(T)$ is an isolated point of $\sigma_{ap}(T)$. Conversely, suppose that $\Delta_a(T) \subseteq iso\,\sigma_{ap}(T)$ and take $\lambda \in \Delta_a(T)$. Then T has the SVEP at λ, since λ is an isolated point of $\sigma_{ap}(T)$, and since $\lambda I - T \in \Phi_+(X)$ the SVEP at λ is equivalent to saying that $p(\lambda I - T) < \infty$, and hence $\lambda I - T \in B_+(X)$. Therefore, $\lambda \in p_{00}^a(T)$, from which we conclude that $\Delta_a(T) = p_{00}^a(T)$.

(ii) $\Rightarrow$ (iii) Obvious.

(iii) $\Rightarrow$ (ii) Suppose that the inclusion $\Delta_a(T) \subseteq \partial\sigma_{ap}(T)$ holds. Let $\lambda_0 \in \Delta_a(T)$ be arbitrarily given. We show that T has the SVEP at λ_0. Let $f : U \to X$ be an analytic function defined on an open disc U of λ_0 which satisfies the equation $(\lambda I - T)f(\lambda) = 0$ for all $\lambda \in U$. Since $\lambda_0 \in \partial\sigma_{ap}(T)$ we can choose $\mu \neq \lambda_0$, $\mu \in U$ such that $\mu \notin \sigma_{ap}(T)$. Consider an open disc W of μ such that $W \subseteq U$. Since T has the SVEP at μ, then $f(\lambda) = 0$ for all $\lambda \in W$. The identity theorem for analytic functions then implies that $f(\lambda) = 0$ for all $\lambda \in U$, hence T has the SVEP at λ_0. Finally, $\lambda_0 I - T \in \Phi_+(X)$, since $\lambda_0 \in \Delta_a(T)$. The SVEP at λ_0 then implies that $\sigma_{ap}(T)$ does not cluster at λ_0, and $\Delta_a(T)$ being a subset of $\sigma_{ap}(T)$ we then conclude that $\lambda_0 \in iso\,\sigma_{ap}(T)$.

(ii) $\Rightarrow$ (iv) Suppose that $\Delta_a(T) \subseteq iso\,\sigma_{ap}(T)$. For every $\lambda_0 \in \Delta_a(T)$ then $\alpha(\lambda_0 I - T) > 0$ and since $\lambda_0 \in iso\,\sigma_{ap}(T)$ there exists an open disc $\mathbb{D}(\lambda_0, \varepsilon)$ such that $\alpha(\lambda I - T) = 0$ for all $\lambda \in \mathbb{D}(\lambda_0, \varepsilon) \setminus \{\lambda_0\}$. Therefore the mapping $\lambda \to \ker(\lambda I - T)$ is not continuous at λ_0.

(iv) $\Rightarrow$ (ii) Let $\lambda_0 \in \Delta_a(T)$ be arbitrary. By the punctured neighborhood theorem there exists an open disc $\mathbb{D}(\lambda_0, \varepsilon)$ such that $\alpha(\lambda I - T)$ is constant as λ ranges over $\mathbb{D}(\lambda_0, \varepsilon) \setminus \{\lambda_0\}$, $\lambda I - T \in \Phi_+(X)$ for all $\lambda \in \mathbb{D}(\lambda_0, \varepsilon)$,

$$\text{ind}(\lambda I - T) = \text{ind}(\lambda_0 I - T) \quad \text{for all} \quad \lambda \in \mathbb{D}(\lambda_0, \varepsilon),$$

and

$$0 \leq \alpha(\lambda I - T) \leq \alpha(\lambda_0 I - T) \quad \text{for all } \lambda \in \mathbb{D}(\lambda_0, \varepsilon).$$

Since the mapping $\lambda \to \ker(\lambda I - T)$ is not continuous at λ_0 it then follows that

$$0 \leq \alpha(\lambda I - T) < \alpha(\lambda_0 I - T) \quad \text{for all } \lambda \in \mathbb{D}(\lambda_0, \varepsilon) \setminus \{\lambda_0\}.$$

We claim that $\alpha(\lambda I - T) = 0$ for all $\lambda \in \mathbb{D}(\lambda_0, \varepsilon) \setminus \{\lambda_0\}$. To see this, suppose that there is a $\lambda_1 \in \mathbb{D}(\lambda_0, \varepsilon) \setminus \{\lambda_0\}$ such that $\alpha(\lambda_1 I - T) > 0$. From $\text{ind}(\lambda_1 I - T) = \text{ind}(\lambda_0 I - T) \leq 0$ we see that $\lambda I - T$ is upper semi-Weyl, and hence $\lambda_1 \in \Delta_a(T)$. Repeating the same reasoning as above we may choose a $\lambda_2 \in \mathbb{D}(\lambda_0, \varepsilon) \setminus \{\lambda_0, \lambda_1\}$ such that

$$0 < \alpha(\lambda_2 I - T) < \alpha(\lambda_1 I - T)$$

and this is impossible since $\alpha(\lambda I - T)$ is constant for all $\lambda \in \mathbb{D}(\lambda_0, \varepsilon) \setminus \{\lambda_0\}$. Therefore $\alpha(\lambda I - T) = 0$ for $\lambda \in \mathbb{D}(\lambda_0, \varepsilon) \setminus \{\lambda_0\}$ and since $(\lambda I - T)(X)$ is closed for all $\lambda \in \mathbb{D}(\lambda_0, \varepsilon)$ we can conclude that $\lambda_0 \in \text{iso } \sigma_{\text{ap}}(T)$, as desired.

(iv) $\Leftrightarrow$ (v) $\Leftrightarrow$ (vi) To show these equivalences observe first that for every $\lambda \in \Delta_a(T)$ the range $(\lambda I - T)(X)$ is closed. The equivalences then follow from Theorem 1.51.

(vi) $\Leftrightarrow$ (vii) If $\lambda_0 \in \Delta_a(T)$ then there exists an open disc $\mathbb{D}(\lambda_0, \varepsilon)$ centered at λ_0 such that $\lambda I - T$ has closed range for all $\lambda \in \mathbb{D}(\lambda, \varepsilon)$. The equivalence (vi) $\Leftrightarrow$ (vii) then easily follows from Theorem 1.51.

(viii) $\Rightarrow$ (vii) Clear.

(vii) $\Rightarrow$ (viii) Suppose that $\Delta_a(T) \subseteq \sigma_{\text{se}}(T)$. If $\lambda_0 \in \Delta_a(T)$ then $\lambda_0 I - T \in \Phi_+(X)$ so $\lambda_0 I - T$ is essentially semi-regular, in particular of Kato-type. By Theorem 1.65 there exists an open disc $\mathbb{D}(\lambda_0, \varepsilon)$ centered at λ_0 such that $\lambda I - T$ is semi-regular for all $\lambda \in \mathbb{D}(\lambda_0, \varepsilon) \setminus \{\lambda_0\}$. But $\lambda_0 \in \sigma_{\text{se}}(T)$, so $\lambda_0 \in \text{iso } \sigma_{\text{se}}(T)$.

(i) $\Leftrightarrow$ (ix) The inclusion $\sigma_{\text{uw}}(T) \cup \text{iso } \sigma_{\text{ap}}(T) \subseteq \sigma_{\text{ap}}(T)$ holds for every $T \in L(X)$, so we need only prove the reverse inclusion. Suppose that a-Browder's theorem holds. If $\lambda \in \sigma_{\text{ap}}(T) \setminus \sigma_{\text{uw}}(T)$ then, by Theorem 5.27, T has the SVEP at λ, and hence, by Theorem 2.97, we deduce that $\lambda \in \text{iso } \sigma_{\text{ap}}(T)$. Therefore $\sigma_{\text{ap}}(T) \subseteq \sigma_{\text{uw}}(T) \cup \text{iso } \sigma_{\text{ap}}(T)$, so the equality (ix) is proved.

Conversely, suppose that $\sigma_{\mathrm{ap}}(T) = \sigma_{\mathrm{uw}}(T) \cup \mathrm{iso}\,\sigma_{\mathrm{ap}}(T)$. Let $\lambda \notin \sigma_{\mathrm{uw}}(T)$. There are two possibilities: $\lambda \in \mathrm{iso}\,\sigma_{\mathrm{ap}}(T)$ or $\lambda \notin \mathrm{iso}\,\sigma_{\mathrm{ap}}(T)$. If $\lambda \in \mathrm{iso}\,\sigma_{\mathrm{ap}}(T)$ then T has the SVEP at λ. In the other case $\lambda \notin \sigma_{\mathrm{uw}}(T) \cup \mathrm{iso}\,\sigma_{\mathrm{ap}}(T) = \sigma_{\mathrm{ap}}(T)$, and hence T has the SVEP at λ. From Theorem 5.27 we then conclude that a-Browder's theorem holds for T.

The second assertion follows by duality, since $\sigma_{\mathrm{s}}(T) = \sigma_{\mathrm{ap}}(T^*)$ and $\sigma_{\mathrm{lw}}(T) = \sigma_{\mathrm{uw}}(T^*)$ for every $T \in L(X)$.   ∎

Remark 5.32 If $T \in \Phi_+(X)$, the property that T is not semi-regular may be expressed by saying that the *jump* $j(T)$ is greater than 0, see Aiena [1, Theorem 1.58], so

$$a\text{-Browder's theorem holds for } T \Leftrightarrow j(\lambda I - T) > 0 \ \text{ for all } \lambda \in \Delta_a(T).$$

Corollary 5.33 *Suppose that T^* has the SVEP. Then $\Delta_a(T) \subseteq \mathrm{iso}\,\sigma(T)$.*

Proof Here we can also suppose that $\Delta_a(T)$ is non-empty. If T^* has the SVEP then a-Browder's theorem holds for T, so by Theorem 5.31 $\Delta_a(T) \subseteq \mathrm{iso}\,\sigma_{\mathrm{ap}}(T)$. By Theorem 2.68, the SVEP for T^* entails that $\sigma_{\mathrm{ap}}(T) = \sigma(T)$ ∎

Corollary 5.34 *Suppose that $T \in L(X)$ has the SVEP and $\mathrm{iso}\,\sigma_{\mathrm{ap}}(T) = \emptyset$. Then*

$$\sigma_{\mathrm{ap}}(T) = \sigma_{\mathrm{uw}}(T) = \sigma_{\mathrm{se}}(T). \tag{5.8}$$

Analogously, if T^* has the SVEP and $\mathrm{iso}\,\sigma_{\mathrm{s}}(T) = \emptyset$, then

$$\sigma_{\mathrm{s}}(T) = \sigma_{\mathrm{lw}}(T) = \sigma_{\mathrm{se}}(T). \tag{5.9}$$

We now give a further characterization of operators satisfying a-Browder's theorem in terms of the quasi-nilpotent part $H_0(\lambda I - T)$.

Theorem 5.35 *For a bounded operator $T \in L(X)$ the following statements are equivalent:*

(i) *a-Browder's theorem holds for T.*
(ii) *$H_0(\lambda I - T)$ is finite-dimensional for every $\lambda \in \Delta_a(T)$.*
(iii) *$H_0(\lambda I - T)$ is closed for every $\lambda \in \Delta_a(T)$.*
(iv) *$\mathcal{N}^\infty(\lambda I - T)$ is finite-dimensional for every $\lambda \in \Delta_a(T)$.*
(v) *$\mathcal{N}^\infty(\lambda I - T)$ is closed for every $\lambda \in \Delta_a(T)$.*

Proof There is nothing to prove if $\Delta_a(T) = \emptyset$. Suppose that $\Delta_a(T) \neq \emptyset$.

(i) $\Leftrightarrow$ (ii) Suppose that T satisfies a-Browder's theorem. By Theorem 5.26 then

$$\Delta_a(T) = p_{00}^a(T) = \sigma_{\mathrm{ap}}(T) \setminus \sigma_{\mathrm{ub}}(T).$$

If $\lambda \in \Delta_a(T)$ then λ is isolated in $\sigma_{\mathrm{ap}}(T)$ and hence T has the SVEP at λ. We also have that $\lambda I - T \in \Phi_+(X)$, so, from Theorem 2.97 we conclude that $H_0(\lambda I - T)$ is finite-dimensional.

Conversely, suppose that $H_0(\lambda I - T)$ is finite-dimensional for every $\lambda \in \Delta_a(T)$. To show that T satisfies a-Browder's theorem it suffices to prove that T has the SVEP at every $\lambda \notin \sigma_{uw}(T)$. Since T has the SVEP at every $\lambda \notin \sigma_{ap}(T)$ we can suppose that $\lambda \in \sigma_{ap}(T) \setminus \sigma_{uw}(T) = \Delta_a(T)$. Since $\lambda I - T \in \Phi_+(X)$ the SVEP at λ then follows by Theorem 2.97.

(ii) $\Leftrightarrow$ (iii) Since $\lambda I - T \in \Phi_+(X)$ for every $\lambda \in \Delta_a(T)$, the equivalence follows from Theorem 2.97.

(ii) $\Rightarrow$ (iv) Clear, since $\mathcal{N}^\infty(\lambda I - T) \subseteq H_0(\lambda I - T)$.

(iv) $\Rightarrow$ (i) Here we also prove that T has the SVEP at every $\lambda \notin \sigma_{uw}(T)$. We can suppose that $\lambda \in \sigma_{ap}(T) \setminus \sigma_{uw}(T) = \Delta_a(T)$, since the SVEP is satisfied at every point $\mu \notin \sigma_{ap}(T)$. By assumption $\mathcal{N}^\infty(\lambda I - T)$ is finite-dimensional and from the inclusion

$$\ker (\lambda I - T)^n \subseteq \ker (\lambda I - T)^{n+1} \subseteq \mathcal{N}^\infty(\lambda I - T) \quad \text{for all } n \in \mathbb{N},$$

it is evident that there exists a $p \in \mathbb{N}$ such that $\ker (\lambda I - T)^p = \ker (\lambda I - T)^{p+1}$. Hence $p(\lambda I - T) < \infty$, so T has the SVEP at λ.

(iv) $\Rightarrow$ (v) Obvious.

(v) $\Rightarrow$ (i) As above it suffices to prove that $p(\lambda I - T) < \infty$ for every $\lambda \notin \sigma_{uw}(T)$. We use a standard argument from the well-known Baire theorem. Suppose $p(\lambda I - T) = \infty$, $\lambda \notin \sigma_{uw}(T)$. By assumption $\mathcal{N}^\infty(\lambda I - T) = \bigcup_{n=1}^\infty \ker T^n$ is closed so it is of second category in itself. Moreover, $\ker (\lambda I - T)^n \neq \mathcal{N}^\infty(\lambda I - T)$ implies that $\ker (\lambda I - T)^n$ is of the first category as a subset of $\mathcal{N}^\infty(\lambda I - T)$ and hence also $\mathcal{N}^\infty(\lambda I - T)$ is of the first category. From this it then follows that $\mathcal{N}^\infty(\lambda I - T)$ is not closed, a contradiction. Therefore $p(\lambda I - T) < \infty$ for every $\lambda \notin \sigma_{uw}(T)$. ∎

Theorem 5.36 *If $K(\lambda I - T)$ is finite-codimensional for all $\lambda \in \Delta_a(T)$ then a-Browder's theorem holds for T.*

Proof We show that T has the SVEP at every $\lambda \notin \sigma_{uw}(T)$. As in the proof of Theorem 5.35 we can suppose that $\lambda \in \sigma_{ap}(T) \setminus \sigma_{uw}(T) = \Delta_a(T)$. By assumption $K(\lambda I - T)$ has finite codimension, and hence, by Theorem 2.98, $q(\lambda I - T) < \infty$, from which it follows that $\mathrm{ind}\,(\lambda I - T) \geq 0$, see Theorem 1.22. On the other hand, $\lambda I - T \in W_+(X)$, so $\mathrm{ind}\,(\lambda I - T) \leq 0$, from which we obtain that $\mathrm{ind}\,(\lambda I - T) = 0$. Again by Theorem 1.22 we conclude that $p(\lambda I - T) < \infty$, and hence T has the SVEP at λ. ∎

The reverse implication of that of Theorem 5.36, in general, does not hold. Later we shall prove that the property of $K(\lambda I - T)$ being finite-codimensional for all $\lambda \in \Delta_a(T)$ is equivalent to property (b), which is stronger than a-Browder's theorem.

We now consider the operators $T \in L(X)$ for which $\sigma_{\mathrm{ubw}}(T) = \sigma_{\mathrm{ld}}(T)$. The operators T which satisfy this property have been said in the literature to satisfy the *generalized a-Browder's theorem*. We shall show that a-Browder's theorem and the generalized a-Browder's theorem are equivalent.

Theorem 5.37 *Let $T \in L(X)$. Then the following statements are equivalent:*

(i) $\sigma_{\mathrm{ubw}}(T) = \sigma_{\mathrm{ld}}(T)$, *i.e., T satisfies the generalized a-Browder's theorem;*
(ii) *T has the SVEP at every $\lambda \notin \sigma_{\mathrm{ubw}}(T)$.*

     Analogously, the following statements are equivalent:
(iii) $\sigma_{\mathrm{lbw}}(T) = \sigma_{\mathrm{ld}}(T)$;
(iv) *T has the SVEP at every $\lambda \notin \sigma_{\mathrm{lbw}}(T)$.*

Proof (i) $\Leftrightarrow$ (ii) If $\sigma_{\mathrm{ld}}(T) = \sigma_{\mathrm{ubw}}(T)$, then, by Theorem 2.97, T has the SVEP at the points $\lambda \notin \sigma_{\mathrm{ubw}}(T)$.

Conversely, assume that T has the SVEP at every point that does not belong to $\sigma_{\mathrm{ubw}}(T)$. If $\lambda \notin \sigma_{\mathrm{ubw}}(T)$ then, by Theorem 2.97, $\lambda \notin \sigma_{\mathrm{ld}}(T)$, so $\sigma_{\mathrm{ld}}(T) \subseteq \sigma_{\mathrm{ubw}}(T)$. On the other hand, by Theorem 1.141 we have $\sigma_{\mathrm{ubw}}(T) \subseteq \sigma_{\mathrm{ld}}(T)$, for all operators $T \in L(X)$, thus $\sigma_{\mathrm{ubw}}(T) = \sigma_{\mathrm{ld}}(T)$.

(iii) $\Leftrightarrow$ (iv) If $\sigma_{\mathrm{ld}}(T) = \sigma_{\mathrm{lbw}}(T)$, then, by Theorem 2.97, T has the SVEP at the points $\lambda \notin \sigma_{\mathrm{ubw}}(T)$. Conversely, assume that T has the SVEP at every point that does not belong to $\sigma_{\mathrm{lbw}}(T)$. If $\lambda \notin \sigma_{\mathrm{lbw}}(T)$ then, again by Theorem 2.97, $\lambda \notin \sigma_{\mathrm{ld}}(T)$, so $\sigma_{\mathrm{ld}}(T) \subseteq \sigma_{\mathrm{lbw}}(T)$.

On the other hand, by Theorem 1.141 we have $\sigma_{\mathrm{lbw}}(T) \subseteq \sigma_{\mathrm{ld}}(T)$, for all operators $T \in L(X)$, thus $\sigma_{\mathrm{lbw}}(T) = \sigma_{\mathrm{ld}}(T)$. ∎

Theorem 5.38 *If $T \in L(X)$ then the following statements are equivalent:*

(i) $\sigma_{\mathrm{uw}}(T) = \sigma_{\mathrm{ub}}(T)$;
(ii) $\sigma_{\mathrm{ubw}}(T) = \sigma_{\mathrm{ld}}(T)$.

Consequently, a-Browder's theorem and the generalized a-Browder's theorem are equivalent.

Proof Suppose that $\sigma_{\mathrm{uw}}(T) = \sigma_{\mathrm{ub}}(T)$. Since, by Theorem 1.141, $\sigma_{\mathrm{ubw}}(T) \subseteq \sigma_{\mathrm{ld}}(T) =$ for all $T \in L(X)$, it suffices to show the opposite inclusion. Assume that $\lambda_0 \notin \sigma_{\mathrm{ubw}}(T)$, i.e. that $\lambda_0 I - T$ is upper semi B-Weyl. By Theorem 1.117, there exists an open disc $\mathbb{D}$ such that $\lambda I - T$ is upper semi-Weyl and hence upper semi-Browder for all $\lambda \in \mathbb{D} \setminus \{\lambda_0\}$. The condition $p(\lambda I - T) < \infty$ implies that T has the SVEP at every $\lambda \in \mathbb{D} \setminus \{\lambda_0\}$, so both T has the SVEP at λ_0. Since every semi B-Fredholm operator has topological uniform descent, by Theorem 2.97, we then have that $\lambda_0 I - T$ is left Drazin invertible, i.e. $\lambda_0 \notin \sigma_{\mathrm{ld}}(T)$. Hence $\sigma_{\mathrm{ubw}}(T) = \sigma_{\mathrm{ld}}(T)$.

Conversely, suppose that $\sigma_{\mathrm{ubw}}(T) = \sigma_{\mathrm{ld}}(T)$. By Theorem 5.37 T has the SVEP at every $\lambda \notin \sigma_{\mathrm{ubw}}(T)$ and in particular T has the SVEP for every $\lambda \notin \sigma_{\mathrm{uw}}(T)$, since $\sigma_{\mathrm{ubw}}(T) \subseteq \sigma_{\mathrm{uw}}(T)$, so, by Theorem 5.27, T satisfies a-Browder's theorem and hence $\sigma_{\mathrm{uw}}(T) = \sigma_{\mathrm{ub}}(T)$. ∎

Let

$$\Pi_a(T) := \sigma_{\mathrm{ap}}(T) \setminus \sigma_{\mathrm{ld}}(T)$$

denote the set of all left poles of the resolvent. By Theorem 4.3 we have $\Pi_a(T) \subseteq$ iso $\sigma_{\mathrm{a}}(T)$ for every $T \in L(X)$.

Define $\Delta_a^g(T) := \sigma_{\mathrm{ap}}(T) \setminus \sigma_{\mathrm{ubw}}(T)$.

Lemma 5.39 *If $T \in L(X)$ then*

$$\Delta_a^g(T) = \{\lambda \in \mathbb{C} : \lambda I - T \text{ is upper semi B-Weyl and } 0 < \alpha(\lambda I - T)\}. \qquad (5.10)$$

Furthermore,

$$\Delta(T) \subseteq \Delta_a^g(T), \quad \Delta(T) \subseteq \Delta^g(T) \subseteq \sigma_{\mathrm{ap}}(T),$$

and $\Pi_a(T) \subseteq \Delta_a^g(T)$.

Proof The inclusion $\supseteq$ in (5.10) is obvious. To show the opposite inclusion, suppose that $\lambda \in \Delta_a^g(T)$. There is no harm in assuming $\lambda = 0$. Then T is upper semi B-Weyl and $0 \in \sigma_{\mathrm{a}}(T)$. Both conditions entail that $\alpha(T) > 0$, otherwise if $\alpha(T) = 0$, and hence $p(T) = 0$, by Corollary 1.101 we would have $T(X)$ closed. Thus, $0 \notin \sigma_{\mathrm{ap}}(T)$, a contradiction. Therefore equality (5.10) holds.

The inclusion $\Delta(T) \subseteq \Delta_a^g(T)$ is evident: if $\lambda \in \sigma(T) \setminus \sigma_{\mathrm{w}}(T)$ then $\lambda I - T$ is Weyl and hence upper semi B-Weyl. Moreover, $\alpha(\lambda I - T) > 0$, otherwise $\alpha(\lambda I - T) = \beta(\lambda I - T) = 0$, in contradiction with the assumption $\lambda \in \sigma(T)$.

The inclusion $\Delta(T) \subseteq \Delta^g(T)$ is clear, since $\sigma_{\mathrm{bw}}(T) \subseteq \sigma_{\mathrm{w}}(T)$. To show the inclusion $\Delta^g(T) \subseteq \sigma_{\mathrm{ap}}(T)$, observe that if $\lambda \in \Delta^g(T)$ then $\lambda I - T$ is B-Weyl. This implies that $\alpha(\lambda I - T) > 0$. Indeed, if $\alpha(\lambda I - T) = 0$, then we would have $p(\lambda I - T) = 0$ and hence T would have the SVEP by part (iii) of Theorem 1.143, and so $\lambda I - T$ would be Drazin invertible. Therefore, $p(\lambda I - T) = q(\lambda I - T) = 0$, so $\lambda \notin \sigma(T)$, which is impossible.

To show the inclusion $\Pi_a(T) \subseteq \Delta_a^g(T)$, let us assume that $\lambda \in \Pi_a(T) = \sigma_{\mathrm{a}}(T) \setminus \sigma_{\mathrm{ld}}(T)$. Then $\lambda \in \sigma_{\mathrm{ap}}(T)$, and $\lambda I - T$ is left Drazin invertible, in particular, upper semi B-Weyl. Therefore $\Pi_a(T) \subseteq \Delta_a^g(T)$. ∎

Obviously, we have $\Delta_a^g(T) = \Pi_a(T)$ exactly when T satisfies Browder's theorem or the generalized Browder's theorem. The equivalence of a-Browder's theorem and the generalized a-Browder's theorem produces a spectral picture of operators satisfying a-Browder's theorem, similar to that established in Theorem 5.31:

Theorem 5.40 *For a bounded operator $T \in L(X)$ the following statements are equivalent:*

(i) *T satisfies the a-generalized Browder's theorem;*
(ii) *Every $\lambda \in \Delta_a^g(T)$ is an isolated point of $\sigma_{\mathrm{ap}}(T)$;*
(iii) *$\Delta_a^g(T) \subseteq \partial\sigma_{\mathrm{ap}}(T)$, $\partial\sigma_{\mathrm{ap}}(T)$ the topological boundary of $\sigma(T)$;*

(iv) int $\Delta_a^g(T) = \emptyset$;
 (v) $\sigma_{ap}(T) = \sigma_{ubw}(T) \cup \partial\sigma_{ap}(T)$;
(vi) $\sigma_{ap}(T) = \sigma_{ubw}(T) \cup \mathrm{iso}\,\sigma_{ap}(T)$.

Proof (i) $\Rightarrow$ (ii) We have $\Delta_a^g(T) = \Pi_a(T) \subseteq \mathrm{iso}\,\sigma_{ap}(T)$.

  (ii) $\Rightarrow$ (iii) Obvious.

  (iii) $\Rightarrow$ (iv) Clear, since int $\partial\sigma_{ap}(T) = \emptyset$.

  (iv) $\Rightarrow$ (v) Suppose that int $\Delta_a^g(T) = \emptyset$. Let $\lambda_0 \in \Delta_a^g(T) = \sigma_a(T) \setminus \sigma_{ubw}(T)$
and suppose that $\lambda_0 \notin \partial\sigma_{ap}(T)$. Then there exists an open disc $\mathbb{D}$ centered at λ_0
contained in $\sigma_{ap}(T)$. Since $\lambda_0 I - T$ is upper semi B-Fredholm there exists, by
Theorem 1.117, a punctured open disc $\mathbb{D}_1$ contained in $\mathbb{D}$ such that $\lambda I - T$ is upper
semi-Fredholm for all $\lambda \in \mathbb{D}_1$. Moreover, $0 < \alpha(\lambda I - T)$ for all $\lambda \in \mathbb{D}_1$. In fact,
if $0 = \alpha(\lambda I - T)$, then $(\lambda I - T)(X)$ being closed, we would have $\lambda \notin \sigma_{ap}(T)$,
a contradiction. By Lemma 5.39 λ_0 belongs to int $\Delta_a^g(T)$, and this contradicts
int $\Delta_a^g(T) = \emptyset$. This shows that $\sigma_{ap}(T) = \sigma_{ubw}(T) \cup \partial\sigma_{ap}(T)$, as desired.

  (v) $\Rightarrow$ (vi) Let $\lambda_0 \in \partial\sigma_{ap}(T)$ and $\lambda_0 \notin \sigma_{ubw}(T)$. Let $\mathbb{D}$ be an open disc centered
at λ_0 and suppose that $(\lambda I - T)f(\lambda) = 0$ for all $\lambda \in \mathbb{D}$. If $\mu \in \mathbb{D}$ and $\mu \notin \sigma_{ap}(T)$
then T has the SVEP at μ, so $f \equiv 0$ in an open disc $\mathbb{U} \subseteq \mathbb{D}$ centered at μ. The
identity theorem for analytic function entails that $f(\lambda) = 0$ for all $\lambda \in \mathbb{D}$, so T has
the SVEP at λ_0. Since $\lambda_0 I - T$ is upper semi B-Fredholm, from Theorem 2.97 it then
follows that $\lambda_0 I - T$ is left Drazin invertible, and hence $\lambda_0 \in \Pi^a(T) \subseteq \mathrm{iso}\,\sigma_{ap}(T)$.
Therefore, $\sigma_{ap}(T) = \sigma_{ubw}(T) \cup \mathrm{iso}\,\sigma_{ap}(T)$.

  (vi) $\Rightarrow$ (i) We show that $\sigma_{ubw}(T) = \sigma_{ld}(T)$. Let $\lambda \notin \sigma_{ubw}(T)$. If $\lambda \notin \sigma_{ap}(T)$
then, since $\sigma_{ld}(T) \subseteq \sigma_{ap}(T)$, $\lambda \notin \sigma_{ld}(T)$. Suppose that $\lambda \in \sigma_{ap}(T)$. Then $\lambda \in$
$\sigma_{ap}(T) \setminus \sigma_{ubw}(T)$ and hence $\lambda \in \mathrm{iso}\,\sigma_{ap}(T)$. This implies that T has the SVEP
at λ. Since $\lambda I - T$ is upper semi B-Fredholm it then follows by Theorem 2.97
that $\lambda I - T$ is left Drazin invertible, so $\lambda \notin \sigma_{ld}(T)$. This proves the inclusion
$\sigma_{ld}(T) \subseteq \sigma_{ubw}(T)$. The opposite inclusion is satisfied by every operator, so we
can then conclude that the equality $\sigma_{ld}(T) = \sigma_{ubw}(T)$ holds and hence T satisfies
the *a*-generalized Browder's theorem.   ∎

Corollary 5.41 *If* $T \in L(X)$ *has the SVEP and* $\mathrm{iso}\,\sigma_{ap}(T) = \emptyset$ *then* $\sigma_{ubw}(T) =$
$\sigma_{ap}(T)$.

Proof T satisfies the generalized *a*-Browder's theorem and hence by Theorem 5.40
$\sigma_{ap}(T) = \sigma_{ubw}(T) \cup \mathrm{iso}\,\sigma_{ap}(T) = \sigma_{ubw}(T)$.   ∎

Theorem 5.42 *For a bounded operator* $T \in L(X)$ *the following statements are
equivalent:*

 (i) *T satisfies the generalized a-Browder's theorem;*
(ii) *For each* $\lambda \in \Delta_a^g(T)$ *there exists a* $\nu := \nu(\lambda) \in \mathbb{N}$ *such that* $H_0(\lambda I - T) =$
     $\ker(\lambda I - T)^\nu$;
(iii) $H_0(\lambda I - T)$ *is closed for all* $\lambda \in \Delta_a^g(T)$;
(iv) *The mapping* $\lambda \to \ker(\lambda I - T)$ *is discontinuous at every* $\lambda \in \Delta_a^g(T)$ *in the
     gap metric.*

Proof (i) $\Rightarrow$ (ii) Assume that T satisfies the generalized a-Browder's theorem, and let $\lambda_0 \in \Delta_a^g(T)$. We may assume that $\lambda_0 = 0$. Then T is upper semi B-Weyl, and by Theorem 5.40 we have $0 \in \mathrm{iso}\,\sigma_{\mathrm{ap}}(T)$, thus T has the SVEP at 0. By Theorem 2.97 it then follows that $H_0(T) = \ker T^\nu$ for some $\nu \in \mathbb{N}$.

(ii) $\Rightarrow$ (iii) Clear.

(iii) $\Rightarrow$ (i) Suppose that $H_0(\lambda I - T)$ is closed for all $\lambda \in \Delta_a^g(T)$. Then T has the SVEP at every $\lambda \in \Delta_a^g(T)$. Now, if $\lambda \notin \sigma_{\mathrm{ubw}}(T)$ there are two possibilities: if $\lambda \notin \sigma_{\mathrm{ap}}(T)$ then T has the SVEP at λ. If $\lambda \in \sigma_{\mathrm{ap}}(T)$ then $\lambda \in \Delta_a^g(T)$, so also in this case T has the SVEP at λ. By Theorem 5.37, T satisfies the generalized a-Browder's theorem.

(i) $\Rightarrow$ (iv) By Theorem 5.40 if T satisfies the generalized Browder's theorem then $\Delta_a^g(T) \subseteq \mathrm{iso}\,\sigma_{\mathrm{ap}}(T)$. If $\lambda_0 \in \Delta_a^g(T)$ then, by Lemma 5.39, $\alpha(\lambda_0 I - T) > 0$ and since $\Delta^a(T) \subseteq \mathrm{iso}\,\sigma_{\mathrm{ap}}(T)$ there exists a punctured open disc $\mathbb{D}(\lambda_0)$ centered at λ_0 such that $\alpha(\lambda_0 I - T) = 0$ for all $\lambda \in \mathbb{D}(\lambda_0)$. Hence $\lambda \to \ker(\lambda I - T)$ is discontinuous at λ_0 in the gap metric.

(iv) $\Rightarrow$ (i) We show that $\Delta_a^g(T)(T) \subseteq \mathrm{iso}\,\sigma_{\mathrm{ap}}(T)$, so Theorem 5.40 applies. Let $\lambda_0 \in \Delta_a^g(T)(T)$ be arbitrary. Then $\lambda_0 I - T$ is upper semi B-Fredholm with $\mathrm{ind}\,(\lambda_0 I - T) \leq 0$. By Theorem 1.117 we know that there exists an open disc $\mathbb{D}(\lambda_0, \varepsilon)$ such that $\lambda I - T$ is upper semi Fredholm for all $\lambda \in \mathbb{D}(\lambda_0, \varepsilon) \setminus \{\lambda_0\}$, $\alpha(\lambda I - T)$ is constant as λ ranges over $\mathbb{D}(\lambda_0, \varepsilon) \setminus \{\lambda_0\}$,

$$\mathrm{ind}(\lambda I - T) = \mathrm{ind}(\lambda_0 I - T) \quad \text{for all} \quad \lambda \in \mathbb{D}(\lambda_0, \varepsilon),$$

and

$$0 \leq \alpha(\lambda I - T) \leq \alpha(\lambda_0 I - T) \quad \text{for all } \lambda \in \mathbb{D}(\lambda_0, \varepsilon).$$

Since the mapping $\lambda \to \ker(\lambda I - T)$ is discontinuous at every $\lambda \in \Delta_a^g(T)$,

$$0 \leq \alpha(\lambda I - T) < \alpha(\lambda_0 I - T) \quad \text{for all } \lambda \in \mathbb{D}(\lambda_0, \varepsilon) \setminus \{\lambda_0\}.$$

We claim that

$$\alpha(\lambda I - T) = 0 \quad \text{for all } \lambda \in \mathbb{D}(\lambda_0, \varepsilon) \setminus \{\lambda_0\}. \tag{5.11}$$

To see this, suppose that there exists a $\lambda_1 \in \mathbb{D}(\lambda_0, \varepsilon) \setminus \{\lambda_0\}$ such that $\alpha(\lambda_1 I - T) > 0$. Clearly, $\lambda_1 \in \Delta^a(T)$, so arguing as for λ_0 we obtain a $\lambda_2 \in \mathbb{D}(\lambda_0, \varepsilon) \setminus \{\lambda_0, \lambda_1\}$ such that

$$0 < \alpha(\lambda_2 I - T) < \alpha(\lambda_1 I - T),$$

and this is impossible since $\alpha(\lambda I - T)$ is constant for all $\lambda \in \mathbb{D}(\lambda_0, \varepsilon) \setminus \{\lambda_0\}$. Therefore (5.11) is satisfied and since $\lambda I - T$ is upper semi-Fredholm for all $\lambda \in \mathbb{D}(\lambda_0, \varepsilon) \setminus \{\lambda_0\}$, the range $(\lambda I - T)(X)$ is closed for all $\lambda \in \mathbb{D}(\lambda_0, \varepsilon) \setminus \{\lambda_0\}$, thus $\lambda_0 \in \mathrm{iso}\,\sigma_{\mathrm{ap}}(T)$, as desired. $\blacksquare$

5.3 Property (*b*)

In this section we introduce a new property which implies *a*-Browder's theorem. We have seen in Theorem 5.27 that $T \in L(X)$ satisfies *a*-Browder's theorem exactly when T has the SVEP at the points $\lambda \in \Delta_a(T)$. In this section we consider those operators for which the dual T^* has the SVEP at the points $\lambda \in \Delta_a(T)$. We shall see that this condition is stronger than the SVEP for T at the points $\lambda \in \Delta_a(T)$.

Define $\Delta_s(T) := \sigma_s(T) \setminus \sigma_{lw}(T)$. We easily have

$$\Delta_s(T) = \{\lambda \in \mathbb{C} : \lambda I - T \in W_-(X), \, 0 < \beta(\lambda I - T)\}.$$

Lemma 5.43 *For all* $T \in L(X)$ *we have*

$$p_{00}(T) \subseteq \Delta_a(T) \cap \Delta_s(T).$$

Furthermore,

$$\Delta_s(T) = \Delta_a(T^*).$$

Proof If $\lambda \in p_{00}(T)$ then $\alpha(\lambda I - T) = \beta(\lambda I - T) > 0$, since $\lambda \in \sigma(T)$. Moreover, $\lambda I - T$ is both upper semi-Weyl and lower semi-Weyl, so $p_{00}(T) \subseteq \Delta_a(T) \cap \Delta_s(T)$. The equality $\Delta_s(T) = \Delta_a(T^*)$ is a clear consequence of the equalities $\sigma_s(T) = \sigma_{ap}(T^*)$ and $\sigma_{uw}(T^*) = \sigma_{lw}(T)$. ∎

The following property was introduced by Berkani and Zariuoh in [74, 75].

Definition 5.44 $T \in L(X)$ satisfies *property* (*b*) if $\Delta_a(T) = p_{00}(T)$.

Property (*b*) may be characterized by the SVEP of T^* at the points of $\Delta_a(T)$. More precisely, we have:

Theorem 5.45 *For a bounded operator* $T \in L(X)$ *the following statements are equivalent:*

(i) *T satisfies property* (*b*);
(ii) *T^* has the SVEP at every* $\lambda \in \Delta_a(T)$;
(iii) $\Delta_a(T) \subseteq \text{iso}\,\sigma(T)$.
  Dually, for every $T \in L(X)$ *the following statements are equivalent:*
(iv) *T^* satisfies property* (*b*);
(v) *T has the SVEP at every* $\lambda \in \Delta_s(T)$;
(vi) $\Delta_s(T) \subseteq \text{iso}\,\sigma(T)$.

Proof (i) $\Leftrightarrow$ (ii) Suppose that T satisfies property (*b*) and let $\lambda \in \Delta_a(T) = p_{00}(T)$ be arbitrarily given. Then λ is an isolated point of $\sigma(T) = \sigma(T^*)$, so T^* has the SVEP at λ.

Conversely, assume (ii) and let $\lambda \in \Delta_a(T)$. Then $\lambda I - T \in W_+(X)$, i.e., $T \in \Phi_+(X)$ and ind $(\lambda I - T) \leq 0$. The SVEP of T^* at λ implies, by Theorem 2.98, that $q(\lambda I - T) < \infty$, and hence ind $(\lambda I - T) \geq 0$, by Theorem 1.22. Therefore, ind $(\lambda I - T) = 0$ and, again by Theorem 1.22, we obtain $p(\lambda I - T) < \infty$. Consequently, $\lambda \in p_{00}(T)$. This shows the inclusion $\Delta_a(T) \subseteq p_{00}(T)$. The opposite inclusion is true for every operator, so that $\Delta_a(T) = p_{00}(T)$.

(i) $\Rightarrow$ (iii) Obvious, since $p_{00}(T) \subseteq \mathrm{iso}\,\sigma(T)$.

(iii) $\Rightarrow$ (ii) Obvious, since T^* has the SVEP at every isolated point of $\sigma(T)$.

The equivalence (iv)–(v)–(vi) concerning property (b) for T^* may be proved by using dual arguments and Lemma 5.43:

(iv) $\Leftrightarrow$ (v) If T^* has property (b) then $p_{00}(T) = p_{00}(T^*) = \Delta_a(T^*) = \Delta_s(T)$, so every $\lambda \in \Delta_s(T)$ is an isolated point of the spectrum of T and, consequently, T has the SVEP at λ. Conversely, suppose (ii). To show that T^* has property (b) we need only to prove $\Delta_a(T^*) \subseteq p_{00}(T^*)$. Let $\lambda \in \Delta_a(T^*) = \Delta_s(T)$. Then $\lambda I - T \in W_-(X)$. Since T has the SVEP at λ, $p(\lambda I - T) < \infty$, by Theorem 2.97, and hence ind $(\lambda I - T) \leq 0$, by Theorem 1.22. But $\lambda I - T \in W_-(X)$, thus ind $(\lambda I - T) = 0$, and this implies that $q(\lambda I - T) < \infty$, again by Theorem 1.22. Thus, $\lambda \in p_{00}(T) = p_{00}(T^*)$.

(iv) $\Leftrightarrow$ (vi) From the equivalence (i) $\Leftrightarrow$ (iii), T^* satisfies property (b) if and only if $\Delta_s(T) = \Delta_a(T^*) \subseteq \mathrm{iso}\,\sigma(T^*) = \mathrm{iso}\,\sigma(T)$.   ■

Corollary 5.46 *If T^* has the SVEP then property (b) holds for T, and analogously if T has the SVEP then property (b) holds for T^*.*

The previous corollary may be extended as follows:

Corollary 5.47 *Suppose that $T, R \in L(X)$ commutes and R is a Riesz operator.*

(i) *If T^* has the SVEP then property (b) holds for $T + R$, and analogously if T has the SVEP then property (b) holds for $T^* + R^*$.*

(ii) *If T^* has the SVEP and $K \in L(X)$ is an algebraic operator which commutes with T, then $T + K$ satisfies property (b). If T has the SVEP, then $T^* + K^*$ satisfies property (b).*

(iii) *If T^* has the SVEP then $f(T)$ satisfies property (b) for every $f \in \mathcal{H}(\sigma(T))$ and analogously, if T has the SVEP then $f(T^*)$ satisfies property (b) for every $f \in \mathcal{H}(\sigma(T))$.*

(iv) *If $T \in L(X)$ has the SVEP and T and $S \in L(Y)$ are intertwined by an injective map $A \in L(X, Y)$ then S satisfies property (b).*

Proof (i) follows from Theorem 2.129 and Corollary 5.46. Statement (ii) follows from Theorem 2.145 and Corollary 5.46, once we observe that K^* is also algebraic. Assertion (iii) immediately follows, since by Theorem 2.86 $f(T)$ or $f(T^*)$ has the SVEP. (iv) The operator S has the SVEP by Lemma 2.141.   ■

Property (*b*) entails *a*-Browder's theorem (or equivalently the generalized *a*-Browder's theorem):

Corollary 5.48 *If property (b) holds for T then a-Browder's theorem holds for T.*

Proof Every isolated point of the spectrum belongs to $\sigma_{ap}(T)$. Consequently, iso $\sigma(T) \subseteq$ iso $\sigma_{ap}(T)$ and hence, combining Theorems 5.45 and 5.31, we conclude that property (*b*) entails *a*-Browder's theorem. ∎

The following example shows that the converse of the result of Corollary 5.48 does not hold in general. This example also shows that the SVEP for T does not ensure, in general, that property (*b*) holds for T.

Example 5.49 Let $R \in L(\ell_2(\mathbb{N}))$ denote the right shift and let $P \in L(\ell_2(\mathbb{N}))$ be the idempotent operator defined by $P(x) := (0, x_2, x_3, \dots)$ for all $x = (x_1, x_2, \dots) \in \ell_2(\mathbb{N})$.

It is easily seen that if $T := R \oplus P$ then $\sigma(T) = \sigma_w(T) = \mathbf{D}(0, 1)$, where $\mathbf{D}(0, 1)$, and $\sigma_{ap}(T) = \Gamma \cup \{0\}$, where Γ denotes the unit circle. Since $\sigma_{uw}(T) = \Gamma$, we then have $\Delta_a(T) = \{0\}$, while $\sigma(T)$ has no isolated points. Therefore, T does not have property (*b*), while T inherits SVEP from R and P, thus, by Theorem 5.27, T satisfies *a*-Browder's theorem.

Now, let us consider the condition that T^* has the SVEP at every $\lambda \notin \sigma_{uw}(T)$. Clearly, $\Delta_a(T) \subseteq \mathbb{C} \setminus \sigma_{uw}(T)$, so, from Theorem 5.45, we obtain:

$$T^* \text{ has the SVEP at every } \lambda \notin \sigma_{uw}(T) \Rightarrow T \text{ has property } (b).$$

Next we give an example of an operator T for which property (*b*) holds, while T^* may fail SVEP at some points $\lambda \notin \sigma_{uw}(T)$.

Example 5.50 Let $T := R \oplus S$, where R is the right shift on $\ell_2(\mathbb{N})$ and $S \in L(\ell_2(\mathbb{N}))$ is defined as

$$S(x_1, x_2, x_3, \dots) := \left(\frac{1}{2}x_2, \frac{1}{3}x_3, \frac{1}{4}x_4, \dots\right) \quad \text{for all } (x_k) \in \ell_2(\mathbb{N}).$$

Then $\sigma(T) = \sigma_w(T) = \mathbf{D}(0, 1)$. This implies that the set of poles is empty, in particular $p_{00}(T) = \emptyset$. On the other hand,

$$\sigma_{ap}(T) = \sigma_{uw}(T) = \Gamma \cup \{0\},$$

Γ the unit circle of $\mathbb{C}$, so T possesses property (*b*). Suppose now that T^* has the SVEP at every $\lambda \notin \sigma_{uw}(T)$. If $\lambda \notin \sigma_{uw}(T)$ then $\lambda I - T$ is upper Weyl, and the SVEP of T^* ensures, by Corollary 2.106, that ind $(\lambda I - T) \leq 0$, hence $\lambda I - T$ is Weyl, and since the inclusion $\sigma_{uw}(T) \subseteq \sigma_w(T)$ is satisfied by every operator it then follows that $\sigma_w(T) = \sigma_{uw}(T)$, and this is not possible. Therefore there exists a $\lambda_0 \notin \sigma_{uw}(T)$ such that T^* fails the SVEP at λ_0. Hence, T^* does not have the SVEP at every point $\lambda \notin \sigma_{uw}(T)$. Now, set $U := T^*$. Clearly $U^* = T$ has property (*b*), but the SVEP fails at $\lambda_0 \notin \sigma_{uw}(T) = \sigma_{uw}(U)$.

Theorem 5.51 *Let* $T \in L(X)$*. Then we have*

(i) T^* *has the SVEP at every* $\lambda \notin \sigma_{\mathrm{uw}}(T)$ *if and only if* T *has property* (b) *and* $\sigma_{\mathrm{uw}}(T) = \sigma_{\mathrm{w}}(T)$*.*
(ii) T *has the SVEP at every* $\lambda \notin \sigma_{\mathrm{lw}}(T)$ *if and only if* T^* *has property* (b) *and* $\sigma_{\mathrm{lw}}(T) = \sigma_{\mathrm{w}}(T)$*.*

Furthermore, if T *satisfies property* (b) *then*

$$\Delta_a(T) = \Delta(T) = p_{00}(T) = p_{00}^a(T).$$

Proof

(i) As observed above the SVEP for T^* at the points $\lambda \notin \sigma_{\mathrm{uw}}(T)$ entails property (b). If $\lambda \notin \sigma_{\mathrm{uw}}(T)$ then $\lambda I - T$ is upper semi-Fredholm with index less than or equal to 0. The SVEP for T^* at λ implies, by Corollary 2.106, that $\mathrm{ind}\,(\lambda I - T) \geq 0$, so $\lambda I - T$ is Weyl and hence $\lambda \notin \sigma_{\mathrm{w}}(T)$. The inclusion $\sigma_{\mathrm{uw}}(T) \subseteq \sigma_{\mathrm{w}}(T)$ holds for every operator, thus $\sigma_{\mathrm{uw}}(T) = \sigma_{\mathrm{w}}(T)$.

Conversely, suppose that T has property (b) and $\sigma_{\mathrm{uw}}(T) = \sigma_{\mathrm{w}}(T)$. Let $\lambda \notin \sigma_{\mathrm{uw}}(T)$. Then $\lambda I - T$ is Weyl. There are two possibilities. If $\lambda \notin \sigma_{\mathrm{ap}}(T)$ then $0 = \alpha(\lambda I - T) = \beta(\lambda I - T)$, so $\lambda \notin \sigma(T) = \sigma(T^*)$, and, trivially, T^* has the SVEP at λ. If $\lambda \in \sigma_{\mathrm{ap}}(T)$ then $\lambda \in \Delta_a(T) \subseteq \mathrm{iso}\,\sigma(T)$, thus T^* has the SVEP at λ.

(ii) Suppose that T has the SVEP at every $\lambda \notin \sigma_{\mathrm{lw}}(T)$. We show first that T^* satisfies property (b). By Theorem 5.45 we need to show that $\Delta_s(T) \subseteq \mathrm{iso}\,\sigma(T)$. Let $\lambda \in \Delta_s(T) = \sigma_s(T) \setminus \sigma_{\mathrm{lw}}(T)$. Then $\lambda I - T$ is lower semi-Weyl, and hence $\mathrm{ind}\,(\lambda I - T) \geq 0$, while the SVEP for T implies, by Corollary 2.106, that $\mathrm{ind}\,(\lambda I - T) \leq 0$. Therefore $\lambda I - T$ is Weyl, and another consequence of the SVEP at λ is that $p(\lambda I - T) < \infty$. By Theorem 1.22 we then conclude that $\lambda I - T$ is Browder, and hence λ is an isolated point of $\sigma(T)$. Therefore, property (b) holds for T^*. The above argument shows that $\sigma_{\mathrm{w}}(T) \subseteq \sigma_{\mathrm{lw}}(T)$, from which it follows that $\sigma_{\mathrm{lw}}(T) = \sigma_{\mathrm{w}}(T)$.

Conversely, suppose that T^* has property (b) and $\sigma_{\mathrm{lw}}(T) = \sigma_{\mathrm{w}}(T)$ and let $\lambda \notin \sigma_{\mathrm{lw}}(T)$. If $\lambda \notin \sigma_s(T)$ then $\lambda I - T$ is onto, and hence $\lambda \notin \sigma_{\mathrm{lw}}(T) = \sigma_{\mathrm{w}}(T)$, consequently $\lambda I - T$ is injective, so $\lambda \notin \sigma(T) = \sigma(T^*)$ and hence T^* has the SVEP at λ. Consider the other case where $\lambda \in \sigma_s(T)$. Then $\lambda \in \Delta_s(T) = p_{00}(T)$, hence λ is an isolated point of $\sigma(T) = \sigma(T^*)$, thus T^* has the SVEP at λ.

The last assertion is clear: property (b) entails a-Browder's theorem, and hence Browder's theorem. ∎

It has been already observed in Theorem 5.36 that the condition that $K(\lambda I - T)$ has finite codimension for all $\lambda \in \Delta_a(T)$ implies that T satisfies a-Browder's theorem. Since the quasi-nilpotent part of an operator is, in a sense, near to be the topological complement of the analytic core, the result of Corollary 5.41 could suggest that the converse of this implication holds. This is not true. Indeed, we next shows that this condition characterizes property (b), which is a formally stronger condition than a-Browder's theorem.

Theorem 5.52 *For an operators* $T \in L(X)$ *the following statements are equivalent:*

(i) *T satisfies property (b);*
(ii) $K(\lambda I - T)$ *has finite codimension for all* $\lambda \in \Delta_a(T)$.
   In this case there exists a natural $\nu := \nu(\lambda)$ *such that* $H_0(\lambda I - T) =$ $\ker(\lambda I - T)^\nu$ *and* $K(\lambda I - T) = (\lambda I - T)^\nu(X)$ *for all* $\lambda \in \Delta_a(T)$.
   Dually, the following statements are equivalent:
(iii) *T* satisfies property (b);*
(iv) $H_0(\lambda I - T)$ *has finite dimension for all* $\lambda \in \Delta_s(T)$.

In this case there exists a natural $\nu := \nu(\lambda)$ *such that* $H_0(\lambda I - T) = \ker(\lambda I - T)^\nu$ *and* $K(\lambda I - T) = (\lambda I - T)^\nu(X)$ *for all* $\lambda \in \Delta_s(T)$.

Proof (i) $\Leftrightarrow$ (ii) Suppose first that T has property (*b*). Then, by Theorem 5.45, every $\lambda \in \Delta_a(T)$ is an isolated point of $\sigma(T)$. Since T satisfies a-Browder's theorem we know, from Theorem 5.35, that $H_0(\lambda I - T)$ is finite dimensional. From the decomposition $X = H_0(\lambda I - T) \oplus K(\lambda I - T)$ we then conclude that $K(\lambda I - T)$ has finite codimension.

Conversely, suppose that $K(\lambda I - T)$ has finite codimension at every point $\lambda \in \Delta_a(T)$. If $\lambda \in \Delta_a(T)$ then $\lambda I - T \in W_+(X)$ and hence, by Theorem 2.98, T^* has SVEP at λ, so Theorem 5.45 implies that T satisfies property (*b*).

The last assertion holds since every $\lambda \in \Delta_a(T) = p_{00}(T)$ is a pole, and if ν is the order of λ then, by Corollary 2.47, $H_0(\lambda I - T) = \ker(\lambda I - T)^\nu$ and $K(\lambda I - T) = (\lambda I - T)^\nu(X)$.

(iii) $\Leftrightarrow$ (iv) Suppose that T^* has property (*b*). From the equivalence (i) $\Leftrightarrow$ (ii) we know that $K(\lambda I - T^*)$ has finite codimension for all $\lambda \in \Delta_a(T^*) = \Delta_s(T)$. Let $\lambda \in \Delta_s(T)$ arbitrary given. Then $\lambda \in \Delta_a(T^*) = p_{00}(T^*) = p_{00}(T)$. Hence λ is a pole of the resolvent of T, as well a pole of the resolvent of T^*, and it is well known that the order of λ as a pole of T coincides with the order of λ as a pole of T^*. Let ν be the order of λ. By Corollary 2.47 we know that $K(\lambda I - T^*)$ is closed and $K(\lambda I - T^*) = (\lambda I - T^*)^\nu(X^*)$, while $H_0(\lambda I - T) = \ker(\lambda I - T)^\nu$. If $M^\perp$ denote the annihilator of $M \subseteq X$, from the classical closed range theorem we have:

$$K(\lambda I - T^*) = (\lambda I - T^*)^\nu(X^*) = [\ker(\lambda I - T)^\nu]^\perp = H_0(\lambda I - T)^\perp,$$

so $H_0(\lambda I - T)^\perp$ has finite codimension in X^*. But from a standard result of functional analysis we know that the quotient $X^*/H_0(\lambda I - T)^\perp$ is isomorphic to $[H_0(\lambda I - T)]^*$, hence the latter space is finite dimensional and, consequently, also $H_0(\lambda I - T)$ is finite dimensional, as desired.

Conversely, suppose that $H_0(\lambda I - T)$ has finite dimension for all $\lambda \in \Delta_s(T)$. Let $\lambda \in \Delta_s(T)$ be arbitrary. Since $\lambda I - T \in W_-(X)$ then T has SVEP at λ, by Theorem 2.97, and hence by Theorem 5.45 T^* satisfies property (*b*).

The last assertion is immediate. Since

$$\Delta_s(T) = \Delta_a(T^*) = p_{00}(T^*) = p_{00}(T),$$

every $\lambda \in \Delta_s(T)$ is a pole, and if ν is the order of λ, then $H_0(\lambda I - T) = \ker (\lambda I - T)^\nu$ while $K(\lambda I - T) = (\lambda I - T)^\nu(X)$, again by Corollary 2.47. ∎

Note that in the statement (iv) of Theorem 5.52 the condition that $H_0(\lambda I - T)$ has finite dimension for all $\lambda \in \Delta_s(T)$ may be replaced by the formally weaker condition that $H_0(\lambda I - T)$ is closed for all $\lambda \in \Delta_s(T)$. This is a consequence of Theorem 2.97.

Theorem 5.53 *If $T \in L(X)$ the following equivalences hold:*

(i) *T satisfies property (b) if and only if $(\lambda I - T)^\infty(X)$ has finite codimension for all $\lambda \in \Delta_a(T)$.*

(ii) *T^* satisfies property (b) if and only if $\mathcal{N}^\infty(\lambda I - T)$ has finite dimension for all $\lambda \in \Delta_s(T)$.*

Proof

(i) Suppose that property (b) holds for T and let $\lambda \in \Delta_a(T)$. Since $K(\lambda I - T) \subseteq (\lambda I - T)^\infty(X)$ it then follows, by Theorem 5.52, that $(\lambda I - T)^\infty(X)$ has finite codimension. Conversely, suppose that $(\lambda I - T)^\infty(X)$ has finite codimension for all $\lambda \in \Delta_a(T)$. From the inclusions $(\lambda I - T)^\infty(X) \subseteq (\lambda I - T)^{n+1}(X) \subseteq (\lambda I - T)^n(X)$, it is evident that $(\lambda I - T)^{\nu+1}(X) = (\lambda I - T)^\nu(X)$ for some $\nu \in \mathbb{N}$, hence $q(\lambda I - T) < \infty$ and this implies that T^* has SVEP at λ, so, by Theorem 5.45, T satisfies property (b).

(ii) We have $\mathcal{N}^\infty(\lambda I - T) \subseteq H_0(\lambda I - T)$, so, if T^* has property (b) and $\lambda \in \Delta_s(T)$, Theorem 5.52 entails that $\mathcal{N}^\infty(\lambda I - T)$ has finite dimension. Conversely, suppose that $\lambda \in \Delta_s(T)$ and that $\mathcal{N}^\infty(\lambda I - T)$ is finite dimensional. From the inclusions $\ker(\lambda I - T)^n \subseteq \ker(\lambda I - T)^{n+1} \subseteq \mathcal{N}^\infty(\lambda I - T)$, we easily see that $p(\lambda I - T) < \infty$, so T has SVEP at λ. By Theorem 5.45 we then conclude that T^* satisfies property (b). ∎

Next we show the exact relation between property (b), Browder's theorem and a-Browder's theorem:

Theorem 5.54 *If $T \in L(X)$ the following statements are equivalent:*

(i) *T satisfies property (b);*

(ii) *T satisfies a-Browder's theorem and $p_{00}^a(T) = p_{00}(T)$;*

(iii) *T satisfies Browder's theorem and $\operatorname{ind}(\lambda I - T) = 0$ for all $\lambda \in \Delta_a(T)$.*

Proof (i) $\Leftrightarrow$ (ii) Suppose (i). By Corollary 5.48 property (b) entails a-Browder's theorem and, by Theorem 5.51, $p_{00}^a(T) = p_{00}(T)$. Conversely, if (ii) holds, then T

satisfies *a*-Browder's theorem and

$$\Delta_a(T) = p_{00}^a(T) = p_{00}(T) \subseteq \text{iso}\,\sigma(T),$$

thus, by Theorem 5.45, T has property (*b*).

(ii) $\Rightarrow$ (iii) *a*-Browder's theorem implies Browder's theorem. Furthermore, $\Delta_a(T) = p_{00}^a(T) = p_{00}(T)$, so $\lambda I - T$ is Browder for all $\lambda \in \Delta_a(T)$. In particular, $\lambda I - T$ has index 0.

(iii) $\Rightarrow$ (i) Since T satisfies Browder's theorem we have $\sigma_w(T) = \sigma_b(T)$. Since $\lambda I - T$ is Weyl for every $\lambda \in \Delta_a(T)$, we have $\Delta_a(T) = \sigma_{ap}(T) \setminus \sigma_w(T)$ and hence

$$\Delta_a(T) \subseteq \sigma(T) \setminus \sigma_w(T) = \sigma(T) \setminus \sigma_b(T) = p_{00}(T) \subseteq \text{iso}\,\sigma(T),$$

and hence T has property (*b*), by Theorem 5.45. ∎

Corollary 5.55 *If* $T \in L(X)$ *is a-polaroid, then property* (*b*) *for* T *is equivalent to a-Browder's theorem for* T.

Proof Property (*b*) entails *a*-Browder's theorem. To prove the converse assume that T satisfies *a*-Browder's theorem. Then, by Corollary 5.41, $\Delta_a(T) \subseteq \text{iso}\,\sigma_{ap}(T) = \text{iso}\,\sigma(T)$, and hence by Theorem 5.45 T has property (*b*). ∎

Remark 5.56 The operator T defined in Example 5.49 shows that the equivalence established in Theorem 5.54 does not hold if the assumption that T is *a*-polaroid is replaced by the weaker assumption that T is polaroid. Indeed, T is polaroid (since $\text{iso}\,\sigma(T) = \emptyset$), and T satisfies *a*-Browder's theorem, since it satisfies SVEP, but not property (*b*).

Property (*b*) is invariant under nilpotent commuting perturbations.

Theorem 5.57 *Let* $T \in L(X)$ *satisfy property* (*b*), *and let* N *be a nilpotent operator which commutes with* T. *Then* $T + N$ *has property* (*b*).

Proof We know that $\sigma(T) = \sigma(T + N)$ and $\sigma_b(T) = \sigma_b(T + N)$, by Corollary 3.9, so $p_{00}(T) = p_{00}(T + N)$. On the other hand, we easily have $\Delta_a(T) = \Delta_a(T + N)$, and property (*b*) for T gives $p_{00}(T + N) = p_{00}(T) = \Delta_a(T) = \Delta_a(T + N)$, hence $T + N$ has property (*b*). ∎

The previous result does not extend to commuting quasi-nilpotent perturbations. A simple counterexample is the following. Let $T = 0$ and Q be the quasi-nilpotent operator defined as

$$Q(x_1, x_2, \dots) := \left(0, \frac{1}{2}x_2, \frac{1}{3}x_3, \dots\right) \quad \text{for all } (x_i) \in \ell_2(\mathbb{N}).$$

Then $\Delta_a(T) = \Pi(T) = \emptyset$, while $\Pi(T + Q) = \{0\} \neq \Delta_a(T + Q)$.

Recall that T is said to be *finitely polaroid* if every $\lambda \in \text{iso}\,\sigma(T)$ is a pole of finite rank, i.e. $\alpha(\lambda I - T) < \infty$.

Theorem 5.58 *Suppose that $T \in L(X)$ is finitely polaroid and has property (b). If $Q \in L(X)$ is a quasi-nilpotent operator which commutes with T, then $T + Q$ has property (b).*

Proof We know that $\sigma_{ap}(T)$ and $\sigma_{uw}(T)$ are invariant under commuting quasi-nilpotent perturbations, so $\Delta_a(T) = \Delta(T + Q)$. We show that $\Pi(T) = \Pi(T + Q)$. Let $\lambda \in \Pi(T) = \sigma(T) \setminus \sigma_d(T)$ be arbitrarily given. Then $\lambda I - T$ has both finite ascent and descent, and since T is finitely polaroid we also have $\alpha(\lambda I - T) < \infty$. From Theorem 1.22 we then have that $\lambda I - T$ is Browder, hence $\lambda I - (T + Q)$ is also Browder, by Theorem 3.8. Since $\sigma(T) = \sigma(T + Q)$, then $\lambda \in \sigma(T + Q) \setminus \sigma_b(T + Q) = p_{00}(T + Q) \subseteq \Pi(T + Q)$. Hence, $\Pi(T) \subseteq \Pi(T + Q)$. To show the opposite inclusion, let $\lambda \in \Pi(T + Q)$. Then $\lambda I - (T + Q)$ is Drazin invertible, and hence $\lambda \in \mathrm{iso}\,\sigma(T + Q) = \mathrm{iso}\,\sigma(T)$. Since T is finitely polaroid then λ is a pole of the resolvent having finite rank. By Theorem 1.22, $\lambda I - T$ is Browder and hence $\lambda I - (T + Q)$ is Browder, by Theorem 3.8. Therefore $\lambda \in \sigma(T + Q) \setminus \sigma_b(T + Q) = \sigma(T) \setminus \sigma_b(T) = p_{00}(T) \subseteq \Pi(T)$. Therefore, $\Pi(T) = \Pi(T + Q)$, and consequently, since T has property (b),

$$\Delta_a(T + Q) = \Delta_a(T) = \Pi(T) = \Pi(T + Q),$$

which shows that $T + Q$ has property (b).   ■

Theorem 5.59 *If $T \in L(X)$ has the SVEP, and $\mathrm{iso}\,\sigma_{ap}(T) = \emptyset$, then $T + Q$ has property (b) for every commuting quasi-nilpotent operator $Q \in L(X)$.*

Proof We have $\sigma_{ap}(T) = \sigma_{ap}(T+Q)$, so $\mathrm{iso}\,\sigma_{ap}(T+Q) = \emptyset$. Clearly, $\Pi(T+Q) = \emptyset$, since every pole of an operator is an eigenvalue.

Suppose now that $\Delta_a(T + Q) \neq \emptyset$, and $\lambda \in \Delta_a(T + Q) = \sigma_{ap}(T + Q) \setminus \sigma_{uw}(T + Q)$. By Theorem 3.8 we know that $T + Q$ has the SVEP, hence $\sigma_{ap}(T + Q)$ does not cluster at λ, by Theorem 2.97. Since $\lambda \in \sigma_{ap}(T + Q)$, λ is an isolated point of $\sigma_{ap}(T + Q)$, and this is impossible. Therefore,

$$\Delta_a(T + Q) = \Pi(T + Q) = \emptyset,$$

so $T + Q$ has property (b).   ■

We have seen that Browder's theorem, as well as a-Browder's theorem are invariant under commuting Riesz perturbations. We now give a sufficient condition for the permanence of property (b) under commuting perturbations K for which some power K^n is finite-dimensional.

Theorem 5.60 *Suppose that $T \in L(X)$ and $K \in L(X)$ is an operator which commutes with T such that K^n is finite-dimensional for some $n \in \mathbb{N}$. If $\mathrm{iso}\,\sigma_{ap}(T) = \mathrm{iso}\,\sigma_{ap}(T + K)$ then property (b) for T implies property (b) for $T + K$.*

Proof The assumption iso $\sigma_{ap}(T) = $ iso $\sigma_{ap}(T + K)$ implies that $\sigma_{ap}(T) = \sigma_{ap}(T + K)$. Indeed, from our assumption and Theorem 3.26 we have

$$\sigma_{ap}(T + K) = \text{iso}\,\sigma_{ap}(T + K) \cup \text{acc}\,\sigma_{ap}(T + K)$$
$$= \text{iso}\,\sigma_{ap}(T) \cup \text{acc}\,\sigma_{ap}(T) = \sigma_{ap}(T).$$

Note that K is a Riesz operator and hence, by Corollary 3.18, $\sigma_{uw}(T) = \sigma_{uw}(T + K)$, from which we conclude that $\Delta_a(T) = \Delta_a(T + K)$. To conclude the proof it suffices to prove $\Pi(T) = \Pi(T + K)$. If $\lambda \in \Pi(T) = \sigma(T) \setminus \sigma_d(T)$ then $\lambda I - T$ is Drazin invertible, and hence, by Theorem 3.78, $\lambda I - (T + K)$ is Drazin invertible. Moreover, since $\lambda I - T$ is Drazin invertible, $\lambda \in \text{iso}\,\sigma_{ap}(T)$, by Theorem 4.3. Hence $\lambda \in \text{iso}\,\sigma_{ap}(T + K)$, and in particular $\lambda \in \sigma(T + K)$. Therefore,

$$\lambda \in \sigma(T + K) \setminus \sigma_d(T + K) = \Pi(T + K),$$

and this shows that $\Pi(T) \subseteq \Pi(T + K)$. A symmetric argument shows that the opposite inclusion is also true, so $\Pi(T) = \Pi(T + K)$. If T has property (b) then

$$\Pi(T) = \Delta_a(T) = \Delta_a(T + K) = \Pi(T + K),$$

thus $T + K$ has property (b). ∎

5.4 Property (gb)

Property (gb), which was also introduced by Berkani and Zariouh in [66], is defined by generalizing property (b) in the sense of the B-Fredholm property. This property, which is formally stronger that property (b), may also be characterized by means of the localized single valued extension property at the point of a certain set.

Definition 5.61 $T \in L(X)$ satisfies *property* (gb) if $\Delta_a^g(T) = \Pi(T)$, i.e., every point of $\Delta_a^g(T)$ is a pole of the resolvent.

Property (gb) may be characterized by means of the localized SVEP as follows.

Theorem 5.62 *For a bounded operator* $T \in L(X)$ *the following statements are equivalent:*

(i) *T satisfies property (gb);*
(ii) $\Delta_a^g(T) \subseteq \text{iso}\,\sigma(T)$;
(iii) $\Delta_a^g(T) \subseteq \text{iso}\,\sigma_s(T)$;
(iv) *T^* has the SVEP at every $\lambda \in \Delta_a^g(T)$;*
(v) $q(\lambda I - T) < \infty$ *for all $\lambda \in \Delta_a^g(T)$.*

Proof The implication (i) $\Rightarrow$ (ii) is obvious, since $\Pi(T) \subseteq \operatorname{iso} \sigma(T)$.

The implication (ii) $\Rightarrow$ (iii) easily follows, once we observe that the isolated points of the spectrum belong to $\sigma_s(T)$.

The implication (iii) $\Rightarrow$ (iv) is clear: for every operator T, its dual T^* has the SVEP at the points $\lambda \in \operatorname{iso} \sigma_s(T)$.

(iv) $\Rightarrow$ (v) If $\lambda \in \Delta_a^g(T)$, then $\lambda I - T$ is upper semi B-Weyl, in particular quasi-Fredholm. By Theorem 1.142 we then have $q(\lambda I - T) < \infty$.

(v) $\Rightarrow$ (i) If $\lambda \in \Delta_a^g(T)$ then $\lambda I - T$ is upper semi B-Weyl and, by Theorem 1.143, the condition $q(\lambda I - T) < \infty$ entails that λ is a pole of the resolvent, thus $\Delta_a^g(T) \subseteq \Pi(T)$. The opposite inclusion is always true for any operator, since for a pole λ we have that $\lambda I - T$ is Drazin invertible, and in particular $\lambda I - T$ is upper semi B-Weyl. Therefore, $\Delta_a^g(T) = \Pi(T)$.   ∎

Corollary 5.63 *If T^* has the SVEP then T satisfies property (gb).*

The next example shows that the SVEP for T does not ensure, in general, that property (gb) holds for T.

Example 5.64 Let $R \in L(\ell_2(\mathbb{N}))$ denote the right shift and let $P \in L(\ell_2(\mathbb{N}))$ be the idempotent operator defined by $P(x) := (0, x_2, x_3, \dots)$ for all $x = (x_1, x_2, \dots) \in \ell_2(\mathbb{N})$. It is easily seen that if $T := R \oplus P$ then $\sigma(T) = \mathbf{D}(0, 1)$, and $\sigma_{ap}(T) = \Gamma \cup \{0\}$, where Γ denotes the unit circle. Since $\sigma_{ubw}(T) = \Gamma$, we then have $\Delta_a^g(T) = \{0\}$, while since $\sigma(T)$ has no isolated points, we have $\Pi(T) = \emptyset$. Therefore, T does not have property (gb), while T inherits SVEP from R and P.

Corollary 5.65 *If $T \in L(X)$ has property (gb) then T satisfies property (b).*

Proof Since $\sigma_{ubw}(T) \subseteq \sigma_{uw}(T)$, we have $\Delta_a(T) \subseteq \Delta_a^g(T)$, hence, by Theorem 5.62, property (gb) entails that $\Delta_a(T) \subseteq \operatorname{iso} \sigma(T)$, and the last inclusion holds precisely when T satisfies property (b), by Theorem 5.45.   ∎

The converse of Corollary 5.65 does not hold in general, as is shown by the following example.

Example 5.66 Let R denote the unilateral right shift on $\ell_2(\mathbb{N})$. Then $\sigma(R) = \mathbf{D}(0, 1)$, and $\sigma_{ap}(R)$ is the unit circle Γ. Moreover, the set of eigenvalues of R is empty, see [260]. Moreover, $\sigma_{uw}(R) = \Gamma$ and obviously, the set $\Pi(R)$ of poles of the resolvent is empty. Define $T := 0 \oplus R$. Then $\ker T = \ell_2(\mathbb{N}) \oplus \{0\}$,

$$\sigma_{uw}(T) = \sigma_{ap}(T) = \Gamma \cup \{0\},$$

and $\sigma_{ubw}(T) = \Gamma$. We also have that the set of left poles $\Pi_a(T) = \{0\}$, and $\Pi(T) = p_{00}^a(T) = \emptyset$. Hence $\Delta_a(T) = \sigma_{ap}(T) \setminus \sigma_{uw}(T) = p_{00}^a(T) = \emptyset$, while $\sigma_{ap}(T) \setminus \sigma_{ubw}(T) = \{0\} \neq \Pi(T)$. Therefore T possesses property (b), but does not have property (gb).

We know that if $\lambda \in \mathrm{iso}\,\sigma(T)$ then $X = H_0(\lambda I - T) \oplus K(\lambda I - T)$, see Theorem 2.45. Consequently, if T has property (*gb*) then

$$X = H_0(\lambda I - T) \oplus K(\lambda I - T) \quad \text{for all } \lambda \in \Delta_a^g(T) \tag{5.12}$$

since $\Delta_a^g(T) \subseteq \Pi(T)$. The following results show that property (*gb*) may be characterized by some conditions that are formally weaker than the one expressed by the decomposition (5.12).

Theorem 5.67 *For an operator* $T \in L(X)$ *the following statements are equivalent:*

(i) *T satisfies property (gb);*
(ii) $X = H_0(\lambda I - T) + K(\lambda I - T)$ *for all* $\lambda \in \Delta_a^g(T)$;
(iii) *there exists a natural* $\nu := \nu(\lambda)$ *such that* $K(\lambda I - T) = (\lambda I - T)^{\nu}(X)$ *for all* $\lambda \in \Delta_a^g(T)$;
(iv) *there exists a natural* $\nu := \nu(\lambda)$ *such that* $(\lambda I - T)^{\infty}(X) = (\lambda I - T)^{\nu}(X)$ *for all* $\lambda \in \Delta_a^g(T)$.

Proof (i) $\Leftrightarrow$ (ii) The implication (i) $\Rightarrow$ (ii) is clear, as observed in (5.12).

To show the implication (ii) $\Rightarrow$ (i) observe that the condition $X = H_0 (\lambda I - T) + K(\lambda I - T)$ is equivalent, by Theorem 2.41, to $\lambda \in \mathrm{iso}\,\sigma_s(T)$. Hence $\Delta_a^g(T) \subseteq \mathrm{iso}\,\sigma_s(T)$ and from Theorem 5.62 it immediately follows that T satisfies property (*gb*).

(i) $\Leftrightarrow$ (iii) If T satisfies property (*gb*) then, by Theorem 5.62, $q := q(\lambda I - T) < \infty$ for all $\lambda \in \Delta_a^g(T)$, so $(\lambda I - T)^{\infty}(X) = (\lambda I - T)^q(X)$. Since $\lambda I - T$ is upper semi B-Fredholm, there exists a $\nu \in \mathbb{N}$ such that $(\lambda I - T)^n(X)$ is closed for all $n \geq \nu$, so $(\lambda I - T)^{\infty}(X)$ is closed. Furthermore, by Theorem 1.79, the restriction $(\lambda I - T)|(\lambda I - T)^{\infty}(X)$ is onto, so

$$(\lambda I - T)((\lambda I - T)^{\infty}(X)) = (\lambda I - T)^{\infty}(X).$$

From Theorem 1.39, part (i), it then follows that $(\lambda I - T)^{\infty}(X) \subseteq K(\lambda I - T)$, and, since the reverse inclusion holds for every operator, we then conclude that

$$(\lambda I - T)^{\infty}(X) = K(\lambda I - T) = (\lambda I - T)^q(X).$$

Conversely, let $\lambda \in \Delta_a^g(T)$ be arbitrarily given and suppose that there exists a natural $\nu := \nu(\lambda)$ such that $K(\lambda I - T) = (\lambda I - T)^{\nu}(X)$. Then we have

$$(\lambda I - T)^{\nu}(X) = K(\lambda I - T) = (\lambda I - T)(K(\lambda I - T)) = (\lambda I - T)^{\nu+1}(X),$$

thus $q(\lambda I - T) \leq \nu$. By Theorem 5.62 we then conclude that property (*gb*) holds for T.

(i) $\Leftrightarrow$ (iv) Suppose that T satisfies property (gb). By Theorem 5.62, $q :=$ $q(\lambda I - T) < \infty$ for all $\lambda \in \Delta_a^g(T)$, and hence $(\lambda I - T)^\infty(X) = (\lambda I - T)^q(X)$ for all $\lambda \in \Delta_a^g(T)$. Conversely, suppose that (v) holds and $\lambda \in \Delta_a^g(T)$. Then

$$(\lambda I - T)^\nu(X) = (\lambda I - T)^\infty(X) \subseteq (\lambda I - T)^{\nu+1}(X),$$

and since $(\lambda I - T)^{n+1}(X) \subseteq (\lambda I - T)^n(X)$ for all $n \in \mathbb{N}$, we then obtain that $(\lambda I - T)^\nu(X) = (\lambda I - T)^{\nu+1}(X)$. Hence $q(\lambda I - T) \le \nu$, so T satisfies property (gb) by Theorem 5.62.   ∎

Property (gb) is related to Browder-type theorems as follows:

Theorem 5.68 *If $T \in L(X)$ the following statements are equivalent:*

 (i) *T satisfies property (gb);*
 (ii) *T satisfies a-Browder's theorem and $\sigma_{\mathrm{bw}}(T) \cap \Delta_a^g(T) = \emptyset$;*
(iii) *T satisfies Browder's theorem and $\sigma_{\mathrm{bw}}(T) \cap \Delta_a^g(T) = \emptyset$;*
(iv) *T has property (b) and $\Pi(T) = \Pi_a(T)$.*

Proof (i) $\Leftrightarrow$ (ii) Assume that T has property (gb), i.e., $\Delta_a^g(T) = \Pi(T)$. Property (gb) implies, by Corollary 5.65, property (b) and hence, by Corollary 5.48, a-Browder's theorem. Suppose that there exists a $\lambda \in \sigma_{\mathrm{bw}}(T) \cap \Delta_a^g(T) = \sigma_{\mathrm{bw}}(T) \cap \Pi(T)$. Then $\lambda I - T$ is Drazin invertible and hence $\lambda \notin \sigma_{\mathrm{d}}(T) = \sigma_{bw}(T)$, since the generalized Browder's theorem holds. This is a contradiction, so $\sigma_{\mathrm{bw}}(T) \cap \Delta_a^g(T) = \emptyset$.

Conversely suppose that (ii) holds. Since T satisfies the generalized a-Browder's theorem, $\sigma_{\mathrm{ubw}}(T) = \sigma_{\mathrm{ld}}(T)$. Let $\lambda \in \Pi(T)$. Then $\lambda I - T$ is Drazin invertible, in particular left Drazin invertible, and hence $\lambda \notin \sigma_{\mathrm{ubw}}(T)$. Since a pole of the resolvent is always an eigenvalue, $\lambda \in \sigma_{\mathrm{ap}}(T) \setminus \sigma_{\mathrm{ubw}}(T) = \Delta_a^g(T)$. This shows the inclusion $\Pi(T) \subseteq \Delta_a^g(T)$. To show the reverse inclusion observe first that if $\lambda \in \Delta_a^g(T)$ then $\lambda \notin \sigma_{\mathrm{bw}}(T)$, by assumption. But $\sigma_{\mathrm{bw}}(T) = \sigma_{\mathrm{d}}(T)$, since Browder's theorem holds for T, or equivalently the generalized Browder's theorem. Therefore, $\lambda \in \sigma_{\mathrm{ap}}(T) \setminus \sigma_{\mathrm{d}}(T) = \Pi(T)$.

The implication (ii) $\Rightarrow$ (iii) is clear. We show the implication (iii) $\Rightarrow$ (i). Let $\lambda \in \Delta_a^g(T)$ be arbitrarily chosen. Then $\lambda I - T$ is B-Weyl, hence $\lambda \notin \sigma_{\mathrm{bw}}(T) = \sigma_{\mathrm{d}}(T)$, since Browder's theorem holds for T. Therefore λ is a pole and hence is an isolated point of $\sigma(T)$. By Theorem 5.62 we conclude that T has property (gb).

(i) $\Rightarrow$ (iv) We know that property (gb) implies property (b), i.e., $\Delta_a^g(T) = \Pi(T)$, and $\Pi(T) \subseteq \Pi_a(T)$. If $\lambda \notin \Pi_a(T)$ then $\lambda I - T$ is left Drazin invertible, or equivalently, by Theorem 3.47, $\lambda I - T$ is upper semi B-Browder. Moreover, by Theorem 4.3, $\lambda \in \sigma_{\mathrm{ap}}(T)$, so $\lambda \in \Delta_a^g(T) = \Pi(T)$. Therefore, $\Pi(T) = \Pi_a(T)$.

Conversely, to show (iv) $\Rightarrow$ (i), assume that T has property (b) and $\Pi(T) = \Pi_a(T)$. Property (b) entails a-Browder's theorem, and this is equivalent to the generalized a-Browder's theorem, by Theorem 5.38. By Theorem 5.38, we have

$$\Delta_a^g(T) = \sigma_{\mathrm{ap}}(T) \setminus \sigma_{\mathrm{ubw}}(T) = \sigma_{\mathrm{ap}}(T) \setminus \sigma_{\mathrm{ld}}(T) = \Pi_a(T).$$

From the assumption $\Pi(T) = \Pi_a(T)$ we then conclude that $\Delta_a^g(T) = \Pi(T)$, thus T possesses property (*gb*).   ∎

From part (iii) of Theorem 5.68 we immediately obtain:

Corollary 5.69 $T \in L(X)$ *has property* (*gb*) *if and only if* T *satisfies Browder's theorem and* ind $(\lambda I - T) = 0$ *for all* $\lambda \in \Delta_a^g(T)$.

Corollary 5.70 *If* $T \in L(X)$ *is a-polaroid then the properties* (*b*), (*gb*), *and a-Browder's theorem for* T, *are equivalent.*

Proof We know that for every $T \in L(X)$ we have (*gb*) $\Rightarrow$ (*b*) $\Rightarrow$ *a*-Browder's theorem, so we have only to prove that *a*-Browder's theorem for T implies property (*gb*) for T. Since *a*-Browder's theorem and the generalized *a*-Browder's theorem are equivalent we have $\sigma_{\mathrm{ubw}}(T) = \sigma_{\mathrm{ld}}(T)$. Therefore $\Delta_a^g(T) = \sigma_{\mathrm{ap}}(T) \setminus \sigma_{\mathrm{ld}}(T) = \Pi^a(T)$. From Theorem 4.3 we know that every left pole λ is an isolated point of $\sigma_{\mathrm{ap}}(T)$. Our assumption that T is *a*-polaroid entails that $\lambda \in \Pi(T)$, and hence $\Delta_a^g(T) \subseteq \Pi(T)$, from which we conclude that property (*gb*) holds for T.   ∎

Corollary 5.71 *Suppose that* $T \in L(X)$ *is a-polaroid and has the SVEP. Then property* (*gb*) *holds for* T.

Proof T satisfies *a*-Browder's theorem.   ∎

We show now that the results of Corollaries 5.71 and 5.70 cannot be extended to polaroid operators.

Example 5.72 Let R denote the unilateral right shift on $\ell_2(\mathbb{N})$. We have $\sigma(R) = \mathbf{D}(0, 1)$, while $\sigma_{\mathrm{ap}}(R)$ is the unit circle Γ. Define $T := 0 \oplus R$. Clearly, T has the SVEP, since T is the direct sum of operators having SVEP, $\sigma(T) = \mathbf{D}(0, 1)$ and

$$\sigma_{\mathrm{ap}}(T) = \sigma_{\mathrm{ap}}(R) \cup \{0\} = \Gamma \cup \{0\}.$$

We show that T is left Drazin invertible. Evidently, $p := p(T) = p(R) + p(0) = 1$. We have $T(X) = \{0\} \oplus R(X)$, so $T(X) = T^2(X)$ is closed, since $R(X)$ is closed. Hence $0 \notin \sigma_{\mathrm{ld}}(T)$ and this implies that $0 \notin \sigma_{\mathrm{ubw}}(T)$, because $\sigma_{\mathrm{ubw}}(T) \subseteq \sigma_{\mathrm{ubb}}(T) = \sigma_{\mathrm{ld}}(T)$, by Corollary 3.49. Therefore, $0 \in \Delta_a^g(T)$ but $0 \notin \mathrm{iso}\,\sigma(T)$, since $\sigma(T)$ has no isolated points. Consequently, T does not satisfy property (*gb*). Observe that T is polaroid, and satisfies *a*-Browder's theorem, since T has the SVEP.

Let us consider the set $E(T)$ of eigenvalues which are isolated points of the spectrum, i.e.,

$$E(T) := \{\lambda \in \mathrm{iso}\,\sigma(T) : \alpha(\lambda I - T) > 0\}.$$

We have seen before that the SVEP for T does not ensure that property (*gb*) holds for T. However, the following result shows that in the case of polaroid operators, property (*gb*) holds for T^*.

Theorem 5.73 *Suppose that $T \in L(X)$ is polaroid. Then we have:*

(i) *T satisfies property (gb) if and only if $\Delta_a^g(T) = E(T)$.*
(ii) *If T has the SVEP then T^* satisfies property (gb).*

Proof Clearly, $\Pi(T) \subseteq E(T)$ for every operator. The opposite inclusion is immediate, since the polaroid condition entails that every $\lambda \in E(T)$ is a pole of the resolvent. Hence $\Pi(T) = E(T)$, from which the equivalence (i) follows.

(ii) The SVEP for T entails a-Browder's theorem for T^*. Moreover, T^* is polaroid and the SVEP for T implies, by Theorem 2.68, that $\sigma(T^*) = \sigma(T) = \sigma_s(T) = \sigma(T^*)$. Hence T^* is a-polaroid, so by Theorem 5.70 T^* has property (gb). ∎

In the sequel we give some results concerning the stability of property (gb) under some commuting perturbations. In order to transfer property (gb) from T to its perturbation $T + Q$, by a commuting quasi-nilpotent operator Q, it is sufficient to assume that T is *finite-polaroid*, i.e., every isolated point of the spectrum $\sigma(T)$ is a pole of finite rank.

Theorem 5.74 *Suppose that T is finite-polaroid and has the SVEP. Then $T^* + Q^*$ satisfies property (gb) for every quasi-nilpotent operator Q commuting with T.*

Proof We prove first that $T + Q$ is polaroid. Let $\lambda \in \text{iso}\,\sigma(T + Q)$. Then $\lambda \in \text{iso}\,\sigma(T)$ and hence is a pole of the resolvent of T (consequently, an eigenvalue of T). Therefore, $p := p(\lambda I - T) = q(\lambda I - T) < \infty$ and since by assumption $\alpha(\lambda I - T) < \infty$ we then have $\alpha(\lambda I - T) = \beta(\lambda I - T)$, by Theorem 1.22, so $\lambda I - T$ is Browder. By Theorem 3.8, we know that the class of Browder operators is stable under quasi-nilpotent commuting perturbations, so $\lambda I - (T + Q)$ is Browder, and hence λ is a pole of the resolvent of $T + Q$. Therefore, $T + Q$ is polaroid.

Now, by Theorem 2.129, the SVEP from T is transmitted to $T + Q$, and this implies, by Theorem 4.15, that $T^* + Q^*$ is a-polaroid. Moreover, the SVEP for $T + Q$ implies that $T^* + Q^*$ satisfies a-Browder's theorem. By Corollary 5.70, we then conclude that property (gb) holds for $T^* + Q^*$. ∎

Theorem 5.75 *Suppose that $T \in L(X)$ has the SVEP and $\text{iso}\,\sigma_{\text{ap}}(T) = \emptyset$. Then we have:*

(i) *$T + Q$ satisfies property (gb) for every commuting quasi-nilpotent operator Q.*
(ii) *$T + K$ satisfies property (gb) for every commuting finite rank operator K.*

Proof

(i) We know that $\sigma_{\text{ap}}(T) = \sigma_{\text{ap}}(T + Q)$ for every commuting quasi-nilpotent operator Q. Therefore $\text{iso}\,\sigma_{\text{ap}}(T + Q) = \emptyset$, from which we conclude that $\Pi(T + Q)$ is empty, since $\Pi(T + Q) \subseteq \text{iso}\,\sigma_{\text{ap}}(T + Q)$.

It remains only to show that $\Delta_a^g(T + Q) = \emptyset$. Suppose that $\Delta_a^g(T + Q)$ is non-empty and let $\lambda \in \Delta_a^g(T + Q) = \sigma_{\text{ap}}(T + Q) \setminus \sigma_{\text{ubw}}(T + Q)$. The SVEP for T is inherited by $T + Q$, and since $\lambda I - (T + Q)$ is upper semi B-Weyl, the SVEP of $T + Q$ at λ implies, by Theorem 2.97, that $\sigma_{\text{ap}}(T + Q)$ does not cluster at λ. But $\lambda \in \sigma_{ap}(T + Q)$, so λ is an isolated point of $\sigma_{ap}(T + Q)$ and this is impossible. Therefore, $\Pi(T + Q) = \Delta_a^g(T + Q) = \emptyset$.

(ii) Observe first that, by Theorem 3.29, we have $\sigma_{ap}(T) = \sigma_{ap}(T + K)$, so $\sigma_{ap}(T + K)$ has no isolated points. Furthermore, the SVEP for T is inherited by $T + K$, by Theorem 2.129. The statement may be proved by using the same arguments as the proof of part (i). ∎

Both the conditions iso $\sigma_{ap}(T) = \emptyset$ and T has the SVEP are satisfied by every non quasi-nilpotent unilateral right shift T on $\ell^p(\mathbb{N})$, with $1 \le p < \infty$. As usual, let $\mathcal{H}(\sigma(T))$ denote the set of all analytic functions defined on a neighborhood of $\sigma(T)$.

Theorem 5.76 *Let $T \in L(X)$ be such that there exists a $\lambda_0 \in \mathbb{C}$ such that*

$$K(\lambda_0 I - T) = \{0\} \quad and \quad \ker(\lambda_0 I - T) = \{0\}. \tag{5.13}$$

Then property (gb) holds for $f(T)$ for all $f \in \mathcal{H}(\sigma(f(T))$.

Proof For all complex $\lambda \ne \lambda_0$ we have $\ker(\lambda I - T) \subseteq K(\lambda_0 I - T)$, so that $\ker(\lambda I - T) = \{0\}$ for all $\lambda \in \mathbb{C}$. Therefore, the point spectrum $\sigma_p(T)$ is empty.

We also show that $\sigma_p(f(T)) = \emptyset$. To see this, let $\mu \in \sigma(f(T))$ and write $\mu - f(\lambda) = p(\lambda)g(\lambda)$, where g is analytic on an open neighborhood $\mathcal{U}$ containing $\sigma(T)$ and without zeros in $\sigma(T)$, p a polynomial of the form $p(\lambda) = \Pi_{k=1}^n (\lambda_k - \lambda)^{\nu_k}$, with distinct roots $\lambda_1, \ldots, \lambda_n$ lying in $\sigma(T)$. Then

$$\mu I - f(T) = \Pi_{k=1}^n (\lambda_k I - T)^{\nu_k} g(T).$$

Since $g(T)$ is invertible, $\sigma_p(T) = \emptyset$ implies that $\ker(\mu I - f(T)) = \{0\}$ for all $\mu \in \mathbb{C}$, so $\sigma_p(f(T)) = \emptyset$.

To prove that property (gb) holds for $f(T)$, observe first that $\Pi(f(T))$ is empty, since each pole is an eigenvalue. So we need only to prove that $\Delta_a^g(f(T))$ is empty. Suppose that there exists a $\lambda \in \Delta_a^g(f(T))$. Then $\lambda \in \sigma_{ap}(f(T))$ and $\lambda I - f(T)$ is upper semi B-Weyl. Since $\sigma_p(f(T)) = \emptyset$ we have that $\lambda I - f(T)$ is injective, hence, by Corollary 1.115, $\lambda I - f(T)$ is bounded below, i.e. $\lambda \notin \sigma_{ap}(f(T))$, a contradiction. Therefore $f(T)$ satisfies property (gb). ∎

The conditions of Theorem 5.76 are satisfied by any injective operator for which the hyper-range $T^\infty(X)$ is $\{0\}$, since $K(T) \subseteq T^\infty(X)$ for all $T \in L(X)$. In particular, the conditions of Theorem 5.76 are satisfied by every semi-shift.

Theorem 5.77 *Let $T \in L(X)$ and let $K \in L(X)$ be a finite rank operator which commutes with T such that $\sigma_{ap}(T) = \sigma_{ap}(T + K)$. If T has property (gb) then $T + K$ also has property (gb).*

Proof Property (gb) entails a-Browder's theorem for T, i.e., $\sigma_{uw}(T) = \sigma_{ub}(T)$. By Corollary 3.18 and Theorem 3.8 we also have $\sigma_{uw}(T) = \sigma_{uw}(T + K)$ and $\sigma_{ub}(T) = \sigma_{ub}(T + K)$, so a-Browder's theorem holds for $T + K$, or equivalently the generalized a-Browder's theorem holds for $T + K$. From Theorem 5.31 it then follows that

$$\Delta_a^g(T + K) \subseteq \text{iso}\,\sigma_{\text{ap}}(T + K) = \text{iso}\,\sigma_{\text{ap}}(T).$$

Let $\lambda \in \Delta_a^g(T + K)$ be arbitrary chosen. Then $T + K$ has the SVEP at λ, since $\lambda \in$ iso $\sigma_{\text{ap}}(T + K)$, and hence, by Theorem 2.97, $\lambda I - (T + K)$ is left Drazin invertible. From Theorem 3.78 it then follows that $\lambda I - (T + K) + K = \lambda I - T$ is also left Drazin invertible, in particular upper semi B-Weyl. Since $\lambda \in \sigma_{\text{ap}}(T)$ we then obtain $\lambda \in \Delta_a^g(T) = \Pi(T)$. Again by Theorem 3.78 we know that $\sigma_d(T) = \sigma_d(T + K)$, so $\Pi(T) = \Pi(T + K)$, and hence $\lambda \in \Pi(T + K)$. This shows the inclusion $\Delta_a^g(T + K) \subseteq \Pi(T + K)$. The opposite inclusion is true for every operator, thus $\Delta_a^g(T + K) = \Pi(T + K)$, and the proof is complete.   ∎

For hereditarily polaroid operators we can say much more:

Theorem 5.78 *Suppose that $T \in L(X)$ is a hereditarily polaroid operator which satisfies SVEP and $K \in L(X)$ an algebraic operator which commutes with T. Then $f(T^* + K^*)$ satisfies property (gb) for every $f \in \mathcal{H}_{nc}(\sigma(T))$.*

Proof Since T is hereditarily polaroid, then $f(T+K)$ is polaroid, by Theorem 4.33. Moreover, $T + K$ satisfies SVEP, by Theorem 2.145, and hence $f(T + K)$ also has the SVEP, by Theorem 2.86. Therefore, the result of Theorem 5.73 applies to $f(T^* + K^*)$.   ∎

5.5 Property (ab)

For every operator $T \in L(X)$ define

$$\Sigma_a(T) := \Delta(T) \cup p_{00}^a(T).$$

Since $\Delta(T) \subseteq \sigma_{\text{ap}}(T)$, $\Sigma_a(T)$ is the set of all points $\lambda \in \sigma_{\text{ap}}(T)$ for which either $\lambda I - T$ is Weyl or λ is a left pole of finite rank.

The following property was introduced by Berkani and Zariuoh in [75].

Definition 5.79 $T \in L(X)$ satisfies *property (ab)* if $\Delta(T) = p_{00}^a(T)$.

Property (ab) also entails Browder's theorem:

Theorem 5.80 *If $T \in L(X)$ satisfies property (ab) then Browder's theorem holds for T.*

Proof Suppose that $\Delta(T) = p_{00}^a(T)$. Let $\lambda \in \Delta(T) = p_{00}^a(T)$. Then $\lambda I - T$ is both Weyl and upper semi-Browder, in particular $p(\lambda I - T) < \infty$. From Theorem 1.22 we deduce that $q(\lambda I - T) < \infty$, so λ is a pole of the resolvent and hence $\lambda \in \text{iso}\,\sigma(T)$. The inclusion $\Delta(T) \subseteq \text{iso}\,\sigma(T)$ is equivalent, by Theorem 5.10, to Browder's theorem for T.   ∎

The next example shows that the converse of Theorem 5.80 does not hold, i.e. property (ab) is stronger than Browder's theorem.

Example 5.81 Let R and L be the right shift and the left shift defined on $\ell_2(\mathbb{N})$, respectively. Define $T : R \oplus L \oplus L$. Recall that L is the adjoint of R. Since $L' = R$ has the SVEP, by Theorem 2.68, $\sigma_{ap}(L) = \sigma(L)$. Since $\sigma(R) = \mathbf{D}(0, 1)$, we have $\sigma(T) = \mathbf{D}(0, 1)$ and

$$\sigma_{ap}(T) = \sigma_{ap}(R) \cup \sigma_{ap}(L) = \mathbf{D}(0, 1).$$

Now, it easily seen that $\alpha(T) = 1$, $\beta(T) = 2$, so T is a Fredholm operator and, consequently has closed range. Since ind $T \le 0$ then $0 \notin \sigma_{uw}(T)$. We also have $p(T) = \infty$, so $0 \in \sigma_{ub}(T)$. Hence T does not satisfies a-Browder's theorem. On the other hand $\sigma_w(T) = \mathbf{D}(0, 1)$, so $\Delta(T) = \emptyset$, and hence T satisfies Browder's theorem. We claim that $p_{00}^a(T)$ is empty. Indeed, suppose that there exists a $\lambda \in p_{00}^a(T) = \sigma_{ap}(T) \setminus \sigma_{ub}(T)$. Then $\lambda \in \sigma_{ap}(T)$ and $\lambda I - T$ is upper semi-Browder. The condition $p(\lambda I - T) < \infty$ is equivalent, by Theorem 2.97, to saying that λ is an isolated point of $\sigma_{ap}(T)$, and this is impossible. Therefore, $\Delta(T) = p_{00}^a(T) = \emptyset$, so T satisfies property (ab).

The precise relationship between property (ab) and Browder's theorem is described by the following theorem.

Theorem 5.82 *Let $T \in L(X)$. Then the following statements are equivalent:*

(i)   *T satisfies property (ab);*
(ii) *T^* has the SVEP at every $\lambda \in p_{00}^a(T)$ and $\Delta(T) \subseteq p_{00}^a(T)$;*
(iii) *Browder's theorem holds for T and every left pole of finite rank of T is a pole of T, i.e. $p_{00}(T) = p_{00}^a(T)$;*
(iv) *Browder's theorem holds for T and $p_{00}^a(T) \subseteq$ iso $\sigma(T)$;*
(v)   *Browder's theorem holds for T and $p_{00}^a(T) \subseteq \partial\sigma(T)$;*
(vi) *$\Sigma_a(T) \subseteq$ iso $\sigma(T)$;*
(vii) *$\Sigma_a(T) \subseteq p_{00}(T)$;*
(viii)*T^* has the SVEP at every point $\lambda \in \Sigma_a(T)$.*

Proof (i) $\Leftrightarrow$ (ii) Suppose that T satisfies (ab), i.e. $\Delta(T) = \sigma(T) \setminus \sigma_w(T) = p_{00}^a(T)$. Let $\lambda \notin p_{00}^a(T)$ be arbitrary. Then $\lambda I - T$ is Weyl and $p(\lambda I - T) < \infty$. By Theorem 1.22, we have $q(\lambda I - T) < \infty$, hence λ is a pole and in particular an isolated point of $\sigma(T) = \sigma(T^*)$. Consequently, T^* has the SVEP at λ.

Conversely, suppose that T^* has the SVEP at every $\lambda \notin p_{00}^a(T)$ and $\Delta(T) \subseteq p_{00}^a(T)$. If $\lambda \in p_{00}^a(T)$ then $\lambda \notin \sigma_{ub}(T)$ and $p(\lambda I - T) < \infty$. Since $\lambda I - T$ is semi-Fredholm the SVEP of T^* at λ entails that $q(\lambda I - T) < \infty$. By Theorem 1.22 we then conclude that $\lambda I - T$ is Browder, hence Weyl, and consequently $p_{00}^a(T) \subseteq \Delta(T)$. Since the reverse inclusion holds by assumption, we then have $\Delta(T) = p_{00}^a(T)$.

(ii) $\Rightarrow$ (iii) We have only to show that $p_{00}(T) = p_{00}^a(T)$. Let $\lambda \in p_{00}^a(T)$. Then $\lambda I - T \in B_+(X)$ and the SVEP of T^* at λ entails that $q(\lambda I - T) < \infty$, by Theorem 2.98. Since $p(\lambda I - T) < \infty$ it then follows that λ is a pole of the resolvent. But $\alpha(\lambda I - T) < \infty$, thus $\lambda I - T \in B(X)$, by Theorem 1.22. Consequently, $p_{00}^a(T) \subseteq p_{00}(T)$, and since the opposite inclusion is always true we then conclude that $p_{00}(T) = p_{00}^a(T)$.

(iii) $\Rightarrow$ (iv) Clear, since $p_{00}(T) \subseteq \mathrm{iso}\,\sigma(T)$.

(iv) $\Rightarrow$ (ii) T^* has the SVEP at every isolated point of $\sigma(T^*) = \sigma(T)$. Browder's theorem implies that $\Delta(T) = p_{00}(T) \subseteq p_{00}^a(T)$.

(iv) $\Rightarrow$ (v) Clear.

(v) $\Rightarrow$ (ii) T^* has the SVEP at every $\lambda \in \partial\sigma(T^*) = \partial\sigma(T)$, and as above Browder's theorem entails that $\Delta(T) \subseteq p_{00}^a(T)$.

(v) $\Leftrightarrow$ (vi) By Theorem 5.10, T satisfies Browder's theorem if and only if $\Delta(T) \subseteq \mathrm{iso}\,\sigma(T)$.

(vi) $\Leftrightarrow$ (vii) If $\lambda \in \Sigma_a(T)$ then $\lambda I - T$ is either Weyl or upper semi-Browder. Since $\Sigma_a(T) \subseteq \mathrm{iso}\,\sigma(T)$, then both T and T^* have the SVEP at λ. This implies, by Theorems 2.97 and 2.98, that $p(\lambda I - T) = q(\lambda I - T) < \infty$. By Theorem 1.22 we then conclude that $\lambda I - T \in B(X)$, i.e., $\Sigma_a(T) \subseteq p_{00}(T)$. The implication (vii) $\Rightarrow$ (vi) is obvious.

(vi) $\Rightarrow$ (viii) T^* has the SVEP at every point $\lambda \in \Sigma_a(T)$ since $\lambda \in \mathrm{iso}\,\sigma(T) = \mathrm{iso}\,\sigma(T^*)$.

(viii) $\Rightarrow$ (ii) Suppose that T^* has the SVEP at every point $\lambda \in \Sigma_a(T)$. Then T^* has the SVEP at the points of $\Delta(T)$, as well as at the points of $p_{00}^a(T)$. It is easily seen that $\Delta(T) \subseteq p_{00}^a(T)$. Indeed, if $\lambda \in \Delta(T)$ then $\lambda I - T$ is Weyl, and the SVEP for T^* at λ entails by Theorem 2.98 that $q(\lambda I - T) < \infty$. By Theorem 1.22, then $\lambda I - T$ is Browder, in particular upper semi-Browder, so $\lambda \in p_{00}^a(T)$.   ∎

Every isolated point of the spectrum belongs to $\sigma_{\mathrm{ap}}(T)$ by Theorem 1.12. Hence $\mathrm{iso}\,\sigma(T) \subseteq \mathrm{iso}\,\sigma_{\mathrm{ap}}(T)$, from which we easily obtain:

Corollary 5.83 *If T has property (ab) then $\Sigma_a(T) \subseteq \mathrm{iso}\,\sigma_{\mathrm{ap}}(T)$.*

Note that T always has the SVEP at the points $\lambda \in p_{00}^a(T)$, since $\lambda I - T \in B_+(X)$, and hence $p(\lambda I - T) < \infty$. The next example shows that the condition $\Delta(T) \subseteq p_{00}^a(T)$ does not ensure that T^* has the SVEP at $\lambda \in p_{00}^a(T)$.

Example 5.84 Let T be the operator defined in Example 5.49. Then $T^* = R^* \oplus P^*$ does not have the SVEP at 0, since R^* is the left shift and this operator fails SVEP at 0. On the other hand, $\Delta(T) = \emptyset$, while $\sigma_{\mathrm{uw}}(T) = \sigma_{\mathrm{ub}}(T) = \Gamma$, since T has the SVEP and hence a-Browder's theorem holds for T, thus $\Delta(T) \subseteq p_{00}^a(T) = \{0\}$.

Corollary 5.85 *If $T \in L(X)$ has property (b) then T has property (ab).*

Proof From Lemma 5.25 we know that $\Sigma_a(T) \subseteq \Delta_a(T)$. The property (b) is equivalent, by Theorem 5.45, to the inclusion $\Delta_a(T) \subseteq \mathrm{iso}\,\sigma(T)$, so, from part (vi) of Theorem 5.82, we deduce that property (b) implies property (ab).   ∎

The following example shows that property (b) is in general stronger than property (ab).

Example 5.86 Let R and L be the unilateral right and left shift, respectively, on $\ell_2(\mathbb{N})$, and define $T = L \oplus R \oplus R$. Then $\alpha(T) = 1$, $\beta(T) = 2$, and $p(T) = \infty$. This implies in particular $0 \notin \sigma_{uw}(T)$. Since $p(T) = \infty$, T does not satisfy a-Browder's theorem. Therefore T does not have property (b), by Corollary 5.48. On the other hand, $\sigma(T) = \mathbf{D}(0, 1)$, hence, the set of left poles is empty, and $\sigma_{bw}(T) = \mathbf{D}(0, 1)$, and hence T has property (ab).

Observe that $\sigma_w(T) \cap \Delta(T)$ is trivially empty. The intersection $\sigma_w(T) \cap \Delta_a(T)$ may be non-empty. For instance, let R and L denote the right shift and the left shift on $\ell_2(\mathbb{N})$, respectively. Let $T := L \oplus R \oplus R$. It is easy to check that $\alpha(T) = 1$, $\beta(T) = 2$, so T is upper Weyl but not Weyl. Evidently, $0 \in \sigma_w(T) \cap \Delta_a(T)$.

The precise relationship between properties (b) and property (ab) is described in the following theorem.

Theorem 5.87 *If $T \in L(X)$ then the following assertions are equivalent:*

(i)   *T has property (b);*
(ii) *T has property (ab) and $\sigma_w(T) \cap \Delta_a(T) = \emptyset$;*
(iii) *T satisfies a-Browder's theorem and $\sigma_w(T) \cap \Delta_a(T) = \emptyset$;*
(iv) *T satisfies Browder's theorem and $\sigma_w(T) \cap \Delta_a(T) = \emptyset$.*

Proof (i) $\Rightarrow$ (ii) Suppose that T has property (b). Then $\Delta_a(T) = p_{00}(T)$ and property (ab) holds for T, by Corollary 5.85. Moreover, a-Browder's theorem holds for T, by Corollary 5.48, and hence Browder's theorem, so that $\sigma_w(T) = \sigma_b(T)$. Therefore, $\sigma_w(T) \cap \Delta_a(T) = \sigma_b(T) \cap p_{00}(T) = \emptyset$.

Conversely, assume (ii). Property (ab) entails Browder's theorem so $\sigma_w(T) = \sigma_b(T)$. Let $\lambda \in \Delta_a(T)$. By assumption then $\lambda \notin \sigma_w(T) = \sigma_b(T)$, hence $\lambda \in p_{00}(T)$. This shows that $\Delta_a(T) \subseteq p_{00}(T)$, and since, by Lemma 5.25, the reverse inclusion holds for every operator it then follows that $\Delta_a(T) = p_{00}(T)$.

The implications (i) $\Rightarrow$ (iii) $\Rightarrow$ (iv) are clear, since property (b) entails a-Browder's theorem and this implies Browder's theorem. To show the implication (iv) $\Rightarrow$ (i), suppose that $\lambda \in \Delta_a(T)$. Then $\lambda \notin \sigma_w(T) = \sigma_b(T)$, hence λ is an isolated point of $\sigma(T)$, so T has property (b), by Theorem 5.45. ∎

The SVEP for T entails Browder's theorem, so we have:

Corollary 5.88 *If T has the SVEP then property (b) holds for T if and only if $\sigma_w(T) \cap \Delta_a(T) = \emptyset$. In this case, properties (b) and (ab), a-Browder's theorem and Browder's theorem for T are equivalent.*

Define

$$p_{00}^s(T) := \sigma_s(T) \setminus \sigma_{lb}(T) = \sigma_{ap}(T^*) \setminus \sigma_{ub}(T^*)$$

and by duality

$$\Sigma_s(T) := p_{00}^s(T) \cup \Delta(T).$$

Evidently, $\Sigma_s(T) = \Sigma_a(T^*)$, since $\Delta(T) = \Delta(T^*)$. The next result shows that property (ab) for T may also be characterized by means of the analytic core $K(\lambda I - T)$.

Theorem 5.89 *For an operator $T \in L(X)$ the following equivalences hold:*

(i) *T satisfies property (ab) $\Leftrightarrow$ $K(\lambda I - T)$ has finite codimension for all $\lambda \in \Sigma_a(T)$. In this case, for all $\lambda \in \Sigma_a(T)$ there exists a $\nu := \nu(\lambda) \in \mathbb{N}$ such that $K(\lambda I - T) = (\lambda I - T)^\nu(X)$ and $H_0(\lambda I - T) = \ker(\lambda I - T)^\nu$.*

(ii) *T^* satisfies property (ab) $\Leftrightarrow$ $K(\lambda I - T)$ has finite dimension for all $\lambda \in \Delta_s(T)$. In this case, for all $\lambda \in \Sigma_s(T)$ there exists a $\nu := \nu(\lambda) \in \mathbb{N}$ such that $K(\lambda I - T) = (\lambda I - T)^\nu(X)$ and $H_0(\lambda I - T) = \ker(\lambda I - T)^\nu$.*

Proof

(i) Suppose that T has property (ab) and let $\lambda \in \Sigma_a(T)$. Then $\lambda \in p_{00}(T)$, by Theorem 5.82, and hence λ is a pole of the resolvent, so that $X = H_0(\lambda I - T) \oplus K(\lambda I - T)$. Moreover, $\lambda \in \Delta(T)$ or $\lambda \in p_{00}^a(T)$. If $\lambda \in \Delta(T)$, then $H_0(\lambda I - T)$ has finite dimension since T satisfies Browder's theorem, by Theorem 5.35. If $\lambda \in p_{00}^a(T)$, then λ is a left pole of finite rank and hence, by Theorem 4.3, $H_0(\lambda I - T)$ has finite dimension. Therefore, $K(\lambda I - T)$ has finite codimension for all $\lambda \in \Sigma_a(T)$.

Conversely, suppose that $K(\lambda I - T)$ has finite codimension for all $\lambda \in \Sigma_a(T) = \Delta(T) \cup p_{00}^a(T)$. Clearly, if $\lambda \in \Sigma_a(T)$ then $\lambda I - T$ is either Weyl or upper semi-Browder and hence, by Theorem 2.98, T^* has the SVEP at λ. In particular, T^* has the SVEP at every $\lambda \in p_{00}^a(T)$. Let $\lambda \in \Delta(T)$. By Theorem 2.98, we have $q(\lambda I - T) < \infty$ and since $\lambda I - T$ is Weyl, by Theorem 1.22 it then follows that $\lambda I - T$ is Browder. In particular, $\lambda \in \sigma_{ap}(T) \setminus \sigma_{ub}(T) = p_{00}^a(T)$ and hence, by Theorem 5.82, T satisfies property (ab). The last assertion follows from Theorem 2.45, since every $\lambda \in \Sigma_a(T)$ is a pole of the resolvent.

(ii) We proceed by duality. Suppose that T^* has property (ab). By part (i) every $\lambda \in \Delta_s(T) = \Delta_a(T^*)$ is a pole of the resolvent of T^*, hence a pole of the resolvent of T, and $K(\lambda I - T^*)$ has finite codimension. Clearly,

$$X^* = H_0(\lambda I - T^*) \oplus K(\lambda I - T^*) = \ker(\lambda I - T^*)^p \oplus (\lambda I - T^*)^p(X^*),$$

and

$$X = H_0(\lambda I - T) \oplus K(\lambda I - T) = \ker(\lambda I - T)^p \oplus (\lambda I - T)^p(X),$$

where p is the order of the pole. By the closed range theorem we have:

$$K(\lambda I - T) = (\lambda I - T)^p(X) = {}^{\perp}[\ker (\lambda I - T^*)^p] = {}^{\perp}H_0(\lambda I - T^*),$$

where ${}^{\perp}M$ denotes the pre-annihilator of $M \subseteq X^*$, so $K(\lambda I - T)$ has finite codimension, since $H_0(\lambda I - T^*)$ has finite dimension.   ∎

Theorem 5.90 *If $T \in L(X)$ we have*

(i) *T satisfies property (ab) $\Leftrightarrow$ $(\lambda I - T)^{\infty}(X)$ has finite codimension for all $\lambda \in \Sigma_a(T)$.*
(i) *T^* satisfies property (ab) $\Leftrightarrow$ $\mathcal{N}^{\infty}(\lambda I - T)$ has finite dimension for all $\lambda \in \Sigma_s(T)$.*

Proof

(i) The proof is analogous to that of Theorem 5.53, just replace $\Delta_a(T)$ with $\Sigma_a(T)$ and use Theorem 5.89. In the part ($\Leftarrow$) of the proof we obtain that $q(\lambda I - T) < \infty$ for all $\lambda \in \Sigma_a(T) = \Delta(T) \cup p_{00}^a(T)$. Consequently, T^* has the SVEP at the points of $p_{00}^a(T)$ and the condition $q(\lambda I - T) < \infty$ at the points of $\Delta(T)$ entails $p(\lambda I - T) = q(\lambda I - T) < \infty$, so $\lambda \in p_{00}(T)$. Property (ab) then follows from Theorem 5.82.
(ii) Analogous to part (ii) of Theorem 5.53.   ∎

5.6   Property (gab)

Property (ab) also admits a generalization in the sense of B-Fredholm theory.

Definition 5.91 $T \in L(X)$ is said to satisfy *property (gab)* if $\Delta^g(T) = \Pi_a(T)$.

Property (gab) entails property (ab):

Theorem 5.92 *Suppose that $T \in L(X)$ has property (gab). Then T has property (ab).*

Proof If T has property (gab) then $\Delta^g(T) = \Pi_a(T)$. If $\lambda \in \Delta(T)$, then $\lambda \in \Delta^g(T)$, hence is a left pole of the resolvent. Since $\alpha(\lambda I - T) < \infty$, we have $\lambda \in p_{00}^a(T)$. This proves the inclusion $\Delta(T) \subseteq p_{00}^a(T)$. Conversely, if $\lambda \in p_{00}^a(T)$, then λ is upper semi-Browder, so $\alpha(\lambda I - T) < \infty$ and $p(\lambda I - T) < \infty$. Since T has property (gab), $\lambda \in \Delta^g(T)$ and $\text{ind}(\lambda I - T) = 0$. Since $p(\lambda I - T) < \infty$, by Theorem 1.22 $\lambda I - T$ is Browder, in particular $\lambda I - T$ is Weyl. Hence $\lambda \in \Delta(T) = \sigma(T) \setminus \sigma_w(T)$. Therefore $\Delta(T) = p_{00}^a(T)$, so T has property (ab).   ∎

The converse of the result of Theorem 5.92 does not hold:

Example 5.93 If R is the unilateral right shift on $\ell_2(\mathbb{N})$, then $\sigma(R) = \mathbf{D}(0, 1)$, and $\sigma_{ap}(R) = \Gamma$, the unit circle, and the set of eigenvalues of R is empty. Moreover,

$\sigma_w(T) = \Gamma$ and $p_{00}(T) = \emptyset$. Define $T = 0 \oplus R$. Then $\sigma(T) = \mathbf{D}(0, 1)$,

$$\ker T = \ell_2(\mathbb{N}) \oplus \{0\}, \ \sigma_{ap}(T) = \Gamma \cup \{0\},$$

and $\sigma_w(T) = \sigma_{bw}(T) = \mathbf{D}(0, 1)$. Since $\Pi_a(T) = \{0\}$, and $p_{00}^a(T) = \emptyset$, we have

$$\Delta(T) = p_{00}^a(T) \text{ and } \Delta^g(T) = \emptyset \neq \Pi_a(T).$$

Therefore, T has property (ab) but not property (gab).

Set

$$\Sigma_a^g(T) := \Delta^g(T) \cup \Pi_a(T).$$

Lemma 5.94 *If* $T \in L(X)$ *then* $\Sigma_a^g(T) \subseteq \Delta_a^g(T)$.

Proof By Lemma 5.39 we have $\Pi_a(T) \subseteq \Delta_a^g(T)$. It remains only to prove that $\Delta^g(T) \subseteq \Delta_a^g(T)$. If $\lambda \in \Delta^g(T)$ then $\lambda \in \sigma_{ap}(T)$, again by Lemma 5.39. On the other hand, we have $\lambda \notin \sigma_{bw}(T)$ and hence $\lambda \notin \sigma_{ubw}(T)$, since $\sigma_{ubw}(T) \subseteq \sigma_{bw}(T)$. ∎

The set $\Sigma_a^g(T)$ may be empty. Indeed, in the case of the unilateral right shift $R \in L(\ell_2(\mathbb{N}))$ it has been observed that $\Delta_a^g(R) = \emptyset$, so, by Lemma 5.94, we have $\Sigma_a^g(T) = \emptyset$

Property (gab) may also be characterized by means of the localized SVEP as follows.

Theorem 5.95 *For a bounded operator* $T \in L(X)$ *the following statements are equivalent:*

  (i) *T satisfies property (gab);*
 (ii) *T^* has the SVEP at every $\lambda \in \Pi_a(T)$ and $\Delta^g(T) \subseteq \Pi_a(T)$;*
(iii) *T^* has the SVEP at every $\lambda \in \Sigma_a^g(T)$;*
 (iv) *Browder's theorem holds for T and $\Pi(T) = \Pi_a(T)$;*
  (v) *Browder's theorem holds for T and $\Pi_a(T) \subseteq \text{iso } \sigma(T)$;*
 (vi) *Browder's theorem holds for T and $\Pi_a(T) \subseteq \partial\sigma(T)$, $\partial\sigma(T)$ the boundary of $\sigma(T)$;*
 (vii) *$\Sigma_a^g(T) \subseteq \text{iso } \sigma(T)$;*
(viii) *$\Sigma_a^g(T) \subseteq \text{iso } \sigma_s(T)$;*
 (vii) *$\Sigma_a^g(T) \subseteq \Pi(T)$.*

Proof To show the equivalence (i) $\Leftrightarrow$ (ii), suppose first that T has property (gab), i.e. $\Delta^g(T) = \Pi_a(T)$. If $\lambda \in \Pi_a(T)$ then $\lambda I - T$ is B-Weyl, in particular lower semi B-Weyl. Since $p(\lambda I - T) < \infty$ then, by Theorem 1.143, $\lambda I - T$ is Drazin invertible, in particular $q(\lambda I - T) < \infty$ and hence T^* has the SVEP at λ. Obviously, $\Delta^g(T) \subseteq \Pi_a(T)$, by assumption.

Conversely, suppose that T^* has the SVEP at every $\lambda \in \Pi_a(T)$ and $\Delta^g(T) \subseteq \Pi_a(T)$. If $\lambda \in \Pi_a(T)$ then λ is a left pole, so $p(\lambda I - T) < \infty$, and $\lambda I - T$ is left Drazin invertible, or equivalently, upper semi B-Browder. Since T^* has the SVEP at λ, we have $q(\lambda I - T) < \infty$. Therefore, $\lambda I - T$ is Drazin invertible and hence $\Pi_a(T) \subseteq \Pi(T)$. The opposite inclusion holds for every operator, so $\Pi_a(T) = \Pi(T)$. If $\lambda \in \Pi_a(T)$ then $\lambda I - T$ is a Drazin invertible operator, and hence B-Weyl. Now, $\lambda \in \sigma(T)$ so we have $\lambda \in \sigma(T) \setminus \sigma_{bw}(T) = \Delta^g(T)$. Thus, $\Pi_a(T) \subseteq \Delta^g(T)$, and since the opposite inclusion holds by assumption we then conclude that $\Pi_a(T) = \Delta^g(T)$.

(ii) $\Rightarrow$ (iii) Since $\Delta^g(T) \subseteq \Pi_a(T)$, we have $\Sigma_a^g(T) = \Delta^g(T) \cup \Pi_a(T) = \Pi_a(T)$, hence T^* has the SVEP at every $\lambda \in \Sigma_a^g(T)$.

(iii) $\Rightarrow$ (iv) Suppose that T^* has the SVEP at every $\lambda \in \Sigma_a^g(T)$. Let $\lambda \in \Pi_a(T)$. Then λ is a left pole and hence $\lambda I - T$ is left Drazin invertible, so $p(\lambda I - T) < \infty$. Since $\Pi_a(T) \subseteq \Sigma_a^g(T)$, the SVEP of T^* at λ implies $q(\lambda I - T) < \infty$, by Theorem 2.98, thus $\lambda \in \Pi(T)$ and consequently $\Pi_a(T) \subseteq \Pi(T)$. The opposite inclusion holds for every $T \in L(X)$, hence $\Pi(T) = \Pi_a(T)$. It remains to prove Browder's theorem for T. Let $\lambda \notin \sigma_w(T)$. Clearly, we can suppose that $\lambda \in \sigma(T)$. Then $\lambda \notin \sigma_{bw}(T)$, since $\sigma_{bw}(T) \subseteq \sigma_w(T)$, hence $\lambda \in \Delta^g(T)$. Since $\lambda I - T$ is B-Weyl, the SVEP of T^* at λ, again by Theorem 2.98, implies that $q(\lambda I - T) < \infty$ and hence, by Theorem 1.143, $\lambda I - T$ is Drazin invertible. But $\alpha(\lambda I - T) < \infty$, so, by Theorem 1.22, $\lambda I - T$ is Browder, hence $\lambda \notin \sigma_b(T)$. Therefore, $\sigma_w(T) = \sigma_b(T)$.

(iv) $\Rightarrow$ (v) If $\lambda \in \Pi_a(T)$ then $\lambda I - T$ is left Drazin invertible, hence upper semi B-Weyl. Since T^* has the SVEP at λ then $q(\lambda I - T) < \infty$, by Theorem 2.98, hence $\lambda I - T$ is Drazin invertible, by Theorem 1.143, and consequently $\lambda \in \Pi(T)$. Therefore, $\Pi_a(T) \subseteq \Pi(T) \subseteq \mathrm{iso}\,\sigma(T)$.

(iv) $\Rightarrow$ (v) Clear, since $\Pi_a(T) = \Pi(T) \subseteq \mathrm{iso}\,\sigma(T)$.

(v) $\Rightarrow$ (ii) T^* has the SVEP at every isolated point of $\sigma(T) = \sigma(T^*)$, so T^* has the SVEP at every $\lambda \in \Pi_a(T)$. Browder's theorem is equivalent to the generalized Browder's theorem, and hence $\Delta^g(T) = \Pi(T) \subseteq \Pi_a(T)$.

(v) $\Rightarrow$ (vi) Obvious, since $\mathrm{iso}\,\sigma(T) \subseteq \partial\sigma(T)$.

(vi) $\Rightarrow$ (ii) T^* has the SVEP at every $\lambda \in \partial\sigma(T) = \partial\sigma(T^*)$ and, as above, Browder's theorem entails that $\Delta^g(T) \subseteq \Pi_a(T)$.

(v) $\Leftrightarrow$ (vii) Assume that T satisfies Browder's theorem, or equivalently the generalized Browder's theorem, and that the inclusion $\Pi_a(T) \subseteq \mathrm{iso}\,\sigma(T)$ holds. By Theorem 5.17 then $\Delta^g(T) \subseteq \mathrm{iso}\,\sigma(T)$, and consequently, $\Sigma_a^g(T) \subseteq \mathrm{iso}\,\sigma(T)$.

Conversely, if $\Sigma_a^g(T) \subseteq \mathrm{iso}\,\sigma(T)$ then $\Pi_a(T) \subseteq \mathrm{iso}\,\sigma(T)$ and $\Delta^g(T) \subseteq \mathrm{iso}\,\sigma(T)$. The last inclusion is equivalent to saying that T satisfies the generalized Browder's theorem, or equivalently Browder's theorem, again by Theorem 5.17. $\blacksquare$

An obvious consequence of Theorem 5.95 is that if T^* has the SVEP then T satisfies property (*gab*). We can say more:

Corollary 5.96 *Suppose that $T, K \in L(X)$ commute and K is a Riesz operator. If T^* has the SVEP then $T + K$ satisfies property (gab).*

Proof The dual of a Riesz operator is also a Riesz operator. The SVEP for T^* is transferred to $T^* + K^* = (T + K)^*$, by Theorem 2.129. ■

Every operator $T \in L(X)$ has the SVEP at the isolated points of the spectrum, and, by Theorem 5.95, property (gab) is equivalent to the inclusion $\Sigma_a^g(T) \subseteq$ iso $\sigma(T)$. Therefore, if T has (gab) then T has the SVEP at every point of $\Sigma_a^g(T)$. The converse is false. Next we give an example of an operator which has the SVEP but the property (gab) fails for T.

Example 5.97 Let T be defined as in Example 5.72. Then T has the SVEP. Let $\lambda \notin \sigma_{\mathrm{bw}}(T)$, and suppose that $\lambda \in \sigma(T)$. By Theorem 1.143, then $\lambda I - T$ is Drazin invertible, and hence λ is a pole of the resolvent of T, in particular an isolated point of $\sigma(T)$, which is impossible. Therefore

$$\sigma_{\mathrm{bw}}(T) = \sigma(T) = \mathbf{D}(0, 1).$$

On the other hand, we know that

$$\sigma_{\mathrm{ap}}(T) = \Gamma \cup \{0\}.$$

We know, see Example 5.72, that T is left Drazin invertible, and because $0 \in \sigma_{\mathrm{ap}}(T)$ we then conclude that 0 is a left pole. Therefore,

$$\Pi_a(T) = \{0\} \neq \Delta^g(T) = \sigma(T) \setminus \sigma_{\mathrm{bw}}(T) = \emptyset,$$

i.e., T does not satisfy property (gab).

In the next theorem we establish the exact relationships between property (gab) and some of the other properties introduced above.

Theorem 5.98 *If $T \in L(X)$ then the following statements are equivalent:*

  (i) *T has property (gab);*
 (ii) *T has property (ab) and $\Pi(T) = \Pi_a(T)$;*
(iii) *T satisfies Browder's theorem and $\Pi(T) = \Pi_a(T)$.*

Proof (i) $\Leftrightarrow$ (ii) If T has property (gab) then $\Delta^g(T) = \Pi_a(T)$ and T has property (ab). By Theorem 5.80, Browder's theorem holds for T, or equivalently, by Theorem 5.38, T satisfies the generalized Browder's theorem, i.e., $\Delta^g(T) = \Pi(T)$. Conversely, assume (ii). Since T satisfies Browder's theorem, or equivalently the generalized Browder's theorem, then $\Delta^g(T) = \Pi(T)$. Since by assumption $\Pi(T) = \Pi_a(T)$ it then follows that $\Delta^g(T) = \Pi_a(T)$, thus property (gab) holds for T.

The implication (ii) $\Rightarrow$ (iii) is clear. We now prove (iii) $\Rightarrow$ (i). Assume that (iii) holds. Since Browder's theorem is equivalent to the generalized Browder's theorem, $\Delta^g(T) = \Pi(T)$ and hence, from our assumption, $\Delta^g(T) = \Pi_a(T)$, so T has property (gab). ■

Property (*gb*) entails property (*gab*) and the precise relationship between these two properties is given in the following theorem.

Theorem 5.99 *If* $T \in L(X)$ *has property* (*gb*) *then* T *has property* (*gab*). T *has property* (*gb*) *precisely when* T *has property* (*gab*) *and* $\mathrm{ind}\,(\lambda I - T) = 0$ *for all* $\lambda \in \Delta_a^g(T)$.

Proof Assume that T has property (*gb*) and $\mathrm{ind}\,(\lambda I - T) = 0$ for all $\lambda \in \Delta_a^g(T)$. Property (*gb*) entails Browder's theorem, by Theorem 5.68, or equivalently the generalized Browder's theorem, so $\Delta^g(T) = \Pi(T)$. Again by Theorem 5.68 we have $\Pi(T) = \Pi_a(T)$, so $\Delta^g(T) = \Pi_a(T)$ and hence T has property (*gab*). To show the second statement, observe that if T has property (*gb*) and $\lambda \in \Delta_a^g(T)$ then $\lambda \in \Pi(T)$. Hence $\lambda I - T$ is B-Weyl, so $\mathrm{ind}\,(\lambda I - T) = 0$. Conversely, if T has property (*gab*) and $\mathrm{ind}\,(\lambda I - T) = 0$ for all $\lambda \in \Delta_a^g(T)$, then, since T satisfies Browder's theorem, from Corollary 5.69 we conclude that T has property (*gb*). ∎

Theorem 5.100 *If* $T \in L(X)$ *the following statements are equivalent:*

  (i) T *has property* (*gb*);
  (ii) T *has property* (*gab*) *and* $\sigma_{\mathrm{bw}}(T) \cap \Delta_a^g(T) = \emptyset$;
  (iii) T *satisfies a-Browder's theorem and* $\sigma_{\mathrm{bw}}(T) \cap \Delta_a^g(T) = \emptyset$;
  (iv) T *satisfies Browder's theorem and* $\sigma_{\mathrm{bw}}(T) \cap \Delta_a^g(T) = \emptyset$.

Proof The equivalence (i) ⇔ (ii) follows from Theorem 5.99. The implications (i) ⇒ (iii) ⇒ (iv) are clear, since property (*gb*) implies *a*-Browder's theorem. To show the implication (iv) ⇒ (i), suppose that $\lambda \in \Delta_a^g(T)$. Then $\lambda \notin \sigma_{\mathrm{bw}}(T)$ by assumption, and since Browder's theorem is equivalent to the generalized Browder's theorem it then follows that $\lambda \notin \sigma_{\mathrm{d}}(T)$, so $\lambda I - T$ is Drazin invertible, and hence $\lambda \in \mathrm{iso}\,\sigma(T)$. The inclusion $\Delta_a^g(T) \subseteq \mathrm{iso}\,\sigma(T)$ is equivalent, by Theorem 5.62, to property (*gb*). ∎

Theorem 5.101 *For an operator* $T \in L(X)$ *the following statements are equivalent:*

  (i) T *satisfies property* (*gab*);
  (ii) $X = H_0(\lambda I - T) + K(\lambda I - T)$ *for all* $\lambda \in \Sigma_a^g(T)$;
  (iii) *there exists a natural* $\nu := \nu(\lambda)$ *such that* $K(\lambda I - T) = (\lambda I - T)^\nu(X)$ *for all* $\lambda \in \Sigma_a^g(T)(T)$;
  (iv) *there exists a natural* $\nu := \nu(\lambda)$ *such that* $(\lambda I - T)^\infty(X) = (\lambda I - T)^\nu(X)$ *for all* $\lambda \in \Sigma_a^g(T)$.

Proof (i) ⇒ (ii) Clear, as observed in (5.12).

(ii) ⇒ (i) By Theorem 2.41 the condition $X = H_0(\lambda I - T) + K(\lambda I - T)$ is equivalent to the inclusion $\lambda \in \mathrm{iso}\,\sigma_{\mathrm{s}}(T)$. Hence $\Sigma_a^g(T) \subseteq \mathrm{iso}\,\sigma_{\mathrm{s}}(T)$. From Theorem 5.95 it immediately follows that T satisfies property (*gab*).

(i) ⇔ (iii) If T satisfies property (*gab*) then, by Theorem 5.95, T^* has the SVEP at every $\lambda \in \Sigma_a^g(T)$. For every $\lambda \in \Sigma_a^g(T)$, $\lambda I - T$ is quasi-Fredholm so, by Theorem 2.98, $q := q(\lambda I - T) < \infty$ for all $\lambda \in \Sigma_a^g(T)$, and hence $(\lambda I - T)^\infty(X) = (\lambda I - T)^q(X)$. Since for every $\lambda \in \Sigma_a^g(T)$ the operator $\lambda I - T$ is upper semi B-Fredholm, there exists a $v \in \mathbb{N}$ such that $(\lambda I - T)^n(X)$ is closed for all $n \geq v$, hence $(\lambda I - T)^\infty(X)$ is closed. As observed above, for every $\lambda \in \Sigma_a^g(T)$, $\lambda I - T$ is quasi-Fredholm and hence has topological uniform descent. Furthermore, by Theorem 1.79, the restriction $(\lambda I - T)|(\lambda I - T)^\infty(X)$ is onto, so $(\lambda I - T)((\lambda I - T)^\infty(X)) = (\lambda I - T)^\infty(X)$. From Theorem 1.39 it then follows that $(\lambda I - T)^\infty(X) \subseteq K(\lambda I - T)$, and, since the reverse inclusion holds for every operator, we then conclude that

$$(\lambda I - T)^\infty(X) = K(\lambda I - T) = (\lambda I - T)^q(X),$$

for all $\lambda \in \Sigma_a^g(T)$.

Conversely, let $\lambda \in \Sigma_a^g(T)$ be arbitrarily given and suppose that there exists a natural $v := v(\lambda)$ such that $K(\lambda I - T) = (\lambda I - T)^v(X)$. Then we have

$$(\lambda I - T)^v(X) = K(\lambda I - T) = (\lambda I - T)(K(\lambda I - T)) = (\lambda I - T)^{v+1}(X),$$

thus $q(\lambda I - T) \leq v$, so T^* has the SVEP at λ, and hence T satisfies (gab), by Theorem 5.95.

(i) $\Leftrightarrow$ (iv) Suppose that T satisfies property (gab). By Theorem 5.95 then T^* has the SVEP at every $\lambda \in \Sigma_a^g(T)$, hence, by Theorem 2.98, $q := q(\lambda I - T) < \infty$ for all $\lambda \in \Sigma_a^g(T)$. Therefore, $(\lambda I - T)^\infty(X) = (\lambda I - T)^q(X)$ for all $\lambda \in \Sigma_a^g(T)$.

Conversely, suppose that (iv) holds and $\lambda \in \Sigma_a^g(T)$. Then

$$(\lambda I - T)^v(X) = (\lambda I - T)^\infty(X) \subseteq (\lambda I - T)^{v+1}(X),$$

and since $(\lambda I - T)^{n+1}(X) \subseteq (\lambda I - T)^n(X)$ holds for all $n \in \mathbb{N}$, we then obtain that $(\lambda I - T)^v(X) = (\lambda I - T)^{v+1}(X)$. Therefore, $q(\lambda I - T) \leq v$, and hence, by Theorem 2.65, T^* has the SVEP at every $\lambda \in \Sigma_a^g(T)$. Consequently, by Theorem 5.95, T^* satisfies property (gab).   ∎

Property (gab) for T^* may be characterized by means of the quasi-nilpotent part as follows:

Theorem 5.102 *If $T \in L(X)$ then T^* has property (gab) if and only if $H_0(\lambda I - T)$ is closed for all $\lambda \in \Sigma_a^g(T^*)$.*

Proof Suppose that T^* has property (gab). By Theorem 5.95 then

$$\Sigma_a^g(T^*) \subseteq \mathrm{iso}\,\sigma(T^*) = \mathrm{iso}\,\sigma(T),$$

so both T and T^* have the SVEP at the points of $\Sigma_a^g(T^*)$. Let $\lambda \in \Sigma_a^g(T^*) = \Delta^g(T^*) \cup \Pi_a(T^*)$. If $\lambda \in \Delta^g(T^*)$ then $\lambda I - T^*$ is B-Weyl, and hence is quasi-Fredholm. By Theorem 1.104 $\lambda I - T$ is also quasi-Fredholm and, since T has the SVEP at λ, Theorem 2.97 entails that $H_0(\lambda I - T)$ is closed. If $\lambda \in \Pi_a(T^*)$ then $\lambda I - T^*$ is left Drazin invertible and hence $\lambda I - T$ is right Drazin invertible, in particular quasi-Fredholm, so the SVEP of T at λ entails, again by Theorem 2.98, that $H_0(\lambda I - T)$ is closed.

Conversely, suppose that $H_0(\lambda I - T)$ is closed for all $\lambda \in \Sigma_a^g(T^*)$. If $\lambda \in \Delta^g(T^*)$ then $\lambda I - T^*$ is B-Weyl, and hence, as above, $\lambda I - T$ is quasi- Fredholm. The condition $H_0(\lambda I - T)$ closed implies, again by Theorem 2.97, that T has the SVEP at λ. By Theorem 1.143 we then conclude that $\lambda I - T$ is Drazin invertible, and hence $\lambda \in \mathrm{iso}\,\sigma(T) = \mathrm{iso}\,\sigma(T^*)$. If $\lambda \in \Pi_a(T^*)$ then $\lambda I - T^*$ is left Drazin invertible, so $\lambda I - T$ is right Drazin invertible. Since $H_0(\lambda I - T)$ is closed, then T has the SVEP at λ, and hence $\lambda I - T$ is Drazin invertible, again by Theorem 1.143. Consequently, $\lambda \in \mathrm{iso}\,\sigma(T) = \mathrm{iso}\,\sigma(T^*)$. Therefore, $\Sigma_a^g(T^*) \subseteq \mathrm{iso}\,\sigma(T^*)$ and hence T^* has property (*gab*) by Theorem 5.95. ∎

Theorem 5.103 *Let $T \in L(X)$ be finitely polaroid. Then T satisfies property (gab) if and only if $K(\lambda I - T)$ has finite codimension for all $\lambda \in \Sigma_a^g(T)$.*

Proof By Theorem 5.95 property (*gab*) entails that $\Sigma_a^g(T) \subseteq \mathrm{iso}\,\sigma(T)$, so, if $\lambda \in \Sigma_a^g(T)$ then $\lambda I - T$ is Browder. Observe that $\beta(\lambda I - T) < \infty$ implies that $\beta(\lambda I - T)^n < \infty$ for every $n \in \mathbb{N}$. Since λ is a pole, then $K(\lambda I - T) = (\lambda I - T)^p(X)$ has finite codimension, where p is the order of the pole.

Conversely, suppose that $K(\lambda I - T)$ has finite codimension for all $\lambda \in \Sigma_a^g(T)$. If $\lambda \in \Sigma_a^g(T)$ then either $\lambda \in \Delta^g(T)$ or $\lambda \in \Pi_a(T)$. If $\lambda \in \Delta^g(T)$, from the inclusion $K(\lambda I - T) \subseteq (\lambda I - T)(X)$ we see that $(\lambda I - T)(X)$ also has finite codimension, hence $\beta(\lambda I - T) < \infty$. Since $\lambda I - T$ is B-Weyl then $\alpha(\lambda I - T) = \beta(\lambda I - T) < \infty$, so $\lambda I - T$ is Weyl. The condition $\mathrm{codim}\,K(\lambda I - T) < \infty$ entails, by Theorem 2.105, that T^* has the SVEP at λ, or equivalently $q(\lambda I - T) < \infty$. By Theorem 1.22 it then follows that λ is a pole, hence $\Delta^g(T) \subseteq \mathrm{iso}\,\sigma(T)$. Consider the other case that $\lambda \in \Pi_a(T)$. Then $p(\lambda I - T) < \infty$ and, as above, the inclusion $K(\lambda I - T) \subseteq (\lambda I - T)(X)$ implies that $\beta(\lambda I - T) < \infty$. Therefore, $\lambda I - T$ is lower semi-Fredholm and hence the condition $K(\lambda I - T)$ has finite codimension implies, again by Theorem 2.105, that $q(\lambda I - T) < \infty$, from which we conclude that $\Pi_a(T) \subseteq \mathrm{iso}\,\sigma(T)$. Consequently, $\Sigma_a^g(T) \subseteq \mathrm{iso}\,\sigma(T)$ and by Theorem 5.95, it then follows that T has property (*gab*). ∎

Theorem 5.104 *Let $T \in L(X)$ be a-polaroid. Then property (gab), property (ab) and Browder's theorem are equivalent for T.*

Proof By Corollary 5.63, in order to show the equivalences we need only to show that Browder's theorem implies property (*gab*). If T satisfies Browder's theorem, or equivalently the generalized Browder's theorem, then $\Delta^g(T) = \Pi(T)$. Since T is a-polaroid, $\Pi_a(T) = \Pi(T)$, so $\Delta^g(T) = \Pi_a(T)$, i.e., T has property (*gab*). ∎

The equivalences of Theorem 5.104 cannot be extended to polaroid operators. Indeed, if T is defined as in Example 5.64, then T is polaroid and satisfies Browder's theorem, since T has the SVEP, while property (*gab*) does not hold for T.

Theorem 5.105 *Suppose that $T, K \in L(X)$ commute and that K^n is a finite rank operator for some $n \in \mathbb{N}$. Furthermore, assume that $\mathrm{iso}\,\sigma_a(T) = \mathrm{iso}\,\sigma_a(T + K)$. If T has property (gab) then $T + K$ also has property (gab).*

Proof We know that $\sigma_d(T) = \sigma_d(T + K)$ and $\sigma_{ld}(T) = \sigma_{ld}(T + K)$. By Theorem 3.27 we then obtain $\Pi(T) = \Pi(T + K)$ and $\Pi_a(T) = \Pi_a(T + K)$.

Now, assume that T has property (gab). Then Browder's theorem holds for T and since K is a Riesz operator, Browder's theorem holds for $T + K$, by Corollary 5.5. Furthermore, property (gab) for T entails, by Theorem 5.95, that $\Pi(T) = \Pi_a(T)$, and hence $\Pi(T) = \Pi(T+K) = \Pi_a(T) = \Pi_a(T+K)$, so, again by Theorem 5.95, $T + K$ has property (gab). ∎

Lemma 5.106 *Suppose that for $T \in L(X)$ we have* iso $\sigma_{ap}(T) = \emptyset$. *If $K \in L(X)$ is such that K^n is finite-dimensional for some $n \in \mathbb{N}$, then* iso $\sigma_{ap}(T + K) = \emptyset$. *Consequently,* $\sigma_{ap}(T + K) = \sigma_{ap}(T)$.

Proof We know that acc $\sigma_{ap}(T) =$ acc $\sigma_{ap}(T + K)$, so

$$\sigma_{ap}(T) = \text{iso}\,\sigma_{ap}\,(T) \cup \text{acc}\,\sigma_{ap}\,(T) = \text{acc}\,\sigma_{ap}\,(T)$$
$$= \text{acc}\,\sigma_{ap}\,(T + K) \subseteq \sigma_{ap}(T + K).$$

On the other hand, $\sigma_{ap}(K)$ is a finite set, say $\sigma_{ap}(K) = \{\lambda_1, \lambda_2, \ldots \lambda_n\}$, so we have

$$\text{iso}\,\sigma_{ap}(T + K) \subseteq \text{iso}\,(\sigma_{ap}(T) + \sigma_{ap}(K)) = \text{iso}\,\bigcup_{k=1}^{n}(\lambda_k + \sigma_{ap}(T)) = \emptyset,$$

hence, by Theorem 3.26, we have

$$\sigma_{ap}(T + K) = \text{iso}\,\sigma_{ap}(T + K) \cup \text{acc}\,\sigma_{ap}(T + K) = \text{acc}\,\sigma_{ap}(T + K)$$
$$= \text{acc}\,\sigma_{ap}(T) = \sigma_{ap}(T),$$

so $\sigma_{ap}(T + K) = \sigma_{ap}(T)$ holds. ∎

Theorem 5.107 *Suppose that $T, K \in L(X)$ commute and that K^n is a finite rank operator for some $n \in \mathbb{N}$. If* iso $\sigma_{ap}(T) = \emptyset$ *and T has (gab) then $T + K$ has (gab).*

Proof The condition iso $\sigma_a(T) = \emptyset$ implies that also iso $\sigma_a(T + K) = \emptyset$. Thus, we are in the situation of Theorem 5.105, hence T transfers property (gab) to $T + K$. ∎

We have seen that the condition that R is Drazin invertible, which means that 0 belongs to the resolvent or it is a pole of the resolvent of its Drazin inverse S, determines the spectral structure (or the local spectral structure) of its Drazin inverse. We conclude this chapter by proving that all Browder-type theorems are transmitted for a Drazin invertible operator R to its Drazin inverse. We begin with a remark.

Remark 5.108 It should be noted that if R is Drazin invertible then

$$R \text{ is upper semi-Weyl} \Leftrightarrow R \text{ is Weyl} \Leftrightarrow R \text{ is Browder.}$$

Theorem 5.109 *Suppose that $R \in L(X)$ is Drazin invertible with Drazin inverse S. Then*

(i) *R satisfies Browder's theorem if and only if S satisfies Browder's theorem.*
(ii) *R satisfies a-Browder's theorem if and only if S satisfies a-Browder's theorem.*

Proof

(i) Suppose that R satisfies Browder's theorem and let $X = Y \oplus Z$, $R = R_1 \oplus R_2$ and $S = 0 \oplus S_2$, where $S_2 = R_2^{-1}$. Let $\lambda \notin \sigma_w(S)$ be arbitrarily given. To prove that Browder's theorem holds for S it suffices to show that S has the SVEP at λ. If $\lambda = 0$ then T has the SVEP at 0, since either S is invertible or 0 is a pole of the resolvent of S (recall that the Drazin inverse S is itself Drazin invertible), and hence an isolated point of the spectrum. If $\lambda \neq 0$ then $1/\lambda \notin \sigma_w(R)$, and since R satisfies Browder's theorem, R has the SVEP at $1/\lambda$. By Theorem 2.184 S has the SVEP at λ. Hence S satisfies Browder's theorem. The converse may be proved by similar arguments.

(ii) Suppose that R satisfies a-Browder's theorem and let $\lambda \notin \sigma_{uw}(S)$. If $\lambda = 0$, since S is Drazin invertible we have $p(S) = q(S) < \infty$, hence S has the SVEP at 0. If $\lambda \neq 0$ then $\frac{1}{\lambda} \notin \sigma_{uw}(R)$, and since R satisfies a-Browder's theorem, R has the SVEP at $1/\lambda$. By Theorem 2.184 S has the SVEP at λ, and hence S satisfies a-Browder's theorem. ∎

Properties (*b*) and (*ab*) are also transmitted from a Drazin invertible operator to its Drazin inverse. To show this we need some preliminary results:

Theorem 5.110 *Suppose that $R \in L(X)$ is Drazin invertible with Drazin inverse S. Then R is Browder if and only if S is Browder.*

Proof If $0 \notin \sigma(R)$ then R is invertible and the Drazin inverse is $S = R^{-1}$ so the assertion is trivial in this case. Suppose that $0 \in \sigma(R)$ and that R is Browder. Then 0 is a pole of the resolvent of R and is also a pole (of the first order) of the resolvent of S. Let $X = Y \oplus Z$ such that $R = R_1 \oplus R_2$, $R_1 = R|Y$ nilpotent and $R_2 = R|Z$ invertible. Observe that

$$\ker R = \ker R_1 \oplus \ker R_2 = \ker R_1 \oplus \{0\}, \tag{5.14}$$

and, analogously, since $S = 0 \oplus S_2$ with $S_2 = R_2^{-1}$, we have

$$\ker S = \ker 0 \oplus \ker S_2 = Y \oplus \{0\}. \tag{5.15}$$

Since R is Browder we have $\alpha(R) = \dim \ker R < \infty$, and from the inclusion $\ker R_1 \subseteq \ker R$ it then follows that $\alpha(R_1) < \infty$. Consequently, $\alpha(R_1^n) < \infty$ for all $n \in \mathbb{N}$. Let $R_1^\nu = 0$. Since $Y = \ker R_1^\nu$ we then conclude that the subspace Y is finite-dimensional and hence $\ker S = Y \oplus \{0\}$ is finite-dimensional, i.e. $\alpha(S) < \infty$. Now, S is Drazin invertible, so $p(S) = q(S) < \infty$ and hence, by Theorem 1.22, $\alpha(S) = \beta(S) < \infty$. Hence S is Browder.

Conversely, suppose that S is Browder. Then $\alpha(S) < \infty$ and hence by (5.15) the subspace Y is finite-dimensional, from which it follows that $\ker R_1 = \ker R|Y$ is finite-dimensional. From (5.14) we then have that $\alpha(R) < \infty$ and since $p(R) = q(R) < \infty$ we then conclude that $\alpha(R) = \beta(R)$, again by Theorem 1.22. Therefore, R is a Browder operator.   ∎

Lemma 5.111 *Let $R \in L(X)$ be Drazin invertible with Drazin inverse S. We have:*

(i) $0 \in p_{00}(R) \Leftrightarrow 0 \in p_{00}(S)$. *If* $\lambda \neq 0$ *then* $\lambda \in p_{00}(R) \Leftrightarrow \frac{1}{\lambda} \in p_{00}(S)$.
(ii) $0 \in p_{00}^a(R) \Leftrightarrow 0 \in p_{00}^a(S)$. *If* $\lambda \neq 0$ *then* $\lambda \in p_{00}^a(R) \Leftrightarrow \frac{1}{\lambda} \in p_{00}^a(S)$.

Proof

(i) Since $0 \in \sigma(R)$ if and only if $0 \in \sigma(S)$ then the first assertion follows from Theorem 5.110. The second assertion is clear from part (ii) of Theorem 3.126.
(ii) The proof is similar to part (i).   ∎

Theorem 5.112 *Suppose that $R \in L(X)$ is Drazin invertible with Drazin inverse S. Then*

(i) *R satisfies property (ab) and only if S satisfies property (ab).*
(i) *R satisfies property (b) and only if S satisfies property (ab).*

Proof

(i) Suppose that R satisfies property (ab). Then R satisfies Browder's theorem and hence S also satisfies Browder's theorem, by Lemma 5.110. Therefore, $\sigma_b(S) = \sigma_w(S)$. Let $\lambda \in \Sigma_a(S)$. By Theorem 5.82 it suffices to show that $\lambda \in \operatorname{iso}\sigma(S)$.
   We distinguish the two cases $\lambda = 0$ and $\lambda \neq 0$.
   If $\lambda = 0$ then $0 \in \operatorname{iso}\sigma(S)$, since S is Drazin invertible. Suppose that $\lambda \neq 0$. Then either $\lambda \in \Delta(S)$ or $\lambda \in p_{00}(S)$. If $\lambda \in \Delta(S) = \sigma(S) \setminus \sigma_w(S) = \sigma(S) \setminus \sigma_b(S) = p_{00}(S)$, then $\lambda I - S$ is Browder, so $\lambda \in \operatorname{iso}\sigma(S)$. If $\lambda \in p_{00}^a(S)$ then, by Lemma 5.111, $\frac{1}{\lambda} \in p_{00}^a(R)$. Property (ab) for R entails, by Theorem 5.82, $\frac{1}{\lambda} \in \operatorname{iso}\sigma(R)$. Consequently, $\lambda \in \operatorname{iso}\sigma(S)$.
   Therefore, S has property (ab). The converse may be proved by using similar arguments.
(ii) Suppose that R satisfies property (b), or equivalently $\Delta_a(R) \subseteq \operatorname{iso}\sigma(R)$. Then R satisfies a-Browder's theorem and hence S also satisfies a- Browder's theorem, by Theorem 5.109, so that $\sigma_{ub}(R) = \sigma_{uw}(SR)$ and $\sigma_{ub}(S) = \sigma_{uw}(S)$. Consequently,

$$\Delta_a(R) = p_{00}^a(R) \quad \text{and} \quad \Delta_a(S) = p_{00}^a(S).$$

To show property (b) for T it suffices to prove, by Theorem 5.82, the inclusion $\Delta_a(S) \subseteq \operatorname{iso}\sigma(S)$. Let $\lambda \in \Delta_a(S)$. If $\lambda = 0$ then 0 is an isolated point of $\sigma(S)$, since S is Drazin invertible. Suppose that $\lambda \neq 0$. Since $\lambda \in \Delta_a(S) = p_{00}^a(S)$

then, by Lemma 5.111 and Theorem 5.82, $\frac{1}{\lambda} \in p_{00}^a(R) = \Delta_a(R) \subseteq \mathrm{iso}\,\sigma(R)$. Consequently, $\lambda \in \mathrm{iso}\,\sigma(S)$. Thus, S has property (b). The converse may be proved in a similar way.   ∎

5.7 Comments

Browder's theorem was introduced by Harte and Lee [173], while the concept of the generalized Browder's theorem was first introduced by Berkani and Koliha in [70]. The equivalence between Browder's theorem and the generalized Browder's theorem was proved by Amouch and Zguitti [48] and was later proved in [36], by using the methods adopted in the first section of this chapter. All the material concerning the characterizations of Browder's theorem and the generalized Browder's theorem, by means of the localized SVEP, as well as by means of the quasi-nilpotent part $H_0(\lambda I - T)$ as λ belongs to certain subsets of $\mathbb{C}$, is modeled after Aiena and Biondi [10], and Aiena and Garcia [13]. However, most of the material of Sect. 5.2 of this chapter is inspired by the works of Aiena et al. [35], and Aiena, and Miller [34].

Property (b), property (gb) and property (ab) were introduced by Berkani and Zariuoh in various articles, for instance [74, 75]. In [73] and [76] Berkani, Sarih and Zariouh established some other results concerning the stability of these properties under commutative finite-rank perturbations, compact perturbations and nilpotent perturbations. Related results may be found in Duggal and Kim [137]. Most of the material concerning the characterizations of the properties by means of the quasi-nilpotent part and the analytic core of $\lambda I - T$ is modeled after Aiena et al. [43]. Property (gab) was introduced by Berkani and Zariouh in [75], but most of the material of this chapter concerning this property may be found in Aiena and Triolo [25]. The results concerning the transmission of Browder-type theorems from a Drazin invertible operator to its Drazin inverse are modeled after Aiena and Triolo [28].

Chapter 6
Weyl-Type Theorems

In the previous chapters we introduced several classes of operators which have their origin in Fredholm theory. We also know that the spectrum of a bounded linear operator T on a Banach space X can be split into subsets in many different ways. In 1908 Weyl [296] proved an important property of self-adjoint operators on Hilbert spaces. He proved that if $T \in L(H)$ is self-adjoint then the spectral points $\lambda \in \sigma(T)$ that do not belong to the intersection of all the spectra $\sigma(T + K)$, where $K \in L(H)$ are compact operators, are exactly the points of the set $\pi_{00}(T)$ of all isolated points λ of $\sigma(T)$ which are eigenvalues of finite multiplicity, i.e., $0 < \alpha(\lambda I - T) < \infty$. In our language, the intersection mentioned above coincides with what we called the Weyl spectrum of T (see Corollary 3.37). Hence, Weyl proved the equality

$$\sigma(T) \setminus \sigma_{\mathrm{w}}(T) = \pi_{00}(T) \tag{6.1}$$

for self-adjoint operators in Hilbert spaces. In 1966 Coburn [97] extended Weyl's result from self-adjoint operators to nonnormal operators, in particular to Toeplitz operators on Hardy spaces, and later this result was extended to several other classes of operators and this, in more recent years, gave rise to an intense line of research in spectral theory. Nowadays, an operator for which the equality (6.1) holds is said to satisfy Weyl's theorem.

Since for a self-adjoint operator T, the SVEP is satisfied by both T and T^* (indeed, T is decomposable), by Theorem 2.68 we then have $\sigma(T) = \sigma_{\mathrm{ap}}(T)$ and, by Theorem 3.44, we also have $\sigma_{\mathrm{uw}}(T) = \sigma_{\mathrm{w}}(T)$. Therefore, for self-adjoint operators we have:

$$\pi_{00}(T) = \sigma(T) \setminus \sigma_{\mathrm{w}}(T) = \sigma_{\mathrm{ap}}(T) \setminus \sigma_{\mathrm{uw}}(T) = \pi_{00}^a(T), \tag{6.2}$$

where $\pi_{00}^a(T)$ is the set of all isolated points of $\sigma_{\mathrm{ap}}(T)$ which are eigenvalues of finite multiplicity. In [262], Rakočević introduced the operators for which the equality $\sigma_{\mathrm{ap}}(T) \setminus \sigma_{\mathrm{uw}}(T) = \pi_{00}^a(T)$ holds. These operators are said to satisfy

© Springer Nature Switzerland AG 2018   419
P. Aiena, *Fredholm and Local Spectral Theory II*, Lecture Notes in Mathematics
2235, https://doi.org/10.1007/978-3-030-02266-2_6

a-Weyl's theorem. From the equalities (6.2) it then follows that a-Weyl's theorem and Weyl's theorem are equivalent for self-adjoint operators. We shall see that this is true only assuming the SVEP for T^*. But for operators which are not self-adjoint, a-Weyl's theorem is in general stronger than Weyl's theorem. Moreover, there are also examples of operators for which the equality $\sigma_{\mathrm{ap}}(T) \setminus \sigma_{\mathrm{uw}}(T) = \pi_{00}(T)$ holds, and in this case we say that T satisfies property (w). We shall see that for any operator a-Weyl's theorem, as well as property (ω), entails Weyl theorem. Furthermore, a-Weyl's theorem and property (ω) are independent.

The Weyl-type theorems mentioned above admit an extension obtained by replacing the classical Fredholm theory by the B-Fredholm theory introduced by Berkani [64]. In the literature the versions of Weyl-type theorems obtained in the framework of B-Fredholm theory are improperly called (because they are stronger versions) *the generalized Weyl's theorem, the generalized a-Weyl's theorem and the generalized property* (ω). Before studying all these Weyl-type theorems we introduce, in the first two sections, the property (R) and the generalized property (gR), which in some sense may be thought of as half property (ω) and property $(g\omega)$, respectively. After a rather detailed study of Weyl-type theorems we shall see that if T is a polaroid-type operator then some of these theorems are equivalent. Weyl-type theorems are also extended from a Drazin invertible operator R to its Drazin inverse.

Weyl-type theorems are satisfied by several classes of operators defined on Banach spaces, for instance Toeplitz operators on Hardy spaces, semi-shifts, and symmetrizable operators. The two conditions of being polaroid and of T, or T^*, having the SVEP provide a useful tool for establishing Weyl-type theorems, but the case of Toeplitz operators provide an example of operators that obey Weyl's theorem, even if neither T and T^* satisfy the SVEP.

We conclude this book by giving, in the last section of this chapter, a very useful and unique theoretical framework from which we can deduce that the Weyl-type theorems hold for many classes of operators which act on Hilbert spaces. This framework is created by introducing the class of *quasi totally hereditarily normaloid* operators and by proving that these operators are *hereditarily polaroid*. Many common classes of operators T on Hilbert spaces are quasi totally hereditarily normaloid, and this fact, together with SVEP, allows us to extend all Weyl-type theorems to the perturbations $f(T + K)$, where K is algebraic and commutes with T and f is an analytic function, defined on an open neighbourhood of the spectrum of $T + K$, such that f is non-constant on each of the components of its domain.

6.1 Property (R)

Recall that by $p_{00}(T) := \sigma(T) \setminus \sigma_{\mathrm{b}}(T)$ we denote the set of all poles of the resolvent having finite rank, while

$$p_{00}^a(T) := \sigma_{\mathrm{ap}}(T) \setminus \sigma_{\mathrm{ub}}(T)$$

denotes the set of all left poles having finite rank. Define

$$\pi_{00}(T) := \{\lambda \in \text{iso}\,\sigma(T) : 0 < \alpha(\lambda I - T) < \infty\},$$

i.e., $\pi_{00}(T)$ is the set of all eigenvalues of T which are isolated points of the spectrum and have finite multiplicity. It is easily seen that

$$p_{00}(T) \subseteq \pi_{00}(T) \quad \text{for all } T \in L(X).$$

Indeed, every $\lambda \in p_{00}(T)$ is a pole of the resolvent and hence an isolated point of the spectrum. Furthermore, $\alpha(\lambda I - T) < \infty$, since $\lambda I - T \in B(X)$, and $\alpha(\lambda I - T) > 0$, otherwise, if $\alpha(\lambda I - T) = 0$ we would have, by Theorem 1.22, $\alpha(\lambda I - T) = \beta(\lambda I - T) = 0$, hence $\lambda \notin \sigma(T)$.

We now consider the operators $T \in L(X)$ on Banach spaces for which the equality $\pi_{00}(T) = p_{00}(T)$ holds. The next theorem shows that this condition may be characterized in several ways:

Theorem 6.1 *For a bounded operator $T \in L(X)$ the following statements are equivalent:*

  (i) $\pi_{00}(T) = p_{00}(T)$;
 (ii) $\sigma_w(T) \cap \pi_{00}(T) = \emptyset$;
(iii) $(\lambda I - T)(X)$ *is closed for all* $\lambda \in \pi_{00}(T)$;
 (iv) $H_0(\lambda I - T)$ *is finite-dimensional for all* $\lambda \in \pi_{00}(T)$;
  (v) $K(\lambda I - T)$ *has finite codimension for all* $\lambda \in \pi_{00}(T)$;
 (vi) $(\lambda I - T)^\infty(X)$ *has finite codimension for all* $\lambda \in \pi_{00}(T)$;
(vii) $\beta(\lambda I - T) < \infty$ *for all* $\lambda \in \pi_{00}(T)$;
(viii) $q(\lambda I - T) < \infty$ *for all* $\lambda \in \pi_{00}(T)$;
 (ix) *The mapping* $\lambda \to \gamma(\lambda I - T)$ *is not continuous at each* $\lambda_0 \in \pi_{00}(T)$, *where* $\gamma(\lambda I - T)$ *denotes the minimal modulus of* $\lambda I - T$.

Proof (i) $\Rightarrow$ (ii) If $p_{00}(T) = \pi_{00}(T)$ then $\pi_{00}(T) \cap \sigma_b(T) = \emptyset$, and hence $\sigma_w(T) \cap \pi_{00}(T) = \emptyset$, since $\sigma_w(T) \subseteq \sigma_b(T)$.

(ii) $\Rightarrow$ (iii) If $\lambda \in \pi_{00}(T)$ then $\lambda I - T$ is Weyl, so $(\lambda I - T)(X)$ is closed.

(iii) $\Rightarrow$ (iv) If $\lambda \in \pi_{00}(T)$ then $\alpha(\lambda I - T) < \infty$, so $\lambda_0 I - T \in \Phi_+(X)$. Since T has the SVEP at every isolated point of $\sigma(T)$, by Theorem 2.105 it then follows that $H_0(\lambda I - T)$ has finite dimension.

(iv) $\Rightarrow$ (v) If $\lambda \in \text{iso}\,\sigma(T)$ then the decomposition $X = H_0(\lambda I - T) \oplus K(\lambda I - T)$ holds, by Theorem 2.45. Consequently, $K(\lambda I - T)$ has finite codimension, since $H_0(\lambda I - T)$ is finite-dimensional.

(v) $\Rightarrow$ (vi) This follows from the inclusion $K(\lambda I - T) \subseteq (\lambda I - T)^\infty(X)$.

(vi) $\Rightarrow$ (vii) Clear, since $(\lambda I - T)^\infty(X) \subseteq (\lambda I - T)(X)$ for every $\lambda \in \mathbb{C}$, and this implies that $\beta(\lambda I - T) < \infty$.

(vii) $\Rightarrow$ (i) For every $\lambda \in \pi_{00}(T)$ we have $\alpha(\lambda I - T) < \infty$, so if $\beta(\lambda I - T) < \infty$ then $\lambda I - T \in \Phi(X)$. Since $\lambda \in \text{iso}\,\sigma(T)$, the SVEP of T and T^* at λ ensures that $p(\lambda I - T)$ and $q(\lambda I - T)$ are both finite, by Theorems 2.97 and 2.98.

Thus, $\pi_{00}(T) \subseteq p_{00}(T)$ and hence, since the opposite inclusion is satisfied by every operator, we may conclude that $\pi_{00}(T) = p_{00}(T)$.

(i) $\Rightarrow$ (viii) Clear.

(viii) $\Rightarrow$ (vii) This is immediate. In fact, by Theorem 1.22, if $q(\lambda I - T) < \infty$ then $\beta(\lambda I - T) \leq \alpha(\lambda I - T) < \infty$ for all $\lambda \in \pi_{00}(T)$.

(iii) $\Leftrightarrow$ (ix) Observe first that if $\lambda_0 \in \pi_{00}(T)$ then there exists a punctured open disc $\mathbb{D}_0$ centered at λ_0 such that

$$\gamma(\lambda I - T) \leq |\lambda - \lambda_0| \quad \text{for all } \lambda \in \mathbb{D}_0. \tag{6.3}$$

In fact, if λ_0 is isolated in $\sigma(T)$ then $\lambda I - T$ is invertible, and hence has closed range in an open punctured disc $\mathbb{D}$ centered at λ_0. Take $0 \neq x \in \ker(\lambda_0 I - T)$. Then

$$\gamma(\lambda I - T) \leq \frac{\|(\lambda I - T)x\|}{\text{dist}\,(x, \ker(\lambda I - T))}] = \frac{\|(\lambda I - T)x\|}{\|x\|}$$

$$= \frac{\|(\lambda I - T)x - (\lambda_0 I - T)x\|}{\|x\|} = |\lambda - \lambda_0|.$$

From the estimate (6.3) it then follows that $\gamma(\lambda I - T) \to 0$ as $\lambda \to \lambda_0$, so the mapping $\lambda \to \gamma(\lambda I - T)$ is not continuous at a point $\lambda_0 \in \pi_{00}(T)$ precisely when $\gamma(\lambda_0 I - T) > 0$, or, equivalently, by Theorem 1.2, when $(\lambda_0 I - T)(X)$ is closed. ∎

Definition 6.2 We say that an operator $T \in L(X)$ satisfies property (R) if the equality $p_{00}^a(T) = \pi_{00}(T)$ holds.

The following example shows that property (R) for an operator T is not transmitted to the dual T^*.

Example 6.3 Let $T \in \ell^2(\mathbb{N})$ be the weighted right unilateral shift defined by

$$T(x_1, x_2, \dots) := \left(0, \frac{x_1}{2}, \frac{x_2}{3}, \dots\right) \quad \text{for all } x = (x_1, x_2, \dots) \in \ell^2(\mathbb{N}).$$

Clearly, T is quasi-nilpotent, $\sigma_a(T) = \sigma_{\text{ub}}(T) = \{0\}$, and $p_{00}^a(T) = \emptyset$, so T satisfies property (R). On the other hand, it is easily seen that T^* does not satisfy property (R).

By duality it is easy to see that T^* satisfies property (R) if and only if $\pi_{00}(T^*)$ coincides with the set of all right poles having finite rank.

Theorem 6.4 *If $T \in L(X)$ satisfies property (R), then $\pi_{00}(T) = p_{00}(T)$. In particular, every left pole of finite rank of T is a pole.*

Proof Observe first that the inclusion $p_{00}(T) \subseteq \pi_{00}(T)$ holds for all $T \in L(X)$, so we need only to show the opposite inclusion. Suppose that T satisfies (R) and let $\lambda \in \pi_{00}(T) = p_{00}^a(T)$. Then $p(\lambda I - T) < \infty$, and since $\lambda \in \text{iso}\,\sigma(T)$ then T^* has the SVEP at λ. By Theorem 2.98, since $\lambda I - T \in B_+(X)$, the SVEP for

T^* at λ is equivalent to saying that $q(\lambda I - T) < \infty$. Moreover, $\alpha(\lambda I - T) < \infty$, since $\lambda \in \pi_{00}(T)$. From Theorem 1.22 it then follows that $\beta(\lambda I - T) < \infty$, so that $\lambda I - T \in B(X)$. Since $\alpha(\lambda I - T) > 0$ we then conclude that $\lambda \in \sigma(T) \setminus \sigma_b(T) = p_{00}(T)$, thus $\pi_{00}(T) = p_{00}(T)$. The last assertion is clear: $p_{00}(T) = p_{00}^a(T)$. ∎

Theorem 6.5 *Let $T \in L(H)$, H a Hilbert space. Then T^* has property (R) if and only if its adjoint T' has property (R).*

Proof By Theorem 3.1 we have

$$p_{00}^a(T^*) = \sigma_{ap}(T^*) \setminus \sigma_{ub}(T^*) = \overline{\sigma_{ap}(T') \setminus \sigma_{ub}(T')} = \overline{p_{00}^a(T')}$$

and obviously, $\pi_{00}(T^*) = \overline{\pi_{00}(T')}$. ∎

The equality $\pi_{00}(T) = p_{00}(T)$ is strictly weaker than property (R) for T. However, we have:

Theorem 6.6 $T \in L(X)$ *satisfies property (R) if and only if the following two conditions hold:*

(i) $p_{00}^a(T) \subseteq \text{iso}\,\sigma(T)$.
(ii) $\pi_{00}(T) = p_{00}(T)$.

Proof If T satisfies property (R) then $p_{00}^a(T) = \pi_{00}(T) \subseteq \text{iso}\,\sigma(T)$ and, by Theorem 6.4, we have $\pi_{00}(T) = p_{00}(T)$. Conversely, suppose that both (i) and (ii) hold. If $\lambda \in p_{00}^a(T) = \sigma_{ap}(T) \setminus \sigma_{ub}(T)$ then $\lambda I - T \in B_+(X)$, hence $\lambda I - T$ has closed range. Since $\lambda \in \sigma_{ap}(T)$, we have $0 < \alpha(\lambda I - T) < \infty$, from which we conclude that $p_{00}^a(T) \subseteq \pi_{00}(T)$. Since $\pi_{00}(T) = p_{00}(T)$ we then have $\pi_{00}(T) \subseteq p_{00}^a(T)$. Therefore $p_{00}^a(T) = \pi_{00}(T)$. ∎

Clearly, every polaroid operator T satisfies $p_{00}(T) = \pi_{00}(T)$.

Theorem 6.7 *Suppose that $T \in L(X)$ is a-polaroid. Then T satisfies property (R).*

Proof If $\lambda \in p_{00}^a(T)$ then λ is a left pole and hence an isolated point of $\sigma_{ap}(T)$. Since T is a-polaroid, λ is a pole of the resolvent of T and hence an isolated point of the spectrum. Clearly, $0 < \alpha(\lambda I - T) < \infty$, thus $\lambda \in \pi_{00}(T)$ and consequently $p_{00}^a(T) \subseteq \pi_{00}(T)$.

To show the opposite inclusion $\pi_{00}(T) \subseteq p_{00}^a(T)$, let $\lambda \in \pi_{00}(T)$ be arbitrarily given. Since $0 < \alpha(\lambda I - T)$, we have $\lambda \in \sigma_{ap}(T)$ and, since $\lambda \in \text{iso}\,\sigma(T)$, we then have $\lambda \in \text{iso}\,\sigma_{ap}(T)$, and hence λ is a pole of the resolvent of T, or equivalently $\lambda I - T$ has both ascent and descent finite. Since $\alpha(\lambda I - T) < \infty$ then $\beta(\lambda I - T) < \infty$, by Theorem 1.22, hence $\lambda I - T \in B(X)$, in particular $\lambda \notin \sigma_{ub}(T)$. Therefore $\lambda \in \sigma_a(T) \setminus \sigma_{ub}(T) = p_{00}^a(T)$, as desired. ∎

By Theorem 4.24 the a-polaroid condition is preserved under commuting perturbations K for which K^n is finite-dimensional for some $n \in \mathbb{N}$, so we have:

Corollary 6.8 *If $T \in L(X)$ is a-polaroid and $K \in L(X)$ commutes with T and K^n is finite-dimensional for some $N \in \mathbb{N}$, then $T + K$ satisfies property (R).*

The next example shows that under the weaker condition of T being polaroid the result of Theorem 6.7 does not hold.

Example 6.9 Let $R \in \ell^2(\mathbb{N})$ be the unilateral right shift and

$$U(x_1, x_2, \ldots) := (0, x_2, x_3, \cdots) \quad \text{for all } (x_n) \in \ell^2(\mathbb{N}).$$

If $T := R \oplus U$ then $\sigma(T) = D(0, 1)$, so iso $\sigma(T) = \pi_{00}(T) = \emptyset$. Therefore, T is polaroid. Moreover, $\sigma_{\mathrm{ap}}(T) = \Gamma \cup \{0\}$, where Γ is the unit circle, so iso $\sigma_a(T) = \{0\}$. Since R is injective and $p(U) = 1$ it then follows that $p(T) = p(R) + p(U) = 1$. Furthermore, $T \in \Phi_+(X)$ and hence T is upper semi-Browder, so $0 \in \sigma_a(T) \setminus \sigma_{\mathrm{ub}}(T) = p_{00}^a(T)$, from which we conclude that $p_{00}^a(T) \neq \pi_{00}(T)$.

The result of Theorem 6.7 may be extended as follows:

Theorem 6.10 *Let $T \in L(X)$ be polaroid and $f \in \mathcal{H}_{nc}(\sigma(T))$.*

(i) *If T^* has the SVEP then property (R) holds for $f(T)$.*
(ii) *If T has the SVEP then property (R) holds for $f(T^*)$.*

Proof

(i) From Theorem 4.19 we know that $f(T)$ is polaroid and, by Corollary 2.89, $f(T^*)$ has the SVEP, hence, by Theorem 4.15, $f(T)$ is a-polaroid. From Theorem 6.29 it then follows that property (R) holds for $f(T)$.
(ii) T^* is also polaroid and hence, again by Theorem 4.19, $f(T^*)$ is polaroid. Moreover, again by Corollary 2.89, $f(T)$ has the SVEP, and hence, by Theorem 4.15, $f(T^*)$ is a-polaroid. From Theorem 6.7 we then conclude that property (R) holds for $f(T^*)$.   ∎

We now investigate the permanence of property (R) under Riesz commuting perturbations. Define

$$\pi_{0f}(T) := \{\lambda \in \text{iso}\,\sigma(T) : \alpha(\lambda I - T) < \infty\}.$$

Obviously, $\pi_{00}(T) \subseteq \pi_{0f}(T)$.

The following result gives useful information on $\pi_{00}(T + K)$.

Theorem 6.11 *Let $T \in L(X)$ and suppose that $R \in L(X)$ is a Riesz operator that commutes with T. Then we have*

(i) $\pi_{0f}(T + R) \cap \sigma(T) \subseteq \text{iso}\,\sigma(T)$.
(ii) $\pi_{00}(T + R) \cap \sigma_{\mathrm{ap}}(T) \subseteq \text{iso}\,\sigma(T)$.

Proof

(i) Assume that $\lambda \in \pi_{0f}(T + K) \cap \sigma(T)$. Since $\lambda \in \text{iso}\,\sigma(T + R)$, by Theorem 2.45 we have

$$X = H_0(\lambda I - (T + R)) \oplus K(T + R - \lambda I).$$

Write $T = T_1 \oplus T_2$ and $R = R_1 \oplus R_2$ with respect to this decomposition. We claim that $\sigma(T_1)$ is a finite set. To show this, let us suppose that $\sigma(T_1)$ is infinite and consider a sequence (λ_n) of distinct scalars in $\sigma(T_1) \setminus \{\lambda\}$. Consider the operator

$$Q := \lambda I - (T_1 + R_1) = (\lambda I - (T + R))|H_0(\lambda I - (T + R)).$$

Evidently, Q is quasi-nilpotent and $\ker Q$ is finite-dimensional, since $\lambda \in \pi_{0f}(T + R)$. Since $\lambda_n - \lambda \neq 0$, we have

$$\lambda I - \lambda I + \lambda_n I - (T_1 + R_1) = \lambda_n I - (T_1 + R_1)$$

is invertible and hence Weyl. Since $R_1 = R|H_0(\lambda I - (T + R))$ is Riesz, by Theorem 3.7, so $\lambda_n I - T_1$ is Weyl and hence $\ker(\lambda_n I - T_1)$ is a non-zero finite-dimensional subspace, because $\lambda_n I - T_1$ is not invertible. From this we then conclude that the restriction of Q to $\ker(\lambda_n I - T_1)$ is nilpotent, so $\ker(\lambda_n I - T_1) \cap \ker Q$ is not trivial and hence it contains a non-zero element x_n. Since each x_n is an eigenvector of T associated to λ_n, and the scalars λ_n are mutually distinct, we can easily check that (x_n) consists of linearly independent vectors. Consequently, since $x_n \in \ker Q$ for every $n \in \mathbb{N}$, the subspace $\ker Q$ is infinite-dimensional, a contradiction. Therefore $\sigma(T_1)$ is finite and hence there exists a deleted neighborhood U_1 of λ such that $U_1 \cap \sigma(T_1) = \emptyset$. On the other hand, since $\lambda I - (T_2 + R_2)$ is invertible, and hence Browder, $\lambda I - T_2$ is Browder, by Theorem 3.8. Consequently there exists a deleted neighborhood U_2 of λ such that $U_2 \cap \sigma(T_2) = \emptyset$. Now, if $U := U_1 \cap U_2$ then $U \cap \sigma(T) = \emptyset$ and since $\lambda \in \sigma(T)$ we then conclude that $\lambda \in \mathrm{iso}\,\sigma(T)$.

(ii) Clearly, we have

$$\pi_{00}(T + R) \cap \sigma_{\mathrm{ap}}(T) \subseteq \pi_{0f}(T + R) \cap \sigma(T),$$

so, from part (i) we deduce that $\pi_{00}(T + R) \cap \sigma_{\mathrm{ap}}(T) \subseteq \mathrm{iso}\,\sigma(T)$. ∎

Theorem 6.12 *Suppose that $T \in L(X)$ has property (R) and $R \in L(X)$ is a Riesz operator for which $TR = RT$. If $\sigma_{\mathrm{ap}}(T) = \sigma_{\mathrm{ap}}(T + R)$ then $\pi_{00}(T) \subseteq \pi_{00}(T + R)$.*

Proof Suppose that T has property (R). By assumption we have $\sigma_{\mathrm{ap}}(T) = \sigma_{\mathrm{ap}}(T + R)$, hence

$$\pi_{00}(T) = \sigma_{\mathrm{ap}}(T) \setminus \sigma_{\mathrm{ub}}(T) = \sigma_{\mathrm{ap}}(T + R) \setminus \sigma_{\mathrm{ub}}(T + R) = p_{00}^a(T + R). \qquad (6.4)$$

Let $\lambda \in \pi_{00}(T)$ be arbitrarily given. Taking into account that $S := T + R$ commutes with R, by part (ii) of Lemma 6.11 and recalling that the isolated point of the spectrum belongs to the approximate point spectrum, we then have

$$\lambda \in \pi_{00}(T) \cap \sigma_{\mathrm{ap}}(T + R) = \pi_{00}(S - R) \cap \sigma_{\mathrm{ap}}(S)$$

$$\subseteq \mathrm{iso}\,\sigma(S) = \mathrm{iso}\,\sigma(T + R) \subseteq \sigma_{\mathrm{ap}}(T + R).$$

Moreover, by Theorem 3.8 we know that $\lambda I - (T + R)$ is upper semi-Browder and hence has closed range. Since $\lambda \in \sigma_{\mathrm{ap}}(T + R)$ it then follows that λ is an eigenvalue, so $0 < \alpha(\lambda I - (T + R)) < \infty$, i.e., $\lambda \in \pi_{00}(T + R)$, as desired.   ∎

Remark 6.13 Recall that $\alpha(T) < \infty$ implies that $\alpha(T^n) < \infty$ for all $n \in \mathbb{N}$. Moreover, if there exists a finite-dimensional operator $S \in L(Z)$ which has finite-dimensional kernel, then the Banach space Z is necessarily finite-dimensional.

Theorem 6.14 *Suppose that $T \in L(X)$ is an isoloid operator for which property (R) holds and let $K \in L(X)$ be a bounded operator commuting with T such that K^n is a finite rank operator for some $n \in \mathbb{N}$. If $\sigma_{\mathrm{ap}}(T) = \sigma_{\mathrm{ap}}(T + K)$ then we have:*

(i) $\pi_{00}(T) = \pi_{00}(T + K)$.
(ii) $T + K$ *has property (R).*

Proof

(i) Observe first that K is a Riesz operator, so, by Theorem 6.12, we need only to prove the inclusion $\pi_{00}(T + K) \subseteq \pi_{00}(T)$.

Let $\lambda \in \pi_{00}(T + K)$. Then λ is an isolated point of $\sigma(T + K)$, and since $\alpha(\lambda I - (T + K)) > 0$ we then have $\lambda \in \sigma_{\mathrm{ap}}(T + K) = \sigma_{\mathrm{ap}}(T)$. Therefore, by Lemma 6.11,

$$\lambda \in \pi_{00}(T + K) \cap \sigma_{\mathrm{ap}}(T) \subseteq \mathrm{iso}\, \sigma(T).$$

Since T is isoloid we then have $\alpha(\lambda I - T) > 0$. We show now that $\alpha(\lambda I - T) < \infty$. Let U denote the restriction of $(\lambda I - (T + K))^n$ to $\ker(\lambda I - T)$. Clearly, if $x \in \ker(\lambda I - T)$ then

$$U x = (-1)^n K^n x \in K^n(X),$$

thus U is a finite rank operator. Moreover, since $\lambda \in \pi_{00}(T + K)$ we have $\alpha(\lambda I - (T + K)) < \infty$ and hence

$$\alpha(U) \leq \alpha(\lambda I - (T + K))^n < \infty.$$

By Remark 6.13 it then follows that $\ker(\lambda I - T)$ is finite-dimensional, as claimed. Therefore, $\lambda \in \pi_{00}(T)$, and, consequently, $\pi_{00}(T + K) \subseteq \pi_{00}(T)$.

(ii) Since K is a Riesz operator we have, by Theorem 3.8, $\sigma_{\mathrm{ub}}(T) = \sigma_{\mathrm{ub}}(T + K)$, thus

$$\pi_{00}(T + K) = \pi_{00}(T) = \sigma_{\mathrm{ap}}(T) \setminus \sigma_{\mathrm{ub}}(T) = \sigma_{\mathrm{ap}}(T + K) \setminus \sigma_{\mathrm{ub}}(T + K),$$

hence $T + K$ satisfies property (R).   ∎

Evidently Theorem 6.14 applies to the case of a commuting nilpotent perturbation N. Indeed, in this case the equality $\sigma_{\mathrm{ap}}(T) = \sigma_{\mathrm{ap}}(T + N)$ is satisfied. The next result shows that in the case of nilpotent perturbations the condition that T is isoloid can be omitted.

Theorem 6.15 *Suppose that $T \in L(X)$ and let $N \in L(X)$ be a nilpotent operator which commutes with T. Then we have:*

(i) $\pi_{00}(T + N) = \pi_{00}(T)$.
(ii) *T satisfies property (R) if and only if $T + N$ satisfies property (R).*

Proof

(i) Suppose that $N^p = 0$. We show first that

$$\ker(\lambda I - (T + N)) \subseteq \ker(\lambda I - T)^p. \tag{6.5}$$

Indeed, if $x \in \ker(\lambda I - (T + N))$ then $(\lambda I - T)x = Nx$, hence

$$(\lambda I - T)^p x = N^p x = 0,$$

so $x \in \ker(\lambda I - T)^p$. It is easily seen that

$$\ker(\lambda I - T) \subseteq \ker(\lambda I - (T + N))^p. \tag{6.6}$$

Indeed, suppose that $x \in \ker(\lambda I - T)$, i.e. $(\lambda I - T)x = 0$. Then for some suitable binomial coefficients $\mu_{j,p}$ we have

$$(\lambda I - (T + N))^p x = \sum_{j=0}^{p} \mu_{j,p}(\lambda I - T)^j N^{p-j} x = N^p x = 0.$$

Finally, suppose that $\lambda \in \pi_{00}(T)$. Then $\lambda \in \operatorname{iso}\sigma(T) = \operatorname{iso}\sigma(T + N)$. Moreover, $\alpha(\lambda I - T) > 0$ entails, by (1.38), that $\alpha(\lambda I - (T + N))^p > 0$ and hence $\alpha(\lambda I - (T + N)) > 0$. From $\alpha(\lambda I - T) < \infty$ we deduce that $\alpha(\lambda I - T)^p < \infty$ and hence, from the inclusion (1.38), we conclude that $\alpha(\lambda I - (T + N)) < \infty$. Therefore $\lambda \in \pi_{00}(T + N)$. To show the opposite inclusion just proceed by symmetry: since N commutes with $T + N$,

$$\pi_{00}(T + N) \subseteq \pi_{00}(T + N - N) = \pi_{00}(T).$$

(ii) Suppose that T has property (R). Then

$$\begin{aligned}
\pi_{00}(T + N) = \pi_{00}(T) &= \sigma_{\mathrm{ap}}(T) \setminus \sigma_{\mathrm{ub}}(T) \\
&= \sigma_{\mathrm{a}}(T + N) \setminus \sigma_{\mathrm{ub}}(T + N) \\
&= p_{00}^a(T + N),
\end{aligned}$$

therefore $T + N$ has property (R). The converse follows by symmetry. ∎

Example 6.16 Generally, property (R) is not transmitted from T to a quasi-nilpotent perturbation $T + Q$. In fact, if $Q \in L(\ell^2(\mathbb{N}))$ is defined by

$$Q(x_1, x_2, \dots) = \left(\frac{x_2}{2}, \frac{x_3}{3}, \dots \right) \quad \text{for all } (x_n) \in \ell^2(\mathbb{N}),$$

then Q is quasi-nilpotent, so $\sigma_{\text{ap}}(Q) = \sigma_{\text{ub}}(Q) = \{0\}$ and hence

$$\{0\} = \pi_{00}(Q) \neq \sigma_{\text{ap}}(Q) \setminus \sigma_{\text{ub}}(Q) = \emptyset.$$

Take $T = 0$. Clearly, T satisfies property (R) but $T + Q = Q$ fails this property.

Recall that $T \in L(X)$ is said to be *finite-isoloid* if every isolated point of $\sigma(T)$ is an eigenvalue of T having finite multiplicity.

Theorem 6.17 *Suppose that $T \in L(X)$ is a finite-isoloid operator which satisfies property (R). If K is a Riesz operator which commutes with T and such that $\sigma_{\text{ap}}(T) = \sigma_{\text{ap}}(T + K)$, then $T + K$ has property (R).*

Proof We show first that $\pi_{00}(T + K) = \pi_{00}(T)$. By Theorem 6.12 it suffices to prove that $\pi_{00}(T + K) \subseteq \pi_{00}(T)$. Let $\lambda \in \pi_{00}(T + K)$ be arbitrarily given. Then $\lambda \in \text{iso } \sigma(T + K)$ and $\alpha(\lambda I - (T + K) > 0$ entails that $\lambda \in \sigma_{\text{ap}}(T + K) = \sigma_{\text{ap}}(T)$. By Lemma 6.11 it then follows that

$$\lambda \in \pi_{00}(T + K) \cap \sigma_{\text{ap}}(T) \subseteq \text{iso } \sigma(T).$$

Since T is finite-isoloid, $0 < \alpha(\lambda I - T) < \infty$, so $\lambda \in \pi_{00}(T)$, and hence $\pi_{00}(T + K) = \pi_{00}(T)$. Property (R) then follows from the following equalities:

$$p_{00}^a(T + K) = \sigma_{\text{ap}}(T + K) \setminus \sigma_{\text{ub}}(T + K) = \sigma_{\text{ap}}(T) \setminus \sigma_{\text{ub}}(T)$$
$$= \pi_{00}(T) = \pi_{00}(T + K).$$

∎

Corollary 6.18 *Suppose that $T \in L(X)$ is a finite-isoloid operator which satisfies property (R). If $Q \in L(X)$ is quasi-nilpotent operator which commutes with T then $T + Q$ has property (R).*

Proof Since $\sigma_{\text{ap}}(T) = \sigma_{\text{ap}}(T + Q)$ the result follows directly from Theorem 6.17. ∎

Theorem 6.19 *If T is a-polaroid and finite-isoloid, and Q is a quasi-nilpotent operator which commutes with T, then $T + Q$ has property (R).*

Proof If $\lambda \in \text{iso } \sigma_{\text{ap}}(T + Q)$ then $\lambda \in \text{iso } \sigma_{\text{ap}}(T)$ and hence, since T is a-polaroid, λ is a pole of the resolvent of T, in particular an isolated point of the spectrum. Therefore, $p := p(\lambda I - T) = q(\lambda I - T) < \infty$ and since by assumption $\alpha(\lambda I - T) <$

∞ we then have $\alpha(\lambda I - T) = \beta(\lambda I - T)$, by Theorem 1.22, so $\lambda I - T$ is Browder. As observed in Chap. 3, Browder operators are invariant under Riesz commuting perturbations, in particular under quasi-nilpotent commuting perturbations, hence $\lambda I - (T + Q)$ is Browder, and consequently λ is a pole of the resolvent of $T + Q$. Therefore $T + Q$ is a-polaroid, thus Theorem 6.7 applies.   ∎

It is natural to ask how to extend the results above to algebraic commuting perturbations. In the following result we give a positive answer in the case when the operator is hereditarily polaroid and has the SVEP.

Theorem 6.20 *Suppose that $T \in L(X)$ and $K \in L(X)$ is an algebraic operator which commutes with T.*

 (i) *If T is hereditarily polaroid then $T^* + K^*$ satisfies property (R).*
(ii) *If T^* is hereditarily polaroid then $T + K$ satisfies property (R).*

Proof The statements are a direct consequence of Theorems 4.32 and 6.7.   ∎

Remark 6.21 In the case of Hilbert space operators, the assertions of Theorem 6.20 are still valid if T^* is replaced with the Hilbert adjoint T'.

The result of Theorem 6.20 may be considerably improved. As usual, let $\mathcal{H}_{nc}(\sigma(T))$ denote the set of all analytic functions, defined on an open neighborhood of $\sigma(T)$, such that f is non-constant on each of the components of its domain. Define, by the classical functional calculus, $f(T)$ for every $f \in \mathcal{H}_{nc}(\sigma(T))$.

Theorem 6.22 *Suppose that $T \in L(X)$ and $K \in L(X)$ is an algebraic operator which commutes with T.*

 (i) *If T is hereditarily polaroid and has the SVEP then $f(T^* + K^*)$ satisfies property (R) for all $f \in \mathcal{H}_{nc}(\sigma(T + K))$.*
(ii) *If T^* is hereditarily polaroid and has the SVEP then $f(T + K)$ satisfies property (R) for all $f \in \mathcal{H}_{nc}(\sigma(T + K))$.*

Proof (i) As in the proof of Theorem 6.20, we have $T^* + K^*$ is polaroid, and hence $f(T^* + K^*)$ is polaroid, by Theorem 4.19. Moreover the SVEP for $T + K$ entails the SVEP for $f(T + K)$, by Corollary 2.89, and hence $\sigma_{ap}(f(T^* + K^*)) = \sigma(f(T^* + K^*))$, by Theorem 2.68. Therefore, $f(T^* + K^*)$ is a-polaroid and property (R) for $f(T^* + K^*)$ then follows from Theorem 6.7.

The proof of part (ii) is analogous.   ∎

In the proof of Theorem 4.32 it is shown that if T is hereditarily polaroid then $T + K$ is polaroid. The polaroid condition is stronger than the isoloid condition. We call $T \in L(X)$ *hereditarily isoloid* if every restriction $T|M$ to a closed invariant subspace of T is isoloid. Obviously, every hereditarily polaroid is hereditarily isoloid. There is some interest in asking if the hereditarily isoloid condition on T entails that $T + K$ is isoloid, where K is algebraic and commutes with T. The answer is positive. To show this we first need to prove the following lemma.

Lemma 6.23 *If $T \in L(X)$ is isoloid and if $N \in L(X)$ is nilpotent and commutes with T then $T + N$ is isoloid.*

Proof As observed in the proof of Theorem 6.15 we have $\ker(\lambda I - T) \subseteq \ker(\lambda I - (T + N))^p$ where $N^p = 0$. If $\lambda \in \operatorname{iso} \sigma(T + N) = \operatorname{iso} \sigma(T)$, then the isoloid condition entails that $\ker(\lambda I - T) \neq \{0\}$. Therefore, we have $\ker(\lambda I - (T + N))^p \neq \{0\}$, and this obviously implies $\ker(\lambda I - (T + N)) \neq \{0\}$. ∎

Theorem 6.24 *If $T \in L(X)$ is hereditarily isoloid and if $K \in L(X)$ is algebraic and commutes with T then $T - K$ is isoloid.*

Proof Using the same denotation of the proof of Theorem 4.32, if $N_i := \lambda_i I - K_i$ we have $\sigma(T_i - K_i) = \sigma(T_i - K_i - N_i) = \sigma(T_i - \lambda_i I)$. Hence,

$$\sigma(T - K) = \bigcup_{i=1}^{n} \sigma(T_i - \lambda_i I).$$

Now, let $\lambda \in \operatorname{iso} \sigma(T - K)$. Then $\lambda \in \operatorname{iso} \sigma(T_j - K_j)$ for some j, and hence $\lambda + \lambda_j \in \operatorname{iso} \sigma(T_j)$. Since T_j is isoloid, $\lambda + \lambda_j$ is an eigenvalue of T_j. If p_j denotes the order of the nilpotent operator N_j then, from the inclusion (1.38), we obtain

$$\{0\} \neq \ker((\lambda + \lambda_j)I - T_j) \subseteq \ker((\lambda + \lambda_j)I - (T_j + N_j))^{p_j}$$
$$= \ker(\lambda I - (T_j - K_j))^{p_j},$$

from which we deduce that λ is an eigenvalue of $T_i - K_j$. Since

$$\ker(\lambda I - (T - K)) = \bigoplus_{j=1}^{n} \ker(\lambda I - (T_j - K_j))$$

it then follows that $\ker(\lambda I - (T - K)) \neq \{0\}$, i.e., λ is an eigenvalue of $T - K$. ∎

6.2 Property $(g R)$

In this section we consider the generalization of property (R) in the sense of B-Fredholm theory. Recall that for every bounded operator $T \in L(X)$ we set:

$$E(T) := \{\lambda \in \operatorname{iso} \sigma(T) : 0 < \alpha(\lambda I - T)\}.$$

Obviously, $\pi_{00}(T) \subseteq E(T)$.

Definition 6.25 An operator $T \in L(X)$ is said to satisfy the *generalized property* (R), abbreviated $(g R)$, if the equality $\sigma_a(T) \setminus \sigma_{\mathrm{ld}}(T) = E(T)$ holds, i.e., $E(T)$ coincides with the set $\Pi_a(T)$ of all left poles of T.

By duality T^* satisfies property (gR) if and only if $E(T^*)$ coincides with the set of right left poles of T.

Theorem 6.26 *Property (gR) implies property (R).*

Proof Let $T \in L(X)$ and suppose that $\lambda \in p_{00}^a(T) = \sigma_{ap}(T) \setminus \sigma_{ub}(T)$. Then $\lambda I - T$ is upper semi-Browder and hence $\alpha(\lambda I - T) < \infty$. Since $\lambda I - T$ has closed range and $\lambda I - T$ is not bounded below then $0 < \alpha(\lambda I - T)$. Trivially, every upper semi-Browder is upper semi B-Browder, or equivalently, by Theorem 3.47, is left Drazin invertible, hence $\lambda \in \sigma_{ap}(T) \setminus \sigma_{ld}(T) = E(T)$. Since $\alpha(\lambda I - T) < \infty$ it then follows that $\lambda \in \pi_{00}(T)$. This shows the inclusion $p_{00}^a(T) \subseteq \pi_{00}(T)$.

Conversely, suppose that $\lambda \in \pi_{00}(T)$. Then $\lambda \in E(T) = \sigma_{ap}(T) \setminus \sigma_{ld}(T)$, so $\lambda I - T$ is left Drazin invertible, and hence $p := p(\lambda I - T) < \infty$ and $(\lambda I - T)^{p+1}(X)$ is closed. Moreover, $\lambda \in \pi_{00}(T)$ entails that $\alpha(\lambda I - T) < \infty$ and hence, $\alpha(\lambda I - T)^{p+1} < \infty$. Consequently, $(\lambda I - T)^{p+1}$ is upper semi-Fredholm and this, by the classical Fredholm theory, implies $\lambda I - T \in \Phi_+(X)$. Therefore, $\lambda I - T$ is upper semi-Browder, hence $\lambda \in p_{00}^a(T)$, and consequently $p_{00}^a(T) = \pi_{00}(T)$. ∎

Theorem 6.27 *Let $T \in L(X)$. Then we have:*

(i) *If T satisfies property (gR) then every left pole of T is a pole of the resolvent of T.*

(ii) *If T^* satisfies property (gR) then every right pole of T is a pole of the resolvent of T.*

Proof

(i) Let $\lambda \in \Pi_a(T)$. Then $\lambda I - T$ is left Drazin invertible and hence $p(\lambda I - T) < \infty$. Since T satisfies property (gR) we have $\lambda \in E(T)$, so λ is an isolated point of $\sigma(T)$. Consequently, T^* has the SVEP at λ and since $\lambda I - T$ is left Drazin invertible, the SVEP of T^* at λ implies, by Theorem 2.98, that $q(\lambda I - T) < \infty$. Therefore, λ is a pole of the resolvent of T and hence $\Pi_a(T) \subseteq \Pi(T)$. Since the opposite inclusion is true for every operator, we then conclude that $\Pi_a(T) = \Pi(T)$.

(ii) Clear, by duality. ∎

An example of an operator with property (R) but not property (gR) may be easily found:

Example 6.28 Let $T := 0 \oplus Q$, where Q is any quasi-nilpotent operator acting on an infinite-dimensional Banach space X such that $Q^n(X)$ is non-closed for all $n \in \mathbb{N}$. Clearly, $\sigma_{ap}(T) = \sigma_{ub}(T) = \{0\}$, so T has property (R), since $p_{00}^a(T) = \pi_{00}(T) = \emptyset$. Since $T^n(X)$ is non-closed for all $n \in \mathbb{N}$, T is not left Drazin invertible, so $\Pi_a(T) = \emptyset$. On the other hand, $E(T) = \{0\}$, so property (gR) does not hold for T.

The next result improves Theorem 6.7.

Theorem 6.29 *If $T \in L(X)$ is a-polaroid then T satisfies property (gR).*

Proof Let $\lambda \in \sigma_a(T) \setminus \sigma_{ld}(T)$. Then $\lambda I - T$ is left Drazin invertible, hence λ is a left pole of T and, consequently, an isolated point of $\sigma_{ap}(T)$. Since T is a-polaroid, λ is a pole of the resolvent of T and hence an isolated point of $\sigma(T)$. From $p(\lambda I - T) = q(\lambda I - T) < \infty$ we deduce that $\alpha(\lambda I - T) = \beta(\lambda I - T)$, see Theorem 1.22, and this excludes that $\alpha(\lambda I - T) = 0$, otherwise $\lambda \notin \sigma(T)$. Therefore, $\lambda \in E(T)$.

Conversely, let $\lambda \in E(T)$. Then $\lambda \in \mathrm{iso}\,\sigma(T)$ and the condition $\alpha(\lambda I - T) > 0$ entails that $\lambda \in \sigma_{ap}(T)$ and hence $\lambda \in \mathrm{iso}\,\sigma_{ap}(T)$. Therefore, λ is a pole, in particular a left pole, so $\lambda \in \Pi_a(T) = \sigma_a(T) \setminus \sigma_{ubb}(T)$.   ∎

Example 6.30 Let $R \in L(\ell^2(\mathbb{N}))$ be the unilateral right shift and

$$U(x_1, x_2, \dots) := (0, x_2, x_3, \cdots) \quad \text{for all } (x_n) \in \ell^2(\mathbb{N}).$$

If $T := R \oplus U$ we have $T \in \Phi_+(X)$ and hence $T^2 \in \Phi_+(X)$, so that $T^2(X)$ is closed. We also have $p(T) = 1$ so that 0 is a left pole of T. Since $\sigma_a(T) = \Gamma \cup \{0\}$, Γ the unit circle, it then follows that T is left polaroid. On the other hand, $q(R) = \infty$, and hence $q(T) = q(R) + q(U) = \infty$, so that T is not a-polaroid. This example also shows that a left polaroid operator in general does not satisfy property (gR).

Theorem 6.31 *Suppose that $T \in L(X)$ is left polaroid.*

(i) *If T satisfies property (R) then $E_a(T) = \Pi_a(T)$.*
(ii) *If T satisfies property (gR) then $E(T) = E_a(T) = \Pi_a(T)$.*

Proof

(i) Trivially, for every left polaroid operator we have $E_a(T) \subseteq \Pi_a(T)$. To show the opposite inclusion, let $\lambda \in \Pi_a(T)$ arbitrarily given. Then $\lambda I - T$ is left Drazin invertible, so $p := p(\lambda I - T) < \infty$ and $(\lambda I - T)^{p+1}(X)$ is closed.

   Suppose now that $\alpha(\lambda I - T) = 0$. Then $\alpha(\lambda I - T)^n = 0$ for all $n \in \mathbb{N}$, and hence $(\lambda I - T)^{p+1}$ is bounded below, in particular upper semi-Browder. From the classical Fredholm theory we then have that $\lambda I - T$ is upper semi-Browder, thus $\lambda \in \sigma_{ap}(T) \setminus \sigma_{ub}(T) = \pi_{00}(T)$, since T has property (R). From the definition of $\pi_{00}(T)$ we then have $\alpha(\lambda I - T) > 0$, a contradiction.

   Therefore, $\alpha(\lambda I - T) > 0$. Now, every left pole of the resolvent is an isolated point of $\sigma_{ap}(T)$. Consequently, $\lambda \in E_a(T)$, and this shows the inclusion $\Pi_a(T) \subseteq E_a(T)$.
(ii) The equality $E_a(T) = \Pi_a(T)$ holds by part (i), since (gR) implies (R). The inclusion $E(T) \subseteq E_a(T)$ holds for every operator, since every isolated point of the spectrum lies in $\sigma_{ap}(T)$, and, as already observed, the left polaroid condition entails $E_a(T) \subseteq \Pi_a(T) = E(T)$.   ∎

Recall that $T \in L(X)$ is said to be *finite a-isoloid* if every isolated point of $\sigma_{ap}(T)$ is an eigenvalue of T having finite multiplicity.

In the sequel we need the following lemma.

Lemma 6.32 *Let $T \in L(X)$ be such that $\alpha(T) < \infty$. Suppose that there exists an injective quasi-nilpotent operator $Q \in L(X)$ such that $TQ = QT$. Then T is injective.*

Proof Set $Y := \ker T$. Clearly, Y is invariant under Q and the restriction $(\lambda I - Q)|Y$ is injective for all $\lambda \neq 0$. By assumption Y is finite-dimensional, so $(\lambda I - Q)|Y$ is also surjective for all $\lambda \neq 0$. Thus $\sigma(Q|Y) \subseteq \{0\}$. On the other hand, from the assumption, we know that $Q|Y$ is injective and hence $Q|Y$ is surjective. Hence $\sigma(Q|Y) = \emptyset$, from which we conclude that $Y = \{0\}$. ∎

Theorem 6.33 *Suppose that $T \in L(X)$ is a finite a-isoloid operator and suppose that there exists an injective quasi-nilpotent operator Q which commutes with T. Then T satisfies property (gR).*

Proof Note first that $E(T)$ is empty. Indeed, suppose that $\lambda \in E(T)$. Then λ is an isolated point of $\sigma(T)$ and hence belongs to $\sigma_{ap}(T)$. Thus $\lambda \in \text{iso}\,\sigma_{ap}(T)$, so that $0 < \alpha(\lambda I - T) < \infty$, since T is finite a-isoloid. But, by Lemma 6.32, we also have $\alpha(\lambda I - T) = 0$, and this is impossible. Therefore, $E(T) = \emptyset$.

In order to show that property (gR) holds for T we need to prove that $\sigma_{ap}(T) \setminus \sigma_{ld}(T)$ is empty. Suppose that $\lambda \in \sigma_{ap}(T) \setminus \sigma_{ld}(T)$. Then $\lambda \in \sigma_{ap}(T)$ and $\lambda I - T$ is left Drazin invertible. By Theorem 2.97 λ is an isolated point of $\sigma_{ap}(T)$, and since T is finite a-isoloid we then have $\alpha(\lambda I - T) < \infty$. Again by Lemma 6.32 we then conclude that $\lambda I - T$ is injective. On the other hand, by Theorem 1.114, we have $\lambda I - T \in \Phi_+(X)$, so $\lambda I - T$ has closed range and hence $\lambda I - T$ is bounded below, i.e. $\lambda \notin \sigma_{ap}(T)$, a contradiction. Therefore, $\sigma_{ap}(T) \setminus \sigma_{ld}(T) = \emptyset$, and consequently T satisfies property (gR).

∎

Theorem 6.34 *Let $T \in L(X)$ and suppose that $N \in L(X)$ is a nilpotent operator which commutes with T. Then we have:*

(i) *T has property (gR) if and only if $T + N$ has property (gR).*
(ii) *T^* has property (gR) if and only if $T^* + N^*$ has property (gR).*

Proof

(i) We show first that $E(T) \subseteq E(T + N)$. Note that $\text{iso}\,\sigma(T) = \text{iso}\,\sigma(T + N)$, so we need only to show that if $\lambda \in E(T)$ then $\alpha(\lambda I - (T + N)) > 0$.

Suppose that $N^p = 0$. It is easily seen that

$$\ker(\lambda I - T) \subseteq \ker(\lambda I - (T + N))^p. \tag{6.7}$$

Indeed, suppose that $x \in \ker(\lambda I - T)$, i.e. $(\lambda I - T)x = 0$. Then for some suitable binomial coefficients μ_j we have

$$(\lambda I - (T + N))^p x = \sum_{j=0}^{p} \mu_j (\lambda I - T)^j N^{p-j} x = N^p x = 0.$$

We know that the condition $\alpha(\lambda I - T) > 0$ entails that $\alpha(\lambda I - (T + N))^p > 0$ and hence $\alpha(\lambda I - (T + N)) > 0$, as desired.

By symmetry we have

$$E(T + N) \subseteq E((T + N) - N) = E(T),$$

from which we conclude that the equality $E(T + N) = E(T)$ holds.

Finally, suppose that T has (gR). Since $\sigma_{\mathrm{ap}}(T)$ is invariant under commuting nilpotent perturbations, by Theorem 3.78 we then have

$$E(T + N) = E(T) = \sigma_{\mathrm{ap}}(T) \setminus \sigma_{\mathrm{ld}}(T) = \sigma_{\mathrm{ap}}(T + N) \setminus \sigma_{\mathrm{ld}}(T + N),$$

so $T + N$ has property (gR). Clearly, the converse implication also holds by symmetry.

(ii) Clearly, $T^* N^* = N^* T^*$ and iso $\sigma(T^*) = $ iso $\sigma(T^* + N^*)$. Proceeding as in the proof of part (i) we then have $\alpha(\lambda I - (T^* + N^*)) > 0$ and hence $E(T^*) \subseteq E(T^* + N^*)$. By symmetry we then have $E(T^*) = E(T^* + N^*)$ and from this it easily follows that T^* has property (gR) if and only if $T^* + N^*$ has the same property. ∎

Example 6.35 Generally, property (gR) is not transmitted from T to a quasi-nilpotent perturbation $T + Q$. In fact, if $Q \in L(\ell^2(\mathbb{N}))$ is the quasi-nilpotent operator defined in Example 6.16, then $\sigma_{\mathrm{ap}}(Q) = \sigma_{\mathrm{ld}}(Q) = \{0\}$ and hence

$$\{0\} = E(Q) \neq \sigma_{\mathrm{ap}}(Q) \setminus \sigma_{\mathrm{ld}}(Q) = \emptyset.$$

Take $T = 0$. Clearly, T satisfies property (gR) but $T + Q = Q$ fails this property.

Theorem 6.36 *If T is a-polaroid and finite-isoloid, Q is a quasi-nilpotent operator which commutes with T, then $T + Q$ has property (gR).*

Proof If $\lambda \in $ iso $\sigma_{\mathrm{ap}}(T + Q)$ then $\lambda \in $ iso $\sigma_{\mathrm{ap}}(T)$ and hence, since T is a-polaroid, λ is a pole of the resolvent of T. Therefore, $p := p(\lambda I - T) = q(\lambda I - T) < \infty$ and hence λ is an isolated point in $\sigma(T)$. Our assumption that T is finite isoloid entails that $\alpha(\lambda I - T) < \infty$, so we have $\alpha(\lambda I - T) = \beta(\lambda I - T)$, by Theorem 1.22, hence $\lambda I - T$ is Browder. Since, by Theorem 3.8, Browder operators are invariant under Riesz commuting perturbations, in particular under quasi-nilpotent commuting perturbations, $\lambda I - (T + Q)$ is Browder, and, consequently, λ is a pole of the resolvent of $T + Q$. Therefore $T + Q$ is a-polaroid, thus Theorem 6.29 applies. ∎

6.3 Weyl-Type Theorems

We observe first that the Weyl spectrum, in some sense, tends to be large. Indeed, the set $\Delta(T) = \sigma(T) \setminus \sigma_{\mathrm{w}}(T)$ is either empty or consists of eigenvalues of finite multiplicity. In fact, if $\lambda \in \Delta(T) = \sigma(T) \setminus \sigma_{\mathrm{w}}(T)$ then $\lambda I - T$ is not invertible, while $0 < \alpha(\lambda I - T) = \beta(\lambda I - T) < \infty$. This implies that $\lambda I - T$ cannot be injective.

In 1981 Coburn [97], while trying to extend to non-normal operators some properties already known for self-adjoint operators on Hilbert spaces, found that there are some other classes of operators, defined on Banach spaces, which satisfy the equality $\Delta(T) := \sigma(T) \setminus \sigma_w(T) = \pi_{00}(T)$. He gave the following definition:

Definition 6.37 Given an operator $T \in L(X)$, where X is a Banach space, we say that *Weyl's theorem holds* for $T \in L(X)$ if

$$\Delta(T) = \pi_{00}(T) = \{\lambda \in \text{iso } \sigma(T) : 0 < \alpha(\lambda I - T) < \infty\}. \tag{6.8}$$

Theorem 6.38 *If a bounded operator $T \in L(X)$ satisfies Weyl's theorem then*

$$p_{00}(T) = \pi_{00}(T) = \Delta(T).$$

Proof Suppose that T satisfies Weyl's theorem. By definition then $\Delta(T) = \pi_{00}(T)$. We show now that the equality $p_{00}(T) = \pi_{00}(T)$ holds. It suffices to prove the inclusion $\pi_{00}(T) \subseteq p_{00}(T)$.

Let λ be an arbitrary point of $\pi_{00}(T)$. Since λ is isolated in $\sigma(T)$, T has the SVEP at λ and from the equality $\pi_{00}(T) = \sigma(T) \setminus \sigma_w(T)$ we know that $\lambda I - T \in W(X)$. Hence $\lambda I - T \in \Phi(X)$, and the SVEP at λ implies, by Theorem 2.97, that $p(\lambda I - T) < \infty$, so $\lambda \in p_{00}(T)$. ∎

From Theorem 5.4 it then follows that

Weyl's theorem for $T \Rightarrow$ Browder's theorem for T.

Example 6.39 It is not difficult to find an example of an operator satisfying Browder's theorem but not Weyl's theorem. For instance, if $T \in L(\ell^2)(\mathbb{N})$ is defined by

$$T(x_0, x_1, \ldots) := \left(\frac{1}{2}x_1, \frac{1}{3}x_2, \ldots\right) \quad \text{for all } (x_n) \in \ell^2(\mathbb{N}),$$

then T is quasi-nilpotent, so T has the SVEP and consequently satisfies Browder's theorem. On the other hand T does not satisfy Weyl's theorem, since $\sigma(T) = \sigma_w(T) = \{0\}$ and $\pi_{00}(T) = \{0\}$.

Let us define

$$\Delta_{00}(T) := \Delta(T) \cup \pi_{00}(T).$$

Weyl's theorem for an operator $T \in L(X)$ may be viewed as the conjunction of two properties. The first one is that Browder's theorem holds for T, and this may be considered, by Theorem 5.4, a local spectral property, and the second one is given by the equality $p_{00}(T) = \pi_{00}(T)$. Indeed, we have:

Theorem 6.40 *Let $T \in L(X)$. Then the following statements are equivalent:*

(i) *T satisfies Weyl's theorem;*
(ii) *T satisfies Browder's theorem and $p_{00}(T) = \pi_{00}(T)$;*
(iii) *the map $\lambda \to \gamma(\lambda I - T)$ is not continuous at every $\lambda \in \Delta_{00}(T)$.*

Proof The implication (i) $\Rightarrow$ (ii) is clear, from Theorem 6.38, while the implication (ii) $\Rightarrow$ (i) follows immediately from Theorem 5.4, so the statements (i) and (ii) are equivalent.

(i) $\Rightarrow$ (iii) By Theorem 6.38 we have that $\Delta_{00}(T) = \Delta(T)$ and T satisfies Browder's theorem. Therefore, by Theorem 5.31, the mapping $\lambda \to \gamma(\lambda I - T)$ is not continuous at every point $\lambda \in \Delta_{00}(T)$.

(iii) $\Rightarrow$ (ii) Suppose that $\lambda \to \gamma(\lambda I - T)$ is not continuous at every $\lambda \in \Delta_{00}(T) = \Delta(T) \cup \pi_{00}(T)$. The discontinuity at the points of $\Delta(T)$ entails, by Theorem 5.31, that T satisfies Browder's theorem, while the discontinuity at the points of $\pi_{00}(T)$ is equivalent, by Theorem 6.1, to saying that $\pi_{00}(T) \cap \sigma_b(T) = \emptyset$. The last equality obviously implies that $\pi_{00}(T) \subseteq p_{00}(T)$, hence $\pi_{00}(T) = p_{00}(T)$. ∎

From Theorems 5.31 and 6.40 we see that Browder's theorem and Weyl's theorem are equivalent to the discontinuity of the mapping $\lambda \to \gamma(\lambda I - T)$ at the points of two sets $\Delta(T)$ and $\Delta_{00}(T)$, respectively, with $\Delta_{00}(T)$ larger than $\Delta(T)$. Note that the discontinuity of the mapping $\lambda \to \ker(\lambda I - T)$ at every $\lambda \in \Delta_{00}(T) = \Delta(T) \cup \pi_{00}(T)$ does not imply Weyl's theorem. In fact, since every point of $\pi_{00}(T)$ is an isolated point of $\sigma(T)$, it is evident that the map $\lambda \to \ker(\lambda I - T)$ is not continuous at every $\lambda \in \pi_{00}^a(T)$ for *all* operators $T \in L(X)$. Therefore, the discontinuity of the mapping $\lambda \to \ker(\lambda I - T)$ at every $\lambda \in \Delta_{00}(T)$ is equivalent to the discontinuity of the same mapping at the points of $\Delta(T)$, i.e., it is equivalent to Browder's theorem for T.

Theorem 6.41 *Let $T \in L(X)$. Then the following statements are equivalent:*

(i) *T satisfies Weyl's theorem;*
(ii) *The quasi-nilpotent part $H_0(\lambda I - T)$ is finite-dimensional for all $\lambda \in \Delta_{00}(T)$;*
(iii) *The analytic core $K(\lambda I - T)$ is finite-codimensional for all $\lambda \in \Delta_{00}(T)$.*

Proof

(i) $\Rightarrow$ (ii) If T satisfies Weyl's theorem then T satisfies Browder's theorem and $\pi_{00}(T) = p_{00}(T) = \Delta(T)$. By Theorem 5.35 then $H_0(\lambda I - T)$ is finite-dimensional for all $\lambda \in \Delta_{00}(T) = \Delta(T)$.

(ii) $\Rightarrow$ (i) Since $\Delta(T) \subseteq \Delta_{00}(T)$, Browder's theorem holds for T, by Theorem 5.35. From the inclusion $\pi_{00}(T) \subseteq \Delta_{00}(T)$ we know that $H_0(\lambda I - T)$ is finite-dimensional for every $\lambda \in \pi_{00}(T)$. Since every $\lambda \in \pi_{00}(T)$ is an isolated point of $\sigma(T)$ we have $X = H_0(\lambda I - T) \oplus K(\lambda I - T)$ and hence $K(\lambda I - T)$ has finite codimension. From the inclusion $K(\lambda I - T) \subseteq (\lambda I - T)(X)$ we deduce that $\beta(\lambda I - T) < \infty$, hence $\lambda I - T$ is Fredholm. But both T and T^* have the

SVEP at λ, hence $p(\lambda I - T) = q(\lambda I - T) < \infty$, by Theorems 2.97 and 2.98, hence $\lambda I - T$ is Browder, by Theorem 1.22. Therefore $\pi_{00}(T) = p_{00}(T)$, so, by Theorem 6.40 T satisfies Weyl's theorem.

(iii) $\Leftrightarrow$ (ii) The equivalence follows from Theorem 5.35 and from the fact that for any point of $\lambda \in \pi_{00}(T)$ we have $X = H_0(\lambda I - T) \oplus K(\lambda I - T)$. ∎

Remark 6.42 In contrast with Theorem 5.35, condition (ii) of Theorem 6.41 cannot be replaced by the formally weaker condition that $H_0(\lambda I - T)$ is closed for all $\lambda \in \Delta_{00}(T)$. Indeed, if T is defined as in Example 6.39 then $0 \in \pi_{00}(T)$, T does not satisfy Weyl's theorem, while $H_0(\lambda I - T) = \ell^2(\mathbb{N})$ is closed, since T is quasi-nilpotent.

Theorem 6.43 *If $T \in L(X)$ is polaroid and either T or T^* has the SVEP then both T and T^* satisfy Weyl's theorem.*

Proof The SVEP of either T or T^* entails Browder's theorem for T, or equivalently Browder's theorem for T^*. The polaroid condition for T entails that $p_{00}(T) = \pi_{00}(T)$, so Weyl's theorem holds for T, by Theorem 6.40. If T is polaroid then even T^* is polaroid and hence $p_{00}(T^*) = \pi_{00}(T^*)$, so Weyl's theorem also holds for T^*. ∎

Later we shall see that the condition that either T or T^* has the SVEP is not a necessary condition for a polaroid operator to satisfy Weyl's theorem. Examples will be given in the framework of Toeplitz operators. The Toeplitz operators are polaroid and satisfy Weyl's theorem, but there are examples for which the SVEP fails for both T and T^*.

Corollary 6.44 *If $T \in L(X)$ is polaroid and either T or T^* has the SVEP then both $f(T)$ and $f(T^*)$ satisfy Weyl's theorem for every $f \in \mathcal{H}_{nc}(\sigma(T))$.*

Proof If T is polaroid then $f(T)$ is polaroid, by Theorem 4.19, and $f(T)$ has the SVEP by Theorem 2.86. So Theorem 6.43 applies. Analogously, if T^* has the SVEP then $f(T^*)$ has the SVEP and $f(T^*)$ is polaroid. ∎

We now give a sufficient condition which ensures that Weyl's theorem is transmitted under commuting perturbations K for which K^n is finite-dimensional.

Theorem 6.45 *Let $T \in L(X)$ be an isoloid operator which satisfies Weyl's theorem. If $K \in L(X)$ commutes with T and K^n is a finite-dimensional operator for some $k \in \mathbb{N}$, then $T + K$ satisfies Weyl's theorem.*

Proof Since K is a Riesz operator and T satisfies Browder's theorem then $T + K$ satisfies Browder's theorem, by Corollary 5.5. Therefore, by Theorem 6.40 it suffices to prove $p_{00}(T + K) = \pi_{00}(T + K)$.

Let $\lambda \in \pi_{00}(T + K)$. If $\lambda I - T$ is invertible then $\lambda I - T + K$ is Browder and hence $\lambda \in p_{00}(T + K)$. Suppose that $\lambda \in \sigma(T)$. From Theorem 6.11 then $\lambda \in \text{iso}\,\sigma(T)$. Furthermore, the restriction

$$(\lambda I - T + K)|\ker(\lambda I - T) = K^n|\ker(\lambda I - T)$$

has both finite-dimensional kernel and range, so $\ker(\lambda I - T)$ is finite-dimensional and consequently $\lambda \in \pi_{00}(T)$, because T is isoloid. Since T satisfies Weyl's theorem we have $\pi_{00}(T) \cap \sigma_w(T) = \emptyset$, so $\lambda I - T$ is Weyl and hence $\lambda I - T + K$ is also Weyl, by Theorem 3.17, and this implies that $\lambda \in p_{00}(T + K)$. This shows that $\pi_{00}(T + K) \subseteq p_{00}(T + K)$, and, since the opposite inclusion is true for every operator, it then follows that $p_{00}(T + K) = \pi_{00}(T + K)$.   ∎

We show that in the preceding theorem it is essential to require that T is isoloid.

Theorem 6.46 *Let $T \in L(X)$ and suppose that there exists a finite rank operator commuting with T which is not nilpotent. If Weyl's theorem holds for $T + K$ for every finite rank operator which commutes with T, then T is isoloid.*

Proof Assume that T is not isoloid and let $\lambda \in \operatorname{iso}\sigma(T)$ be such that $\lambda I - T$ is injective. By hypothesis there exists a finite rank operator K that is not nilpotent and commutes with T. Observe that K is not quasi-nilpotent, otherwise its restriction $K|K(X)$ is nilpotent and hence K is also nilpotent. The spectrum of K is finite and contains 0, so $X = H_0(T) \oplus K(T)$, by Theorem 2.45. Since K is not quasi-nilpotent and the restriction $K|K(T)$ is an invertible operator of finite rank it then follows that $K(T)$ is a non-zero subspace of X having finite dimension. Let $T = T_1 \oplus T_2$ be the decomposition of T with $X = H_0(T) \oplus K(T)$ and let $\mu \in \mathbb{C}$ such that $\lambda - \mu \in \sigma(T_2)$. Clearly, $\lambda - \mu$ is an eigenvalue of T_2. The operator defined by $S := 0 \oplus \mu I$ is a finite rank operator which commutes with T and

$$\sigma(T + S) = \sigma(T_1) \cup \sigma(\mu I + T_2).$$

Since $\lambda \in \operatorname{iso}\sigma(T)$ and $\lambda I - T$ is injective we then have $\lambda \notin \sigma(T_2)$ and $\lambda \in \operatorname{iso}\sigma(T_1) \subseteq \operatorname{iso}\sigma(T + S)$. Moreover, $\ker(\lambda I - T + S) = \ker((\lambda - \mu)I - T_2)$ is a non-trivial subspace having finite dimension, so $\lambda \in \pi_{00}(T + S)$. On the other hand, $\lambda \notin p_{00}(T)$, so $\lambda I - T$ is not Browder, and hence $\lambda I - T + S$ is not Browder, by Theorem 3.8, and this implies that $\lambda \notin p_{00}(T + S)$. Consequently, $T + S$ does not satisfy Weyl's theorem, a contradiction with our assumptions.   ∎

Remark 6.47 Suppose that every finite rank operator commuting with T is nilpotent. This means precisely that X does not admit a decomposition $X = X_1 \oplus X_2$, X_1 and X_2 two closed T-invariant subspaces such that one of them has finite-dimension. In particular, $p_{00}(T) = \emptyset$. If, in addition, T satisfies Weyl's theorem then, by Theorem 6.40, we have $p_{00}(T) = \pi_{00}(T) = \emptyset$. Therefore, the hypotheses introduced in Theorem 6.46 are fulfilled by every operator for which $p_{00}(T)$ is non-empty, i.e., whenever $\sigma_b(T)$ is properly contained in $\sigma(T)$.

Theorem 6.48 *Let $T \in L(X)$ be a finite-isoloid operator which satisfies Weyl's theorem. If $R \in L(X)$ is a Riesz operator that commutes with T then $T + R$ satisfies Weyl's theorem.*

Proof Since T satisfies Browder's theorem, $T + R$ satisfies Browder's theorem, by Corollary 5.5, so it suffices to prove, by Theorem 6.40, the equality $p_{00}(T + R) = \pi_{00}(T + R)$. Let $\lambda \in \pi_{00}(T + R)$. If $\lambda I - T$ is invertible then $\lambda I - (T + R)$ is Browder and hence $\lambda \in p_{00}(T + R)$. If $\lambda \in \sigma(T)$ then, by Theorem 6.11, $\lambda \in \mathrm{iso}\,\sigma(T)$ and, since T is finite-isoloid, we also have $\lambda \in \pi_{00}(T)$. But Weyl's theorem for T entails that $\pi_{00}(T) \cap \sigma_b(T) = \emptyset$, therefore $\lambda I - T$ is Browder and hence $\lambda \in p_{00}(T + R)$. The other inclusion is trivial, so $T + R$ satisfies Weyl's theorem. ∎

In the next result we give a necessary and sufficient condition under which Weyl's theorem is transmitted from T to the perturbation $T + R$ by a commuting Riesz operator R.

Theorem 6.49 *Let $T \in L(X)$ be an operator which satisfies Weyl's theorem and let $R \in L(X)$ be a Riesz operator that commutes with T. Then the following conditions are equivalent:*

  (i) *$T + R$ satisfies Weyl's theorem;*
  (ii) *$\pi_{00}(T + R) = p_{00}(T + R)$;*
  (iii) *$\pi_{00}(T + R) \cap \sigma(T) \subseteq \pi_{00}(T)$.*

Proof (i) $\Rightarrow$ (ii) is clear by Theorem 6.40. The reverse implication is again an easy consequence of Theorem 6.40, once it is observed that Weyl's theorem implies Browder's theorem for T and hence for $T + R$.

(ii) $\Rightarrow$ (i) Let $\lambda_0 \in \pi_{00}(T+R) \cap \sigma(T)$ be arbitrary. Then $\lambda_0 \in p_{00}(T+R) \cap \sigma(T)$, so $\lambda_0 \in \sigma(T)$ and $\lambda_0 \notin \sigma_b(T + R) = \sigma_b(T)$, by Corollary 3.9. Hence $\lambda_0 \in p_{00}(T) = \pi_{00}(T)$, since T satisfies Weyl's theorem.

(iii) $\Rightarrow$ (ii) The inclusion $p_{00}(T + R) \subseteq \pi_{00}(T + R)$ is clear. To show the reverse inclusion, let $\mu_0 \in \pi_{00}(T + R)$ be arbitrary. Then $\mu_0 \notin \sigma(T)$ or $\mu_0 \in \sigma(T)$. If $\mu_0 \notin \sigma(T)$ then $\mu_0 \notin \sigma_b(T) = \sigma_b(T + R)$, and since $\mu_0 \in \sigma(T + R)$ it then follows that $\mu_0 \in p_{00}(T + R)$. In the case where $\mu_0 \in \sigma(T)$ we have

$$\mu_0 \in \pi_{00}(T + R) \cap \sigma(T) \subseteq \pi_{00}(T) = p_{00}(T),$$

since T satisfies Weyl's theorem. Therefore $\mu_0 \notin \sigma_b(T) = \sigma_b(T + R)$. Since $\mu_0 \in \sigma(T + R)$ then $\mu_0 \in p_{00}(T + R)$. Hence the inclusion $\pi_{00}(T + R) \subseteq p_{00}(T + R)$ is proved. ∎

An obvious consequence of Theorem 6.48 is that for a finite-isoloid operator $T \in L(X)$, Weyl's theorem from T is transmitted to $T + Q$, where $Q \in L(X)$ is a quasi-nilpotent operator which commutes with T. We now show that this result remains true if we suppose only the weaker condition $\sigma_p(T) \cap \mathrm{iso}\,\sigma(T) \subseteq \pi_{00}(T)$, where $\sigma_p(T)$ is the point spectrum of T.

Theorem 6.50 *Let $T \in L(X)$ be such that $\sigma_p(T) \cap \mathrm{iso}\,\sigma(T) \subseteq \pi_{00}(T)$. If T satisfies Weyl's theorem and $Q \in L(X)$ is a quasi-nilpotent operator which commutes with T, then $T + Q$ satisfies Weyl's theorem.*

Proof Since $\sigma(T + Q) = \sigma(T)$ and $\sigma_b(T) = \sigma_b(T + Q)$ we have $p_{00}(T) = p_{00}(T + Q)$. By Theorem 6.40 we have $p_{00}(T) = \pi_{00}(T)$, so it suffices to prove that $\pi_{00}(T) = \pi_{00}(T + Q)$. Let $\lambda \in \pi_{00}(T) = p_{00}(T)$. Then $\lambda I - T$ is Browder and hence $\lambda \in \operatorname{iso} \sigma(T) = \operatorname{iso} \sigma(T + Q)$. By Theorem 3.8 $\lambda I - (T + Q)$ is Browder and hence $\lambda \in \pi_{00}(T + Q)$.

Conversely, suppose that $\lambda \in \pi_{00}(T + Q)$. Since Q is quasi-nilpotent and commutes with T, the restriction of $\lambda I - T$ to the finite-dimensional subspace $\ker(\lambda I - (T + Q))$ is not invertible, hence $\ker(\lambda I - T)$ is non-trivial. Therefore, $\lambda \in \sigma_{\mathrm{p}}(T) \cap \operatorname{iso} \sigma(T) \subseteq \pi_{00}(T)$, which completes the proof. ∎

We now show that Weyl's theorem is extended from T to $f(T)$, where $f \in \mathcal{H}(\sigma(T))$, in the case when T is isoloid and either T or T^* has the SVEP. First we need the following lemma.

Lemma 6.51 *For every $T \in L(X)$, X a Banach space, and $f \in \mathcal{H}(\sigma(T))$ we have*

$$\sigma(f(T)) \setminus \pi_{00}(f(T)) \subseteq f(\sigma(T) \setminus \pi_{00}(T)). \qquad (6.9)$$

Furthermore, if T is isoloid then

$$\sigma(f(T)) \setminus \pi_{00}(f(T)) = f(\sigma(T) \setminus \pi_{00}(T)). \qquad (6.10)$$

Proof To show the inclusion (6.9) suppose that $\lambda_0 \in \sigma(f(T)) \setminus \pi_{00}(f(T))$. We distinguish two cases:

Case I λ_0 is not an isolated point of $f(\sigma(T))$. In this case there is a sequence $(\lambda_n) \subseteq f(\sigma(T))$ such that $\lambda_n \to \lambda_0$ as $n \to \infty$. Since $f(\sigma(T)) = \sigma(f(T))$, there exists a sequence (μ_n) in $\sigma(T)$ such that $f(\mu_n) = \lambda_n \to \lambda_0$. The sequence (μ_n) contains a convergent subsequence and we may assume that $\lim_{n \to \infty} \mu_n = \mu_0$. Hence

$$\lambda_0 = \lim_{n \to \infty} f(\mu_n) = f(\mu_0).$$

Since $\mu_0 \in \sigma(T) \setminus \pi_{00}(T)$ it then follows that $\lambda_0 \in f(\sigma(T) \setminus \pi_{00}(T))$.

Case II λ_0 is an isolated point of $\sigma(f(T))$, so either λ_0 is not an eigenvalue of $f(T)$ or it is an eigenvalue for which $\alpha(\lambda_0 - f(T)) = \infty$. Set $g(\lambda) := \lambda_0 - f(\lambda)$. The function $g(\lambda)$ is analytic and has only a finite number of zeros in $\sigma(T)$, say $\{\lambda_1, \dots, \lambda_k\}$. Write

$$g(\lambda) = p(\lambda)h(\lambda) \quad \text{with} \quad p(\lambda) := \prod_{i=1}^{k} (\lambda_i - \lambda)^{n_i},$$

where n_i denotes the multiplicity of λ_i for every $i = 1, \dots, k$. Clearly,

$$\lambda_0 I - f(T) = g(T) = p(T)h(T),$$

and $h(T)$ is invertible.

Now, suppose that λ_0 is not an eigenvalue of $f(T)$. Then none of the scalars $\lambda_1, \ldots, \lambda_k$ can be an eigenvalue of T, hence $\lambda_0 \in f(\sigma(T) \setminus \pi_{00}(T))$.

Consider the other possibility, i.e., λ_0 is an eigenvalue of T of infinite multiplicity. Then at least one of the scalars $\lambda_1, \ldots, \lambda_k$, say λ_1, is an eigenvalue of T of infinite multiplicity. Consequently, $\lambda_1 \in \sigma(T) \setminus \pi_{00}(T)$ and $f(\lambda_1) = \lambda_0$, so $\lambda_0 \in f(\sigma(T) \setminus \pi_{00}(T))$. Thus, the inclusion (6.9) is proved.

To prove the equality (6.10) suppose that T is isoloid. We need only to prove the inclusion

$$f(\sigma(T) \setminus \pi_{00}(T)) \subseteq \sigma(f(T)) \setminus \pi_{00}(f(T)). \qquad (6.11)$$

Let $\lambda_0 \in f(\sigma(T) \setminus \pi_{00}(T))$. From $f(\sigma(T)) = \sigma(f(T))$ we know that $\lambda_0 \in \sigma(f(T))$. If possible, let $\lambda_0 \in \pi_{00}(f(T))$, in particular, λ_0 is an isolated point of $\sigma(f(T))$. As above we can write $\lambda_0 I - f(T) = p(T)h(T)$, with

$$p(T) = \prod_{i=1}^{k} (\lambda_i I - T)^{n_i}. \qquad (6.12)$$

From the equality (6.12) it follows that any of the scalars $\lambda_1, \ldots, \lambda_k$ must be an isolated point of $\sigma(T)$, hence an eigenvalue of T, since by assumption T is isoloid. Moreover, λ_0 is an eigenvalue of finite multiplicity, so any scalar λ_i must be an eigenvalue of finite multiplicity, and hence $\lambda_i \in \pi_{00}(T)$. This contradicts that $\lambda_0 \in f(\sigma(T) \setminus \pi_{00}(T))$. Therefore, $\lambda_0 \notin \pi_{00}(f(T))$, which completes the proof of the equality (6.10). ∎

Theorem 6.52 *Suppose that $T \in L(X)$ is isoloid. Then $f(T)$ satisfies Weyl's theorem for every $f \in \mathcal{H}(\sigma(T))$ if and only if T satisfies Weyl's theorem and T is of stable sign index on $\rho_f(T)$.*

Proof Suppose that T satisfies Weyl's theorem and T is of stable sign index on $\rho_f(T)$. Since T is isoloid, by Lemma 6.51 we have

$$\sigma(f(T)) \setminus \pi_{00}(f(T)) = f(\sigma(T) \setminus \pi_{00}(T)).$$

Since T is of stable sign index on $\rho_f(T)$ we also have $f(\sigma_w(T)) = \sigma_w(f(T))$ for every $f \in \mathcal{H}(\sigma(T))$, by Theorem 3.119. Weyl's theorem for T then entails $\sigma(T) \setminus \pi_{00}(T) = \sigma_w(T)$, hence

$$\sigma(f(T)) \setminus \sigma_w(f(T)) = f(\sigma(T) \setminus \pi_{00}(T)) = f(\sigma_w(T)) = \sigma_w(f(T)),$$

from which we see that Weyl's theorem holds for $f(T)$.

Conversely, if $f(T)$ satisfies Weyl's theorem then $f(T)$, for every $f \in \mathcal{H}(\sigma(T))$, satisfies Browder's theorem, by Theorem 6.40, and hence, by Theorem 5.7, $f(\sigma_w(T)) = \sigma_w(f(T))$ for every $f \in \mathcal{H}(\sigma(T))$. By Theorem 3.119 T is of stable sign index on $\rho_f(T)$. Furthermore, by taking $f(\lambda) := \lambda$, we deduce that T satisfies Weyl's theorem. ∎

The next example shows that the transmission of Weyl's theorem from T to $f(T)$ may fail if T is not isoloid, also if the equality $f(\sigma_w(T)) = \sigma_w(f(T))$ is satisfied.

Example 6.53 Define on $\ell^2(\mathbb{N})$ the operators:

$$T_1(x_1, x_2, \dots) := \left(x_1, 0, \frac{x_2}{2}, \frac{x_3}{2}, \dots\right) \quad \text{for all } (x_n) \in \ell^2(\mathbb{N}),$$

and

$$T_2(x_1, x_2, \dots) := \left(0, \frac{x_1}{2}, \frac{x_2}{3}, \dots\right) \quad \text{for all } (x_n) \in \ell^2(\mathbb{N}).$$

Let $T := T_1 \oplus T_2$. Then

$$\sigma(T) = \{1\} \cup \mathbf{D}(0, \frac{1}{2}) \cup \{-1\},$$

and

$$\sigma_w(T) = \mathbf{D}(0, \frac{1}{2}) \cup \{-1\}.$$

Evidently, T satisfies Weyl's theorem. Set $f(t) = t^2$. Then

$$\sigma(T^2) = \sigma_w(T^2) = \mathbf{D}(0, \frac{1}{4}) \cup \{1\},$$

while $\pi_{00}(T^2) = \{1\}$, so Weyl's theorem does not hold for T^2. Note that $[(\sigma_w(T))]^2 = \sigma_w(T^2)$, and T is not isoloid.

Theorem 6.54 *Suppose that $T \in L(X)$ is isoloid and either T or T^* has the SVEP. Then $f(T)$ satisfies Weyl's theorem for every $f \in \mathcal{H}(\sigma(T))$ if and only if T satisfies Weyl's theorem.*

Proof If either T or T^* has the SVEP then $f(\sigma_w(T)) = \sigma_w(f(T))$, by Corollary 3.120, or equivalently T is of stable sign index on $\rho_f(T)$. Hence Theorem 6.52 applies. ∎

From Corollary 3.120, if $f \in \mathcal{H}(\sigma(T))$ is injective, then $f(\sigma_w(T)) = \sigma_w(f(T))$. Thus we have:

Corollary 6.55 *Let $T \in L(X)$ be isoloid and suppose that $f \in \mathcal{H}(\sigma(T))$ is injective on $\sigma_w(T)$. If Weyl's theorem holds for T then Weyl's theorem holds for $f(T)$.*

The following example shows that without SVEP the result of Corollary 6.54 does not hold.

Example 6.56 Let R be the right unilateral shift on $\ell_2(\mathbb{N})$ and define T in $\ell_2(\mathbb{N}) \oplus \ell_2(\mathbb{N})$ by

$$T = \begin{pmatrix} R + I & 0 \\ 0 & R' - I \end{pmatrix},$$

where R' is the adjoint of R, i.e., the left unilateral shift. We have $\sigma(T) = \sigma_w(T)$ and $\pi_{00}(T) = \emptyset$, so T is isoloid. Weyl's theorem holds for T, $1 \notin \sigma_w(T^2)$, while $1 \in \sigma(T)$ and $1 \notin \sigma_w(T^2) \cup \pi_{00}(T^2)$, thus Weyl's theorem does not hold for T^2.

It should be noted that if $f(T)$ satisfies Weyl's theorem for every $f \in \mathcal{H}(\sigma(T))$ then T is isoloid. To see this we first need the following lemma:

Lemma 6.57 *Suppose that $f(T)$ satisfies Weyl's theorem for every $f \in \mathcal{H}(\sigma(T))$. If $p_{00}(T) \neq \emptyset$ then $\mathrm{iso}\,\sigma(T) \subseteq \sigma_p(T)$.*

Proof Suppose that the implication $p_{00}(T) \neq \emptyset \Rightarrow \mathrm{iso}\,\sigma(T) \subseteq \sigma_p(T)$ is not satisfied. We show that there exists a $g \in \mathcal{H}(\sigma(T))$ for which Weyl's theorem does not hold for $g(T)$. We can choose $\lambda_1 \in p_{00}(T)$ and $\lambda_2 \in \mathrm{iso}\,\sigma(T)$ with $\lambda_2 \notin \sigma_p(T)$. Set $\sigma := \sigma(T) \setminus \{\lambda_1, \lambda_2\}$ and write $X = P_1(X) \oplus P_2(X) \oplus P_\sigma(X)$, where P_1, P_2 and P_σ are the spectral projections associated with $\{\lambda_1\}$, $\{\lambda_2\}$, and σ. If we set $T_i := T|P_i(X)$, $i = 1, 2$, and $T_3 := T|P_\sigma(X)$ then $T = T_1 \oplus T_2 \oplus T_3$, with $\sigma(T_i) = \{\lambda_i\}$ $i = 1, 2$, and $0 < \dim P_1(X) < \infty$, by Theorem 3.2. Note that $\lambda_1, \lambda_2 \notin \sigma(T_3) = \sigma$. Since $\lambda_2 \notin p_{00}(T)$, we have also $\dim P_2(X) = \infty$. If we define $g(\lambda) := (\lambda_1 - \lambda)(\lambda_2 - \lambda)$ we then have, with respect to the decomposition $X = P_1(X) \oplus P_2(X) \oplus P_\sigma(X)$,

$$g(T) = g(T_1) \oplus g(T_2) \oplus g(T_3).$$

It is easily seen that:

(a) $\sigma(g(T_1)) = g(\sigma(T_1)) = \{0\} = g(\sigma(T_2)) = \sigma(g(T_2))$ and $0 \notin g((\sigma(T_3)) = \sigma(g(T_3))$,
(b) $0 \in \mathrm{iso}\,\sigma(g(T_3))$, $P_\sigma(T) = P_1(X) \oplus P_2(X)$,
(c) $0 < \alpha(g(T_3) < \infty, \alpha(g(T_2)) = 0 = \alpha(g(T_3))$.

Hence $\alpha(g(T)) = \alpha(g(T_1)) + \alpha(g(T_2)) + \alpha(g(T_3)) = \alpha(g(T_1))$ and $0 \in \pi_{00}(g(T))$. But $0 \notin p_{00}(g(T))$, since the spectral projection associated with $g(T)$ and $\{0\}$ is infinite-dimensional. Therefore, $g(T)$ does not satisfy Weyl's theorem, by Theorem 6.38. ∎

Theorem 6.58 *If $f(T)$ satisfies Weyl's theorem for every $f \in \mathcal{H}(\sigma(T))$, then T is isoloid.*

Proof If $\lambda \in \mathrm{iso}\,\sigma(T)$ then either $\lambda \in p_{00}(T)$ or $\lambda \notin p_{00}(T)$. If $\lambda \in p_{00}(T)$, then $p_{00}(T) \neq \emptyset$, so, by Lemma 6.57, $\lambda \in \sigma_p(T)$. If $\lambda \notin p_{00}(T) = \sigma(T) \setminus \sigma_b(T)$ then $\lambda I - T$ is Browder and $\lambda \in \sigma(T)$. This implies $\alpha(\lambda I - T) = \beta(\lambda I - T) > 0$, otherwise $\lambda \notin \sigma(T)$, so also in this case we have $\lambda \in \sigma_p(T)$. ∎

Weyl's theorem also admits a generalization by considering B-Fredholm theory. The following concept was introduced by Berkani [65]. Recall that $\Delta^g(T)$ denotes the set $\sigma(T) \setminus \sigma_{bw}(T)$.

Definition 6.59 An operator $T \in L(X)$ is said to satisfy the *generalized Weyl's theorem* if $\Delta^g(T) = E(T)$.

The generalized Weyl's theorem entails Weyl's theorem:

Theorem 6.60 *If $T \in L(X)$ satisfies the generalized Weyl's theorem then T satisfies Weyl's theorem.*

Proof Suppose that the generalized Weyl's theorem holds for T and let $\lambda \in \sigma(T) \setminus \sigma_w(T)$. Then $\lambda I - T$ is Weyl and hence B-Weyl, so $\lambda \in \sigma(T) \setminus \sigma_{bw}(T) = E(T)$, and hence $\alpha(\lambda I - T) > 0$. Moreover, λ is an isolated point of $\sigma(T)$ and since $\alpha(\lambda I - T) < \infty$ we then have $\lambda \in \pi_{00}(T)$. Therefore, $\sigma(T) \setminus \sigma_w(T) \subseteq \pi_{00}(T)$. To prove the opposite inclusion, let $\lambda \in \pi_{00}(T)$. Obviously, $\pi_{00}(T) \subseteq E(T) = \sigma(T) \setminus \sigma_{bw}(T)$, so $\lambda I - T$ is B-Weyl. Since T has the SVEP at λ, $p(\lambda I - T) > \infty$, by Theorem 2.97, and hence from Theorem 1.143 we conclude that $\lambda I - T$ has both ascent and descent finite. Since $\alpha(\lambda I - T) < \infty$ we then have, by Theorem 1.22, that $\beta(\lambda I - T) < \infty$, so $\lambda \notin \sigma_w(T)$, and hence $\lambda \in \sigma(T) \setminus \sigma_w(T)$. Therefore, $\sigma(T) \setminus \sigma_w(T) = \pi_{00}(T)$.   ∎

In the following example we show that there exist operators which satisfy Weyl's theorem but not the generalized Weyl's theorem.

Example 6.61 Let $T \in L(X)$, where $X := \ell^1(\mathbb{N})$, be defined for each $x := (x_k) \in \ell^1(\mathbb{N})$) as

$$T(x_1, x_2, \ldots, x_k, \ldots) := (0, \alpha_1 x_1, \alpha_2 x_2, \ldots, \alpha_{k-1} x_{k-1}, \ldots),$$

where the sequence of complex numbers (α_k) is chosen such that $0 < |\alpha_k| \le 1$ and $\sum_{k=1}^{\infty} |\alpha_k| < \infty$. We have $T^n(X) \ne \overline{T^n(X)}$ for all $n = 1, 2, \ldots$. Indeed, for a given $n \in \mathbb{N}$, let $x_k^{(n)} := (1, \ldots, 1, 0, 0, \ldots)$ with the first $(n + k)$-terms equal to 1. Then $y^{(n)} := \lim_{k \to \infty} T x_k^{(n)}$ exists and lies in $\overline{T^n(X)}$. On the other hand there is no element $x^{(n)} \in \ell^1(\mathbb{N})$ which satisfies the equation $T^n x^{(n)} = y^{(n)}$, since the algebraic solution to this equation is $(1, 1, \ldots) \notin \ell^1(\mathbb{N})$.

Define $S := T \oplus 0$ on $Y := \ell^1(\mathbb{N}) \oplus \ell^1(\mathbb{N})$. Then $\ker S = \{0\} \oplus \ell^1(\mathbb{N})$, $\sigma(S) = \sigma_{ap}(S) = \{0\}$, and $E(T) = \{0\}$. Since $S^n(Y) = T^n(X) \oplus \{0\}$, the subspaces $S^n(Y)$ are not closed for all n, so S is not B-Weyl and $\sigma_{bw}(S) = \{0\}$. Furthermore, S is not upper semi-Weyl and $\sigma_{uw}(S) = \{0\}$. Hence $E(S) \ne \sigma(S) \setminus \sigma_{bw}(S)$, from which we see that S does not satisfy the generalized Weyl's theorem. It is easily seen that S satisfies Weyl's theorem. Indeed, S^* is quasi-nilpotent and hence has the SVEP. By Theorem 3.44 we then have $\sigma_w(S) = \sigma_{uw}(S) = \{0\}$, so $\sigma(S) \setminus \sigma_w(S) = \emptyset$. Also, $\pi_{00}(S)$ is empty, since 0 is the unique isolated point of $\sigma(S)$ and $\alpha(S) = \infty$.

Define

$$\Delta_1^g(T) := \Delta^g(T) \cup E(T),$$

where, as usual, $\Pi(T)$ denotes the set of all poles of T.

Theorem 6.62 *For a bounded operator* $T \in L(X)$ *the following statements are equivalent:*

(i) *T satisfies the generalized Weyl's theorem;*
(ii) *T satisfies the generalized Browder's theorem and $E(T) = \Pi(T)$;*
(iii) *For every $\lambda \in \Delta_1^g(T)$ there exists a $p := p(\lambda) \in \mathbb{N}$ such that $H_0(\lambda I - T) = \ker(\lambda I - T)^p$.*

Proof (i) $\Rightarrow$ (ii) The generalized Weyl's theorem entails Weyl's theorem and hence Browder's theorem, or, equivalently, the generalized Browder's theorem. To show the equality $E(T) = \Pi(T)$ it suffices to prove the inclusion $E(T) \subseteq \Pi(T)$. Let $\lambda \in E(T)$. Clearly, both T and T^* have the SVEP at λ, since $\lambda \in \mathrm{iso}\,\sigma(T)$. Moreover, $\lambda I - T$ is B-Weyl, and hence, by Theorems 2.97 and 2.98, we have that $\lambda I - T$ is Drazin invertible, i.e., $\lambda \in \Pi(T)$. Therefore, $E(T) = \Pi(T)$.

(ii) $\Rightarrow$ (i) By Theorem 5.17 we have $\Delta^g(T) = \Pi(T) = E(T)$.

(ii) $\Rightarrow$ (iii) By Theorem 5.22 we have only to show that there exists a $p := p(\lambda) \in \mathbb{N}$ such that $H_0(\lambda I - T) = \ker(\lambda I - T)^p$ for all $\lambda \in E(T)$. If $\lambda \in E(T) = \Pi(T)$ then λ is a pole of the resolvent so, by Theorem 2.45 $H_0(\lambda_0 I - T) = \ker(\lambda_0 I - T)^p$ where $p =: p(\lambda_0 I - T) = q(\lambda_0 I - T)$.

(iii) $\Rightarrow$ (ii) Suppose that $\lambda \notin \sigma_{\mathrm{bw}}(T)$. To show the generalized Browder's theorem for T it suffices to prove, by Theorem 5.14, that T has the SVEP at λ. If $\lambda \notin \sigma(T)$ there is nothing to prove. If $\lambda \in \sigma(T)$ then $\lambda \in \Delta^g(T) \subseteq \Delta_1^g(T)$, so $H_0(\lambda I - T)$ is closed, and hence T has the SVEP at λ, by Theorem 2.39. Therefore, the generalized Browder's theorem holds for T. To show the equality $E(T) = \Pi(T)$ we have only to prove that $E(T) \subseteq \Pi(T)$. Consider a point $\lambda \in E(T)$. Then $\lambda \in \Delta_1^g(T)$ and hence $H_0(\lambda I - T) = \ker(\lambda I - T)^p$ for some $p \in \mathbb{N}$. Since λ is an isolated point of $\sigma(T)$, by Theorem 2.45 we have

$$X = H_0(\lambda I - T) \oplus K(\lambda I - T) = \ker(\lambda I - T)^p \oplus K(\lambda I - T).$$

Hence

$$(\lambda I - T)^p(X) = (\lambda I - T)^p(K(\lambda I - T)) = K(\lambda I - T).$$

This shows that $X = \ker(\lambda I - T)^p \oplus (\lambda I - T)^p(X)$, so by Theorem 1.35, $\lambda \in \Pi(T)$. Consequently, $E(T) \subseteq \Pi(T)$, as desired. ∎

We conclude this section by giving a perturbation result concerning the generalized Weyl's theorem.

Theorem 6.63 *Suppose that $T \in L(X)$ is polaroid and that $K \in L(X)$ is such that $TK = KT$ and K^n is a finite rank operator for some $n \in \mathbb{N}$. If T satisfies the generalized Weyl's theorem then $T + K$ satisfies the generalized Weyl's theorem.*

Proof T satisfies the generalized Browder's theorem, or equivalently Browder's theorem, hence, by Corollary 5.5, $T + K$ satisfies Browder's theorem, or equivalently, the generalized Browder's theorem. By Theorem 6.62 it suffices to prove that $E(T + K) = \Pi(T + K)$. Since the inclusion $\Pi(T + K) \subseteq E(T + K)$ holds for every operator we need only to prove the reverse inclusion. Let $\lambda \in E(T + K)$. Then $\lambda \in \text{iso}\,\sigma(T + K)$ and since, by Theorem 4.24, $T + K$ is polaroid, then $\lambda \in \Pi(T + K)$. Therefore, $T + K$ satisfies the generalized Weyl's theorem.   ∎

We now show that the generalized Weyl's theorem is transmitted from T to $f(T)$, where $f \in \mathcal{H}(\sigma(T))$, in the case where T is isoloid and either T or T^* has the SVEP. First we need the following lemma.

Lemma 6.64 *Suppose that $T \in L(X)$ and either T or T^* has the SVEP, and $f \in \mathcal{H}(\sigma(T))$. Then we have*

$$\sigma(f(T)) \setminus E(f(T)) \subseteq f(\sigma(T) \setminus E(T)). \tag{6.13}$$

Furthermore, if T is isoloid then

$$\sigma(f(T)) \setminus E(f(T)) = f(\sigma(T) \setminus E(T)). \tag{6.14}$$

Proof To show the inclusion (6.13) suppose that $\lambda_0 \in \sigma(f(T)) \setminus \pi_{00}(f(T))$. We distinguish two cases:

Case (I) λ_0 is not an isolated point of $f(\sigma(T))$. In this case there is a sequence $(\lambda_n) \subseteq f(\sigma(T))$ such that $\lambda_n \to \lambda_0$ as $n \to \infty$. Since $f(\sigma(T)) = \sigma(f(T))$, there exists a sequence (μ_n) in $\sigma(T)$ such that $f(\mu_n) = \lambda_n \to \lambda_0$. The sequence (μ_n) contains a convergent subsequence and we may assume that $\lim_{n\to\infty} \mu_n = \mu_0$. Hence $\lambda_0 = \lim_{n\to\infty} f(\mu_n) = f(\mu_0)$. Since $\mu_0 \in \sigma(T) \setminus E(T)$ it then follows that $\lambda_0 \in f(\sigma(T) \setminus E(T))$.

Case (II) λ_0 is an isolated point of $\sigma(f(T))$. Since $\lambda \notin E(f(T))$, λ is not an eigenvalue of $f(T)$. Write $g(\lambda) := \lambda_0 - f(\lambda)$. The function $g(\lambda)$ is analytic and has only a finite number of zeros in $\sigma(T)$, say $\{\lambda_1, \ldots, \lambda_k\}$. Write

$$g(\lambda) = p(\lambda)h(\lambda) \quad \text{with} \quad p(\lambda) := \prod_{i=1}^{k} (\mu_i - \lambda)^{n_i},$$

where n_i denotes the multiplicity of μ_i for every $i = 1, \ldots, k$. Clearly, $\lambda_0 I - f(T) = g(T) = p(T)h(T)$ and $h(T)$ is invertible. As $\lambda_0 \notin E(f(T))$, none of $\mu_1, \ldots, \mu_k$ can be an eigenvalue of T. Since $\lambda_0 I - f(T)$ is not invertible, there exists a $\mu \in \{\mu_1, \ldots, \mu_k\}$ such that $\mu I - T$ is not invertible. Hence $f(\mu) = \lambda_0$ and $\lambda_0 \in f(\sigma(T) \setminus E(T))$.

To prove the equality (6.14) suppose that T is isoloid. We need only to prove the inclusion

$$f(\sigma(T) \setminus E(T)) \subseteq \sigma(f(T)) \setminus E(f(T)). \tag{6.15}$$

Let $\lambda_0 \in f(\sigma(T) \cap E(f(T))$. Then we can write $\lambda_0 I - f(T) = p(T)h(T)$, where $p(T) = \prod_{i=1}^{k}(\mu_i I - T)^{n_i}$, $\mu_i \neq \mu_j$ for $i \neq j$, and $h(T)$ is invertible. Since $\lambda_0 I - f(T)$ is not invertible, there exists a $\mu \in \{\mu_1, \ldots, \mu_k\}$ such that $\mu \in \sigma(T)$. Since $\lambda_0 \in \operatorname{iso}\sigma(f(T))$, μ is isolated in $\sigma(T)$. Hence $\lambda_0 = f(\mu) \notin f(\sigma(T) \setminus E(T))$, from which we obtain the inclusion (6.15), and the proof is complete. ∎

Theorem 6.65 *Suppose that $T \in L(X)$ is isoloid and either T or T^* has the SVEP. If T satisfies the generalized Weyl's theorem then the generalized Weyl's theorem holds for $f(T)$ for every $f \in \mathcal{H}(\sigma(T))$.*

Proof The generalized Weyl's theorem for T implies that $\sigma(T) \setminus E(T) = \sigma_{bw}(T)$. Since T is isoloid, by Lemma 6.64 we have

$$f(\sigma_{bw}(T) \setminus E(f(T))) = \sigma(f(T) \setminus E(f(T))).$$

But either T or T^* has the SVEP, so, by Corollary 3.123, $f(\sigma_{bw}(T)) = \sigma_{bw}(f(T))$. Thus, $\sigma(f(T)) \setminus E(f(T)) = \sigma_{bw}(f(T))$. ∎

In the next example we show that the condition that T is isoloid is crucial in Theorem 6.65. First we need a preliminary result.

Theorem 6.66 *If $T \in L(X)$ has no eigenvalues then T satisfies the generalized Weyl's theorem.*

Proof Clearly, $E(T) = \emptyset$. We show that $\sigma(T) = \sigma_{bw}(T)$. Let $\lambda \in \sigma(T)$. We can assume $\lambda = 0$. If $0 \notin \sigma_{bw}(T)$ then T is B-Weyl, and since $\ker T = \{0\}$ we have, by Theorem 1.143, $p(T) = q(T) = 0$, i.e., T is invertible, a contradiction. Hence, $\sigma(T) = \sigma_{bw}(T)$ and T satisfies the generalized Weyl's theorem. ∎

It should be noted that in the proof of Theorem 6.66 it has been proved that if T is injective and non-invertible, then T cannot be B-Weyl.

Example 6.67 Let I_1 and I_2 denote the identities on $\mathbb{C}$ and ℓ_2, respectively. Define S_1 and S_2 on $\ell_2(\mathbb{N})$ as follows:

$$S_1(x_1, x_2, \ldots) := \left(0, \frac{1}{3}x_1, \frac{1}{3}x_2, \ldots\right) \quad \text{for all } (x_n) \in \ell_2(\mathbb{N}),$$

and

$$S_2(x_1, x_2, \ldots) := \left(0, \frac{1}{2}x_1, \frac{1}{3}x_2, \ldots\right) \quad \text{for all } (x_n) \in \ell_2(\mathbb{N}).$$

Let $T_1 := I_1 \oplus S_1$ and $T_2 := S_2 - I_2$. Since T_2 has no eigenvalue, then, see the proof of Theorem 6.66, $\sigma(T_2) = \sigma_{\mathrm{bw}}(T_2) = \{-1\}$. Evidently,

$$\sigma(T_1) = \left\{\lambda \in \mathbb{C} : |\lambda| \leq \frac{1}{3}\right\} \cup \{1\}.$$

We claim that

$$\sigma_{\mathrm{bw}}(T_1) = \left\{\lambda \in \mathbb{C} : |\lambda| \leq \frac{1}{3}\right\}. \tag{6.16}$$

Since

$$T_1 - (I_1 \oplus I_2) = 0 \oplus (S_1 - I_2)$$

and $S_1 - I_2$ is invertible then, by Theorem 1.119, $T_1 - (I_1 \oplus I_2)$ is B-Weyl. Thus, $1 \notin \sigma_{\mathrm{bw}}(T_1)$. If there exists some $0 \leq |\lambda| \leq \frac{1}{3}$ such that $T_1 - \lambda(I_1 \oplus I_2)$ is B-Weyl then, by Theorem 1.127, we deduce that $S_1 - \lambda I_2$ is B-Weyl, which is a contradiction, since $S_1 - \lambda I_2$ has no eigenvalue. Therefore, the equality (6.16) is proved. Now, let $T := T_1 \oplus T_2$. Then T has the SVEP and

$$\sigma_{\mathrm{bw}}(T) = \{-1\} \cup \left\{\lambda \in \mathbb{C} : |\lambda| \leq \frac{1}{3}\right\}.$$

Since $E(T) = \{1\}$ we deduce that the generalized Weyl's theorem holds for T. We also have

$$\sigma_{\mathrm{bw}}(T^2) = \{1\} \cup \left\{\lambda \in \mathbb{C} : |\lambda| \leq \frac{1}{9}\right\},$$

which is equal to $\sigma(T^2)$. Since $E(T^2) = \{1\}$, the generalized Weyl's theorem does not hold for T^2.

6.4 a-Weyl's Theorem

Let $\lambda \in \Delta_a(T) := \sigma_{\mathrm{ap}}(T) \setminus \sigma_{\mathrm{uw}}(T)$. Then $\lambda I - T \in W_+(X)$, so $\alpha(\lambda I - T) < \infty$ and $(\lambda I - T)(X)$ is closed. Since $\lambda I - T$ is not bounded below, necessarily $0 < \alpha(\lambda I - T)$. Therefore, for every operator we have:

$$\Delta_a(T) \subseteq \{\lambda \in \sigma_{\mathrm{ap}}(T) : 0 < \alpha(\lambda I - T) < \infty\}.$$

The following approximate point version of Weyl's theorem has been introduced by Rakočević [262]. Set

$$\pi_{00}^a(T) := \{\lambda \in \text{iso } \sigma_{\text{ap}}(T) : 0 < \alpha(\lambda I - T) < \infty\}.$$

Definition 6.68 An operator $T \in L(X)$ is said to satisfy a-Weyl's theorem if

$$\Delta_a(T) := \sigma_{\text{ap}}(T) \setminus \sigma_{\text{uw}}(T) = \pi_{00}^a(T).$$

Theorem 6.69 *If a bounded operator $T \in L(X)$ satisfies a-Weyl's theorem then*

$$p_{00}^a(T) = \pi_{00}^a(T) = \Delta_a(T).$$

In particular, a-Weyl's theorem for T implies a-Browder's theorem for T.

Proof Suppose that T satisfies a-Weyl's theorem. By definition then $\Delta_a(T) = \pi_{00}^a(T)$. We show now the equality $p_{00}^a(T) = \pi_{00}^a(T)$. It suffices to prove the inclusion $\pi_{00}^a(T) \subseteq p_{00}^a(T)$. Let λ be an arbitrary point of $\pi_{00}^a(T)$. Since λ is isolated in $\sigma_{\text{ap}}(T)$ then T has the SVEP at λ and, from the equality $\pi_{00}^a(T) = \sigma_{\text{ap}}(T) \setminus \sigma_{\text{uw}}(T)$, we know that $\lambda I - T \in W_+(X)$. The SVEP at λ by Theorem 2.97 implies that $p(\lambda I - T) < \infty$, so $\lambda \in p_{00}^a(T)$.

The last assertion follows from Theorem 5.26. ∎

It is not difficult to find an example of an operator satisfying a-Browder's theorem but not a-Weyl's theorem. For instance, if $Q \in L(\ell^2)$ is defined by

$$Q(x_0, x_1, \dots) := \left(\frac{1}{2}x_1, \frac{1}{3}x_2, \dots\right) \quad \text{for all } (x_n) \in \ell^2,$$

then Q is quasi-nilpotent, so has the SVEP and consequently satisfies a-Browder's theorem. On the other hand, Q does not satisfy a-Weyl's theorem, since $\sigma_{\text{ap}}(Q) = \sigma_{\text{uw}}(Q) = \{0\}$ and $\pi_{00}^a(Q) = \{0\}$.

Note that the condition $\Delta_a(T) = \emptyset$ does not ensure that a-Weyl's theorem holds for T. To describe the operators which satisfy a-Weyl's theorem, let us define

$$\Gamma_a(T) := \Delta_a(T) \cup \pi_{00}^a(T).$$

Clearly, if $\Gamma(T) = \emptyset$ then a-Weyl's theorem holds for T. In the following theorem we shall exclude the trivial case $\Gamma(T) = \emptyset$. As usual, $\gamma(T)$ denotes the reduced minimal modulus of T.

Theorem 6.70 *Let $T \in L(X)$. Then the following statements are equivalent:*

(i) *T satisfies a-Weyl's theorem;*
(ii) *T satisfies a-Browder's theorem and $p_{00}^a(T) = \pi_{00}^a(T)$;*
(iii) *$\Delta_a(T)$ is a discrete set, i.e., all points of $\Delta_a(T)$ are isolated, and $\pi_{00}^a(T) \subseteq \rho_{\text{sf}}(T)$;*
(iv) *a-Browder's theorem holds for T and $(\lambda I - T)(X)$ is closed for all $\lambda \in \pi_{00}^a(T)$;*
(v) *the map $\lambda \to \gamma(\lambda I - T)$ is not continuous at every $\lambda \in \Gamma_a(T)$.*

Proof (i) $\Leftrightarrow$ (ii) The implication (i) $\Rightarrow$ (ii) is clear, from Theorem 6.69. The implication (ii) $\Rightarrow$ (i) follows immediately from Theorem 5.26.

(ii) $\Rightarrow$ (iii) Evidently, $\pi_{00}^a(T) = p_{00}^a(T) = \sigma(T) \setminus \sigma_{\mathrm{ub}}(T) \subseteq \rho_{\mathrm{sf}}(T)$, while, by Theorem 5.31, $\Delta_a(T) \subseteq \mathrm{iso}\,\sigma_{\mathrm{ap}}(T)$, thus $\Delta_a(T)$ is discrete.

(iii) $\Rightarrow$ (ii) Let $\lambda \in \pi_{00}^a(T)$. Then $\lambda I - T \in \Phi_\pm(X)$ and, since λ is isolated in $\sigma_{\mathrm{ap}}(T)$, the ascent $p(\lambda I - T)$ is finite, by Theorem 2.97. From Theorem 1.22 it then follows that $\alpha(\lambda I - T) \le \beta(\lambda I - T)$, from which we obtain $\lambda I - T \in \Phi_+(X)$. Consequently, $\lambda I - T$ is upper semi-Browder, and hence $\lambda \in p_{00}^a(T)$. The reverse inclusion $p_{00}^a(T) \subseteq \pi_{00}^a(T)$ holds for every operator, hence $\pi_{00}^a(T) = p_{00}^a(T)$. To show a-Weyl's theorem for T it suffices to prove that T satisfies a-Browder's theorem, or equivalently, by Theorem 5.31, $\Delta_a(T) \subseteq \mathrm{iso}\,\sigma_{\mathrm{ap}}(T)$. Let $\lambda_0 \in \Delta_a(T)$. Then, by Lemma 5.25, $\lambda_0 I - T \in W_+(X)$ and hence, for some $\delta_1 > 0$, we have $\lambda I - T \in W_+(X)$ for all $\lambda \in \mathbb{D}(\lambda_0, \delta_1)$. In particular, $\lambda I - T$ has closed range for all $\lambda \in \mathbb{D}(\lambda_0, \delta_1)$. But, by Lemma 5.25, we have

$$\Delta_a(T) = [\rho_{\mathrm{sf}}^-(T) \cup \rho_{\mathrm{w}}(T)] \cap \sigma_{\mathrm{p}}(T),$$

from which the assumption that $\Delta_a(T)$ is discrete yields that there exists a $\delta_2 > 0$ for which $\alpha(\lambda I - T) = 0$ for $\lambda \in \mathbb{D}(\lambda_0, \delta_2)$. If $\delta := \min\{\delta_1, \delta_2\}$ then $\lambda I - T$ is bounded below for every $\lambda \in \mathbb{D}(\lambda_0, \delta)$, $\lambda \ne \lambda_0$. Thus, $\lambda_0 \in \mathrm{iso}\,\sigma_{\mathrm{ap}}(T)$.

(i) $\Rightarrow$ (iv) If T satisfies a-Weyl's theorem then T obeys a-Browder's theorem. Furthermore, $\pi_{00}^a(T) = p_{00}^a(T)$ by Theorem 6.69, so $\lambda I - T \in B_+(X)$ for all $\lambda \in \pi_{00}^a(T)$, and hence $(\lambda_0 I - T)(X)$ is closed.

(iv) $\Rightarrow$ (i) The condition $(\lambda_0 I - T)(X)$ closed for all $\lambda \in \pi_{00}^a(T)$ entails that for these values of λ we have $\lambda I - T \in \Phi_+(X)$. Now, T has the SVEP at every isolated point of $\sigma_{\mathrm{ap}}(T)$, and in particular T has the SVEP at every point of $\pi_{00}^a(T)$. By Theorem 2.97 it then follows that $p(\lambda I - T) < \infty$ for all $\lambda \in \pi_{00}^a(T)$, from which we deduce that $\pi_{00}^a(T) = p_{00}^a(T)$.

(i) $\Rightarrow$ (v) By Theorem 6.69 we have that $\Gamma_a(T) = \Delta_a(T)$ and T satisfies a-Browder's theorem, hence, by Theorem 5.31, the map $\lambda \to \gamma(\lambda I - T)$ is not continuous at every $\lambda \in \Gamma_a(T)$.

(v) $\Rightarrow$ (iv) Suppose that $\lambda \to \gamma(\lambda I - T)$ is not continuous at every $\lambda \in \Gamma_a(T) = \Delta_a(T) \cup \pi_{00}^a(T)$. The discontinuity at the points of $\Delta_a(T)$ entails, by Theorem 5.31, that T satisfies a-Browder's theorem. We show now that the discontinuity at a point λ_0 of $\pi_{00}^a(T)$ implies that $(\lambda_0 I - T)(X)$ is closed. In fact, if $\lambda_0 \in \pi_{00}^a(T)$ then $\lambda_0 \in \mathrm{iso}\,\sigma_{\mathrm{ap}}(T)$ and $0 < \alpha(\lambda_0 I - T) < \infty$. Clearly, $\lambda I - T$ is injective in an open punctured disc $\mathbb{D}$ centered at λ_0. Take $0 \ne x \in \ker(\lambda_0 I - T)$. If $\lambda \in \mathbb{D}$ then

$$\gamma(\lambda I - T) \le \frac{\|(\lambda I - T)x\|}{\mathrm{dist}\,(x, \ker(\lambda I - T))} = \frac{\|(\lambda I - T)x\|}{\|x\|}$$

$$= \frac{\|(\lambda I - T)x - (\lambda_0 I - T)x\|}{\|x\|} = |\lambda - \lambda_0|.$$

From this estimate it follows that

$$\lim_{\lambda \to \lambda_0} \gamma(\lambda I - T) = 0 \neq \gamma(\lambda_0 I - T),$$

so $(\lambda_0 I - T)(X)$ is closed. ∎

From Theorems 5.31 and 6.70 we see that *a*-Browder's theorem and *a*-Weyl's theorem are equivalent to the discontinuity of the mapping $\lambda \to \gamma(\lambda I - T)$ at the points of the two sets $\Delta_a(T)$ and $\Gamma_a(T)$, respectively, where $\Gamma_a(T)$ is larger than $\Delta_a(T)$. Comparing Theorems 6.70 and 5.31 one might expect that the discontinuity of the mapping $\lambda \to \ker(\lambda I - T)$ at every $\lambda \in \Gamma_a(T) = \Delta_a(T) \cup \pi_{00}^a(T)$ is equivalent to *a*-Weyl's theorem for T. This does not work. In fact, by definition of $\pi_{00}^a(T)$ the map $\lambda \to \ker(\lambda I - T)$ is not continuous at every $\lambda \in \pi_{00}^a(T)$ for *all* operators $T \in L(X)$, since every $\lambda \in \pi_{00}^a(T)$ is an isolated point of $\sigma_{\mathrm{ap}}(T)$. From this it obviously follows that the discontinuity of the mapping $\lambda \to \ker(\lambda I - T)$ at every $\lambda \in \Gamma_a(T)$ is equivalent to *a*-Browder's theorem for T.

Analogously, the inclusion $\Gamma_a(T) \subseteq \sigma_{\mathrm{ap}}(T)$ is equivalent to *a*-Browder's theorem for T, since the inclusion $\pi_{00}^a(T) \subseteq \mathrm{iso}\,\sigma_{\mathrm{ap}}(T)$ holds by definition of $\pi_{00}^a(T)$.

Theorem 6.71 *Let $T \in L(X)$ and let N be a nilpotent operator commuting with T. If a-Weyl's theorem holds for T, then it also holds for $T + N$.*

Proof First we prove that $\pi_{00}^a(T) = \pi_{00}^a(T + N)$. It is enough to prove that $0 \in \pi_{00}^a(T)$ if and only if $0 \in \pi_{00}^a(T+N)$. Suppose that $0 \in \pi_{00}^a(T)$, so $0 < \alpha(T) < \infty$. We prove that $\alpha(T + N) < \infty$. Let $x \in \ker(T + N)$. Then $(T + N)x = 0$, so $Tx = -Nx$, and hence

$$T^k x = (-1)^k N^k x \quad \text{for } n \in \mathbb{N}.$$

Let ν be such that $N^\nu = 0$. Then $x \in \ker T^\nu$, and hence $\ker(T + N) \subseteq \ker T^\nu$. From $\alpha(T^\nu) < \infty$ we then have $\alpha(T + N) < \infty$. We prove that $\alpha(T + N) > 0$. Since $\alpha(T) > 0$, there exists an $x \neq 0$ such that $Tx = 0$. Then $(T + N)^\nu x = 0$, so $x \in \ker(T + N)^\nu$ and $\alpha(T + N)^\nu > 0$. This, obviously, implies that $\alpha(T + N) > 0$. Hence $0 \in \pi_{00}^a(T + N)$, thus

$$\pi_{00}^a(T) \subseteq \pi_{00}^a(T + N).$$

The reverse inclusion follows by symmetry. To show that *a*-Weyl's theorem holds for $T + N$, observe that

$$\pi_{00}^a(T + N) = \pi_{00}^a(T) = \sigma_{\mathrm{ap}}(T) \setminus \sigma_{\mathrm{uw}}(T) = \sigma_{\mathrm{ap}}(T + N) \setminus \sigma_{\mathrm{uw}}(T + N),$$

so *a*-Weyl's theorem holds for $T + N$. ∎

In the sequel we shall need the following elementary lemma

Lemma 6.72 *Let* $Q \in L(X)$ *be a quasi-nilpotent operator having finite-dimensional kernel. If* $R \in L(X)$ *is a Riesz operator that commutes with* Q *then* $Q + R$ *has a finite spectrum.*

Proof Suppose on the contrary that there exists a sequence $\{\lambda_n\}$ of distinct numbers in $\sigma(Q + R) \setminus \{0\}$. Then $\lambda_n I - Q$ is invertible for all $n \in \mathbb{N}$, and since R commutes with Q we have, by Theorem 3.8, that $\lambda_n I - (Q + R)$ is Browder. Therefore, ker $(\lambda_n I - (Q + R))$ is a non-zero finite-dimensional subspace, because $\lambda_n I - (Q+R)$ is non-invertible and hence the restriction of Q to ker $(\lambda_n I - (Q+R))$ is nilpotent. Consequently, ker $(\lambda_n I - (Q + R)) \cap$ ker Q is non-trivial and contains a non-zero element x_n. Every x_n is an eigenvector of $Q+R$ associated to λ_n and the numbers λ_n are mutually distinct, and it is easily seen that $\{x_n\}$ consists of linearly independent vectors of ker Q. Hence ker Q has infinite dimension, a contradiction. ∎

For a bounded operator $T \in L(X)$ we set

$$\pi_f^a(T) := \{\lambda \in \sigma_{\mathrm{ap}}(T) : \alpha(\lambda I - T) < \infty\}.$$

Evidently,

$$p_{00}(T) \subseteq \pi_{00}(T) \subseteq \pi_f^a(T).$$

Theorem 6.73 *If* $T \in L(X)$ *and* $R \in L(X)$ *is a Riesz operator that commutes with* T, *then*

$$\pi_f^a(T + R) \cap \sigma_{\mathrm{ap}}(T) \subseteq \mathrm{iso}\,\sigma_{\mathrm{ap}}(T).$$

Proof Let $\lambda \in \pi_f^a(T + R)$. Then there exists a deleted neighborhood U of λ such that $\mu I - (T + R)$ is bounded below for all $\mu \in U$. In particular, $\mu I - (T + R)$ is upper semi-Browder and hence $\mu I - T$ is also upper semi-Browder, by Theorem 3.8. By Theorem 2.97 it then follows that $H_0(\mu I - T) \cap K(\mu I - T) = \{0\}$ for all $\mu \in U$. On the other hand, the closed subspaces $H_0(\mu I - T) + K(\mu I - T) = H_0(\mu I - T) \oplus K(\mu I - T)$ are constant on U, by Corollary 2.121. Let Z denote one of them and set $T_0 := T|Z$ and $R_0 := R|Z$. We claim that λ is not an accumulation point of T_0. To see this, let $\mu \in U$. Since the restriction $(\mu I - T)|K(\mu I - T)$ is invertible, then the restriction $(\mu I - (T + R))|K(\mu I - T)$ is Browder, again by Theorem 3.8, and hence $\mu I - (T_0 + R_0)$ is also Browder, since $H_0(\mu I - T)$ is finite-dimensional. Moreover, $\mu I - (T + R)$ is injective, from which we deduce that $\mu I - (T_0 + R_0)$ is invertible and this shows that $\lambda \notin \mathrm{acc}\,(T_0 + R_0)$. Consequently,

$$Z = H_0(\lambda I - (T_0 + R_0)) \oplus K(\lambda I - (T_0 + R_0)). \tag{6.17}$$

Now, write $T_0 = T_1 + T_2$ and $R_0 = R_1 + R_2$ with respect to the decomposition (6.17). Since $\lambda I - (T_1 + R_1)$ is a quasi-nilpotent operator having a finite-dimensional kernel, from Lemma 6.72 we obtain that $\sigma(T_1)$ is finite, hence there exists a deleted neighborhood V_1 of λ such that $V_1 \cap \sigma(T_1) = \emptyset$. Moreover, since $\lambda I - (T_2 + R_2)$ is invertible, $\lambda I - T_2$ is Browder, by Theorem 3.8, and hence λ is an isolated point of $\sigma(T_2)$, so there exists a deleted neighborhood V_2 of λ such that $V_2 \cap \sigma(T_2) = \emptyset$. Set $V := V_1 \cap V_2 \cap U$. Clearly, $V \cap \sigma(T_0) = \emptyset$. Finally, since

$$\ker(\mu I - T) \subseteq H_0(\mu I - T) \subseteq Z,$$

we have

$$\ker(\mu I - T) = \ker(\mu I - T_0) = \{0\} \quad \text{for } \mu \in V.$$

But $\mu I - T$ is semi-Fredholm, hence $\mu I - T$ is bounded below. This completes the proof. ∎

Theorem 6.74 *Suppose that $T \in L(X)$ is a-isoloid and that it satisfies a-Weyl's theorem. If $K \in L(X)$ is an operator that commutes with T and such that there exists an $n \in \mathbb{N}$ such that K^n is a finite rank operator then a-Weyl's theorem holds for $T + K$.*

Proof Since K is a Riesz operator and T satisfies a-Browder's theorem, $T + K$ satisfies a-Browder's theorem, by Corollary 5.29. By Theorem 6.70 it suffices to prove the equality $p_{00}^a(T + K) = \pi_{00}^a(T + K)$. Let $\lambda \in \pi_{00}^a(T + K)$ be arbitrarily given. If $\lambda I - T$ is bounded below, in particular upper semi-Browder, then $\lambda I - (T + K)$ is upper semi-Browder, by Theorem 3.8, so $\lambda \in p_{00}^a(T + K)$. Consider the other case that $\lambda \in \sigma_{\mathrm{ap}}(T)$. By Theorem 6.73, $\lambda \in \mathrm{iso}\,\sigma_{\mathrm{ap}}(T)$. Since the restriction of the operator $\lambda I - (T + K)$ to $\ker(\lambda I - T)$ has both finite-dimensional range and kernel, $\ker(\lambda I - T)$ is also finite-dimensional, so that $\lambda \in \pi_{00}^a(T)$, since T is a-isoloid. On the other hand, a-Weyl's theorem for T entails that $\pi_{00}^a(T) = \pi_{00}^a(T)$, so $\pi_{00}^a(T) \cap \sigma_{\mathrm{ub}}(T) = \emptyset$. Consequently, $\lambda I - T$ is upper semi-Browder, and hence $\lambda I - (T + K)$ is also upper semi-Browder, by Theorem 3.8, from which we conclude that $\lambda \in p_{00}^a(T + K)$. This shows the inclusion $\pi_{00}^a(T + K) \subseteq p_{00}^a(T + K)$ and since the opposite inclusion is trivial we then conclude that $\pi_{00}^a(T + K) = p_{00}^a(T + K)$, so the proof is complete. ∎

Notice that in the preceding result it is essential to require that T is a-isoloid. Indeed, we have:

Theorem 6.75 *Let $T \in L(X)$ and suppose that there exists a finite rank operator commuting with T which is not nilpotent. If a-Weyl's theorem holds for $T + K$ for every finite rank operator which commutes with T, then T is a-isoloid.*

Proof Assume that T is not a-isoloid and let $\lambda \in \mathrm{iso}\,\sigma_{\mathrm{ap}}(T)$ such that $\lambda I - T$ is injective. By hypothesis there exists a finite rank operator K that is not nilpotent and commutes with T. The operator K cannot be quasi-nilpotent, otherwise its restriction $K|K(X)$ is nilpotent and hence K is also nilpotent. The spectrum of K is finite and contains 0, so $X = H_0(T) \oplus K(T)$, by Theorem 2.45. Since K is not quasi-nilpotent and the restriction $K|K(T)$ is an invertible operator of finite rank, it then follows that $K(T)$ is a non-zero subspace of X having finite dimension. Represent $T = T_1 \oplus T_2$ with respect to the decomposition $X = X_1 \oplus X_2$, where $X_1 = H_0(K)$ and X_2 is the analytic core of K. Since K is not quasi-nilpotent then $K|X_2$ is an invertible operator having finite rank, and this implies that X_2 is a non-zero subspace of finite dimension. Let $T := T_1 \oplus T_2$ with respect to $X = X_1 \oplus X_2$, and let $\mu \in \mathbb{C}$ such that $\lambda - \mu \in \sigma(T_2)$. Clearly, $\lambda - \mu$ is an eigenvalue of T_2. The operator $S := 0 \oplus \mu I$ is a finite-rank operator which commutes with T and

$$\sigma_{\mathrm{ap}}(T + S) = \sigma_{\mathrm{ap}}(T_1) \cup \sigma_{\mathrm{ap}}(\mu I + T_2).$$

Since $\lambda \in \mathrm{iso}\,\sigma_{\mathrm{ap}}(T)$ and $\lambda I - T$ is injective we then have $\lambda \notin \sigma_{\mathrm{ap}}(T_2)$ and $\lambda \in \mathrm{iso}\,\sigma_{\mathrm{ap}}(T_1) \subseteq \mathrm{iso}\,\sigma_{\mathrm{ap}}(T + S)$. Moreover,

$$\ker(\lambda I - T + S) = \ker((\lambda - \mu)I - T_2)$$

is a non-trivial subspace having finite dimension, so $\lambda \in \pi_{00}^a(T + S)$. On the other hand, $\lambda \notin p_{00}^a(T)$, so $\lambda I - T$ is not upper semi-Browder, and hence $\lambda I - T + S$ is not upper semi-Browder, by Theorem 3.8, and this implies that $\lambda \notin p_{00}^a(T + S)$. Consequently, $T + S$ does not satisfy a-Weyl's theorem, contradicting our assumptions.   ∎

Theorem 6.76 *Let $T \in L(X)$ be a finite a-isoloid operator which satisfies a-Weyl's theorem. If $R \in L(X)$ is a Riesz operator that commutes with T then $T + R$ satisfies a-Weyl's theorem.*

Proof Since T satisfies a-Browder's theorem, $T + R$ satisfies a-Browder's theorem, by Corollary 5.29, so it suffices to prove, by Theorem 6.70, the equality $p_{00}^a(T + R) = \pi_{00}^a(T + R)$. Let $\lambda \in \pi_{00}^a(T + R)$. If $\lambda I - T$ is bounded below then $\lambda I - (T + R)$ is upper semi-Browder and hence $\lambda \in p_{00}^a(T + R)$. If $\lambda \in \sigma_{\mathrm{ap}}(T)$ then, by Theorem 6.73, $\lambda \in \mathrm{iso}\,\sigma_{\mathrm{ap}}(T)$, and since T is finite a-isoloid it then follows that $\lambda \in \pi_{00}^a(T)$. But a-Weyl's theorem for T entails that $\pi_{00}^a(T) \cap \sigma_{\mathrm{ub}}(T) = \emptyset$, therefore $\lambda I - T$ is upper semi-Browder and hence $\lambda \in p_{00}^a(T + R)$. The other inclusion is trivial, so $T + R$ satisfies a-Weyl's theorem.   ∎

The proof of the following theorem is omitted, since it is very similar to the proof of Theorem 6.49, just take into account Theorem 6.70 and that $\sigma_{\mathrm{ub}}(T)$ is invariant under Riesz commuting perturbations.

Theorem 6.77 *Let $T \in L(X)$ be an operator which satisfies a-Weyl's theorem and let $R \in L(X)$ be a Riesz operator that commutes with T. Then the following conditions are equivalent:*

(i) $T + R$ *satisfies a-Weyl's theorem;*
(ii) $\pi_{00}^a(T + R) = p_{00}^a(T + R)$;
(iii) $\pi_{00}^a(T + R) \cap \sigma_{\mathrm{ap}}(T) \subseteq \pi_{00}^a(T)$.

An obvious consequence of that for a finite a-isoloid operator $T \in L(X)$ a-Weyl's theorem from T is transmitted to $T + Q$, where $Q \in L(X)$ is a quasi-nilpotent operator which commutes with T. We now show that this result remains true if we suppose only the following weaker condition

$$\sigma_p(T) \cap \mathrm{iso}\,\sigma_{\mathrm{ap}}(T) \subseteq \pi_{00}^a(T).$$

Theorem 6.78 *Let $T \in L(X)$ be such that $\sigma_p(T) \cap \mathrm{iso}\,\sigma_{\mathrm{ap}}(T) \subseteq \pi_{00}(T)$. If T satisfies a-Weyl's theorem and $Q \in L(X)$ is a quasi-nilpotent operator which commutes with T, then $T + Q$ satisfies a-Weyl's theorem.*

Proof We already know that $\sigma_{\mathrm{ap}}(T + Q) = \sigma_{\mathrm{ap}}(T)$ and $\sigma_b(T) = \sigma_b(T + Q)$ so $p_{00}^a(T) = p_{00}^a(T + Q)$. Since $p_{00}^a(T) = \pi_{00}^a(T)$, it then suffices to prove the equality $\pi_{00}^a(T) = \pi_{00}^a(T + Q)$. Let $\lambda \in \pi_{00}^a(T) = p00^a(T)$. Then $\lambda I - T$ is upper semi-Browder and hence $\lambda \in \mathrm{iso}\,\sigma_{\mathrm{ap}}(T) = \mathrm{iso}\,\sigma_{\mathrm{ap}}(T + Q)$. Theorem 3.8 then entails that $\lambda I - (T + Q)$ is upper semi-Browder, and hence $\lambda \in \pi_{00}^a(T + Q)$.

Conversely, suppose that $\lambda \in \pi_{00}^a(T + Q)$. Since Q is quasi-nilpotent and commutes with T, the restriction of $\lambda I - T$ to the finite-dimensional subspace $\ker(\lambda I - (T + Q))$ is not invertible, hence $\ker(\lambda I - T)$ is non-trivial. Therefore, $\lambda \in \sigma_p(T) \cap \mathrm{iso}\,\sigma_{\mathrm{ap}}(T) \subseteq \pi_{00}^a(T)$, which completes the proof. ∎

In the sequel we need the following lemma.

Lemma 6.79 *For every $T \in L(X)$, X a Banach space, and $f \in \mathcal{H}(\sigma(T))$ we have*

$$\sigma_{\mathrm{ap}}(f(T)) \setminus \pi_{00}^a(f(T)) \subseteq f\left(\sigma_{\mathrm{ap}}(T) \setminus \pi_{00}^a(T)\right). \tag{6.18}$$

Furthermore, if T is a-isoloid then

$$\sigma_{\mathrm{ap}}(f(T)) \setminus \pi_{00}^a(f(T)) = f\left(\sigma_{\mathrm{ap}}(T) \setminus \pi_{00}^a(T)\right). \tag{6.19}$$

Proof Suppose that $\lambda \in \sigma_{\mathrm{ap}}(f(T)) \setminus \pi_{00}^a(f(T)) \subseteq f(\sigma_{\mathrm{ap}}(T))$. We consider three cases.

Case (I) Suppose that $\lambda \notin \mathrm{iso}\,f(\sigma_{\mathrm{ap}}(T))$. Then there is a sequence (μ_n) which converges to a scalar $\mu_0 \in \sigma_{\mathrm{ap}}(T)$ such that $f(\mu_n) \to \lambda$. Hence, $\lambda = f(\mu_0) \in f\left(\sigma_{\mathrm{ap}}(T) \setminus \pi_{00}^a(T)\right)$.

Case (II) Suppose that $\lambda \in \mathrm{iso}\,f(\sigma_{\mathrm{ap}}(T))$ and λ is not an eigenvalue of $f(T)$. Write

$$\lambda I - f(T) = (\mu_1 I - T) \cdots (\mu_n I - T)g(T), \tag{6.20}$$

where $\mu_1 \in \sigma_{\mathrm{ap}}(T)$, $\mu_2, \ldots, \mu_n \in \sigma(T)$, and $g(T)$ is invertible. Obviously, the operators on the right-hand side of (6.20) mutually commute. Since λ is not an eigenvalue of $f(T)$, none of the scalars $\mu_1, \ldots, \mu_n$ can be an eigenvalue of T. Hence $\lambda = f(\mu_1) \in f\left(\sigma_{\mathrm{ap}}(T) \setminus \pi_{00}^a(T)\right)$.

Case (III) Suppose that λ is an eigenvalue of $f(T)$ having infinite multiplicity. According to the equality (6.20) there exists some μ_i such that μ_i is an eigenvalue of T having infinite multiplicity. Then $\mu_i \in \sigma_{\mathrm{ap}}(T) \setminus \pi_{00}^a(T)$ and $\lambda \in f\left(\sigma_{\mathrm{ap}}(T) \setminus \pi_{00}^a(T)\right)$.

To show the equality (6.19), let

$$\lambda \in f\left(\sigma_{\mathrm{ap}}(T) \setminus \pi_{00}^a(T)\right) \subseteq f(\sigma_{\mathrm{ap}}(T)) = \sigma_{\mathrm{ap}}(f(T)).$$

Suppose that $\lambda \in \pi_{00}^a(f(T))$. Then $\lambda \in \mathrm{iso}\,\sigma_{\mathrm{ap}}(f(T))$ and, in the equality (6.20), we see that if some μ_i belongs to $\sigma_{\mathrm{ap}}(T)$, then μ_i is isolated in $\sigma_{\mathrm{ap}}(T)$, and hence an eigenvalue of T. Since λ is an eigenvalue of finite multiplicity, then all $\mu_i \in \sigma_a(T)$ are eigenvalues of T of finite multiplicity. Therefore all $\mu_i \in \sigma_{\mathrm{ap}}(T)$ are in $\pi_{00}^a(T)$ and this contradicts our assumption that $\lambda \in f\left(\sigma_{\mathrm{ap}}(T) \setminus \pi_{00}^a(T)\right)$. ∎

Theorem 6.80 *Let $T \in L(X)$ be a-isoloid and suppose that either T or T^* has the SVEP. If T satisfies a-Weyl's theorem then $f(T)$ satisfies a-Weyl's theorem $f \in \mathcal{H}(\sigma(T))$.*

Proof If T or T^* has the SVEP then the spectral mapping theorem holds for $\sigma_{\mathrm{uw}}(T)$, by Corollary 3.120. Since T satisfies a-Weyl's theorem we have $\sigma_{\mathrm{uw}}(T) = \sigma_{\mathrm{ap}}(T) \setminus \pi_{00}^a(T)$. By Lemma 6.79 we then have

$$\sigma_{\mathrm{uw}}(f(T)) = f(\sigma_{\mathrm{uw}}(T)) = f(\sigma_{\mathrm{ap}}(T) \setminus \pi_{00}^a(T)) = \sigma_{\mathrm{ap}}(f(T)) \setminus \pi_{00}^a(f(T)),$$

thus a-Weyl's theorem holds for $f(T)$. ∎

a-Weyl's theorem also admits a generalization in the sense of B-Fredholm theory. Define first

$$E^a(T) := \{\lambda \in \mathrm{iso}\,\sigma_a(T) : 0 < \alpha(\lambda I - T)\},$$

and recall that by $\Delta_a^g(T)$ we denote the set $\sigma_{\mathrm{ap}}(T) \setminus \sigma_{\mathrm{ubw}}(T)$.

Definition 6.81 An operator $T \in L(X)$ is said to satisfy the *generalized a-Weyl's theorem* if $\Delta_a^g(T) = E_a(T)$.

The generalized a-Weyl's theorem entails a-Weyl's theorem and the generalized Weyl's theorem:

Theorem 6.82 *Suppose that $T \in L(X)$ satisfies the generalized a-Weyl's theorem. Then we have:*

(i) *T satisfies a-Weyl's theorem.*
(ii) *T satisfies the generalized Weyl's theorem.*

Proof

(i) Because T satisfies the generalized a-Weyl's theorem we have $\sigma_{\mathrm{ap}}(T) \setminus \sigma_{\mathrm{ubw}}(T) = E_a(T)$. Let $\lambda \in \sigma_{\mathrm{ap}}(T) \setminus \sigma_{\mathrm{bw}}(T)$. Then $\lambda I - T$ is not upper semi B-Weyl and hence $\lambda \in \sigma_{\mathrm{ap}}(T) \setminus \sigma_{\mathrm{ubw}}(T) = E_a(T)$. Since $\alpha(\lambda I - T) < \infty$, we have $\lambda \in \pi_{00}^a(T)$. Conversely, suppose that $\lambda \in \pi_{00}^a(T)$. Obviously,

$\lambda \in E_a(T) = \sigma_{ap}(T) \setminus \sigma_{ubw}(T)$, so $\lambda I - T$ is upper semi B-Weyl. Since $\alpha(\lambda I - T) < \infty$, $\lambda I - T$ is upper semi-Weyl, by Theorem 1.114, hence $\lambda \in \sigma_{ap}(T) \setminus \sigma_{uw}(T)$. Therefore, $\sigma_{ap}(T) \setminus \sigma_{uw}(T) = \pi_{00}^a(T)$, and T satisfies a-Weyl's theorem.

(ii) Suppose that $\lambda \in \sigma(T) \setminus \sigma_{bw}(T)$. Then $\lambda I - T$ is B-Weyl, in particular upper semi B-Weyl, and $\lambda \in \sigma(T)$. Observe that $\lambda \in \sigma_{ap}(T)$, otherwise if $\lambda I - T$ were bounded below we would have, by Theorem 1.143, $p(\lambda I - T) = q(\lambda I - T) = 0$, hence $\lambda \notin \sigma(T)$, a contradiction. Therefore, $\lambda \in \sigma_{ap}(T) \setminus \sigma_{ubw}(T) = E_a(T)$ and consequently $\mu I - T$ is injective in a deleted neighborhood U_1 of $\lambda \in \sigma_{ap}(T)$. Since $\lambda I - T$ is B-Weyl, by Theorem 1.117 then $\mu I - T$ is Weyl in a deleted neighborhood U_2 of λ. If $U =:= U_1 \cap U_2$ then $\mu I - T$ is invertible for all $\mu \in U$, so $\lambda \in \mathrm{iso}\,\sigma(T)$. Consequently, $\lambda \in E(T)$ and hence $\sigma(T) \setminus \sigma_{bw}(T) \subseteq E(T)$.

Conversely, if $\lambda \in E(T)$ then λ is an isolated point of $\sigma_{ap}(T)$, since the boundary of the spectrum belongs to the approximate-point spectrum. Hence, $\lambda \in E_a(T) = \sigma_{ap}(T) \setminus \sigma_{ubw}(T)$ and $\lambda I - T$ is upper semi B-Weyl. Because $\lambda \in \mathrm{iso}\,\sigma(T)$, both T and T^* have the SVEP at λ, so, by Theorems 2.97 and 2.98, $\lambda I - T$ is Drazin invertible. By Theorem 1.141 it then follows that $\lambda \notin \sigma_{bw}(T)$, so $\sigma(T) \setminus \sigma_{bw}(T) = E(T)$, as desired. ∎

We show now that in general, a-Weyl's theorem does not imply the generalized Weyl's theorem, nor does it imply the generalized a-Weyl's theorem.

Example 6.83 Let $S \in L(X)$ as in Example 6.61. Then $\pi_{00}^a(S) = \sigma_{ap}(S) \setminus \sigma_{uw}(S) = \emptyset$, so S satisfies a-Weyl's theorem, but does not satisfy the generalized Weyl's theorem. Furthermore, by Theorem 6.82, S does not satisfy the generalized a-Weyl's theorem.

Recall that $\Delta_a^g(T) = \sigma_{ap}(T) \setminus \sigma_{ubw}(T)$. Define

$$\Delta_2^g(T) := \Delta_a^g(T) \cup E_a(T).$$

By Theorem 6.82, the generalized a-Weyl's theorem entails a-Weyl's theorem and hence a-Browder's theorem, or equivalently, the generalized a-Browder's theorem. The exact relationship between the generalized a-Weyl's theorem and the generalized a-Browder's theorem is described in the following theorem.

Theorem 6.84 *For a bounded operator $T \in L(X)$ the following statements are equivalent:*

(i) *T satisfies the generalized a-Weyl's theorem;*
(ii) *T satisfies the generalized a-Browder's theorem and $E_a(T) = \Pi_a(T)$;*
(iii) *For every $\lambda \in \Delta_2^g(T)$ there exists a $p := p(\lambda) \in \mathbb{N}$ such that $H_0(\lambda I - T) = \ker(\lambda I - T)^p$ and $(\lambda I - T)^n(X)$ is closed for all $n \geq p$.*

Proof (i) $\Rightarrow$ (ii) We have only to show the equality $E_a(T) = \Pi_a(T)$. Suppose that $\Delta_a^g(T) = E_a(T)$. To prove the equality $E_a(T) = \Pi_a(T)$ it suffices to prove the inclusion $E_a(T) \subseteq \Pi_a(T)$.

Let $\lambda \in E_a(T)$. Clearly, T has the SVEP at λ, because $\lambda \in \operatorname{iso}\sigma_{ap}(T)$. Since $\sigma_{ap}(T) \setminus \sigma_{ubw}(T) = E_a(T)$, $\lambda I - T$ is upper semi B-Weyl and hence, by Theorem 2.97, $\lambda I - T$ is left Drazin invertible, i.e., $\lambda \in \Pi_a(T)$. Therefore, $E_a(T) = \Pi_a(T)$.

(ii) $\Rightarrow$ (i) The generalized a-Browder's theorem entails that $\Delta_a^g(T) = \Pi_a(T) = E_a(T)$, so T satisfies the generalized a-Weyl's theorem.

(ii) $\Rightarrow$ (iii) If T satisfies the generalized a-Browder's theorem then $\Delta_a^g(T) = \Pi_a(T)$ and from the assumption $E_a(T) = \Pi_a(T)$ we have $\Delta_a^g(T) = \Delta_a(T) = \Pi_a(T)$. Let $\lambda \in \Delta_2^g(T)$. From Theorem 5.42 there exists an $m \in \mathbb{N}$ such that $H_0(\lambda I - T) = \ker(\lambda I - T)^m$. Clearly, $p := p(\lambda I - T)$ is finite, and $H_0(\lambda I - T) = \ker(\lambda I - T)^p$. Since $\lambda \in \Pi_a(T)$ then $\lambda I - T$ is left Drazin invertible, thus $(\lambda I - T)^{p+1}(X)$ is closed and hence, by Corollary 1.101, $(\lambda I - T)^n(X)$ is closed for all $n \geq p$.

(iii) $\Rightarrow$ (ii) Since $\Delta_a^g(T) \subseteq \Delta_2^g(T)$, from Theorem 5.42 we know that T satisfies the generalized a-Browder's theorem. To show that $E_a(T) = \Pi_a(T)$ it suffices to prove that $E_a(T) \subseteq \Pi_a(T)$. Suppose that $\lambda \in E_a(T)$. Then there exists a $\nu \in \mathbb{N}$ such that $H_0(\lambda I - T) = \ker(\lambda I - T)^\nu$ and this implies that $\lambda I - T$ has ascent $p = p(\lambda I - T) \leq \nu$. Thus, from our assumption, we obtain that $\lambda I - T)^{p+1}(X)$ is closed and hence $\lambda \in \Pi_a(T)$. $\blacksquare$

Corollary 6.85 *Suppose that T is left polaroid and either T or T^* has the SVEP. Then T satisfies the generalized a-Weyl's theorem.*

Proof T satisfies the generalized a-Browder's theorem and the left polaroid condition entails that $E_a(T) = \Pi_a(T)$. $\blacksquare$

Theorem 6.86 *If T^* has the SVEP then the generalized a-Weyl's theorem holds for T if and only if the generalized Weyl's theorem holds for T.*

Proof Suppose that T satisfies the generalized Weyl's theorem. Since T^* has the SVEP, by Theorem 2.68 $\sigma(T) = \sigma_{ap}(T)$, and hence $E(T) = E_a(T)$. By Theorem 3.44 we also have $\sigma_{ubw}(T) = \sigma_{bw}(T)$, so that

$$E(T) = \sigma(T) \setminus \sigma_{bw}(T) = \sigma_{ap}(T) \setminus \sigma_{ubw}(T) = E^a(T),$$

and hence the generalized a-Weyl's theorem holds for T. $\blacksquare$

6.5 Property (ω)

The following variant of Weyl's theorem was introduced by Rakočević in a short note [261] and studied extensively by Aiena and Peña in [22]. Evidently, for any operator $T \in L(X)$ we have $\Delta(T) \subseteq \{\lambda \in \sigma(T) : 0 < \alpha(\lambda I - T) < \infty\}$.

Definition 6.87 A bounded operator $T \in L(X)$ is said to satisfy property (ω) if

$$\Delta_a(T) = \pi_{00}(T) = \{\lambda \in \text{iso } \sigma(T) : 0 < \alpha(\lambda I - T) < \infty\}.$$

Property (ω) entails a-Browder's theorem:

Theorem 6.88 *Suppose that $T \in L(X)$ satisfies property (ω). Then a-Browder's holds for T and*

$$\sigma_{\text{ap}}(T) = \sigma_{\text{uw}}(T) \cup \text{iso } \sigma_{\text{ap}}(T).$$

Proof By Theorem 5.27 it suffices to show that T has the SVEP at every $\lambda \notin \sigma_{\text{uw}}(T)$. Let $\lambda \notin \sigma_{\text{uw}}(T)$. If $\lambda \notin \sigma_{\text{ap}}(T)$ then T has the SVEP at λ, while if $\lambda \in \sigma_{\text{ap}}(T)$ then $\lambda \in \Delta_a(T) = \pi_{00}(T)$ and hence λ is an isolated point of $\sigma(T)$, so also in this case T has the SVEP at λ.

The inclusion $\sigma_{\text{uw}}(T) \cup \text{iso } \sigma_{\text{ap}}(T) \subseteq \sigma_{\text{ap}}(T)$ holds for every operator, since $\sigma_{\text{uw}}(T) \subseteq \sigma_{\text{ap}}(T)$. To show the opposite implication, suppose that $\lambda \in \sigma_{\text{ap}}(T)$. If $\lambda \notin \sigma_{\text{uw}}(T)$ then $\lambda \in \sigma_{\text{ap}}(T) \setminus \sigma_{\text{uw}}(T) = \Delta_a(T) = \pi_{00}(T)$ and hence $\lambda \in \text{iso } \sigma(T)$, in particular $\lambda \in \text{iso } \sigma_{\text{ap}}(T)$, since every isolated point of the spectrum belongs to the approximate point spectrum. Therefore, $\sigma_{\text{ap}}(T) \subseteq \sigma_{\text{uw}}(T) \cup \text{iso } \sigma_{\text{ap}}(T)$, as desired. ∎

The next result shows that, roughly speaking, property (R) may be thought of as half of the property (ω):

Theorem 6.89 *If $T \in L(X)$ the following statements are equivalent:*

(i) *T satisfies property (ω);*
(ii) *a-Browder's theorem holds for T and T has property (R).*

Proof (i) $\Rightarrow$ (ii) Suppose that T has property (ω). By Theorem 6.88 we need only to show that property (R) holds for T, i.e. $p_{00}^a(T) = \pi_{00}(T)$. If $\lambda \in \pi_{00}(T) = \sigma_{\text{ap}}(T) \setminus \sigma_{\text{uw}}(T)$ then $\lambda \in \text{iso } \sigma(T)$ and $\lambda I - T$ is upper semi-Weyl. The SVEP for T at λ is equivalent to saying that $p(\lambda I - T) < \infty$, by Theorem 2.97. Therefore $\lambda I - T$ is upper semi-Browder and hence $\lambda \in p_{00}^a(T)$. This shows the inclusion $\pi_{00}(T) \subseteq p_{00}^a(T)$.

To show the opposite inclusion, suppose that $\lambda \in p_{00}^a(T)$. Since T satisfies a-Browder's theorem, by Theorem 6.88, then $\sigma_{\text{ub}}(T) = \sigma_{\text{uw}}(T)$, and hence $\lambda \in \sigma_{\text{ap}}(T) \setminus \sigma_{\text{uw}}(T) = \pi_{00}(T)$, so the equality $p_{00}^a(T) = \pi_{00}(T)$ is proved.

(ii) $\Rightarrow$ (i) If $\lambda \in \Delta_a(T) = \sigma_{\text{ap}}(T) \setminus \sigma_{\text{uw}}(T)$, a-Browder's theorem entails that $\lambda \in \sigma_{\text{ap}}(T) \setminus \sigma_{\text{ub}}(T) = p_{00}^a(T)$. Since T has property (R), it then follows that $\lambda \in \pi_{00}(T)$, so $\Delta_a(T) \subseteq \pi_{00}(T)$. On the other hand, if $\lambda \in \pi_{00}(T)$ then property (R) entails that $\lambda \in p_{00}^a(T) = \sigma_{\text{ap}}(T) \setminus \sigma_{\text{ub}}(T) = \sigma_{\text{ap}}(T) \setminus \sigma_{\text{uw}}(T)$. Therefore, $\Delta_a(T) = \pi_{00}(T)$. ∎

It is not surprising that without the SVEP the equivalence observed in the previous theorem fails. Indeed, the next example shows that property (R) is weaker than property (ω).

Example 6.90 Let $R \in L(\ell^2(\mathbb{N}))$ denote the classical unilateral right shift, let Q denote a quasi-nilpotent operator. Define $T := R \oplus R' \oplus Q$, R' the Hilbert adjoint of R. It is well-known that R' is a unilateral left shift. Clearly, $\sigma_{ap}(T) = \sigma_{ub}(T) = \mathbf{D}(0,1)$, where $\mathbf{D}(0,1)$ denotes the closed unit disc. Since $\pi_{00}(T) = \emptyset$, T satisfies property (R), while T does not satisfy property (ω), since $\sigma_{uw}(T) = \Gamma \cup \{0\}$, where Γ denotes the unit circle of $\mathbb{C}$, so $\sigma_{ap}(T) \setminus \sigma_{uw}(T) \neq \emptyset = \pi_{00}(T)$.

Property (ω) entails Weyl's theorem:

Theorem 6.91 *If $T \in L(X)$ satisfies property (ω) then Weyl's theorem holds for T.*

Proof T satisfies a-Browder's theorem by Theorem 6.88, and hence Browder's theorem. To show Weyl's theorem holds for T it suffices to prove that $\pi_{00}(T) = p_{00}(T)$. If $\lambda \in \pi_{00}(T)$ then $\lambda \in \sigma_{ap}(T)$, since $\alpha(\lambda I - T) > 0$, and from $\lambda \in \mathrm{iso}\,\sigma(T)$ we know that both T and T^* have the SVEP at λ. Since $\pi_{00}(T) = \sigma_{ap}(T) \setminus \sigma_{uw}(T)$, $\lambda I - T$ is upper semi-Weyl and hence, by Theorems 2.97 and 2.98, we have $p(\lambda I - T) = q(\lambda I - T) < \infty$. Since $\alpha(\lambda I - T) < \infty$, from Theorem 1.22 we then obtain that $\lambda I - T$ is Browder, i.e. $\lambda \in p_{00}(T)$. Hence, $\pi_{00}(T) \subseteq p_{00}(T)$ and since the reverse inclusion holds for every operator we conclude that $\pi_{00}(T) = p_{00}(T)$. ∎

The following examples show that property (ω) is not intermediate between Weyl's theorem and a-Weyl's theorem. The first example provides an operator satisfying property (w) but not a-Weyl's theorem.

Example 6.92 Let $T \in L(\ell_2(\mathbb{N}))$ denote the unilateral right shift and $Q \in L(\ell_2(\mathbb{N}))$ the quasi-nilpotent operator defined as

$$Q(x_1, x_2, \ldots) := \left(\frac{x_2}{2}, \frac{x_3}{3}, \ldots \right) \quad \text{for all } (x_k) \in \ell_2(\mathbb{N}).$$

Consider on $X := \ell_2(\mathbb{N}) \oplus \ell_2(\mathbb{N})$ the operator $S := T \oplus Q$. Then $\sigma(S) = \sigma_w(S) = \mathbf{D}(0,1)$, while $\sigma_{ap}(S) = \sigma_{uw}(S) = \Gamma \cup \{0\}$, Γ the unit circle. Moreover, $\pi_{00}^a(S) = \{0\}$ and since $\pi_{00}(S) = \emptyset$ we then see that $\sigma_{ap}(S) \setminus \sigma_{uw}(S) = \pi_{00}(S)$, so S satisfies property (ω). On the other hand, $\sigma_{ap}(S) \setminus \sigma_{uw}(S) = \emptyset \neq \pi_{00}^a(S) = \{0\}$, so S does not satisfy a-Weyl's theorem.

The following example provides an operator that satisfies a-Weyl theorem but not property (ω).

Example 6.93 Let $R \in \ell^2(\mathbb{N})$ be the unilateral right shift and

$$U(x_1, x_2, \ldots) := (0, x_2, x_3, \cdots) \quad \text{for all } (x_n) \in \ell^2(\mathbb{N}).$$

If $T := R \oplus U$ then $\sigma(T) = \mathbf{D}(0,1)$ so $\mathrm{iso}\,\sigma(T) = \pi_{00}(T) = \emptyset$. Moreover, $\sigma_{ap}(T) = \Gamma \cup \{0\}$ and $\sigma_{uw}(T) = \Gamma$, from which we see that T does not satisfy property (ω), since $\Delta_a(T) = \{0\}$. On the other hand, we also have $\pi_{00}^a(T) = \{0\}$, so T satisfies a-Weyl's theorem.

Define

$$\Lambda(T) := \{\lambda \in \Delta_a(T) : \text{ind}\,(\lambda I - T) < 0\}.$$

Clearly,

$$\Delta_a(T) = \Delta(T) \cup \Lambda(T) \quad \text{and} \quad \Lambda(T) \cap \Delta(T) = \emptyset. \tag{6.21}$$

The next result relates Weyl's theorem and property (w).

Theorem 6.94 *If $T \in L(X)$ satisfies property (w) then $\Lambda(T) = \emptyset$. Furthermore, the following statements are equivalent:*

(i) *T satisfies property (ω);*
(ii) *T satisfies Weyl's theorem and $\Lambda(T) = \emptyset$;*
(iii) *T satisfies Weyl's theorem and $\Delta_a(T) \subseteq \text{iso}\,\sigma(T)$;*
(iv) *T satisfies Weyl's theorem and $\Delta_a(T) \subseteq \partial\sigma(T)$, $\partial\sigma(T)$ the topological boundary of $\sigma(T)$.*

Proof Suppose that T satisfies property (ω) and that $\Lambda(T)$ is non-empty. Let $\lambda \in \Lambda(T)$. Then $\lambda \in \Delta_a(T) = \pi_{00}(T)$, so $\lambda \in \text{iso}\,\sigma(T)$ and hence T^* has the SVEP at λ. Since $\lambda \notin \sigma_{\text{uw}}(T)$, $\lambda I - T$ is upper semi-Weyl, so ind $(\lambda I - T) \leq 0$. By Corollary 2.106 the SVEP for T^* implies that ind $(\lambda I - T) \geq 0$, hence ind $(\lambda I - T) = 0$, and this contradicts our assumption that $\lambda \in \Lambda(T)$.

To show the equivalence (i) $\Leftrightarrow$ (ii) observe first that the implication (i) $\Rightarrow$ (ii) is clear from the first part of the proof and from Theorem 6.91. Conversely, if $\Lambda(T) = \emptyset$ and T satisfies Weyl's theorem then $\Delta_a(T) = \Delta(T) = \pi_{00}(T)$, so property (ω) holds.

(iii) $\Rightarrow$ (ii) Suppose that T satisfies Weyl's theorem. If $\Delta_a(T) \subseteq \text{iso}\,\sigma(T)$ then both T and T^* have the SVEP at λ and, as above, this implies that ind $(\lambda I - T) = 0$ for every $\lambda \in \Delta_a(T)$, so $\Lambda(T) = \emptyset$. Hence property (ω) holds for T.

(i) $\Rightarrow$ (iii) If property (ω) holds then $\Delta_a(T) = \pi_{00}(T) \subseteq \text{iso}\,\sigma(T)$.

(iii) $\Rightarrow$ (iv) Obvious.

(iv) $\Rightarrow$ (ii) Both T and T^* have the SVEP at the points $\lambda \in \partial\sigma(T) = \partial\sigma(T^*)$. If $\lambda \in \Delta_a(T)$ then, by Corollary 2.106, ind $(\lambda I - T) = 0$, hence $\Lambda(T) = \emptyset$. ∎

We give now two sufficient conditions under which a-Weyl's theorem for T (respectively, T^*) implies property (ω) for T (respectively, T^*). Observe that these conditions, see Theorem 5.48, are a bit stronger than the assumption that T satisfies a-Browder's theorem.

Theorem 6.95 *If $T \in L(X)$ then the following statements hold:*

(i) *If T has property (b) and T satisfies a-Weyl's theorem then property (ω) holds for T.*
(ii) *If T^* has property (b) and T^* satisfies a-Weyl's theorem then property (ω) holds for T^*.*

Proof

(i) If T has property (b) then $\Delta_a(T) \subseteq \mathrm{iso}\,\sigma(T)$, by Theorem 5.45. Consequently,
$\Delta_a(T) \subseteq \sigma_{\mathrm{ap}}(T)$. Let $\lambda \in \Delta_a(T)$. Then $\lambda I - T$ is upper semi-Weyl, so $\alpha(\lambda I - T) < \infty$ and $\lambda I - T$ has closed range. Since $\lambda \in \sigma_{\mathrm{ap}}(T)$ it then follows that
$\alpha(\lambda I - T) > 0$, so $\lambda \in \pi_{00}(T)$. To show the inclusion $\pi_{00}(T) \subseteq \Delta_a(T)$ it
suffices to observe that $\pi_{00}(T) \subseteq \pi_{00}^a(T) = \Delta_a(T)$, since T satisfies a-Weyl's
theorem. Hence, $\pi_{00}(T) = \Delta_a(T)$.
(ii) If T^* has property (b) then

$$\Delta_s(T) = \sigma_s(T) \setminus \sigma_{\mathrm{lw}}(T) \subseteq \mathrm{iso}\,\sigma(T) = \mathrm{iso}\,\sigma(T^*),$$

again by Theorem 5.45, from which we easily obtain that $\Delta_s(T) \subseteq \sigma_s(T)$. Let
$\lambda \in \Delta_s(T) = \Delta_a(T^*)$. Then $\beta(\lambda I - T) = \alpha(\lambda I - T^*) > 0$ and since $\lambda I - T$
is lower semi-Weyl we also have $\beta(\lambda I - T) = \alpha(\lambda I - T^*) < \infty$. Therefore,
$\lambda \in \pi_{00}(T^*)$. To show the reverse inclusion $\pi_{00}(T^*) \subseteq \Delta_a(T^*)$, observe that
since T^* satisfies a-Weyl's theorem then $\pi_{00}(T^*) \subseteq \pi_{00}^a(T^*) = \Delta_a(T^*)$, so
$\Delta_a(T^*) = \pi_{00}^a(T^*)$, as desired.   ∎

Theorem 6.96 *If $T \in L(X)$, then the following equivalences holds:*

(i) *If T^* has the SVEP, property (ω) holds for T if and only if Weyl's theorem holds
for T, and this is the case if and only if a-Weyl's theorem holds for T.*
(ii) *If T has the SVEP, property (ω) holds for T^* if and only if Weyl's theorem holds
for T^*, and this is the case if and only if a-Weyl's theorem holds for T^*.*

Proof

(i) We know that if T^* has the SVEP then property (b) holds for T, by Corollary 5.46. From part (i) of Theorems 6.95 and 6.91, the following implications
hold for T:

$$\text{a-Weyl} \Rightarrow (\omega) \Rightarrow \text{Weyl.} \tag{6.22}$$

Assume now that T satisfies Weyl's theorem. The SVEP for T^* implies, by
Theorem 2.68, that $\sigma(T) = \sigma_{\mathrm{ap}}(T)$, so $\pi_{00}^a(T) = \pi_{00}(T) = \sigma(T) \setminus \sigma_{\mathrm{w}}(T)$. By
Theorem 3.44 we also have that $\sigma_{\mathrm{w}}(T) = \sigma_{\mathrm{ub}}(T)$, from which we obtain that

$$\pi_{00}^a(T) = \sigma_{\mathrm{ap}}(T) \setminus \sigma_{\mathrm{ub}}(T) = p_{00}^a(T).$$

Since the SVEP for T^* entails a-Browder's theorem for T, by Theorem 5.27,
then a-Weyl's theorem holds for T, by Theorem 6.70.
(ii) The argument is similar to that used in the proof of part (i). If T has the SVEP,
the implications (6.22) hold for T^*, again by Theorems 6.95 and 6.91. If T has
the SVEP then $\sigma(T^*) = \sigma_{\mathrm{ap}}(T^*)$, by Theorem 2.68, and hence $\pi_{00}^a(T^*) = \pi_{00}(T^*)$. By Theorem 3.44 we also have that $\sigma_{\mathrm{w}}(T^*) = \sigma_{\mathrm{w}}(T) = \sigma_{\mathrm{lb}}(T) = \sigma_{\mathrm{ub}}(T^*)$, from which it easily follows that $\pi_{00}^a(T^*) = p_{00}^a(T^*)$. The SVEP

for T also implies that T^* satisfies a-Browder's theorem, by Theorem 5.27, so a-Weyl's theorem holds for T^*, by Theorem 6.70. ∎

Theorem 6.97 *Suppose that $T \in L(H)$, H a Hilbert space. Then property (ω) holds for T^* if and only if property (ω) holds for the Hilbert adjoint T'.*

Proof From Theorem 3.1 we have $\sigma_{uw}(T') = \overline{\sigma_{uw}(T^*)}$, so

$$\sigma_{ap}(T') \setminus \sigma_{uw}(T') = \overline{\sigma_{ap}(T^*)} \setminus \overline{\sigma_{uw}(T^*)} = \overline{\sigma_{ap}(T^*) \setminus \sigma_{uw}(T^*)}$$

$$= \overline{\pi_{00}(T^*)} = \pi_{00}(T'),$$

so T' satisfies property (ω). The opposite implication follows in a similar way. ∎

Remark 6.98 The operator T considered in Example 6.92 shows that in the statement (i) of Theorem 6.96 the SVEP for T^* cannot be replaced by the SVEP for T. Indeed, $S = T \oplus Q$ has the SVEP, since every right shift operator has the SVEP and Q has the SVEP, since it is quasi-nilpotent. But S satisfies property (w) while a-Weyl's theorem does not hold for S.

Analogously, in the statements (ii) of Theorem 6.96 the assumption that T has the SVEP cannot be replaced by the assumption that T^* has the SVEP. Indeed, let us consider the left shift $L \in L(\ell^2(\mathbb{N}))$, and let U' be the adjoint of the quasi-nilpotent operator U defined in Example 6.93. We have $L' = R$, R the unilateral right shift. If we define $S := L \oplus U'$ then, as observed in Example 6.93, the Hilbert adjoint $S' = T = R \oplus U$ has the SVEP (and hence the dual S^* has the SVEP). We also know that $\sigma(S) = \overline{\sigma(S')} = \mathbf{D}(0, 1)$, and $S' = T$ does not have property (ω), or equivalently, S^* does not have property (ω).

Theorem 6.99 *Suppose that T is a-polaroid. Then a-Weyl's theorem holds for T if and only if T satisfies property (ω).*

Proof If T is a-polaroid then $\pi_{00}^a(T) = p_{00}(T)$. Indeed, if $\lambda \in \pi_{00}^a(T)$ then $\lambda I - T$ has both ascent and descent finite. Since $\alpha(\lambda I - T) < \infty$ then, by Theorem 1.22, $\lambda I - T$ is Browder, and hence $\lambda \in p_{00}(T)$. The reverse inclusion is obvious, so $\pi_{00}^a(T) = p_{00}(T)$. Now, if T satisfies a-Weyl's theorem then $\Delta_a(T) = \pi_{00}^a(T) = p_{00}(T)$. Moreover, since a-Weyl's theorem entails Weyl's theorem, by Theorem 6.40 we then have $p_{00}(T) = \pi_{00}(T)$, thus $\Delta_a(T) = \pi_{00}(T)$, and hence T has property (ω).

Conversely, if T has property (ω) then $\Delta_a(T) = \pi_{00}(T)$. Property ($\omega$) entails that T satisfies Weyl's theorem. By Theorem 6.40 then $p_{00}(T) = \pi_{00}(T) = \pi_{00}^a(T)$, thus T satisfies a-Weyl's theorem. ∎

The operator defined in Example 6.30 shows that a similar result to that of Theorem 6.99 does not hold for polaroid operators, i.e. if $T \in L(X)$ is polaroid Weyl's theorem for T and property (ω) for T in general are not equivalent.

Theorem 6.100 *Suppose that $T \in L(X)$ is polaroid. Then the following statements hold:*

(i) *If T^* has the SVEP then $f(T)$ has property (ω) for all $f \in \mathcal{H}_{nc}(\sigma(T))$.*
(ii) *If T has the SVEP then $f(T^*)$ has property (ω) for all $f \in \mathcal{H}_{nc}(\sigma(T))$.*

Proof

(i) If T is polaroid then $f(T)$ is polaroid, by Theorem 4.19 and $f(T^*) = (f(T))^*$ has the SVEP, by Theorem 2.86. From Theorem 4.15 then $f(T)$ is a-polaroid. Since, by Theorem 6.43, $f(T)$ satisfies Weyl's theorem it then follows, by Theorem 6.96, that $f(T)$ has property (ω).

(ii) The proof is analogous: $f(T)$ is polaroid and hence if $f(T^*$ is polaroid. The SVEP for T is transmitted to $f(T)$, so $f(T^*)$ is a-polaroid. By Theorem 6.43 then $f(T^*)$ satisfies Weyl's theorem and hence, by Theorem 6.96, $f(T^*)$ satisfies property (ω). ∎

We consider now the transmission of property (ω) from $T \in L(X)$ to $T + K$, where K is a suitable commuting perturbation.

Theorem 6.101 *Suppose that $T \in L(X)$ is isoloid and property (ω) holds for T. Let $K \in L(X)$ be such that $TK = KT$ and K^n is a finite rank operator for some $n \in \mathbb{N}$. If $\mathrm{iso}\,\sigma_{\mathrm{ap}}(T) = \mathrm{iso}\,\sigma_{\mathrm{ap}}(T + K)$ then $T + K$ has property (ω).*

Proof By Theorem 3.27 we have $\sigma_{\mathrm{ap}}(T) = \sigma_{\mathrm{ap}}(T + K)$. Suppose that T has property (ω). Then T satisfies a-Browder's theorem and property (R). By Theorem 6.14 $T + K$ has property (R) and a-Browder's theorem holds for $T + K$, by Corollary 5.29. Hence, $T + K$ has property (ω). ∎

Generally, property (ω) is not transmitted from T to a quasi-nilpotent perturbation $T + Q$. In fact, if $Q \in L(\ell^2(\mathbb{N}))$ is defined by

$$Q(x_1, x_2, \dots) = \left(\frac{x_2}{2}, \frac{x_3}{3}, \dots \right) \quad \text{for all } (x_n) \in \ell^2(\mathbb{N}),$$

then Q is quasi-nilpotent and

$$\{0\} = \pi_{00}(Q) \neq \sigma_{\mathrm{ap}}(Q) \setminus \sigma_{\mathrm{uw}}(Q) = \emptyset.$$

Take $T = 0$. Clearly, T satisfies property (ω) but $T + Q = Q$ fails this property. Note that Q is not injective.

Theorem 6.102 *Suppose that for $T \in L(X)$ there exists an injective quasi-nilpotent Q operator commuting with T. Then both T and $T + Q$ satisfy property (ω), a-Weyl's and Weyl's theorem.*

Proof We show first property (ω) for T. It is evident, by Lemma 6.32, that $\pi_{00}(T)$ is empty.

Suppose that $\sigma_{\mathrm{ap}}(T) \setminus \sigma_{\mathrm{uw}}(T)$ is not empty and let $\lambda \in \sigma_{\mathrm{ap}}(T) \setminus \sigma_{\mathrm{uw}}(T)$. Since $\lambda I - T \in W_+(X)$, $\alpha(\lambda I - T) < \infty$ and $\lambda I - T$ has closed range. Since $\lambda I - T$

commutes with Q it then follows, by Lemma 2.189, that $\lambda I - T$ is injective, so $\lambda \notin \sigma_{\mathrm{ap}}(T)$, a contradiction. Hence $\sigma_{\mathrm{a}}(T) \setminus \sigma_{\mathrm{uw}}(T)$ is also empty, and property (ω) holds for T.

To show that a-Weyl's theorem holds for T observe that by Lemma 6.32, $\pi_{00}^a(T)$ is also empty, hence

$$\sigma_{\mathrm{ap}}(T) \setminus \sigma_{\mathrm{uw}}(T) = \pi_{00}^a(T) = \emptyset.$$

Analogously, a-Weyl's theorem also holds for $T + Q$, since the operator $T + Q$ commutes with Q. Weyl's theorem is obvious: property (ω), as well as a-Weyl's theorem, entails Weyl's theorem. Property (ω), as well as a-Weyl's theorem and Weyl's theorem, for $T + Q$ is clear, since $T + Q$ commutes with Q. ∎

Obviously, by Theorem 6.102 any injective quasi-nilpotent operator satisfies property (ω).

Example 6.103 In Theorem 6.102 the condition *quasi-nilpotent* cannot be replaced by the condition of being compact. For example, consider the following operators $T := U \oplus I$ and $K := V \oplus Q$ on $\ell_2(\mathbb{N}) \oplus \ell_2(\mathbb{N})$, where, $Q \in L(\ell_2(\mathbb{N}))$ is an injective compact quasi-nilpotent operator,

$$Ux := \left(0, \frac{x_2}{2}, \frac{x_3}{3}, \dots\right), \quad x := (x_n)_{n=1,2,\dots} \in \ell_2(\mathbb{N}),$$

and

$$Vx := \left(1, -\frac{x_2}{2}, -\frac{x_3}{3}, \dots\right), \quad x := (x_n)_{n=1,2,\dots} \in \ell_2(\mathbb{N}).$$

The operator U is compact, so $T = U \oplus I$ and T^* have the SVEP since both operator have discrete spectrum. Consequently, by Theorem 2.68, $\sigma_{\mathrm{ap}}(T) = \sigma(T)$ and $\sigma_{\mathrm{uw}}(T) = \sigma_{\mathrm{w}}(T) = \{0, 1\}$. Clearly,

$$\sigma_{\mathrm{ap}}(T) \setminus \sigma_{\mathrm{uw}}(T) = \sigma(T) \setminus \sigma_{\mathrm{w}}(T) = \pi_{00}(T) = \left\{\frac{1}{n} : n = 2, 3, \dots\right\},$$

thus property (ω) holds for T. Note that K is an injective compact operator, $TK = KT$ and

$$\sigma(T + K) = \sigma_{\mathrm{w}}(T + K) = \{0, 1\} \quad \text{and} \quad \pi_{00}(T + K) = \{1\}. \tag{6.23}$$

Clearly, $T^* + K^*$ has the SVEP, since it has finite spectrum. Moreover,

$$\sigma(T + K) = \sigma_{\mathrm{ap}}(T + K) \quad \text{and} \quad \sigma_{\mathrm{uw}}(T + K) = \sigma_{\mathrm{w}}(T + K),$$

from which we deduce that property (ω) does not hold for $T + K$.

From Lemma 2.189 we deduce that if $0 < \alpha(T) < \infty$ then there exists no injective quasi-nilpotent operator Q which commutes with T. Theorem 6.102 has the following interesting consequence:

Corollary 6.104 *Suppose that T does not satisfy Browder's theorem. Then there exists no injective quasi-nilpotent operator commuting with T.*

Proof If T does not satisfy Browder's theorem then T does not satisfy Weyl's theorem, so the assertion follows from Theorem 1.70.   ∎

Remark 6.105 For a finite rank operator K and any $T \in L(X)$, it is known that $\alpha(T) = \infty$ if and only if $\alpha(T + K) = \infty$.

Theorem 6.106 *Suppose that $Q \in L(X)$ is quasi-nilpotent and $K \in L(X)$ is a finite rank operator commuting with Q. If Q satisfies property (ω) then $Q + K$ satisfies property (w).*

Proof If Q is injective then $Q+K$ satisfies property (ω) by Theorem 6.102. Suppose that Q is non-injective and satisfies property (ω). Clearly, $\{0\} = \sigma_{uw}(Q) = \sigma_{ap}(Q)$ since both $\sigma_{uw}(Q)$ and $\sigma_{ap}(Q)$ are non-empty, and from Corollary 6.120 and Theorem 3.17, we know that

$$\{0\} = \sigma_{uw}(Q) = \sigma_{uw}(Q + K)$$

and $\sigma_{ap}(Q + K) = \sigma_{ap}(K)$, so $\sigma_{ap}(Q + K) \setminus \sigma_{uw}(Q + K)$ is the set of all non-zero eigenvalues of K. Say $\lambda_1, \ldots, \lambda_n$.

We show that $\pi_{00}(Q + K) = \{\lambda_1, \ldots, \lambda_n\}$. Since Q satisfies property (ω) we have

$$\emptyset = \sigma_{ap}(Q) \setminus \sigma_{uw}(Q) = \pi_{00}(Q),$$

and since $\alpha(Q) > 0$ this implies that $\alpha(Q) = \infty$. As observed in Remark 6.105, this implies that $\alpha(Q + K) = \infty$, so that $0 \notin \pi_{00}(Q + K)$. Therefore,

$$\pi_{00}(Q + K) \subseteq \sigma_{ap}(Q + K) = \sigma_{ap}(K) \setminus \{0\} = \{\lambda_1, \ldots, \lambda_n\}.$$

We show the opposite inclusion. For every $i = 1, \ldots, n$ the operators $\lambda_i I - Q$ are invertible, in particular Fredholm operators, so $\lambda_i I - (Q + K)$ is a Fredholm operator. Therefore, $\alpha(\lambda_i I - (Q + K)) < \infty$ and $\lambda_i I - (Q + K)$ has closed range.

Now, suppose that $\alpha(\lambda_i I - (Q + K)) = 0$. Then $\lambda_i \notin \sigma_{ap}(Q + K) = \sigma_{ap}(K)$, hence $\lambda_i I - K$ is injective. Since K is a finite-rank operator it then follows that

$$\alpha(\lambda_i I - K) = \beta(\lambda_i I - K) = 0,$$

i.e. $\lambda_i \notin \sigma(K)$, a contradiction. Therefore $\lambda_i \in \pi_{00}(Q + K)$, and consequently property (ω) holds for $Q + K$.   ∎

Property (ω) may also be extended in the sense of B-Fredholm theory.

Definition 6.107 An operator $T \in L(X)$ is said to satisfy the *generalized property* (ω), in symbols $(g\omega)$, if $\Delta_a^g = E(T)$.

Property $(g\omega)$ is stronger than property (ω):

Theorem 6.108 *If $T \in L(X)$ has property $(g\omega)$ then T has property (ω).*

Proof Assume property $(g\omega)$ for T and let $\lambda \in \Delta_a(T) = E(T)$. Clearly, $\lambda \in \Delta_a^g(T)$, because $\sigma_{ubw}(T) \subseteq \sigma_{uw}(T)$, hence $0 < \alpha(\lambda I - T)$ and since $\lambda I - T$ is Weyl we then have $\alpha(\lambda I - T) < \infty$, so $\lambda \in \pi_{00}(T)$.

Conversely, if $\lambda \in \pi_{00}(T)$ then $\lambda \in E(T) = \Delta_a^g(T)$ and hence $\lambda I - T$ is upper semi B-Fredholm. But $\alpha(\lambda I - T) < \infty$ and this implies, by Theorem 1.114, that $\lambda I - T$ is upper semi-Fredholm. Therefore, $\lambda \in \sigma_{ap}(T) \setminus \sigma_{uw}(T) = \Delta_a(T)$. Thus, property (ω) holds for T. ∎

The next example shows that the converse of the previous theorem does not hold in general.

Example 6.109 Denote by Q any quasi-nilpotent operator, acting on an infinite-dimensional Banach space, for which all the ranges $Q^n(X)$ are non-closed for all $n \in \mathbb{N}$. Let $T := 0 \oplus Q$. Since $T^n(X) = Q^n(X)$ is non-closed for all $n \in \mathbb{N}$, T is not semi B-Fredholm, so $\sigma_{ubw}(T) = \{0\}$. Since $\sigma_{ap}(T) = \{0\}$ and $E(T = \{0\}$, T does not satisfy property $(g\omega)$, while, since $\pi_{00}(T) = \emptyset$ and $\sigma_{uw}(T) = \{0\}$, T satisfies property (ω).

Property $(g\omega)$ is related to the generalized Weyl's theorem as follows:

Theorem 6.110 *Let $T \in L(X)$. Then the following statements are equivalent:*

(i) *T satisfies property $(g\omega)$;*
(ii) *T satisfies the generalized Weyl's theorem and $\mathrm{ind}\,(\lambda I - T) = 0$ for all $\lambda \in \Delta_a^g(T)$.*

Proof (i) $\Rightarrow$ (ii) Suppose that T satisfies property $(g\omega)$ and let $\lambda \in \Delta^g(T) = \sigma(T) \setminus \sigma_{bw}(T)$. Then $\lambda \notin \sigma_{ubw}(T)$. If $\alpha(\lambda I - T) = 0$ then $\lambda I - T$ is invertible, and this is impossible. Hence $\alpha(\lambda I - T) > 0$ and $\lambda \in \sigma_{ap}(T)$. Since T has property $(g\omega)$ it then follows that $\lambda \in E(T)$, from which we obtain that $\Delta^g(T) \subseteq E(T)$. To show the converse inclusion, let $\lambda \in E(T)$ be arbitrary. Since T has property $(g\omega)$, $\lambda \notin \sigma_{ubw}(T)$ and hence $\mathrm{ind}\,(\lambda I - T) \leq 0$. On the other hand, since $\lambda \in E(T)$ then $\lambda \in \mathrm{iso}\,\sigma(T)$, hence T^* has the SVEP at λ and, consequently, $\mathrm{ind}\,(\lambda I - T) \geq 0$, from which we obtain $\mathrm{ind}\,(\lambda I - T) = 0$ and $\lambda \notin \sigma_{bw}(T)$. Hence $\Delta^g(T) = E(T)$ and $\mathrm{ind}\,(\lambda I - T) = 0$ for all $\lambda \in \Delta_a^g(T)$.

(ii) $\Rightarrow$ (i) Conversely, assume that T satisfies the generalized Weyl's theorem $\mathrm{ind}\,(\lambda I - T) = 0$ for all $\lambda \in \Delta_a^g(T)$ and $\mathrm{ind}\,(\lambda I - T) = 0$ for all $\lambda \in \Delta_a^g(T)$. If $\lambda \in \Delta_a^g(T)$, then $\lambda I - T$ is upper semi B-Weyl, and since $\mathrm{ind}\,(\lambda I - T) = 0$ we then have that $\lambda I - T$ is B-Weyl. But T satisfies the generalized Weyl's theorem, so $\lambda \in E(T)$ and hence $\Delta_a^g(T) \subseteq E(T)$. To show the reverse inclusion, let $\lambda \in E(T)$. Then $\lambda I - T$ is B-Weyl and $\alpha(\lambda I - T) > 0$, since $\lambda \in \sigma(T)$. Thus $\lambda \in \Delta_a^g(T)$, and consequently T satisfies property $(g\omega)$. ∎

In the following theorem we give the exact relationship between property (ω) and property $(g\omega)$.

Theorem 6.111 *Let $T \in L(X)$. Then the following statements are equivalent:*

 (i) *T satisfies property $(g\omega)$;*
 (ii) *T satisfies property (ω) and has property (gR);*
(iii) *T satisfies Weyl's theorem and has property (gR);*
(iv) *T satisfies Browder's theorem holds and has property (gR).*

Proof (i) $\Rightarrow$ (ii) Assume that T has property $(g\omega)$. By Theorem 6.108 T has property (ω) and, by Theorem 6.110, T satisfies the generalized Weyl's theorem. Hence $E(T) = \Pi(T)$. If $\lambda \in \Pi_a(T)$ then $\lambda I - T$ is left Drazin invertible and since T^* has the SVEP at λ it then follows that $q(\lambda I - T) < \infty$, by Theorem 2.98, so $\lambda \in \Pi(T)$. This shows that $\Pi(T) \subseteq \Pi_a(T)$ and hence $\Pi_a(T) = E(T)$, i.e., T has property (gR).

(ii) $\Rightarrow$ (iii) Clear, since property (ω) entails Weyl's theorem for T, by Theorem 6.91.

(iii) $\Rightarrow$ (iv) Clear: indeed Weyl's theorem entails Browder's theorem.

(iv) $\Rightarrow$ (i) Suppose that T satisfies Browder's theorem, or equivalently the generalized Browder's theorem, and suppose that $\Pi_a(T) = E(T)$. Then $\Delta_a(T) = \Pi_a(T) = E(T)$, so T has property $(g\omega)$. ∎

Theorem 6.112 *Let $T \in L(X)$. If T^* has the SVEP then the following statements are equivalent.*

 (i) *T has property $(g\omega)$;*
 (ii) *T satisfies the generalized a-Weyl's theorem;*
(iii) *T satisfies the generalized Weyl's theorem.*
        Dually, if T has the SVEP then the following statements are equivalent.
(iv) *T^* has property $(g\omega)$;*
 (v) *T^* satisfies the generalized Weyl's theorem;*
(vi) *T^* satisfies the generalized a-Weyl's theorem.*

Proof The implications (i) $\Rightarrow$ (ii) is clear, by Theorem 6.110. To show that (ii) $\Leftrightarrow$ (iii), observe first that, by Theorem 2.68, the SVEP for T^* entails that $\sigma(T) = \sigma_{\mathrm{ap}}(T)$, so $E_a(T) = E(T)$. By Theorem 3.53 we also have $\sigma_{\mathrm{bw}}(T) = \sigma_{\mathrm{ubw}}(T)$, hence $\Delta_a^g(T) = \Delta^g(T)$, so the statements (ii) and (iii) are equivalent. To show the implication (ii) $\Rightarrow$ (i), suppose that T satisfies the generalized Weyl's theorem and let us consider $\lambda \in \Delta_a^g(T)$. Then $\lambda I - T$ is upper semi B-Weyl, so ind $(\lambda I - T) \leq 0$. On the other hand, the SVEP for T^* ensures, by Theorem 2.98, that $q(\lambda I - T) < \infty$ and hence, by Theorem 1.22, ind $(\lambda I - T) \geq 0$. Therefore, ind $(\lambda I - T) = 0$ for all $\lambda \in \Delta_a^g(T)$. By Theorem 6.110 then T has property $(g\omega)$.

The proof of the equivalences (iv) $\Leftrightarrow$ (v) $\Leftrightarrow$ (vi) is analogous, just use Theorem 2.97 instead of Theorem 2.98. ∎

Theorem 6.113 *Let $T \in L(X)$ be polaroid and $f \in \mathcal{H}_{nc}(\sigma(T))$.*

(i) *If T^* has the SVEP, then $f(T)$ has property ($g\omega$).*
(ii) *If T has the SVEP, then $f(T^*)$ has property ($g\omega$).*

Proof

(i) $f(T^*)$ has the SVEP, by Theorem 2.86, and $f(T)$ is polaroid by Theorem 4.19. The SVEP for $f(T^*)$ entails that Browder's theorem, or equivalently the generalized Browder's theorem, holds for $f(T)$. Since $f(T)$ is polaroid then $E(f(T)) = \Pi(f(T))$, so $f(T)$ satisfies the generalized Weyl's theorem. By Theorem 6.112 we then conclude that $f(T)$ satisfies property ($g\omega$).

Since T^* has the SVEP, Browder's theorem holds for T, or equivalently the generalized Browder's theorem holds for T, by Theorem 5.15. Since T is polaroid, $E(T) = \Pi(T)$, so T satisfies the generalized Weyl's theorem. Since T^* has the SVEP, by Theorem 6.112 then property ($g\omega$) holds for T.

(ii) The proof is analogous to the proof of part (i), taking into account that T is polaroid if and only if T^* is polaroid. ∎

Recall that, by Theorem 3.78, if $K \in L(X)$ commutes with $T \in L(X)$ and K^n is a finite rank operator for some $n \in \mathbb{N}$ then $\sigma_{ubw}(T + K) = \sigma_{ubw}(T)$.

Theorem 6.114 *Suppose that $T \in L(X)$ is polaroid and that $K \in L(X)$ is such that $TK = KT$ and K^n is a finite rank operator for some $n \in \mathbb{N}$. If T satisfies property ($g\omega$) then $T + K$ satisfies property ($g\omega$) if and only if $\Pi_a(T + K) \subseteq E(T + K)$.*

Proof Assume that T has property ($g\omega$) and $\Pi_a(T + K) \subseteq E(T + K)$. Property ($g\omega$) entails property ($w$) and hence a-Browder's theorem. Then $T + K$ satisfies a-Browder's theorem, by Corollary 5.29, or equivalently, T satisfies the generalized a-Browder's theorem by Theorem 5.38. Hence $\Delta_a^g(T) = \Pi_a(T + K)$, and from the assumption we then have $\Delta_a^g(T) \subseteq E(T + K)$. Conversely, suppose that $\lambda \in E(T + K)$. Then $\lambda \in \mathrm{iso}\,\sigma(T + K)$, and since $T + K$ is polaroid, by Theorem 4.24, we then have

$$\lambda \in \Pi(T + K) \subseteq \Pi_a(T + K) = \Delta_a^g(T).$$

Thus, $\Delta_a^g(T) = E(T + K)$ and hence $T + K$ has property ($g\omega$).

Conversely, if $T + K$ has property ($g\omega$) then $\Delta_a^g(T + K) = E(T + K)$. Since $\Pi_a(T + K) \subseteq \Delta_a^g(T + K)$, we have $\Pi_a(T + K) \subseteq E(T + K)$. ∎

We consider now the special case of nilpotent perturbations. Recall that $\sigma_{ubw}(T + N) = \sigma_{ubw}(N)$ for every nilpotent operator N that commutes with T.

Theorem 6.115 *Suppose that $T \in L(X)$ is polaroid and that $N \in L(X)$ is a nilpotent operator that commutes with T. If T satisfies property ($g\omega$) then $T + N$ satisfies property ($g\omega$).*

Proof First note that

$$\Delta_a^g(T) = \sigma_{ap}(T) \setminus \sigma_{ubw}(T) = \sigma_{ap}(T + N) \setminus \sigma_{ubw}(T + N) = \Delta_a^g(T + N).$$

Now, if $\lambda \in E(T)$ then $\lambda \in \text{iso}\,\sigma(T) = \text{iso}\,\sigma(T + N)$, and since $T + N$ is polaroid, by Theorem 4.24, then λ is a pole of the resolvent of $T + N$. In particular, $\lambda \in E(T + N)$. This shows that $E(T) \subseteq E(T + N)$ and the converse of this inclusion may be obtained by a symmetric argument. Hence $E(T) = E(T + N)$. Finally, since T satisfies $(g\omega)$,

$$\Delta_a^g(T + N) = \Delta_a^g(T) = E(T) = E(T + N),$$

so $T + N$ has property $(g\omega)$.   ∎

Lemma 6.116 *Suppose that for a bounded operator $T \in L(X)$ there exists a $\lambda_0 \in \mathbb{C}$ such that $K(\lambda_0 I - T) = \{0\}$ and $\ker(\lambda_0 I - T) = \{0\}$. Then $\sigma_p(T) = \emptyset$.*

Proof For all complex $\lambda \neq \lambda_0$ we have $\ker(\lambda I - T) \subseteq K(\lambda_0 I - T)$, so that $\ker(\lambda I - T) = \{0\}$ for all $\lambda \in \mathbb{C}$.   ∎

Theorem 6.117 *Let $T \in L(X)$ be such that there exists a $\lambda_0 \in \mathbb{C}$ such that*

$$K(\lambda_0 I - T) = \{0\} \quad and \quad \ker(\lambda_0 I - T) = \{0\}. \tag{6.24}$$

Then property $(g\omega)$ holds for $f(T)$ for all $f \in \mathcal{H}(\sigma(f(T)))$.

Proof We know from Lemma 6.116 that $\sigma_p(T) = \emptyset$, so T has the SVEP. We show that also $\sigma_p(f(T)) = \emptyset$. Let $\mu \in \sigma(f(T))$ and write $\mu - f(\lambda) = p(\lambda)g(\lambda)$, where g is analytic on an open neighborhood $\mathcal{U}$ containing $\sigma(T)$ and without zeros in $\sigma(T)$, p a polynomial of the form $p(\lambda) = \Pi_{k=1}^n (\lambda_k - \lambda)^{\nu_k}$, with distinct roots $\lambda_1, \ldots, \lambda_n$ lying in $\sigma(T)$. Then

$$\mu I - f(T) = \Pi_{k=1}^n (\lambda_k I - T)^{\nu_k} g(T).$$

Since $g(T)$ is invertible, $\sigma_p(T) = \emptyset$ implies that $\ker(\mu I - f(T)) = \{0\}$ for all $\mu \in \mathbb{C}$, hence $\sigma_p(f(T)) = \emptyset$ and, consequently, $f(T)$ has the SVEP. This implies that a-Browder's theorem holds for $f(T)$, or equivalently, the generalized a-Browder's theorem holds for $f(T)$, i.e., $\sigma_{ubw}(f(T)) = \sigma_{ld}(f(T))$. Clearly, $E_a(f(T)) = E(f(T)) = \emptyset$. Now, suppose that $\lambda \notin \sigma_{ubw}(f(T))$, so $\lambda I - f(T)$ is upper semi B-Weyl. Since $\alpha(\lambda I - f(T)) = 0$ then, by Corollary 1.115, $\lambda I - f(T)$ is bounded below, so $\lambda \notin \sigma_a(f(T))$. Therefore, $\sigma_a(f(T)) \subseteq \sigma_{ubw}(f(T))$, and since the opposite inclusion holds for every operator, we then have $\sigma_{ap}(f(T)) = \sigma_{ubw}(f(T))$, so $\Delta_a^g(T) = \emptyset$. Hence $f(T)$ has property $(g\omega)$.   ∎

In the last part of this section we show that Weyl-type theorems are transferred from a Drazin invertible operator to its Drazin inverse.

Lemma 6.118 *Let $R \in L(X)$ be Drazin invertible with Drazin inverse S. Then we have:*

(i) $0 \in \pi_{00}(R) \Leftrightarrow 0 \in \pi_{00}(S)$. *If $\lambda \neq 0$ then $\lambda \in \pi_{00}(R) \Leftrightarrow \frac{1}{\lambda} \in \pi_{00}(S)$.*
(ii) $0 \in \pi_{00}^{a}(R) \Leftrightarrow 0 \in \pi_{00}^{a}(S)$. *If $\lambda \neq 0$ then $\lambda \in \pi_{00}^{a}(R) \Leftrightarrow \frac{1}{\lambda} \in \pi_{00}^{a}(S)$.*

Proof

(i) Suppose first that $0 \in \pi_{00}(R)$. Then $0 \in \text{iso}\,\sigma(R)$, and this implies $0 \in \text{iso}\,\sigma(S)$. Obviously, with respect to the usual decomposition $X = Y \oplus Z$, $R = R_1 \oplus R_2$ we have $\alpha(R) = \alpha(R_1) + \alpha(R_2) = \alpha(R_1)$, since R_2 is invertible. Since $\alpha(R) < \infty$, $\alpha(R_1) < \infty$ and hence $\alpha(R_1^{\nu}) < \infty$, where $R_1^{\nu} = 0$. But $\ker R_1^{\nu} = Y$, so Y is finite-dimensional.

On the other hand, $\ker(S) = \ker 0 \oplus \ker S_2 = Y \oplus \{0\}$, thus $\alpha(S) < \infty$. It remains to prove that $0 < \alpha(S)$. By assumption we have $\alpha(R) = \alpha(R_1) > 0$, so $\alpha(R_1^{\nu}) = \dim Y > 0$, since $\ker R_1 \subseteq \ker R_1^{\nu}$ and hence $\alpha(S) = \dim Y > 0$. Therefore, $0 \in \pi_{00}(S)$. Analogous arguments show the reverse implication.

The second assertion follows from the equality $\ker(\lambda I - R) = \ker(\frac{1}{\lambda}I - S)$.
(ii) If $0 \in \pi_{00}^{a}(R)$ then $0 \in \text{iso}\,\sigma_a(R)$ and hence $0 \in \text{iso}\,\sigma_a(S)$. To show that $0 < \alpha(S) < \infty$, proceed as in the proof part (i). An analogous reasoning shows that if $0 \in \pi_{00}^{a}(S)$ then $0 \in \pi_{00}^{a}(R)$. The second assertion follows from the equality $\ker(\lambda I - R) = \ker(\frac{1}{\lambda}I - S)$. $\blacksquare$

Theorem 6.119 *Let $R \in L(X)$ be Drazin invertible with Drazin inverse S. Then*

(i) *R satisfies Weyl's theorem if and only if S satisfies Weyl's theorem.*
(ii) *R satisfies a-Weyl's theorem if and only if S satisfies a-Weyl's theorem.*
(iii) *R satisfies property (ω) if and only if S satisfies property (ω).*

Proof

(i) Suppose that R satisfies Weyl's theorem. Then R satisfies Browder's theorem and hence, by part (i) of Theorem 5.109, Browder's theorem holds for S. Let $\lambda \in \pi_{00}(S)$. If $\lambda = 0$ then, by Lemma 6.118, $0 \in \pi_{00}(R) = p_{00}(R)$, hence, by Lemma 5.111, $0 \in p_{00}(S)$. If $\lambda \neq 0$ then $\frac{1}{\lambda} \in \pi_{00}(R) = p_{00}(R)$, hence $\lambda \in p_{00}(S)$. Therefore, $\pi_{00}(S) \subseteq p_{00}(S)$ and since the opposite inclusion holds for every operator we then conclude that $\pi_{00}(S) = p_{00}(S)$, thus, by Theorem 6.40 S satisfies Weyl's theorem. In a similar way Weyl's theorem for S implies Weyl's theorem for R.
(ii) If R satisfies a-Weyl's theorem, then R satisfies a-Browder's theorem and $p_{00}^{a}(R) = \pi_{00}^{a}(R)$. From part (ii) of Theorem 5.109 we know that a-Browder's theorem holds for S. To show that S satisfies a-Weyl's theorem then it suffices, by Theorem 6.40, to prove that $\pi_{00}^{a}(S) = p_{00}^{a}(S)$.

Let $\lambda \in \pi_{00}^{a}(S)$. If $\lambda = 0$ then, by Lemma 6.118, $0 \in \pi_{00}^{a}(R) = p_{00}^{a}(R)$. Hence, by Lemma 5.111, $0 \in p_{00}^{a}(S)$. If $\lambda \neq 0$ then, by Lemma 6.118, $\frac{1}{\lambda} \in \pi_{00}^{a}(R) = p_{00}^{a}(R)$, and hence, by Lemma 5.111, we have $\lambda \in p_{00}^{a}(S)$. Therefore, $\pi_{00}^{a}(S) \subseteq p_{00}^{a}(S)$. The opposite inclusion holds for every operator, so $\pi_{00}^a(S) = p_{00}^a(S)$, thus S satisfies a-Weyl's theorem. In a similar way a-Weyl's theorem for S implies a-Weyl's theorem for R.

(iii) If R satisfies property (ω), then, by Theorem 6.89, R satisfies a-Browder's theorem and $p_{00}^a(R) = \pi_{00}(R)$. By part (ii) of Theorem 5.109, a- Browder's theorem holds for S, so, in order to show property (ω) for S, it suffices to prove, by Theorem 6.89, that $\pi_{00}(S) = p_{00}^a(S)$.

Let $\lambda \in \pi_{00}(S)$. If $\lambda = 0$ then, by Lemma 6.118, $0 \in \pi_{00}(R) = p_{00}^a(R)$. Hence, by Lemma 5.111, $0 \in p_{00}^a(S)$. Suppose the other case, $\lambda \neq 0$. Then $\frac{1}{\lambda} \in \pi_{00}(R) = p_{00}^a(R)$, and hence $\lambda \in p_{00}^a(S)$. Therefore, $\pi_{00}(S) \subseteq p_{00}^a(S)$. It remains to prove that $p_{00}^a(S) \subseteq \pi_{00}(S)$. Let $\lambda \in p_{00}^a(S)$. If $\lambda = 0$ then S is Browder, see Lemma 5.111, so 0 is an isolated point of $\sigma(S)$. Clearly, $0 < \alpha(T)$, since 0 is an eigenvalue and $S(X)$ is closed. Moreover, $\alpha(S) < \infty$, so $0 \in \pi_{00}(S)$.

If $\lambda \neq 0$ then, again by Lemma 6.118, $\frac{1}{\lambda} \in p_{00}^a(R) = \pi_{00}(R)$, and hence, by Lemma 5.111, $\lambda \in \pi_{00}(S)$, from which we conclude that $\pi_{00}(S) = p_{00}^a(S)$, thus S satisfies property (ω). In a similar way, property (ω) for S implies property (ω) for R.   ∎

6.6   Weyl-Type Theorems for Polaroid Operators

We have seen that assuming the SVEP for T^* (respectively, for T) the Weyl-type theorems are equivalent for T (respectively, for T^*). In this section we show that also assuming polaroid-type conditions then some of the Weyl-type theorems are equivalent.

In the following diagram we resume the relationships between all Weyl-type theorems, generalized or not, proved in the previous section. We shall use the symbols (W) and (gW) for Weyl's theorem and the generalized Weyl's theorem, respectively. By (aW) and (gaW) we shall denote a-Weyl's theorem and the generalized a-Weyl's theorem, respectively. We have

$$(g\omega) \;\Rightarrow\; (\omega) \;\Rightarrow\; (W)$$

$$(gaW) \Rightarrow (aW) \Rightarrow (W).$$

Furthermore,

$$(g\omega) \;\Rightarrow\; (gW) \Rightarrow (W)$$

$$(gaW) \Rightarrow (gW) \Rightarrow (W).$$

Theorem 6.120 *Let $T \in L(X)$. Then we have:*

(i) *If T is left-polaroid then (aW) and (gaW) for T are equivalent. If T is right-polaroid then (aW) and (gaW) for T^* are equivalent.*

(ii) *If T is polaroid then (W) and (gW) for T are equivalent. Analogously, (W) and (gW) for T^* are equivalent.*

Proof

(i) We know that (gaW) entails property (aW) without any assumption on T. To show the converse, suppose that property (aW) holds for T, i.e., $\Delta_a(T) = \sigma_{ap}(T) \setminus \sigma_{uw}(T) = \pi_{00}^a(T)$. We have only to prove that $\sigma_{ap}(T) \setminus \sigma_{ubw}(T) = E_a(T)$. We show first that the inclusion $\sigma_a(T) \setminus \sigma_{ubw}(T) \subseteq E_a(T)$ holds without any assumption on T.

Let $\lambda \in \sigma_{ap}(T) \setminus \sigma_{ubw}(T)$. We can suppose that $\lambda = 0$. Therefore, $0 \in \sigma_{ap}(T)$ and T is upper semi B-Weyl. By Theorem 1.117, there exists an $\varepsilon > 0$ such that $\mu I - T$ is upper semi-Weyl for all $0 < |\mu| < \varepsilon$. We claim that T has the SVEP at every μ. If $\mu \notin \sigma_{ap}(T)$ this is obvious. Suppose that $\mu \in \sigma_{ap}(T)$. Then $\mu \in \sigma_{ap}(T) \setminus \sigma_{uw}(T) = \pi_{00}^a(T)$, so μ is an isolated point of $\sigma_{ap}(T)$ and hence T has the SVEP at μ. The following argument shows that T has the SVEP at 0. Let $f : \mathbb{D}_0 \to X$ be an analytic function defined on an open disc $\mathbb{D}_0$ centered at 0 for which the equation $(\lambda I - T) f(\lambda) = 0$ for all $\lambda \in \mathbb{D}_0$. Take $0 \neq \mu \in \mathbb{D}_0$ and let $\mathbb{D}_1$ be an open disc centered at μ contained in $\mathbb{D}_0$. The SVEP of T at μ implies that $f \equiv 0$ on $\mathbb{D}_1$ and hence, from the identity theorem for analytic functions, it then follows that $f \equiv 0$ on $\mathbb{D}_0$, so T has the SVEP at 0. But T is upper semi B-Fredholm, so, by Theorem 2.97, $0 \in \mathrm{iso}\, \sigma_{ap}(T)$.

Suppose that $\alpha(T) = 0$. By Theorem 1.114 then $T \in \Phi_+(X)$, so the range $T(X)$ is closed and, consequently, $0 \notin \sigma_{ap}(T)$, a contradiction. Therefore $\alpha(T) > 0$, from which we conclude that $0 \in E_a(T)$ and hence $\sigma_{ap}(T) \setminus \sigma_{ubw}(T) \subseteq E_a(T)$.

Suppose now that T is left polaroid and let $\lambda \in E_a(T)$. Then λ is an isolated point of $\sigma_{ap}(T)$, and hence by the left polaroid condition λ is a left pole of T. In particular, $\lambda I - T$ is left Drazin invertible. Since $\sigma_{ubw}(T) \subseteq \sigma_{ld}(T)$ we then have $\lambda \in \sigma_{ap}(T) \setminus \sigma_{ld}(T) \subseteq \sigma_{ap}(T) \setminus \sigma_{ubw}(T)$. Therefore, $\sigma_{ap}(T) \setminus \sigma_{ubw}(T) = E_a(T)$, and the generalized a-Weyl's theorem holds for T.

The assertion concerning right-polaroid operators is obvious by Theorem 1.144.

(ii) We have only to show that Weyl's theorem entails the generalized Weyl's theorem. Suppose first that $\lambda_0 \in E(T)$. Since T is polaroid then λ_0 is a pole of T, hence $0 < p(\lambda_0 I - T) = q(\lambda_0 I - T) < \infty$. Therefore, $\lambda_0 I - T$ is Drazin invertible or equivalently, by Theorem 3.47, $\lambda_0 I - T$ is B-Browder and hence B-Weyl. Consequently, $\lambda_0 \in \sigma(T) \setminus \sigma_{bw}(T)$ and hence $E(T) \subseteq \sigma(T) \setminus \sigma_{bw}(T)$.

Conversely, assume that $\lambda_0 \in \sigma(T) \setminus \sigma_{\mathrm{bw}}(T)$. Then $\lambda_0 I - T$ is B-Weyl and hence, again by Theorem 1.117, there exists an $\varepsilon > 0$ such that $\lambda I - T$ is Weyl for all $0 < |\lambda - \lambda_0| < \varepsilon$. By Theorem 1.118 we know that $\lambda I - T$ is semi-regular in a punctured open disc centered at λ_0, so we can assume that

$$\ker(\lambda I - T) \subseteq \mathcal{N}^\infty(\lambda I - T) \subseteq (\lambda I - T)^\infty(X) \quad \text{for all } 0 < |\lambda - \lambda_0| < \varepsilon.$$

Since Weyl's theorem for T entails Browder's theorem for T, we have $\sigma_{\mathrm{w}}(T) = \sigma_{\mathrm{b}}(T)$. Therefore, $\lambda I - T$ is Browder for all $0 < |\lambda - \lambda_0| < \varepsilon$ and, consequently, $p(\lambda I - T) = q(\lambda I - T) < \infty$. By Lemma 1.19 we then have

$$\ker(\lambda I - T) = \ker(\lambda I - T) \cap (\lambda I - T)^\infty(X) = \{0\},$$

thus $\alpha(\lambda I - T) = 0$ and since $\lambda I - T$ is Weyl we then conclude that also $\beta(\lambda I - T) = 0$, so $\lambda I - T$ is invertible for all $0 < |\lambda - \lambda_0|$ and hence $\lambda_0 \in \mathrm{iso}\,\sigma(T)$.

To show that $\lambda_0 \in E(T)$ it remains to prove that $\alpha(\lambda_0 I - T) > 0$. Suppose that $\alpha(\lambda_0 I - T) = 0$. Since $\lambda_0 I - T$ is B-Weyl then, by Lemma 1.114, $\lambda_0 I - T$ is Weyl and since $\alpha(\lambda_0 I - T) = 0$ it then follows that $\lambda_0 I - T$ is invertible, a contradiction since $\lambda_0 \in \sigma(T)$. Therefore, $\lambda_0 \in E(T)$, so the generalized Weyl's theorem holds for T.

The last assertion is clear: T^* is also polaroid.   ∎

Corollary 6.121 *If $T \in L(X)$ is a-polaroid then (aW), (gaW), (ω), $(g\omega)$ for T are equivalent.*

Proof Every a-polaroid operator is left polaroid so, by part (i) of Theorem 6.120, (aW) and (gaW) are equivalent. Property (ω) and (aW) are equivalent, since $\pi_{00}(T) = \pi_{00}^a(T)$. We also have $E(T) = E_a(T)$, from which it easily follows that (gaW) and $(g\omega)$ are equivalent.   ∎

In the following example we show that the result of Corollary 6.121 does not hold if we replace the condition of being a-polaroid by the weaker conditions of being left polaroid or polaroid.

Example 6.122 Let R and U be defined as in Example 6.30. As observed before $T := R \oplus U$ is both left polaroid and polaroid. Moreover, $\sigma_{\mathrm{ap}}(T) = \Gamma \cup \{0\}$, Γ the unit circle, and $\mathrm{iso}\,\sigma(T) = \pi_{00}(T) = \emptyset$, so $\sigma_{\mathrm{ap}}(T) \setminus \sigma_{\mathrm{uw}}(T) = \{0\} \neq \pi_{00}(T)$, i.e., T does not satisfy property (ω). On the other hand, we have $\pi_{00}^a(T) = \{0\}$, hence T satisfies a-Weyl's theorem.

In the following result we show that if T is polaroid and T^* has the SVEP, (respectively, T has the SVEP), we can say much more: all Weyl-type theorems, generalized or not, are equivalent and hold for T, (respectively, for T^*).

Theorem 6.123 *Let* $T \in L(X)$ *be polaroid. Then we have*

(i) *If* T^* *has the SVEP then* (W), (aW), (ω), (gW), (gaW) *and* (gw) *hold for* T. *Moreover,* T^* *satisfies* (gW).

(ii) *If* T *has the SVEP then* (W), (aW), (ω), (gW), (gaW) *and* $(g\omega)$ *hold for* T^*. *Moreover,* T *satisfies* (gW).

Proof

(i) T satisfies (W) by Theorem 6.44. The first statement is then proved if we show that (W) is equivalent to each one of the other Weyl-type theorems for T, generalized or not. The SVEP for T^* ensures, by Theorem 6.96, that (W) and (aW) for T are equivalent. By Theorem 4.15 T is a-polaroid hence, by Theorem 6.29, (aW), (gaW), (ω), $(g\omega)$ for T are equivalent. Finally, from part (ii) of Theorem 6.120, (W) and (gW) for T are equivalent. By Theorem 6.44 T^* satisfies (W) and since T^* is polaroid then, by part (ii) of Theorem 6.120, (gW) holds for T^*.

(ii) T^* satisfies (W) by Theorem 6.44, so it suffices to prove for T^* that (W) is equivalent to each one of the other Weyl-type theorems, generalized or not. Because, by Theorem 4.15, T^* is a-polaroid, by Theorem 6.29 it then follows that (aW), (gaW), (ω), $(g\omega)$ are equivalent for T^*. The SVEP for T entails, by Theorem 6.96, that (W) and (aW) are equivalent for T^*, while (W) and (gW) for T^* are equivalent by part (ii) of Theorem 6.120. By Theorem 6.44 T satisfies (W) and since T is polaroid this is equivalent to (gW) for T, by part (ii) of Theorem 6.120. ∎

The result of Theorem 6.123 may be considerably extended as follows

Theorem 6.124 *Let* $T \in L(X)$ *be polaroid and suppose that* $f \in \mathcal{H}(\sigma(T))$ *is not constant on each of the components of its domain. Then we have*

(i) *If* T^* *has the SVEP then* (W), (aW), (ω), (gW), (gaW) *and* $(g\omega)$ *hold for* $f(T)$.

(ii) *If* T *has the SVEP then* (W), (aW), (ω), (gW), (gaW) *and* $(g\omega)$ *hold for* $f(T^*)$.

Proof

(i) If T^* has the SVEP then $f(T)^* = f(T^*)$ has the SVEP, by Theorem 2.86. Moreover, T is left polaroid by Theorem 4.15, so $f(T)$ is left polaroid, as observed after Theorem 4.19. Again by Theorem 4.19, $f(T)$ is polaroid, hence Theorem 6.123 applies to $f(T)$.

(ii) Argue as in the proof of part (i), just replace T with T^*. ∎

Remark 6.125 Obviously, in the case of Hilbert space operators, the condition that T^* has the SVEP in Theorem 6.124 may be replaced by the SVEP of the Hilbert adjoint T'.

Theorem 6.126 *Suppose that* $T \in L(X)$ *has the SVEP and* $f \in \mathcal{H}_{nc}(\sigma(T))$. *Then we have*

(i) *If $T \in L(X)$ is left-polaroid, then (aW) holds for $f(T)$, or equivalently (gaW) holds for $f(T)$.*

(ii) *If $T \in L(X)$ is polaroid, then (W) holds for $f(T)$, or equivalently (gW) holds for $f(T)$.*

Proof

(i) As observed above, if T is left polaroid then $f(T)$ is left polaroid. Since $f(T)$ has the SVEP, by Theorem 2.86, then Corollary 6.85 applies to $f(T)$. The equivalence of (aW) and (gaW) follows from Theorem 6.120.

(ii) If T is polaroid then $f(T)$ is polaroid and has the SVEP, so Theorem 6.44 applies to $f(T)$. The equivalence of (W) and (gW) follows from Theorem 6.120.   ∎

We now apply the results obtained in this section in order to produce a general framework for establishing Weyl type theorems for perturbations of hereditarily polaroid operators

Theorem 6.127 *Suppose $K \in L(X)$ is an algebraic operator commuting with $T \in L(X)$ and let $f \in \mathcal{H}_{nc}(\sigma(T+K))$. Then we have*

(i) *If T is hereditarily polaroid then $f(T+K)$ satisfies (gW), while $f(T^*+K^*)$ satisfies every Weyl-type theorem (generalized or not).*

(ii) *If T^* is hereditarily polaroid then $f(T^*+K^*)$ satisfies (gW), while $f(T+K)$ satisfies every Weyl-type theorem (generalized or not).*

Proof

(i) Every hereditarily polaroid has the SVEP, by Theorem 4.31. Then $T+K$ has the SVEP by Theorem 2.145, and hence $f(T+K)$ also has the SVEP, by Theorem 2.86. Moreover, by Theorem 4.32, $T+K$ is polaroid, and consequently, $f(T+K)$ is polaroid, by Theorem 4.19. From Theorem 6.43 it then follows that $f(T+K)$ satisfies Weyl's theorem and this, by Theorem 6.126, is equivalent to the generalized Weyl's theorem.

To show the second assertion, observe that the SVEP for $T+K$ entails, by Theorem 4.15, that T^*+K^* is a-polaroid, in particular left polaroid. According to Remark 4.20 then $f(T^*+K^*)$ is left polaroid. By Theorem 6.126 it then follows that (gaW) holds for $f(T^*+K^*)$ and this, by part (ii) of Theorem 6.120, is equivalent to property $(g\omega)$.

(ii) The proof is analogous.   ∎

6.7   Weyl-Type Theorems Under Compact Perturbations

In this section we study Browder-type and Weyl-type theorems for operators $T+K$, where $K \in L(X)$ is a (not necessarily commuting) compact operator. We have already seen that in general, Weyl-type theorems, such as Weyl's theorem, a-Weyl's theorem and property (w), are not preserved under compact, also commuting, perturbations K. In Theorems 6.45, 6.74 and 6.101 we have seen that this happens only under some special conditions on T, or on K. Thus, it is not surprising that the permanence of Weyl-type theorems under compact perturbations requires some rather restrictive conditions.

Lemma 6.128 *If $T \in L(X)$, X a Banach space, then* iso $\sigma(T) \subseteq \sigma_{sf}(T) \cup p_{00}(T)$.

Proof Let $\lambda \in$ iso $\sigma(T)$. Then either $\lambda \in \sigma_{sf}(T)$ or $\lambda \notin \sigma_{sf}(T)$. If $\lambda \notin \sigma_{sf}(T)$ then $\lambda I - T \in \Phi_\pm(X)$, and since both T and T^* have the SVEP at λ, $\lambda I - T$ is Browder, by Theorem 2.100, so $\lambda \in p_{00}(T) = \sigma(T) \setminus \sigma_b(T)$. ∎

Recall that given a compact set $\sigma \subset \mathbb{C}$, a *hole* of σ is a bounded component of the complement $\mathbb{C} \setminus \sigma$. Since $\mathbb{C} \setminus \sigma$ always has an unbounded component, $\mathbb{C} \setminus \sigma$ is connected precisely when σ has no holes.

Lemma 6.129 *Let $T \in L(X)$.*

(i) *If $\rho_w(T)$ is connected then $\rho_w(T) = \rho(T) \cup p_{00}(T)$. Furthermore, $\Delta(T) = p_{00}(T)$.*
(ii) *If $\rho_{uw}(T)$ is connected then $\rho_{uw}(T) = \rho_{ap}(T) \cup p_{00}^a(T)$. Furthermore, $\Delta_a(T) = p_{00}^a(T)$.*

Proof

(i) Let Ω be the unique component of $\rho_w(T)$. Clearly, $\Omega \supseteq \rho(T)$. Since $\rho_w(T) \subseteq \rho_{sf}(T)$ there is a component Ω_1 of $\rho_{sf}(T)$ which contains Ω and hence $\rho_w(T)$. Now, Ω_1 contains the resolvent $\rho(T)$. Trivially both T and T^* have the SVEP at every point of $\rho(T)$, so, see Remark 2.119, both T and T^* have the SVEP at every point of Ω_1, and in particular have the SVEP at every point of Ω. Now, let $\lambda \in \rho_w(T)$, so $\lambda I - T \in W(X)$. The SVEP of T and T^* at λ implies, see Theorems 2.97 and 2.98, that $p(\lambda I - T) = q(\lambda I - T) < \infty$. We have either $\lambda \notin \rho(T)$ or $\lambda \notin \rho(T)$. If $\lambda \notin \rho(T)$ then $\lambda \in \sigma(T)$ and hence $\lambda \in \sigma(T) \setminus \sigma_b(T) = p_{00}(T)$. Therefore, the equality $\Delta(T) = p_{00}(T)$ holds.
(ii) The proof is analogous to that of part (i): let Ω be the unique component of $\rho_{uw}(T)$. Clearly $\rho_{ap}(T) \subseteq \Omega$. Since $\rho_{uw}(T) \subseteq \rho_{sf}(T)$ there is a component Ω_1 of $\rho_{sf}(T)$ which contains Ω. Obviously, T has the SVEP at every point of Ω, by Remark 2.119. Suppose that $\lambda \in \rho_{uw}(T)$. The SVEP at λ entails, by Theorem 2.97, that $p(\lambda I - T) < \infty$, i.e., $\lambda \notin \sigma_{ub}(T)$. We have either $\lambda \in \rho_{ap}(T)$ or $\lambda \notin \rho_{ap}(T)$. If $\lambda \notin \rho_{ap}(T)$ then $\lambda \in \sigma_{ap}(T)$, so $\lambda \in \sigma_{ap}(T) \setminus \sigma_{ub}(T) = p_{00}^a(T)$. ∎

From Lemma 6.129 we see that if $\rho_w(T)$ is connected then T satisfies Browder's theorem, and, analogously, if $\rho_{uw}(T)$ is connected then T satisfies a-Browder's theorem. Set $\rho_{ap}(T) =: \mathbb{C} \setminus \sigma_{ap}(T)$.

Theorem 6.130 *Let $T \in L(X)$. Then*

(i) *$\rho_w(T)$ is connected if and only if $\rho(T)$ is connected and T satisfies Browder's theorem.*

(ii) $\rho_{uw}(T)$ *is connected if and only if* $\rho_{ap}(T)$ *is connected and* T *satisfies a-Browder's theorem.*

Proof

(i) Suppose that $\rho_w(T)$ is connected. By Lemma 6.129 then T satisfies Browder's theorem. Suppose that $\rho(T)$ is not connected. Then there is a hole of $\sigma(T)$, i.e. a bounded component Ω of $\rho(T)$. On the other hand, $\rho_w(T)$ has a unique (unbounded) component. Since $\Omega \subseteq \rho_w(T)$ we have a contradiction. Therefore, $\rho(T)$ is connected.

Conversely, assume that $\rho(T)$ is connected and that T satisfies Browder's theorem. Suppose that $\rho_w(T)$ is not connected. Then there is a bounded component Ω of $\rho_w(T)$. Then either $\Omega \subset \sigma(T)$ or $\Omega \setminus \sigma(T) \neq \emptyset$. If $\Omega \subset \sigma(T)$ then $\Omega \subseteq \sigma(T) \setminus \sigma_w(T) \subseteq \text{iso}\,\sigma(T)$, where the last inclusion follows since T satisfies Browder's theorem, and this is impossible. If $\Omega \setminus \sigma(T) \neq \emptyset$, then there exists a $\lambda_0 \in \Omega$ such that $\lambda_0 \notin \sigma(T)$. Since $\lambda_0 \in \rho(T)$ then there exists a component Ω_0 of $\rho(T)$ such that $\lambda_0 \in \Omega_0$. The set Ω_0 is an open connected set contained in $\rho_w(T)$. Furthermore, Ω and Ω_0 has a common point, so $\Omega_0 \subseteq \Omega$. Consequently, Ω_0 is a bounded component of $\rho(T)$, i.e. $\rho(T)$ is not connected, a contradiction.

(ii) The proof is similar to that of part (i). Replace in the proof of part (i) the set $\rho(T)$ by $\rho_{ap}(T)$. ∎

In the sequel by $\mathcal{K}(X)$ we shall denote the two-sided ideal of all *compact operators* in $L(X)$.

Corollary 6.131 *Let* $T \in L(X)$ *be such that* $\rho_w(T)$ *is connected and* $\text{int}\,\sigma_w(T) = \emptyset$. *Then both* $T + K$ *and* $T^* + K^*$ *have SVEP for all* $K \in \mathcal{K}(X)$.

Proof $\rho_w(T + K) = \rho_w(T^* + K^*) = \rho_w(T)$ is connected, so both $T + K$ and T^* satisfy Browder's theorem. Since $\text{int}\,\sigma_w(T) = \text{int}\,\sigma_w(T + K) = \text{int}\,\sigma_w(T^* + K^*) = \emptyset$, then, by Theorem 5.6, $T + K$ and $T^* + K^*$ have SVEP for all $K \in \mathcal{K}(X)$. ∎

Lemma 6.132 *If* $T \in L(X)$ *and* $\rho_{uw}(T)$ *is connected then* $\rho_w(T) = \rho_{uw}(T)$.

Proof We know that $\rho_{uw}(T) = \rho_{sf}^-(T) \cup \rho_{uw}(T)$ and both sets $\rho_{sf}^-(T)$ and $\rho_{uw}(T)$ are open subsets of $\mathbb{C}$. Since $\rho_{uw}(T)$ is non-empty and $\rho_{sf}^-(T) \cap \rho_{uw}(T) = \emptyset$, the assumption that $\rho_{uw}(T)$ is connected entails $\rho_{sf}^-(T) = \emptyset$. Thus, $\rho_w(T) = \rho_{uw}(T)$. ∎

We now state a result for Hilbert space operators. The proof involves some rather technical result on Hilbert spaces. For a proof we refer to [188, Proposition 3.2].

Lemma 6.133 *Let* $T \in L(H)$, H *a Hilbert space. If* Ω *is a bounded component of* $\sigma_{sf}(T)$ *then there exists a* $K \in \mathcal{K}(H)$ *such that* $\Omega \subset \sigma_p(T + K)$.

The condition that $\rho_w(T)$ is connected entails Browder's theorem for T, by part (i) of Theorem 6.130. The converse is true for Hilbert space operators:

Theorem 6.134 *Let* $T \in L(X)$ *be such that* $\rho_w(T)$ *is connected. Then* $T + K$ *satisfies Browder's theorem for all* $K \in \mathcal{K}(X)$.

Furthermore, if $T \in L(H)$, H *a Hilbert space, then the following statements are equivalent:*

(i) $\rho_w(T)$ *is connected;*
(ii) *Browder's theorem holds for* $T + K$ *for every* $K \in \mathcal{K}(H)$.

Proof (i) $\Rightarrow$ (ii) Let $S := T + K$, with $K \in K(X)$ arbitrarily given, and suppose that $\rho_w(T)$ is connected. To show that Browder's theorem holds for S we need to prove the equality $\Delta(S) = p_{00}(S)$. Observe first that

$$\rho_w(T) = \mathbb{C} \setminus \sigma_w(T) = \mathbb{C} \setminus \sigma_w(T + K) = \rho_w(S).$$

Thus, $\rho_w(S)$ is connected. By Lemma 6.129 we then have $\Delta(S) = p_{00}(S)$, so Browder's theorem holds for $S = T + K$.

To show (ii) $\Rightarrow$ (i), suppose that $T \in L(H)$, and $\rho_w(T)$ is not connected. Then $\rho_w(T)$ has a bounded component Ω, and by Lemma 6.133 there exists a $K \in \mathcal{K}(H)$ for which $\Omega \subseteq \sigma_p(T + K)$. Then

$$\Omega \subseteq \Delta(T + K) = \sigma(T + K) \setminus \sigma_w(T + K),$$

and hence $\Delta(T + K)$ cannot be contained in $\mathrm{iso}\,\sigma(T + K)$. Consequently, by Theorem 5.10, $T + K$ does not satisfy Browder's theorem. This shows that (ii) $\Rightarrow$ (i). ∎

An analogous result holds for property (*b*):

Theorem 6.135 *Let* $T \in L(X)$. *If* $\rho_{uw}(T)$ *is connected then property* (*b*) *holds for* $T + K$ *for every* $K \in \mathcal{K}(X)$.

Proof Let $K \in K(X)$ be arbitrarily given and set $S := T + K$. To prove $\Delta_a(S) = p_{00}(S)$ it suffices to prove the inclusion $\Delta_a(S) \subseteq p_{00}(S)$, since the opposite inclusion is true for every operator. Let $\lambda_0 \in \Delta_a(S) = \sigma_{ap}(S) \setminus \sigma_{uw}(S)$. Then $\lambda_0 \in \mathbb{C} \setminus \sigma_{ap}(S) = \rho_{uw}(S)$. By Lemmas 6.132 and 6.129 we have $\rho_{uw}(S) = \rho_w(S) = \rho(S) \cup p_{00}(S)$. Therefore, either $\lambda_0 \in \rho(S)$ or $\lambda_0 \in p_{00}(S)$. But $\lambda_0 \in \rho(S)$ is impossible, since $\lambda_0 \in \sigma(S)$. Hence $\lambda_0 \in p_{00}(S)$, and consequently $\Delta_a(S) \subseteq p_{00}(S)$. ∎

In the case of Hilbert space operators the result of Theorem 6.135 may be reversed:

Theorem 6.136 *Let* $T \in L(H)$, H *a Hilbert space. Then the following statements are equivalent:*

(i) $\rho_{uw}(T)$ *is connected;*
(ii) *a-Browder's theorem holds for* $T + K$ *for every* $K \in \mathcal{K}(H)$;
(iii) *property* (*b*) *holds for* $T + K$ *for every* $K \in \mathcal{K}(H)$.

Proof (i) $\Rightarrow$ (iii) has been proved in Theorem 6.135. The implication (iii) $\Rightarrow$ (ii) holds, because property (b) for T entails that T satisfies a-Browder's theorem. It suffices to prove the implication (ii) $\Rightarrow$ (i). Suppose that $\rho_{uw}(T)$ is not connected, so there exists a bounded component Ω of $\rho_{uw}(T)$ and obviously, $\Omega \subseteq \rho_{sf}(T)$. By Lemma 6.133, there exists a $K \in K(H)$ such that $\Omega \subseteq \sigma_p(T + K)$, and $\Omega \subseteq \rho_{sf}(T + K)$. Hence,

$$\Omega \subseteq \rho_{sf}(T + K) \cap \sigma_p(T + K) = \Delta_a(T + K),$$

and hence $\Delta_a(T + K)$ is not contained in $\sigma_{ap}(T + K)$, so a-Browder's theorem does not hold for $T + K$.   ∎

Remark 6.137 The proof of the equivalence of (i), (ii) and (iii) in Theorem 6.136 is based on Lemma 6.133. In the absence of a similar result for perturbed Banach space operators, a corresponding result does not seem to be possible for Banach space operators.

Lemma 6.138 *Let $T \in L(X)$, X a Banach space.*

(i) *If* iso $\sigma_w(T) = \emptyset$ *then* $\pi_{00}(T) \subseteq \rho_{sf}(T)$ *and* $p_{00}(T) = \pi_{00}(T)$.
(ii) *If* iso $\sigma_{uw}(T) = \emptyset$ *then* $\pi_{00}^a(T) \subseteq \rho_{sf}(T)$ *and* $p_{00}^a(T) = \pi_{00}^a(T)$.

Proof

(i) If $\lambda_0 \in \pi_{00}(T)$ then $\lambda \in$ iso $\sigma(T)$. Suppose that $\lambda_0 \notin \rho_{sf}(T)$. Then $\lambda_0 \in \sigma_{sf}(T) \subseteq \sigma_w(T)$, hence $\lambda_0 \in$ iso $\sigma_w(T) \neq \emptyset$, a contradiction.

To show the equality $p_{00}(T) = \pi_{00}(T)$, it suffices to show that $\pi_{00}(T) \subseteq p_{00}(T)$. Let $\lambda \in \pi_{00}(T)$. Then $\alpha(\lambda I - T) < \infty$, $\lambda \in$ iso(T), and $\lambda \in \rho_{sf}(T)$ by the first part, so $\lambda I - T \in \Phi_\pm(X)$. By Theorems 2.97 and 2.98 we have $p(\lambda I - T) = q(\lambda I - T) < \infty$ and hence $\alpha(\lambda I - T) = \beta(\lambda I - T) < \infty$, by Theorem 1.22. Thus, $\lambda \in p_{00}(T)$.

(ii) If $\lambda_0 \in \pi_{00}^a(T)$ then $\lambda_0 \in$ iso $\sigma_{ap}(T)$, so there exists a $\delta > 0$ such that $\lambda I - T$ is bounded below for all $0 < |\lambda| < \delta$, in particular $\lambda I - T \in W_+(X)$ for all $0 < |\lambda| < \delta$. Suppose that $\lambda_0 \notin \rho_{sf}(T)$. Then $\lambda_0 \in \sigma_{sf}(T) \subseteq \sigma_{uw}(T)$, hence $\lambda_0 \in$ iso $\sigma_{uw}(T) \neq \emptyset$, a contradiction.

To show the equality $p_{00}^a(T) = \pi_{00}^a(T)$, it suffices to show that $\pi_{00}^a(T) \subseteq p_{00}^a(T)$. Let $\lambda \in \pi_{00}^a(T)$. Then $\alpha(\lambda I - T) < \infty$, $\lambda \in$ iso $\sigma_{ap}(T)$, and $\lambda \in \rho_{sf}(T)$, so $\lambda I - T \in \Phi_+(X)$. By Theorem 2.97 we then have $p(\lambda I - T) < \infty$, thus $\lambda \in p_{00}^a(T)$.   ∎

We now state another result for Hilbert space operators. Its proof depends upon the following lemma, whose rather technical proof may be found in [188].

Lemma 6.139 *Let $T \in L(H)$, H a Hilbert space.*

(i) *If $\lambda \in$ iso $\sigma_w(T)$ then there exists a compact operator $K \in K(H)$ for which $\lambda \in \pi_{00}(T + K)$.*
(ii) *If $\lambda \in$ iso $\sigma_{uw}(T)$ then there exists a compact operator $K \in K(H)$ for which $\lambda \in \pi_{00}^a(T + K)$.*

Theorem 6.140 *Let $T \in L(X)$, X a Banach space, be such that $\rho_w(T)$ is connected and* iso $\sigma_w(T) = \emptyset$. *Then $T + K$ satisfies Weyl's theorem for every $K \in \mathcal{K}(X)$.*

Furthermore, if $T \in L(H)$, H a Hilbert space, then the following statements are equivalent:

(i) $\rho_w(T)$ *is connected and* iso $\sigma_w(T) = \emptyset$;

(ii) *Weyl's theorem holds for $T + K$ for every $K \in \mathcal{K}(H)$.*

Proof Suppose that $\rho_w(T)$ is connected and iso $\sigma_w(T) = \emptyset$. Let $S := T + K$, where $K \in K(X)$ is arbitrarily given. Since $\sigma_w(T) = \sigma_w(S)$, $\rho_w(T) = \rho_w(S)$ is connected, so, by Lemmas 6.129 and 5.25, $\Delta(S) = p_{00}(S)$. On the other hand iso $\sigma_w(T + K) =$ iso $\sigma_w(T) = \emptyset$, so by Lemma 6.138, we also have $\pi_{00}(S) = p_{00}(S) = \Delta(S)$. Thus, Weyl's theorem holds for $S = T + K$.

We show now that if $T \in L(H)$, H a Hilbert space, then (ii) $\Rightarrow$ (i). Suppose that Weyl's theorem holds for $T + K$ for every $K \in \mathcal{K}(H)$. Then

$$\pi_{00}(T + K) = p_{00}(T + K) \quad \text{for all } K \in K(H). \tag{6.25}$$

Assume that iso $\sigma_w(T) \neq \emptyset$ and choose $\lambda \in$ iso $\sigma_w(T)$. By Lemma 6.139 there exists a $K_0 \in \mathcal{K}(H)$ such that $\lambda \in \pi_{00}(T + K_0)$, and hence $\lambda \in$ iso $\sigma(T + K_0)$. Since $\lambda \in$ iso $\sigma_w(T)$, by Theorem 3.58 we have $\lambda \in \sigma_{\mathrm{usf}}(T)$, and since $\sigma_{\mathrm{usf}}(T)$ is stable under compact perturbations, $\lambda \in \sigma_{\mathrm{usf}}(T + K_0) \subseteq \sigma_b(T + K_0)$. Consequently,

$$\lambda \notin \sigma(T + K_0) \setminus \sigma_b(T + K_0) = p_{00}(T + K_0).$$

Therefore $\pi_{00}(T + K_0) \neq p_{00}(T + K_0)$, contradicting the equality (6.25). Thus, iso $\sigma_w(T) = \emptyset$ for every $K \in \mathcal{K}(H)$.

On the other hand, assume that $\rho_w(T)$ is not connected, and let Ω be a bounded component of $\rho_w(T)$. Evidently, Ω is also a bounded component of $\rho_{\mathrm{sf}}(T)$ and by Lemma 6.133 there exists a $K_0 \in K(H)$ such that $\Omega \subset \sigma_p(T)$. Then $\Omega \subseteq \Delta(T + K_0)$, and since Weyl's theorem for $T + K_0$ entails that $T + K_0$ satisfies Browder's theorem, by Theorem 5.10 we then have $\Omega \subseteq$ iso $\sigma(T + K_0)$, which is impossible. Thus $\rho_w(T)$ is connected. ∎

Lemma 6.141 *If* iso $\sigma_w(T) = \emptyset$ *then T is polaroid and hence isoloid.*

Proof Let $\lambda \in$ iso $\sigma(T)$. Then either $\lambda \in \sigma_w(T)$ or $\lambda \notin \sigma_w(T)$. If $\lambda \in \sigma_w(T)$ then $\lambda \in$ iso $\sigma_w(T)$ and this is impossible. Therefore, $\lambda \notin \sigma_w(T)$, so $\lambda I - T$ is Weyl and, since both T and T^* have the SVEP at every isolated point of the spectrum, it then follows, from Theorem 2.100, that $p(\lambda I - T) = q(\lambda I - T) < \infty$, i.e., λ is a pole of the resolvent. ∎

Trivially, by Theorem 6.140, if $\rho_w(T)$ is connected and iso $\sigma_w(T) = \emptyset$ then T satisfies Weyl's theorem. Assuming SVEP and $KT = TK$, the condition that $\rho_w(T)$ is connected may be omitted:

Corollary 6.142 *Assume that T or T^* has the SVEP. If* iso $\sigma_{\mathrm{w}}(T) = \emptyset$ *then $f(T)$ satisfies Weyl's theorem for every $f \in \mathcal{H}_{nc}(\sigma(T))$. Additionally, if $KT = TK$, $K \in \mathcal{K}(X)$, then Weyl's theorem holds for $f(T + K)$ for every $f \in \mathcal{H}_{nc}(\sigma(T + K))$.*

Proof As observed above the condition iso $\sigma_{\mathrm{w}}(T) = \emptyset$ entails that T is polaroid and hence $f(T)$ is polaroid for every $f \in \mathcal{H}_{nc}(\sigma(T))$, by Theorem 4.19. The SVEP for T or T^* entails the SVEP for $f(T)$, or $f(T^*)$, by Theorem 2.86. Hence, by Theorem 6.43, Weyl's theorem holds for $f(T)$.

Additionally, assuming $TK = KT$, if T has the SVEP then $T + K$ has the SVEP, by Theorem 2.129, and, by Lemma 6.141, $T + K$ is polaroid since iso $\sigma_{\mathrm{w}}(T + K) =$ iso $\sigma_{\mathrm{w}}(T) = \emptyset$. Then $f(T + K)$ is polaroid for every $f \in \mathcal{H}_{nc}(\sigma(T + K))$, again by Theorem 4.19, and hence, again by Theorem 6.43, Weyl's theorem holds for $f(T + K)$ for every $f \in \mathcal{H}_{nc}(\sigma(T + K))$.   ∎

Theorem 6.143 *Let $T \in L(X)$, X a Banach space, be such that $\rho_{\mathrm{uw}}(T)$ is connected and* iso $\sigma_{\mathrm{uw}}(T) = \emptyset$. *Then $T + K$ satisfies both property (w) and a-Weyl's theorem, for every $K \in \mathcal{K}(X)$.*

If $T \in L(H)$, H a Hilbert space, then the following statements are equivalent:

(i) *$\rho_{\mathrm{w}}(T)$ is connected and* iso $\sigma_{\mathrm{w}}(T) = \emptyset$;
(ii) *property (ω) holds for $T + K$ for every $K \in \mathcal{K}(H)$;*
(iii) *a-Weyl's theorem holds for $T + K$ for every $K \in \mathcal{K}(H)$.*

Proof Let $S := T + K$, where $K \in \mathcal{K}(X)$. The condition that $\rho_{\mathrm{uw}}(T)$ is connected entails a-Browder's theorem, by Theorem 6.136, so, to show property (ω) for S it suffices to prove $\pi_{00}^a(S) = p_{00}(S)$. Since $p_{00}(S) \subseteq \pi_{00}^a(S)$ holds for every operator, it suffices to prove the inclusion $\pi_{00}^a(S) \subseteq p_{00}(S)$. Observe that iso $\sigma_{\mathrm{w}}(S) =$ iso $\sigma_{\mathrm{w}}(T) = \emptyset$, so, by Lemma 6.138, we have $\pi_{00}^a(S) \subseteq \rho_{\mathrm{sf}}(S)$. If $\lambda_0 \in \pi_{00}^a(S)$ then $\lambda_0 \in$ iso $\sigma_{\mathrm{ap}}(T)$ and $\lambda_0 I - S \in \Phi_{\pm}(X)$. By Theorem 2.97 we have $p(\lambda_0 I - T) < \infty$ and hence $\alpha(\lambda_0 I - T) \le \beta(\lambda_0 I - T)$, by Theorem 1.22. This implies $\lambda_0 I - S \in \Phi_+(X)$ and ind $(\lambda_0 I - T) \le 0$, so $\lambda_0 \in \rho_{\mathrm{uw}}(T)$.

By Lemma 6.132 we have $\rho_{\mathrm{uw}}(S) = \rho_{\mathrm{w}}(S)$, and from Lemma 6.129, it then follows that $\rho_{\mathrm{uw}}(S) = \rho(S) \cup p_{00}(S)$. Since $\lambda \notin \rho(S)$, we then conclude that $\lambda_0 \in p_{00}(S)$, hence $\pi_{00}^a(S) \subseteq p_{00}(S)$. Therefore, $\pi_{00}^a(S) = p_{00}(S)$ and hence property (ω) holds for S.

To show that a-Weyl's theorem holds for S, note that $\pi_{00}^a(S) = p_{00}(S) \subseteq p_{00}^a(S)$, by Lemma 5.25. Since the inclusion $p_{00}^a(S) \subseteq \pi_{00}^a(S)$ holds for every operator, it then follows that $p_{00}^a(S) = \pi_{00}^a(S)$ and taking into account that a-Browder's theorem holds for S it then follows, from Theorem 6.70, that a-Weyl's theorem holds for S.

(ii) $\Rightarrow$ (i) Suppose that $T \in L(H)$ and property (ω) holds for $T + K$ for every $K \in \mathcal{K}(H)$. In particular, by Theorem 6.89 property (R) holds for $T + K$, i.e.,

$$\pi_{00}^a(T + K) = p_{00}(T + K) \text{ for every } K \in \mathcal{K}(H). \tag{6.26}$$

We show that if one of the conditions (i) is not satisfied then there exists a compact operator $K_0 \in \mathcal{K}(H)$ such that $T + K_0$ does not satisfy property (ω). If $\rho_{\mathrm{uw}}(T)$ is not connected then, by Theorem 6.136, there exists a $K_0 \in K(H)$ such that $T + K_0$ does not satisfy a-Browder's theorem and hence property (ω) does not hold for $T + K_0$, since property (ω) entails a-Browder's theorem. Suppose the other case, iso $\sigma_{\mathrm{uw}}(T) \neq \emptyset$, and choose $\lambda \in$ iso $\sigma_{\mathrm{uw}}(T)$. Then, by Lemma 6.139 there exists a $K_0 \in \mathcal{K}(H)$ such that $\lambda \in \pi_{00}^a(T + K_0)$. Evidently, $\lambda \in$ iso $\sigma_{\mathrm{uw}}(T + K_0) =$ iso $\sigma_{\mathrm{uw}}(T)$. From Theorem 3.58, we know that $\lambda \in \sigma_{\mathrm{usf}}(T + K_0) \subseteq \sigma_{\mathrm{b}}(T + K_0)$, so

$$\lambda \notin \sigma(T + K_0) \setminus \sigma_{\mathrm{b}}(T + K_0) = p_{00}(T + K_0),$$

contradicting (6.26). Thus iso $\sigma_{\mathrm{uw}}(T) = \emptyset$.

(i) $\Rightarrow$ (iii) follows by the first part of the proof.

(iii) $\Rightarrow$ (i) Proceed as before: Suppose that a-Weyl's theorem holds for $T + K$ for every $K \in \mathcal{K}(H)$. If $\rho_{\mathrm{uw}}(T)$ is not connected then, by Theorem 6.136, there exists a $K_0 \in \mathcal{K}(H)$ such that $T + K_0$ does not satisfy a-Browder's theorem and hence a-Weyl's theorem does not hold for $T + K_0$, a contradiction. Hence $\rho_{\mathrm{uw}}(T)$ is connected, so property (b) holds for every $T + K$, by Theorem 6.136, and hence

$$\Delta_a(T + K) = p_{00}(T + K) \text{ for all } K \in \mathcal{K}(H).$$

Now, if iso $\sigma_{\mathrm{uw}}(T) \neq \emptyset$ then, as above, there exists a $K_0 \in \mathcal{K}(H)$ such that

$$\pi_{00}^a(T + K_0) \neq p_{00}(T + K_0) = \Delta_a(T + K_0),$$

so a-Weyl's theorem does not hold for $T + K_0$ ∎

Lemma 6.144 *Every operator $T \in L(X)$ for which* iso $\sigma_{\mathrm{uw}}(T) = \emptyset$ *is both left polaroid and finite a-isoloid.*

Proof Suppose that $\lambda_0 \in$ iso $\sigma_{\mathrm{ap}}(T)$ is arbitrary. Then there is an $\varepsilon > 0$ such that $\lambda I - T$ is bounded below for all $0 < |\lambda - \lambda_0| < \varepsilon$, and hence $\lambda I - T \in W_+(X)$ for all $0 < |\lambda - \lambda_0| < \varepsilon$. Now, we have either $\lambda_0 \in \sigma_{\mathrm{uw}}(T)$ or $\lambda_0 \notin \sigma_{\mathrm{uw}}(T)$. If $\lambda_0 \in \sigma_{\mathrm{uw}}(T)$ then we would have $\lambda_0 \in$ iso $\sigma_{\mathrm{uw}}(T) \neq \emptyset$, a contradiction. Therefore, $\lambda_0 \notin \sigma_{\mathrm{uw}}(T)$, i.e. $\lambda_0 I - T \in W_+(X)$. Since T has the SVEP at every isolated point of $\sigma_{\mathrm{ap}}(T)$, it then follows, by Theorem 2.97, that $p(\lambda_0 I - T) < \infty$, hence $\lambda_0 I - T \in B_+(X)$, so λ_0 is a left pole of the resolvent, hence T is left polaroid. Since $(\lambda_0 I - T)(X)$ is closed, the condition $\lambda_0 \in \sigma_{\mathrm{ap}}(T)$ entails that $\alpha(\lambda_0 I - T) > 0$, thus T is a-isoloid. But $\alpha(\lambda_0 I - T) < \infty$, since $\lambda_0 I - T \in B_+(X)$, so T is finite a-isoloid. ∎

Assuming that a Riesz operator R commutes with T, the result of Theorem 6.143 may be extended to Riesz operators:

Corollary 6.145 *Let $T \in L(X)$ and suppose $\rho_{\mathrm{uw}}(T)$ is connected and* iso $\sigma_{\mathrm{uw}}(T) = \emptyset$. *Then $T + R$ satisfies a-Weyl's theorem, or equivalently $T + R$ satisfies the generalized a-Weyl's theorem, for every Riesz operator R which commutes with T.*

Proof By Corollary 3.18 $\sigma_{uw}(T + R) = \sigma_{uw}(T)$ is connected and

$$\text{iso } \sigma_{uw}(T + R) = \text{iso } \sigma_{uw}(T) = \emptyset,$$

so a-Weyl's theorem holds for $T + R$, and this is equivalent, by Theorem 6.120, to $T + R$ satisfying the generalized a-Weyl's theorem.   ∎

Theorem 6.146 *Let* $T \in L(X)$ *be such that* $\rho_{uw}(T)$ *is connected and* iso $\sigma_{uw}(T) = \emptyset$. *If* $K \in \mathcal{K}(X)$ *we have:*

(i) *The generalized a-Weyl's theorem holds for* $T + K$.

(ii) *If* T *or* T^* *has the SVEP,* R *is a Riesz operator for which* $RT = TR$, *and* $f \in \mathcal{H}(\sigma(T + R))$, *where* f *is injective and non-constant on the components of* $\sigma(T + R)$, *then the generalized a-Weyl's theorem holds for* $f(T + R)$.

(iii) *If* T^* *has the SVEP,* R *a Riesz operator for which* $RT = TR$, *and* $f \in \mathcal{H}(\sigma(T + R))$ *is injective and non-constant on the components of* $\sigma(T + R)$, *then generalized property* (ω) *holds for* $f(T + R)$.

Proof

(i) By Theorem 6.143, $T + K$ satisfies a-Weyl's theorem for all $K \in \mathcal{K}(X)$. According to Lemma 6.144, the condition iso $\sigma_{uw}(T) = \text{iso } \sigma_{uw}(T + K) = \emptyset$ implies that $T + K$ is left polaroid. This implies, by Theorem 6.120, that a-Weyl's theorem holds for $T + K$ if and only if the generalized a-Weyl's theorem holds for $T + K$.

(ii) The SVEP for T or T^* is extended to $T + R$ or $T^* + R^*$, by Theorem 2.129. This ensures that the spectral mapping theorem holds for $\sigma_{uw}(T + R)$, by Corollary 3.120, and that a-Browder's theorem holds for T as well as for $T + R$, by Corollary 5.5. Thus

$$\text{iso } \sigma_{uw}(T + R) = \text{iso } \sigma_{ub}(T + R) = \text{iso } \sigma_{ub}(T) = \text{iso } \sigma_{uw}(T) = \emptyset$$

and analogously $\rho_{uw}(T + R)$ is connected. Thus $T + R$ is a-isoloid, by Lemma 6.144, and $T + R$ satisfies a-Weyl's theorem, by Theorem 6.145. By Theorem 6.80, $f(T + R)$ satisfies a-Weyl's theorem. Since $T + R$ is left polaroid $f(T + R)$ is also left polaroid, by Theorem 4.21. Therefore, as in part (i), the generalized a-Weyl's theorem holds for $f(T + R)$.

(iii) $T + R$ is polaroid and has the SVEP, so the generalized property (ω) for $f(T + R)$, by Theorem 6.100.   ∎

6.8   Weyl's Theorem for Toeplitz Operators

The results of the previous section apply in particular to Toeplitz analytic operators on Hardy spaces. As observed in Chap. 3, if $\phi \in C(\mathbf{T})$, $\sigma_w(T_\phi)$ consists of $\mathbf{\Gamma} := \phi(\mathbf{T})$ and every Toeplitz operator with continuous symbol is polaroid. The Toeplitz operators defined in Example 4.101 shows that the SVEP may fail for both T_ϕ and T'_ϕ, so the result of Theorem 6.43 cannot be applied to deduce Weyl's theorem for T_ϕ. However, we know, see Corollary 4.98, that $\sigma(T_\phi) = \sigma_w(T_\phi)$, and hence $\Delta(T_\phi) = \sigma(T_\phi) \setminus \sigma_w(T_\phi) = \emptyset$. Moreover, if ϕ is non-constant, by Theorem 4.99 we have iso $\sigma(T_\phi) = $ iso $\sigma_w(T_\phi)$ and hence $\pi_{00}(T) = \emptyset$.

Theorem 6.147 *If $\phi \in C(\mathbf{T})$ is non-constant then T_ϕ satisfies Weyl's theorem.*

The next Example 6.154 provides an example of a Toeplitz operator T_ϕ which satisfies Weyl's theorem, but whose symbol ϕ is not continuous.

We also know that $\sigma_w(T_\phi)$ consists of Γ and those holes with respect to which the winding number of ϕ is nonzero, while $\sigma_{uw}(T_\phi)$ consists of Γ and those holes with respect to which the winding number of ϕ is negative. Therefore, $\rho_w(T_\phi)$ is connected (respectively, $\rho_{uw}(T_\phi)$ is connected) precisely when the winding number of ϕ with respect to each hole of Γ is nonzero (respectively, the winding number of ϕ with respect to each hole of Γ is negative). From Theorems 6.134, 6.136, and 6.140 we then obtain:

Theorem 6.148 *Let $0 \neq \phi \in C(\mathbf{T})$, and T_ϕ be the corresponding Toeplitz operator on $L^2(\mathbf{T})$. Then we have*

(i) *$T_\phi + K$ satisfies Browder's theorem for any compact operator K on $L^2(\mathbf{T})$ if and only if the winding number of ϕ with respect to each hole of $\phi(\mathbf{T})$ is nonzero. $T_\phi + K$ satisfies Weyl's theorem, for any compact operator K on $L^2(\mathbf{T})$, if and only if ϕ is non-constant and the winding number of ϕ with respect to each hole of $\phi(\mathbf{T})$ is nonzero.*

(ii) *$T_\phi + K$ satisfies a-Browder's theorem for any compact operator K on $L^2(\mathbf{T})$ if and only if the winding number of ϕ with respect to each hole of $\phi(\mathbf{T})$ is negative. $T_\phi + K$ satisfies both a-Weyl's theorem and property (w), for any compact operator K on $L^2(\mathbf{T})$, if and only if ϕ is non-constant and the winding number of ϕ with respect to each hole of $\phi(\mathbf{T})$ is negative.*

We now consider the special case when $\phi \in H^\infty(\mathbf{T})$.

Theorem 6.149 *If $\phi \in H^\infty(\mathbf{T})$ then Weyl's theorem holds for $f(T_\phi)$, for every $f \in \mathcal{H}(\sigma(T_\phi))$. If K is an algebraic operator on $L^2(\mathbf{T})$ which commutes with T_ϕ, then Weyl's theorem holds for $f(T_\phi + K)$ for every $f \in \mathcal{H}_{nc}(\sigma(T_\phi + K))$.*

Proof If $\phi \in H^\infty(\mathbf{T})$ then, by Theorem 4.91 T_ϕ is hereditarily polaroid and has the SVEP. By Corollary 6.44 then Weyl's theorem holds for $f(T_\phi)$. The last assertion follows from Theorem 6.127. ∎

In the case when $\phi \in C(\mathbf{T})$ the argument used in the proof of Theorem 6.149 does not work, since, as already noted, both T_ϕ and T'_ϕ may fail SVEP. However, if ϕ satisfies condition (i) of Theorem 4.100, or the condition of Theorem 4.102, then T_ϕ has the SVEP and is also polaroid. From Corollary 6.44 we then conclude that Weyl's theorem holds for $f(T_\phi)$.

Lemma 6.150 *Let* $\phi \in C(\mathbf{T})$ *and let* $f \in \mathcal{H}(\sigma(T_\phi))$. *Then* $T_{f \circ \phi} - f(T_\phi)$ *is compact.*

Proof We know that T_ϕ is invertible if and only if T_ϕ is Fredholm. Therefore, for every $\lambda \notin \sigma(T_\phi)$ both the functions $\phi - \lambda$ and $\overline{\phi - \lambda}$ are invertible in $C(\mathbf{T})$ and $(\phi - \lambda)^{-1} \in C(\mathbf{T})$. Using this fact and Theorem 4.93 we have, for $\psi \in L^\infty(\mathbf{T})$ and $\lambda, \mu \in \mathbb{C}$

$$T_{\phi - \mu} T_\psi T_{\phi - \lambda}^{-1} - T_{(\phi - \mu)\psi(\phi - \lambda)^{-1}} \in K(L^2(\mathbf{T})) \text{ whenever } \lambda \notin \sigma(T_\phi).$$

The argument used above also works for rational functions to yield: If r is any rational function with all its poles outside of $\sigma(T_\phi)$, then $r(T_\phi) - T_{r \circ \phi}$ is compact on $L^2(\mathbf{T})$. Suppose that $f \in \mathcal{H}(\sigma(T_\phi))$. By Runge's theorem (see Appendix A) there exists a sequence of rational functions r_n for which the poles of each r_n lie outside of $\sigma(T_\phi)$, and $r_n \to f$ uniformly on $\sigma(T_\phi)$. Thus, $r_n(T_\phi) \to f(T_\phi)$ in the norm topology of $L(L^2(\mathbf{T}))$. Since $r_n \circ \phi \to f \circ \phi$ uniformly, it then follows that $T_{r_n \circ \phi}$ converges to $T_{f \circ \phi}$. Therefore $T_{f \circ \phi} - f(T_\phi)$ is the limit of the compact operators $T_{r_n \circ \phi} - r_n(T_\phi)$, hence is a compact operator.   ∎

Corollary 6.151 *If* $\phi \in C(\mathbf{T})$ *and* $f \in \mathcal{H}(\sigma(T_\phi))$, *then*

$$\sigma_{\mathrm{w}}(f(T_\phi)) = \sigma(T_{f \circ \phi}).$$

Proof The Weyl spectrum is stable under compact perturbations, so, by Lemma 6.150

$$\sigma_{\mathrm{w}}(f(T_\phi)) = \sigma_{\mathrm{w}}(T_{f \circ \phi}) = \sigma(T_{f \circ \phi}).$$

   ∎

We now characterize the operators T_ϕ for which $f(T_\phi)$ satisfies Weyl's theorem.

Theorem 6.152 *Let* $\phi \in C(\mathbf{T})$ *and* $f \in \mathcal{H}(\sigma(T_\phi))$. *Then*

$$\sigma(T_{f \circ \phi}) \subseteq f(\sigma(T_\phi)).$$

$f(T_\phi)$ *satisfies Weyl's theorem if and only if* $\sigma(T_{f \circ \phi}) = f(\sigma(T_\phi))$.

Proof By Corollary 6.151 we have

$$\sigma(T_{f \circ \phi}) \subseteq \sigma_{\mathrm{w}}(f(T_\phi)) \subseteq \sigma(f(T_\phi)) = f(\sigma(T_\phi)).$$

Since $\sigma(T_\phi)$ is connected, $f(\sigma(T_\phi)) = f(\sigma(T_\phi))$ is connected. Therefore, $\pi_{00}(f(T_\phi)) = \emptyset$. By Corollary 6.151 we then have $\sigma_{\mathrm{w}}(f(T_\phi)) = \sigma(T_{f \circ \phi})$. Consequently, $\sigma(f(T_\phi)) \setminus \sigma_{\mathrm{w}}(f(T_\phi)) = \emptyset$ if and only if $\sigma(T_{f \circ \phi}) = f(\sigma(T_\phi))$.   ∎

We now show that there exists a continuous function for which $\sigma(T_{\phi^2}) \neq [\sigma(T_\phi)]^2$.

Example 6.153 Let ϕ be defined as in Example 4.101. We know that $\sigma(T_\phi) = \text{conv}\,\phi(\mathbf{T})$ and $\sigma_e(T_\phi) = \phi(\mathbf{T})$. A straightforward calculation shows that $\phi^2(\mathbf{T})$ is the cardioid Γ having equation $\rho = 2(1 + \cos\theta)$. In particular, $\phi^2(\mathbf{T})$ traverses Γ once in a counterclockwise direction, and then traverses Γ once in clockwise direction. Thus, $wn(\phi^2, \lambda) = 0$ for each λ in the hole of $\phi^2(\mathbf{T})$. This shows that $\sigma(T_{\phi^2})$ is the cardioid Γ, and since

$$[\sigma(T_\phi)]^2 = \text{conv}\,\phi(\mathbf{T}) = \{(\rho, \theta) : \rho \leq 2(1 + \cos\theta)\},$$

it then follows that $\sigma(T_{\phi^2}) \neq [\sigma(T_\phi)]^2$.

The Toeplitz operator defined in the previous example satisfies Weyl's theorem, while, by Theorem 6.152, T_ϕ^2 does not satisfy Weyl's theorem.

Example 6.154 If ϕ is not continuous it is possible that Weyl's theorem holds for $f(T_\phi)$ without $\sigma(T_{f \circ \phi})$ being equal to $f(\sigma(T_\phi))$. Define

$$\phi(e^{i\theta}) := e^{\frac{i\theta}{3}}, \quad \text{where} \quad 0 \leq \theta \leq 2\pi.$$

Evidently, ϕ is a piecewise continuous function. The operator T_ϕ is invertible, while T_{ϕ^2} is not invertible, so $0 \in \sigma(T_{\phi^2}) \setminus [\sigma(T_\phi)]^2$. However, we have $\sigma(T_\phi) = \sigma_w(T_\phi)$, see the following figure:

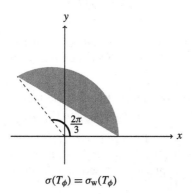

$$\sigma(T_\phi) = \sigma_w(T_\phi)$$

Because the set $\pi_{00}(T_\phi)$ is empty, Weyl's theorem holds for T_ϕ. Note that the equality $\sigma(T_\phi^2) = \sigma_w(T_\phi^2)$ is still true, see the following figure,

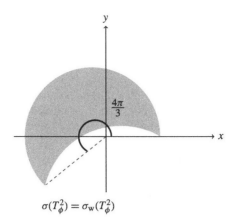

$$\sigma(T_\phi^2) = \sigma_{\mathrm{w}}(T_\phi^2)$$

and since $\pi_{00}(T_\phi^2)$ is also empty it then follows that Weyl's theorem holds for T_ϕ^2.

Corollary 6.155 *Let* $\phi \in C(T)$ *and suppose that either* T_ϕ *or* $T\phi'$ *has the SVEP. Then Weyl's theorem holds for* $f(T_\phi)$ *and* $f(T_\phi')$ *for every* $f \in \mathcal{H}(\sigma(T_\phi))$. *Furthermore,* $\sigma(T_{f \circ \phi}) = f(\sigma(T_\phi))$ *for every* $f \in \mathcal{H}(\sigma(T_\phi))$.

Proof T_ϕ is polaroid and either T_ϕ or $T\phi'$ has the SVEP. By Corollary 6.44, Weyl's theorem holds for $f(T_\phi)$ for every $f \in \mathcal{H}(\sigma(T_\phi))$. By Theorem 6.152 then $\sigma(T_{f \circ \phi}) = f(\sigma(T_\phi))$ for every $f \in \mathcal{H}(\sigma(T_\phi))$. ∎

Since for $\phi \in H^\infty(T)$, T_ϕ has the SVEP, since it is hyponormal, we also have:

Corollary 6.156 *If* $\phi \in H^\infty(T)$ *then* $\sigma(T_{f \circ \phi}) = f(\sigma(T_\phi))$ *for every* $f \in \mathcal{H}(\sigma(T_\phi))$.

In particular, by Theorem 4.100 the result of Corollary 6.155 applies to the case where the orientation of the curve $\phi(T)$ traced out by ϕ is counterclockwise or clockwise.

Remark 6.157 If ϕ is not continuous it is possible that Weyl's theorem holds for some $f(T_\phi)$ without $\sigma(T_{f \circ \phi})$ being equal to $f(\sigma(T_\phi))$.

Example 6.158 Let ϕ have the form $p(\frac{a}{z} + bz)$, where $a, b \in \mathbb{R}$ and $p(z)$ is any polynomial. Then Weyl's theorem holds for $f(T_\phi)$ for every $f \in \mathcal{H}(\sigma(T_\phi))$. Indeed, if $a = b$ then T_ϕ is self-adjoint and the assertion is evident, since T_ϕ has the SVEP. If $a \neq b$ set $\psi = \frac{a}{z} + bz$. Then

$$\psi(T) = \left\{ (u, v) \in \mathbb{C} : \left(\frac{u}{b + a} \right)^2 + \left(\frac{v}{b - a} \right)^2 = 1 \right\},$$

which is a circle or an ellipse. Then

$$\phi(T) = (p \circ \psi)(T) = p(\psi)(T),$$

and hence $\phi(\mathbf{T})$ has no holes, or exactly one hole (in this case it is oriented counterclockwise), because polynomials map continuous curves onto continuous curves and open sets onto open sets. Therefore, Weyl's theorem holds for $f(T_\phi)$ for every $f \in \mathcal{H}(\sigma(T_\phi))$.

Theorem 6.159 *If $\phi \in C(\mathbf{T})$ is such that $\sigma(T_\phi)$ has planar Lebesgue measure zero then Weyl's theorem holds for $f(T_\phi)$ and $f(T'_\phi)$ for every $f \in \mathcal{H}(\sigma(T_\phi))$.*

Proof T_ϕ is polaroid and has the SVEP, by Theorem 4.102, so Theorem 6.43 applies. ∎

It should be noted that if $\phi \in C(\mathbf{T})$ has planar Lebesgue measure zero then T_ϕ is the sum of a normal operator and a compact operator. Indeed, T_ϕ is essentially normal, and as observed after Theorem 4.93, T_ϕ is the sum of a normal operator and a compact operator since $\lambda I - T$ is Weyl for every $\lambda \notin \sigma_e(T_\phi)$.

6.9 Weyl's Theorem for Isometries and Weighted Shift Operators

In this section we exhibit some other classes of operators for which the resolvent sets $\rho_w(T)$ and $\rho_{uw}(T)$ are connected, and iso $\sigma_w(T) = \emptyset$, or iso $\sigma_{uw}(T) = \emptyset$.

(a) Let $T \in L(X)$, X an infinite-dimensional Banach space, be non-invertible and $i(T) = r(T)$, where $r(T)$ denotes the spectral radius of T and

$$i(T) := \lim_{n \to \infty} k(T^n)^{1/n}$$

has been defined in Chap. 3.

We have seen in Chap. 3 that $\sigma(T) = \sigma_w(T) = \mathbf{D}(0, r(T))$, while $\sigma_{ap}(T) = \sigma_{sf}(T) = \partial\mathbf{D}(0, r(T))$. Since

$$\sigma_{sf}(T) \subseteq \sigma_{uw}(T) \subseteq \sigma_{ap}(T),$$

it then follows that $\sigma_{uw}(T) = \partial\mathbf{D}(0, r(T))$. Suppose that T is not quasi-nilpotent, i.e. $r(T) > 0$. Then $\rho_w(T)$ is connected and iso $\sigma_w(T) = \emptyset$, so by Theorem 6.140, $T + K$ satisfies Weyl's theorem for all $K \in K(H)$. The condition $i(T) = r(T)$ entails that T has the SVEP, so, by Corollary 6.142, $f(T)$ satisfies Weyl's theorem for all $f \in \mathcal{H}(\sigma(T))$. The SVEP for T also entails that a-Browder's theorem holds for T. Since $\Delta_a(T) = \pi_{00}^a(T) = \emptyset$, T satisfies a-Weyl's theorem. Of course, if T is a Hilbert space operator, since $\rho_{uw}(T) = \mathbb{C} \setminus \partial\mathbf{D}(0, r(T))$ is not connected, there exists a $K \in \mathcal{K}(H)$ for which $T + K$ does not satisfy a-Weyl's theorem.

(b) **Weyl-type theorems for isometries** Let $T \in L(X)$ be an isometry on a Banach space X. Then $i(T) = r(T) = 1$, and by Corollary 4.71, T always has the SVEP, so T satisfies Browder's theorem. Suppose first that T is non-invertible. By Theorem 4.72, every non-invertible isometry is a-polaroid, and, by part (ii) of

Theorem 4.68, $\sigma(T) = \mathbf{D}(0, 1)$, and $\sigma_{\mathrm{ap}}(T) = \Gamma$, Γ the unit circle of $\mathbb{C}$. Therefore $\rho(T) = \mathbb{C} \setminus \mathbf{D}(0, 1)$ is connected, and by part (ii) of Theorem 6.130 we deduce that $\rho_{\mathrm{w}}(T)$ is connected. Actually, we have $\sigma_{\mathrm{w}}(T) = \mathbf{D}(0, 1)$. Indeed, suppose that $\lambda \notin \sigma_{\mathrm{w}}(T)$ and $\lambda \in \sigma(T)$. Then $\lambda I - T \in W(X)$, and the SVEP for T implies, by Theorem 2.97, that $p(\lambda I - T) < \infty$ and hence, by Theorem 1.22, $q(\lambda I - T) < \infty$, so λ is a pole of T, in particular an isolated point of $\sigma(T)$, and this is impossible. Therefore iso $\sigma_{\mathrm{w}}(T) = \emptyset$. Combining Theorem 6.140, part (ii) of Theorem 6.126, and part (ii) of Theorem 6.123, we then obtain:

Theorem 6.160 *If T is a non-invertible isometry on a Banach space X then $T + K$ satisfies Weyl's theorem for all $K \in \mathcal{K}(X)$. Moreover, for every $f \in \mathcal{H}_{nc}(\sigma(T))$, we have:*

(i) *The generalized Weyl's theorem holds for $f(T)$.*
(ii) *The generalized a-Weyl's theorem and the generalized property (ω) hold for $f(T^*)$.*

Consider the case when $T \in L(X)$ is an invertible isometry. As observed in Theorem 4.72, T is hereditarily polaroid. From Theorem 6.127 we then have:

Theorem 6.161 *Suppose that T is an invertible isometry on a Banach space X and $K \in K(X)$ is an algebraic operator which commutes with T. If $f \in \mathcal{H}_{nc}(\sigma(T))$, then $f(T + K)$ satisfies the generalized Weyl's theorem, while $f(T^* + K^*)$ satisfies every Weyl-type theorem (generalized or not),*

In the case of invertible isometries for which $\sigma(T) = \Gamma$, Γ the unit circle, the result of Theorem 6.140 cannot be applied to compact perturbations $T + K$. In fact, in this case $\rho(T)$ is not connected, T has the SVEP and hence Browder's theorem holds for T. By Theorem 6.130, $\rho_{\mathrm{w}}(T)$ is not connected. Therefore, for Hilbert space invertible isometries $T \in L(H)$, there exist a compact operator $K \in K(H)$ for which Weyl's theorem fails.

(c) **Weyl-type theorems for weighted shift operators** We first consider the case of weighted right shift operators T defined on $\ell^p(\mathbb{N})$. Recall that $\sigma_{\mathrm{ap}}(T)$ is the possibly degenerate annulus $\{\lambda \in \mathbb{C} : i(T) \leq |\lambda| \leq r(T)\}$, while, by Theorems 4.79 and 4.83, we have $\sigma_{\mathrm{w}}(T) = \mathbf{D}(0, r(T))$.

Theorem 6.162 *Let T be a non-quasinilpotent weighted right shift operators T on $\ell^p(\mathbb{N})$. Then we have:*

(i) *$f(T)$ and $f(T^*)$ satisfy Weyl's theorem for all $f \in \mathcal{H}_{nc}(\sigma(T))$.*
(ii) *$T + K$ satisfies Weyl's theorem for all compact operators $K \in \mathcal{K}(\ell^p(\mathbb{N}))$.*

Proof

(i) Obviously, T is polaroid, since iso $\sigma(T) = \emptyset$, and has the SVEP. Then $f(T)$ is polaroid, by Theorem 4.19, and $f(T)$ has the SVEP, by Theorem 2.88. From Theorem 6.43 we then conclude that both $f(T)$ and $f(T^*)$ satisfy Weyl's theorem.

(ii) $\rho_w(T) = \mathbb{C} \setminus \mathbf{D}(0, r(T))$ is connected and $\mathrm{iso}\,\sigma_w(T) = \emptyset$, so the assertion follows from Theorem 6.140.   ∎

Concerning a-Weyl's theorem, or property (ω), we distinguish the two situations $i(T) = 0$ and $i(T) > 0$.

Theorem 6.163 *Let T be a non-quasinilpotent unilateral weighted right shift on $\ell^p(\mathbb{N})$, $1 \le p \le \infty$. Then we have:*

(i) *If $i(T) = 0$ then $f(T)$ satisfies the generalized a-Weyl's theorem for every $f \in \mathcal{H}_c^i(\sigma(T))$.*

(ii) *If $i(T) = 0$, $T + K$ satisfies Weyl's theorem for all compact operators K in $\ell^p(\mathbb{N})$.*

(iii) *If $i(T) > 0$, for $p = 2$ there exists a compact operator K in $\ell^p(\mathbb{N})$ for which a-Weyl's theorem does not hold.*

Proof

(i) We have $\sigma_{ap}(T) = \sigma(T) = \mathbf{D}(0, r(T))$, so T is left polaroid, since $\mathrm{iso}\,\sigma_{ap}(T) = \emptyset$, and hence $f(T)$ is left polaroid, by Theorem 4.21. Moreover, T has the SVEP, hence $f(T)$ has the SVEP, by Theorem 2.88. From Corollary 6.85 it then follows that $f(T)$ satisfies the generalized a-Weyl's theorem.

(ii) Since T has the SVEP and $\rho(T)$ is connected then $\rho_w(T)$ is connected. Moreover, it is easily seen that $\sigma_w(T) = \mathbf{D}(0, r(T))$. Indeed, if $\lambda \notin \sigma_w(T)$ and $\lambda \in \mathbf{D}(0, r(T))$, then $\lambda I - T \in W(X)$. The SVEP for T at λ implies, by Theorem 2.97, that $p(\lambda I - T) < \infty$ and hence, by Theorem 1.22, $q(\lambda I - T) < \infty$, so λ is an isolated point of $\mathbf{D}(0, r(T))$, a contradiction. Therefore, $\mathrm{iso}\,\sigma_w(T) = \emptyset$, so the assertion follows from Theorem 6.140.

(iii) If $i(T) > 0$ then $\rho_{ap}(T)$ is disconnected, so $\rho_{uw}(T)$ is disconnected, by Theorem 6.130. Since $p = 2$, the assertion follows from Theorem 6.143.   ∎

We conclude this section with some remarks on the classical bilateral shift T on $\ell^2(\mathbb{Z})$. It is well-known that $\sigma(T)$ is the unit circle Γ. This implies that both T and T^* have the SVEP. Furthermore, $\sigma_w(T) = \Gamma$. This follows by using the same argument as part (ii) of Theorem 6.163. Hence $\rho_w(T)$ is not connected.

Theorem 6.164 *Let T be a bilateral shift on $\ell^2(\mathbb{Z})$. Then, both $f(T)$ and $f(T^*)$ satisfy all generalized Weyl-type theorems for every $f \in \mathcal{H}_{nc}(\sigma(T))$. Moreover, there exists a compact operator on $\ell^2(\mathbb{Z})$ such that Weyl's theorem fails for $T + K$.*

Proof Since $\mathrm{iso}\,\sigma(T) = \emptyset$, T is polaroid and hence $f(T)$ is polaroid. Since T and T^* have the SVEP, $f(T)$ and $f(T)^*$ also have the SVEP. By Theorem 6.113 it then follows that the generalized property (ω) holds for $f(T)$ and $f(T^*)$, and this is equivalent to saying that $f(T)$ and $f(T^*)$ satisfy the generalized a-Weyl's theorem, by Theorem 6.112. The last assertion is clear from Theorem 6.140, since $\rho_w(T)$ is not connected.   ■

(d) **Weyl's theorem for $H(p)$-operators** Let $T \in L(X)$ be a $H(p)$-operator, i.e. $H_0(\lambda I - T) = \ker(\lambda I - T)^p$ for all $\lambda \in \mathbb{C}$ and some $p \in \mathbb{N}$. By Theorem 4.37, T is a hereditarily polaroid operator, so, by Theorem 6.127, we have:

Theorem 6.165 *Let $T \in L(X)$ be an $H(p)$-operator and $K \in K(X)$ an algebraic operator which commutes with T. If $f \in \mathcal{H}_{nc}(\sigma(T))$, then $f(T + K)$ satisfies the generalized Weyl's theorem, while $f(T^* + K^*)$ satisfies every Weyl-type theorem (generalized or not).*

In particular, Theorem 6.165 applies to convolution operators T_μ on the group algebra $L^1(G)$, G a locally compact abelian group, and more generally to every multiplier defined on a commutative semi-simple Banach algebra, since these are $H(1)$-operators. Since every paranormal operator is hereditarily polaroid, by Theorem 4.56, another consequence of Theorem 6.127 is the following result:

Theorem 6.166 *Let $T \in L(X)$ be paranormal and $K \in K(X)$ an algebraic operator which commutes with T. If $f \in \mathcal{H}_{nc}(\sigma(T))$, then $f(T + K)$ satisfies the generalized Weyl's theorem, while $f(T^* + K^*)$ satisfies every Weyl-type theorem (generalized or not).*

(e) **Weyl's theorem for operators reduced by each of its finite-dimensional eigenspaces** We outline in this section some results due to Berberian [60]. Let $T \in L(H)$, H a Hilbert space, and suppose that T is reduced by each of its finite-dimensional eigenspaces. If $\pi_{0f} = \{\lambda \in \mathbb{C} : 0 < \alpha(\lambda I - T) < \infty\}$, set

$$\mathcal{M} := \bigvee_{\lambda \in \pi_{0f}} \ker(\lambda I - T).$$

It is easily seen that $\mathcal{M}$ reduces T. Set $T_1 := T|\mathcal{M}$ and $T_2 := T|\mathcal{M}^\perp$. Then we have, see [60, Proposition 4.1]:

(i) T_1 is normal with pure point spectrum (this means that the space $\mathcal{M}$ is the closed linear span of the eigenvalues of T_1).
(ii) $\pi_0(T_1) = \pi_{0f}(T)$, where $\pi_0(T_1)$ is the set of all eigenvalues of infinite multiplicity.
(iii) $\sigma(T_1) = \overline{\sigma_p(T_1)}$.
(iv) $\pi_0(T_2) = \pi_0(T) \setminus \pi_{0f} = \pi_0(T)$.

In this case, Berberian [60, Definition 5.4] defined the set $\tau(T) := \sigma(T_2) \cup \pi_{0f}(T)$ and showed that $\tau(T)$ is a non-empty compact subset of $\sigma(T)$. The set $\tau(T)$ is called the *Berberian spectrum* of T. Berberian has also shown that if T is reduced by each of its finite-dimensional subspaces then $\tau(T) = \sigma_w(T) = \sigma_b(T)$, so T satisfies Browder's theorem, see also Lee [223].

$$\sigma(T) \setminus \sigma_w(T) = \sigma(T) \setminus \sigma_b(T) \subseteq \pi_{00}(T). \tag{6.27}$$

The inclusion (6.27) easily implies that if $\text{iso}\,\sigma(T) = \emptyset$ and T is reduced by each of its finite-dimensional subspaces then Weyl's theorem holds for T, as well as for T^*, and $\sigma(T) = \sigma_w(T)$. The following result was proved in [60].

Theorem 6.167 *If $T \in L(H)$ is reduced by each of its finite-dimensional subspaces and every restriction of T to any reducing subspace is isoloid then Weyl's theorem holds for T.*

Theorem 6.167 applies to hyponormal operators and p-hyponormal operators, since these are reduced by each of its finite-dimensional subspaces, see Curto [104]. We can say much more: hyponormal operators and p-hyponormal operators are $H(p)$-operators, hence have the SVEP and are hereditarily polaroid. From Theorem 6.44, $f(T)$ and $f(T^*)$ satisfies Weyl's theorem for every $f \in \mathcal{H}(\sigma(T))$. Since, by Theorem 4.15, the SVEP for T entails that (aW), (ω), (gW), (gaW) and $(g\omega)$ hold for $f(T^*)$.

An operator $T \in L(X)$ is said to have property (G_1) if

$$\|(\lambda I - T)^{-1}\| \leq \frac{1}{\text{dist}\,(\lambda, \sigma(T))}.$$

Note that the right-hand side of the inequality above is equal to the spectral radius of $(\lambda I - T)^{-1}$, so that we actually have equality, thus, T satisfies (G_1) if and only if $(\lambda I - T)^{-1}$ is normaloid for all $\lambda \notin \sigma(T)$. In [288] Stampfli showed that the condition (G_1) entails that T is isoloid. In [183] Istrătescu proved that if the restriction of T to any invariant subspace satisfies property (G_1) and each point of $\sigma(T)$ is a bare point (that is, it lies on the circumference of some closed disc that contains $\sigma(T)$), then Weyl's theorem holds for T. Berberian in [60] has shown that if T is reduced by each of its finite-dimensional eigenspaces and the restriction of T to any reducing subspace has property (G_1) then Weyl's theorem holds for T.

(f) **Weyl's theorem for Cesàro operators** Let C_p denote the *Cesàro operator* on the Hardy space $H^p(\mathbb{D})$ for $1 \leq p < \infty$, defined as

$$(C_p f)(\lambda) := \frac{1}{\lambda} \int_0^\lambda \frac{f(\mu)}{1 - \mu} d\mu \quad \text{for all } f \in H^p(\mathbb{D}) \, \lambda \in \mathbb{D}.$$

The spectral properties of Cesàro operators have been studied by Siskasis [286]. In [240] Miller et al. proved that C_p has property (β) whenever $1 < p < \infty$. In particular, C_p has the SVEP for $1 < p < \infty$. The spectrum of C_p is connected and has no isolated points, since

$$\sigma(C_p) = \left\{ \lambda \in \mathbb{C} : |\lambda - \frac{p}{2}| < \frac{p}{2} \right\}.$$

Furthermore, $\sigma_{\mathrm{ap}}(C_p) = \partial \sigma(C_p)$, so $\rho_{\mathrm{ap}}(T)$ is not connected. From Theorem 6.130 we then deduce that $\rho_{\mathrm{uw}}(T)$ is not connected. Therefore, Weyl's theorem holds for $C_p + K$ for every compact operator on $H^p(\mathbb{D})$, while this is not true for a-Weyl's theorem, or property (ω).

6.10   Weyl's Theorem for Symmetrizable Operators

Let X be a complex infinite-dimensional Banach space and suppose that X is a subspace of another Banach space Y. Assume that the embedding of X into Y is continuous, i.e. there is a constant $k > 0$ such that

$$\|x\|_Y \leq k \|x\|_X \quad \text{for all } x \in X.$$

Let $T \in L(X)$ and denote by $\overline{T} \in L(Y)$ an extension of T to Y. In general, very few things can be said concerning the relationship between the spectral theory and Fredholm theory of T and $\overline{T}$, see Example 1 and Example 2 of Barnes [56]. The spectral theory and Fredholm theory of T and $\overline{T}$ are almost the same if we assume:

A) X is dense in Y and $\overline{T}(Y) \subseteq X$.

In [220] some aspects of Fredholm theory have been studied where we assume:

B) Y is a Hilbert space and T is symmetrizable (see later for the definition).

In this section we are mainly concerned with the transmission of Weyl-type theorems from $\overline{T}$ to T in both cases (A) and (B). In the sequel we always assume that X and Y are Banach spaces with X a proper subspace of Y. Suppose that $T \in L(X)$ admits an extension $\overline{T} \in L(Y)$ and set

$$\mathcal{M}(X) := \{T \in L(X) : \overline{T}(Y) \subseteq X\}.$$

It is easily seen that $\mathcal{M}(X)$ is a left ideal of $L(X)$, i.e., if $T \in \mathcal{M}$ and $S \in L(X)$ then $ST \in \mathcal{M}(X)$. If $T \in \mathcal{M}(X)$, $\sigma(T)$ and $\sigma(\overline{T})$ may differ only by 0. More precisely, we have:

Theorem 6.168 *If $T \in \mathcal{M}(X)$ then*

(i) $\ker(\lambda I - \overline{T}) = \ker(\lambda I - T) \quad$ *for all $\lambda \neq 0$.*
(ii) $\sigma(T) \setminus \{0\} = \sigma(\overline{T}) \setminus \{0\}$.
(iii) $\sigma_{\mathrm{w}}(T) \setminus \{0\} = \sigma_{\mathrm{w}}(\overline{T}) \setminus \{0\}$.
(iv) $\sigma_{\mathrm{b}}(T) \setminus \{0\} = \sigma_{\mathrm{b}}(\overline{T}) \setminus \{0\}$.

Proof To show (i), note first that $\ker(\lambda I - T) = \ker(\lambda I - \overline{T}) \cap X$ for all $\lambda \in \mathbb{C}$. Suppose that $\lambda \neq 0$ and $y \in \ker(\lambda I - \overline{T})$. Then $y = \frac{1}{\lambda}\overline{T}y \in \overline{T}(Y) \subset X$, which proves assertion (i). A direct proof of the assertions (ii) and (iii) can be found in

[55], but it is possible to prove these by using an argument of [56]. Let $S \in L(X, Y)$ denote the canonical embedding of X into Y and define $R \in L(Y, X)$ by $Ry := \overline{T}y$ for all $y \in Y$. Then $T = RS$ and $\overline{T} = SR$, and hence the assertions (ii) and (iii) follow from [56, Theorem 6], while (iv) follows from [56, Theorem 6 and Proposition 10].   ∎

Remark 6.169 Note that since X is dense in Y, $T \in \mathcal{M}(X)$ if and only if there exists a $c > 0$ such that $\|Tx\|_X \leq c\|x\|_Y$ for all $x \in X$ [62].

Remark 6.170 Since in the notation of the proof of Theorem 6.168 we have $T = RS$ and $\overline{T} = SR$, $\overline{T}$ has the SVEP if and only if T has the SVEP, by Lemma 2.158.

Theorem 6.171 *Suppose that X is dense in Y and $T \in \mathcal{M}(X)$. Then $0 \in \sigma_w(T) \cap \sigma_w(\overline{T}) \subseteq \sigma(T) \cap \sigma(\overline{T})$. Consequently, $\sigma(T) = \sigma(\overline{T})$ and $\sigma_w(T) = \sigma_w(\overline{T})$.*

Proof Suppose that $0 \notin \sigma_w(\overline{T})$. Then $\overline{T} \in \Phi(Y)$, so $\overline{T}(Y)$ has finite codimension in Y and hence has finite codimension in X. Therefore there exists a finite-dimensional subspace Z such that $X = \overline{T}(Y) \oplus Z$. But $\overline{T}(Y)$ is closed in Y, hence X is a closed subspace of Y. Since X is assumed to be dense in Y, it then follows that $X = \overline{X} = Y$, contradicting our assumption that X is a proper subspace of Y.

Suppose now that $0 \notin \sigma_w(T)$. Then $T \in W(X)$, hence there exists an invertible operator $U \in L(X)$ and a finite-dimensional operator $K \in L(X)$ such that $T = U - K \in \mathcal{M}(X)$, see Theorem 3.35. From this we obtain

$$U^{-1}(U - K) = I - U^{-1}K = I - K_0 \in \mathcal{M}(X),$$

where $K_0 := U^{-1}K$ is a finite-dimensional operator. By Remark 6.170 then ker K_0 is closed in Y. Since ker K_0 has finite codimension in X, there is a finite-dimensional subspace N such that $X = \ker K_0 \oplus N$. Therefore X is closed in Y and this implies $X = Y$, again contradicting the assumption that X is a proper subspace of Y.

The last assertion is clear by Theorem 6.168.   ∎

Corollary 6.172 *Suppose that X is dense in Y and $T \in \mathcal{M}(X)$. Then T satisfies Browder's theorem if and only if $\overline{T}$ satisfies Browder's theorem.*

Proof By Theorem 6.171 we have $0 \in \sigma_w(T) \cap \sigma_w(\overline{T}) \subseteq \sigma_b(T) \cap \sigma_b(\overline{T})$. Therefore, by Theorem 6.168, $\sigma_b(T) = \sigma_b(\overline{T})$, and hence $\sigma_w(T) = \sigma_b(T)$ if and only if $\sigma_w(\overline{T}) = \sigma_b(\overline{T})$.   ∎

The equivalence of Weyl's theorem for T and $\overline{T}$ requires a very special condition on the range of T.

Theorem 6.173 *Suppose that X is dense in Y, $T \in \mathcal{M}(X)$ and $T(X)$ is closed in X. Then $\overline{T}$ satisfies Weyl's theorem if and only if T satisfies Weyl's theorem. In particular, this equivalence holds if $\beta(T) < \infty$.*

Proof Suppose that Weyl's theorem holds for $\overline{T}$. By Theorem 6.171 then $0 \notin \sigma(\overline{T}) \setminus \sigma_w(\overline{T}) = \pi_{00}(\overline{T})$. If $\lambda \in \pi_{00}(\overline{T})$ then $\lambda \neq 0$ so, by part (i) of Theorem 6.171, we have $\alpha(\lambda I - T) = \alpha(\lambda I - \overline{T})$. Since $\lambda \in \mathrm{iso}\,\sigma(\overline{T}) = \mathrm{iso}\,\sigma(T)$, it then follows that $\lambda \in \pi_{00}(T)$. Therefore, $\pi_{00}(\overline{T}) \subseteq \pi_{00}(T)$.

We now show that the reverse inclusion also holds. We claim that $\alpha(T) = \infty$. To see this, suppose $\alpha(T) < \infty$. Then $\ker T$ is complemented, since it is finite-dimensional, so there exists a closed subspace M of X such that $X = \ker T \oplus M$. The restriction $T|M : M \to T(X)$ admits an inverse $(T|M)^{-1}$. Define $V \in L(X)$ by

$$V : x \in X \to (T|M)^{-1}Tx \in X.$$

Clearly, $V(\ker T) = \{0\}$ and $Vm = m$ for all $m \in M$. Consequently, $I - V$ is finite-dimensional. We show that $V \in \mathcal{M}(X)$. Since $T \in \mathcal{M}(X)$ there exists a $c > 0$ such that $\|Tx\|_X \le c\|x\|_X$. Therefore,

$$\|Vx\|_X = \|(T|M)^{-1}Tx\|_X \le \|(T|M)^{-1}\|\|Tx\|_X$$
$$\le c\|(T|M)^{-1}\|\|x\|_Y,$$

from which we conclude that $V \in \mathcal{M}(X)$. Now, $\ker(I - V)$ is closed in Y. Indeed, let (x_n) be a sequence of elements of $\ker(I - V) \subset X$ such that $\|x_n - x_0\|_Y \to 0$ for some $x_0 \in Y$. Then

$$\|x_n - x_m\|_X = \|V(x_n - x_m)\|_X \le c\|(T|M)^{-1}\|\|x - x - x_m\|_Y \to 0,$$

so (x_n) is a Cauchy sequence in X. Since X is a Banach space, there exists a $z \in X$ such that $\|x_n - z\|_X \to 0$. Therefore, for some $c' > 0$ we have

$$\|x_n - z\|_Y \le c'\|x_n - z\|_X \to 0,$$

and $z = x_0$. Consequently, $\|x_n - x_0\|_X \to 0$ which shows that $\ker(I - V)$ is closed in Y, as desired.

Since $I - V$ is finite-dimensional, we have $X = \ker(I - V) \oplus N$, with N finite-dimensional, and hence X is closed in Y. Hence $X = \overline{X} = Y$, a contradiction.

Therefore, $\alpha(T) = \infty$ and hence $0 \notin \pi_{00}(T)$. Consequently, by part (i) of Theorem 6.168, we have $\alpha(\lambda I - T) = \alpha(\lambda I - \overline{T})$ for all $\lambda \in \pi_{00}(T)$, so $\pi_{00}(T) \subseteq \pi_{00}(\overline{T})$. Therefore, $\pi_{00}(T) = \pi_{00}(\overline{T})$. Finally

$$\sigma(T) \setminus \sigma_{\mathrm{w}}(T) = \sigma(\overline{T}) \setminus \sigma_{\mathrm{w}}(\overline{T}) = \pi_{00}(\overline{T}) = \pi_{00}(T),$$

so T satisfies Weyl's theorem.

Suppose that T satisfies Weyl's theorem. Then $0 \notin \sigma(T) \setminus \sigma_{\mathrm{w}}(T) = \pi_{00}(T)$ and hence $\pi_{00}(T) \subseteq \pi_{00}(\overline{T})$. Suppose that $\alpha(\overline{T}) < \infty$. Then $\alpha(T) < \infty$ and as it has been proved before this is impossible. Therefore, $\alpha(\overline{T}) = \infty$ and hence $0 \notin \pi_{00}(\overline{T})$. As above, it then follows that $\pi_{00}(T) = \pi_{00}(\overline{T})$ and hence $\sigma(\overline{T}) \setminus \sigma_{\mathrm{w}}(\overline{T}) = \pi_{00}(\overline{T})$.

The last assertion is obvious: every finite codimensional subspace is closed. ∎

In the next corollary we consider the case when X is a dense subspace of a Hilbert space.

Corollary 6.174 *Suppose that X is dense in a Hilbert space H and let $T \in \mathcal{M}(X)$ be such that $T(X)$ is closed in X. If $\overline{T}$ is self-adjoint then T satisfies a-Weyl's theorem.*

Proof If $\overline{T}$ is self-adjoint then $\overline{T}$ is decomposable, hence the dual $\overline{T}^*$ (or equivalently, the Hilbert adjoint of $\overline{T}$) has the SVEP. In the notation of the proof of Theorem 6.168 we have $\overline{T}^* = R^* S^*$ and $T^* = S^* R^*$, and this implies that T^* also has the SVEP. By Theorem 6.173 we know that T satisfies Weyl's theorem and the SVEP of T^* implies that a-Weyl's theorem also holds for T, while T^* satisfies the generalized Weyl's theorem.   ∎

Note that instead of assuming that $\overline{T}$ is self-adjoint we can assume that $\overline{T}$ is generalized scalar. Indeed, every generalized scalar operator is decomposable and hence its dual has the SVEP, so the argument of Corollary 6.174 still works.

Theorem 6.175 *Suppose that X is dense in Y and $T \in \mathcal{M}(X)$.*

(i) *If T is polaroid and $\overline{T}$ satisfies Weyl's theorem then T satisfies the generalized Weyl's theorem.*

(ii) *If $\overline{T}$ is polaroid and T satisfies Weyl's theorem then $\overline{T}$ satisfies the generalized Weyl's theorem.*

Proof

(i) Proceeding as in the first part of the proof of Theorem 6.173 we see that $\pi_{00}(\overline{T}) \subseteq \pi_{00}(T)$. The polaroid condition on T entails that $0 \notin \pi_{00}(T)$. Indeed if $0 \in \pi_{00}(T)$ then 0 is a pole of the resolvent and hence $p(T) = q(T) < \infty$. By definition of $\pi_{00}(T)$ we also have $\alpha(T) < \infty$, so $\beta(T) = \alpha(T)$, hence $0 \notin \sigma_w(T)$, which is impossible by Theorem 6.171. By part (i) of Theorem 6.168 we then conclude that $\pi_{00}(T) \subseteq \pi_{00}(\overline{T})$. Therefore, $\pi_{00}(T) = \pi_{00}(\overline{T})$ and as in the proof of Theorem 6.171 this implies that T satisfies Weyl's theorem. Since T is polaroid, T satisfies the generalized Weyl's theorem, by Theorem 6.44.

(ii) The proof is analogous to that of part (i).   ∎

We now show that Weyl's theorem holds for symmetrizable operators. Indeed, in this very special situation the assumptions of Corollary 6.174 can be simplified. To see this, assume that the Banach space X is a subspace of a Hilbert space H and assume that the embedding of X into H is continuous and X is dense in H. Following Lax [220] $T \in L(X)$ is said to be *symmetrizable* if T is symmetric with respect to the inner product $\langle \cdot, \cdot \rangle$ induced by H on X, i.e.,

$$\langle Tx, y \rangle = \langle x, Ty \rangle \quad \text{for all } x, y \in X.$$

Note that every *quasi-hermitian* operator in the sense of Dieudonné [109] is symmetrizable (a bounded operator T on a Hilbert space is said to be quasi-hermitian if it satisfies a relation of the form $ST = T^* S$, where S is a metric operator, i.e., a strictly positive self-adjoint operator, see for further information also Antoine and Trapani [50]). Applications of symmetrizable operators to partial differential equations may be found in Lax [220] and Gohkberg and Zambitski [154].

The proof of the following important properties of symmetrizable operators $T \in L(X)$ may be found in Lax [220]:

(a) T is bounded with respect to the Hilbert norm. Moreover, the natural extension $\overline{T}$ of T to H is a bounded self-adjoint operator.
(b) $\sigma(\overline{T}) \subseteq \sigma(T)$. Clearly, since $\overline{T}$ is a self-adjoint operator then $\sigma(\overline{T}) \subset \mathbb{R}$. This inclusion may be strict, since $\sigma(T)$ may contain non-real points, see the next Example 6.180.
(c) If $\lambda I - T \in W(X)$ then $\lambda I - \overline{T} \in W(H)$. In this case $\ker(\lambda I - T) = \ker(\lambda I - \overline{T})$. Moreover, $\sigma_w(\overline{T}) \subseteq \sigma_w(T)$.

Lemma 6.176 *If T is symmetrizable and λ_0 is an eigenvalue of T then $\lambda_0 \in \mathbb{R}$. Furthermore, if λ_0 is an isolated eigenvalue of T then λ_0 is an isolated eigenvalue of $\overline{T}$.*

Proof Clearly, every eigenvalue λ of T is an eigenvalue of $\overline{T}$. Since $\overline{T}$ is self-adjoint then $\lambda \in \mathbb{R}$. If λ_0 is an isolated eigenvalue of T then there exists a punctured open disc $\mathbb{D}_0$ centered at λ_0 such that $\lambda \notin \sigma(T)$ of all $\lambda \in \mathbb{D} \setminus \{\lambda_0\}$, and hence by (b) we have $\lambda \notin \sigma(\overline{T})$, from which we deduce that λ_0 is an isolated eigenvalue of $\overline{T}$. ∎

Lemma 6.177 *Every symmetrizable operator T has the SVEP.*

Proof Observe first that since $\overline{T}$ is self-adjoint then $\overline{T}$ has the SVEP. This entails that T also has the SVEP. In fact, let $\lambda \in \mathbb{C}$ be arbitrarily given. Since every analytic function $f : U \to X$ defined on an open disc U centered at λ remains analytic when considered as a function from U to H, it is clear that T inherits the SVEP at λ. Observe that the SVEP is also an immediate consequence of Theorem 6.176, since $\sigma_p(T)$ has an empty interior. ∎

Theorem 6.178 *If $T \in L(X)$ is symmetrizable then Weyl's theorem holds for T.*

Proof Since T has the SVEP, T satisfies Browder's theorem. By Theorem 6.40, in order to prove that T satisfies Weyl's theorem, it suffices to show that $\pi_{00}(T) = p_{00}(T)$. For this suppose that $\lambda_0 \in \pi_{00}(T)$. Then λ_0 is an isolated eigenvalue of finite multiplicity in $\sigma(T)$ and hence, by Lemma 6.176, it follows that λ_0 is also an isolated eigenvalue of $\overline{T}$. Since $\overline{T}$ is self-adjoint, λ_0 is a pole of first order of the resolvent of $\overline{T}$, see [179, Proposition 70.5]. Therefore, $p(\lambda_0 I - \overline{T}) = q(\lambda_0 I - \overline{T}) = 1$. If P_0 denotes the spectral projection of $\overline{T}$ associated with λ_0, then, by (c),

$$H_0(\lambda_0 I - \overline{T}) = P_0(H) = \ker(\lambda_0 I - \overline{T}) = \ker(\lambda_0 I - T).$$

Therefore, $(\lambda_0 I - T)P_0 x = 0$ for all $x \in H$.

On the other hand, the restriction of P_0 to X coincides with the spectral projection of T associated with the isolated point λ_0 of $\sigma(T)$. For all $x \in X$ then

$(\lambda_0 I - T) P_0 x = 0$, which implies

$$H_0(\lambda_0 I - T) = R(P_0|X) \subseteq \ker(\lambda_0 I - T),$$

where $R(P_0|X)$ is the range of $P_0|X$. This implies that $H_0(\lambda_0 I - T) = \ker(\lambda_0 I - T)$ for all $\lambda \in \pi_{00}(T)$. From the decomposition

$$X = H_0(\lambda_0 I - T) \oplus K(\lambda_0 I - T) = \ker(\lambda_0 I - T) \oplus K(\lambda_0 I - T),$$

we then deduce that $(\lambda I - T)(X) = K(\lambda I - T)$, so $X = \ker(\lambda_0 I - T) \oplus (\lambda I - T)(X)$ and this implies, by Theorem 1.35, that $p(\lambda_0 I - T) = q(\lambda_0 I - T) = 1$. By the definition of $\pi_{00}(T)$ we know that $\alpha(\lambda_0 I - T) < \infty$ and this implies by Theorem 1.22 that $\beta(\lambda_0 I - T)$ is also finite. Since $\lambda_0 I - T$ has both ascent and descent finite then $\lambda_0 I - T$ is Browder, so $\lambda_0 \in p_{00}(T)$ and hence $\pi_{00}(T) \subseteq p_{00}(T)$. Since the opposite inclusion holds for every operator we then conclude that $p_{00}(T) = \pi_{00}(T)$.   ∎

In the sequel we need the following elementary lemma:

Lemma 6.179 *Let X and Y be vector spaces and T a linear injective mapping from x onto Y. If M is a linear subspace of X for which $Y/T(M)$ is finite-dimensional then X/M is also finite-dimensional, with $\dim(Y/T(M)) = \dim(X/M)$.*

Proof Let $\hat{x} := x + M$ and define $\phi : X/M \to Y/T(M)$ by $\phi(\hat{x}) := Tx + T(M)$, where $x \in \hat{x}$. Clearly, since $Y/T(M)$ is finite-dimensional and ϕ maps in a one-to-one way X/M onto $Y/T(M)$, the two quotients X/M and $Y/T(M)$ have the same dimension.   ∎

Example 6.180 Let R and L denote the right shift and the left shift in $\ell^p(\mathbb{N})$, $1 \leq p > \infty$, respectively. Let $U \in L(\ell^p(\mathbb{N}))$ be defined as

$$U := aR + bI + cL, \quad a, b, c \in \mathbb{R}. \tag{6.28}$$

Some spectral properties of U have been described by Gohkberg and Zambitski in [154]. In particular,

1) If $|a| \leq |c|$ then $\sigma(U) = \{\lambda = x + iy\}$ inside and on the ellipse having equation

$$\frac{(x-b)^2}{(a+c)^2} + \frac{y^2}{(a-c)^2} = 1.$$

If λ belongs to the interior of this ellipse then $(\lambda I - U)(X)$ is closed and

$$\alpha(\lambda I - U) := \begin{cases} 1 \text{ if } |a| > |c|, \\ 0 \text{ if } |a| < |c|, \end{cases}$$

and

$$\beta(\lambda I - U) := \begin{cases} 0 \text{ if } |a| > |c|, \\ 1 \text{ if } |a| < |c|. \end{cases}$$

2) When $a = c$, $\sigma(U)$ is the closed interval $[-2c + b, 2|c| + b]$.
3) In the case $a = -c$, $\sigma(U) = \{\lambda = i\mu + b : -2|c| \le \mu \le 2|c|\}$.

Furthermore, in the case (2) and (3) we have $\alpha(\lambda I - U) = 0$ and $\overline{(\lambda I - U)(x)} = \ell^p(\mathbb{N})$.

Let now X be the space of all complex sequences $x := (x_1, x_2, \dots, x_n, \dots)$ for which $\sum_{j=1}^{\infty} 2^j |x_j| < \infty$. The space X provided with the norm

$$\|x\|_X = \sum_{j=1}^{n} 2^j |x_j|$$

is a Banach space which is dense in the Hilbert space $\ell^2(\mathbb{N})$, and $\|x\|_2 \le \|x\|_X$. Define a map $\Psi : X \to \ell^1$ by

$$\Psi(x) := (2x_1, 2^2 x_2, \dots, 2^j x_j, \dots).$$

Evidently, Ψ is an isometry from X onto $\ell^1(\mathbb{N})$. Let S be a bounded operator in $\ell^1(\mathbb{N})$ defined by $S := \frac{1}{2} a R + bI + 2cL$ and U defined in X as in (6.28). Then

$$\Psi(U(x)) = S(\Psi(x)) \quad \text{for all } x \in X.$$

In particular, $\sigma(U) = \sigma(S)$.

Let us take $H := \ell^2(\mathbb{N})$ and $T := R + 2I + L$. Then T is symmetrizable over X. Indeed, T is a bounded linear operator on X, and

$$\langle Tx, y \rangle = \langle x, Ty \rangle \quad \text{for all } x, y \in X,$$

where $\langle \cdot, \cdot \rangle$ is the inner product in $\ell^2(\mathbb{N})$. Therefore, T satisfies Weyl's theorem. We can say much more:

(i) $\sigma_w(\overline{T}) = \sigma(\overline{T}) = [0, 4]$.
(ii) $\sigma_w(T) = \sigma(T)$ consists of the points $\lambda := x + iy$ which lie inside or on the ellipse Γ having equation

$$\frac{(x - 2)^2}{\frac{25}{4}} + \frac{y^2}{\frac{9}{4}} = 1.$$

In particular, $\sigma_w(T)$ is connected
(iii) $\sigma_e(T) = \sigma_{ap}(T) = \Gamma$.

To show (i), consider above the case $a = c$. Then $\sigma(\overline{T}) = [0, 4]$. On the other hand, since the only points λ for which ind$(\lambda I - \overline{T})$ are the points in the resolvent of T, $\sigma_w(\overline{T}) = \sigma(\overline{T})$. To prove (ii) and (iii) observe that $\sigma(T)$ coincides with the spectrum of $S = \frac{1}{2}aR + bI + 2cL$ on $\ell^1(\mathbb{N})$, which according to part (1) above, consists of all $\lambda = x + iy$ inside and on the ellipse Γ. Furthermore, for $\lambda \in \sigma(S)$ inside the ellipse Γ we have $\alpha(\lambda I - S) = 0$ and $\beta(\lambda I - S) = 1$. Using the fact that the mapping Ψ defined above is an injective linear map from X onto $\ell^1(\mathbb{N})$ with $(\lambda I - S)(\ell^1) = \Psi[(\lambda I - T)(X)]$, one obtains by Lemma 6.179,

$$\alpha(\lambda I - T) = 0 \text{ and } \beta(\lambda I - T) = 1, \text{ for any } \lambda \text{ inside the ellipse } \Gamma.$$

The following argument shows that $\sigma_e(T) = \Gamma$. The inclusion $\sigma_e(T) \subseteq \Gamma$ is clear, since inside the ellipse $\lambda I - T$ is Fredholm.

Let $\lambda \in \Gamma$ and suppose that $\lambda \notin \sigma_e(T)$. Then $\lambda I - T$ is Fredholm and since Γ is the border of $\sigma(T)$ then both T and T^* have the SVEP at λ. This implies, by Theorems 2.97 and 2.98, that $p(\lambda I - T) = q(\lambda I - T) < \infty$, hence, by Theorem 1.22, $\lambda I - T$ is Weyl. Consequently, there exists an open disc $\mathbb{D}(\lambda, \delta)$ centered at λ for which $\mu I - T$ is Weyl for all $\mu \in \mathbb{D}(\lambda, \delta)$, and this is impossible, since ind$(\mu I - T) = -1$ inside Γ. Therefore, $\sigma_e(T) = \Gamma$. Evidently, $\lambda I - T \in W_+(X)$ inside Γ, so $\sigma_{uw}(T) \subseteq \Gamma$. The opposite inclusion is also true, since, as above, for every $\lambda \in \Gamma$ we cannot have $\lambda I - T \in W_+(X)$, otherwise from the SVEP for T and T^* at λ we would obtain ind$(\mu I - T) = 0$ in an open disc centered at λ. From $\sigma_{uw}(T) \subseteq \sigma_{ap}(T) \subseteq \Gamma$ we then obtain $\sigma_{ap}(T) = \sigma_{uw}(T) = \Gamma$, in particular we see that $\rho_{uw}(T)$ is not connected.

The operator T defined in Example 6.180 is polaroid, since iso$\sigma(T) = \emptyset$ and has the SVEP, so by Theorem 6.44 both $f(T)$ and $f(T^*)$ satisfy Weyl's theorem for every $f \in \mathcal{H}(\sigma(T))$. Since T has the SVEP, $f(T^*)$ satisfies a-Weyl's theorem and property (ω). Moreover, since $\rho_w(T)$ is connected and iso$\sigma_w(T) = \emptyset$, from Theorem 6.140 we then deduce that $T + K$ satisfies Weyl's theorem for every compact operator $K \in K(X)$. Since $\rho_{uw}(T)$ is not connected there exists a $K \in K(H)$ such that a-Weyl's theorem does not hold for $T + K$.

The results above may be improved as follows.

Definition 6.181 Let $T \in L(X)$, X a Banach space with norm $\|\cdot\|$. A linear operator K is said to be T-*compact* if for any sequence $\{x_n\}$ satisfying $\|x_n\| + \|Tx_n\| \le c$, where $c > 0$, the sequence $\{Kx_n\}$ has a convergent subsequence.

Clearly, every compact operator is T-compact for any $T \in L(X)$ and each T-compact operator is bounded. This can be shown by establishing first the inequality $\|Kx\| \le M(\|x\| + \|Tx\|)$ and then using the boundedness of T.

Corollary 6.182 Let $T \in L(X)$ be symmetrizable on X and suppose that K is a T-compact operator in X, symmetric with respect to the inner product induced by the Hilbert space H on X. Then Weyl's theorem holds for $T + K$.

Proof Since K is bounded, K is symmetrizable, and hence $T + K$ too. By Theorem 6.178 then $T + K$ satisfies Weyl's theorem.   ∎

Example 6.183 Let Ω be a compact set of $\mathbb{R}^n$, and $C(\Omega)$ be the Banach space of all continuous function on Ω. Let T be the following convolution operator

$$(Tf)(x) := \int_\Omega \frac{f(y)}{|x - y|^{n-\alpha}} \mathrm{d}y \quad 0 < \alpha < n, \ n > 2.$$

T is a bounded linear operator on $C(\Omega)$. This follows from the fact that if $f \in C(\Omega)$, then $Tf \in C(\Omega)$ (see the section on equations of elliptic type in Hellwig [176]), and since the integral

$$\int_\Omega \frac{\mathrm{d}y}{|x - y|^{n-\alpha}}$$

is bounded. Taking $X = C(\Omega)$ and $H = L^2(\Omega)$, we see that T is symmetric with respect to the inner product

$$\langle f, g \rangle = \int_\Omega f(x)\overline{g(x)}\mathrm{d}x.$$

Therefore, T satisfies Weyl's theorem.

6.11   Quasi-$\mathcal{THN}$ Operators

In this section, in order to give a general framework for Weyl-type theorems, we introduce a new class of operators which properly contain the class $\mathcal{THN}$ of all totally hereditarily normaloid operators, already studied in Chap. 3.

Remark 6.184 It is rather easy to see that if $T \in L(X)$ is $\mathcal{THN}$ and M is a T-invariant closed subspace of X, then the restriction $T|M$ is also $\mathcal{THN}$.

Definition 6.185 An operator $T \in L(X)$, X a Banach space, is said to be *k-quasi totally hereditarily normaloid*, k a nonnegative integer, if the restriction $T|\overline{T^k(X)}$ is $\mathcal{THN}$.

Evidently, every $\mathcal{THN}$-operator is quasi-$\mathcal{THN}$, and if $T^k(X)$ is dense in X then a quasi-$\mathcal{THN}$ operator T is $\mathcal{THN}$. In the sequel by $\overline{Y}$ we denote the closure of $Y \subseteq X$.

Lemma 6.186 *If $T \in L(X)$ is quasi-$\mathcal{THN}$ and M is a closed T-invariant subspace of X, then $T|M$ is quasi-$\mathcal{THN}$.*

Proof Let k be a nonnegative integer such that $T_k := T|\overline{T^k(X)}$ is $\mathcal{THN}$. Let T_M denote the restriction $T|M$. Clearly, $\overline{T_M{}^k(M)} \subseteq \overline{T^k(X)}$, so $T_M{}^k(M)$ is T_k-invariant subspace of $\overline{T^k(X)}$. By Remark 6.184 it then follows that

$$T_M | \overline{T_M{}^k(M)} = T_k | \overline{T_M{}^k(M)}$$

is a $\mathcal{THN}$-operator.   ∎

We recall now some elementary algebraic facts. Suppose that $T \in L(X)$ and $X = M \oplus N$, with M and N closed subspaces of X, M invariant under T. With respect to this decomposition of X it is known that T may be represented by an upper triangular operator matrix $\begin{pmatrix} A & B \\ 0 & C \end{pmatrix}$, where $A \in L(M)$, $C \in L(N)$ and $B \in L(N, M)$. It is easily seen that for every $x = \begin{pmatrix} x \\ 0 \end{pmatrix} \in M$ we have $Tx = Ax$, so $A = T|M$. Let us consider now the case of operators T acting on a Hilbert space H, and suppose that $T^k(H)$ is not dense in H. In this case we can consider the nontrivial orthogonal decomposition

$$H = \overline{T^k(H)} \oplus \overline{T^k(H)}^\perp, \tag{6.29}$$

where $\overline{T^k(H)}^\perp = \ker(T^*)^k$, T' the adjoint of T. Note that the subspace $\overline{T^k(H)}$ is T-invariant, since

$$T(\overline{T^k(H)}) \subseteq \overline{T(T^k(H))} = \overline{T^{k+1}(H)} \subseteq \overline{T^k(H)}.$$

Thus we can represent, with respect to the decomposition (6.29), T as an upper triangular operator matrix

$$\begin{pmatrix} T_1 & T_2 \\ 0 & T_3 \end{pmatrix}, \tag{6.30}$$

where $T_1 = T|\overline{T^k(H)}$. Moreover, T_3 is nilpotent. Indeed, if $x \in \overline{T^k(X)}^\perp$, an easy computation yields $T^k x = T \begin{pmatrix} 0 \\ x \end{pmatrix} = T_3{}^k x$. Hence $T_3{}^k x = 0$, since $T^k x \in \overline{T^k(H)} \cup \overline{T^k(H)}^\perp = \{0\}$. Therefore we have:

Theorem 6.187 *Suppose that* $T \in L(H)$ *and* $T^k(H)$ *non-dense in* H. *Then, according to the decomposition* (6.29), $T = \begin{pmatrix} T_1 & T_2 \\ 0 & T_3 \end{pmatrix}$ *is quasi-*$\mathcal{THN}$ *if and only if* T_1 *is* $\mathcal{THN}$. *Furthermore,*

$$\sigma(T) = \sigma(T_1) \cup \sigma(T_3) = \sigma(T_1) \cup \{0\}.$$

Proof The first assertion is clear, since $T_1 = T|\overline{T^k(H)}$. The second assertion follows from the following general result: if $T := \begin{pmatrix} A & C \\ 0 & B \end{pmatrix}$ is an upper triangular operator matrix acting on some direct sum of Banach spaces and $\sigma(A) \cap \sigma(B)$ has no interior points, then $\sigma(T) = \sigma(A) \cup \sigma(B)$; see [222] for details. ∎

Paranormal operators, and in particular hyponormal operators, are obvious examples of quasi-$\mathcal{THN}$ operators, since, as has been observed in Chap. 3, these operators are $\mathcal{THN}$. In the sequel we give some other examples of operators which are quasi totally hereditarily normaloid, that generalize these classes.

(iv) The class of quasi-paranormal operators may be extended as follows: $T \in L(H)$ is said to be (n, k)-*quasiparanormal* if

$$\|T^{k+1}x\| \leq \|T^{1+n}(T^k x)\|^{\frac{1}{1+n}} \|T^k x\|^{\frac{n}{1+n}} \quad \text{for all } x \in H.$$

The class of $(1, k)$-quasiparanormal operators has been studied in [238]. The $(1, 1)$-quasiparanormal operators have been studied in Cao et al. [86]. If $T^k(H)$ is not dense then, in the triangulation $T = \begin{pmatrix} T_1 & T_2 \\ 0 & T_3 \end{pmatrix}$, $T_1 = T|\overline{T^k(H)}$ is n-quasiparanormal, and hence $\mathcal{THN}$, see Yuan and Ji [299].

(v) An extension of class A operators is given by the class of all k-*quasiclass A operators*, where $T \in L(H)$, H a separable infinite-dimensional Hilbert space, is said to be a k-quasiclass A operator if

$$T^{*k}(|T|^2 - |T|^2)T^k \geq 0.$$

Every k-quasiclass A operator is quasi-$\mathcal{THN}$. Indeed, if T has dense range then T is a class A operator and hence paranormal. If T does not have dense range then T with respect the decomposition $H = \overline{T^k(H)} \oplus \ker T^{*k}$ may be represented as a matrix $T = \begin{pmatrix} T_1 & T_2 \\ 0 & T_3 \end{pmatrix}$, where $T_1 := T|\overline{T^k(H)}$ is a class A operator, and hence $\mathcal{THN}$, see Tanahashi [290].

As has been observed in Duggal and Jeon [135, Example 0.2], a quasi-class A operator (i.e. $k = 1$), need not be normaloid. This shows that, in general, a quasi-$\mathcal{THN}$ operator is not normaloid, so the class of quasi-$\mathcal{THN}$ operators properly contains the class of $\mathcal{THN}$ operators.

(vi) An operator $T \in L(H)$, H a separable infinite-dimensional Hilbert space, is said to be k-*quasi $*$-paranormal*, $k \in \mathbb{N}$, if

$$\|T^* T^k x\|^2 \leq \|T^{k+2}x\| \|T^k x\| \quad \text{for all unit vectors } x \in H.$$

This class of operators contains the class of all quasi- $*$-paranormal operators (which corresponds to the value $k = 1$). Every k-quasi $*$-paranormal operator is quasi-$\mathcal{THN}$. Indeed, if T^k has dense range then T is $*$-paranormal and hence

$\mathcal{THN}$. If T^k does not have dense range then T may be decomposed, according the decomposition $H = \overline{T^k(H)} \oplus \ker T^{*k}$, as $T = \begin{pmatrix} T_1 & T_2 \\ 0 & T_3 \end{pmatrix}$, where $T_1 = T|\overline{T^k(H)}$ is *-paranormal, hence $\mathcal{THN}$, see [239, Lemma 2.1].

(vii) An extension of p-quasi-hyponormal operators is defined as follows: an operator $T \in L(H)$ is said to be (p, k)-*quasihyponormal* for some $0 < p \le 1$ and $k \in \mathbb{N}$, if

$$T^{*k}|T^*|^{2p}T^k \le T^{*k}|T|^{2p}T^k.$$

Every (p, k)-quasihyponormal operator T with respect to the decomposition $H = \overline{T^k(H)} \oplus \ker T^{*k}$ may be represented as a matrix $T = \begin{pmatrix} T_1 & T_2 \\ 0 & 0 \end{pmatrix}$, where $T_1 := T|\overline{T^k(H)}$ is k-hyponormal (hence paranormal) and consequently $\mathcal{THN}$, see Kim [198].

The next result generalizes the result of Theorem 4.60.

Theorem 6.188 *Suppose that $T \in L(H)$, H a Hilbert space, is analytically quasi-$\mathcal{THN}$ and quasi-nilpotent. Then T is nilpotent.*

Proof Suppose first that T is quasi-nilpotent and k-quasi $\mathcal{THN}$. If $T^k(H)$ is dense then T is $\mathcal{THN}$, so T is nilpotent by Theorem 4.60. Suppose that $T^k(H)$ is not dense and write $T = \begin{pmatrix} T_1 & T_2 \\ 0 & T_3 \end{pmatrix}$, where T_1 is $\mathcal{THN}$, $T_3{}^k = 0$, and $\sigma(T) = \sigma(T_1) \cup \{0\}$. Since $\sigma(T) = \{0\}$ and $\sigma(T_1)$ is not empty, we then have $\sigma(T_1) = \{0\}$, thus T_1 is a quasi-nilpotent $\mathcal{THN}$ operator and hence $T_1 = 0$. Therefore $T = \begin{pmatrix} 0 & T_2 \\ 0 & T_3 \end{pmatrix}$. An easy computation yields that

$$T^{k+1} = \begin{pmatrix} 0 & T_2 \\ 0 & T_3 \end{pmatrix}^{k+1} = \begin{pmatrix} 0 & T_2 T_3^k \\ 0 & T_3^{k+1} \end{pmatrix}^{k+1} = 0,$$

so that T is nilpotent.

Finally, suppose that T is quasi-nilpotent and analytically k-quasi $\mathcal{THN}$. Let $h \in \mathcal{H}_{nc}(\sigma(T))$ be such that $h(T)$ is quasi-$\mathcal{THN}$. We claim that $h(T)$ is nilpotent. If $h(T)^k$ has dense range then $h(T)$ is $\mathcal{THN}$ and hence, by Theorem 4.60, $h(T)$ is nilpotent. Suppose that $h(T)^k$ does not have dense range. Then with respect to the decomposition $X = \overline{h(T)^k(H)} \oplus \overline{h(T)^k(H)}^\perp$, the operator $h(T)$ has a triangulation $h(T) = \begin{pmatrix} A & B \\ 0 & C \end{pmatrix}$ such that $A = h(T)|\overline{h(T)^k(H)}$ is $\mathcal{THN}$ and

$$\sigma(h(T)) = \sigma(A) \cup \{0\}.$$

By the spectral mapping theorem we have

$$\sigma(h(T)) = h(\sigma(T)) = \{h(0)\}.$$

Consequently, $0 \in \{h(0)\}$, i.e. $h(0) = 0$, and therefore $h(T)$ is quasi-nilpotent. Since $h(T)$ is quasi-$\mathcal{THN}$, by the first part of the proof it then follows that $h(T)$ is nilpotent. Now, $h(0) = 0$ so we can write

$$h(\lambda) = \mu \lambda^n \prod_{i=1}^{n} (\lambda_i I - T)^{n_i} g(\lambda),$$

where $g(\lambda)$ has no zeros in $\sigma(T)$ and $\lambda_i \neq 0$ are the other zeros of g with multiplicity n_i. Hence

$$h(T) = \mu\, T^n \prod_{i=1}^{n} (\lambda_i I - T)^{n_i} g(T),$$

where all $\lambda_i I - T$ and $g(T)$ are invertible. Since $h(T)$ is nilpotent, T is also nilpotent.   ∎

Theorem 6.189 *If $T \in L(H)$ is an analytically quasi $\mathcal{THN}$ operator, then T is polaroid.*

Proof We show that for every isolated point λ of $\sigma(T)$ we have $p(\lambda I - T) = q(\lambda I - T) < \infty$. Let λ be an isolated point of $\sigma(T)$, and denote by P_λ the spectral projection associated with $\{\lambda\}$. Then $M := K(\lambda I - T) = \ker P_\lambda$ and $N := H_0(\lambda I - T) = P_\lambda(X)$, and $H = H_0(\lambda I - T) \oplus K(\lambda I - T)$, see Theorem 2.45. Furthermore, the restriction $\lambda I - T|N$ is quasi-nilpotent, while $\lambda I - T|M$ is invertible. Since $\lambda I - T|N$ is analytically quasi $\mathcal{THN}$, Lemma 6.188 implies that $\lambda I - T|N$ is nilpotent. In other words, $\lambda I - T$ is an operator of Kato Type, and hence has uniform topological descent by Theorem 1.83.

Now, both T and the dual T^* have the SVEP at λ, since λ is isolated in $\sigma(T) = \sigma(T^*)$, and this implies, by Theorems 2.97 and 2.98, that both $p(\lambda I - T)$ and $q(\lambda I - T)$ are finite. Therefore, λ is a pole of the resolvent.   ∎

Theorem 6.190 *If $T \in L(H)$ is analytically quasi $\mathcal{THN}$, then T is hereditarily polaroid and hence has the SVEP.*

Proof Let $f \in \mathcal{H}_{nc}(\sigma(T))$ such that $f(T)$ is quasi $\mathcal{THN}$. If M is a closed T-invariant subspace of X, we know that $f(T)|M$ is quasi $\mathcal{THN}$, by Lemma 6.186, and $f(T)|M = f(T|M)$, so $f(T|M)$ is polaroid, by Theorem 6.189, and consequently, $T|M$ is polaroid, by Theorem 4.19.   ∎

Corollary 6.191 *If $T \in L(H)$ is the direct sum $T = S \oplus N$, where S is $\mathcal{THN}$ and N is nilpotent, then T is hereditarily polaroid.*

Proof If $T = S \oplus N$, where S is $\mathcal{THN}$ and N is nilpotent, then T is quasi $\mathcal{THN}$, since T admits a triangulation $T = \begin{pmatrix} S & 0 \\ 0 & N \end{pmatrix}$, with respect to a suitable decomposition. ∎

Theorem 6.192 *Let* $T \in L(H)$ *be an analytically quasi* $\mathcal{THN}$ *operator on a Hilbert space* H, *and let* $K \in L(H)$ *be an algebraic operator commuting with* T. *Then both* $f(T + K)$ *and* $f(T' + K')$ *satisfy* (gW) *for every* $f \in \mathcal{H}_{nc}(\sigma(T + K))$.

Proof Suppose that $T \in L(H)$ is analytically quasi $\mathcal{THN}$, and let $f \in \mathcal{H}_{nc}(\sigma(T))$ be such that $f(T)$ is quasi $\mathcal{THN}$. Since T has the SVEP, by Theorem 4.31, $f(T)$ has the SVEP. Now, by Theorem 6.190 T is hereditarily polaroid, and hence the results of Theorem 6.127 apply. ∎

6.12 Comments

The properties (R), (gR) were introduced by Aiena et al. in [41, 43] and [44] and the section concerning these properties is completely modeled after these papers. The crucial Theorem 6.11 is due to Oudghiri [251]. The perturbation result concerning hereditarily polaroid operators of Theorem 4.31 is taken from Aiena and Aponte [8]. The definition of Weyl's theorem was introduced by Coburn [97], which started from the work of Weyl [296], which studied the spectra of all the compact perturbations $T + K$ of a self-adjoint operator T acting on a Hilbert space and showed that $\lambda \in \mathbb{C}$ belongs to the Weyl spectrum precisely when λ is not an isolated point of finite multiplicity in $\sigma(T)$. Later Coburn [97] extended Weyl's theorem to Toeplitz operators, and Berberian extended Weyl's theorem to some other classes of operators in [60] and [61]. Later, Weyl's theorem was studied in connection with the single-valued extension property, by Curto and Han [105]. In particular they showed that if T or T^* has the SVEP then the spectral theorem holds for the Weyl spectrum. Later, Weyl's theorem was studied and extended to other classes of operators by several authors. We cite some of these articles [16, 54, 57, 105, 106, 119, 129–131, 166, 168, 173, 183, 224]. The characterizations of Weyl's theorem of Theorems 6.40 and 6.41 are taken from [10, 32, 105], and Weyl's theorem for polaroid operators was investigated by Duggal et al. [138]. Lemma 6.51 and Theorem 6.52 is modeled after Oberai [250].

The generalized version of Weyl's theorem was introduced by Berkani and Koliha [70] and investigated in several papers by Berkani et al. [9, 64, 65] and [69]. However, the characterizations of the generalized Weyl's theorem established in Theorem 6.62 is taken from Aiena and Garcia [13]. Theorem 6.64 and Example 6.67 is modeled after Zguitti [306].

a-Weyl's theorem for operators was introduced by Rakočević [262] and has been studied by several other authors [2, 35, 105, 111, 252]. In particular, Theorem 6.70 is modeled after [35]. Theorems 6.73 and 6.75 are due to Oudghiri

[252], while Theorem 6.80 is an adaptation of a result of Djordjević and Djordjević [113].

The generalized version of a-Weyl's theorem was introduced by Berkani and Koliha [70], and studied by several other authors. Theorem 6.84 may be found in Aiena and Miller [47].

Property (w) was introduced in a short article by Rakočević [261] and later studied in several other articles by Aiena et al. [4, 11, 22, 37, 42]. Property (gw) was introduced and studied by Amouch and Berkani [67, 68]. The section concerning the equivalence between Weyl-type theorems for polaroid-type operators is modeled after Aiena et al. [38], but pertinent results may be found in Duggal and Djordjević [129–133]. The section concerning Weyl-type theorems for Drazin invertible operators is modeled after Aiena and Triolo [28].

A streamlined study of Toeplitz operators may be found in [223]. The result that a Toeplitz operator satisfies Weyl's theorem was first established by Coburn [97]. The section concerning Weyl-type theorems under compact perturbations is modeled after Duggal and Kim [137], Aiena and Triolo [29], and Jia and Feng [188]. Theorems 6.150 and 6.152 and Example 6.154 are taken from Farenick and Lee [145].

The results concerning Weyl-type theorems for extensions of bounded linear operators and symmetrizable operators are modeled after [40] and Nieto [248], who first proved Theorem 6.178 by using different methods. The class of quasi $\mathcal{THN}$ operators, which give a general framework for Weyl-type theorems for operators on Hilbert spaces, was introduced in Aiena et al. [45]. Upper triangular operator matrices have been studied by many authors, see for instance Han et al. [167] and Cao et al. [86, 118, 305].

Appendix A

In this appendix we collect some of the basic definitions and results from the theory of semi-Fredholm operators acting between Banach spaces, and some other classes of operators related to them. We are mainly concerned with the algebraic and topological structure of this class, as well as with some perturbation properties.

A.1 Basic Functional Analysis

Let M be a subset of a Banach space X. The *annihilator* of M is the closed subspace of X^* defined by

$$M^\perp := \{f \in X^* : f(x) = 0 \text{ for every } x \in M\},$$

while the *pre-annihilator* of a subset N of X^* is the closed subspace of X defined by

$$^\perp N := \{x \in X : f(x) = 0 \text{ for every } f \in N\}.$$

An application of the Hahn–Banach theorem shows that if M is a linear subspace of X then $^\perp(M^\perp)$ is the norm closure of M, so, if M is closed then $^\perp(M^\perp) = M$. If N is a linear subspace of X^*, then $(^\perp N)^\perp$ is the weak-star closure of N, see [269, Chap. 4].

The following classical result is essentially due to Kato [196, Chapter IV], its proof may be found in several other standard books on functional analysis:

Theorem A.1 (Sum Theorem) *Let M and N be two closed subspaces of a Banach space X. Then the following assertions are equivalent:*

(i) *$M + N$ is closed;*
(ii) *$M^\perp + N^\perp$ is closed;*

© Springer Nature Switzerland AG 2018
P. Aiena, *Fredholm and Local Spectral Theory II*, Lecture Notes in Mathematics 2235, https://doi.org/10.1007/978-3-030-02266-2

(iii) $M^{\perp} + N^{\perp} = (M \cap N)^{\perp}$;
(iv) $M + N = {}^{\perp}(M^{\perp} \cap N^{\perp})$.

Let $L(X, Y)$ denote the space of all continuous linear operators from the Banach space X into the Banach space Y. The following duality relationships between the kernels and ranges of a bounded operator $T \in L(X)$, on a Banach space X, and its dual T^* are well known, (the reader can find the proofs, for instance, in Heuser [179, p. 135], or Goldberg [156]:

$$\ker T = {}^{\perp}\overline{T^*(X^*)} \quad \text{and} \quad {}^{\perp}\ker T^* = \overline{T(X)}, \tag{A.1}$$

and

$$T(X)^{\perp} = \ker T^* \quad \text{and} \quad \overline{T^*(X^*)} \subseteq \ker T^{\perp}. \tag{A.2}$$

Note that the last inclusion is, in general, strict. However, a classical consequence of the *closed range theorem* establishes that the equality holds precisely when T has closed range, see Kato [196, Theorem 5.13, Chapter IV].

In the next theorem we establish some well-known basic isomorphisms (for a proof, see [269]).

Theorem A.2 (Annihilator Theorem) *Let M be a closed subspace of a Banach space X. Then M^* is isometrically isomorphic to the quotient $X^*/M^{\perp}$, while $(X/M)^*$ is isometrically isomorphic to $M^{\perp}$. Moreover, if N is a weak-star closed linear subspace of X^* then X^*/N is isometrically isomorphic to $({}^{\perp}N)^*$ and $(X/{}^{\perp}N)^*$ is isometrically isomorphic to N.*

Definition A.3 Let X be a Banach space. A subspace M of X is said to be *paracomplete, or paraclosed*, if there exists a $T \in L(X)$ such that $T(X) = M$.

A subspace M of a Banach space X is paracomplete if and only if there is a complete norm $|||\cdot|||$ on M which is greater than the original norm $\|\cdot\|$. Every closed subspace is paracomplete but the opposite is not true, see [147].

The following lemma due to Neubauer (see [208, Prop. 2.1.1]), gives sufficient conditions under which paracomplete subspaces are closed.

Lemma A.4 (Neubauer Lemma) *Let X be a Banach space, and M and N paracomplete subspaces of X. If $M \cap N$ and $M + N$ are closed, then both M and N are closed.*

Proofs of the following two basic principles of operator theory may be found in several standard books on functional analysis, for instance Rudin [269].

Theorem A.5 (Open Mapping Theorem) *Let $T \in L(X, Y)$ be surjective. Then T is open, i.e. there is a constant $c > 0$ such that, for every $y \in Y$, there exists an element $x \in X$ for which $Tx = y$ and $\|x\| \le c\|y\|$.*

Theorem A.6 *If $T \in L(X, Y)$, X and Y Banach spaces, is such that $T(X)$ is of second category in X, then T is surjective and hence open.*

Let D be a linear subspace of X and T a linear map from D into Y. The *graph* of T is the set $G_T := \{(x, Tx) : x \in D\}$. If X and Y are Banach spaces, the operator T is said to be *closed* if G_T is a closed subset of $X \times Y$, where the product $X \times Y$ is provided with the norm $\|(x, y)\| := \|x\| + \|y\|$.

Theorem A.7 (Closed Graph Theorem) *Let $T \in L(X, Y)$ be closed. Then T is continuous.*

It is obvious that the sum $M + N$ of two linear subspaces M and N of a vector space X is again a linear subspace. If $M \cap N = \{0\}$ then this sum is called the *direct sum* of M and N and will be denoted by $M \oplus N$. In this case for every $z = x + y$ in $M + N$ the components x, y are uniquely determined. If $X = M \oplus N$ then N is called an *algebraic complement* of M. In this case the (Hamel) basis of X is the union of the basis of M with the basis of N. It is obvious that every subspace of a vector space admits at least one algebraic complement. The *codimension* of a subspace M of X is the dimension of every algebraic complement N of M, or equivalently the dimension of the quotient X/M. Note that codim $M = \dim M^{\perp}$. Indeed, by Theorem A.2 we have:

$$\text{codim } M = \dim X/M = \dim (X/M)^* = \dim M^{\perp}.$$

A particularly important class of endomorphisms are the so-called *projections*. If $X = M \oplus N$ and $x = y + z$, with $x \in M$ and $y \in N$, define $P : X \to M$ by $Px := y$. The linear map P projects X onto M along N. Clearly, $I - P$ projects X onto N along M and we have

$$P(X) = \ker (I - P) = M, \quad \ker P = (I - P)(X) = N, \quad \text{with } P^2 = P,$$

i.e. P is an idempotent operator. Suppose now that X is a Banach space. If $X = M \oplus N$ and the projection P is continuous then M is said to be *complemented* and N is said to be a *topological complement* of M. Note that each complemented subspace is closed, but the converse is not true, for instance c_0, the Banach space of all sequences which converge to 0, is a non-complemented closed subspace of ℓ_∞, where ℓ_∞ denotes the Banach space of all bounded sequences, see [246]. It is well known that if $X = M \oplus N$ then both M and N are invariant under T (i.e. T is *reduced* by the pair (M, N)) if and only if T commutes with the projection P of X onto M along N.

Definition A.8 $T \in L(X, Y)$, X a Banach space, is said to be *relatively regular* if there exists an operator $S \in L(Y, X)$ for which

$$T = TST \quad \text{and} \quad STS = S.$$

In this case S is called a *pseudo-inverse* of T.

There is no loss of generality if we require in the definition above only $T = TST$. In fact, if $T = TST$ holds then the operator $S' := STS$ will satisfy both the equalities

$$T = TS'T \quad \text{and} \quad S' = S'TS'.$$

In general, a relatively regular operator admits infinite pseudo-inverses. In fact, if S is a pseudo-inverse then all operators of the form

$$STS + U - STUTS \quad \text{with } U \in L(X) \text{ arbitrary}$$

are pseudo-inverses of T, see [87, Theorem 2]. We now establish a basic result.

Theorem A.9 *A bounded operator* $T \in L(X, Y)$ *is relatively regular if and only if* $\ker T$ *and* $T(X)$ *are complemented.*

Proof If $T = TST$ and $STS = S$ then $P := TS \in L(Y)$ and $Q := ST \in L(X)$ are idempotents, hence projections. Indeed

$$(TS)^2 = TSTS = TS \quad \text{and} \quad (ST)^2 = STST = ST.$$

Moreover, from the inclusions

$$T(X) = (TST)(X) \subseteq (TS)(Y) \subseteq T(X),$$

and

$$\ker T \subseteq \ker(ST) \subseteq \ker(STS) = \ker T,$$

we obtain $P(Y) = T(X)$ and $\ker Q = (I_X - Q)(X) = \ker T$.

Conversely, suppose that $\ker T$ and $T(X)$ are complemented in X and Y, respectively. Write $X = \ker T \oplus U$ and $Y = T(X) \oplus V$ and let us denote by P the projection of X onto $\ker T$ along U and by Q_0 the projection of Y onto $T(X)$ along V. Define $T_0 : U \to T(X)$ by $T_0 x = Tx$ for all $x \in U$. Clearly T_0 is bijective. Put $S := T_0^{-1} Q_0$. If we represent an arbitrary $x \in X$ in the form $x = y + z$, with $y \in \ker T$ and $z \in U$, we obtain

$$ST x = T_0^{-1} Q_0 T(y + z) = T_0^{-1} Q_0 T z$$
$$= T_0^{-1} T z = z = x - y = x - Px.$$

Similarly one obtains $TS = Q_0$. If $Q := I_Y - Q_0$ then

$$ST = I_X - P \quad \text{and} \quad TS = I_Y - Q. \tag{A.3}$$

If we multiply the first equation in (A.3) from the left by T we obtain $TST = T$, and analogously multiplying the second equation in (A.3) from the left by S we obtain $STS = S$.   ∎

The left, or right, invertible operators may be characterized as follows:

Theorem A.10 *Let* $T \in L(X, Y)$, *X and Y Banach spaces.*

(i) T *is injective and* $T(X)$ *is complemented if and only if there exists an* $S \in L(Y, X)$ *such that* $ST = I_X$.

(ii) T *is surjective and* ker T *is complemented if and only if there exists an* $S \in L(Y, X)$ *such that* $TS = I_Y$.

Proof

(i) If $S \in L(Y, X)$ and $ST = I_X$ then $TST = T$, thus T is relatively regular and hence has complemented range, by Theorem A.9. Clearly, T is injective. Conversely, if T is bounded below and P is a projection of X onto $T(X)$, let $S_0 : T(X) \to X$ be the inverse of T. If $S := S_0 P$ then $ST = I_X$.

(ii) If $TS = I_Y$ then $TST = T$, so T has complemented kernel by Theorem A.9 and, as it is easy to see, T is onto. Conversely, if T is onto and $X = \ker T \oplus N$ then $T|N : N \to Y$ is bijective. Let $J_N : N \to X$ be the natural embedding and set $S := J_N(T|N)^{-1}$. Clearly, $TS = I_Y$.   ∎

A.2   Compact Operators

In this section we reassume some of the basic properties of compact linear operators. Many of the results of this section are classical and are contained in standard texts of functional analysis. For this reason, we do not give the proofs of many of these results.

Definition A.11 A bounded operator T from a normed space X into a normed space Y is said to be *compact* if for every bounded sequence (x_n) of elements of X the corresponding sequence (Tx_n) contains a convergent subsequence. This is equivalent to saying that the closure of $T(B_X)$, B_X the closed unit ball of X, is a compact subset of Y.

Denote by $\mathcal{F}(X, Y)$ the set of all continuous finite-dimensional operators, i.e. those operators T for which $T(X)$ is finite-dimensional. Let $\mathcal{K}(X, Y)$ be the set of all compact operators. In the sequel we list some basic properties of these sets (see [179, §13]).

Theorem A.12 *Let X, Y and Z be Banach spaces. Then*

(i) $\mathcal{F}(X, Y)$ *and* $\mathcal{K}(X, Y)$ *are linear subspaces of* $L(X, Y)$. *Moreover,* $\mathcal{F}(X, Y) \subseteq \mathcal{K}(X, Y)$.

(ii) *If $T \in \mathcal{F}(X, Y)$, $S \in L(Y, Z)$, $U \in L(Z, X)$ then $ST \in \mathcal{F}(X, Z)$ and $TU \in \mathcal{F}(Z, Y)$. Analogous statements hold for $T \in \mathcal{K}(X, Y)$.*

(iii) *If (T_n) is a sequence of $\mathcal{K}(X, Y)$ which converges to T then $T \in \mathcal{K}(X, Y)$. Consequently, $\mathcal{K}(X, Y)$ is a closed subspace of $L(X, Y)$.*

Note that by Theorem A.12, part (ii), the integral operator K defined in Sect. 1.1 is compact. A famous counter-example of Enflo [144] shows that not every compact operator is the limit of finite-dimensional operators.

Note that $\mathcal{K}(X, Y)$ may coincide with $L(X, Y)$. This for instance is the case when $X = \ell^q$ or $X = c_0$, $Y = \ell^p$, with $1 \le p < q < \infty$, see [155].

Let us now consider the case $X = Y$ and set $\mathcal{F}(X) := \mathcal{F}(X, X)$, $\mathcal{K}(X) := \mathcal{K}(X, X)$. Recall that a subset J of a Banach algebra $\mathcal{A}$ is said to be a (two-sided) *ideal* if J is a linear subspace of $\mathcal{A}$ and for every $x \in J$, $a \in \mathcal{A}$, the products xa and ax lie in J. From Theorem A.12 we then deduce that $\mathcal{F}(X)$ as well as $\mathcal{K}(X)$ are ideals of the Banach algebra $L(X)$. Note that if X is infinite-dimensional then $I \notin K(X)$, otherwise any bounded sequence would contain a convergent subsequence and by the Bolzano–Weierstrass theorem this is not possible. It is known that we can define on the quotient algebra $L(X)/\mathcal{K}(X)$ the *quotient norm*:

$$\|\hat{T}\| := \inf_{T \in \hat{T}} \|T\|, \quad \text{where } \hat{T} := T + \mathcal{K}(X).$$

Since $\mathcal{K}(X)$ is closed, the quotient algebra $\hat{\mathcal{L}} := L(X)/\mathcal{K}(X)$, with respect to the quotient norm defined above, is a Banach algebra, known in the literature as the *Calkin algebra*. Also $L(X)/\mathcal{F}(X)$ is an algebra, but in general is not a Banach algebra.

Compactness is preserved by duality, see Proposition 42.2, Proposition 42.3 of [179]:

Theorem A.13 (Schauder's Theorem) *If $T \in L(X, Y)$, X and Y Banach spaces, then T is compact if and only if T^* is compact.*

We now establish some important properties of compact endomorphisms.

Theorem A.14 *Let $T \in \mathcal{K}(X)$, X a Banach space. Then $\alpha(\lambda I - T) < \infty$, $\beta(\lambda I - T) < \infty$ and $\mathrm{ind}(\lambda I - T) = 0$ for all $\lambda \neq 0$.*

Theorem A.15 *If $T \in \mathcal{K}(X)$, X a Banach space, then $p(\lambda I - T) = q(\lambda I - T) < \infty$ for all $\lambda \neq 0$.*

Theorem A.16 *If X and Y are Banach spaces and $T \in L(X, Y)$ is a compact operator having closed range then T is finite-dimensional.*

Proof Suppose first that T is compact and bounded below, namely T has closed range and is injective. Let (y_n) be a bounded sequence of $T(X)$ and let (x_n) be a sequence of X for which $Tx_n = y_n$ for all $n \in \mathbb{N}$. Since T is bounded below there exists a $\delta > 0$ such that $\|y_n\| = \|Kx_n\| \ge \delta \|x_n\|$ for all $n \in \mathbb{N}$, so (x_n) is bounded. The compactness of T then implies that there exists a subsequence (x_{n_k})

of (x_n) such that $Tx_{n_k} = y_{n_k}$ converges as $k \to \infty$. Hence every bounded sequence of $T(X)$ contains a convergent subsequence and by the Stone–Weierstrass theorem this implies that $T(X)$ is finite-dimensional.

Assume now the more general case when $T \in \mathcal{K}(X, Y)$ has closed range. If $T_0 : X \to T(X)$ is defined by $T_0 x := Tx$ for all $x \in X$, then T_0 is a compact operator from X onto $T(X)$. Therefore the dual $T_0^* : T(X)^* \to X^*$ is a compact operator by Schauder's theorem. Moreover, T_0 is onto, so T_0^* is bounded below by Theorem 1.10. The first part of the proof then gives that T_0^* is finite-dimensional, and hence $T(X)^*$ is finite-dimensional because it is isomorphic to the range of T_0^*. From this it follows that $T(X)$ is also finite-dimensional. ∎

If $C[a, b]$ is the Banach space of all continuous functions on the interval $[a, b]$, then an example of a compact operator is given by the classical *Fredholm integral equation* (of the second kind),

$$x(s) - \int_a^b k(s, t)x(t)\mathrm{d}t = y(s),$$

where the kernel of the equation $k(s, t)$ is continuous on the square $[a, b] \times [a, b]$, the function y is continuous on the interval $[a, b]$, and we look for solutions $x \in C[a, b]$. If we define $K : C[a, b] \to C[a, b]$ as

$$(Kx)(s) := \int_a^b k(s, t)x(t)\mathrm{d}t$$

then K is compact on $C[a, b]$, see [179].

A.3   Semi-Fredholm Operators

We now introduce some important classes of operators.

Definition A.17 Given two Banach spaces X and Y, the set of all *upper semi-Fredholm operators* is defined by

$$\Phi_+(X, Y) := \{T \in L(X, Y) : \alpha(T) < \infty \text{ and } T(X) \text{ is closed}\},$$

while the set of all *lower semi-Fredholm operators* is defined by

$$\Phi_-(X, Y) := \{T \in L(X, Y) : \beta(T) < \infty\}.$$

The set of all *semi-Fredholm operators* is defined by

$$\Phi_\pm(X, Y) := \Phi_+(X, Y) \cup \Phi_-(X, Y).$$

The class $\Phi(X, Y)$ of all *Fredholm operators* is defined by

$$\Phi(X, Y) = \Phi_+(X, Y) \cap \Phi_-(X, Y).$$

At first glance the definitions of semi-Fredholm operators seems to be asymmetric, but this is not the case since the condition $\beta(T) < \infty$ entails by Corollary 1.7 that $T(X)$ is closed.

We shall set

$$\Phi_+(X) := \Phi_+(X, X) \quad \text{and} \quad \Phi_-(X) := \Phi_-(X, X),$$

while

$$\Phi(X) := \Phi(X, X) \quad \text{and} \quad \Phi_\pm(X) := \Phi_\pm(X, X).$$

If $T \in \Phi_\pm(X, Y)$ the *index* of T is defined by $\operatorname{ind} T := \alpha(T) - \beta(T)$. Clearly, if T is bounded below then T is upper semi-Fredholm with index less than or equal to 0, while any surjective operator is lower semi-Fredholm with index greater than or equal to 0. Clearly, if $T \in \Phi_+(X, Y)$ is not Fredholm then $\operatorname{ind} T = -\infty$, while if $T \in \Phi_-(X, Y)$ is not Fredholm then $\operatorname{ind} T = \infty$.

Observe that in the case $X = Y$ the class $\Phi(X)$ is non-empty since the identity trivially is a Fredholm operator. This is a substantial difference from the case in which X and Y are different. In fact, if $T \in \Phi(X, Y)$ for some infinite-dimensional Banach spaces X and Y then there exist two subspaces M and N such that $X = \ker T \oplus M$ and $Y = T(X) \oplus N$, with M and $T(X)$ closed infinite-dimensional subspaces of X and Y, respectively. The restriction of T to M clearly has a bounded inverse, so the existence of a Fredholm operator from X into a different Banach space Y implies the existence of isomorphisms between some closed infinite-dimensional subspaces of X and Y. For this reason, for certain Banach spaces X, Y no bounded Fredholm operator from X to Y may exist, i.e., $\Phi(X, Y) = \varnothing$. This is, for instance, the case for $X := L^p[0, 1]$ and $Y = L^q[0, 1]$ with $0 < p < q < \infty$. Other examples may be found in Aiena [1, Chapter 7].

It can be proved that each finite-dimensional subspace as well as every closed finite-codimensional subspace is complemented, see Proposition 24.2 of Heuser [179]. Therefore, by Corollary 1.7 and Theorem A.9 we have

Theorem A.18 *Every Fredholm operator $T \in \Phi(X, Y)$ is relatively regular.*

Fredholm operators may be characterized as follows, see Heuser [179].

Theorem A.19 (Atkinson's Theorem) *If $T \in L(X)$ then the following statements are equivalent:*

  (i) *T is a Fredholm operator;*
 (ii) *The class residue $T + \mathcal{F}(X)$ is invertible in $L(X)/\mathcal{F}(X)$;*
(iii) *The class residue $T + \mathcal{K}(X)$ is invertible in $L(X)/\mathcal{K}(X)$.*

Theorem A.20 *Suppose that X,Y and Z are Banach spaces.*

(i) *If $T \in \Phi_-(X, Y)$ and $S \in \Phi_-(Y, Z)$ then $ST \in \Phi_-(X, Z)$.*
(ii) *If $T \in \Phi_+(X, Y)$ and $S \in \Phi_+(Y, Z)$ then $ST \in \Phi_+(X, Z)$.*
(iii) *If $T \in \Phi(X, Y)$ and $S \in \Phi(Y, Z)$ then $ST \in \Phi(X, Z)$.*
    In particular, if T belongs to one of the classes $\Phi_-(X, Y)$, $\Phi_+(X, Y)$, $\Phi(X, Y)$ then T^n belongs to the same class for all $n \in \mathbb{N}$.

Theorem A.21 *If $T \in \Phi_\pm(X)$ then $p(T) = q(T^*)$ and $q(T) = p(T^*)$.*

Theorem A.22 *Suppose that X, Y and Z are Banach spaces, $T \in L(X, Y)$, and $S \in L(Y, Z)$.*

(i) *If $ST \in \Phi_-(X, Z)$ then $S \in \Phi_-(Y, Z)$.*
(ii) *If $ST \in \Phi_+(X, Z)$ then $T \in \Phi_+(X, Y)$.*
(iii) *If $ST \in \Phi(X, Z)$ then $T \in \Phi_+(X, Y)$ and $S \in \Phi_-(Y, Z)$.*

Theorem A.23 *Let $T \in \Phi_-(X, Y)$ and $S \in \Phi_-(Y, X)$ (or $T \in \Phi_+(X, Y)$ and $S \in \Phi_+(Y, X)$), then $\mathrm{ind}\,(ST) = \mathrm{ind}\,S + \mathrm{ind}\,T$.*

If $T \in \Phi_+(X, Y)$ the range $T(X)$ need not be complemented, and analogously if $T \in \Phi_-(X, Y)$ the kernel may be not complemented. The study of the following two classes of operators was initiated by Atkinson [52].

Definition A.24 If X and Y are Banach spaces then $T \in L(X, Y)$ is said to be *left Atkinson* if $T \in \Phi_+(X, Y)$ and $T(X)$ is complemented in X. The operator $T \in L(X, Y)$ is said to be *right Atkinson* if $T \in \Phi_-(X, Y)$ and $\ker(T)$ is complemented in Y. The class of left Atkinson operators and right Atkinson operators will be denoted by $\Phi_l(X, Y)$ and $\Phi_r(X, Y)$, respectively.

Clearly,

$$\Phi(X, Y) \subseteq \Phi_l(X, Y) \subseteq \Phi_+(X, Y),$$

and

$$\Phi(X, Y) \subseteq \Phi_r(X, Y) \subseteq \Phi_-(X, Y).$$

Moreover,

$$\Phi(X, Y) = \Phi_l(X, Y) \cap \Phi_r(X, Y).$$

Theorem A.25 *Let X, Y, and Z be Banach spaces and $T \in L(X, Y)$. Then the following assertions are equivalent:*

(i) *$T \in \Phi_l(X, Y)$;*
(ii) *there exists an $S \in L(Y, X)$ such that $I_X - ST \in \mathcal{F}(X)$;*
(iii) *there exists an $S \in L(Y, X)$ such that $I_X - ST \in \mathcal{K}(X)$.*

Analogously, the following assertions are equivalent:
(iv) $T \in \Phi_r(X, Y)$;
 (v) *there exists an $S \in L(Y, X)$ such that $I_Y - TS \in \mathcal{F}(Y)$;*
(vi) *there exists an $S \in L(Y, X)$ such that $I_Y - TS \in \mathcal{K}(Y)$.*

Theorem A.26 *If $T \in L(X, Y)$ then the following statements are equivalent:*

 (i) $T \in \Phi(X, Y)$;
 (ii) *there exists an $S \in L(Y, X)$ such that $I_X - ST \in \mathcal{F}(X)$ and $I_Y - TS \in \mathcal{F}(Y)$;*
(iii) *there exists an $S \in L(Y, X)$ such that $I_X - ST \in \mathcal{K}(X)$ and $I_Y - TS \in \mathcal{K}(Y)$.*

Theorem A.27 *Suppose that $T \in L(X)$. Then*

 (i) $T \in \Phi_r(X)$ *if and only if the class residue $\hat{T} = T + \mathcal{K}(X)$ is right invertible in the Calkin algebra $L(X)/\mathcal{K}(X)$.*
(ii) $T \in \Phi_l(X)$ *if and only if the class residue $\hat{T} = T + \mathcal{K}(X)$ is left invertible in $L(X)/\mathcal{K}(X)$.*

It should be noted that if X is an infinite-dimensional complex Banach space then $\lambda I - T \notin \Phi(X)$ for some $\lambda \in \mathbb{C}$. This follows from the classical result that the spectrum of an arbitrary element of a complex infinite-dimensional Banach algebra is always non-empty. In fact, by the Atkinson characterization of Fredholm operators, $\lambda I - T \notin \Phi(X)$ if and only if $\hat{T} := T + \mathcal{K}(X)$ is non-invertible in the Calkin algebra $L(X)/\mathcal{K}(X)$. An analogous result holds for semi-Fredholm operators on infinite-dimensional Banach spaces: if X is an infinite-dimensional Banach space and $T \in L(X)$ then $\lambda I - T \notin \Phi_+(X)$ (respectively, $\lambda I - T \notin \Phi_-(X)$) for some $\lambda \in \mathbb{C}$.

In the following theorem we list some other properties of Atkinson operators, the proof is left to the reader, see Problems IV.13 of Lay and Taylor [221].

Theorem A.28 *If X, Y and Z are Banach spaces we have:*

 (i) *If $T \in \Phi_l(X, Y)$ and $S \in \Phi_l(Y, Z)$ then $ST \in \Phi_l(X, Z)$. Analogously, if $T \in \Phi_r(X, Y)$ and $S \in \Phi_r(Y, Z)$ then $ST \in \Phi_r(X, Z)$. The sets $\Phi_l(X)$, $\Phi_r(X)$ and $\Phi(X)$ are semi-groups in $L(X)$.*
(ii) *Suppose that $T \in L(X, Y)$, $S \in L(Y, Z)$ and $ST \in \Phi_l(X, Z)$. Then $S \in \Phi_l(Y, Z)$. Analogously, suppose that $T \in L(X, Y)$, $S \in L(Y, Z)$ and $ST \in \Phi_r(X, Z)$. Then $T \in \Phi_r(X, Y)$.*

A.4 Some Perturbation Properties of Semi-Fredholm Operators

The next theorem shows that the classes $\Phi_+(X, Y)$, $\Phi_-(X, Y)$ and $\Phi(X, Y)$ are stable under compact perturbations.

Theorem A.29 *If X and Y are Banach spaces the following statements hold:*

(i) *If $\Phi_+(X, Y) \neq \varnothing$ then $\Phi_+(X, Y) + \mathcal{K}(X, Y) \subseteq \Phi_+(X, Y)$.*
(ii) *If $\Phi_-(X, Y) \neq \varnothing$ then $\Phi_-(X, Y) + \mathcal{K}(X, Y) \subseteq \Phi_-(X, Y)$.*
(iii) *If $\Phi(X, Y) \neq \varnothing$ then $\Phi(X, Y) + \mathcal{K}(X, Y) \subseteq \Phi(X, Y)$.*

Theorem A.30 *Let $T \in \Phi_\pm(X, Y)$. Then $\operatorname{ind}(T + K) = \operatorname{ind} T$ for all $K \in \mathcal{K}(X, Y)$.*

We show now that $\Phi(X, Y)$ is stable under *small* perturbations.

Theorem A.31 *For every $T \in \Phi(X, Y)$, X and Y Banach spaces, there exists a $\rho := \rho(T) > 0$ such that for all $S \in L(X, Y)$ with $\|S\| < \rho$ then $T + S \in \Phi(X, Y)$ and $\operatorname{ind}(T + S) = \operatorname{ind} T$. The set $\Phi(X, Y)$ is open in $L(X, Y)$.*

Also the classes of semi-Fredholm operators are stable under small perturbations.

Theorem A.32 *Suppose that $T \in L(X, Y)$. Then we have*

(i) *If $T \in \Phi_+(X, Y)$ then there exists an $\varepsilon > 0$ such that for every $S \in L(X, Y)$ for which $\|S\| < \varepsilon$ we have $T + S \in \Phi_+(X, Y)$. Moreover, $\alpha(T + S) \leq \alpha(T)$ and $\operatorname{ind}(T + S) = \operatorname{ind} T$.*
(ii) *If $T \in \Phi_-(X, Y)$ then there exists an $\varepsilon > 0$ such that for every $S \in L(X, Y)$ for which $\|S\| < \varepsilon$ we have $T + S \in \Phi_-(X, Y)$. Moreover, $\beta(T + S) \leq \beta(T)$ and $\operatorname{ind}(T + S) = \operatorname{ind} T$.*

An immediate consequence of Theorem A.32 is that the sets $\Phi_+(X, Y)$, $\Phi_-(X, Y)$ and $\Phi(X, Y)$ are open subsets of $L(X, Y)$. Now, if Ω is a connected open component of $\Phi_\pm(X)$ the index function $T \to \operatorname{ind} T$ is continuous on Ω. If we fix $T_0 \in \Omega$, then the set $\{T \in \Omega; \operatorname{ind} T = \operatorname{ind} T_0$ is both open and closed so it coincides with Ω. Hence we have:

Theorem A.33 *The sets $\Phi_+(X, Y)$, $\Phi_-(X, Y)$ and $\Phi(X, Y)$ are open subsets of $L(X, Y)$. The index is constant on every connected component of $\Phi_\pm(X, Y)$.*

If the perturbation of $T \in \Phi_\pm(X)$ is caused by a multiple of the identity we have:

Theorem A.34 (Punctured Neighborhood Theorem) *Let $T \in \Phi_+(X)$. Then there exists an $\varepsilon > 0$ such that*

$$\alpha(\lambda I + T) \leq \alpha(T) \quad \text{for all } |\lambda| < \varepsilon,$$

and $\alpha(\lambda I - T)$ is constant for all $0 < |\lambda| < \varepsilon$.
Analogously, if $T \in \Phi_-(X)$ then there exists an $\varepsilon > 0$ such that

$$\beta(\lambda I + T) \leq \beta(T) \quad \text{for all } |\lambda| < \varepsilon,$$

and $\beta(\lambda I + T)$ is constant for all $0 < |\lambda| < \varepsilon$.

In the sequel, we establish some basic relationships between the ascent and the descent of a bounded operator $T \in L(X)$ on a Banach space X in the case of semi-Fredholm operators.

Suppose that $T \in \Phi_\pm(X)$. Then $T^n \in \Phi_\pm(X)$, and hence the range of T^n is closed for all n. Analogously, $T^{\star n}$ also has closed range, and therefore for every $n \in \mathbb{N}$,

$$\ker T^{n\star} = T^n(X)^\perp, \quad \ker T^n = {}^\perp T^{n\star}(X^\star) = {}^\perp T^{\star n}(X^\star).$$

Obviously these equalities imply that $p(T^\star) = q(T)$ and $p(T) = q(T^\star)$. Note that these equalities hold in the Hilbert space sense: in the case of Hilbert space operators $T \in \Phi_\pm(H)$ the equalities $p(T^\star) = q(T)$ and $p(T) = q(T^\star)$ hold for the adjoint $T^\star$.

A.5 The Riesz Functional Calculus

Let A be a complex algebra with identity u. For every $a \in A$ and every analytic function $f : U \to \mathbb{C}$ defined on some open neighborhood U of the spectrum $\sigma(a)$, define

$$f(a) := \frac{1}{2\pi i} \int_\Gamma f(\lambda)(\lambda u - a)^{-1} d\lambda,$$

where Γ is a contour that surrounds $\sigma(a)$ in U (this means that Γ is a finite system of positively oriented closed rectifiable curves in $U \setminus \sigma(a)$ such that $\sigma(a)$ is contained in the inside of Γ and $\mathbb{C} \setminus U$ in the outside of Γ. By using Cauchy's theorem and the Hahn–Banach theorem it can be shown that $f(a)$ is independent of the choice of the contour that surrounds $\sigma(a)$ in U. Let $\mathcal{H}(\sigma(a))$ denote the space of all analytic functions on some open neighborhood U of the spectrum $\sigma(a)$. Then the mapping $f \in \mathcal{H}(\sigma(a)) \to f(a)$ is a continuous algebra homomorphism of $\mathcal{H}(\sigma(a))$ into A and is called the *analytic functional calculus* for the element a, see [179, Section 48]. Moreover, the following fundamental property holds.

Theorem A.35 (Spectral Mapping Theorem) *If* $a \in A$ *then the equality* $\sigma(f(a)) = f(\sigma(a))$ *holds for all* $f \in \mathcal{H}(\sigma(a))$.

In the particular case $A = L(X)$, where X is a non-trivial complex Banach space, the analytic functional calculus is commonly called the *Riesz functional calculus*. If $T \in L(X)$ consider a *spectral subset* σ_1 (possibly empty) of $\mathbb{C}$, i.e. such that σ_1 and $\sigma_2 := \sigma(T) \setminus \sigma_1$ are closed in $\mathbb{C}$. Let Γ_1 and Γ_2 be two closed positively oriented contours in $\rho(T) = \mathbb{C} \setminus \sigma(T)$ which contain σ_1 and σ_2, respectively, but do not contain any other subset of $\sigma(T)$. Then the operators

$$P_k := \frac{1}{2\pi i} \int_{\Gamma_k} (\lambda I - T)^{-1} d\lambda, \quad k = 1, 2$$

are idempotent, i.e. $P_k{}^2 = P_k$. Moreover, $P_1 P_2 = P_2 P_1 = 0$ and $P_1 + P_2 = I$. This result may be extended as follows: let $\sigma_1, \ldots, \sigma_n$ be spectral subsets such that $\sigma_i \cap \sigma_j = \emptyset$ for $i \neq j$ and $\sigma(T) = \sigma_1 \cup \ldots \sigma_n$. Let $P_1, \ldots, P_n$ be the corresponding spectral projections. Then $P_i P_j = 0$ if $i \neq j$, while $I = \sum_{k=1}^{n} P_k$ and $X = P_1(X) \oplus \cdots \oplus P_n(X)$.

For a spectral subset σ (possibly empty) the operator

$$P_\sigma := \frac{1}{2\pi i} \int_{\Gamma_\sigma} (\lambda I - T)^{-1} d\lambda, \quad k = 1, 2,$$

where Γ_σ is a contour as above, is a projection and is called the *spectral projection* associated with σ.

Theorem A.36 (Spectral Decomposition Theorem) *If $T \in L(X)$ and σ is a spectral subset of T then $X = M_\sigma \oplus N_\sigma$, where $M_\sigma = P_\sigma(X)$ and $N_\sigma = \ker P_\sigma$. Both subspaces M_σ and N_σ are invariant under every $f(T)$, with $f \in \mathcal{H}(\sigma(T))$. Moreover,*

$$\sigma(T|M_\sigma) = \sigma \quad and \quad \sigma(T|N_\sigma) = \sigma(T) \setminus \sigma.$$

Suppose that for $T \in L(X)$ for clopen subsets σ_1 and σ_2 of $\sigma(T)$ we have $\sigma(T) = \sigma_1 \cup \sigma_2$ and $\sigma_1 \cap \sigma_2 = \emptyset$. Then $X = P_{\sigma_1} \oplus P_{\sigma_2}$ and T with respect to this decomposition of X may be represented as a matrix $T = \begin{pmatrix} T_1 & 0 \\ 0 & T_2 \end{pmatrix}$, where $\sigma(T_k) = \sigma_k$, $(k = 1, 2)$, see [260, Theorem 2.10].

Let us now consider the case when λ_0 is an isolated point of the spectrum $\sigma(T)$. Then the analytic resolvent function $R_\lambda : \lambda \to (\lambda I - T)^{-1}$ admits a *Laurent expansion* on a punctured disc $0 < |\lambda - \lambda_0| < r$ centered at λ_0 with radius r, i.e. R_λ may be represented by the sum of series

$$R_\lambda = \sum_{n=1}^{\infty} \frac{P_n}{(\lambda - \lambda_0)^n} + \sum_{n=0}^{\infty} Q_n (\lambda - \lambda_0)^n \quad \text{for } 0 < |\lambda - \lambda_0| < r,$$

where the coefficients are calculated according to the formulas

$$P_n = \frac{1}{2\pi i} \int_\Gamma (\lambda - \lambda_0)^{n-1} R_\lambda d\lambda,$$

and

$$Q_n = \frac{1}{2\pi i} \int_\Gamma \frac{R_\lambda}{(\lambda - \lambda_0)^{n+1}} d\lambda,$$

where Γ is a sufficiently small positively oriented circle around λ_0. If P_0 denotes the spectral projection associated with the spectral set $\{\lambda_0\}$, it is easily seen that

$P_1 = P_0$ and $P_n = (T - \lambda_0 I)^{n-1} P_0$, for $n = 1, 2, \ldots$. These equations show that either $P_n \neq 0$ for all n (in this case λ_0 is said to be an *essential singularity* of R_λ), or that there exists a $p \in \mathbb{N}$ such that $P_n \neq 0$ for $n = 1, \ldots, p$ but $P_n = 0$ for $n > p$. In this last case λ_0 is said to be a *pole* of order p of the resolvent. The proof of the following important characterization of the poles may be found in Heuser [179, Proposition 50.2].

Theorem A.37 *If $T \in L(X)$ then $\lambda_0 \in \sigma(T)$ is a pole of R_λ if and only if $0 < p(\lambda_0 I - T) = q(\lambda_0 I - T) < \infty$. Moreover, if $p := p(\lambda_0 I - T) = q(\lambda_0 I - T)$ then p is the order of the pole. In this case λ_0 is an eigenvalue of T, and if P_0 is the spectral projection associated with $\{\lambda_0\}$ then*

$$P_0(X) = \ker(\lambda_0 I - T)^p, \quad \ker P_0 = (\lambda_0 I - T)^p(X).$$

A.6   Vector-Valued Analytic Functions

For ease of reference we give in this section some notions and few results concerning vector-valued analytic functions. However, we refer to Section III.14 of the classical monograph of Dunford and Schwartz [143], or to the book of Rudin [269, Chap. 3].

Let X be a complex Banach space and Δ an open subset of $\mathbb{C}$. A vector-valued function $f : \Delta \to X$ is said to be *analytic* if the composition $\phi \circ f : \Delta \to \mathbb{C}$ is analytic for every $\phi \in X^*$. Analytic functions are complex differentiable, i.e., there exists the norm limit

$$f'(\lambda_0) := \lim_{\lambda \to \lambda_0} \frac{f(\lambda) - f(\lambda_0)}{\lambda - \lambda_0)}$$

for every $\lambda_0 \in \Delta$. As in the classical scalar-valued case, Cauchy's integral formula holds for every X-valued analytic function.

Theorem A.38 (Cauchy's Integral Formula) *If $f : \Delta \to X$ is analytic and if Γ_1 and Γ_2 are two integration paths with the same initial and final points, which can be deformed into each other continuously in Δ, then*

$$\int_{\Gamma_1} f(\lambda) d\lambda = \int_{\Gamma_2} f(\lambda) d\lambda.$$

In particular, $\int_\Gamma f(\lambda) d\lambda = 0$ if Γ is a closed curve whose interior contains only points of Δ.

For every X-valued analytic function, the higher derivative of f at λ_0 is given by

$$f^{(n)}(\lambda_0) := \frac{n!}{2\pi i} \int_\Gamma (\lambda - \lambda_0)^{-1} f(\lambda) d\lambda$$

for all $n = 0, 1, 2, \ldots, \lambda_0 \in \Delta$, where Γ is a positively oriented closed rectifiable curve in Δ for which λ_0 belongs to the inside of Γ and $\mathbb{C} \setminus \Delta$ to the outside of Γ. Furthermore, f can be expanded into a power series

$$f(\lambda) = \sum_{k=0}^{n} a_k (\lambda - \lambda_0)^k, \quad \text{with } a_k \in X,$$

around every point $\lambda_0 \in \Delta$, and the series converges at least in the largest open disk around λ_0 which contains only points of Δ. The coefficients a_k are given by $a_k = \frac{f^{(n)}(\lambda_0)}{n!}$.

The next two theorems follow by combining classical results concerning scalar-valued analytic functions and the Hahn–Banach theorem:

Theorem A.39 (Identity Theorem for an Analytic Function) *Suppose that Y is a closed linear subspace of a Banach space X, and $f : U \to X$ an analytic function defined on a connected open subset $U \subseteq \mathbb{C}$. If there exists a set $W \subseteq U$ that clusters in U such that $f(W) \subseteq Y$, then $f(U) \subseteq Y$. In particular, $f \equiv 0$ on U if f vanishes on a set that clusters in U.*

Theorem A.40 (Liouville's Theorem) *If f is analytic on all of $\mathbb{C}$ and is bounded then f is constant.*

We conclude with the classical Runge's theorem.

Theorem A.41 (Runge's Theorem) *Let K be a compact subset of an open set $U \subseteq \mathbb{C}$. Suppose that each bounded component of $\mathbb{C} \setminus K$ contains some points of $\mathbb{C} \setminus U$. If A is a set containing at least one complex number from every bounded connected component, then every analytic function on some neighbourhood of K may be approximated uniformly on K by a sequence of rational functions r_n which converges uniformly to f on K and such that all the poles of the functions r_n are in A.*

A.7 Operators on Hilbert Spaces

In this section we reassume some of the basic properties of Hilbert space operators. We refer to the books of Rudin [269], Heuser [179], and Furuta [151] for details and proofs. Let H be a complex Hilbert space with an inner product $(\cdot, \cdot)$. The inner product satisfies the *Schwarz inequality*, i.e., $|(x, y)| \leq \|x\| \|y\|$ for all $x, y \in H$. The dual of a Hilbert space is described by the following theorem.

Theorem A.42 (Frechét–Riesz Representation Theorem) *For each fixed element $z \in H$ the map $f : x \in H \to (x, z)$ defines a continuous linear form on H. Conversely, for every continuous linear form f on H there exists a vector $z \in H$ such that $f(x) = (x, z)$ for all $x \in H$. Furthermore, $\|f\| = \|z\|$.*

A consequence of this theorem is that every Hilbert space is isometrically isomorphic to its dual. If $T \in L(H)$, for a fixed $y \in H$ define $f(x) := (Tx, y)$. According to the Frechét–Riesz representation theorem, there exists a unique element $z \in H$ such that $f(x) = (Tx, y) = (x, z)$ for all $x \in H$. The adjoint operator $T' \in L(H)$ is defined by $(Tx, y) = (x, z) = (x, T'y)$. If U is the conjugate-linear isometry that associates to each $y \in H$ the linear form $f(x) = (x, y)$, then the dual $(\lambda I - T)^*$ and the adjoint $(\lambda I - T)'$ are related by the following equality

$$\bar{\lambda} I - T' = (\lambda I - T)' = U^{-1}(\lambda I - T)^* U \quad \text{for all } \lambda \in \mathbb{C}. \tag{A.4}$$

An operator $T \in L(H)$ is said to be *self-adjoint* if $T = T'$, i.e. $(Tx, y) = (x, Ty)$ for all $x, y \in H$. If $TT' = T'T$ then T is said to be a *normal* operator, and if $TT' = T'T = I$ then T is said to be *unitary*. On a Hilbert space H over $\mathbb{C}$, T is normal if and only if $\|Tx\| = \|T'x\|$, for all $x \in H$, and T is self-adjoint if and only if (Tx, x) is real for all $x \in H$. Moreover, T is unitary if and only if $\|Tx\| = \|T'x\| = \|x\|$ for all $x \in H$, so every unitary operator is an isometry. Eigenvectors x, y of a normal operator T corresponding to distinct eigenvalues are orthogonal to each other, i.e. $(x, y) = 0$.

Given two self-adjoint operator $T, S \in L(H)$ the symbol $T \geq S$ means that $(Tx, x) \geq (Sx, x)$ for all $x \in H$ and we say that a self-adjoint operator T is positive if $T \geq 0$, i.e., $(Tx, x) \geq 0$ for all $x \in H$. Obviously, $T \geq S$ if and only if $T - S \geq 0$. If $T \geq S$ and $U \geq 0$ then $T + U \geq S + U$, and the product of positive commuting operators is always positive.

For any positive operator $T \in L(H)$ there exists a unique operator S such that $S^2 = T$. S is called the *square root* of T and denoted by $T^{\frac{1}{2}}$.

Let M be a closed subspace of H. Then $H = M \oplus M^{\perp}$, where $M^{\perp}$ is the orthogonal complement of M, i.e. $M^{\perp} := \{y \in H : (x, y) = 0 \text{ for all } x \in M\}$. The projection P_M of H onto M along $M^{\perp}$ is called the *orthogonal projection* from H onto M. A projection P is orthogonal if and only if P is self-adjoint. Every orthogonal projection P_M has norm equal to 1, moreover $0 \leq P_M \leq I$.

An operator $U \in L(H)$ is said to be a *partial isometry* if there exists a closed subspace M such that

$$\|Ux\| = \|x\| \text{ for any } x \in M, \text{ and } Ux = 0 \text{ for any } x \in M.$$

The subspace M is said to be the *initial space* of U, while the range $N := U(H)$ is said to be the *final space* of U. Evidently, U is an isometry if and only if U is a partial isometry and $M = H$, while U is unitary if and only if U is a partial isometry and $M = N = H$.

Theorem A.43 *Let $U \in L(H)$ be a partial isometry with initial space M and final space N. Then we have*

(i) $U P_M = U$ and $U'U = P_M$.

(ii) N is a closed subspace of H.

(iii) The adjoint U' is a partial isometry with initial space N and final space M.

Note that an operator $U \in L(H)$ is a partial isometry if and only if U' is a partial isometry, and in this case UU' and $U'U$ are projections. Set $|T| := (T'T)^{\frac{1}{2}}$. It is easily seen that $\ker T = \ker |T|$.

Theorem A.44 (Polar Decomposition) *For every $T \in L(H)$ there exists a partial isometry U such that $T = U|T|$. The initial space of U is $M := \overline{|T|(H)} = \overline{T'(H)}$, the final space is $N := \overline{T(H)}$. Moreover, $\ker U = \ker |T|$ and $U'U|T| = |T|$.*

If U is as in Theorem A.44 the product $T = U|T|$ is called the *polar decomposition* of T. The partial isometry U is uniquely determined. If $T = U|T|$ is the polar decomposition of T then $T' = U'|T'|$ is the polar decomposition of T'. Some important properties are transmitted from T to U, for instance if T is normal then U is normal, if T is self-adjoint then U is self-adjoint, and if T is positive then U is positive.

Bibliography

1. P. Aiena, *Fredholm and Local Spectral Theory, with Application to Multipliers* (Kluwer Acadamic Publishers, Dordrecht, 2004)
2. P. Aiena, Classes of operators satisfying a-Weyl's theorem. Stud. Math. **169**, 105–122 (2005)
3. P. Aiena, Quasi Fredholm operators and localized SVEP. Acta Sci. Math. (Szeged), **73**, 251–263 (2007)
4. P. Aiena, Property (w) and perturbations II. J. Math. Anal. Appl. **342**, 830–837 (2008)
5. P. Aiena, Algebraically paranormal operators on Banach spaces. Banach J. Math. Anal. **7**(2), 136–145 (2013)
6. P. Aiena, Fredholm theory and localized SVEP. Funct. Anal. Approx. Comput. **7**(2), 9–58 (2015)
7. P. Aiena, *Polaroid Operators and Weyl Type Theorems*. Applied Mathematics in Tunisia, Springer Proceedings in Mathematics and Statistics, vol. 131 (Springer, Cham, 2015), pp. 1–20
8. P. Aiena, E. Aponte, Polaroid type operators under perturbations. Stud. Math. **214**(2), 121–136 (2013)
9. P. Aiena, M. Berkani, Generalized Weyl's theorem and quasi-affinity. Stud. Math. **198**(2), 105–120 (2010)
10. P. Aiena, M.T. Biondi, Browder's theorems through localized SVEP. Mediterr. J. Math. **2**, 137–151 (2005)
11. P. Aiena, M.T. Biondi, Property (w) and perturbations. J. Math. Anal. Appl. **336**, 683–692 (2007)
12. P. Aiena, C. Carpintero, Single-valued extension property and semi-Browder spectra. Acta Sci. Math. (Szeged) **70**(1–2), 265–278 (2004)
13. P. Aiena, O. Garcia, Generalized Browder's theorem and SVEP. Mediterr. J. Math. **4**, 215–228 (2007)
14. P. Aiena, M. González, On the Dunford property (C) for bounded linear operators RS and SR. Integr. Equ. Oper. Theory **70**(4), 561–568 (2011)
15. P. Aiena, M. González, Local spectral theory for operators R and S satisfying $RSR = R^2$. Extracta Math. **31**(1), 37–46 (2016)
16. P. Aiena, J.R. Guillén, Weyl's theorem for perturbations of paranormal operators. Proc. Am. Math. Soc. **35** 2433–2442 (2007)
17. P. Aiena, T.L. Miller, On generalized a-Browder's theorem. Stud. Math. **180**(3), 285–300 (2007)
18. P. Aiena, O. Monsalve, The single valued extension property and the generalized Kato decomposition property. Acta Sci. Math. (Szeged) **67**, 461–477 (2001)

© Springer Nature Switzerland AG 2018

P. Aiena, *Fredholm and Local Spectral Theory II*, Lecture Notes in Mathematics 2235, https://doi.org/10.1007/978-3-030-02266-2

19. P. Aiena, O. Monsalve, Operators which do not have the single valued extension property. J. Math. Anal. Appl. **250**(2), 435–448 (2003)
20. P. Aiena, V. Muller, The localized single-valued extension property and Riesz operators. Proc. Am. Math. Soc. **143**(5), 2051–2055 (2015)
21. P. Aiena, M.M. Neumann, On the stability of the localized single-valued extension property under commuting perturbations. Proc. Am. Math. Soc. **141**(6), 2039–2050 (2013)
22. P. Aiena, P. Peña, A variation on Weyl's theorem. J. Math. Anal. Appl. **324**, 566–579 (2006)
23. P. Aiena, E. Rosas, The single valued extension property at the points of the approximate point spectrum, J. Math. Anal. Appl. **279** (1), 180–188 (2003)
24. P. Aiena, J.E. Sanabria, On left and right poles of the resolvent. Acta Sci. Math. (Szeged) **74**, 669–687 (2008)
25. P. Aiena, S. Triolo, Property (gab) through localized SVEP. Moroccan J. Pure Appl. Anal. **1**(2), 91–107 (2015)
26. P. Aiena, S. Triolo, Some perturbation results through localized SVEP. Acta Sci. Math. (Szeged), **82**(1–2), 205–219 (2016)
27. P. Aiena, S. Triolo, Local spectral theory for Drazin invertible operators. J. Math. Anal. Appl. **435**(1), 414–424 (2016)
28. P. Aiena, S. Triolo, Fredholm spectra and Weyl type theorems for Drazin invertible operators. Mediterr. J. Math. **13**(6), 4385–4400 (2016)
29. P. Aiena, S. Triolo, Weyl type theorems on Banach spaces under compact perturbations. Mediterr. J. Math. **15**, 126 (2018). https://doi.org/10.1007/s00009-018-1176-y
30. P. Aiena, S. Triolo, Projections and isolated points of parts of the spectrum. Adv. Oper. Theory (2018). https://doi.org/10.15352/aot.1804-1348 ISSN: 2538-225X (electronic) https://projecteuclid.org/aot
31. P. Aiena, F. Villafãne, Components of resolvent sets and local spectral theory. Contemp. Math. **328**, 1–14 (2003)
32. P. Aiena, F. Villafãne, Weyl's theorem for some classes of operators. Integr. Equ. Oper. Theory **53**, 453–466 (2005)
33. P. Aiena, M.L. Colasante, M. González, Operators which have a closed quasi-nilpotent part. Proc. Am. Math. Soc. **130**(9), 2701–2710 (2002)
34. P. Aiena, T.L. Miller, M.M. Neumann, On a localized single-valued extension property. Math. Proc. R. Ir. Acad. **104A**(1), 17–34 (2004)
35. P. Aiena, C. Carpintero, E. Rosas, Some characterization of operators satisfying a-Browder theorem. J. Math. Anal. Appl. **311**, 530–544 (2005)
36. P. Aiena, M.T. Biondi, C. Carpintero, On Drazin invertibility. Proc. Am. Math. Soc. **136**, 2839–2848 (2008)
37. P. Aiena, M.T. Biondi, F. Villafãne, Property (w) and perturbations III. J. Math. Anal. Appl. **353**, 205–214 (2009)
38. P. Aiena, E. Aponte, E. Bazan, Weyl type theorems for left and right polaroid operators. Integr. Equ. Oper. Theory **66**, 1–20 (2010)
39. P. Aiena, M. Chō, M. González, Polaroid type operator under quasi-affinities. J. Math. Anal. Appl. **371**(2), 485–495 (2010)
40. P. Aiena, M. Chō, L. Zhang, Weyl's theorems and extensions of bounded linear operators. Tokyo J. Math. **35**(2), 279–289 (2012)
41. P. Aiena, E. Aponte, J. Guillén, P. Peña, Property (R) under perturbations. Mediterr. J. Math. **10**(1), 367–382 (2013)
42. P. Aiena, J. Guillén, P. Peña, Property (w) for perturbation of polaroid operators. Linear Algebra Appl. **4284**, 1791–1802 (2008)
43. P. Aiena, J. Guillén, P. Peña, Property (R) for bounded linear operators. Mediterr. J. Math. **8**, 491–508 (2011)
44. P. Aiena, J. Guillén, P. Peña, Property (gR) and perturbations. Acta Sci. Math. (Szeged) 78(3–4), 569–588 (2012)
45. P. Aiena, J. Guillén, P. Peña, A unifying approach to Weyl type theorems for Banach space operators. Integr. Equ. Oper. Theory **77**(3), 371–384 (2013)

46. P. Aiena, J. Guillén, P. Peña, Localized SVEP, property (b) and property (ab). Mediterr. J. Math. **10**(4), 1965–1978 (2013)
47. A. Aluthge, On p-hyponormal operators for $1 < p < 1$. Integr. Equ. Oper. Theory **13**, 307–315 (1990)
48. M. Amouch, H. Zguitti, On the equivalence of Browder's and generalized Browder's theorem. Glasgow Math. J. **48**, 179–185 (2006)
49. T. Ando, Operators with a norm condition, Acta Sci. Math. (Szeged) **33**, 169–178 (1972)
50. J.P. Antoine, C. Trapani, Some remarks on quasi-Hermitian operators J. Math. Phys. v. **55**(1), 1–20 (2014)
51. C. Apostol, The reduced minimum modulus. Mich. Math. J. **32**, 279–294 (1984)
52. F.V. Atkinson, On relatively regular operators. Acta Sci. Math. (Szeged) **15**, 38–56 (1953)
53. B. Aupetit, *A Primer on Spectral Theory* (Springer, New York, 1991)
54. B.A. Barnes, Some general conditions implying Weyl's theorem. Rev. Roum. Math. Pures Appl. **8**, 187–192 (1971)
55. B.A. Barnes, The spectral and Fredholm theory of extension of bounded linear operators. Proc. Am. Math. Soc. **105**(4), 941–949 (1989)
56. B.A. Barnes, Common operator properties of the linear operators RS and SR. Proc. Am. Math. Soc. **126**(4), 1055–1061 (1998)
57. B.A. Barnes, Riesz points and Weyl's theorem. Integr. Equ. Oper. Theory **34**, 187–192 (1999)
58. B.A. Barnes, Spectral and Fredholm theory involving the diagonal of bounded linear operators. Acta Math. (Szeged) **73**, 237–250 (2007)
59. R.G. Bartle, C.A. Kariotis, Some localizations of the spectral mapping theorem. Duke Math. J. **40**, 651–660 (1973)
60. S.K. Berberian, An extension of Weyl's theorem to a class of not necessarily normal operators. Mich. Math. J. **16**, 273–279 (1969)
61. S.K. Berberian, The Weyl spectrum of an operator. Indiana Univ. Math. J. **20**, 529–544 (1970)
62. M. Berkani, On a class of quasi-Fredholm operators. Int. Equ. Oper. Theory **34**(1), 244–249 (1999)
63. M. Berkani, Restriction of an operator to the range of its powers. Stud. Math. **140**(2), 163–175 (2000)
64. M. Berkani, Index of B-Fredholm operators and generalization of a Weyl's theorem. Proc. Am. Math. Soc. **130**(6), 1717–1723 (2001)
65. M. Berkani, B-Weyl spectrum and poles of the resolvent. J. Math. Anal. Appl. **272**, 596–603 (2002)
66. C. Benhida, E.H. Zerouali, Local spectral theory of linear operators RS and SR. Integr. Equ. Oper. Theory **54**, 1–8 (2006)
67. M. Berkani, M. Amouch, On the property (gw). Mediterr. J. Math. **5**(3), 371–378 (2008)
68. M. Berkani, M. Amouch, Preservation of property (gw) under perturbations. Acta Sci. Math. (Szeged) **74**(3–4), 769–781 (2008)
69. M. Berkani, A. Arroud, Generalized Weyl's theorem and hyponormal operators. J. Aust. Math. Soc. **76**, 1–12 (2004)
70. M. Berkani, J.J. Koliha, Weyl type theorems for bounded linear operators. Acta Sci. Math. (Szeged) **69**(1–2), 359–376 (2003)
71. M. Berkani, M. Sarih, On semi B-Fredholm operators. Glasgow Math. J. **43**, 457–465 (2001)
72. M. Berkani, M. Sarih, An Atkinson-type theorem for *B-Fredholm operators*. Stud. Math. **148**(3), 251–257 (2001)
73. M. Berkani, M. Sarih, M. Zariouh, *Browder-type theorems and SVEP*. Mediterr. J. Math. **8**(3), 399–409 (2011)
74. M. Berkani, H. Zariuoh, Extended Weyl type theorems. Math. Bohem. **134**(4), 369–378 (2009)
75. M. Berkani, H. Zariuoh, New extended Weyl type theorems. Mat. Vesn. **62**(2), 145–154 (2010)
76. M. Berkani, H. Zariouh, Extended Weyl type theorems and perturbations. Math. Proc. R. Ir. Acad. **110A**(1), 73–82 (2010)

77. M. Berkani, H. Zariuoh, B-Fredholm spectra and Riesz perturbations. Mat. Vesn. **67**(3), 155–165 (2015)
78. M. Berkani, N. Castro, S.V. Djordjević, Single valued extension property and generalized Weyl's theorem. Math. Bohem. **131**(1), 29–38 (2006)
79. E. Bishop, A duality theorem for an arbitrary operator. Pac. J. Math. **9**, 379–397 (1959)
80. F.F. Bonsall, J. Duncan, *Complete Normed Algebras* (Springer, Berlin, 1959)
81. A. Bourhim, V.G. Miller, The single valued extension property is not preserved under sums and products of commuting operators. Glasgow Math. J. **49**, 99–104 (2007)
82. L. Brown, P.R. Halmos, Algebraic properties of Toeplitz operators. J. Reine Angew. Math. **213**, 89–102 (1963/1964)
83. L. Brown, R. Douglas, P. Fillmore, Unitary equivalence modulo the compact operators and extensions of C*-algebras, in *Proceedings of the Conference on Operator Theory*, Halifax, NS. Lecture Notes in Mathematics, vol. 3445 (Springer, Berlin, 1973)
84. J.J. Buoni, J.D. Faires, Ascent, descent, nullity and defect of product of operators. Indiana Univ. Math. J. **25**(7), 703–707 (1976)
85. M. Burgos, A. Kaidi, M. Mbekhta, M. Oudghiri, The descent spectrum and perturbations. J. Oper. Theory **56**, 259–271 (2006)
86. X. Cao, M. Guo, B. Meng, Weyl's theorem for upper triangular operator matrices. Linear Algebra Appl. **402**, 61–73 (2005)
87. S.R. Caradus, *Operator Theory of the Pseudo Inverse.* Queen's Papers in Pure and Applied Mathematics, vol. 38 (Queen's University, Kingston, 1974)
88. S.R. Caradus, W.E. Pfaffenberger, B. Yood, *Calkin Algebras and Algebras of Operators in Banach Spaces* (Dekker, New York, 1974)
89. C. Carpintero, O. Garcia, E. Rosas, J. Sanabria, B-Browder spectra and localized SVEP. Rend. Circ. Mat. Palermo (2) **57**(2), 239–254 (2008)
90. M. Chō, Spectral properties of p-hyponormal operators. Glasgow Math. J. **436**, 117–122 (1994)
91. M. Chō, J.I. Lee, p-hyponormality is not translation-invariant. Proc. Am. Math. Soc. **131**(10), 3109–3111 (2003)
92. M. Chō, H. Jeon, J.I. Lee, Spectral and structural properties of log-hyponormal operators. Glasgow Math. J. **42**, 345–350 (2000)
93. M.D. Choi, C. Davis, The spectral mapping theorem for joint approximate point spectrum. Bull. Am. Math. Soc. **80**, 317–321 (1974)
94. N.N. Chourasia, P.B. Ramanujan, Paranormal operators on Banach spaces. Bull. Austral. Math. Soc. **21**, 161–168 (1980)
95. S. Clary, Equality of spectra of quasisimilar operators. Proc. Am. Math. Soc. **53**, 88–90 (1975)
96. R.E. Cline, An application of representation for the generalized inverse of a matrix. MRC Technical Report **592** (1965)
97. L.A. Coburn, Weyl's theorem for nonnormal operators. Mich. Math. J. **13**(3), 285–288 (1966)
98. I. Colojoară, C. Foiaş, *Theory of Generalized Spectral Operators* (Gordon and Breach, New York, 1968)
99. J.B. Conway, Subnormal operators. Mich. Math. J. **20**, 529–544 (1970)
100. J.B. Conway, *A Course in Functional Analysis*, 2nd edn. (Springer, New York, 1990)
101. J.B. Conway, *The Theory of Subnormal Operators.* Mathematical Survey and Monographs, vol. 36 (American Mathematical Society/Springer, Providence/New York, 1992)
102. G. Corach, B.P. Duggal, R.E. Harte, Extensions of Jacobson lemma. Commun. Algebra **41**, 520–531 (2013)
103. C. Cowen, Hyponormality of Toeplitz Operators. Proc. Am. Math. Soc. **103**, 809–812 (1988)
104. R.E. Curto *Normality, Operator Theory: Operator Algebras and Applications*, Durham, NH, 1988. Proceedings of Symposia in Pure Mathematics, ed. by W.B. Arveson, R.G. Douglas, vol. 51, part II (American Mathematical Society, Providence, 1990)
105. R.E. Curto, Y.M. Han, Weyl's theorem, a-Weyl's theorem, and local spectral theory. J. Lond. Math. Soc. (2) 67, 499–509 (2003)

106. R.E. Curto, Y.M. Han, Generalized Browder's theorem and Weyl's theorem for Banach spaces operators. J. Math. Anal. Appl. (2) **67**, 1424–1442 (2007)
107. J. Daneš, On local spectral radius. Č. Pěst Mat. **112**, 177–178 (1987)
108. A. Devinatz, Toeplitz operators on H^2 spaces. Trans. Am. Math. Soc. **112**, 304–317 (1964)
109. J. Dieudonné, Quasi-hermitian operators. *Proceedings of the International Symposium on Linear Spaces*, Jerusalem (1961), pp. 115–122
110. J. Dieudonné, *Élements d'analyse*. Tome II (Gauthiers-Villars, Paris, 1968)
111. D.S. Djordjević, Operators obeying a-Weyl's theorem. Publicationes Math. Debrecen **55**(3), 283–298 (1999)
112. D.S. Djordjević, S.V. Djordjević, Weyl's theorems: continuity of the spectrum and quasi-hyponormal operators. Acta Sci. Math. (Szeged) **64**(3), 259–269 (1998)
113. D. Djordjević, S.V. Djordjević, On a-Weyl's theorem. Rev. Roumaine Math. Pures Appl. **44**(3), 361–369 (1999)
114. D.S. Djordjević, V. Rakočević, *Lectures on Generalized Inverse*. Faculty of Science and Mathematics (University of Niš, Niš, 2008)
115. S.V. Djordjević, B.P. Duggal, Spectral properties of linear operators through invariant subspaces. Funct. Anal. Approx. Comput. **1**(1), 19–29 (2009)
116. S.V. Djordjević, B.P. Duggal, Drazin invertibility of the diagonal of an operator. Linear Multilinear Algebra **60**, 65–71 (2012)
117. S.V. Djordjević, Y.M. Han, Browder's theorems and spectral continuity. Glasgow Math. J. **42**, 479–486 (2000)
118. S.V. Djordjevíc, H. Zguitti, Essential point spectra of operator matrices through local spectral theory. J. Math. Anal. Appl. **338**, 285–291 (2008)
119. S.V. Djordjević, I.H. Jeon, E. Ko, Weyl's theorem through local spectral theory. Glasgow Math. J. **44**, 323–327 (2002)
120. R.G. Douglas, On extending commutative semigroups of isometries. Bull. Lond. Math. Soc. **1**, 157–159 (1969)
121. R.G. Douglas, *Banach Algebra Techniques in Operator Theory*. Graduate Texts in Mathematics, vol. 179, 2nd edn. (Springer, New York, 1998)
122. H.R. Dowson, *Spectral Theory of Linear Operators* (Academic Press, London, 1978)
123. M.P. Drazin, Pseudoinverse in associative rings and semigroups. Am. Math. Mon. **65**, 506–514 (1958)
124. B.P. Duggal, Quasi-similar p-hyponormal operators. Integr. Equ. Oper. Theory **26**, 338–345 (1996)
125. B.P. Duggal, Hereditarily polaroid operators, SVEP and Weyl's theorem. J. Math. Anal. Appl. **340**, 366–373 (2008)
126. B.P. Duggal, Operator equations $ABA = A^2$ and $BAB = B^2$. Funct. Anal. Approx. Comput. **3**(1), 9–18 (2011)
127. B.P. Duggal, Compact perturbations and consequent hereditarily polaroid operatos. Stud. Math. **221**, 175–929 (2014)
128. B.P. Duggal, Generalized Drazin invertibility and local spectral theory. Monatsh. Math. **182**(4), 841–849 (2017)
129. B.P. Duggal, S.V. Djordjević, Weyl's theorems and continuity of the spectra in the class of p-hyponormal operators. Stud. Math. **143**(1), 23–32 (2000)
130. B.P. Duggal, S.V. Djordjević, Weyl's theorems in the class of algebraically p-hyponormal operators. Comment. Math. Prace Mat. **40**(1), 49–56 (2000)
131. B.P. Duggal, S.V. Djordjević, Dunford's property (C) and Weyl's theorems. Integr. Equ. Oper. Theory **43**(3), 290–297 (2002)
132. B.P. Duggal, S.V. Djordjević, Generalised Weyl's theorem for a class of operators satisfying a norm condition. Math. Proc. R. Ir. Acad. **104A**(1), 75–81 (2004)
133. B.P. Duggal, S.V. Djordjević, Generalised Weyl's theorem for a class of operators satisfying a norm condition II. Math. Proc. R. Ir. Acad. **106A**(1), 1–9 (2006)
134. B.P. Duggal, H. Jeon, Remarks on spectral properties of p-hyponormal and log-hyponormal operators. Bull. Korean Math. Soc. **42**, 541–552 (2005)

135. B.P. Duggal, H. Jeon, On p-quasi hyponormal operators. Linear Algebra Appl. **422**, 331–340 (2007)
136. B.P. Duggal, H. Jeon, H. Kim, Upper triangular operator matrices, asymptotic intertwining and Browder, Weyl theorems. J. Inequl. Appl. **268**, 1–12 (2013)
137. B.P. Duggal, H. Kim, Generalized Browder, Weyl spectra and the polaroid property under compact perturbations. J. Korean Math. Soc. **54**, 281–302 (2017)
138. B.P. Duggal, R.E. Harte, H. Jeon, Polaroid operators and Weyl's theorem. Proc. Am. Math. Soc. **132**, 1345–1349 (2004)
139. B.P. Duggal, D.S. Djordjević, R.E. Harte, S.Č. Živkovic-Zlatanović, Polynomially meromorphic operators. Math. Proc. R. Ir. Acad. **116A**(1), 83–98 (2016)
140. N. Dunford, Spectral theory II. Resolution of the identity. Pac. J. Math. **2**, 559–614 (1952)
141. N. Dunford, Spectral operators. Pac. J. Math. **4**, 321–354 (1954)
142. N. Dunford, A survey of the theory of spectral operators. Bull. Am. Math. Soc. **64**, 217–274 (1958)
143. N. Dunford, J.T. Schwartz, *Linear Operators.* Part I (1958), Part II (1963), Part III (1971) (Wiley, New York)
144. P. Enflo, A counterexample to the approximation problem in Banach spaces. Acta Math. **130**, 309–317 (1973)
145. D.R. Farenick, W.Y. Lee, Hyponormality and spectra of Toeplitz operators. Trans. Am. Math. Soc. **348**(10), 4153–4174 (1996)
146. L.A. Fialkow, A note on quasisimilarity of operators. Acta Sci. Math. **39**, 67–85 (1977)
147. P.A. Filmore, J.P. Williams, On operator ranges. Adv. Math. **7**, 254–281 (1971)
148. J.K. Finch, The single valued extension property on a Banach space. Pac. J. Math. **58**, 61–69 (1975)
149. O.B.H. Fredj, Essential descent spectrum and commuting compact perturbations. Extracta Math. **21**(3), 261–271 (2006)
150. O.B.H. Fredj, M. Burgos, M. Oudghiri, Ascent spectrum and essential ascent spectrum. Stud. Math. **187**(1), 59–73 (2008)
151. T. Furuta, *Invitation to Linear Operators* (Taylor and Francis, London, 2001)
152. T. Furuta, M. Ito, T. Yamazaki, A subclass of paranormal operators including class of log-hyponormal and several related classes. Scientiae Mathematicae **1**, 389–403 (1998)
153. A.M. Gleason, The abstract Theorem of Cauchy–Weyl. Pac. J. Math. **12**(2), 511–525 (1962)
154. I.C. Gohberg, M.K. Zambitski, On the theory of linear operators in spaces with two norms. Pukrain Mat. Z **18**(1), 11–23 (1966, Russian)
155. I.C. Gohberg, A.S. Markus, I.A. Fel'dman, Normally solvable operators and ideals associated with them. Bul. Akad. Stiince RSS Moldoven **10**(76), 51–70 (1960). (translated in Amer. Math. Soc. Transl.) (2), (1967) **61**, 63–84
156. S. Goldberg, *Unbounded Linear Operators, Theory and Applications* (McGraw-Hill, New York, 1966)
157. M. González, The fine spectrum of the Cesàro operator in ℓ_p, $(1 < p < \infty)$. Arch. Math. **44**, 355–358 (1985)
158. M. González, M. Mbekhta, M. Oudghiri, On the isolated points of the surjective spectrum of a bounded operator. Proc. Am. Math. Soc. **136**(10), 3521–3528 (2008)
159. S. Grabiner, Ranges of products of operators. Can. J. Math. **26**, 1430–1441 (1974)
160. S. Grabiner, Ascent, descent and compact perturbations. Proc. Am. Math. Soc. **71**(1), 79–80 (1978)
161. S. Grabiner, Spectral consequences of the existence of intertwining operators. Comment. Math. Prace Mat. **22**, 227–238 (1980–81)
162. S. Grabiner, Uniform ascent and descent of bounded operators. J. Math. Soc. Jpn. **34**(2), 317–337 (1982)
163. S. Grabiner, J. Zemanek, Ascent, descent, and ergodic properties of linear operators. J. Oper. Theory **48**, 69–81 (2002)
164. B. Gramsch, D.C. Lay, Spectral mapping theorems for essential spectra. Math. Ann. **192**, 17–32 (1971)

165. P.R. Halmos, Ten problems in Hilbert space. Bull. Am. Math. Soc. **76**, 887–933 (1970)
166. Y.M. Han, W.Y. Lee, Weyl's theorem holds for algebraically hyponormal operators. Proc. Am. Math. Soc. **128**(8), 2291–2296 (2000)
167. Y.K. Han, H.Y. Lee, W.Y. Lee, Invertible completions of 2×2 upper triangular operator matrices. Proc. Am. Math. Soc. **129**, 119–123 (2001)
168. Y.M. Han, J.I. Lee, D. Wang, Riesz idempotent and Weyl's theorem for w-hyponormal operator. Integr. Equ. Oper. Theory **53**, 51–60 (2005)
169. R.E. Harte, Fredholm, Weyl and Browder theory. Proc. R. Ir. Acad. **85A**(2), 151–176 (1985)
170. R.E. Harte, *Invertibility and Singularity for Bounded Linear Operators*, vol. 109 (Marcel Dekker, New York, 1988)
171. R.E. Harte, Fredholm, Weyl and Browder theory II. Proc. R. Ir. Acad. **91A**(1), 79–88 (1991)
172. R.E. Harte, Taylor exactness and Kaplansky's Lemma. J. Oper. Theory **25**(2), 399–416 (1991)
173. R.E. Harte, W.Y. Lee, *Another note on Weyl's theorem*. Trans. Am. Math. Soc. **349**(1), 2115–2124 (1997)
174. R.E. Harte, H. Raubenheimer, Fredholm, Weyl and Browder Theory III. Proc. R. Ir. Acad. **95A**(1), 11–16 (1995)
175. R.E. Harte, W.Y. Lee, L.L. Littlejohn, On generalized Riesz points. J. Oper. Theory **47**(1), 187–196 (2002)
176. G. Hellwig, *Partial Differential Equations* (Blaisdell Publishing Company, New York, 1964)
177. D.A. Herrero, Economical compact perturbations II, Filling in the holes. J. Oper. Theory **19**(1), 25–42 (1988)
178. D.A. Herrero, *Approximation of Hilbert Space Operators*, vol. 1, 2nd edn., Pitman Research Notes in Mathematics Series, vol. 224 (Longman Scientific and Technical, Harlow, 1989)
179. H. Heuser, *Functional Analysis* (Wiley Interscience, Chichester, 1982)
180. K. Hofmann, *Banach Spaces of Analytic Functions* (Prentice-Hall, Englewood Cliffs, 1962)
181. J.R. Holub, On shifts operators. Can. Math. Bull. **31**, 85–94 (1988)
182. T.B. Hoover, Quasisimilarity of operators. Ill. J. Math. **16**, 678–686 (1972)
183. V.I. Istrătescu, *Weyl's theorem for a class of operators*. Rev. Roum. Math. Pures et Appl. **8**, 1103–1105 (1968)
184. V.I. Istrătescu, *Introduction to Linear Operator Theory*, vol. 65 (Marcel Dekker, New York, 1981)
185. R.C. James, Orthogonality and linear functionals in normed linear spaces. Trans. Am. Math. Soc. **61**, 265–292 (1947)
186. A. Jeribi, *Spectral Theory and Applications of Linear Operators and Block Operator Matrices* (Springer, Cham, 2015)
187. Y.Q. Ji, Quasi-triangular + small compact = strongly irreducible. Trans. Am. Math. Soc. **351**(11), 4657–4673 (1999)
188. B. Jia, Y. Feng, Weyl type theorems under compact perturbations. Mediterr. J. Math. **15**(3) (2018). https://doi.org/10.1007/s00009-017-1051-2
189. C.L. Jiang, Z.Y. Wang, *Structure of Hilbert Space Operators* (World Scientific Publishing Co. Pte. Ltd., Hackensack, 2006)
190. Q. Jiang, H. Zhong, Generalized decomposition, single-valued extension property and approximate point spectrum. J. Math. Anal. Appl. **356**, 322–327 (2009)
191. Q. Jiang, H. Zhong, Topological uniform descent and localized SVEP J. Math. Anal. Appl. **390**, 355–361 (2012)
192. Q. Jiang, H. Zhong, S. Zhang, Components of generalized Kato resolvent set and single-valued extension property. Front. Math. China **7**(4), 695–702 (2012)
193. Q. Jiang, H. Zhong, S. Zhang, Components of topological uniform descent resolvent set and local spectral theory. Linear Algebra Appl. **438**(3), 1149–1158 (2013)
194. M.A. Kaashoek, M.A. Lay, Ascent, descent, and commuting perturbations. Trans. Am. Math. Soc. 169, **72**, 35–47 (1972)
195. T. Kato, Perturbation theory for nullity, deficiency and other quantities of linear operators. J. Anal. Math. **6**, 261–322 (1958)
196. T. Kato, *Perturbation Theory for Linear Operators* (Springer, New York, 1966)

197. R.V. Kadison, J.R. Ringrose, *Fundamentals of the Theory of Operator Algebras*, vol. I (Academic Press, New York, 1983)
198. I.H. Kim, On (p, k)-quasihyponormal operators. Math. Inequal. Appl. 7, **4**, 629–638 (2004)
199. J. Kleinecke, Almost finite, compact and inessential operators. Proc. Am. Math. Soc. **14**, 863–868 (1963)
200. E. Ko, On p-hyponormal operators. Proc. Am. Math. Soc. **128**(3), 775–780 (2000)
201. J.J. Koliha, A generalized Drazin inverse. Glasgow Math. J. **38**, 367–381 (1996)
202. J.J. Koliha, P. Patricio, Elements of rings with equal spectral idempotents. J. Aust. Math. Soc. **72**, 137–152 (2002)
203. V. Kordula, V. Müller, On the axiomatic theory of spectrum. Stud. Math. **119**, 109–128 (1996)
204. V. Kordula, V. Müller, The distance from the Apostol spectrum. Proc. Am. Math. Soc. **124**, 3055–3061 (1996)
205. V. Kordula, V. Müller, V. Rakočević, On the semi-Browder spectrum. Stud. Math. **123**, 1–13 (1997)
206. V. Kordula, M. Mbektha, V. Müller, P.W. Poon, Corrigendum and Addendum: "On the axiomatic theory of spectrum II." Stud. Math. **130**, 193–199 (1998)
207. M.G. Krein, Integral equations on the half line with kernel depending upon the difference of the arguments. Usp. Mat. Nauk **13**, 3–120 (1958, Russian). Am. Math. Soc. Transl. **22**, (2), (1962), 163–288
208. J.P. Labrousse, Les opérateurs quasi-Fredholm. Rend. Circ. Mat. Palermo, XXIX **2**, 161–258 (1980)
209. J. Lambek, *Lectures on Rings and Modules* (Blaisdell, Waltham, 1966)
210. R. Larsen, *An Introduction to the Theory of Multipliers* (Springer, New York, 1979)
211. K.B. Laursen, Operators with finite ascent. Pac. J. Math. **152**, 323–336 (1992)
212. K.B. Laursen, The essential spectra through local spectral theory. Proc. Am. Math. Soc. **125**, 1425–1434 (1997)
213. K.B. Laursen, The Browder spectrum through local spectral theory, in *Proceedings of the International Conference on Banach Algebras*, July–August 1997 (de Gruyter, Blaubeuren, 1998)
214. K.B. Laursen, M.M. Neumann, Asymptotic intertwining and spectral inclusions on Banach spaces. Czechoslov. Math. J. **43**(118), 483–497 (1993)
215. K.B. Laursen, M.M. Neumann, Local spectral theory and spectral inclusions. Glasgow Math. J. **36**, 331–343 (1994)
216. K.B. Laursen, M.M. Neumann, *An Introduction to Local Spectral Theory*. London Mathematical Society Monographs, vol. 20 (Clarendon Press, Oxford, 2000)
217. K.B. Laursen, P. Vrbová, Intertwiners and automatic continuity. J. Lond. Math. Soc. **2**(114), 149–155 (1991)
218. K.B. Laursen, P. Vrbová, Some remarks on the surjectivity spectrum of linear operators. Czechoslov. Math. J. **39**(114), 730–739 (1993)
219. K.B. Laursen, V.G. Miller, M.M. Neumann, Local spectral properties of commutators. Proc. Edinb. Math. Soc. **38**(2), 313–329 (1995)
220. P. Lax, Symmetrizable linear transformations. Commun. Pure Appl. Math. **7**, 633–647 (1954)
221. D. Lay, A. Taylor, *Introduction to Functional Analysis* (Wiley, New York, 1980)
222. W.Y. Lee, Weyl'spectra of operator matrices. Proc. Am. Math. Soc. **129**, 131–138 (2001)
223. W.Y. Lee, *Lecture Notes in Operator Theory*. Graduate Text (Seoul National University, Seoul, 2008)
224. S.H. Lee, W.Y. Lee, A spectral mapping theorem for the Weyl spectrum. Glasgow Math. J. **38**, 61–64 (1996)
225. M.Y. Lee, S.H. Lee, Some generalized theorems on p-quasihyponormal operators for $0 < p < 1$. Nihonkai Math. J. **8**, 109–115 (1997)
226. J. Leiterer, Banach coherent analytic Fréchet sheaves. Math. Nachr. **85**, 91–109 (1978)
227. C.G. Li, S. Zhu, Polaroid type operators and compact perturbations. Stud. Math. **221**, 175–192 (2014)

228. C. Lin, Y. Ruan, Z. Yan, w-hyponormal operators are subscalar. Integr. Equ. Oper. Theory **50**, 165–168 (2004)
229. M. Mbekhta, Généralisation de la décomposition de Kato aux opérateurs paranormaux et spectraux. Glasgow Math. J. **29**, 159–175 (1987)
230. M. Mbekhta, Résolvant généralisé et théorie spectrale. J. Oper. Theory **21**, 69–105 (1989)
231. M. Mbekhta, Sur l'unicité de la décomposition de Kato. Acta Sci. Math. (Szeged) **54**, 367–377 (1990)
232. M. Mbekhta, Sur la théorie spectrale locale et limite des nilpotents. Proc. Am. Math. Soc. **110**, 621–631 (1990)
233. M. Mbekhta, Local spectrum and generalized spectrum. Proc. Am. Math. Soc. **112**, 457–463 (1991)
234. M. Mbekhta, On the generalized resolvent in Banach spaces. J. Math. Anal. Appl. **189**, 362–377 (1995)
235. M. Mbekhta, V. Müller, On the axiomatic theory of spectrum II. Stud. Math. **119**, 129–147 (1996)
236. M. Mbekhta, A. Ouahab, Opérateur s-régulier dans un espace de Banach et théorie spectrale. Acta Sci. Math. (Szeged) **59**, 525–543 (1994)
237. M. Mbekhta, A. Ouahab, *Perturbation des opérateurs s-réguliers*. Topics in Operator Theory, Operator Algebras and Applications, Timisoara (Institute of Mathematics of the Romanian Academy, Bucharest, 1994), pp. 239–249
238. S. Mecheri, Bishop's property (β) and Riesz idempotent for k-quasi-paranormal operators. Banach J. Math. Anal. **6**, 147–154 (2012)
239. S. Mecheri, On a new class of operators and Weyl type theorems. Filomat **27**(4), 629–636 (2013)
240. T.L. Miller, V.G. Miller, An operator satisfying Dunford's condition (C) but without Bishop's property (β). Glasgow Math. J. **40**, 427–430 (1998)
241. T.L. Miller, V.G. Miller, R.C. Smith, Bishop's property (β) and the Césaro operator. J. Lond. Math. Soc. (2) **58**, 197–207 (1998)
242. T.L. Miller, V.G. Miller, M.M. Neumann, Operators with closed analytic core. Rend. Circolo Mat. Palermo **51**(3), 495–502 (2003)
243. V. Müller, On the regular spectrum. J. Oper. Theory **31**, 363–380 (1994)
244. V. Müller, Axiomatic theory of spectrum III-Semiregularities. Stud. Math. **142**, 159–169 (2000)
245. V. Müller, *Spectral Theory of Linear Operators and Spectral Systems on Banach Algebras*. Operator Theory, Advances and Applications, 2nd edn. (Birkhäuser, Berlin, 2007)
246. F.J. Murray, On complementary manifold and projections in spaces L_p and l_p. Trans. Am. Math. Soc. **41**, 138–152 (1937)
247. M.M. Neumann, Banach algebras and local spectral theory. Suppl. Rend. Circ. Matem. Palermo **52**, 649–661 (1998)
248. J.I. Nieto, On the essential spectrum of symmetrizable operators. Math. Ann. **178**, 145–153 (1968)
249. E.A. Nordgren, Composition operators. Can. J. Math. **20**, 422–449 (1968)
250. K.K. Oberai, On the Weyl spectrum II. Ill. J. Math. **21**, 84–90 (1977)
251. M. Oudghiri, Weyl's and Browder's theorem for operators satisfying the SVEP. Stud. Math. **163**(1), 85–101 (2004)
252. M. Oudghiri, a-Weyl's and perturbations. Stud. Math. **173**(2), 193–201 (2004)
253. M. Oudghiri, Weyl's theorem and perturbations. Integr. Equ. Oper. Theory **53**(4), 535–545 (2005)
254. M. Oudghiri, K. Souilah, The perturbation class of algebraic operators and applications. Ann. Funct. Anal. **9**(3), 426–434 (2018)
255. P. Patricio, R.E. Hartwig, Some additive results on Drazin inverse. Appl. Math. Comput. **215**, 530–538 (2009)
256. A. Pełczyński, On strictly singular and strictly cosingular operators I, II. Bull. Acad. Polon. Sci. Sér. Sci. Math. Astronom. Phys. **13**, 31–36, 37–41 (1965)

257. P. Porcelli, *Linear Spaces of Analytic Functions* (Rand McNally, Chigago, 1966)
258. M. Putinar, Hyponormal operators are subscalar. J. Oper. Theory **12**, 385–395 (1984)
259. M. Putinar, Quasi-similarity of tuples with Bishop's property (β). Integr. Equ. Oper. Theory **15**, 1047–1052 (1992)
260. H. Radjavi, P. Rosenthal, *Invariant Subspaces* (Springer, Berlin, 1973)
261. V. Rakočević, On a class of operators. Mat. Vesn. **37**, 423–426 (1985)
262. V. Rakočević, Operators obeying a-Weyl's theorem. Rev. Roumaine Math. Pures Appl. **34**(10), 915–919 (1989)
263. V. Rakočević, Generalized spectrum and commuting compact perturbation. Proc. Edinb. Math. Soc. **36**(2), 197–209 (1993)
264. V. Rakočević, Semi-Fredholm operators with finite ascent or descent and perturbations. Proc. Am. Math. Soc. **123**(12), 3823–3825 (1995)
265. V. Rakočević, Semi-Browder operators and perturbations. Stud. Math. **122**, 131–137 (1997)
266. B.E. Rhoades, Spectra of some Hausdorff operators. Acta Sci. Math. (Szeged) **32**, 91–100 (1971)
267. S. Roch, B. Silbermann, Continuity of generalized inverses in Banach algebras. Stud. Math. **136**, 197–227 (1999)
268. M. Rosenblum, On the operator equation $BX - XA = Q$. Duke Math. J. **23**, 263–269 (1956)
269. W. Rudin, *Functional Analysis*, 2nd edn. (McGraw-Hill, New York, 1991)
270. P. Saphar, Contribution á l'étude des applications linéaires dans un espace de Banach. Bull. Soc. Math. Fr. **92**, 363–384 (1964)
271. M. Schechter, R. Withley, Best Fredholm perturbation theorems. Stud. Math. **90**(1), 175–190 (1988)
272. C. Schmoeger, Ein Spektralabbildungssatz. Arch. Math. Basel **55**, 484–489 (1990)
273. C. Schmoeger, On isolated points of the spectrum of a bounded operator. Proc. Am. Math. Soc. **117**, 715–719 (1993)
274. C. Schmoeger, On operators of Saphar type. Port. Math. **51**, 617–628 (1994)
275. C. Schmoeger, On a class of generalized Fredholm operators I. Demonstratio Math. **30**, 829–842 (1997)
276. C. Schmoeger, On operators T such that Weyl's theorem holds for $f(T)$. Extracta Math. **13**(1), 27–33 (1998)
277. C. Schmoeger, On a class of generalized Fredholm operators II. Demonstratio Math. **31**, 705–722 (1998)
278. C. Schmoeger, On a class of generalized Fredholm operators III. Demonstratio Math. **31**, 723–733 (1998)
279. C. Schmoeger, On a class of generalized Fredholm operators IV. Demonstratio Math. **32**, 581–594 (1999)
280. C. Schmoeger, On a class of generalized Fredholm operators V. Demonstratio Math. **32**, 595–604 (1999)
281. C. Schmoeger, On a class of generalized Fredholm operators VI. Demonstratio Math. **32**(4), 811–822 (1999)
282. C. Schmoeger, On the operator equations $ABA = A^2$ and $BAB = B^2$. Publ. de L'Inst. Math (N.S), **78**(92), 127–133 (2005)
283. C. Schmoeger, Common spectral properties of linear operators A and B such that $ABA = A^2$ and $BAB = B^2$. Publ. de L'Inst. Math (N.S), **79**(93), 109–114 (2006)
284. A.L. Shields, Weighted shift operators and analytic function theory, in *Topics in Operator theory, Mathematical Survey*, ed. by C. Pearcy, vol. 105 (American Mathematical Society, Providence, 1974), pp. 49–128
285. A.M. Sinclair, Eigenvalues in the boundary of the numerical range. Pac. J. Math. **81**, 231–234 (1970)
286. A. Siskakis, Composition semigroups and the Cesàro operator on H^p. J. Lond. Math. Soc. (2) **36**, 153–164 (1987)
287. R.C. Smith, Local spectral theory for invertible composition operators on H^p. Integr. Equ. Oper. Theory **25**, 329–335 (1996)

288. G.J. Stampfli, Hyponormal operators and spectral density. Trans. Am. Math. Soc. **117**, 469–476 (1965). Errata: ibid. **115**, (1965), 550
289. G.J. Stampfli, Quasi-similarity of operators. Proc. R. Ir. Acad. **81**, 109–119 (1981)
290. K. Tanahashi, On log-hyponormal operators. Integr. Equ. Oper. Theory **34**, 364–372 (1999)
291. A.E. Taylor, D.C. Lay, *Introduction to Functional Analysis* (Wiley, New York, 1980)
292. F.H. Vasilescu, *Analytic Functional Calculus and Spectral Decompositions* (Editura Academiei/D. Reidel Publishing Company, Bucharest/Dordrecht, 1982)
293. I. Vidav, On idempotent operators in a Hilbert space. Publ. de L'Inst. Math (NS), **4**(18), 157–163 (1964)
294. P. Vrbová, On local spectral properties of operators in Banach spaces. Czechoslov. Math. J. **23**, 483–492 (1973a)
295. Z. Wang, J.L. Chen, Pseudo Drazin inverses in associative rings and Banach algebras. Linear Algebra Appl. **437**(2), 1332–1345 (2012)
296. H. Weyl, Uber beschrankte quadratiche Formen, deren Differenz vollsteig ist. Rend. Circ. Mat. Palermo **27**, 373–392 (1909)
297. H. Widom, On the spectrum of Toeplitz operators. Pac. J. Math. **14**, 365–375 (1964)
298. D. Xia, *Spectral Theory of Hyponormal Operators* (Birkhäuser, Boston, 1993)
299. J.T. Yuan, G.X. Ji, On (n, k)-quasi paranormal operators. Stud. Math. **209**, 289–301 (2012)
300. Q. Zeng, H. Zhong, New results on common properties of the bounded linear operators RS and SR. Acta Math. Sin (Engl. Ser.) **29**, 1871–1884 (2013)
301. Q. Zeng, H. Zhong, New results on common properties of bounded linear operators AC and BA: spectral theory. Math. Nachr. **287**, 717–725 (2014)
302. Q. Zeng, H. Zhong, Common properties of bounded linear operators AC and BA: local spectral theory. J. Math. Anal. Appl. **414**(2), 553–560 (2014)
303. Q. Zeng, H. Zhong, New results on common properties of the products AC and BA. J. Math. Anal. Appl. **427**(2), 830–840 (2015)
304. Q. Zeng, Q. Jianf, H. Zhong, Spectra originating from semi B-Fredholm theory and commuting perturbations. Stud. Math. **219**(1), 1–18 (2013)
305. Q. Zeng, H. Zhong, Q. Jiang, Localized SVEP and the components of quasi-Fredholm resolvent set. Glasnik Mat. **50**(3), 429–440 (2015)
306. H. Zguitti, A note on generalized Weyl's theorem. J. Math. Anal. Appl. **316**, 373–381 (2006)
307. S. Zhu, C.G. Li, SVEP and compact perturbations. J. Math. Anal. Appl. **380**, 69–75 (2011)

Index

© Springer Nature Switzerland AG 2018

P. Aiena, *Fredholm and Local Spectral Theory II*, Lecture Notes in Mathematics 2235, https://doi.org/10.1007/978-3-030-02266-2

LECTURE NOTES IN MATHEMATICS

 **Springer**

Editors in Chief: J.-M. Morel, B. Teissier;

Editorial Policy

1. Lecture Notes aim to report new developments in all areas of mathematics and their applications – quickly, informally and at a high level. Mathematical texts analysing new developments in modelling and numerical simulation are welcome.

   Manuscripts should be reasonably self-contained and rounded off. Thus they may, and often will, present not only results of the author but also related work by other people. They may be based on specialised lecture courses. Furthermore, the manuscripts should provide sufficient motivation, examples and applications. This clearly distinguishes Lecture Notes from journal articles or technical reports which normally are very concise. Articles intended for a journal but too long to be accepted by most journals, usually do not have this "lecture notes" character. For similar reasons it is unusual for doctoral theses to be accepted for the Lecture Notes series, though habilitation theses may be appropriate.

2. Besides monographs, multi-author manuscripts resulting from SUMMER SCHOOLS or similar INTENSIVE COURSES are welcome, provided their objective was held to present an active mathematical topic to an audience at the beginning or intermediate graduate level (a list of participants should be provided).

   The resulting manuscript should not be just a collection of course notes, but should require advance planning and coordination among the main lecturers. The subject matter should dictate the structure of the book. This structure should be motivated and explained in a scientific introduction, and the notation, references, index and formulation of results should be, if possible, unified by the editors. Each contribution should have an abstract and an introduction referring to the other contributions. In other words, more preparatory work must go into a multi-authored volume than simply assembling a disparate collection of papers, communicated at the event.

3. Manuscripts should be submitted either online at www.editorialmanager.com/lnm to Springer's mathematics editorial in Heidelberg, or electronically to one of the series editors. Authors should be aware that incomplete or insufficiently close-to-final manuscripts almost always result in longer refereeing times and nevertheless unclear referees' recommendations, making further refereeing of a final draft necessary. The strict minimum amount of material that will be considered should include a detailed outline describing the planned contents of each chapter, a bibliography and several sample chapters. Parallel submission of a manuscript to another publisher while under consideration for LNM is not acceptable and can lead to rejection.

4. In general, **monographs** will be sent out to at least 2 external referees for evaluation.

   A final decision to publish can be made only on the basis of the complete manuscript, however a refereeing process leading to a preliminary decision can be based on a pre-final or incomplete manuscript.

   Volume Editors of **multi-author works** are expected to arrange for the refereeing, to the usual scientific standards, of the individual contributions. If the resulting reports can be forwarded to the LNM Editorial Board, this is very helpful. If no reports are forwarded or if other questions remain unclear in respect of homogeneity etc, the series editors may wish to consult external referees for an overall evaluation of the volume.

5. Manuscripts should in general be submitted in English. Final manuscripts should contain at least 100 pages of mathematical text and should always include

   – a table of contents;
   – an informative introduction, with adequate motivation and perhaps some historical remarks: it should be accessible to a reader not intimately familiar with the topic treated;
   – a subject index: as a rule this is genuinely helpful for the reader.
   – For evaluation purposes, manuscripts should be submitted as pdf files.

6. Careful preparation of the manuscripts will help keep production time short besides ensuring satisfactory appearance of the finished book in print and online. After acceptance of the manuscript authors will be asked to prepare the final LaTeX source files (see LaTeX templates online: https://www.springer.com/gb/authors-editors/book-authors-editors/manuscriptpreparation/5636) plus the corresponding pdf- or zipped ps-file. The LaTeX source files are essential for producing the full-text online version of the book, see http://link.springer.com/bookseries/304 for the existing online volumes of LNM). The technical production of a Lecture Notes volume takes approximately 12 weeks. Additional instructions, if necessary, are available on request from lnm@springer.com.

7. Authors receive a total of 30 free copies of their volume and free access to their book on SpringerLink, but no royalties. They are entitled to a discount of 33.3 % on the price of Springer books purchased for their personal use, if ordering directly from Springer.

8. Commitment to publish is made by a *Publishing Agreement*; contributing authors of multiauthor books are requested to sign a *Consent to Publish form*. Springer-Verlag registers the copyright for each volume. Authors are free to reuse material contained in their LNM volumes in later publications: a brief written (or e-mail) request for formal permission is sufficient.

Addresses:
Professor Jean-Michel Morel, CMLA, École Normale Supérieure de Cachan, France
E-mail: moreljeanmichel@gmail.com

Professor Bernard Teissier, Equipe Géométrie et Dynamique,
Institut de Mathématiques de Jussieu – Paris Rive Gauche, Paris, France
E-mail: bernard.teissier@imj-prg.fr

Springer: Ute McCrory, Mathematics, Heidelberg, Germany,
E-mail: lnm@springer.com

Printed in the United States
By Bookmasters